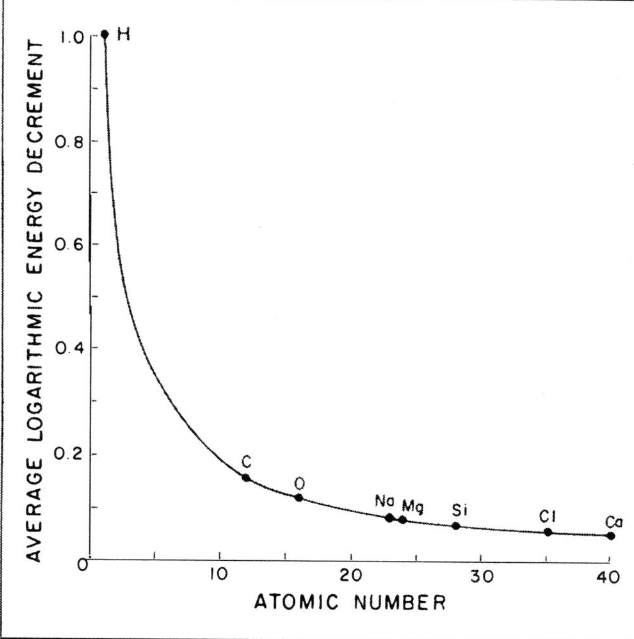

Fig. 2.16—ξ vs. A for element encountered in water-bearing sandstone, limestone, and dolomite.

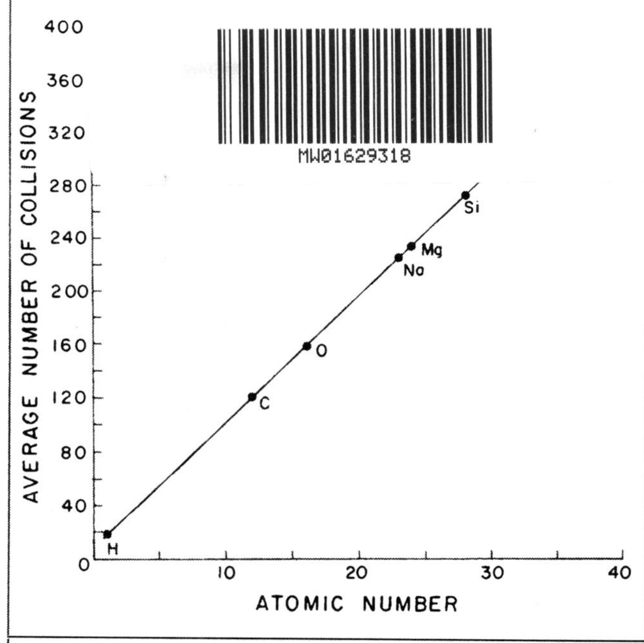

Fig. 2.17—Average number of collisions to thermalize a 4-MeV neutron, $u=18.89$.

The average number of collisions required to reduce a neutron energy from E_0 to E can then be defined by

$$\bar{n}=\frac{u}{\xi}=\frac{1}{\xi}\ln\frac{E_0}{E}. \qquad (2.60)$$

Eqs. 2.57, 2.58, and 2.60 indicate that hydrogen is the most effective neutron moderator. Because of its atomic mass, $A=1$, which is almost equal to the neutron mass, hydrogen exhibits the highest average energy decrement per collision ($\xi=1$). Subsequently, fewer collisions with hydrogen nuclei are needed to moderate neutrons. In fact, a neutron loses all its energy in one direct (head-on) collision with a hydrogen atom. Elastically scattered neutrons will continue the scattering process until they are absorbed.

Example 2.10. A geologic environment made up of water-bearing sandstone, limestone, and dolomite contains H, C, O, Na, Mg, Si, Cl, and Ca.

a. Calculate the average logarithmic energy decrement, ξ, for each of the aforementioned elements.

b. Calculate the average number of collisions with each of the above elements required to thermalize a neutron produced by α-Be reaction.

Solution.

a. For all elements in this environment except hydrogen, $A>10$. ξ can then be calculated from Eq. 2.58. For hydrogen, $\xi=1$. **Table 2.9** gives values of ξ. **Fig. 2.16** is a plot of ξ vs. A.

b. $E_0=4$ MeV (average energy produced by α-Be reaction) and

$E=0.025$ eV (energy of thermal neutron).

From Eqs. 2.59 and 2.60,

$u=\ln(4\times10^6/0.025)=18.891$

and $\bar{n}=18.891/\xi$.

Values of $\bar{n}$ are listed in Table 2.9. **Fig. 2.17** also shows a plot of $\bar{n}$ vs. A.

2.10.3 Radiation Capture and Activation. Neutrons at any energy level can be captured by a nucleus. However, the capture cross section, σ_c, varies inversely with neutron energy. This reaction is particularly important for thermal neutrons. **Table 2.10** lists the cross sections of thermal neutron capture for common earth elements. Of these elements, chlorine (Cl) exhibits the highest capture cross section. Neutron capture by a nucleus results in a heavier isotope in an excited state. The excited nucleus releases the excess energy in the form of gamma photons referred to as gamma rays of capture.

Neutron capture reactions for hydrogen and chlorine are

$${}_1H^1+{}_0n^1\rightarrow{}_1H^2+\gamma \qquad (2.61)$$

and $${}_{17}Cl^{35}+{}_0n^1\rightarrow{}_{17}Cl^{36}+\gamma. \qquad (2.62)$$

Energy levels of capture gamma rays, E_γ, are characteristic of the capturing nucleus. Table 2.10 gives typical energy levels of capture gamma rays emitted by common earth elements. **Fig. 2.18** is an example of a neutron capture gamma ray spectra in an artificial oil-bearing sandstone.

In the activation process, the neutron capture results in a radioactive isotope. This isotope usually decays by emitting a charged particle. In certain cases, characteristic gamma rays are emitted. This emission occurs, however, much later than capture gamma ray emission.

2.11 Neutron Diffusion

A group of fast neutrons introduced into a medium undergoes a series of interactions described in the previous section. First, the neutrons' energy is moderated by inelastic and elastic scattering. They slow down and go through the intermediate, epithermal, and thermal energy levels. Neutrons scatter at random and diffuse through the medium. Thermal neutrons are eventually captured by the medium nuclei.

The neutron population present at any point in the medium depends on the capture and diffusion processes. Capture reduces the neutron population. Diffusion, however, can reduce or increase the neutron population, depending on the overall distribution. The Boltzmann transport equation[10] is used to describe the population of neutrons in a unit volume. It takes into consideration the population increase resulting from source production and scattering from other regions of space. It also considers the population decrease caused by scattering out and absorption effects.

For a monoenergetic point source located at the origin of an infinite, homogeneous, and isotropic medium, the epithermal neutron distribution can be expressed analytically as[10,14]

$$\Psi_e(r)=\frac{N_N}{4\pi D_e}\frac{e^{-r/L_e}}{r}, \qquad (2.63)$$

TABLE 2.10—THERMAL NEUTRON CROSS SECTION, σ_c, OF EARTH ELEMENTS AND ENERGY LEVELS OF CAPTURE GAMMA RAY (AFTER REF. 13)

Element	σ_c (mbarns)	Number of Gammas Emitted per 100 Captures	E_γ (Capture)* (MeV)
Oxygen	0.2	Very small	—
Hydrogen	330	100	2.23
Carbon	3.3	30	4.9 (1)
			3.7 (2)
Calcium	430	94	6.4 (2)
			4.4 (4)
			2.0 (3)
			1.94(1)
Silicon	130	9	6.4 (5)
			4.9 (1)
			4.2 (4)
			3.5 (3)
			2.7 (2)
Chlorine	32,000	18	7.7 (5)
			7.4 (4)
			6.6 (3)
			6.12(2)
			3.0 (6)
			1.97(1)
Magnesium	63		3.9 (1)
			2.8 (2)
			1.87(3)
Aluminum	230		7.7 (1)
			4.8 (4)
			2.9 (2)
Sulfur	490	84	5.4 (1)
			4.8 (4)
			3.0 (2)
			2.4 (3)
Sodium	505	61	6.4 (3)
			3.9 (2)
			3.6 (1)
			2.2 (4)
Iron	2,430	36	7.64(5)
			7.27(4)
			6.43(3)
			5.92(2)
			5.5 (1)

*These energy levels represent groups of gamma rays in some cases. The gamma rays listed are not all those present; only the most predominant are shown. The numbers in parentheses refer to the relative intensity of the gamma ray groups, with (1) being the largest, and so on.

where $\Psi_e(r)$=epithermal neutron flux, neutrons/cm^2/s at a distance r from the source; N_N=source strength, neutrons/s; and L_e=slowing-down length, cm. L_e is a measure of the rectified ("crow flight") distance traveled by a neutron in its zig-zag path between the source and the point where it reached the epithermal energy level. L_e depends on the lethargy and the properties of the medium.

D_e is the epithermal diffusion coefficient defined by

$$D_e=\xi L_e^2\Sigma_e, \qquad (2.64)$$

where ξ is the logarithmic energy decrement per collision. ξ depends on the physical properties of the medium and is defined by Eq. 2.56, and Σ_e is the macroscopic scattering cross section of the medium (in cm^2/cm^3), which depends on the final neutron energy and the properties of the medium.

Eq. 2.63 shows that the concentration of epithermal neutrons decreases exponentially with distance from the source. At a given distance, the flux is a function of source strength and the medium properties. The experimental count rate of epithermal neutrons at a certain distance from a source placed in an 8-in. water-filled borehole is shown in **Fig. 2.19** as a function of slowing-down length for water-saturated sandstone, limestone, and dolomite.

The flux of thermal neutrons, $\Psi_t(r)$, can also be expressed by use of a similar theoretical approach[10]:

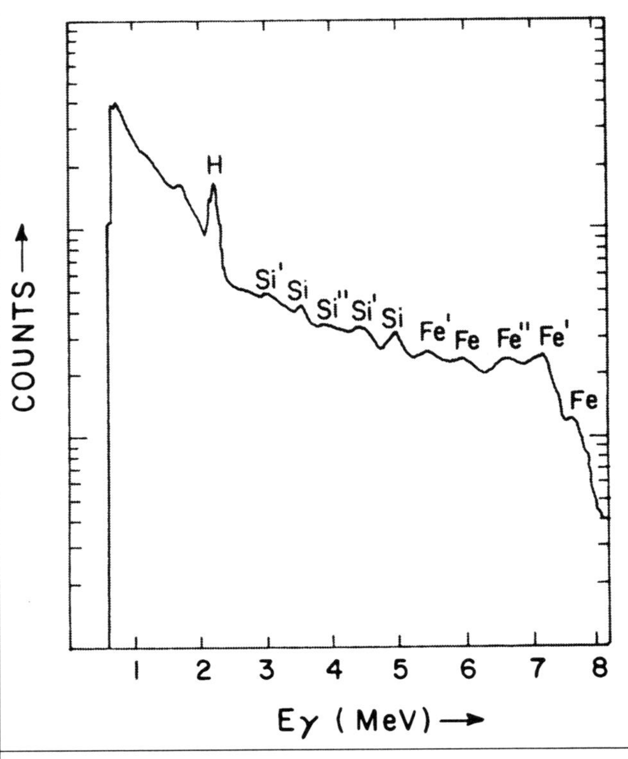

Fig. 2.18—Example of neutron capture gamma ray spectra in artificial oil-bearing sand (from Ref. 12).

$$\Psi_t(r)=\frac{N_N}{4\pi D_t}\frac{L_t^2}{(L_e^2-L_t^2)}\frac{e^{-r/L_e}-e^{-r/L_t}}{r}, \qquad (2.65)$$

where L_t and D_t are the diffusion length and diffusion coefficient of thermal neutrons, respectively. L_t is a measure of the rectified distance a thermal neutron travels before its capture. D_t is expressed as

$$D_t=L_t^2\Sigma_t, \qquad (2.66)$$

where Σ_t is the macroscopic capture cross section of the medium.

The parameters Σ_t, L_t, and D_t depend on the properties of the medium. As in the case of epithermal neutrons, the count rate of thermal neutrons at a given distance from the source can be related to the formation porosity.

Example 2.11.

a. Prepare a plot of neutron flux vs. distance from a monoenergetic point source placed in an infinite medium of water. The source strength is 10^6 neutrons/s. D_e and L_e for water are 0.14 and 7.4 cm, respectively.

b. Repeat Part a, replacing the water medium by an infinite nonporous rock medium. D_e and L_e for the rock are 0.47 and 25 cm, respectively.

c. Repeat Part b for the same rock, this time with a 20% water-filled porosity.

d. Repeat Part c, this time for a porosity value of 40%.

e. Comment on the relationship between porosity and neutron flux.

f. Plot flux vs. porosity on a semilog grid. Show that the relationship can be approximated by a straight line to porosities up to 40%. Give the equation of this line.

Solution.

a. Using Eq. 2.63 gives

$$\Psi_e(r)=\frac{10^6}{4\pi(0.14)}\frac{e^{-r/7.4}}{r}$$

$$=5.6841\times10^5(e^{-r/7.4}/r).$$

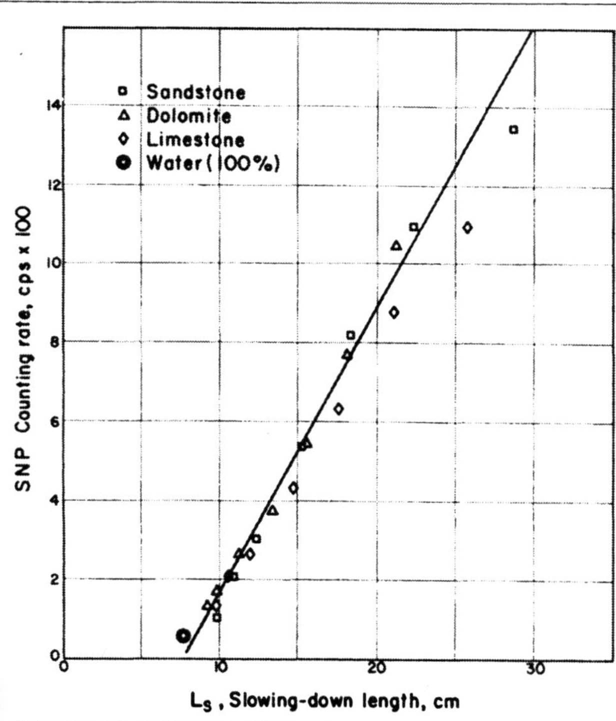

Fig. 2.19—Epithermal-neutron count rate as a function of slowing-down length for water-bearing laboratory formations (after Ref. 15).

TABLE 2.11—$\Psi_e(r)$ VALUES, EXAMPLE 2.11 NEUTRON FLUX (NEUTRONS/$cm^2 \cdot s$)

r (cm)	water	40%-Porosity Formation	20%-Porosity Formation	0%-Porosity (i.e., Matrix)
5	57,843	35,645	31,214	27,724
10	14,716	13,492	12,366	11,349
20	1,905	3,866	3,882	3,804
30	329	1,477	1,625	1,700
40	64	635	765	855
50	13	291	384	458

Values of $\Psi_e(r)$ are listed in **Table 2.11** and plotted vs. r in **Fig. 2.20** (for Parts a through d).

b. $\Psi_e(r)=1.6931\times10^5(e^{-r/25}/r)$.

c. To calculate D_e and L_e of the porous water-bearing formation, a mixing rule based on volumetric fractions is used:

$$D_e=0.2(0.14)+0.8(0.47)=0.404 \text{ cm},$$

$$L_e=0.2(7.4)+0.8(25)=21.48 \text{ cm},$$

and $\Psi_e(r)=1.9697\times10^5(e^{-r/21.48}/r)$.

d. $D_e=0.338$ cm, $L_e=17.96$ cm,

and $\Psi(r)=2.3544\times10^5\ (e^{-r/17.96}/r)$.

e. Fig. 2.20 clearly shows a dependence of neutron flux on porosity. However, the plot can be subdivided into three ranges: near, far, and crossover.

Near the source, the neutron flux increases with porosity (i.e., hydrogen content) because, the higher the hydrogen atom content, the faster the neutrons are slowed down and the higher the concentration of slowed-down (epithermal) neutrons near the source.

Far from the source, the neutron flux decreases with porosity, or hydrogen content. In the case of high hydrogen content, which results in the generation of a high number of epithermal neutrons near the source, these epithermal neutrons will be thermalized and captured at a high rate because of the same high hydrogen content. Fewer neutrons remain to diffuse away from the source, resulting in the observed low epithermal-neutron flux.

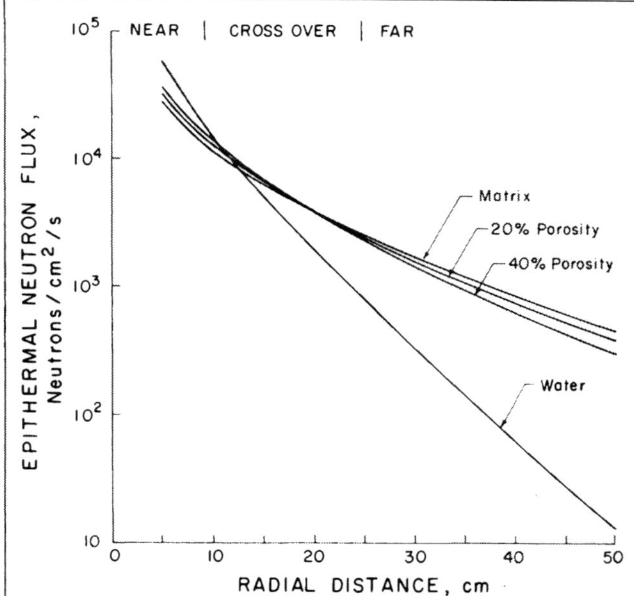

Fig. 2.20—Epithermal-neutron flux around a point source placed in an infinite water-bearing formation of different porosities.

The inverse is true in the case of low porosity. Few fast neutrons will reach the epithermal level near the source, leaving more of them to diffuse away from the source. These neutrons reach the epithermal level far from the source, resulting in the observed low epithermal-neutron flux.

Because epithermal neutrons are the source of thermal neutrons and thermal neutrons are ultimately the source of capture γ rays, the flux of thermal neutrons and capture γ rays assumes the same distribution as shown in Fig. 2.20.

The curves cross (Fig. 2.20) at an intermediate distance from the source. Over this "crossover" range, the flux/porosity dependence is not quite clear.

Although the equation used to develop the curves of Fig. 2.20 ignores the actual geometry involved in neutron logging, it clearly shows that the source-detector spacing has to be chosen carefully to obtain a predictable relationship between neutron log response and porosity. A spacing of 40 to 60 cm is commonly used in neutron porosity logging.

f. **Fig. 2.21** is a plot of Ψ (40) values calculated for porosity values ranging from 0% to 100%. For porosities below 40%, the data can be approximated by a straight line. The equation of this line is

$$\phi=9.025-3.067 \log \Psi.$$

2.12 Neutron Logging Methods

Several phenomena occur during the moderation and capture of neutrons. Physical parameters associated with these phenomena are measurable and are empirically related to rock properties.

Emission of scattering gamma rays characteristic of the medium accompanies inelastic scattering. Fig. 2.15 shows an example of an inelastic scattering γ ray spectrum. Measurement of the spectra of these gamma rays allows the determination of carbon/oxygen content ratio. This ratio, in turn, is used to obtain oil content.

Because hydrogen is responsible for most of the slowing-down effect, measuring the concentration of epithermal neutrons (see the discussion of neutron porosity tools in Chap. 9) indicates the hydrogen concentration in the material. In shale-free, water-bearing formations, the hydrogen concentration reflects the formation porosity. **Fig. 2.22** shows the laboratory results of Fig. 2.19 as a function of laboratory formations' porosities. Fig.2.22 clearly indicates the dependence of epithermal-neutron concentration on porosity and lithology. Because epithermal neutrons are the source of thermal neutrons and thermal neutrons are eventually the source of capture

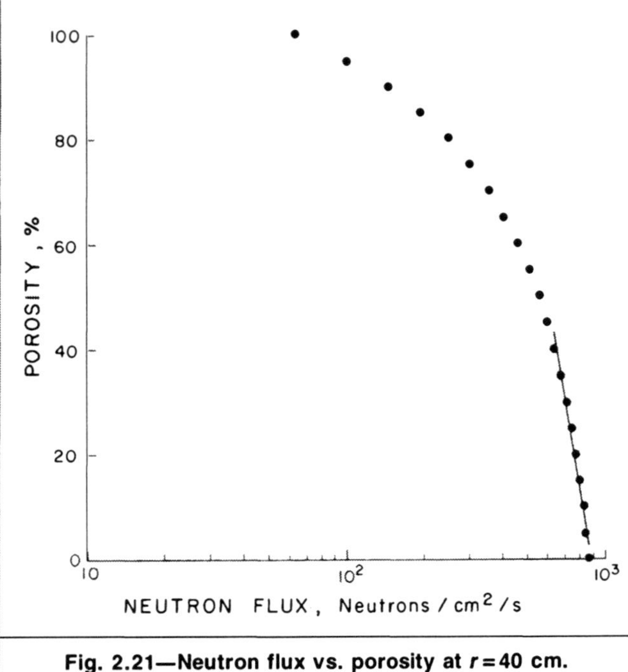

Fig. 2.21—Neutron flux vs. porosity at $r = 40$ cm.

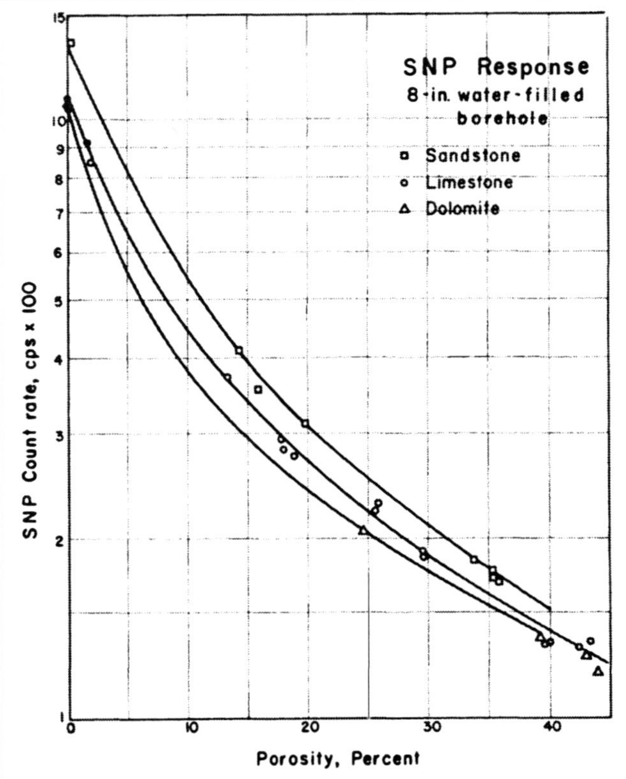

Fig. 2.22—Epithermal-neutron count rate as a function of porosity for water-bearing laboratory formations (after Ref. 15).

gamma rays, thermal and capture gamma ray counts also indicate porosity.

Capture gamma rays also are used for other determinations. The decay-rate measurement of capture γ rays (pulsed neutron tools, Chap. 9) yields decay-time constants and the capture cross sections of the formations. The neutron decay rate indicates the chlorine content of a formation because chlorine has the highest capture cross section of common formation elements. In clean formations, chlorine content is used to indicate water content and, by complement, oil content.

Capture gamma ray energy is also characteristic of the capture nucleus. An example of a capture γ ray spectrum is shown in Fig. 2.18. The measurements of the spectral contribution of iron, silicon, and calcium are used in lithology determination. Chlorine and hydrogen contents are used to investigate salinity changes.[16]

Review Questions

1. State the symbol, mass, and electrical charge of the electron.
2. State the symbol, relative mass, and relative electrical charge of the proton and the neutron.
3. What are isotopes?
4. What are the names, symbols, compositions, atomic numbers, and atomic masses of hydrogen isotopes? What proportions of these isotopes exist in nature?
5. Explain the meaning of nuclei ground and excited states.
6. What is a photon? What is the relationship between its wavelength and frequency?
7. What is an electron-volt, and what is its relation to the Joule?
8. What is the energy range of gamma photons?
9. What are the α and β particles, and what are their masses and charges?
10. What are the γ rays? How penetrating are they? How does their penetrating ability compare to those of the α and β particles?
11. In what ways do radioactive atoms disintegrate? Give an example of each.
12. How many protons and neutrons are in the nucleus of chlorine-34 and chlorine-37?
13. Why are neutrons so penetrating?
14. By what mechanism are photons absorbed by matter?
15. How do the three linear absorption coefficients for aluminum vary with photon energy?
16. At what gamma photon energy is absorption caused almost entirely by Compton scattering?
17. How does the mass absorption coefficient for hydrogen compare to those of other elements?
18. What nuclei are responsible for nearly all the natural radioactivity in the earth? What are the relative abundances, half-lives, and final decay products of these nuclei?
19. What are the comparative radioactivity levels of sedimentary rocks?
20. What are the different tools that measure natural radioactivity of rocks? What information do they provide?
21. Define the following parameters: (a) absorption cross section, (b) linear absorption coefficient, (c) mass absorption coefficient, (d) mean free path, and (e) half-thickness value.
22. Define the following parameters: (a) decay-time constant and (b) half-life time.
23. What element plays the major role in slowing down neutrons by elastic collisions in sedimentary rocks?
24. Why are thermal neutrons called that?
25. Why do neutrons seldom exist in the free state in nature?
26. Describe, in detail, the interactions that a high-energy, fast neutron undergoes between its creation and its capture.
27. What element plays the major role in capturing thermal neutrons in sedimentary rocks?
28. Compare the probability of hitting a 1-ft-diameter wastebasket with a wad of paper to that of hitting a 2-ft-diameter basket.
29. Why was paraffin used to shield fixed neutron sources on logging tools?
30. What information is obtained from the spectrum of scattering and capture gamma rays?
31. What information is obtained from the concentration at a distance from a source of epithermal neutrons, thermal neutrons, and capture γ rays?
32. What information is obtained from the decay rate of thermal neutrons?
33. Explain the principle of the neutron fixed sources and accelerators. What is the advantage of accelerators over the fixed sources?
34. Explain why, at a short distance from a fast neutron source, the thermal-neutron density is higher in 40% porous and water-

filled rock than in 10% porosity water-filled rock. Why is the reverse true further from the source?

35. What generalizations can be stated for the changes that occur when an atom emits an alpha particle and a beta particle?

36. Write the equation which describes each of the following nuclear reactions: (a) bombardment of beryllium by alpha particles and (b) capture of neutrons by hydrogen.

Problems

2.1 Calculate the energy equivalent of a proton mass.

2.2 When a uranium-235 nucleus captures a neutron, the nucleus becomes unstable and splits into a tin nucleus and a molybdenum nucleus. This typical fission reaction is expressed as

$$_{92}U^{235} + {_0n^1} \rightarrow {_{50}Sn^{124}} + {_{42}Mo^{100}} + 12\ {_0n^1} + \text{energy}.$$

Calculate the energy released per gram of uranium. (Atomic masses of U-235, Sn-124, and Mo-100 are 235.116, 118.69, and 95.94 amu, respectively.)

2.3 Visible light has a range of wavelengths from about 4,000 Å at the blue end to about 8,000 Å at the red end. What is the frequency and energy of a "green" photon whose wavelength is 6,000 Å?

2.4 What is the frequency and energy of a 0.008-Å gamma photon?

2.5 Show analytically that the average lifetime, τ, of a radionuclide is 1.44 $t_{1/2}$. [Hint $\tau=(1/\int_0^\infty)t(-dN)$.]

2.6 The average radioactivity of black shale is 26.1×10^{-12} radium equivalent/g. Express this activity in Bequerels. The radium equivalent is defined as the number of grams of radium per grams of inert rock required to duplicate the rock's gamma radiation.

2.7 Radioactive cobalt-60 is used in density logging tools. How much activity will a CO^{60} source lose in 1 month? The half-life of cobalt-60 is 5.26 years.

2.8 If the activity of 1 g of pure radium-226 is measured to be 3.67×10^{10} integrations/sec, calculate the half-life time and decay constant.

2.9 Uranium-238 disintegrates and transforms to stable lead-206. The half-life of this reaction is 7.6 billion years. Estimate the geologic age of a rock that contains 0.05 and 0.0005 wt% U^{238} and Pb^{206}, respectively. List the assumptions implied in the estimate.

2.10 Illustrate the decay process of K^{40} on a Z/A and Z/N plot.

2.11 Calculate the velocity of the photoelectrons ejected from tin interacting with 0.1-Å gamma rays. The energy required to remove an electron from the K shell of a tin atom is 29.201 keV.

2.12 What are the energies of the electrons involved in the following successive processes?

a. A 2-MeV gamma photon is absorbed by pair production.

b. The annihilation radiation is Compton scattered at 60° and 90° angles.

c. The scattered photons are absorbed in barium.

2.13 Using the following data obtained in an experiment designed to determine γ ray absorption properties of lead, determine the linear and mass absorption coefficients, half-thickness value, and energy of the photons used. Also, calculate the absorption cross section and mean free path.

Thickness (cm)	Relative Intensity I/I_0
0	1
2	0.90
4	0.81
8	0.66

2.14 Derive an equation that expresses rock porosity in terms of the mass absorption coefficient of the matrix, α_{ma}; the fluid, α_f; and the bulk rock, α_b.

2.15 How many electrons are there in 1 g of carbon $_6C^{12}$, oxygen $_8O^{16}$, silicon $_{14}Si^{28}$, sulfur $_{16}S^{34}$, and calcium $_{20}Ca^{40}$?

2.16 How many electrons are there in 1 g of hydrogen?

2.17 a. What is the velocity of a 2-MeV neutron

b. If a 2-MeV neutron has a direct (head-on) elastic collision with a hydrogen atom, what is the velocity of the neutron and hydrogen nucleus after collision?

c. If a 2-MeV neutron has a direct elastic collision with a carbon atom, what is the final velocity and kinetic energy of the neutron?

2.18 a. On average, how many collisions must a 2-MeV neutron make with hydrogen nuclei to reach a thermal energy level of 0.025 eV?

b. Repeat Part a for carbon.

2.19 After passing through 0.5 cm of cobalt-59, the intensity of a beam of neutrons was reduced from 4,000 to 780 neutrons/s·cm². Calculate the mean free path and cross section of the absorption of cobalt. Cobalt density is 8.71 g/cm³.

2.20 Calculate the thermal neutron macroscopic capture cross section, Σ, of freshwater.

2.21 Calculate the Σ of oil (C_nH_{2n}) and methane (CH_4) at reservoir conditions where oil and gas densities are 0.8 and 0.1 g/cm³, respectively.

2.22 Calculate the macroscopic capture cross section of the following sedimentary rocks.

Rock	Composition	Density (g/cm³)
Silica	SiO_2	2.65
Calcite	$CaCO_3$	2.71
Dolomite	$CaMg(CO_3)$	2.87
Anhydrite	$CaSO_4$	2.97
Salt	$NaCl$	2.17

2.23 a. Show that Σ_w for formation water increases almost linearly with salinity, n, and can be approximated by

$$\Sigma_w = 22.2 + 3.75\times10^{-4}n$$

where Σ_w is in c.u. and n in ppm.

b. What is Σ_w for 100,000-ppm formation water?

2.24 Using data from the previous problems, calculate Σ for a 15% porosity limestone formation partially saturated with oil. Water saturation is 30%.

2.25 a. Calculate the intensity of a 1000 photons/s·cm², 1.25-MeV γ ray beam after it passes through 10 cm of liquid water.

b. Repeat Part a for ice. Ice density is 0.92 g/cm³.

c. Repeat Part a for high-pressure steam of 0.12 g/cm³ density.

2.26 Cobalt-60 is used as a source of gamma rays in density logging tools. How much will the calibration of the tool change in 1 year? Cobalt-60 half-life is 5.2 years.

2.27 Consider 50 mCuries of a radioactive material with a half-life of 20 days.

a. What is the decay constant for this material in days^{-1} and seconds^{-1}.

b. How many millicuries are left after 40 days? After 1 year?

c. How many radioactive atoms were present initially and then after 1 year?

2.28 If K^{40} has a half-life of 1.4×10^9 years and only 10% of the original K^{40} remains, how old is the earth?

2.29 A sample of pitchblende was found to contain 51.16% U^{238}, 2.24% Pb^{206}, and 0.26% Pb^{207} and Pb^{208}. The half-life of U^{238} is 4.5×10^9 years and its ultimate decay product is Pb^{206}. Assuming no preferential leaching of either the U^{238} mineral or the Pb^{206}, calculate the age of the rock.

2.30 Experiments show that samples of new wood uniformly give 12.5 counts per minute per gram of carbon. Beta ray activity of wood from an Egyptian mummy case is only 9.4 counts per minute per gram of carbon. If the half-life of carbon-14 is 5,720 years, how old is the wood?

2.31 After passing through 10 cm of a brine-saturated pure calcite, a beam of 1.25-MeV gammas is reduced from 10,000 to 2,730 counts/min. If the density of the calcite is 2.71 g/cm³, and the density of the "brine" is 1.146 g/cm³, what is the porosity of the calcite formation? Mass absorption coefficients of brine and calcite are 0.055 and 0.062 cm²/g, respectively.

2.32 What mass of cobalt-60 ($t_{½}$ =5.2 years) will have an activity of 1 Curie?

2.33 The nucleus of platinum ($_{78}Pt^{192}$) decays to osmium ($_{76}Os^{188}$). The atomic masses of Pt-192 and Os-188 are 191.9614 and 187.9560 amu, respectively. How much energy is released in this reaction?

2.34 a. An incident beam of 2.0-MeV gammas has an intensity of 1,200 photons/s·cm^2. What is its intensity after it passes through 8.0 cm of 40% porosity steel wool? Solid steel with a density of 7.85 g/cm^3 has a mass absorption coefficient of 0.044 cm^2/g for 2.0-MeV gammas. (Hint: air contribution is negligible.)

b. If the steel wool is saturated with water whose mass absorption coefficient is 0.049 cm^2/g, what is the linear and mass absorption coefficient of the mixture?

c. To what value will the water-saturated steel wool reduce the incident beam of 1,200 photon/s·cm^2 after it passes through 8.0 cm of it?

2.35 Calculate the number of hydrogen atoms per cubic centimeter for the following matters: (a) fresh water, (b) oil (C_nH_{2n}, 0.8 g/cm^3), and (c) methane (CH_4, 0.01 g/cm^3).

Nomenclature

a = instantaneous decay rate or activity, Bq
A = mass number (number of protons and neutrons)
A = area, m^2
A = atomic mass, amu
C = speed of light, m/s
C_d = decay constant of a specific element
C_K = Boltzmann's constant, J/°K
D = diffusion coefficient, m
e^+ = positron
e^- = electron
E = energy, J
f = fraction
f = frequency, Hz
h = mean free path, m
$h_{½}$ = half-thickness value, m
I = intensity of incident photons, photons/s·cm^2
I_{sh} = shale index, fraction
L = slowing-down length, m
m = mass, g
M = molecular mass, amu
n = neutron
N = number
$\bar{N}$ = mean number
N_a = number of atoms per unit volume, atoms/cm^3
N_A = Avogadro's number, mol/gmol
N_e = number of electrons per unit volume, electrons/cm^3
N_N = source strength, neutrons/s
p^+ = proton
P_e = effective photoelectric index
$\mathcal{P}_N$ = probability of N events
$\mathcal{P}_0$ = probability for zero events
Q = electric charge, C
t = time, seconds
$t_{½}$ = half-life, seconds
T_a = absolute temperature, °K
u = neutron lethargy
v = velocity, m/s
V = volume, m^3
x = distance, m
Z = atomic number (number of protons)
α_l = linear absorption coefficient, m^{-1}
α_m = mass absorption coefficient, cm^2/g or m^2/kg
θ = scattering angle, degrees
λ = wavelength, m
ξ = average logarithmic energy decrement per collision
ρ = bulk density, g/cm^3
σ = absorption cross section, barns
σ = standard deviation
Σ = macroscopic capture cross section, cm^2/cm^3
τ = decay-time constant, seconds
ν = neutrino
ϕ = porosity, fraction
Ψ = neutron flux, neutrons/cm^2·s

Subscripts

a = atom
b = bulk
c = capture
cs = clean sand
Cs = Compton scattering
e = epithermal
e = electron
f = fluid
H = hydrogen
ine = inelastic
k = kinetic
log = logging tool response
m = mass
ma = matrix
mp = most probable
O = oxygen
pe = photoelectric effect
pp = pair production
sh = shale
ss = sandstone
t = thermal
V = volumetric
w = water
0 = time zero
α = alpha particles
β = beta particles
γ = gamma rays

Superscript

$^-$ = average

References

1. Lapp, R.E. and Andrews, H.L.: *Nuclear Radiation Physics,* third edition, Prentice-Hall Inc., Englewood Cliffs, NJ (1964).
2. Enge, H.: *Introduction to Nuclear Physics,* Addison-Wesley Publishing Co., Reading, MA (1966).
3. *Radioactivity Well Logging Handbook,* Lane-Wells Co. (1958).
4. *Log Interpretation Principles/Applications,* Schlumberger, Houston (1987).
5. Evans, R.D.: *The Atomic Nucleus,* McGraw-Hill Book Co. Inc., New York City (1967).
6. Barnes, W.E.: *Basic Physics of Radiotracers,* CRC Press, Boca Raton, FL (1983) **I, II**.
7. Attix, G.H. and Roesch, W.C.: *Radiation Dosimetry,* Academic Press, New York City (1968).
8. Bertozzi, W., Ellis, D.V., and Wahl, J.S.: "The Physical Foundation of Formation Lithology Logging with Gamma Rays," *Geophysics* (1981) **46**, No. 10.
9. Gardner, J.S. and Dumanoir, J.L.: "Litho-Density Log Interpretation," *Trans.*, SPWLA 21st Symposium, Lafayette, LA (1980).
10. Tittman, J.: *Physics of Wireline Measurements,* Schlumberger Educational Services, Houston (1986).
11. *Log Review 1,* Dresser Atlas, Houston (1971).
12. Hertzog, R.C.: "Laboratory and Field Evaluation of an Inelastic Neutron Scattering and Capture Gamma Ray Spectrometry Tool," *SPEJ* (Oct. 1980) 327–40.
13. Owen, J.D.: "A Review of Fundamental Nuclear Physics Applied to Gamma Ray Spectral Logging," *Trans.*, SPWLA First Symposium, Tulsa (1960).
14. Tittle, C.W.: "Theory of Neutron Logging 1," *Geophysics* (Feb. 1961) **25**, No. 1, 27–39.
15. Edmundson, H. and Raymer, L.L.: "Radioactive Logging Parameters for Common Minerals," *Trans.*, SPWLA 20th Symposium, Tulsa (1979).
16. Ralson, W.R.: "Chlorine Detection by the Spectral Log," *Pet. Eng.* (March 1959).

Chapter 3
Acoustic Properties of Rocks

3.1 Introduction

Acoustic logging is an important part of formation evaluation. This type of logging uses the propagation of acoustic waves within and around the borehole. Acoustic properties measured in well logging are compressional- and shear-wave velocities, compressional- and shear-wave attenuation, and amplitude of reflected waves.

The measurement of wave velocity can be used to evaluate formation porosity, lithology, and bulk and pore compressibilities. Cement-bond-quality determination and fractured-zone identification are based on measurements of wave attenuation. The measurement of reflected-wave amplitude is used to locate vugs and fractures, to determine fracture orientation, and to inspect casing.[1]

Acoustic logging in open holes (uncased boreholes) consists mainly of acoustical velocity measurement. This measurement, usually called a sonic log, is a record of the time required for an acoustic wave to travel a given distance through the formation that surrounds a borehole. This parameter is referred to as acoustic transit time, Δt, and is usually expressed in microseconds per foot. Velocity, v, and transit time, Δt, are related by

$$\Delta t=10^6/v, \quad (3.1)$$

where Δt is in μsec/ft and v is in ft/sec.

Sonic logging was originally developed to aid in the evaluation of seismic surveying. Advances in electronics and downhole tool design extended its usefulness to formation evaluation, especially porosity determination.

Because acoustic velocity depends on the elastic properties of rocks, this chapter treats basic concepts of elasticity and elastic wave propagation in fluid-filled boreholes and in porous media. The various parameters that affect acoustic velocity and the most useful models used to relate acoustic velocity to porosity are also discussed.

3.2 Basic Concepts of Elasticity

3.2.1 Stresses and Strains. "Elasticity" refers to the relationship between the external forces applied to a body and the resulting changes in its shape and size. Force is quantitatively described by stress—i.e., force per unit area. A force applied perpendicularly to a rod of length L and diameter d and away from the body on which it acts results in tensile stress. Tensile stress causes rod elongation by ΔL and a decrease in diameter by Δd (**Fig. 3.1a**). When the perpendicular force is applied into the rod (**Fig. 3.1b**), the result is a compressive stress that causes a shortening of the rod by ΔL and an increase in its diameter by Δd. If the force is applied tangentially to the area (**Fig. 3.1c**), it is referred to as shear stress. Shear stress causes deformation by displacement without a change in volume.

Deformation and displacements that result from stresses are described as strains. Strains resulting from compressive and tensile stresses are called longitudinal strains, ϵ_ℓ, and transverse strains, ϵ_t, defined by

$$\epsilon_\ell=\Delta L/L \quad (3.2)$$

and $\epsilon_t=\Delta d/d$. (3.3)

Shear stress results in a shear strain, ϵ_s, defined by

$$\epsilon_s=\Delta L/L=\tan\theta, \quad (3.4)$$

and when the strain is small,

$$\epsilon_s\cong\theta, \quad (3.5)$$

where θ is the deformation angle.

3.2.2 Elastic Constants. The elastic properties of a matter are described by certain elastic constants. The elastic constants are defined below for strains within the elastic limit; i.e., the body returns to its original condition if the force causing the strain is removed.

Young's modulus, E, is the ratio of tensile or compressive stress to the corresponding strain so that

$$E=(F/A)/(\Delta L/L). \quad (3.6)$$

For most rocks, E ranges[2] from 10^{10} to 10^{11} Pa.

Shear modulus, G, describes the ratio of shear stress to shear strain. G is defined as

$$G=(F/A)/\theta. \quad (3.7)$$

For most rocks, G is about one-third to one-half as great as E.

Bulk modulus, K, is a measure of the stress/strain ratio when a body is subjected to uniform compressive stress. The stress or, in this case, pressure, p, is related to volume change, ΔV, by

$$K=p/(\Delta V/V). \quad (3.8)$$

The bulk modulus is the reciprocal of the compressibility.

Poisson's ratio, μ, is a measure of the geometric change of shape under stress. It is defined as the ratio of transverse to longitudinal strains:

$$\mu=\epsilon_t/\epsilon_\ell. \quad (3.9)$$

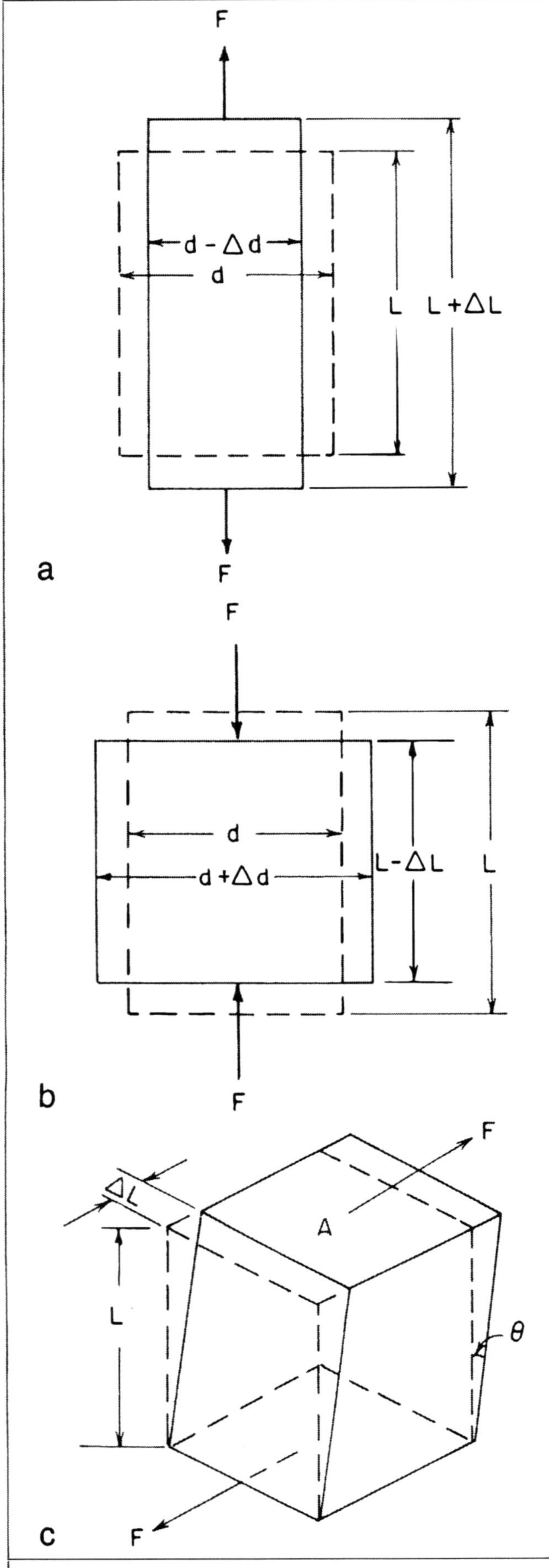

Fig. 3.1—(a) Longitudinal and transverse strains caused by tensile stress; (b) longitudinal and transverse strains caused by compressive stress; (c) shearing strain resulting from shearing stress.

In the case of a cylinder, μ is expressed (using the symbols in Fig. 3.1) as

$$\mu=(\Delta d/d)/(\Delta L/L). \quad (3.10)$$

Poisson's ratio for rocks ranges from 0.05 to 0.40, averaging about 0.25 for sedimentary rocks.[2]

The four elastic constants—E, G, K, and μ—are dependent parameters. Any one of these constants can be expressed in terms of two others. The most useful relationships among them are

$$G=E/[2(1+\mu)] \quad (3.11)$$

and $K=E/[3(1-2\mu)]$. (3.12)

3.2.3 Elastic Body Waves. An elastic body is instantaneously compressed when subjected to sudden stress or pressure. The region where the particles of the body are most compressed will propagate away from the point of impact. The compressions are transmitted to the whole body by a series of compressions and rarefactions (release of compression) in a wave-like form.

Wave propagation is expressed mathematically by the following set of equations:

$$\sigma=A_o \cos 2\pi[(ft-(x/\lambda)], \quad (3.13)$$

$$v=\lambda f, \quad (3.14)$$

and $f=1/t$, (3.15)

where σ=stress at any time t and at a distance x within an elastic wave; A_o=amplitude of the stress at the source; λ=wavelength, or the distance between successive maximum compressions or rarefactions at any time; τ=period, or the time interval between successive maximum compressions or rarefactions at any point; f=frequency of cycles of compression and rarefaction; and v=velocity of propagation. **Fig. 3.2** illustrates this wave motion.

Elastic waves are attenuated by absorption. Because of attenuation, the amplitude, A, of a wave at a distance x from the source can be expressed by

$$A=A_o e^{-\alpha x}, \quad (3.16)$$

where α is the absorption coefficient. The coefficient α depends on the characteristics of the medium through which the wave is traveling. Including attenuation effects, Eq. 3.13 becomes

$$\sigma=Ae^{-\alpha x} \cos 2\pi[(ft-(x/\lambda)]. \quad (3.17)$$

Elastic waves can be classified as body waves and boundary waves, also known as guided waves. Body waves propagate in unbounded media and are distinguished from boundary waves that arise because of the presence of such boundaries as a borehole wall. The two main types of body waves are compressional and shear.

Compressional waves, also known as longitudinal waves or P (primary) waves, are those where the particle motion is in the direction of wave propagation, as shown in **Fig. 3.3a**. The velocity of compression propagation, v_p, depends on the body's elastic properties. It can be derived from the equation of motion and is expressed as

$$v_p=[(K+\tfrac{4}{3})G/\rho]^{1/2} \quad (3.18)$$

$$=\{(E/\rho)(1-\mu)/[(1-2\mu)(1+\mu)]\}^{1/2}, \quad (3.19)$$

where ρ is the density of the medium.[2]

Shear waves, also known as transverse or S (secondary) waves, are those where the particle motion is perpendicular to the direction of wave propagation, as shown in **Fig. 3.3b**. The velocity of shear waves, v_s, can be derived from the equation of motion and is expressed as[2]

$$v_s=(G/\rho)^{1/2} \quad (3.20)$$

$$=[(E/\rho)/2(1+\mu)]^{1/2}. \quad (3.21)$$

The presence of shear waves requires the medium to possess shear strength. Hence, shear waves can propagate only in solids.

Using Eqs. 3.18 through 3.21 to compare compression- and shear-wave velocities yields

$$v_p/v_s=[(\tfrac{4}{3})+(K/G)]^{1/2} \quad (3.22)$$

$$=[2(1-\mu)/(1-2\mu)]^{1/2}. \quad (3.23)$$

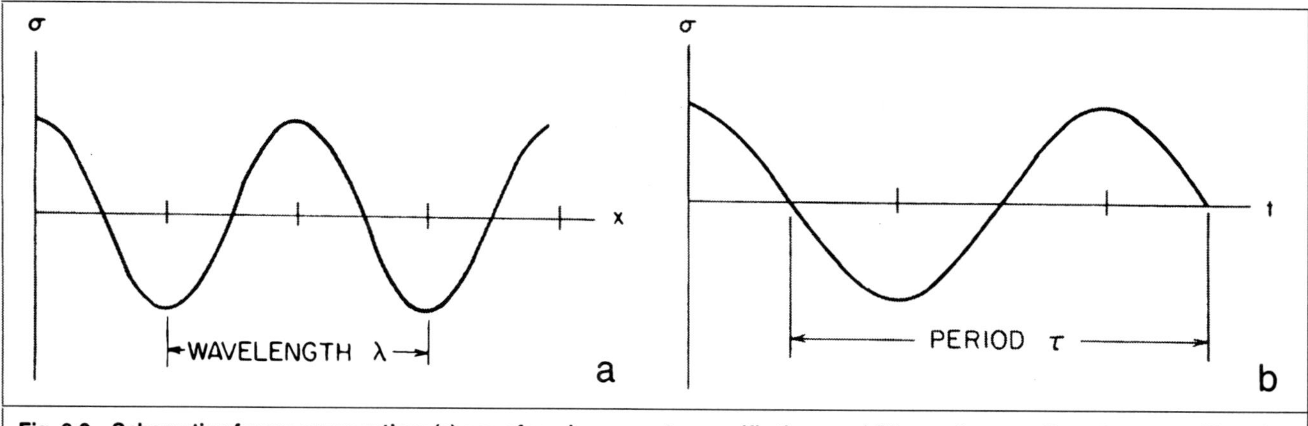

Fig. 3.2—Schematic of wave propagation: (a) waveform in space at a specific time; and (b) waveform vs. time at any specific point.

Because $G>0$ and $K>0$, then $v_p>v_s$, and because $0<\mu<0.5$, then

$$v_p>\sqrt{2}\,v_s, \quad (3.24)$$

and in terms of transit time, Δt,

$$\Delta t_s>\sqrt{2}\,\Delta t_p, \quad (3.25)$$

where Δt_p and Δt_s are the transit times of primary and secondary waves, respectively. Eqs. 3.24 and 3.25 indicate that compressional waves propagate faster than shear waves in an elastic medium. As an example, **Table 3.1** gives the velocities for quartz and calcite, two common minerals.

3.2.4 Reflection and Refraction of Elastic Waves. Acoustic (elastic) waves undergo interference, diffraction, reflection, and refraction. Reflection and refraction occur when a wave encounters a boundary separating two media with different elastic properties. Part of the energy of the incident wave is reflected and part is refracted. The incident wave may be converted into other types of vibrations upon reflection or refraction. This phenomenon is called mode conversion. For example, upon refraction, an incident compressional wave is partially converted into a wave or waves of another type, such as shear wave.

Fig. 3.4 schematically shows the geometry of the rays along which the acoustic waves propagate. Fig. 3.4 shows a wave with velocity v_1, incident at angle α_1, on a plane boundary separating two me-

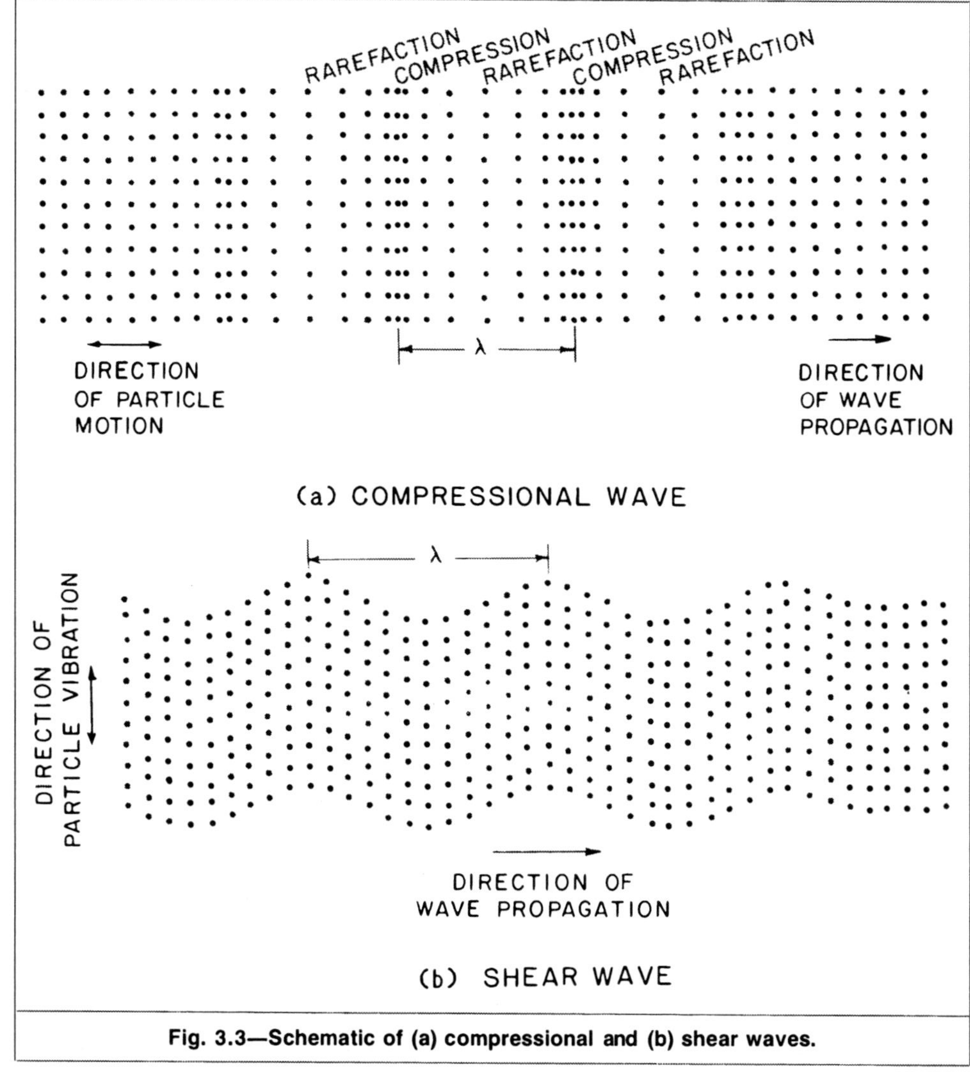

Fig. 3.3—Schematic of (a) compressional and (b) shear waves.

TABLE 3.1—ACOUSTIC PROPERTIES OF QUARTZ AND CALCITE

	Quartz	Calcite
Density, g/cm^3	2.649	2.712
v_p, ft/sec*	19,849	21,424
v_s, ft/sec*	13,419	11,024
v_p/v_s	1.479	1.943
Δt_p, μsec/ft**	50.4	46.7
Δt_s, μsec/ft**	74.5	90.7

*From Ref. 3.
**Using Eq. 3.1.

dia of different elastic characteristics. A wave with velocity v_2 is refracted into the second medium at angle α_2. A third wave with velocity v_3 is reflected back into the first medium at angle α_3. v_1, v_2, and v_3 are characteristic of the media and the types of waves. According to Snell's law,

$$\sin\alpha_1/v_1=\sin\alpha_2/v_2=\sin\alpha_3/v_3. \quad (3.26)$$

If the reflected wave is of the same type as the incident wave, then $v_1=v_3$ and subsequently $\alpha_1=\alpha_3$. The angle of refraction α_2 is always different from α_1 as $v_1 \neq v_2$. The angle α_2 is expressed as

$$\sin\alpha_2=(v_2/v_1)\sin\alpha_1. \quad (3.27)$$

For the special case where

$$\sin\alpha_1=v_1/v_2=\sin\alpha_c, \quad (3.28)$$

$\sin\alpha_2=1$ and $\alpha_2=90°$, as shown in **Fig. 3.5**. Angle α_c is the critical angle of refraction. The refracted wave does not penetrate the second medium but travels along the interface at velocity v_2. This critical refracted wave, called the head wave, propagates energy back into the first medium as it travels along the boundary.

If the incident angle is greater than the critical angle, no refraction will occur and the wave is totally reflected.

A compressional wave traveling in Medium 1 at a velocity v_{p1} will generate a compressional head wave in Medium 2 if its angle of incidence is critical. This critical angle, α_{pc}, is defined according to Eq. 3.28 as

$$\sin\alpha_{pc}=v_{p1}/v_{p2}. \quad (3.29)$$

A compressional wave traveling in Medium 1 will generate a shear head wave if its angle of incidence is critical. This critical angle, α_{sc}, is also defined according to Eq. 3.28 as

$$\sin\alpha_{sc}=v_{p1}/v_{s2}. \quad (3.30)$$

Combining Eqs. 3.30 and 3.29 results in

$$\sin\alpha_{sc}/\sin\alpha_{pc}=v_{p2}/v_{s2}. \quad (3.31)$$

Because v_{p2} is always greater than v_{s2} (Eq. 3.24), then

$$\alpha_{sc}>\alpha_{pc}. \quad (3.32)$$

Fig. 3.6 illustrates this geometry.

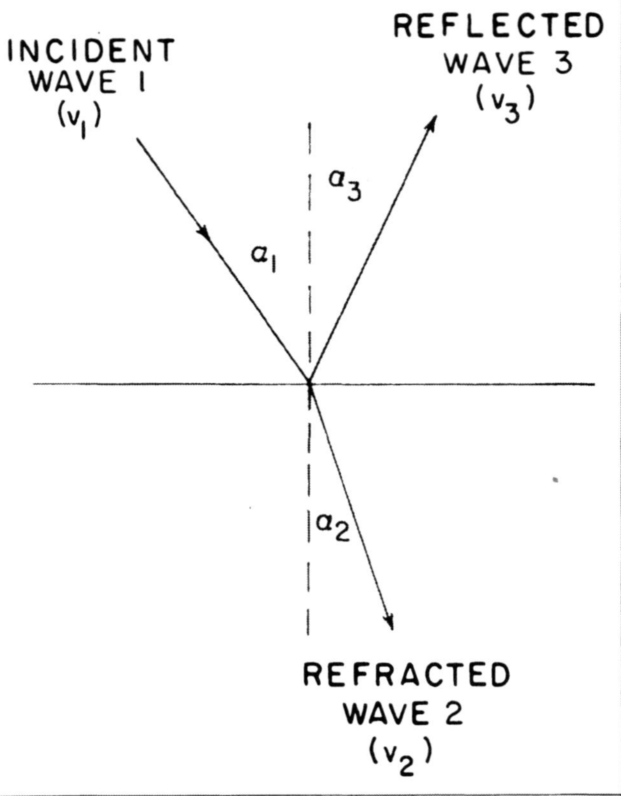

Fig. 3.4—Reflection and refraction of elastic waves at interface between two media of different properties.

Example 3.1. A compressional wave is traveling in a fluid at a velocity of 1830 m/s. The wave encounters a surface of a solid medium at an 18° angle. Part of the wave energy is reflected and part is refracted. The refracted wave is partially converted to a shear wave.

a. Sketch the geometry of the rays along which the reflected and refracted waves are propagated.

b. Calculate the angles of reflection and refraction if the velocities of the compressional wave and the shear wave in the solid are 4570 and 2740 m/s, respectively.

c. Calculate the angle of incidence that would result in a compressional head wave.

d. Calculate the angle of incidence that would result in a shear head wave.

Solution.

a. **Fig. 3.7** shows the geometry of the rays that pertain to this situation.

b. Angle of reflection α_{p3}: the reflected compressional wave has the same velocity as the incident wave, so the angle of reflection $\alpha_{p3}=\alpha_{p1}=18°$.

Angle of the refracted compressional wave α_{p2}: according to Snell's law,

$\sin\alpha_{p1}/v_{p1}=\sin\alpha_{p2}/v_{p2}$,

$\sin\alpha_{p2}=(4{,}570/1{,}830)\sin(18)=0.772$, and $\alpha_{p2}=50.5°$.

Angle of the refracted shear wave α_{s2}: according to Snell's law,

$\sin\alpha_{p1}/v_{p1}=\sin\alpha_{s2}/v_{s2}$,

$\sin\alpha_{s2}=(2{,}740/1{,}830)\sin(18)=0.463$, and $\alpha_{s2}=27.6°$.

c. Critical angle of refraction for compressional head wave, α_{pc}: using Eq. 3.29,

$\sin\alpha_{pc}=1{,}830/4{,}570=0.4$ and $\alpha_{pc}=23.6°$.

d. Critical angle of refraction for shear head wave, α_{sc}: using Eq. 3.30,

$\sin\alpha_{sc}=1{,}830/2{,}740=0.668$ and $\alpha_{sc}=41.9°$

3.3 Acoustic-Wave Propagation in Fluid-Filled Borehole

An exact treatment of acoustic-wave propagation in a fluid-filled borehole is quite involved and complex. The complexity arises from the relatively involved geometry, the presence of a tool in the borehole, and the porous nature of the media surrounding the borehole. The basic concept of acoustic-wave propagation is presented here in a simplified form. The treatment, however, will provide enough insight into acoustic logging physics to suit the purpose of identifying the components of an acoustic wave recorded in a borehole. Thorough and detailed treatment of the phenomenon is available in the literature.[4-9]

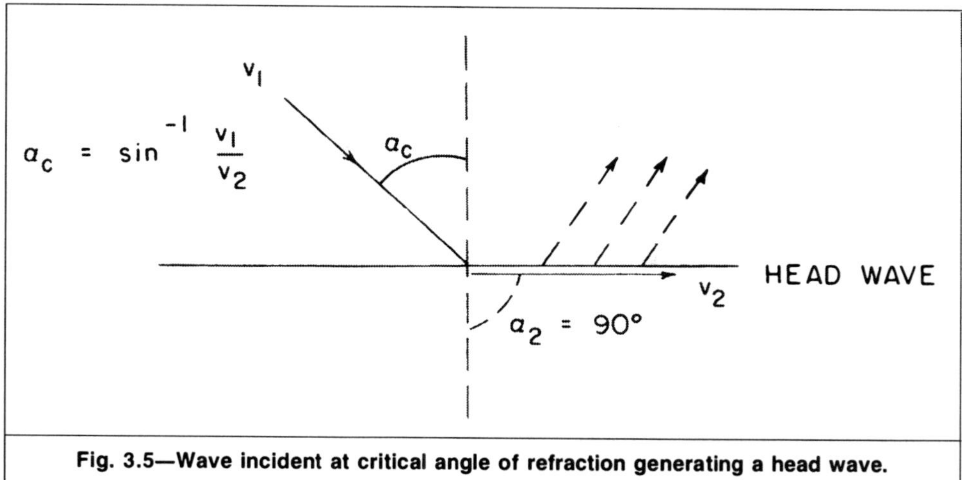

Fig. 3.5—Wave incident at critical angle of refraction generating a head wave.

The source of the acoustic signal is usually centered in the borehole, as illustrated in **Fig. 3.8**. Because fluids cannot sustain a shear stress, the source (transmitter, T) is a pressure transducer that generates compressional waves in the borehole fluid. The receiver, R, is another pressure transducer used to detect acoustic waves. Transducers are piezoelectric or magnetostrictive. Piezoelectric transducers are crystals of quartz or barium titanate that change sizes, thus generating an elastic wave when subjected to an electric field. They generate an electric signal when subjected to a stress. Magnetostrictive transducers are metal alloys, such as nickel/cobalt/iron, that change sizes when subjected to a magnetic field and generate a magnetic field when subjected to a stress. A transmitter transducer excited by an electric or magnetic pulse emits an acoustic wave in the borehole. A receiver transducer excited by an acoustic wave generates a voltage or a magnetic field that is converted to voltage and then transmitted to the surface by cable.

The received signal is a composite of several acoustic waves called a waveform. The main components of the waveform are, in order of their arrival at the receiver, (1) the compressional head wave, (2) the shear head wave, (3) pseudo-Rayleigh or conical waves, and (4) Stoneley or tube waves.

3.3.1 Compressional Head Wave. The first wave to arrive at the receiver is the compressional head wave, which is critically refracted at the borehole wall. Fig. 3.8 shows its path. The wave begins as a compressional wave generated in the borehole by the transmitter transducer. It encounters the borehole wall at an angle α_{pc} and is critically refracted into the formation as a compressional head wave. The head wave travels along the borehole wall with the formation compressional velocity, v_p. As the head wave travels along the borehole fluid/formation interface, energy is continuously radiated back into the borehole as compressional vibrations in the borehole fluid. These vibrations, also propagating at a critical angle, α_{pc}, are sensed by the receiver centered in the borehole.

The compressional wave arrives at the receiver first because of a combination of velocity and path effects. Three velocities play roles in wave propagation in a fluid-filled borehole: the formation compressional-wave velocity, v_p; the formation shear-wave velocity, v_s; and the fluid compressional-wave velocity, v_f. According to Eq. 3.24, the compressional wave propagates the fastest in the formation, and according to Eq. 3.32, this wave also has the shortest path through the drilling fluid, where acoustic velocity is the slowest.

3.3.2 Shear Head Wave. The shear head wave also begins as a compressional wave in the borehole fluid. It encounters the wall at the critical refraction angle, α_{sc}, and thus refracts into the formation as a shear wave (Fig. 3.8). Like the compressional head wave, the shear head wave travels along the borehole wall at a velocity v_s and continuously sheds compressional waves into the borehole fluid at the critical angle α_{sc}. Because $v_s < v_p$ and because it has a longer path in the borehole fluid, the shear head wave arrives at the receiver behind the compressional head wave. If $v_f > v_s$, such as in extremely unconsolidated formations, shear waves cannot be critically refracted along the borehole wall and the shear head wave is not generated.

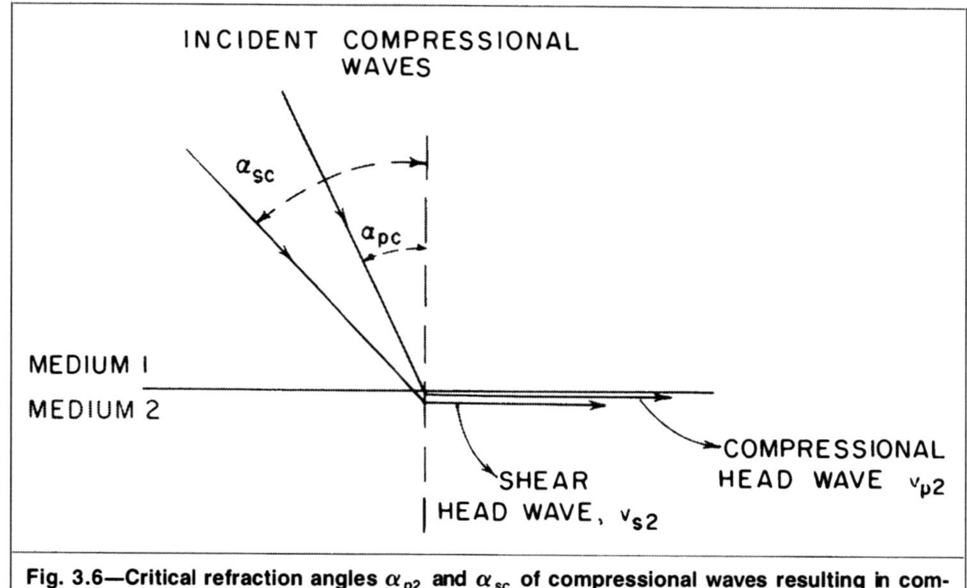

Fig. 3.6—Critical refraction angles α_{p2} and α_{sc} of compressional waves resulting in compressional and shear head waves.

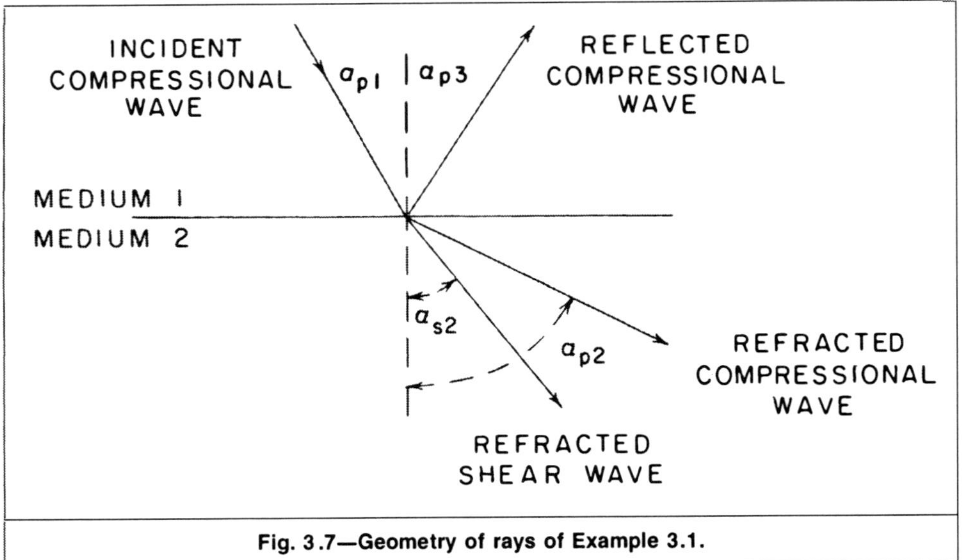

Fig. 3.7—Geometry of rays of Example 3.1.

Under certain geometry and velocity conditions, another type of wave, known as the "hybrid" or "leaky" mode, is observed between the compressional- and shear-wave arrivals.[9] The leaky mode arises from compressional waves incident at the borehole wall with angles between α_{pc} and α_{sc}. These compressional waves undergo both reflection and conversion into refracted shear waves, as **Fig. 3.9** shows. The reflected wave propagates in the borehole fluid as a conical wave. The amplitude of this mode decreases with distance from the transmitter as energy is leaked to the formation at each reflection. The spacing between the transmitter and receiver could be designed long enough so that the remaining energy of the mode at the receiver is too low to be detectable.

Example 3.2. An acoustic transmitter and receiver are placed 5 ft apart on the axis of a cylindrical borehole 8 in. in diameter. Calculate the times of arrival of the compressional and shear head waves if the acoustic velocities are as follows: formation $v_p = 4570$ m/s, formation $v_s = 2740$ m/s, and fluid $v_f = 1830$ m/s.

Solution. *Calculation of interval travel times.* Converting velocities from m/s to ft/s gives

$v_p = 4570(3.28084) = 14{,}993$ ft/sec,

$v_s = 2740(3.28084) = 8{,}990$ ft/sec,

and $v_f = 1830(3.28084) = 6{,}004$ ft/sec.

From Eq. 3.1,

$\Delta t_p = 10^6/14{,}993 = 66.7$ μsec/ft,

$\Delta t_s = 10^6/8{,}990 = 111.2$ μsec/ft,

and $\Delta t_f = 10^6/6{,}004 = 166.6$ μsec/ft.

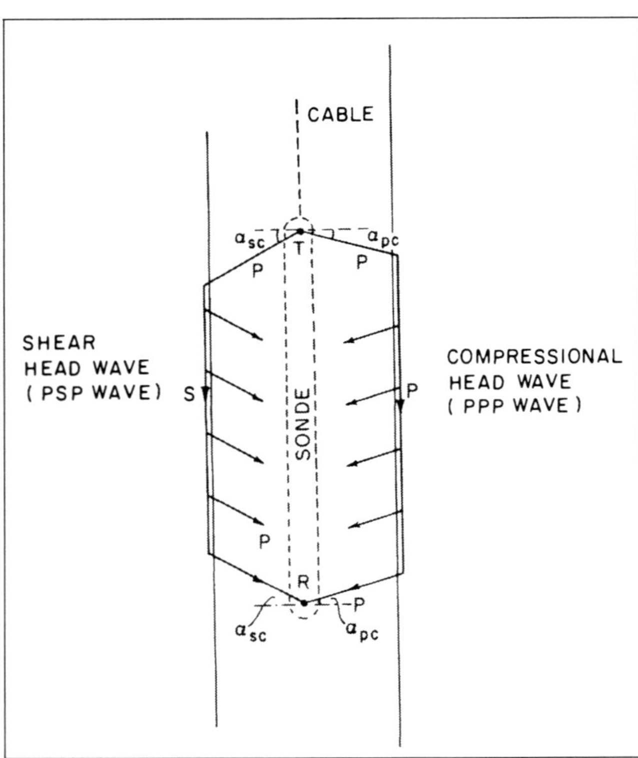

Fig. 3.8—Ray path of compressional- and shear-wave propagation within a fluid-filled borehole.

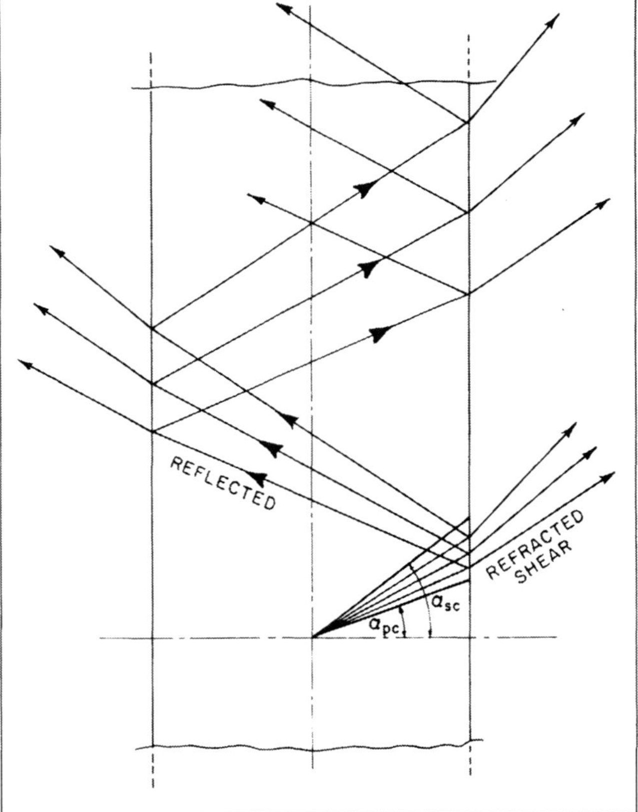

Fig. 3.9—Schematic showing ray paths of the "leaky" mode (after Ref. 9).

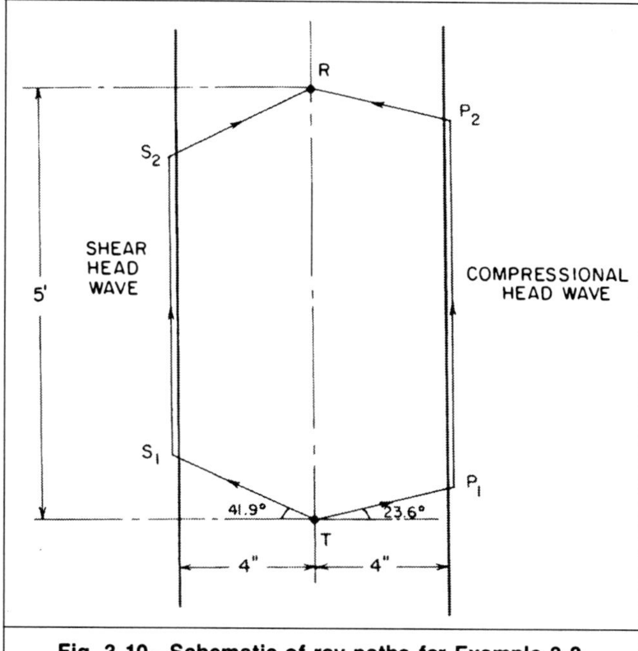

Fig. 3.10—Schematic of ray paths for Example 3.2.

Wave path geometry (see **Fig. 3.10**). From Example 3.1, $\alpha_{pc}=23.6°$ and $\alpha_{sc}=41.9°$.

$\overline{TP_1}=\overline{P_2R}=(4/12)/\cos(23.6)=0.364$ ft,

$\overline{P_1P_2}=5-2(4/12)\tan(23.6)=4.709$ ft,

$\overline{TS_1}=\overline{S_2R}=(4/12)/\cos(41.9)=0.448$ ft,

and $\overline{S_1S_2}=5-2(4/12)\tan(41.9)=4.402$ ft.

Arrival time of compressional head wave along Path TP_1P_2R:

$\Delta t_p=2(0.364)(166.6)+4.709(66.7)=435$ μsec.

Arrival time of shear head wave along Path TS_1S_2R:

$\Delta t_s=2(0.448)(166.6)+4.402(111.2)=639$ μsec.

3.3.3 Pseudo-Rayleigh Waves. Pseudo-Rayleigh waves arise from the portion of the wave incident on the borehole wall at angles greater than the critical angle of refraction of shear wave, α_{sc}. This wave undergoes total reflection and propagates within the borehole as a conical wave, as illustrated in **Fig. 3.11**. This conical wave is called the pseudo-Rayleigh wave by analogy to the Rayleigh waves that propagate along a plane interface.[2] This wave type is also referred to as trapped mode, guided wave, and normal mode. Because no energy is lost to refraction outside the cylindrical surface, the pseudo-Rayleigh waves exhibit relatively large amplitudes.

The velocity of pseudo-Rayleigh waves varies with frequency. The velocity ranges between about nine-tenths of the fluid velocity at high frequency and the formation shear velocity at low frequency. The variation of velocity with frequency is known as dispersion. Because different frequencies travel at different speeds, a dispersive wave appears as a train of events with successive cycles that have increasing or decreasing periods. A cutoff frequency exists below which pseudo-Rayleigh waves will not be generated.

3.3.4 Stoneley Waves. The Stoneley wave is a surface wave at the fluid/formation interface. This wave type is named by analogy to a surface wave discovered by Stoneley on the plane interface between two semi-infinite elastic solids.[9] The Stoneley waves generated in a borehole are also called tube waves and guided waves.

Stoneley waves are also dispersive, but the dispersion is much milder than in pseudo-Rayleigh waves. There is no cutoff frequency, and the velocity is always less than the borehole fluid velocity. If $v_s>v_f$, Stoneley wave velocity is about nine-tenths of the fluid velocity. This fact has resulted in the general erroneous identification of the Stoneley wave as the fluid or mud wave.[10]

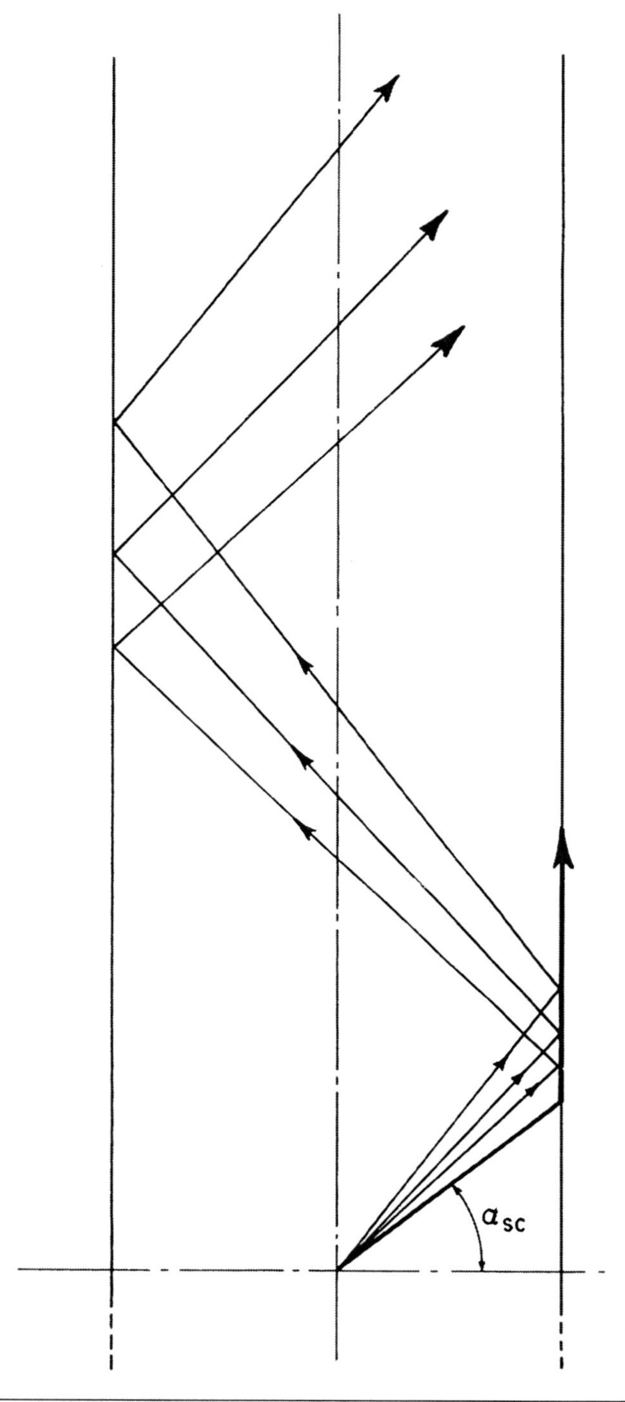

Fig. 3.11—Schematic showing ray paths of the pseudo-Rayleigh wave (after Ref. 9).

3.3.5 Full-Wave Acoustic Signal. The four main waves discussed previously interact in a complicated manner as they propagate from the transmitter to the receiver where they produce the observed acoustic signal. **Fig. 3.12** is a schematic of a typical wave train observed in acoustic logging. Point P marks the energy that is first sensed by the receiver. The first to arrive is the compressional, or primary, wave, followed by the leaky mode if it is present in the wave train. Point S marks the first arrival of shear, or secondary, waves. The guided waves (i.e., pseudo-Rayleigh and Stoneley waves) follow the arrival of shear waves. The characteristic of the wave train depends on transmitter frequency, tool design, borehole radius, and formation and fluid properties. A pseudo-Rayleigh wave propagating at cutoff frequency arrives simultaneously with the shear wave, and its amplitude dominates the shear wave's amplitude. Because of its dispersive nature, several pseudo-Rayleigh waves are

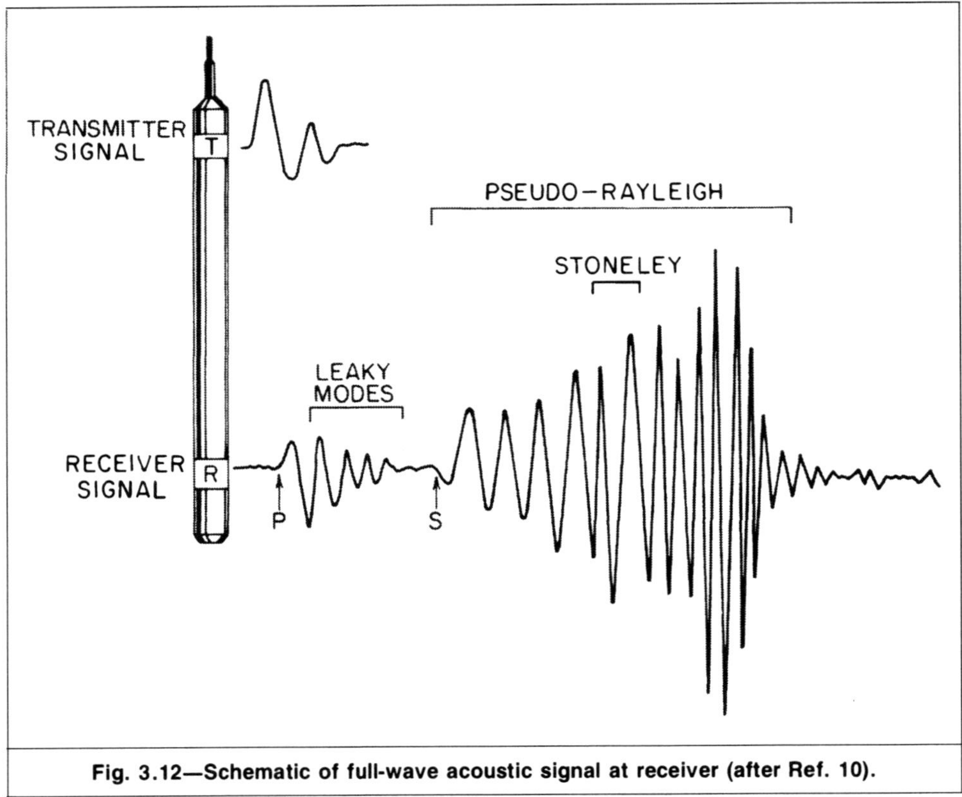

Fig. 3.12—Schematic of full-wave acoustic signal at receiver (after Ref. 10).

present in the wave train. The Stoneley wave appears within the pseudo-Rayleigh train of modes. Because of its low dispersion, all frequencies of a Stoneley wave propagate at the same velocity. Consequently, it usually appears as an impulse.[10] All waves are present in the receiver signal if the formation has a shear velocity higher than the fluid velocity. If the formation shear velocity is less than the borehole fluid velocity, only the primary and Stoneley waves are present.

The different features of the acoustic wave train can be used to determine the acoustic velocity, attenuation coefficient, amplitude, and frequency of the different waves. Different logging tools were developed to record one or more of these parameters. In turn, one

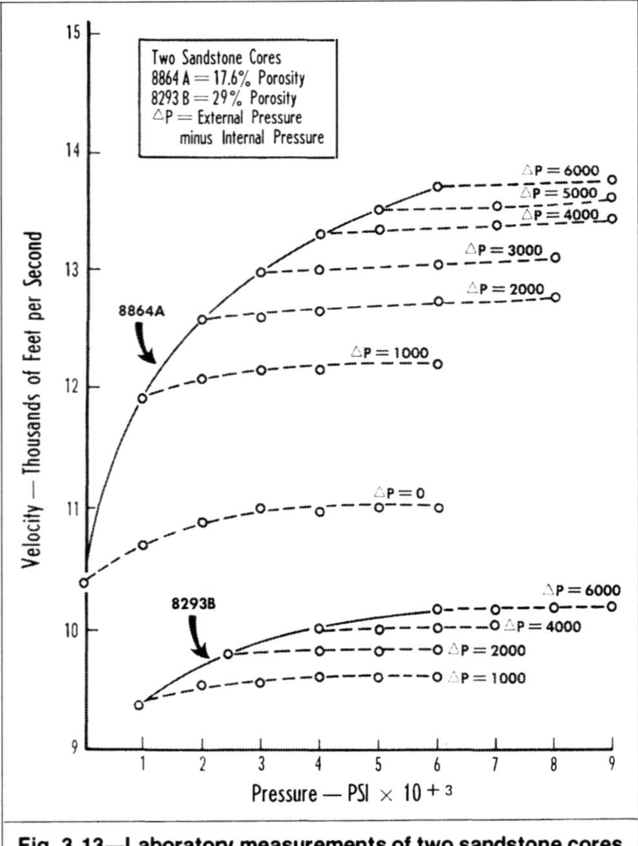

Fig. 3.13—Laboratory measurements of two sandstone cores showing acoustic velocity as a function of the difference between external and internal pressure (from Ref. 20).

TABLE 3.2—ACOUSTIC COMPRESSIONAL VELOCITIES AND TRANSIT TIMES IN ROCK MATRICES OF INTEREST IN WELL LOGGING (from Ref. 14)

Material	v_p (ft/sec)	Δt (μsec/ft)
Sandstone	18,000 to 19,500	55.5 to 51.0
Limestone	21,000 to 23,000	47.6 to 43.5
Dolomite	23,000	43.5
Anhydrite	20,000	50
Shale	5,900 to 17,000	170 to 60
Salt	15,000	66.7

TABLE 3.3—ACOUSTIC COMPRESSIONAL VELOCITIES AND TRANSIT TIMES FOR FLUIDS OF INTEREST IN WELL LOGGING (from Refs. 17 and 18)

Fluid	v_p (ft/sec)	Δt (μsec/ft)
Water		
200 kppm, 15 psia	5,540	180.5
150 kppm, 15 psia	5,375	186.0
100 kppm, 15 psia	5,200	192.3
Pure	4,380	207.0
Drilling mud (26°C)	4,870	205.3
Drilling-mud cake (26°C)	4,980	200.8
Oil	4,200	238.0
Methane (15 psia)	1,600	626.0
Air (15 psia)	1,088	919.0
Ethane (ρ = 0.00125 g/cm^3)	1,010	989.6
Carbon dioxide (ρ = 0.0019776 g/cm^3)	850	1,176.5

parameter or a combination of parameters is used to estimate different formation properties, such as porosity value, porosity type, lithology, and mechanical properties.

The conventional sonic log tool discussed in Chap. 10 measures an interval travel time, Δt, that is the reciprocal of the velocity of compressional wave, v_p. The interval transit time is derived from the time of the first arrival at the receiver.

3.4 Acoustic-Wave Propagation in Rocks

Theoretical expressions of v_p and v_s given by Eqs. 3.18 and 3.20 were derived assuming an infinite, isotropic, homogeneous, elastic medium. Application of these relationships to sedimentary rocks is complicated by the presence of liquid-filled pores, fractures, and/or vugs. Gassman made the first attempt to formulate a theoretical expression for acoustic velocities in porous media.[11] Gassmann's model considers the elastic properties of (1) the matrix, which is taken to be a homogeneous, isotropic, elastic solid material; (2) the dry rock consisting of the matrix and the excavated pore space; and (3) the pore fluid. Gassmann's theory assumes that, at low frequency, relative motion between fluid and rock matrix can be ignored.

The fluid-filled composite rock elastic properties are expressed in terms of the model's components by[9]

$$G_b = G_d, \quad \text{(3.33)}$$

$$\rho_b = (1-\phi)\rho_{ma} + \phi\rho_f, \quad \text{(3.34)}$$

$$\text{and } K_b = K_d + \frac{[1-(K_d/K_{ma})]^2}{(\phi/K_f)+[(1-\phi)/K_{ma}]-(K_d/K_{ma}^2)}, \quad \text{(3.35)}$$

where the subscripts *b*, *ma*, *f*, and *d* refer to the bulk, matrix, fluid, and dry rock, respectively. Note that dry-rock properties G_d and K_d are functions of ϕ.

Substituting Eqs. 3.33 through 3.35 into Eq. 3.18 yields the following expressions for compressional- and shear-wave velocities:

$$v_p^2 = \frac{1}{\rho_b}\left[K_d + \frac{[1-(K_d/K_{ma})]^2}{(\phi/K_f)+[(1-\phi)/K_{ma}]-(K_d/K_{ma}^2)} + \tfrac{4}{3}G_d\right] \quad \text{(3.36)}$$

$$\text{and } v_s^2 = G_d/\rho_b. \quad \text{(3.37)}$$

In Eq. 3.36, replacing the bulk modulus constants with the corresponding compressibility yields

$$v_p^2 = \frac{1}{\rho_b}\left[\frac{\beta}{c_{ma}} + \frac{(1-\beta)^2}{\phi c_f + (1-\phi-\beta)c_{ma}} + \tfrac{4}{3}G_d\right], \quad \text{(3.38)}$$

$$\text{where } \beta = K_d/K_{ma} = c_{ma}/c_d \quad \text{(3.39)}$$

and c_{ma}, c_d, and c_f are the compressibilities of the matrix material, dry bulk rock, and pore fluid, respectively.

Combining Eqs. 3.11 and 3.12 yields

$$G_d = \frac{3K_d(1-2\mu)}{2(1+\mu)} = \frac{3(1-2\mu)}{2c_d(1+\mu)}, \quad \text{(3.40)}$$

which, by taking an average value of Poisson's ratio $\mu=0.2$, reduces to

$$G_d = 3K_d/4 \quad \text{(3.41)}$$

and can be written in terms of compressibility as

$$G_d = 3/4c_d = 3\beta/4c_{ma}. \quad \text{(3.42)}$$

Substituting Eq. 3.42 into Eq. 3.38 gives

$$v_p^2 = \frac{1}{\rho_b}\left[\frac{2\beta}{c_{ma}} + \frac{(1-\beta)^2}{\phi c_f + (1-\phi-\beta)c_{ma}}\right]. \quad \text{(3.43)}$$

Similarly, v_s can be expressed as

$$v_s^2 = 3/(4c_d\rho_b). \quad \text{(3.44)}$$

Biot[12,13] developed a more comprehensive theory of elastic wave propagation in fluid-saturated porous solids. This theory computes v_p and v_s at all frequencies. Below a certain critical frequency, the difference in velocities given by Biot's expression and by Eqs. 3.36 and 3.37 is negligible. Conventional logging frequencies[9] usually fall below this critical limit.

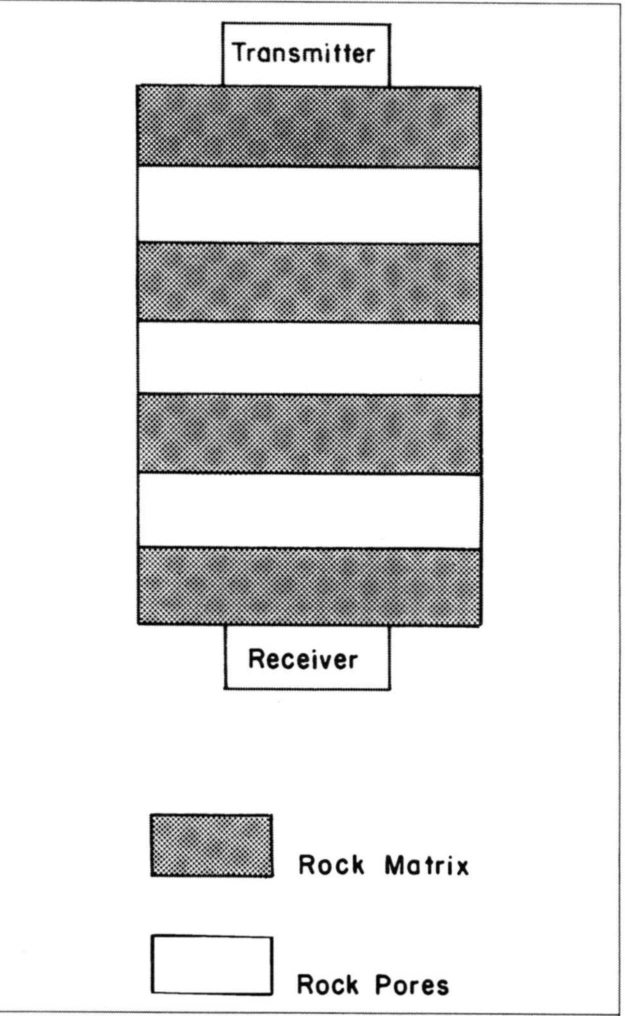

Fig. 3.14—Idealized model used to derive the time-average equation (after Ref. 20).

Compressional velocity of acoustic waves in porous rocks depends on lithology of rock type, differential pressure, pore-fluid type, and porosity. As indicated by Eq. 3.36, the acoustic velocity depends on the elastic properties of the matrix. **Table 3.2** gives acoustic compressional velocities and transit times in rock matrices and fluids of interest in well logging. Values are derived from field observation and thus do not refer to pure minerals. Most of the entries are presented as ranges because of dependence on composition, degree of compaction, and pressure.

A rock is subjected to external (overburden) and internal (pore fluid) pressures. Laboratory measurements,[15,16] such as those in **Fig. 3.13**, have shown that the acoustic velocity, v_p, is affected by the effective stress, which is the difference between external and internal pressures, Δp. Fig. 3.13 indicates that the velocity increases as Δp increases. The velocity is not affected by the absolute value of the external pressure or the internal pressure. Because Δp determines the degree of rock compaction and its bulk modulus, it follows that acoustic velocity depends on compaction. The more compacted the rock, the higher its acoustic velocity is.

The velocity of acoustic waves depends also on the fluid density and bulk modulus. **Table 3.3** lists acoustic compressional velocities and transit times for fluids of interest in well logging.

Acoustic velocity in water depends on temperature, salinity, and pressure. The variation of velocity caused by these factors is about 500 ft/sec under normal well-logging conditions. Acoustic velocity in oil is somewhat less than that in water; it depends on oil composition, gas in solution, temperature, and pressure. The velocity in natural gas is considerably less than in oil and water. The acoustic

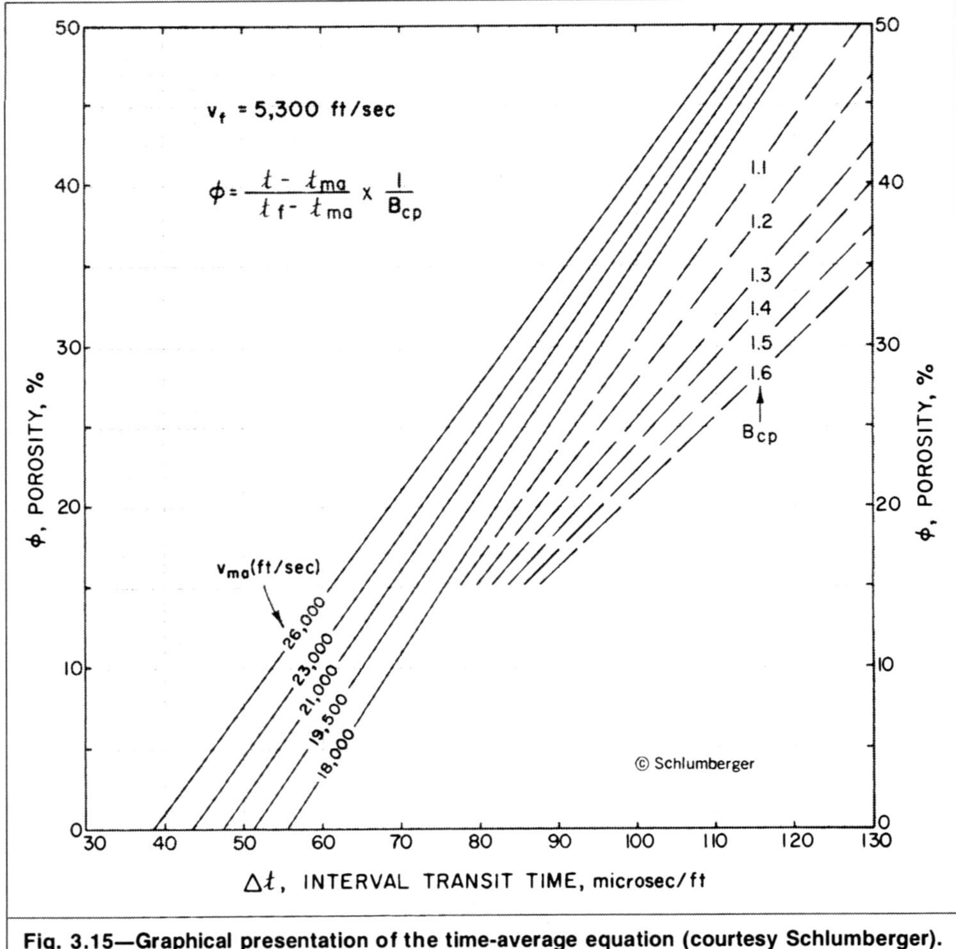

Fig. 3.15—Graphical presentation of the time-average equation (courtesy Schlumberger).

velocity in dry natural gas approaches that in air, which is about 1,100 ft/sec.

The dependence of acoustic velocity on porosity is self-evident: the greater the porosity, the lower the velocity because the amount of low-velocity fluids will be greater and the matrix path through which the wave must travel will be more tortuous. The porosity appears explicitly and implicitly in Eq. 3.36; it controls the value of K_d and G_d.

3.5 Porosity/Transit-Time Relationships

3.5.1 Time-Average Relation in Compacted Formations. A good correlation often exists between porosity and acoustic interval travel time. The time-average equation advocated by Wyllie *et al.*[18] and suggested earlier by Hughes and Jones[19] has been very popular with log analysts. **Fig. 3.14** illustrates the idealized model required by the time-average equation. The model consists of a layered system of parallel slices alternating a solid and a liquid and then crossed by a wave path perpendicular to the interfaces. According to this model, the total travel time is equal to the sum of the signal's travel time through the pore-fluid fraction plus the travel time through the rock-solid fraction. The equation is written as

total travel time = travel time in liquid fraction + travel time in matrix fraction,

$$\frac{1}{v_b}=\frac{\phi}{v_f}+\frac{1-\phi}{v_{ma}} \text{ or}$$

$$\Delta t=\Delta t_f\phi+\Delta t_{ma}(1-\phi). \quad (3.45)$$

Solving for porosity yields

$$\phi=(\Delta t-\Delta t_{ma})/(\Delta t_f-\Delta t_{ma}). \quad (3.46)$$

The matrix and fluid interval transit times commonly used in this formula are as follows.

Material	v_{ma} (ft/sec)	Δt_{ma} (μsec/ft)
Sandstone	18,000	55.5
Limestone	21,000	47.5
Dolomite	23,000	43.5
Fluid	5,300	189

Fig. 3.15 is a graphical presentation of the time-average relationship.

Unfortunately, the time-average model is based on the wrong physical picture. It suggests that only rock matrix and fluid properties influence wave velocity. Nevertheless, the use of this equation in clean, compacted, and consolidated sandstones yields representative porosity values.[21] The equation is unsuitable, however, for use in uncompacted formations and carbonates.

3.5.2 Time-Average Relation in Uncompacted Formations. Application of Eq. 3.46 to uncompacted formations where differential pressure is <4,000 psia yields porosity values that are too high. Lack of compaction is usually indicated when the interval transit time of adjacent shales, Δt_{sh}, exceeds 100 μsec/ft. Tixier *et al.*[22] suggested the introduction of a compaction correction factor, B_{cp}, in Eq. 3.46:

$$\phi=[(\Delta t-\Delta t_{ma})/(\Delta t_f-\Delta t_{ma})]/B_{cp}. \quad (3.47)$$

Fig. 3.15 graphically shows Eq. 3.47. The compaction correction factor can be estimated from the empirical relation

$$B_{cp}=\Delta t_{sh}/100, \quad (3.48)$$

where Δt_{sh} is the transit time in adjacent shales. The value of B_{cp} can also be determined in a given interval by comparing a porosity value derived in a clean-water formation from other logs to that obtained from Eq. 3.46.

3.5.3 Geertsma's Model. Eqs. 3.36 through 3.39 display the effect of the properties of rock matrix, rock bulk material, and pore

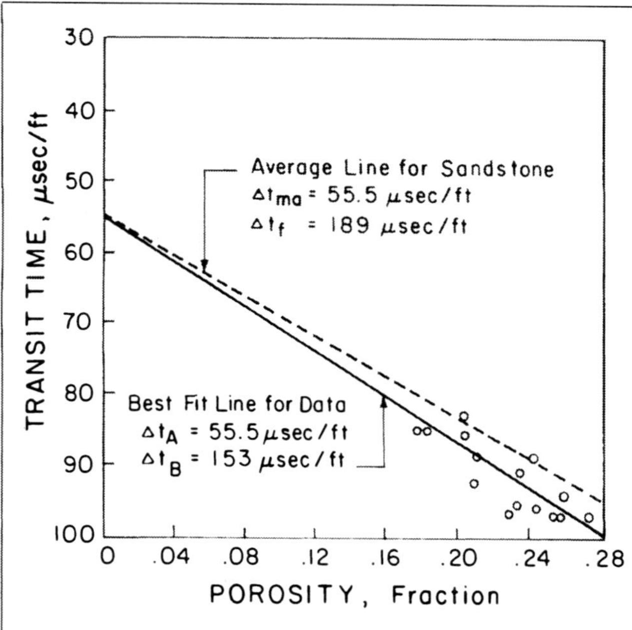

Fig. 3.16—Transit time vs. porosity obtained for Delaware sand (after Ref. 20).

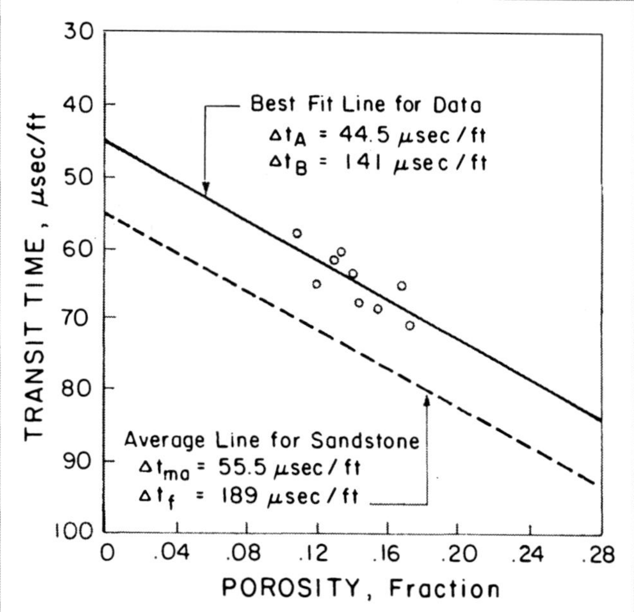

Fig. 3.17—Transit time vs. porosity obtained in siliceous sand (after Ref. 20).

fluid on the wave velocity. Because rock bulk compressibility is a function of porosity, pore-size distribution, composition of the cementing material, type of sedimentary rock, and effective stress, these equations present only a partial solution for the problem of acoustic velocity interpretation in terms of porosity.

Using the classic theory of elasticity, Geertsma[21] suggested a relationship between Δt and ϕ. Replacing E expressed by Eq. 3.12 in Eq. 3.19 results in

$$v_b=\{(3K_b/\rho_b)[(1-\mu_b)/(1+\mu_b)]\}^{1/2}, \quad (3.49)$$

and $$v_{ma}=\{(3K_{ma}/\rho_{ma})[(1-\mu_{ma})/(1+\mu_{ma})]\}^{1/2}, \quad (3.50)$$

where v_b and v_{ma} are the acoustic velocities in bulk formation and matrix material, respectively. Assuming that Poisson's ratio is independent of porosity (i.e., $\mu_{ma} \cong \mu_b$), v_b can be expressed in terms of v_{ma} as

$$v_b=v_{ma}\left(\frac{\rho_{ma}K_b}{\rho_bK_{ma}}\right)^{1/2}=v_{ma}\left(\frac{\rho_{ma}c_{ma}}{\rho_bc_b}\right)^{1/2} \quad (3.51)$$

Using Van der Knapp's[23] definitions of bulk and pore compressibility for a model of a solid with sufficiently isolated holes distributed at random yields

$$c_b=c_{ma}+\phi c_p \quad (3.52)$$

$$=c_{ma}+C\phi/(1-\phi), \quad (3.53)$$

where c_p is the pore compressibility and C is a proportionality constant. Substituting Eq. 3.53 into Eq. 3.51 and expressing ρ_b in terms of ρ_{ma} gives

$$\rho_b=\rho_{ma}(1-\phi). \quad (3.54)$$

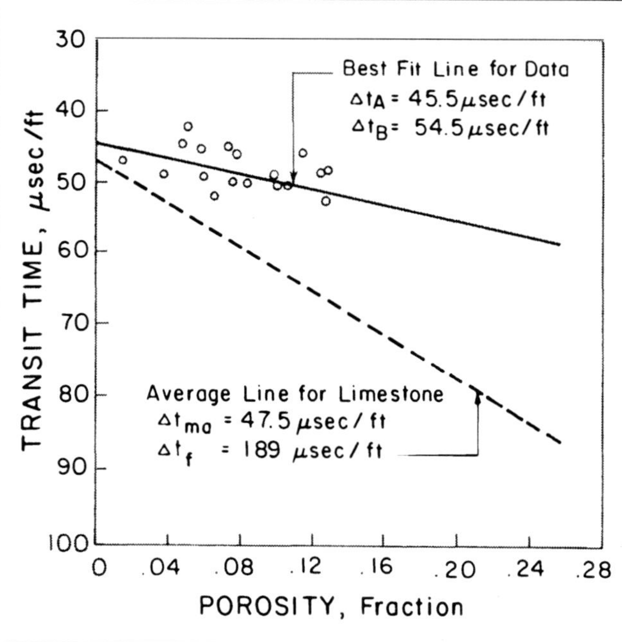

Fig. 3.18—Transit time vs. porosity obtained in a Smackover oolite limestone (after Ref. 20).

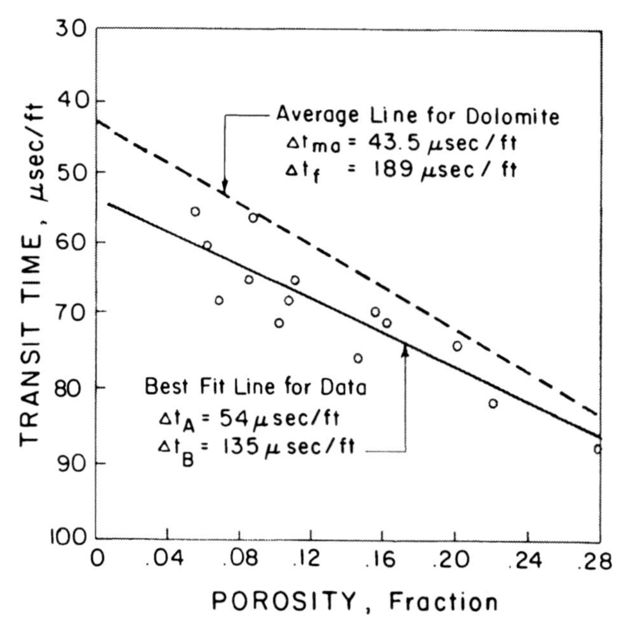

Fig. 3.19—Transit time vs. porosity obtained in a dolomite formation after (Ref. 20).

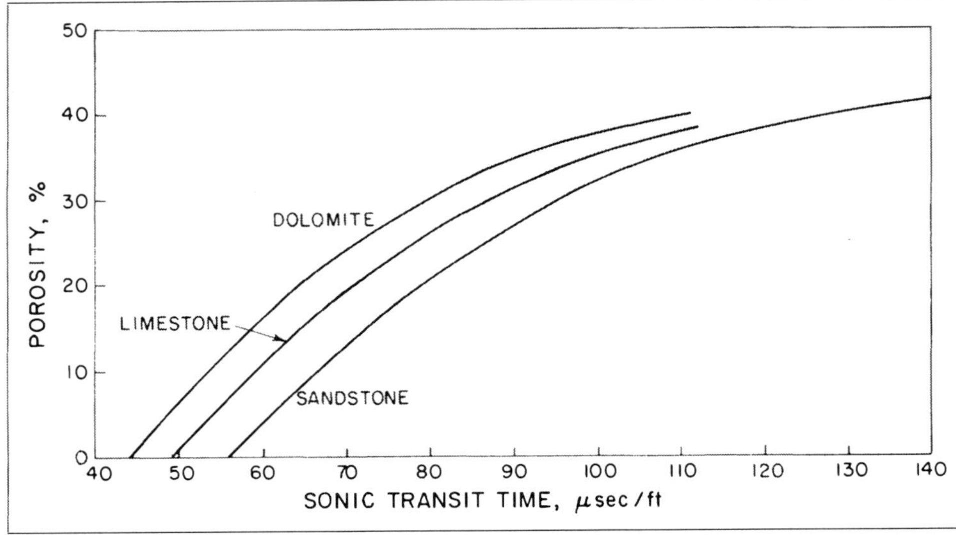

Fig. 3.20—Raymer-Hunt $\Delta t/\phi$ transform showing response in sandstone, limestone, and dolomite (after Ref. 24).

This results in

$$\frac{1}{v_b}=\frac{1}{v_{ma}}\left[1+\left(\frac{C}{c_{ma}}-1\right)\phi\right]^{1/2}, \quad (3.55)$$

which, for small porosity values, can be further approximated as

$$\frac{1}{v_b}=\frac{1}{v_{ma}}\left[1+1/2\left(\frac{C}{c_{ma}}-1\right)\phi\right]. \quad (3.56)$$

This equation is of the form

$$\Delta t=\Delta t_A+\Delta t_B\phi. \quad (3.57)$$

The coefficients Δt_A and Δt_B do not correspond to well-defined physical parameters. They have to be determined empirically for each formation of interest. If the rock matrix and saturated rock are assumed to have the same μ, then $\Delta t_A=\Delta t_{ma}$. The coefficient Δt_B depends strongly on the value of pore compressibility but is not tied to the presence of a fluid in the pores. Because of implied assumptions, this model is most suited to vuggy carbonate rocks.

Laboratory results from experiments on different rock types, however, can be better explained by Eq. 3.57, as shown by **Figs. 3.16 through 3.19**. Fig. 3.16 shows data from the Delaware sand. This sand is friable and very fine grained with little cementing material. Observed transit time deviates from the time-average equation line for sandstone—i.e., $\Delta t_{ma}=55.5$ μsec/ft and $\Delta t_f=189$ μsec/ft. An equation of the form proposed by Geertsma where $\Delta t_A=55.5$ μsec/ft and $\Delta t_B=153$ μsec/ft fits the data better. In this case, $\Delta t_A=\Delta t_{ma}$ for sands—i.e., 55 μsec/ft.

Fig. 3.17 shows data from an offshore Louisiana deep Miocene sand. The rock is very clean and very well cemented with siliceous material. As can be seen, the transit time deviates considerably from the time-average equation. A best fit is obtained from Eq. 3.57 with $\Delta t_A=44.5$ μsec/ft and $\Delta t_B=141$ μsec/ft. In this case, $\Delta t_A\neq 55.5$ μsec/ft. Figs. 3.18 and 3.19 show similar data and results in a Smackover well-cemented oolitic limestone and a dolomite formation, respectively.

The time-average relation given by Eq. 3.45 can also be considered an approximation of the general theory that led to Eq. 3.57. Eq. 3.45 can be written as

$$\Delta t=\Delta t_{ma}+\phi(\Delta t_f-\Delta t_{ma}), \quad (3.58)$$

which is of the same form as Eq. 3.57, where

$$\Delta t_A=\Delta t_{ma} \quad (3.59)$$

and $\Delta t_B=\Delta t_f-\Delta t_{ma}$. (3.60)

Eq. 3.47, which is a time-average relation modified to account for compaction variations, is also of the same form as Eq. 3.57, where Δt_A is given by Eq. 3.59 and Δt_B is expressed as

$$\Delta t_B=B_{cp}(\Delta t_f-\Delta t_{ma}). \quad (3.61)$$

3.5.4 Raymer-Hunt Transform. Raymer and Hunt[24] proposed an empirical transform of Δt to ϕ (**Fig. 3.20**) based on extensive field observations. Although totally empirical in origin, the transform does not contradict theoretical acoustic-wave propagation considerations.

In the range of porosities germane to well-log interpretations, the transform can be expressed by the algorithm

$$\Delta t=\{[(1-\phi)^2/\Delta t_{ma}]+(\phi/\Delta t_f)\}^{-1}. \quad (3.62)$$

The matrix and fluid properties suggested for use in this equation are given below.

Material	v (ft/sec)	Δt (μsec/ft)
Sandstone	17,850	56
Limestone	20,500	49
Dolomite	22,750	44
Fluid	5,300	189

Fig. 3.21 compares the time-average equation and the Raymer-Hunt transform. Because the transform covers part of an uncompacted formation range, it is suggested that it can be used for uncompacted formations without a correction factor.[24] **Fig. 3.22** indicates, however, that Gulf of Mexico unconsolidated sands do not conform to this transform.[25]

Example 3.3. An interval transit time of 70 μsec/ft is measured in a zone of interest in the siliceous sand of Fig. 3.17.

a. Calculate the porosity of the zone using the time-average equation, the Raymer-Hunt transform, and Geertsma's model.

b. Comment on the deviation between the porosity values obtained in Part a, knowing that m, n, R_w, and R_t values of the zone are 2, 2, 0.45 Ω·m, and 3.7 Ω·m, respectively.

Solution.

a. Using the time-average equation yields

$\phi=(70-55.5)/(189-55.5)=0.109$ (11%).

From the Raymer-Hunt transform and Fig. 3.20, $\phi=13\%$. From Fig. 3.17, the empirical correlation of the form $\Delta t=\Delta t_A+\Delta t_B\phi$ is

$\Delta t=44.5+141\phi$;

$\phi=(\Delta t-44.5)/141=(70-44.5)/141=0.18$ (18%).

b. Using each of the three porosity values to estimate water saturation from Eq. 1.44 yields

$S_w=100\%$ (using $\phi=11\%$),

$S_w=85\%$ (using $\phi=13\%$),

and $S_w=61\%$ (using $\phi=18\%$).

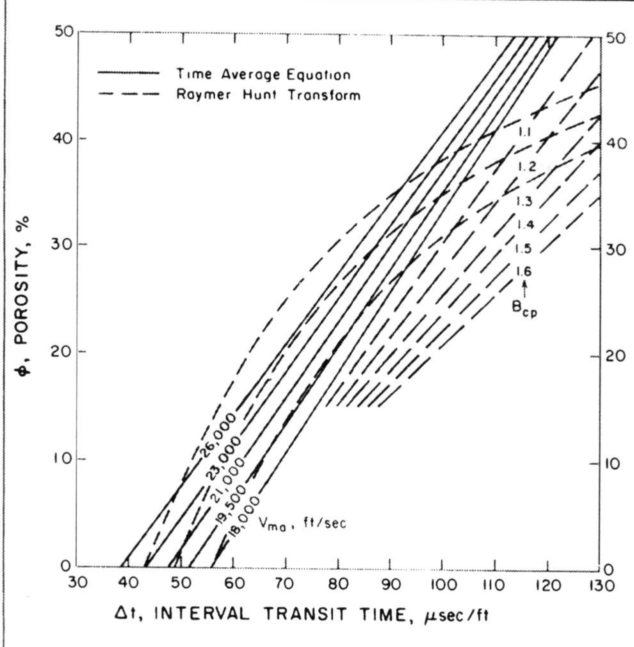

Fig. 3.21—Raymer-Hunt transform compared with time-average equation (courtesy Schlumberger).

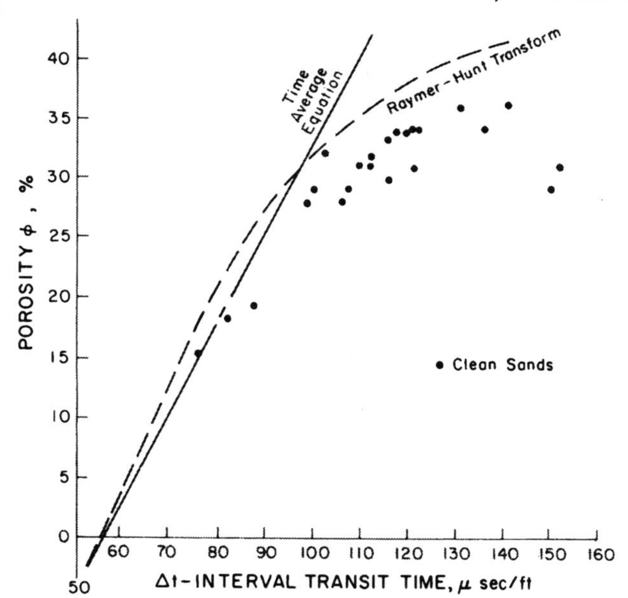

Fig. 3.22—Time-average and Raymer-Hunt correlations compared with Gulf of Mexico data obtained from moderately consolidated to unconsolidated Pleistocene to Miocene sands (after Ref. 25).

The difference between the porosity values has a critical impact on the qualification of the zone's potential. A value of 100% water saturation indicates no potential; a value of 61% water saturation indicates a hydrocarbon zone of some interest.

The most representative porosity and water saturation values of this zone are 18% and 61%, respectively. The porosity is estimated with a correlation developed specifically for the type of formation under consideration.

Example 3.4. The deviation of the best-fit line from the time-average line, shown by Fig. 3.16, could be a lack of compaction.

a. Can this information be represented by a correlation of the form given by Eq. 3.47? What is the B_{cp} value required?

b. Can this information be represented adequately by the Raymer-Hunt transform without a special correction factor?

Solution.

a. The equation of the best fit of the form given by Eq. 3.57 is

$$\Delta t = 55.5 + 153\phi,$$

where $\Delta t_A = 55.5$ μsec/ft and $\Delta t_B = 153$ μsec/ft. Because $\Delta t_A = \Delta t_{ma} = 55.5$ μsec/ft, Δt_B can be expressed by Eq. 3.61; hence,

$$B_{cp} = 153/(189 - 55.5) = 1.15.$$

b. The Raymer-Hunt transform given by Eq. 3.62 can be expressed as

$$\Delta t = \{[(1-\phi)^2/56] + (\phi/189)\}^{-1}.$$

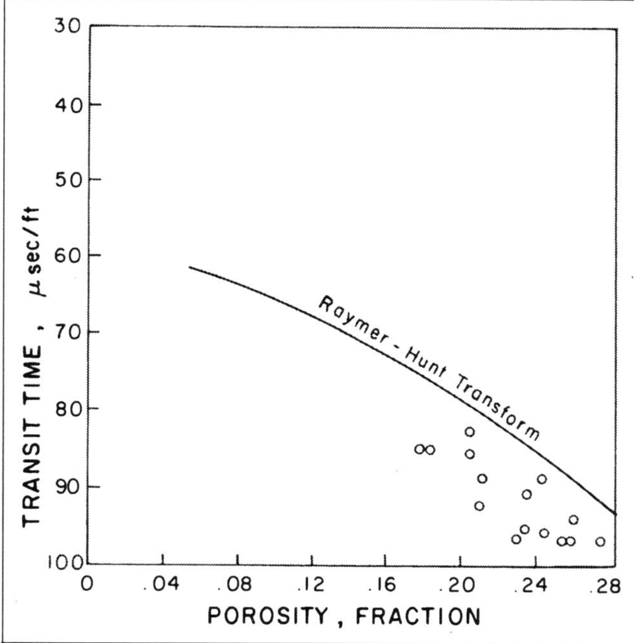

Fig. 3.23—Comparison between Raymer-Hunt correlation and data from the Delaware sandstone.

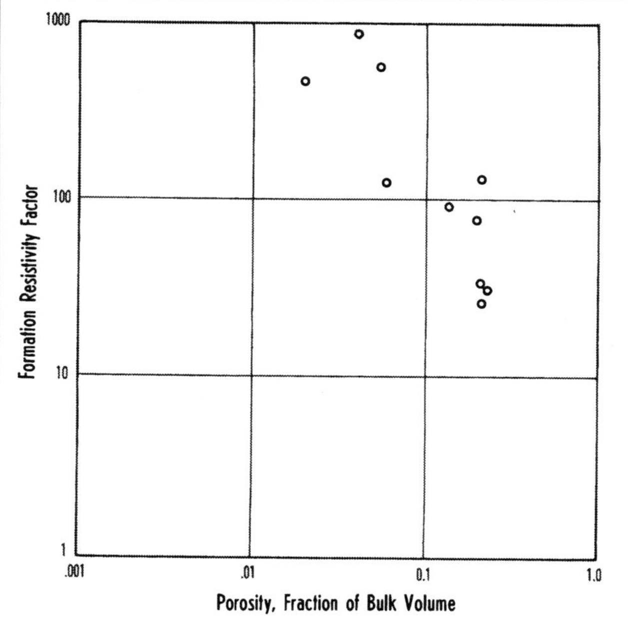

Fig. 3.24—Porosity vs. formation resistivity factor data for Edward formation (from Ref. 20).

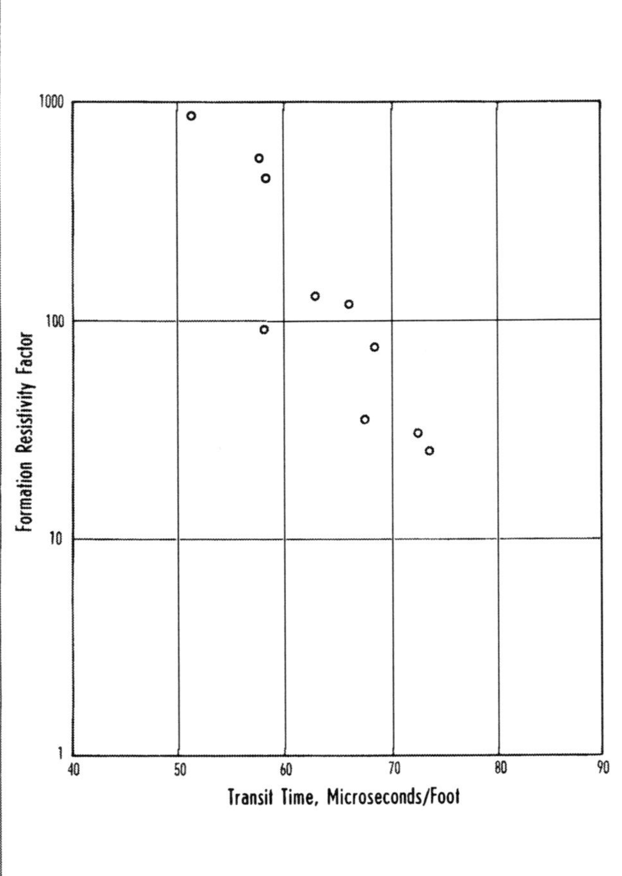

Fig. 3.25—Acoustic transit time vs. formation resistivity factor for Edwards formation (from Ref. 20).

Fig. 3.23 shows the comparison between this correlation and the data of Fig. 3.16. As shown, these data cannot be represented adequately by a Raymer-Hunt transform.

Review Questions

1. What are the different types of stresses to which a body can be subjected? What type of strain results from each of these stresses?
2. What is the physical significance of the different elastic constants?
3. What is a guided wave? How does it differ from a body wave?
4. How do vibrations associated with compressional waves differ from those associated with shear waves?
5. Do the two body waves propagate at the same velocity? Explain.
6. Can a compressional wave be totally or partially converted to a shear wave? Explain.
7. In what respect are light and acoustic waves similar?
8. What are the main components of the waveform that reach a receiver of an acoustic sonde placed in a fluid-filled borehole? Discuss the path from the transmitter to the receiver taken by each of these components.
9. Explain why the compressional head wave arrives first at the receiver of a sonic sonde placed in fluid-filled borehole.
10. Why does the amplitude of the pseudo-Rayleigh wave dominate that of the shear wave?
11. What are the factors affecting acoustic velocities in porous rocks?
12. List the advantages and shortcomings of the following relationships between Δt and ϕ: the time-average equation, Geertsma's model, and the Raymer-Hunt transform. Which of these three relationships would provide the most representative porosity values? Why?

Problems

3.1 A compressional wave traveling in Solid Medium a encounters the boundary of Solid Medium b at an angle of incidence θ. If the wave is partially converted to a shear wave, sketch the geometry of rays for the following three angles of incidence: 60.5°, 43°, 25.5°. The two media have the following velocities: v_{pa}=2,670 m/s, v_{sa}=1,120 m/s, v_{pb}=6,220 m/s, and v_{sb} = 3080 m/s.

3.2 An acoustic wave travels in water at a velocity of 1,500 m/s. Calculate the angle of incidence on a solid medium that results in a compressional and a shear head wave. Body wave velocities in the solid medium are 3,350 and 5,850 m/s.

3.3 Show that the total transit time, Δt_p, of a compressional head wave can be expressed as

$$\Delta t_p=(L/v_p)+2s[(1/v_f^2)-(1/v_p^2)]^{1/2}, \quad \ldots\ldots\ldots\ldots (3.63)$$

where L is the transmitter-receiver spacing and s is the distance from the transducer surfaces to the borehole wall.

3.4 A 5-ft-spacing acoustic device is centered next to a thick limestone formation in a 10-in. borehole filled with drilling mud. Calculate the first arrival time of the compressional and shear waves. The velocities are v_p=6,100 m/s, v_s=3,000 m/s, and v_f=1,800 m/s.

3.5 Calculate the acoustic longitudinal and transverse velocities in a solid medium having the following typical properties: ρ=2.65 g/cm^3, ϕ=20%, μ=0.2, $c_{ma}=2.5\times10^{-5}$ m^2/MN, $c_d=2.75\times10^{-4}$ m^2/MN, and $c_f=4.35\times10^{-4}$ m^2/MN.

3.6 **Figs. 3.24 and 3.25** show the formation resistivity factor and transit time vs. porosity data obtained experimentally for the Edwards formation.

a. Prepare a plot of Δt vs. ϕ.

b. Determine the coefficients Δt_A and Δt_B of Geertsma's model that best fit the data.

c. On the same graph, plot the time-average and Raymer-Hunt correlations.

d. Can this formation be adequately represented by these correlations?

e. What is the cementation exponent of this formation?

Nomenclature

A = amplitude
A = area, ft^2
B_{cp} = compaction correction factor
c = compressibility
C = proportionality constant
d = diameter, ft
E = Young's modulus, Pa
f = frequency, Hz
F = force, lbf
G = shear modulus
K = bulk modulus
L = length, ft
m = cementation exponent
n = saturation exponent
p = pressure, psia
R = resistivity, Ω·m
s = spacing
S_w = water saturation, fraction
Δt = transit time, μsec/ft
$\Delta t_A,\Delta t_B$ = coefficients in Geertsma's model
v = velocity, ft/sec
V = volume, ft^3
α = coefficient of absorption, ft^{-1}
α = angle
β = given by Eq. 3.39
ϵ = strain
θ = deformation angle
λ = wavelength, m

μ = Poisson's ratio
ρ = density, g/cm^3
σ = stress, Pa
τ = period, seconds
ϕ = porosity, fraction

Subscripts

b = bulk
c = critical
d = dry bulk rock
f = fluid
ℓ = longitudinal
ma = matrix
o = original
p = compressional
P = primary
s = shear
S = secondary
sh = shale
t = traverse
t = true
w = water

Superscript

$\bar{}$ = average

References

1. Jennings, H.Y. Jr. and Timur, A.: "Significant Contributions in Formation Evaluation and Well Testing," *JPT* (Dec. 1973) 1432–46.
2. Dobrin, M.B.: *Introduction to Geophysical Prospecting,* McGraw-Hill Book Co. Inc., New York City (1952).
3. Carmichael, R.S.: *Handbook of Physical Properties of Rocks,* CRC Press, Boca Raton, FL (1982) **II**.
4. Biot, M.A.: "Propagation of Elastic Waves in a Cylindrical Bore Containing a Fluid," *J. Appl. Phys.* (1952) **23**, 997–1005.
5. White, J.E.: "Elastic Waves Along a Cylindrical Bore," *Geophysics* (1962) **27**, 327–33.
6. Liu, O.Y.: "Stoneley Wave-Derived At Shear Log," paper presented at the 1984 SPWLA Annual Logging Symposium, June 10–13.
7. Cheng, C.H. and Toksoz, M.N.: "Elastic Wave Propagation in a Fluid Filled Borehole and Synthetic Acoustic Logs," *Geophysics* (1981) **46**, 1042–53.
8. Paillet, F.L.: "Predicting the Frequency Content of Acoustic Waveforms in Boreholes," paper SS presented at the 1981 SPWLA Annual Logging Symposium, June.
9. Tittman, J.: "Physics of Wireline Measurements," Schlumberger Educational Services, Houston (1986).
10. Minear, J.W. and Fletcher, C.R.: "Full-Wave Acoustic Logging," paper EE presented at the 1983 SPWLA Annual Logging Symposium, June.
11. White, J.E.: *Underground Sound,* Elsevier Science Publishers, Amsterdam (1983).
12. Biot, M.A.: "Theory of Propagation of Elastic Waves in a Fluid-Saturated Porous Solid: I—Low Frequency Range," *J. Acoustical Soc. America* (1956) **28**, 168–78.
13. Biot, M.A.: "Theory of Propagation of Elastic Waves in a Fluid-Saturated Porous Solid: II—High Frequency Range," *J. Acoustical Soc. America* (1956) **28**, 179–91.
14. *Log Interpretation, Vol. I: Principles,* Schlumberger, Houston (1972).
15. Hicks, W.G. and Berry, J.E.: "Application of Continuous Velocity Logs to Determination of Fluid Saturation of Reservoir Rocks," *Geophysics* (1956) **21**, 739–54.
16. Wyllie, M.R.J., Gregory, A.R., and Gardner, G.H.F.: "An Experimental Investigation of Factors Affecting Elastic Wave Propagation in Porous Media," *Geophysics* (July 1958) **23**, No. 3, 459–93.
17. *Formation Evaluation Data Handbook,* Gearhart Industries Inc. (1978).
18. Wyllie, M.R.J., Gregory, A.R., and Gardner, G.H.F.: "Elastic Wave Velocity in Heterogeneous and Porous Media," *Geophysics* (Jan. 1956) **21**, No. 1, 41–70.
19. Hughes, D.S. and Jones, H.J.: "Variation of Elastic Moduli of Igneous Rocks with Pressure and Temperature," *Bull.*, Geological Soc. of America (Aug. 1950) **61**, 843.
20. *Fundamentals of Core Analysis,* Core Laboratories Inc., Houston (1982).
21. Geertsma, J.: "Velocity-Log Interpretation: The Effect of Rock Bulk Compressibility," *SPEJ* (Dec. 1961) 235–48; *Trans.*, AIME, **222**.
22. Tixier, M.P., Alger, R.P., and Doll, C.A.: "Sonic Logging," *Pet. Trans.*, AIME (1959) **216**, 106–14.
23. Van der Knapp, W.: "Nonlinear Behavior of Elastic Porous Media," *Pet. Trans.*, AIME (1959) **216**, 179–87.
24. Raymer, L.L. and Hunt, E.R.: "An Improved Sonic Transit Time-to-Porosity Transform," paper P presented at the 1980 SPWLA Annual Logging Symposium, July.
25. Hartley, K.B.: "Factors Affecting Sandstone Acoustic Compressional Velocities and an Examination of Empirical Correlations Between Velocities and Porosities," paper PP presented at the 1981 SPWLA Annual Logging Symposium, June.

Chapter 4
Measurement Environment

4.1 Measurement Environment Effects

Most theoretical concepts of well-logging techniques were developed with the assumption of an infinite, homogeneous, and isotropic medium. When the borehole is considered, it is taken to be a regular cylinder of known diameter, filled with homogeneous fluid of known properties. When stratification is taken into account, beds are usually homogeneous and isotropic.

These assumptions of ideal measurement environment are also extended to the development of equations for quantitative interpretation of log responses. Data obtained in an actual measurement environment must be corrected before use in the interpretation equations. These corrections consist of removing that part of the signal caused by deviation of the actual environment from the ideal.

A logging tool is usually designed to function best in a certain environment. Use of the tool in an appreciably different environment results in a low-quality or totally nonrepresentative log.

Choosing the tool to be run in a certain wellbore, judging the log quality, and extracting quantitative information from the measurement require knowledge of the actual measurement environment. Significant parameters are the borehole diameter and shape, the properties of the drilling fluid filling the borehole, the borehole and formation temperatures, and the radial variation of formation properties.

4.2 Borehole Diameter and Shape

Measurement of borehole diameter with caliper logging has indicated clearly that the actual borehole diameter often differs from the bit size used to drill it. The difference is considerable in some cases. **Fig. 4.1** shows an actual profile of a borehole drilled with a 12-in. bit. The section gauge log was used to estimate the hole volume. It clearly shows that the hole is far from being a regular cylinder with uniform diameter.

The borehole's actual diameter and shape depend on the formation drilled. The upper part of the hole appears to be smooth, with a diameter equal to the bit size. In fact, Section A is drilled to gauge, which is usually the case in hard, consolidated, and impermeable formations.

The diameter of Section B is actually smaller than the bit diameter. This is usually the case in permeable formations drilled with mud that contains solids. Drilling safety usually requires that the hydrostatic head of the mud column be kept above formation pressure. This pressure difference causes the mud to flow into permeable formations. The solid particles that exceed the pore size are retained at the formation face. Their buildup forms a plaster-like layer of very low permeability called a "mudcake." The mudcake thickness depends on mud properties and ranges from a thin film to 1 in. thick in most cases. In permeable formations, the borehole drilled diameter is reduced by twice the mudcake thickness. The mudcake properties, such as resistivity and density, differ considerably from those of the surrounding formation.

In Section C of Fig. 4.1, the borehole's actual diameter is enlarged by as much as 3½ in. Such enlargement occurs in soft, unconsolidated formations because of the scouring effect of drilling muds. Enlargement also occurs in water-soluble formations, such as salt, as a result of leaching, and in naturally fractured formations owing to a weakening of their mechanical integrity.

Borehole enlargements are most commonly observed in shales and shaly formations.[1] Because of their electrochemical properties, clay minerals absorb water, causing the shale formation to swell. The swelling weakens the formation, and the shale sloughs and caves. The intensity of sloughing and caving, which results in borehole enlargement, depends on the physical properties of the clay and drilling fluid. Different clay minerals have different affinities to water. Freshwater-based mud causes more caving than saltwater-based mud. Caving is usually absent when oil-based mud is used.

Formation properties responsible for borehole enlargement vary, causing erratic borehole enlargement, which in turn results in a rugose borehole. Borehole rugosity is apparent in Section D of Fig. 4.1.

Borehole enlargement causes mechanically centered tools to be situated farther from the formation investigated. The space between the centered tool and the formation is occupied by drilling mud, which typically exhibits physical properties very different from those of the formation. Logging tools are designed to investigate a prescribed volume; they can usually accommodate a small fraction of the volume to be occupied by the mud. As that fraction increases with borehole diameter, the tool response cannot be attributed entirely to the formation. In extreme borehole enlargement, the tool response corresponds entirely to the drilling fluid.

When pad-type tools are pressed against the formation face, the sensors are separated from the formation by mudcake and/or pockets of drilling mud (**Figs. 4.2 and 4.3**). Pad-type tools (e.g., microresistivity and density tools) have a relatively small radius of investigation. Consequently, the mudcake zone and mud pockets can considerably affect the tool response.

Adequate analyses of certain log measurements require knowledge of borehole size and shape. To determine borehole geome-

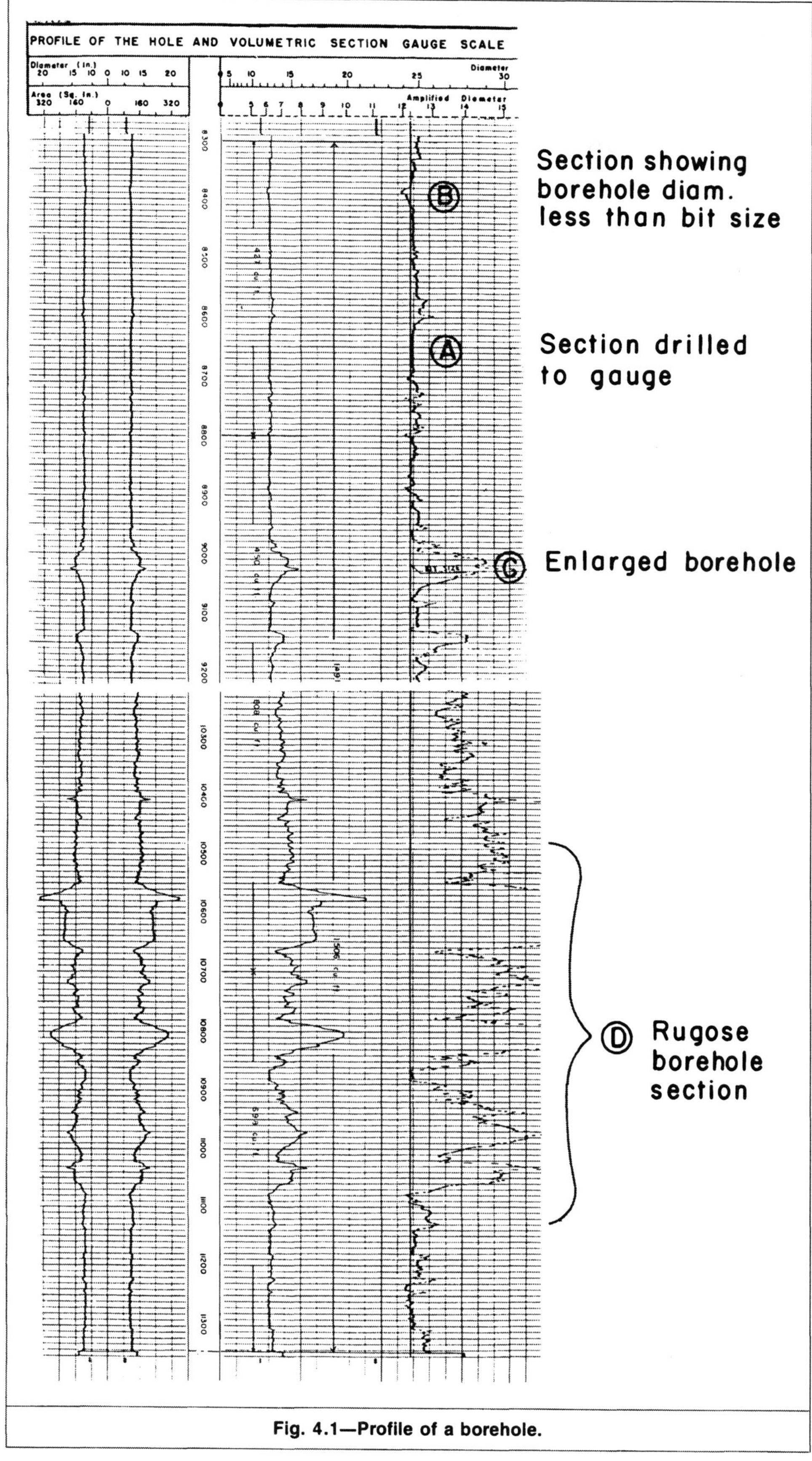

Fig. 4.1—Profile of a borehole.

try, a caliper log is usually run with microresistivity, density, sidewall neutron, sonic, and dipmeter logs. **Fig. 4.4** is a schematic of the microresistivity log caliper. Two pads are pushed against the formation by two opposing arms that work together to keep the main body of the tool centered in the hole. A spring-like mechanism extends the arms so that the pads are in continuous contact with the borehole wall. The outward stretch of the arms that follow the borehole diameter is converted to an electric signal calibrated to yield the borehole diameter. The tool is calibrated at the surface by placing it in metallic rings of standard diameters, usually 8 and 12 in.

The contact pads of the microresistivity tools are at least 6 in. long. Therefore, small hole irregularities cannot be detected. Maximum tool stretch is usually limited to 16 or 18 in. The distance between the pad faces when the tool is collapsed is usually 6 in.

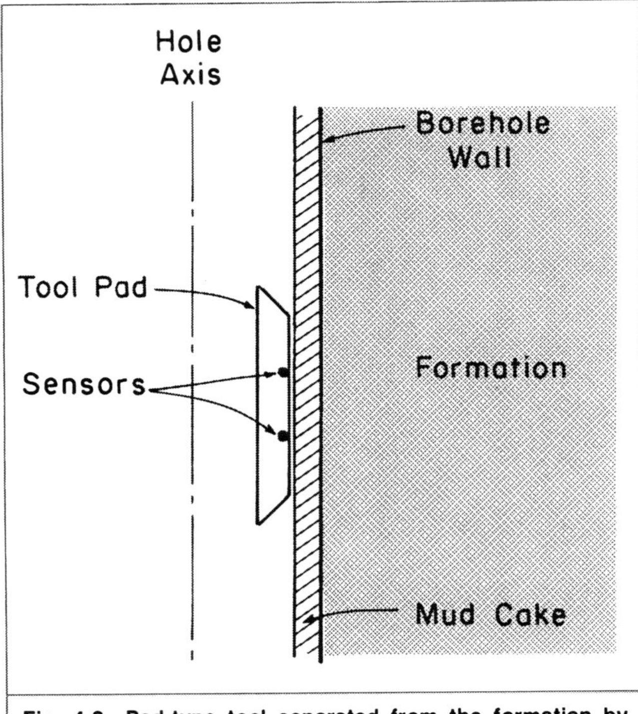

Fig. 4.2—Pad-type tool separated from the formation by mudcake.

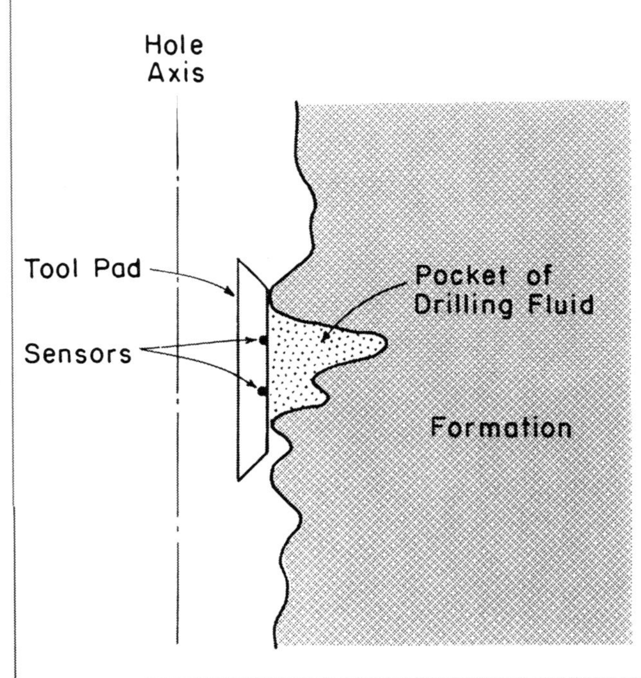

Fig. 4.3—Pad-type tool trapping drilling-mud pockets in rugose borehole.

The range of caliper measurement is then 6 to 16 or 18 in. The pressure exerted on the pads is low, so in permeable zones, they ride over the mudcake. The log reading is then the hole drilled diameter minus twice the mudcake thickness.

The density- and pad-type neutron tool calipers are obtained from an arm attached to the body of the tool (**Fig. 4.5**). The arm is pushed hard against the borehole wall to keep the tool in good contact with the formation. The measured caliper is the distance between the metallic skid attached to the arm and the tool face.

The rigid, relatively long tool cannot move in and out of a borehole enlargement that is shorter than the tool length. The borehole rugosity seen on the log reflects the path of the tool's extended arm on only one side of the borehole. The other side is being smoothed by the longer tool. Because of the pressure exerted on the tool, the arm tends to cut through the mudcake. The log reading is roughly equal to the borehole drilled diameter minus one mudcake thickness.

The sonic log tool is centralized in the hole with three equally spaced bow springs (**Fig. 4.6**). Hole diameter is derived from the stretch of these three arms, which usually work together (i.e., they open and close equally).

Anisotropic formation mechanical properties result in an oval or egg-shaped cross section.[2] An enlarged borehole usually assumes

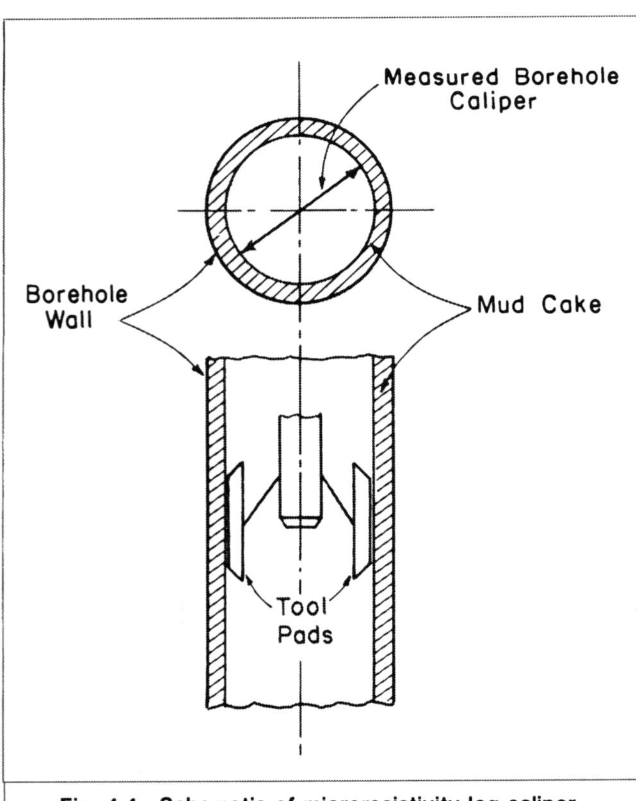

Fig. 4.4—Schematic of microresistivity log caliper.

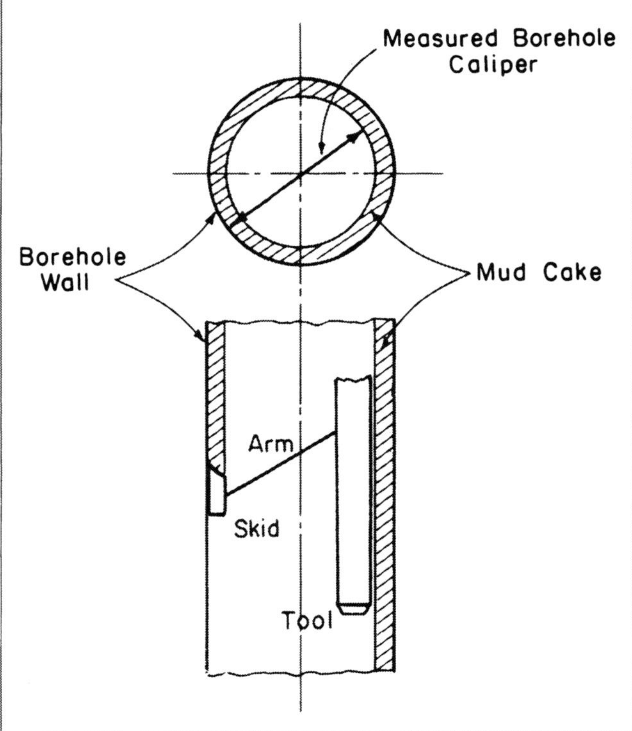

Fig. 4.5—Schematic of density and pad-type neutron log caliper.

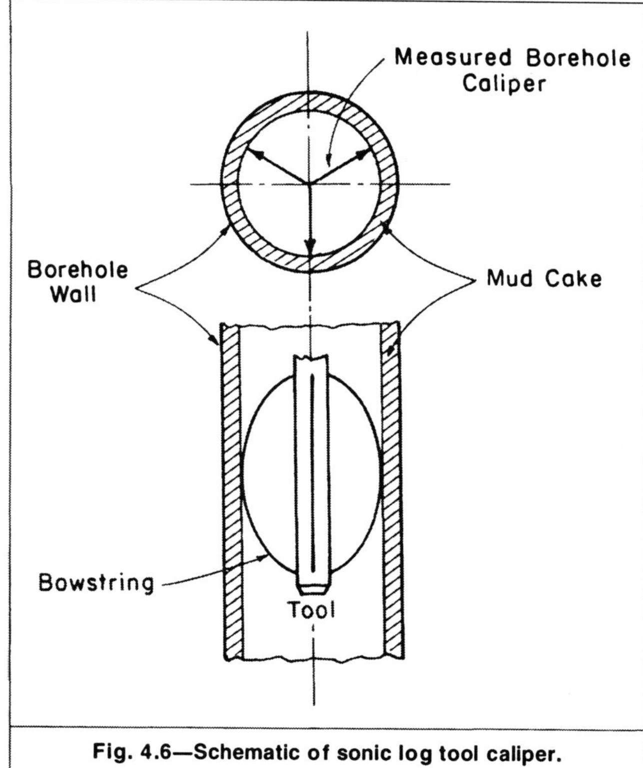

Fig. 4.6—Schematic of sonic log tool caliper.

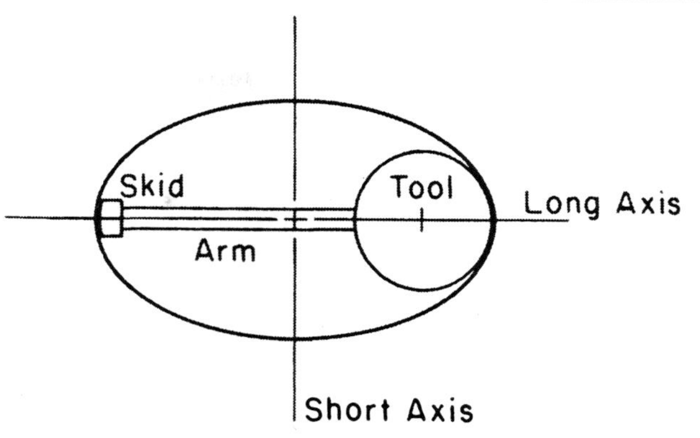

Fig. 4.7—Schematic of elliptical borehole showing the preferential position of a pad-type tool.

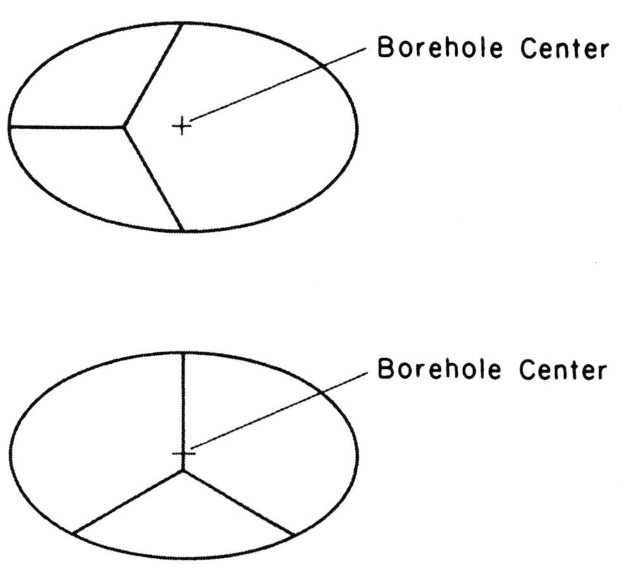

Fig. 4.8—Most probable positions assumed by a three-arm caliper in an elliptical borehole (from Ref. 1).

a noncircular cross section. Because of the pressure exerted on the pads, the tool rotates, seeking the position of least potential energy. This is usually the long axis for an elliptical cross section of the borehole (**Fig. 4.7**). Microresistivity, density, and neutron log calipers generally measure the long axes of oval holes.

In an elliptical hole, a three-arm caliper, such as the sonic caliper, most probably assumes one of the positions shown in **Fig. 4.8.** The indicated diameter is less than the maximum hole axis. In addition, the reading is reduced by twice the mudcake thickness. Note that the three dependent bow springs fail to centralize the tool properly in a noncircular borehole.

The tool design and the borehole shape and rugosity determine the caliper's vertical and horizontal resolution. Different calipers obtained in the same borehole yield different results. **Fig. 4.9** compares the microlog and a density log caliper. Major disagreements are observed over Sections X and Y. Section X is a borehole enlargement. Unable to fit closely in this geometry, the density tool smooths out one side of the hole. This results in a reading smaller than that of the microlog. Section Y is a permeable interval that exhibits mudcake buildup. The density-tool skid cuts through the mudcake while the microlog pads ride over it. The microlog-caliper reading is one mudcake thickness less than that of the density caliper.

Fig. 4.10 compares two calipers: one obtained by a two-arm pad device (e.g., a microlog) and the other by a three-arm bow-spring device (a sonic log). The noncircular hole shape causes the marked disagreement. The two-arm device reads the long axis of the oval hole; the three-arm device reads the short axis.

A four-arm caliper gives a better definition of a noncircular borehole. This tool configuration usually accompanies the dipmeter log where definition of the borehole circumference is essential. The four-arm device has two independent two-arm calipers placed at right angles. The tool provides two caliper logs. In circular boreholes, the two calipers are equal. They separate in noncircular holes as one caliper reads the long axis and the other reads the short axis. **Fig. 4.11** shows an example of the four-arm caliper output. The approximate borehole cross sections at Levels A, B, and C are also shown.

Qualitative and quantitative information can be extracted from the caliper reading. The borehole diameter or mudcake thickness is needed when a correction for the effect of drilling fluids on log measurements is warranted. The presence of mudcake itself is proof of the presence of permeability. Mudcake buildup, borehole enlargement, and borehole rugosity can be used with other log data to mark changes in lithology. Proper evaluation of pad-type tools (density or sidewall neutron) is enhanced by knowledge of the extent of the borehole rugosity.

In well-completion operations, the caliper log can help estimate required cement volumes. It also helps select the optimum location for packer seats. The caliper log is indispensable in dipmeter interpretation because it provides the position of readings' pickup points on the dipping plane.

Example 4.1. Fig. 4.12A shows an interval of a microcaliper (i.e., a two-arm caliper measurement) that accompanies the microlog. The interval logged consists of sands and shales. The vertical line on the log represents the nominal borehole diameter, which is taken to be the bit size.

a. What bit size was used to drill this hole?

b. Does the log show mudcake buildup? What is the thickness of the permeable zone?

c. What is the best estimate of the mudcake thickness?

d. Does the log show severe borehole enlargement? What is the maximum diameter recorded by the log?

e. What is the lithology in the enlarged interval?

f. What is the vertical resolution of this tool? How does it compare with the pad length (about 6 in.)? How accurate is the thickness value obtained in Part b?

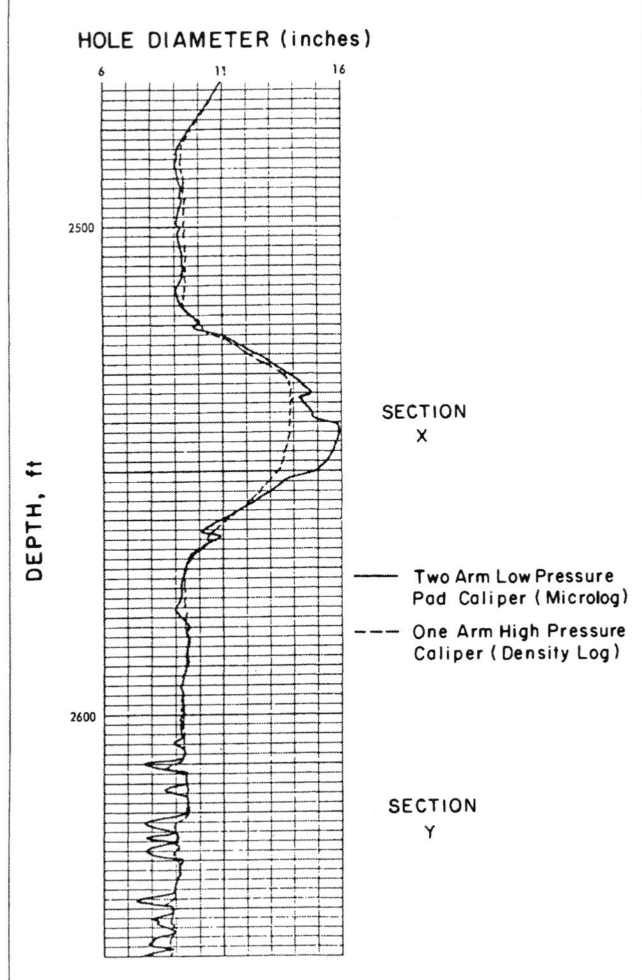

Fig. 4.9—Comparison of microlog caliper and density caliper (from Ref. 2).

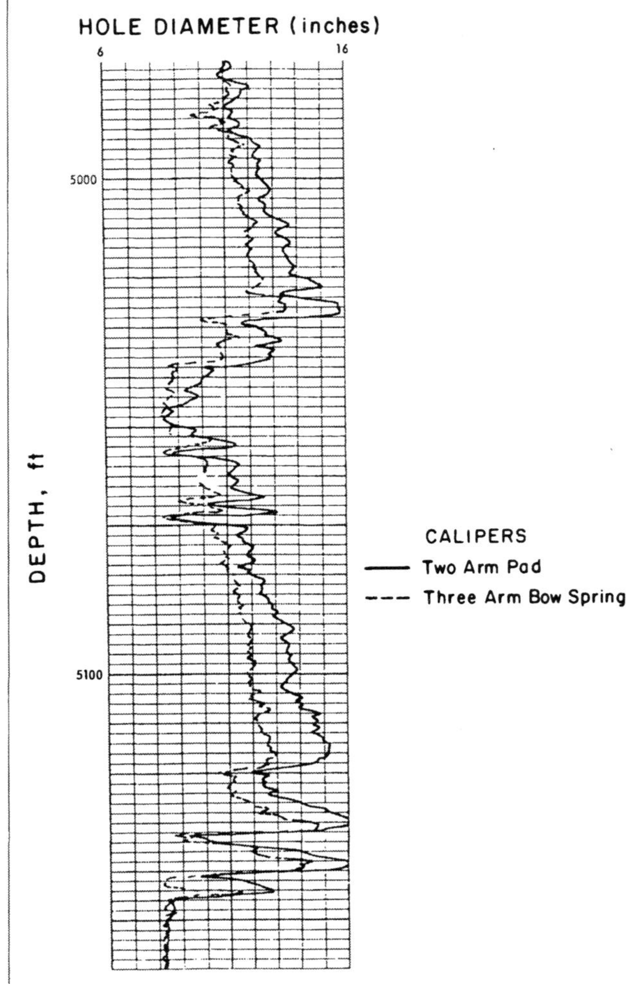

Fig. 4.10—Comparison of two-arm and three-arm bow-spring calipers (from Ref. 2).

g. Using log information, can you guess the tool diameter in the collapsed position?

Solution. **Fig. 4.12B** illustrates the answers to the above questions.

a. The vertical line indicates that a 6¾-in. bit was used to drill this hole.

b. Yes, the actual hole diameter between 10,494 and 10,533 ft is less than the bit size. This indicates that there is mudcake buildup against a 39-ft permeable zone.

c. Two-arm pad-type calipers read the actual hole diameter minus twice the mudcake thickness, h_{mc}. The caliper reads 6-in., so $2h_{mc}=6¾-6=¾$ in. and $h_{mc}=⅜$ in.

The calculation of a ⅜-in. mudcake thickness assumes that the hole was drilled to gauge. If the drilled-hole diameter is larger than the bit size to start with, ⅜ in. is the minimum thickness value. Note that if the drilled-hole diameter is considerably larger than the bit size, mudcake buildup can go undetected.

d. The log shows a severe borehole enlargement in the interval of 10,449 to 10,484 ft. The maximum recorded diameter is 11¾ in., indicating a 5-in. borehole enlargement.

e. Because, as stated, only sands and shales are present in the interval shown in Fig. 4.12A, the lithology in the enlarged interval is shale.

f. The log shows diameter variation over thin zones ½ to 1 ft. thick. The tool's vertical resolution is then on the same order of magnitude as the pad length. Thus, the estimated permeable-zone thickness of 39 ft in Part B is a very good estimate.

g. From Fig. 4.12B, the pads first established contact with the borehole wall at 10,598 ft. The tool reading below that depth corresponds to the collapsed position of the tool. It is about 5.4 in.

4.3 Mud, Mud-Filtrate, and Mudcake Properties

Well-logging tools are usually run in boreholes filled with a drilling fluid. Possible drilling fluids are air, water, or a slurry made up of liquid and solid phases. Slurries, the most frequently used drilling fluid, are called drilling muds. The principal functions of drilling muds are to remove the drilled solids (i.e., cuttings), to prevent formation fluids from flowing into the borehole, to prevent the borehole walls from caving, and to cool the bit. Drilling mud also plays an important role in logging operations, especially electric logging. In effect, a conductive drilling mud is needed to provide electric continuity between the electric electrode-type tools and the formation.

The drilling mud affects, generally adversely, the response of logging tools, depending on its type and properties. Water-based mud is the most common type. It is made of a continuous liquid phase of water in which clay or clay-like material is in suspension. Barite is usually added to increase mud density. Chemicals are also added to control the rheological properties of the fluid. The water used to mix the mud is usually fresh. Some drilling applications however, call for salt water.

The water is usually replaced by liquid hydrocarbons for hot, deep formations, water-soluble formations, and water-sensitive potential producing formations. This type of mud is known as oil-based mud because the liquid phase consists of diesel, weathered crude, refined, or mineral oil. Oil-based muds are nonconductive and limit the use of resistivity logs to induction-type devices. It is generally accepted that oil-based mud minimizes borehole environmental effects. Recent studies[3] show, however, that this is not always the case. As the use of oil-based muds increases, our understanding of their effect on the logging environment improves.

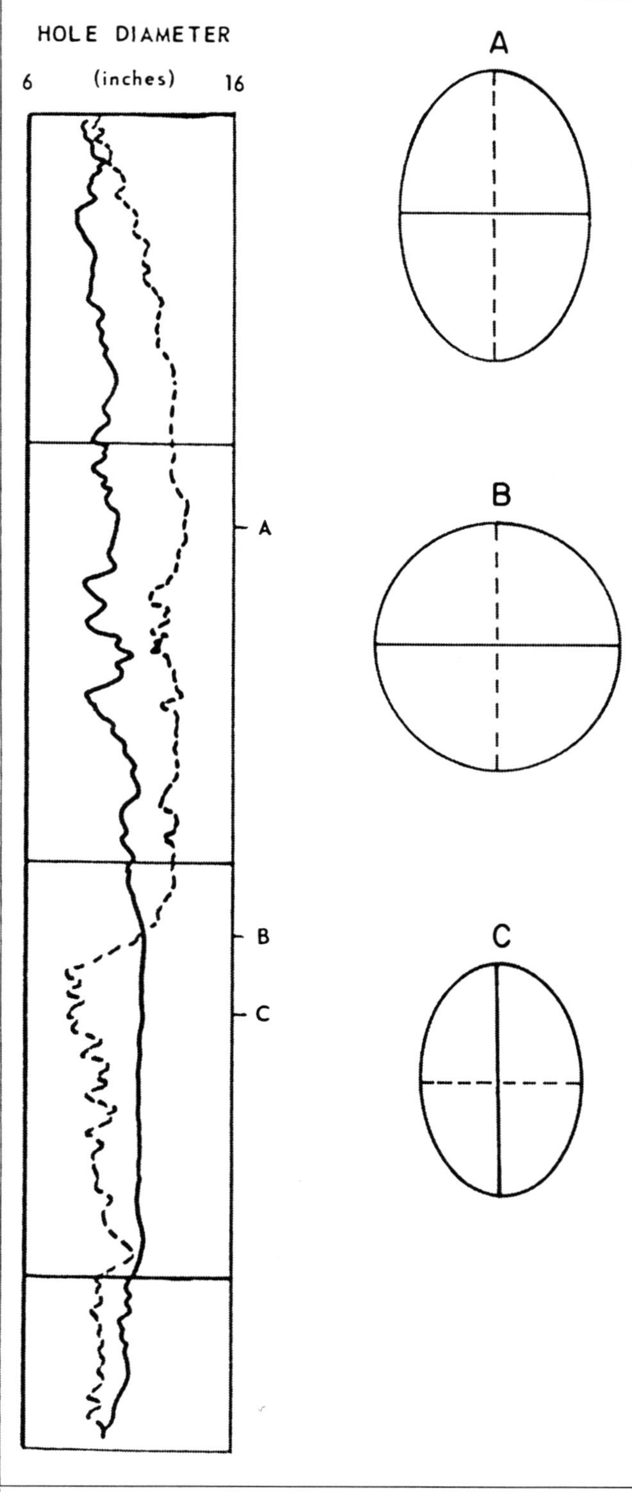

Fig. 4.11—Example of four-arm caliper output (from Ref. 2).

The rest of this section focuses on the more common water-based mud. As mentioned, the drilling mud flows into a permeable formation because of a pressure differential between the wellbore and the formation. Large particles relative to pore size are screened out and coat the formation face with a mudcake. The liquid phase that invades the formation is called the mud filtrate. As **Fig. 4.13** illustrates, the mud, mudcake, the zone of the formation invaded by mud filtrate, and the uninvaded zone of the formation contribute to the tool measurement. In addition to the properties of the formation of interest, the tool response is affected by mud, mudcake, and mud-filtrate properties. The degree of these effects depends on the tool design, physical properties of the zones involved, borehole size, mudcake thickness, and depth of invasion.

Drilling-mud data pertinent to logging operations are type; density; viscosity; pH; fluid loss; and mud, mudcake, and mud-filtrate resistivities.

4.3.1 Mud Type. Water-based muds usually are chemically treated and are designed according to the type of treatment. Commonly used muds can be classified as[4] (1) natural muds (untreated), (2) phosphate muds, (3) organically treated muds (lignate, quebracho, and chrome lignosulfonates), (4) calcium-treated muds (lime, calcium chloride, and gypsum), (5) saltwater muds (seawater and saturated saltwater muds), and (6) oil-emulsion muds (oil-in-water). The mud type reflects the predominant elements in the mud. Knowing the mud type aids in the analysis of nuclear logs because their response is affected by the atomic composition of the surrounding medium.

4.3.2 Density. In field work, mud density is called mud weight. It is expressed in field units of pounds per gallon. It is also referred to in terms of its pressure gradient, which is expressed in pounds per square inch per foot. The mud balance generally is used to measure mud density.[5] The mud density affects the filtration process because it reflects the solid content of the mud. It also determines the magnitude of the pressure differential between the mud and the formation, which in turn determines the severity of mud-filtrate invasion. The response of gamma ray devices (gamma-ray logs, density logs) depends on the mud density because the attenuation of gamma rays depends heavily on the density of the surrounding medium.

4.3.3 Viscosity. A relative measure of mud viscosity is easily obtained on the rig with the Marsh funnel. The Marsh funnel viscosity is the time in seconds needed for 1 qt of mud to flow through the funnel.[5] For reference, the Marsh funnel viscosity of fresh water is about 26 seconds at 70°F.

Viscosity is an indicator of mud quality, or solid content. However, no strong relationship exists between mud viscosity and other mud parameters, like filtration, that are pertinent to logging operations.

4.3.4 pH. The pH, which is the abbreviation for potential hydrogen ion, reflects the relative acidity or alkalinity of mud. pH values range from 0 to 14. Pure water, which is neutral (not acidic or alkaline) has a pH of 7. pH values less than 7 indicate acidity, and values greater than 7 indicate alkalinity. The pH reflects mud chemical quality and is used for mud quality control. In log interpretation, it can be used to indicate qualitatively the presence of certain ions. For example, mud alkalinity usually results from the presence of bicarbonates (HCO_3), carbonates (CO_3), and hydroxyls (OH). Knowledge of mud ionic composition is important in some log analyses [e.g., the self-potential (SP) log].

Mud pH is determined by the use of paper test strips. The paper strip, which is impregnated with dyes sensitive to the solution pH, is wetted by the mud liquid phase. The color of the wet paper is compared with a standard color chart to determine the pH value. pH values can also be determined by use of a glass-electrode pH meter.[5]

4.3.5 Fluid Loss. Fluid loss is a measure of the relative amount of filtrate lost during filtration. It qualitatively indicates the relative severity of mud-filtrate invasion into a permeable formation.

Fluid loss is determined with a filter press. A representative mud sample, usually obtained from the return flowline, is poured into a mud cell with a standard filter paper at its bottom. A 100-psi pressure differential is applied for 30 minutes. The fluid loss is the amount of filtrate measured in cubic centimeters that collects in a graduated cylinder during those 30 minutes.[5] The amount of fluid loss is very useful for mud conditioning, but might not accurately represent filtration under dynamic conditions.[6]

A sample of the mud, mud filtrate, and mudcake deposited on the filter paper is usually retained for resistivity measurements.

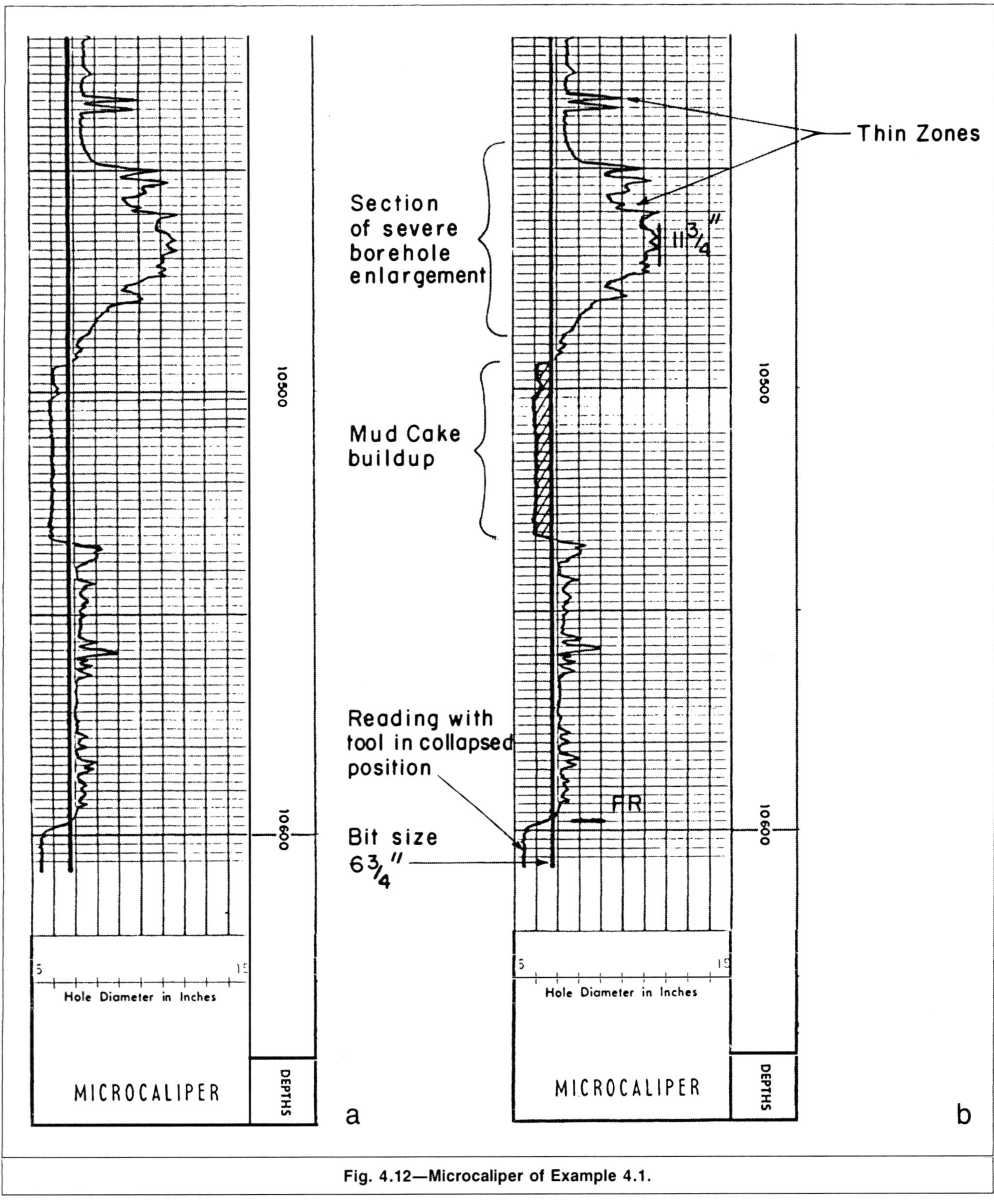

Fig. 4.12—Microcaliper of Example 4.1.

4.3.6 Mud, Mud-Filtrate, and Mudcake Resistivities. Mud, mud-filtrate, and mudcake resistivities—R_m, R_{mf}, and R_{mc}, respectively—are the drilling-fluid properties most pertinent to log analyses, primarily because electric properties of the mud differ drastically from those of the formation and formation fluids, which causes a considerable resistivity contrast between the borehole medium and the rock. This contrast controls the log quality. The value of R_m, is needed to remove the borehole signal from the total tool response. Nuclear and acoustic properties of the borehole medium also differ from formation properties. Nuclear and acoustic logging devices, however, were designed with knowledge gained from earlier electric tools. They were designed to minimize the effect of mud and mudcake on the tool response. The effect of mud and mudcake is also removed (compensated for) automatically.

The contrast of chemical activities between mud filtrate and formation water originates and determines the magnitude of the SP measured in logging operations.

Mud, mud filtrate, and mudcake samples, collected in conjunction with filter-press tests, are placed successively in a resistivity meter, which provides a direct resistivity reading. The temperature of the sample has to be measured because resistivity is strongly dependent on temperature. R_m and R_{mf} values provided by the meter are usually representative, especially if the measurement procedure recommended by API[5] is followed. R_{mc} values are usually suspect

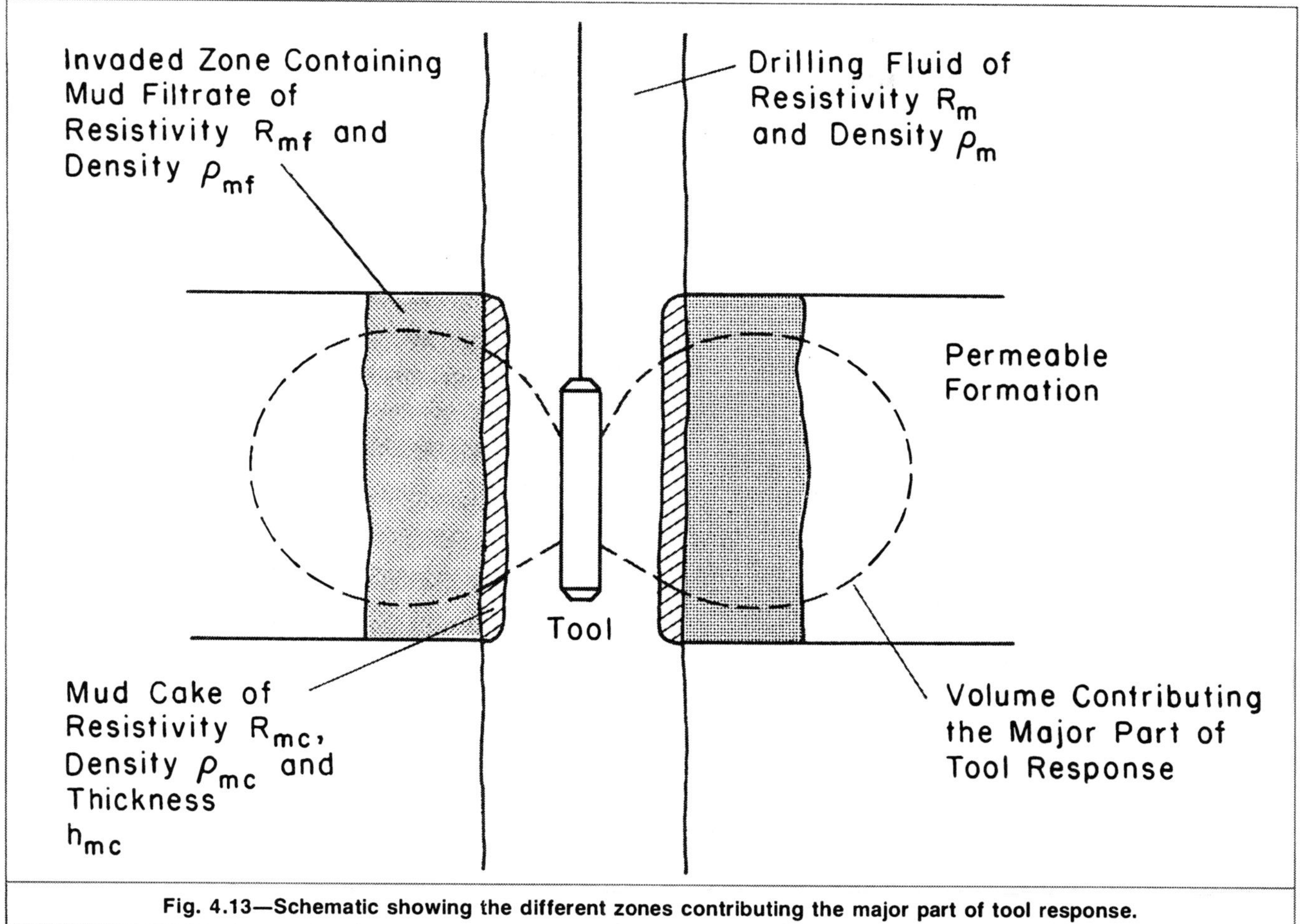

Fig. 4.13—Schematic showing the different zones contributing the major part of tool response.

TABLE 4.1—K_m VALUES AS A FUNCTION OF MUD WEIGHT

Mud Weight (lbm/gal)	Mud Weight kg/m³	K_m
10	1200	0.847
11	1320	0.708
12	1440	0.584
13	1560	0.488
14	1680	0.412
16	1920	0.380
18	2160	0.350

because the manner in which the mudcake is packed in the cell dictates the resistivity value indicated by the meter.

4.3.7 Correlation of Mud-Filtrate and Mudcake Resistivities to Mud Resistivity. In early practice, only the drilling-mud resistivity was measured. Even in present practice, some mud logging units measure R_m only periodically. Also, values of R_m are provided by some measurement-while-drilling (MWD) systems.[7] In this case, R_{mf} is estimated from empirical correlation to R_m. Because of the difficulty associated with measuring R_{mc}, even if a measured value is available, it is usually estimated through empirical correlations.

The following empirical equation was derived from data taken from 94 field muds[8]:

$$R_{mf}=K_m(R_m)^{1.07}, \quad (4.1)$$

where K_m is a coefficient that varies with mud weight. **Table 4.1** gives K_m values as a function of mud weight. It was also found that[8]

$$R_{mc}=0.69R_{mf}(R_m/R_{mf})^{2.65}. \quad (4.2)$$

Eqs. 4.1 and 4.2 are presented graphically in **Fig. 4.14.** The use of the correlation in Eq. 4.1 was restricted to nonlignosulfate muds, which were not used at the time that the correlation was developed. A recent study[7] shows that the use of the correlation can be expanded to today's widely used lignosulfonate muds. The same study also proposed the following correlation for all types of freshwater muds:

$$\log(R_{mf}/R_m)=0.396-0.0475\rho_m, \quad (4.3)$$

where ρ_m is the mud density in lbm/gal.

Other statistical correlations that are valid only for low-weight, predominantly sodium chloride (NaCl) muds are[9]

$$R_{mf}=0.75R_m \quad (4.4)$$

and $R_{mc}=1.5R_m$. (4.5)

Empirical correlations for specific mud types—such as lime, gypsum, and calcium lignite/calcium lignosulfonate muds—are available in Ref. 10.

4.3.8 Effect of Temperature on Mud, Mud-Filtrate, and Mudcake Resistivities. The mud, mud-filtrate, and mudcake resistivities are usually measured at surface temperature. Quantitative interpretation of electric logs requires representative values of R_m, R_{mf}, and R_{mc} under borehole conditions.

As in the case of rocks, the conductivity of the mud results mainly from the liquid phase. Drilling mud's conductivity and that of its filtrate and cake increase with temperature. For predominantly NaCl muds, the nomograph of Fig. 1.7 or Eq. 1.11 can be used to convert surface resistivity values to borehole values. **Fig. 4.15** graphically presents NaCl solution resistivity vs. temperature. Salinity obtained from Fig. 1.7 or 4.15 is approximate and not true salinity because the graphs are for solids-free NaCl solutions.

Several studies[10,11] showed that the effect of temperature on field drilling muds is not the same as NaCl solution. Over the temperature range of 75 to 225°F, the difference may be as much as 10% for R_m and R_{mf} and 25% for R_{mc}. If a certain ionic content

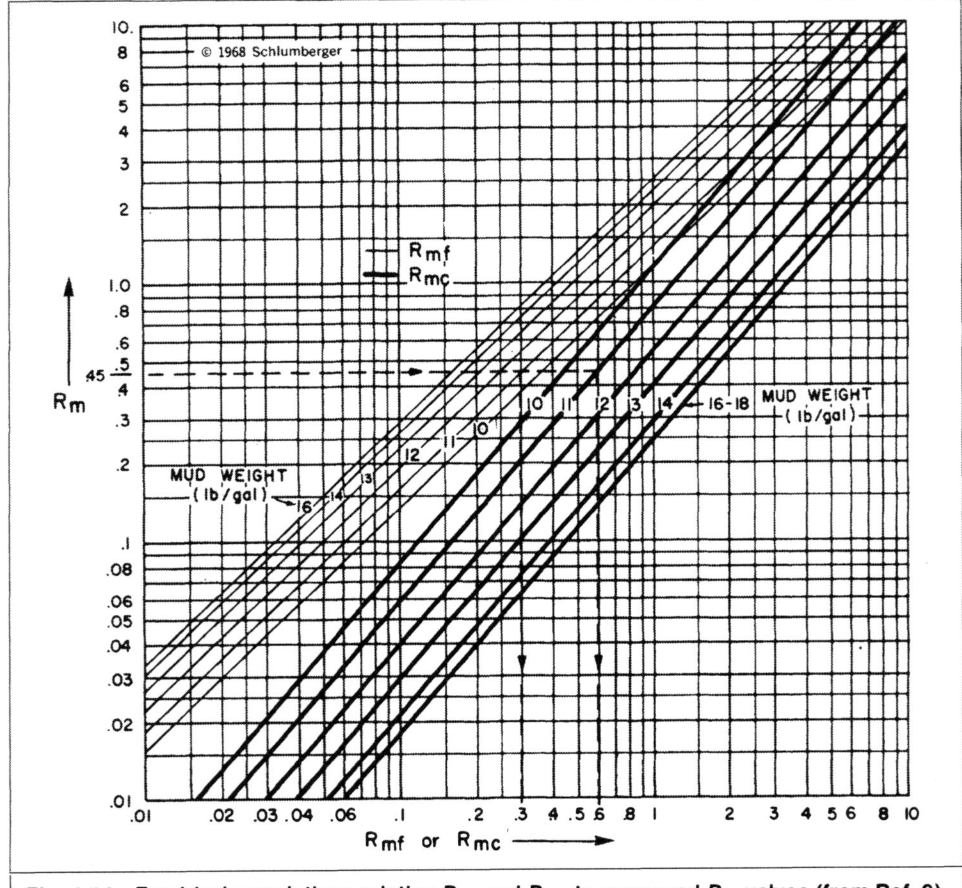

Fig. 4.14—Empirical correlations relating R_{mf} and R_{mc} to measured R_m values (from Ref. 9).

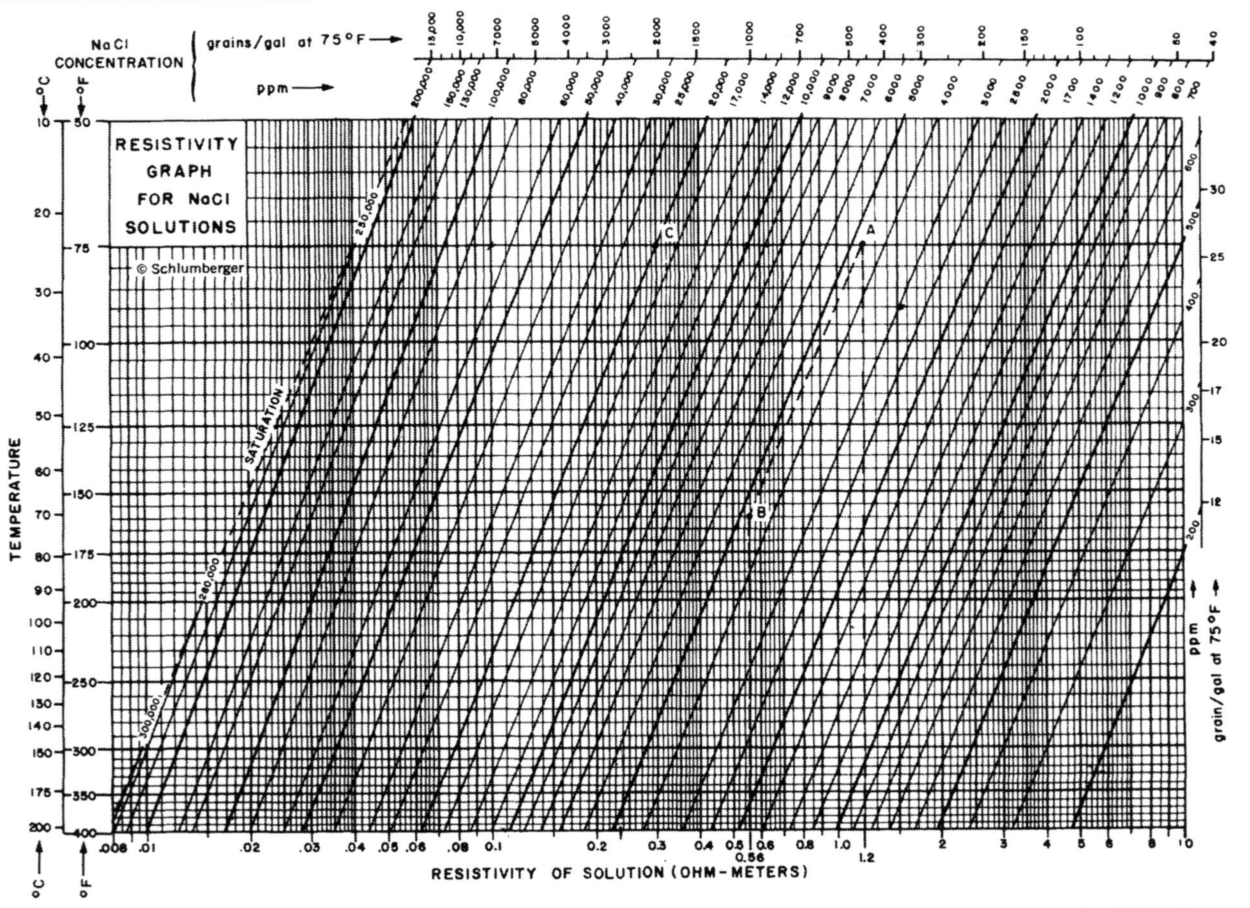

Fig. 4.15—Resistivity vs. temperature for NaCl solutions (from Ref. 9).

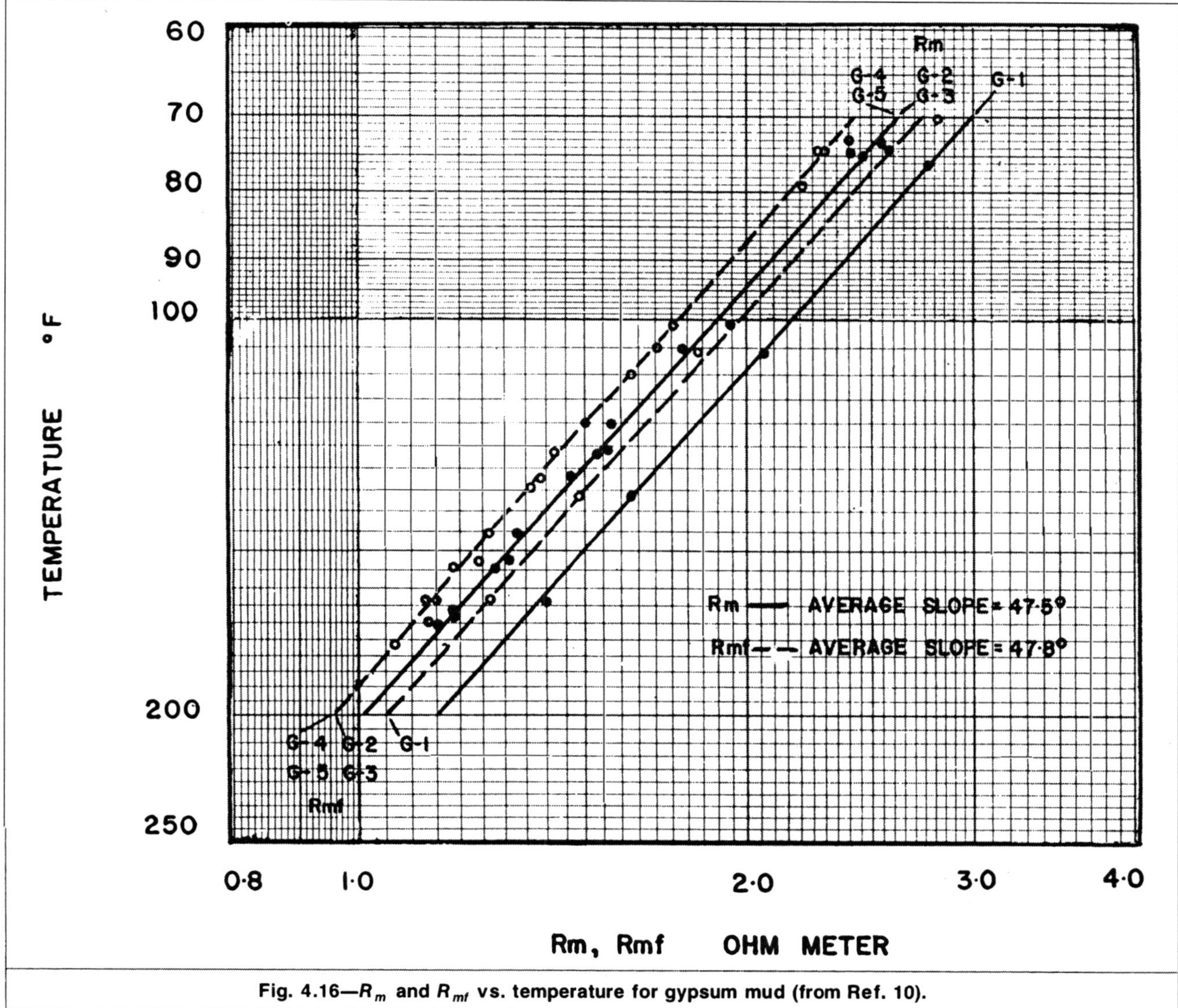

Fig. 4.16—R_m and R_{mf} vs. temperature for gypsum mud (from Ref. 10).

is such that it cannot be approximated by NaCl solution, the effect of temperature on this mud type should be investigated. **Fig. 4.16** shows the results of such an investigation for gypsum muds.

Bottomhole mud resistivity values can be measured in situ if the microlog (Chap. 5) is run. With its arms collapsed, the microlog sonde is lowered into the borehole. In boreholes with diameters exceeding 8 in., the reading of the device in the collapsed mode is determined mostly by the mud. A recording of the reading vs. depth yields a "mud log." The lowest resistivity, usually given by the microinverse electrode array (solid curve), corresponds to the in-situ value of R_m. This value can be used to check the surface R_m measurement. The mud log has other potential applications, including detection of mud system changes and possible downhole water flow.[12]

Fig. 4.17 is an example of a mud log. Note that the caliper curve is absent. The solid vertical line shows an automatically posted value that corresponds to bit size. Note also that the depth scale increases from the bottom of the log to the top. The reason is that this log is run going into the hole; other logs are run going out of the hole.

Example 4.2.

a. Give the in-situ mud resistivity indicated by the mud log in Fig. 4.17.

b. How does this value compare with a surface measurement of 0.9 Ω·m at 80°F, if in-situ temperature is 218°F?

Solution.

a. The minimum value indicated by the mud log at the bottom of the hole is 0.32 Ω·m.

b. Using Eq. 1.11 gives

$$R_m = 0.90 \frac{80+6.77}{218+6.77} = 0.35\ \Omega\cdot\text{m}.$$

The two values compare reasonably well despite the assumption implied in Eq. 1.11 that mud is a predominantly NaCl solution.

4.3.9 Drilling-Mud Resistivity Variation. Mud properties typically are measured on a sample collected when the well is logged. The openhole section to be logged is usually several thousand feet long. This section is drilled over several days and sometimes over several weeks; **Fig. 4.18** illustrates the complications that might arise in this situation. When the top formation is drilled, if the hole contains drilling mud of resistivity R_{m1}, then the resistivity of mud filtrate that invades the formation is R_{mf1} because most of the filtration occurs within a few hours of drilling.[13] When the target depth is reached and the well is logged, if the borehole contains a different mud of resistivity R_{m2}, then the sampled mud-filtrate resistivity is R_{mf2}. The sample value corresponds to the mud filtrate that invaded the lower zone but is different from the mud filtrate that invaded the top formation.

Short-term variations in drilling-mud resistivity are linked to chemical treatment of mud conducted during drilling operations.[14,15] **Fig. 4.19** depicts variations in mud resistivity that occurred during the drilling of a well in Brazoria County, TX. Unless R_m and R_{mf} measurements are made routinely during the drilling phase, the exact R_{mf} value that corresponds to the mud filtrate that

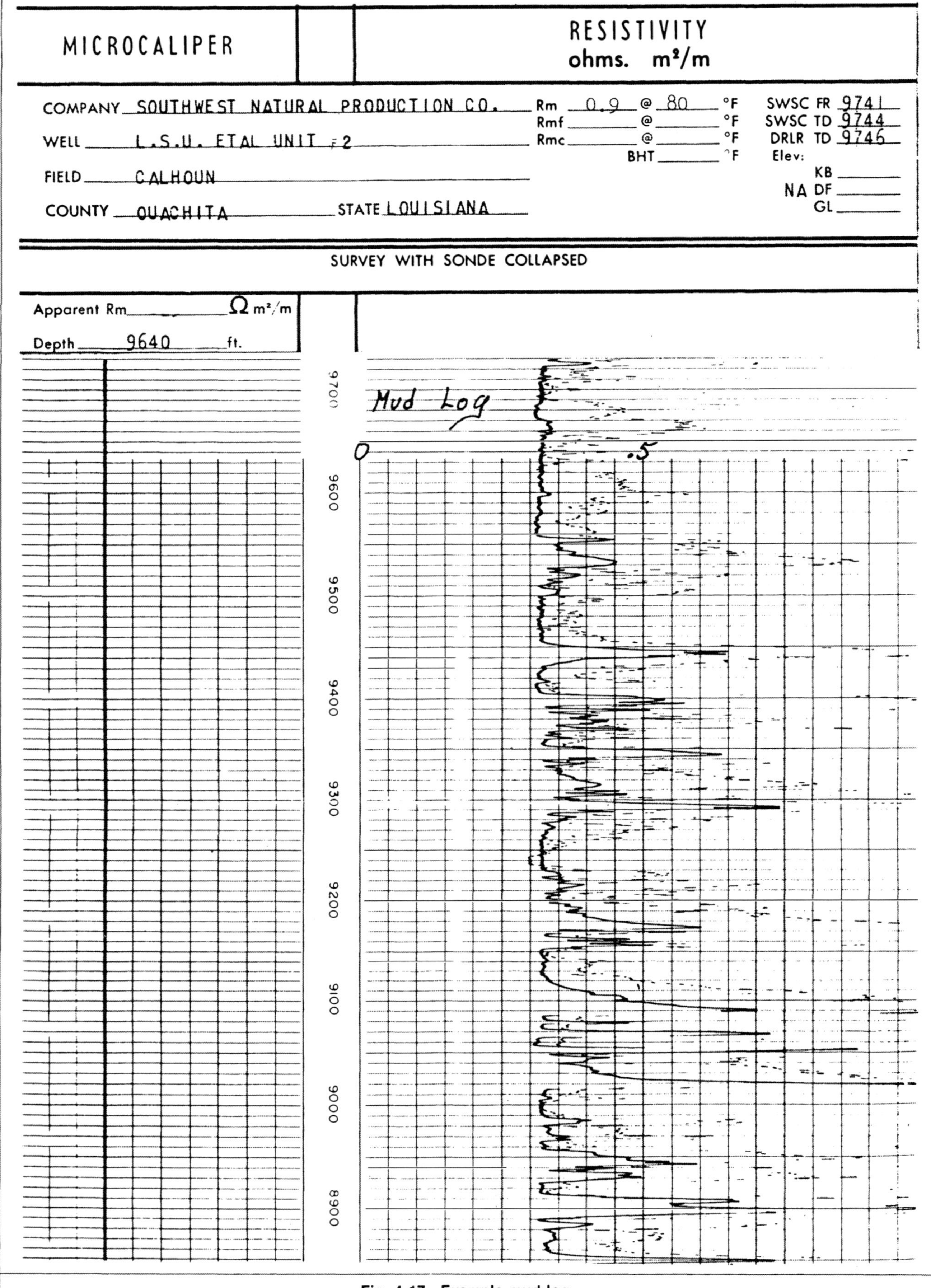
MICROCALIPER | RESISTIVITY ohms. m²/m

COMPANY SOUTHWEST NATURAL PRODUCTION CO. Rm 0.9 @ 80 °F SWSC FR 9741
WELL L.S.U. ETAL UNIT #2 Rmf @ °F SWSC TD 9744
FIELD CALHOUN Rmc @ °F DRLR TD 9746
COUNTY OUACHITA STATE LOUISIANA BHT °F Elev: NA KB DF GL

SURVEY WITH SONDE COLLAPSED

Apparent Rm Ω m²/m
Depth 9640 ft.

Fig. 4.17—Example mud log.

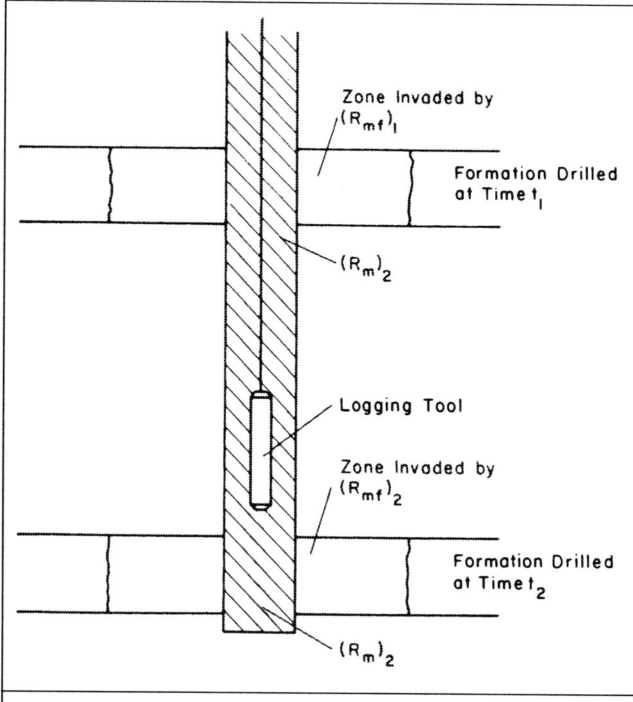

Fig. 4.18—Schematic showing effect of variation of drilling-mud properties.

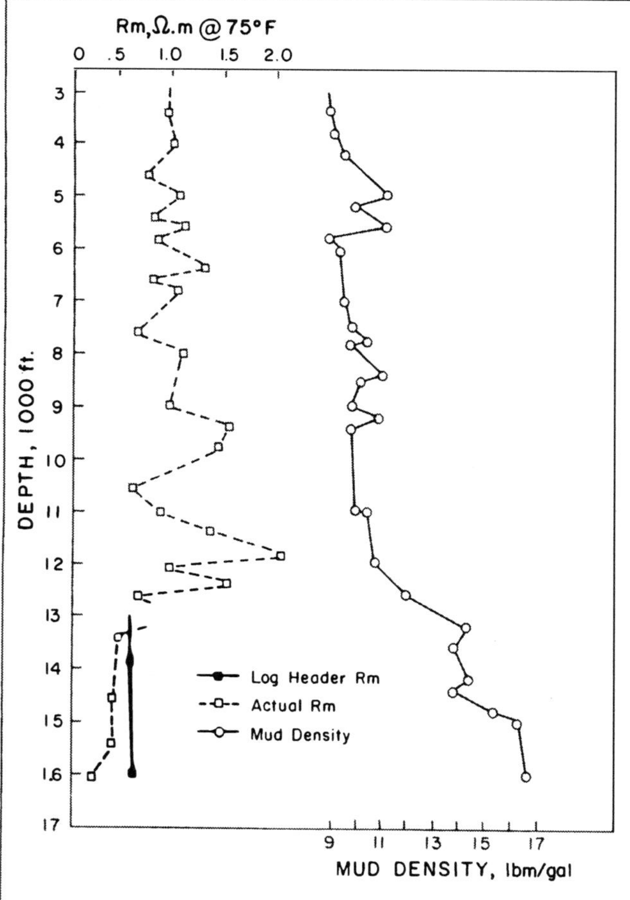

Fig. 4.19—Log Header, R_m, and short-term variation of R_m, and ρ_m vs. depth for a Brazoria County, TX, well (from Ref. 14).

actually invaded the formation remains unknown. This lack of data could affect the interpretation of both the SP and the shallow resistivity logs.

4.4 Invasion Profile

As mud filtrate invades a permeable formation, it displaces formation fluids and mixes with formation water. The invasion alters fluid distribution around the wellbore and subsequently creates zones of resistivities that differ considerably from the formation's true resistivity. Knowing what invasion profile to expect in different formations greatly helps with the interpretation of logs, especially resistivity logs.

The mud-filtrate volume to invade a permeable formation is determined by the fluid loss of the mud, differential pressure between borehole and formation, mudcake permeability, and length of time that the formation has been exposed to the mud.[16,17] The filtrate volume that invades a formation increases as these parameters increase. Note, however, that most of the invasion occurs within a few hours after drilling because filtration is accompanied by mudcake buildup. The low-permeability cake eventually reaches a thickness across which the available pressure drop cannot sustain a significant flow rate.

The formation porosity usually determines the depth of invasion. For a given fluid loss, differential pressure, mudcake permeability, and exposure time, the invasion in a low-porosity formation is deeper than in a high-porosity formation.

4.4.1 Step Profile. Assuming piston-like displacement (i.e., only the filtrate is moving behind the invasion front) results in the ideal step profile of resistivity shown in **Fig. 4.20**. R_{xo} and R_t are the resistivities of the flushed zone and uninvaded formation, respectively. R_t and R_{xo} can be expressed according to Eq. 1.43 as

$$R_t = FR_w/S_w{}^n \quad (4.6)$$

and $$R_{xo} = FR_{mf}/S_{xo}{}^n, \quad (4.7)$$

where S_{xo} is the saturation of mud filtrate in the flushed zone. Because mud filtrate and formation water are miscible, S_{xo} in a water-bearing formation is unity and in an oil-bearing formation is $(1-S_{or})$, with S_{or} being the residual oil saturation. Assuming a step profile yields the simplest possible geometry and subsequently is often retained in the analysis of log measurements.

In water-bearing zones and in the case of $R_{mf}>R_w$, R_{xo} is always greater than R_o. In an oil-bearing formation, the relation of R_{xo} to R_t depends on the R_{mf}/R_w contrast; the values of oil saturation, S_o, in the uninvaded zone; and the residual oil saturation, S_{or}, in the flushed zone. Dividing Eq. 4.7 by Eq. 4.6 yields

$$R_{xo}/R_t = (R_{mf}/R_w)(S_w/S_{xo})^n \quad (4.8)$$

and $$R_{xo}/R_t = (R_{mf}/R_w)[(1-S_o)/(1-S_{or})]^n. \quad (4.9)$$

Because $R_{mf}>R_w$, $R_{mf}/R_w>1$, and because $S_o>S_{or}$, $S_{xo}>S_w$ and $S_w/S_{xo}<1$. The value of the ratio R_{xo}/R_t relative to unity cannot be determined *a priori*. In most cases, however, R_{mf}/R_w is large enough to offset the contribution of S_w/S_{xo} and $R_{xo}>R_t$.

Example 4.3. A given fluid loss, differential pressure, mudcake permeability, and exposure time indicate that 20 L of mud filtrate invaded a porous, permeable oil-bearing formation. If the formation thickness, h, is 10 ft, $S_{or}=25\%$, and borehole diameter, d_h, is 8 in., calculate the diameter of invasion, d_i, if (a) $\phi=40\%$ and (b) $\phi=10\%$.

Solution. Assuming piston-like displacement, the mud filtrate volume, V_{mf}, in the invaded zone is

$$V_{mf} = (\pi/4)(d_i^2 - d_h^2)h\phi S_{xo}. \quad (4.10)$$

Solving for d_i yields

$$d_i = [(4V_{mf}/\pi h\phi S_{xo}) + d_h^2]^{1/2} \quad (4.11)$$

$$= \left\{\frac{4[20{,}000/144(30.48)^3]}{\pi(10)\phi(0.75)} + 8^2\right\}^{1/2} = \left(\frac{17.266}{\phi} + 64\right)^{1/2} \text{ in.}$$

(a) For $\phi=40\%$, $d_i=10$ in.;
(b) for $\phi=10\%$, $d_i=15$ in.

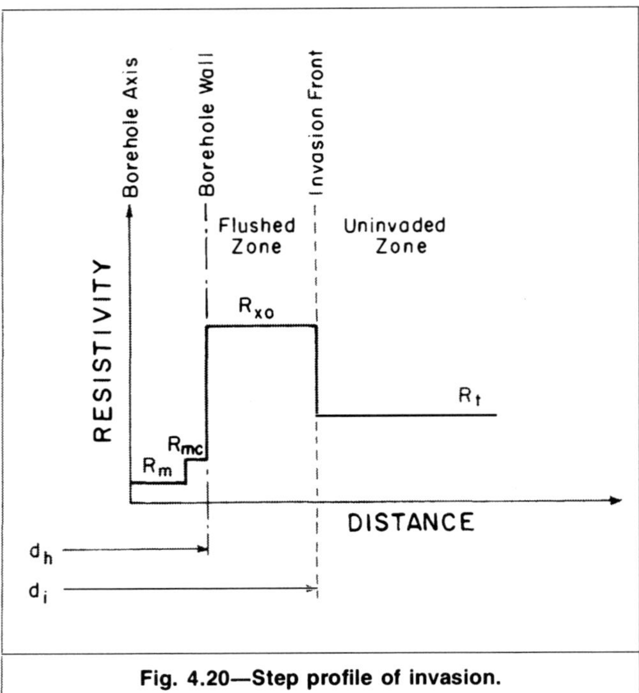

Fig. 4.20—Step profile of invasion.

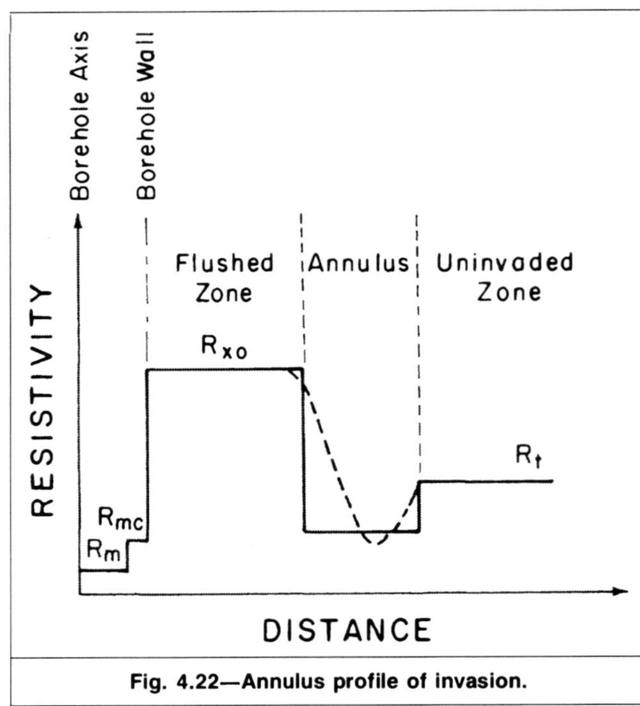

Fig. 4.22—Annulus profile of invasion.

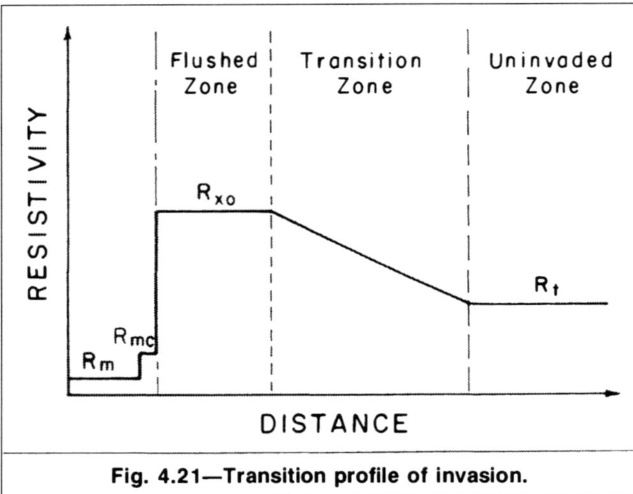

Fig. 4.21—Transition profile of invasion.

4.4.2 Transition Profile. Formation water and mud filtrate mix by diffusion. A salinity gradient exists between the uninvaded, where the salinity corresponds to that of formation water, and the flushed zone, where the salinity equals that of the mud filtrate. The salinity gradient generates, in turn, a resistivity gradient. **Fig. 4.21** shows the resistivity profile for $R_{mf} > R_w$ with the assumption of a linear gradient. The zone over which the resistivity gradually changes from R_{xo} to R_t is called the transition zone.

In an oil-bearing formation where formation water saturation is irreducible, a resistivity profile similar to that of Fig. 4.21 is also developed. In certain cases, however, R_t can be greater than R_{xo}, as previously explained.

4.4.3 Annulus Profile. In an oil-bearing formation where water saturation exceeds the irreducible level, Gondouin and Heim[18] showed experimentally that a formation water bank is formed at the displacement front. The water bank is first followed by a zone of mixed waters, then the flushed zone near the wellbore. When $R_{mf} > R_w$, this fluid distribution results in a resistivity profile that can be approximated by the solid line in **Fig. 4.22**. In reality, capillarity and mixing result in a rounded profile shown by the dashed line.

The presence of a low-resistivity annulus was confirmed by the running of a combination of several resistivity tools with different radii of investigation. However, an annulus does not form in all oil zones because its formation requires favorable water saturation, fluid viscosities, and relative permeability characteristics. The annulus was also found to dissipate with time. An annulus could form and disappear before the well is logged. If an annulus is detected by resistivity tools, it indicates the presence of oil. It also indicates that water is mobile. Subsequently, a completion in a zone that exhibits an annulus produces both oil and water.[19]

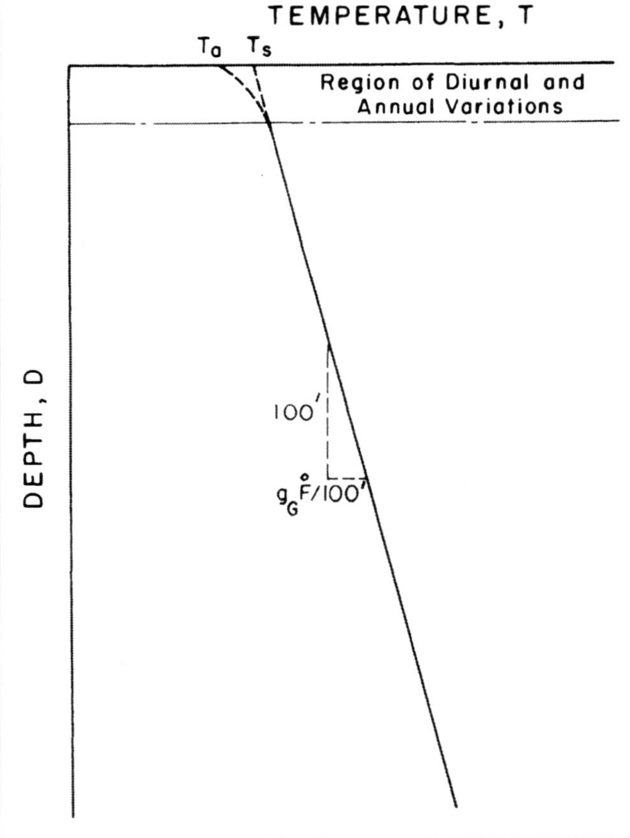

Fig. 4.23—Schematic of linear distribution of subsurface temperature.

TABLE 4.2—THERMAL CONDUCTIVITY OF SELECTED MATERIAL (10^{-3} cal/(sec-cm-°C) (after Ref. 22)

Material	Conductivity
Shale	2.8 to 5.6
Sand	3.5 to 7.7
Porous limestone	4 to 7
Dense limestone	6 to 8
Dolomite	9 to 13
Quartzite	13
Gypsum	3.1
Anhydrite	13
Salt	12.75
Sulphur	0.6
Steel	110
Cement	0.7
Water	1.2 to 1.4
Air	0.06
Gas	0.065
Oil	0.35

4.5 Formation Temperature

4.5.1 Subsurface Temperature Distribution. The prevailing temperature in a wellbore determines the resistivity value of the mud and its derivatives in the region surrounding the tool. The temperature also affects most formation properties (e.g., formation water resistivity). Temperature distribution in a wellbore must be known for proper log analysis. The electronics and sensors of the tool are sensitive to temperature, so the maximum temperature encountered in the wellbore dictates tool design and tool selection.

Earth temperature increases with depth because the core is extremely hot. Approximating the Earth's crust that surrounds a wellbore by an infinite and homogeneous slab of constant thermal conductivity, k_h, we can express heat conduction by using Fourier's law:

$$\dot{Q}=k_h A(\mathrm{d}T/\mathrm{d}z), \quad (4.12)$$

where $\dot{Q}$=heat flow rate, dT/dz=vertical temperature gradient, and A=area at a right angle to the direction of flow.

For heat flow through the Earth's crust, the amount of heat flow through A is independent of time[20]; i.e., $\dot{Q}$ is a constant. Then,

$$\mathrm{d}T/\mathrm{d}z=g_G, \quad (4.13)$$

where g_G is a constant called the geothermal gradient. Integrating between the surface and depth, D, gives

$$\int_{T_s}^{T_f}\mathrm{d}T=g_G\int_0^D \mathrm{d}z,$$

$$T_f-T_s=g_G D,$$

and $T_f=T_s+g_G D$. (4.14)

Eq. 4.13 and thus Eq. 4.14 are not valid close to the surface. Down to about 100 ft, the heat flow and temperature are affected by diurnal and annual variations in air temperature.

According to Eq. 4.14, subsurface temperature varies linearly with depth (**Fig. 4.23**). Several observations indicate that the mean annual temperature of air, T_a, at the ground surface is less than T_s that results from the extrapolation of the linear temperature trend.[21]

The thermal conductivity of formations usually penetrated by a well is not constant. **Table 4.2** lists typical values for selected materials encountered in the environment surrounding a wellbore. The ideal linear temperature distribution curve is modulated by the change of thermal conductivity from one bed to another (**Fig. 4.24**). Neglecting these local modulations has been an acceptable practice in logging applications.

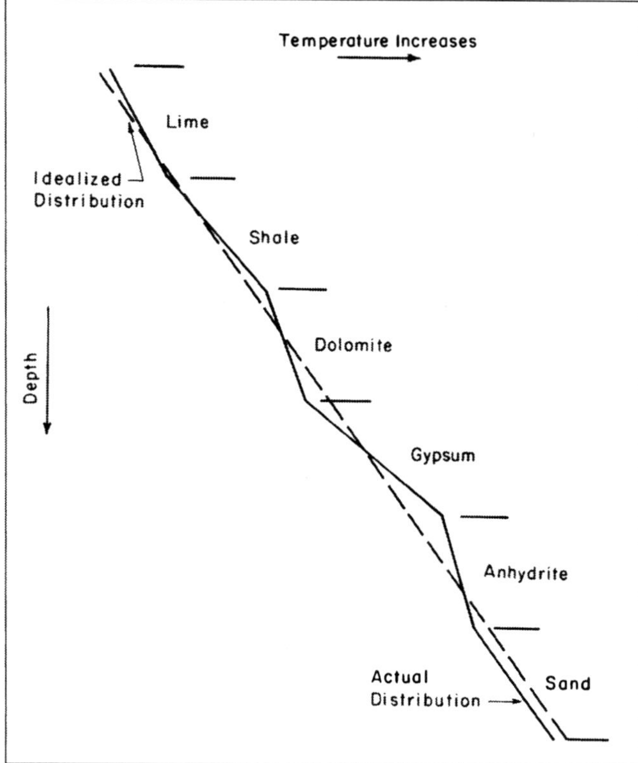

Fig. 4.24—Actual and idealized subsurface temperature distribution.

4.5.2 Temperature Log. Tools are available that provide a continuous measurement of temperature in a wellbore. The temperature sonde consists mainly of a platinum wire exposed to the borehole fluid. The resistivity of the wire, which is measured with a wheatstone bridge, varies with temperature according to a simple, well-known relationship. The resolution of this platinum-resistance thermometer is about 0.5°F.

A newer generation of temperature tools uses a semiconductor-type thermometer called a thermistor. Electric properties of thermistors vary with and can be calibrated to indicate temperature. Thermistors have a resolution of 0.005°F and respond to temperature changes much faster than the resistance thermometer.

Continuous temperature measurement plays an important role in production operations. Temperature logs are very useful in locating cement tops and detecting gas entry into the wellbore, gas and water channels behind the casing, etc. The aforementioned situations are accompanied by temperature anomalies. The detection of such anomalies, which can be quite small, is enhanced by recording the temperature gradient vs. depth. This recording usually, is called the differential log. The differential log is obtained by electronically memorizing the temperature at a certain depth, then subtracting it from a subsequent reading obtained at a known distance from the first reading point. **Fig. 4.25** shows a section of a temperature log. It exhibits a linear trend that increases with depth. The average rate of temperature change is the geothermal gradient in this well. The log shows localized variations in slope caused by changes in thermal conductivity of different beds. These variations are clearly shown by the differential curve. This temperature survey was run inside the tubing in a completed well. The tic marks on the left of the log at about 30-ft intervals correspond to tubing joints. These joints are detected by a special sensor called the casing collar locator. These sensors are usually incorporated into tools run in cased holes.

4.5.3 Maximum-Indicating Thermometer. In an open (uncased) hole, minimizing logging time is desirable. The longer the hole is open, the higher is the risk of drilling complications that affect safety and cost. The practice is to reduce the number of logging runs. Several tools can be run simultaneously. The amount of data that can be collected during one run is limited, however, by the electrical and mechanical properties of the logging cable. For this reason, it is always desirable to collect, if possible, some of the logging data without cable transmission.

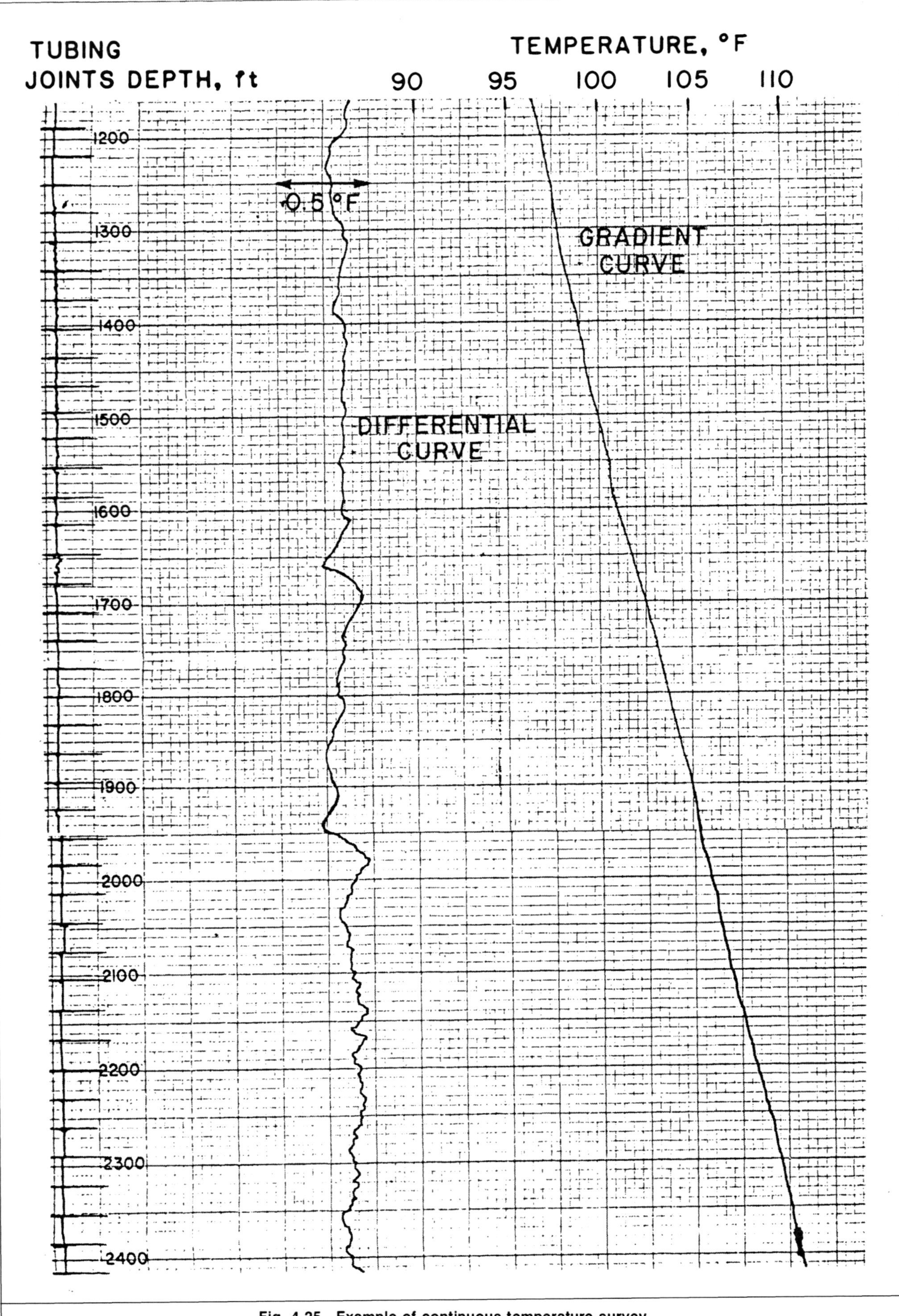

Fig. 4.25—Example of continuous temperature survey.

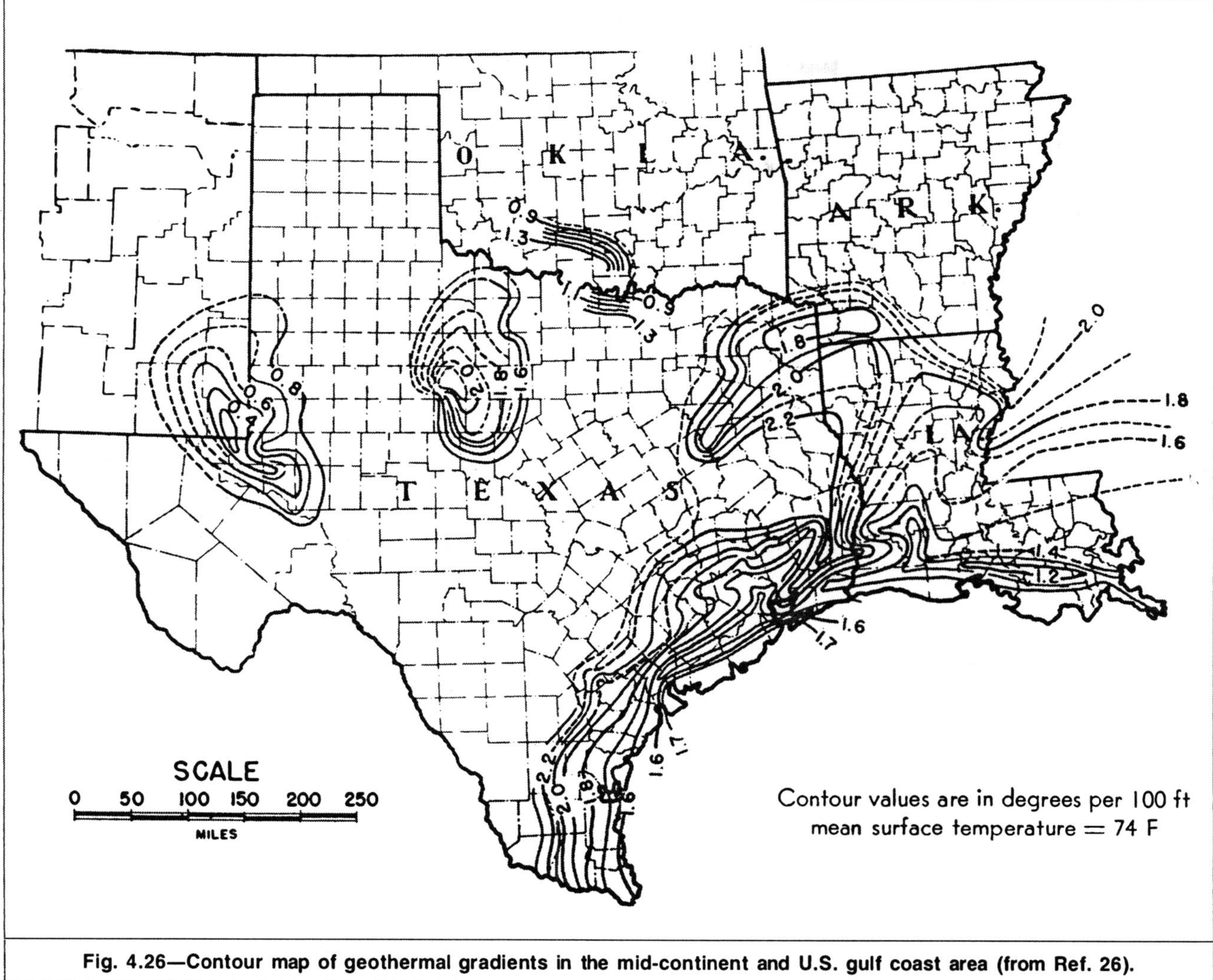

Fig. 4.26—Contour map of geothermal gradients in the mid-continent and U.S. gulf coast area (from Ref. 26).

Downhole temperature is an example of the data collected without cable use. With linear temperature distribution assumed, only temperature values at two different depths are needed to define the temperature gradient and hence the temperature at any desired depth. One of these values is the temperature near the surface, T_s, which can easily be determined *a priori*. Only one temperature value actually needs to be measured in the hole. This value is determined with a maximum-indicating thermometer.

A maximum-indicating thermometer is similar to a medical thermometer. It consists of a capillary tube attached to a mercury-filled glass reservoir. A restriction in the capillary tube exists immediately above the mercury reservoir. The mercury expands and rises in the capillary tube, which is scaled to read temperature. Because the restriction prevents the mercury from receding into the reservoir, the thermometer indicates the highest temperature to which it has been exposed. The thermometer must be shaken vigorously for the mercury to recede. A maximum-indicating thermometer is always run in open holes. The thermometer is enclosed in a steel housing and placed in a special slot in the tool wall. The thermometer reading after a logging run is taken to be the bottomhole temperature (BHT), T_{bh}. Note that few effects can cause the thermometer to indicate a nonrepresentative value. Mercury thermometers have some lag and require a few minutes to reach thermal equilibrium with the surroundings.[23] In a high-pressure environment, the mercury reservoir is compressed to a degree that causes erroneously high readings.[24] If hole conditions cause the tool to jar and vibrate, the mercury in the capillary tube could recede, causing erroneously low readings. To check for any of these effects, it is customary to run more than one thermometer.[25]

4.5.4 Gradient and Formation Temperature Calculations. The maximum recorded temperature, T_{bh}, is used to calculate the geothermal gradient, g_G. Eq. 4.14 can be arranged so that

$$g_G=[(T_{bh}-T_s)/D_{bh}]100, \quad (4.15)$$

where D_{bh}=total depth (TD) of the logged borehole and g_G=average geothermal gradient practically expressed in °F/100 ft or °C/100 m, depending on the units used.

In most cases, the temperature near the surface, T_s, falls between 60 and 80°F. An average value of 70°F can be assumed. The formation temperature, T_f, at any depth, D, is calculated from Eq. 4.14, which is expressed as

$$T_f=T_s+g_G(D/100) \quad (4.16)$$

to account for the practical unit of g_G.

The geothermal gradient varies from region to region. Its value is determined by the subsurface geologic structure. Contour maps can be used to show the geographical variation in geothermal gradient. **Fig. 4.26** is an early example of such a map prepared for the midcontinent and U.S. gulf coast areas.[26] This type of map must be updated as more temperature/depth information becomes available. Fig. 4.26 has been updated once[27] and could be updated again and again as more data are collected.

Example 4.4. A maximum temperature of 216°F is measured in a well drilled to 14,600 ft TD.

a. Calculate the geothermal gradient.

b. Calculate the temperature of a formation situated at 12,000 ft.

c. Investigate the calculated gradient's and the formation temperature's sensitivities to surface temperatures between 60 and 80°F.

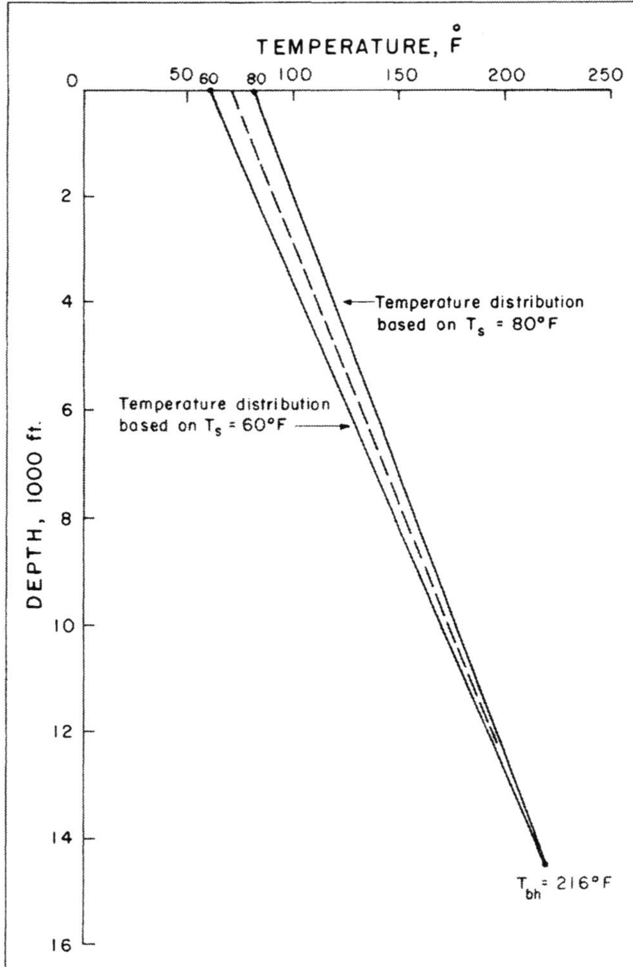

Fig. 4.27—Sensitivity to assumed surface temperature in the 60 to 80°F range for Example 4.4.

Solution.

a. Using Eq. 4.15 and assuming a surface temperature of 70°F gives

$$g_G = \frac{216-70}{14{,}600} 100 = 1\,°F/100 \text{ ft}.$$

b. Using Eq. 4.16 yields

$$T_f = 70 + 1\frac{12{,}000}{100} = 190°F.$$

c. Geothermal gradients of 1.068 and 0.932°F/100 ft and formation temperatures of 188 and 192°F are calculated with surface temperatures of 60 and 80°F, respectively.

The deviation in calculated gradient is only ±0.068°F/100 ft or ±6.8%. The deviation in the calculated formation temperature is ±2°F, or only ±1.1%. The sensitivity to assumed surface temperature in the range of 60 to 80°F is very small in this case. It generally decreases as the TD increases. The sensitivity to assumed surface temperature is illustrated graphically in **Fig. 4.27**. The separation between the two illustrated trends is very small in the lower part of the hole, which is usually the part of practical interest.

4.5.5 Static Formation Temperature. During the drilling of a borehole, mud is pumped from mud tanks down the drillpipe. It returns to the surface by moving up the annulus. When drilling-fluid circulation starts, the mud temperature is relatively low. Starting at the bit level, heat flows from the relatively hot formation to the mud. The mud continues to be heated by the formation as it travels up the annulus. At a certain depth, the mud temperature is higher than the formation temperature, and heat flows from the mud to the formation. This cycle of heating and cooling the drilling mud continues during borehole drilling and cleaning. The cleaning phase starts when the bit reaches the target depth and drilling is stopped. The mud is circulated for a few hours to remove all cuttings from the wellbore. When the borehole is judged to be "clean" (i.e., reasonably free of cuttings), circulation is stopped, the drillstring is removed, and the logging tool is run in the borehole.

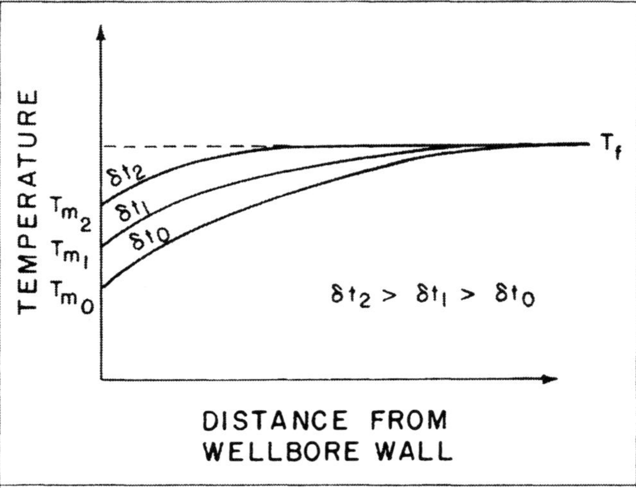

Fig. 4.28—Schematic of temperature distribution around a wellbore at different time intervals since circulation stopped.

When mud circulation is stopped, the mud temperature, T_m, at the bottom of the borehole is lower than the surrounding formation temperature, T_f. **Fig. 4.28** is a schematic of temperature radial distribution around the wellbore for different values of the time elapsed since circulation has stopped, δt. **Fig. 4.29** shows that the maximum-indicating thermometer records the mud temperature, which is lower than the formation temperature. How far the maximum recorded temperature is from T_f depends on several factors: the time δt at which the thermometer reaches the bottom of the hole after circulation has stopped; the mud temperature at $\delta t=0$ (i.e., the time circulation first stopped); and the volume of mud to be heated, which is proportional to the borehole diameter.[28]

Assuming that temperature buildup is similar to pressure buildup allows mud temperature vs. time to be expressed as[29]

$$T_m(\delta t) = T_f - C \log\frac{\delta t}{t+\delta t}, \quad \text{(4.17)}$$

where t=mud circulation time, δt=time elapsed since circulation stopped, $T_m(\delta t)$=mud temperature at time δt, T_f=static formation temperature, and C=constant.

A plot of $T_m(\delta t)$ vs. $(\delta t/t+\delta t)$ on semilog paper (**Fig. 4.30**) is a straight line. Because lim $(t+\delta t)/\delta t = 1$, the value of T_m read at $(t+\delta t)/\delta t = 1$ corresponds to T_f.

This plot, known in pressure-buildup analysis as the Horner plot, can be used to determine T_f, provided that maximum-indicating thermometer readings are available at different δt's and that mud circulating time, t, is known. More than one $T_m(\delta t)$ value is usually available when more than one logging run is used to obtain the needed data.

The circulating time is not part of the standard data collected and recorded by the logging engineer. Fortunately, as Example 4.6 illustrates, the estimation of T_f with the Horner-type plot is not very sensitive to t, which can be obtained from drilling records. If drilling records are not easily accessible, t can be estimated from the well depth. The estimate is based on the time range necessary for one or two complete circulations of the mud.

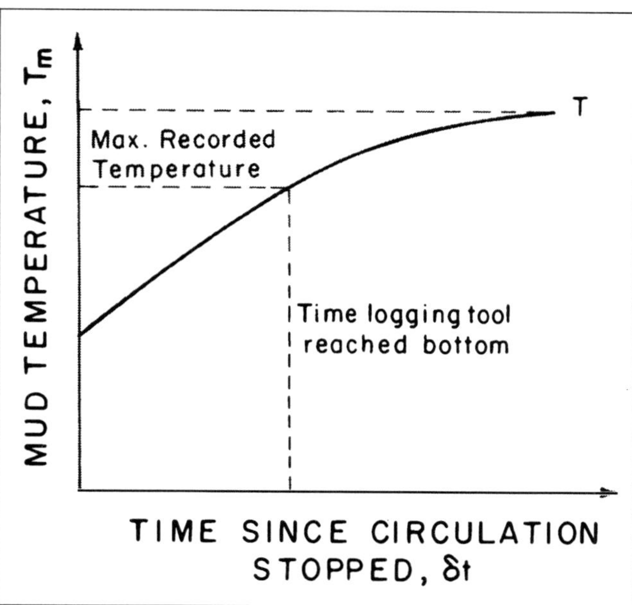

Fig. 4.29—Mud temperature vs. time elapsed since circulation stopped.

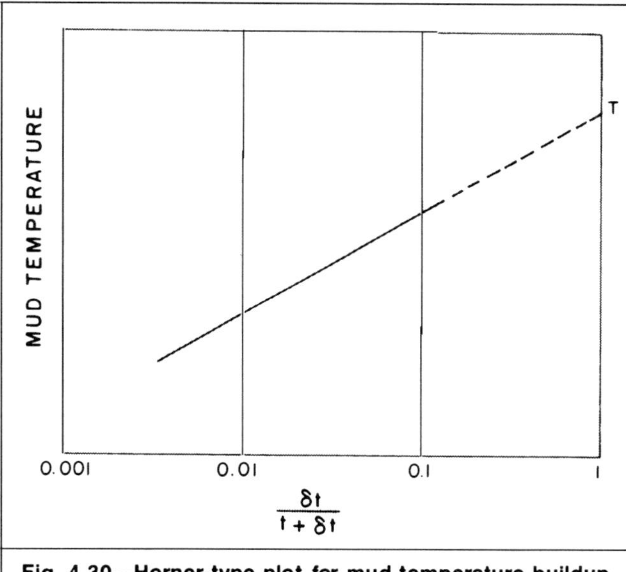

Fig. 4.30—Horner-type plot for mud temperature buildup.

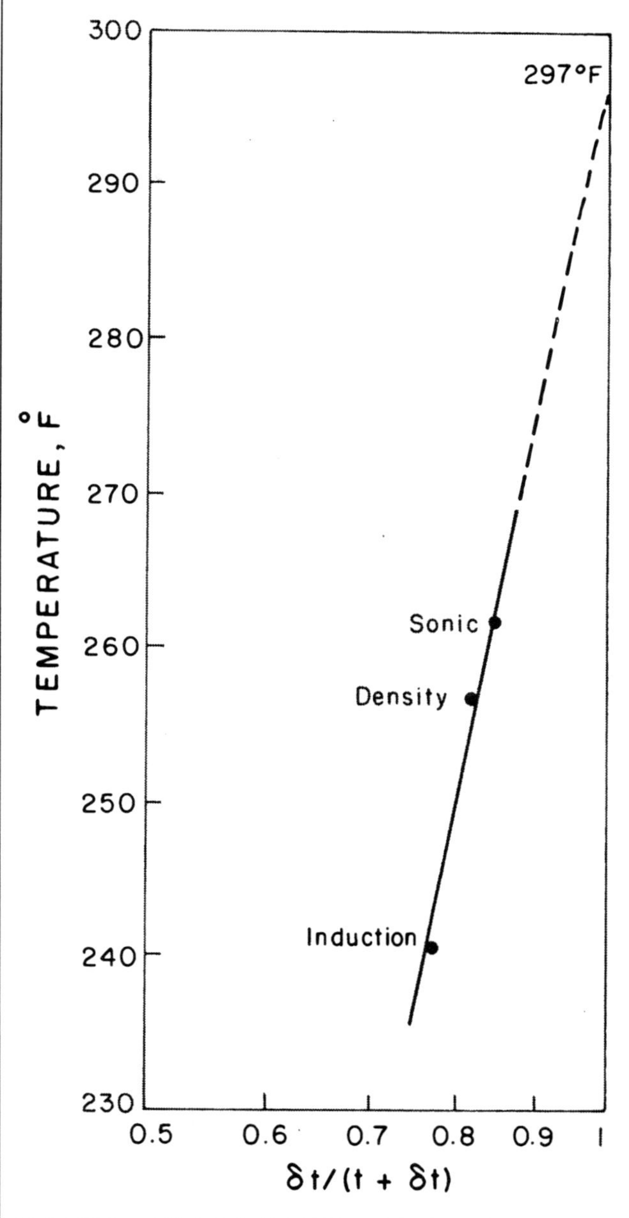

Fig. 4.31—Horner-type plot for temperature buildup of Example 4.5.

Example 4.5. Drilling of a 16,200-ft borehole was finished at 1:30 a.m. The mud was then circulated until 4:00 a.m. Three logging tools were run to TD successively. The tool type, the time the tool moved off bottom, and the maximum recorded temperatures are shown below.[30]

Tool	Time Off Bottom	Maximum Recorded Temperature (°F)
Induction	12:15	241
Density	15:00	257
Neutron	17:30	262

a. Estimate T_f at 16,200 ft.
b. Calculate g_G.
c. If the formation-water salinity is 100,000 ppm, determine the error in the hydrocarbon saturation calculation that is based on 241°F instead of the static temperature estimated in Part a.

Solution.
a. t=2.5 hours.

Tool	T_m (°F)	δt (hours)	$\delta t/(t+\delta t)$
Induction	241	8.25	0.767
Density	257	11.0	0.815
Sonic	262	13.50	0.844

Fig. 4.31 shows a plot of T_m vs. $\log[\delta t/(t+\delta t)]$. Extrapolating the best fit through the three data points to $\delta t/(t+\delta t)=1$ yields T_f=297°F.

b. Assuming T_s=70°F and using Eq. 4.15 gives

$$g_G=\frac{297-70}{16{,}200}\times 100=1.4\text{°F}/100 \text{ ft.}$$

c. Let S_{w1} and S_{w2} be the water saturation values calculated with 241°F and 297°F, respectively. Because the formation temperature used affects only the calculated value of R_w, then

$$S_{w1}=(FR_{w1}/R_t)^{1/2}$$

and $S_{w2}=(FR_{w2}/R_t)^{1/2}$,

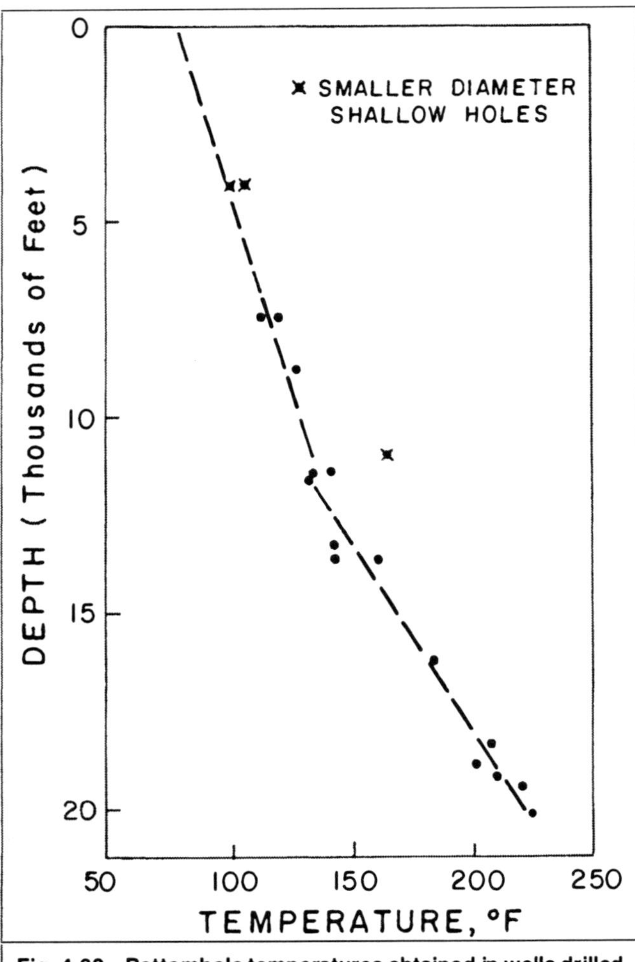

Fig. 4.32—Bottomhole temperatures obtained in wells drilled in Pecos County, TX (from Ref. 23).

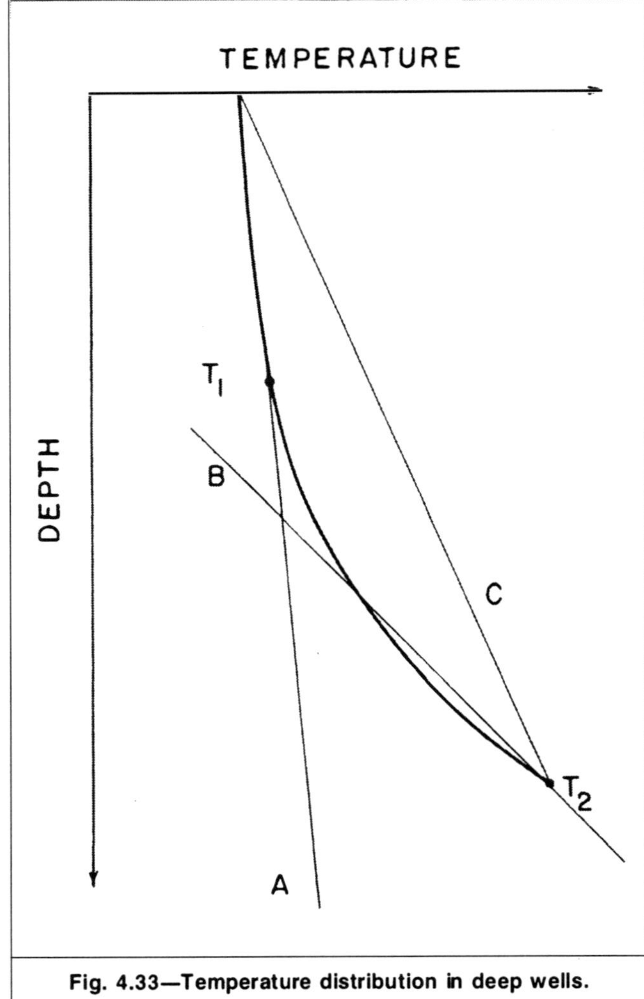

Fig. 4.33—Temperature distribution in deep wells.

and the relative variation of water saturation can be expressed as

$$\frac{\Delta S_w}{S_w}=\frac{S_{w1}-S_{w2}}{S_{w2}}=\frac{R_{w1}{}^{1/2}-R_{w2}{}^{1/2}}{R_{w2}{}^{1/2}}$$

From Fig. 4.15, $R_{w1}=0.0255\ \Omega\cdot\text{m}$, $R_{w2}=0.0210\ \Omega\cdot\text{m}$, and $\Delta S_w/S_w=0.10$ or 10%.

A theoretical study[30] showed that a Horner-type analysis of temperature buildup is not mathematically correct. The Horner plot results in a value of T_f that is still less than the actual value. However, for the short circulating times that prevail in most practical field conditions, the method yields reasonable estimates of T_f.

4.5.6 Temperature Distribution in Deep Wells. Deep wells are drilled in several sections. After a section is drilled, it is logged and cased before drilling begins on the following section, which is, of course, drilled with a smaller bit. In addition to log runs performed at the target depth of each section, logs are run at intermediate depths to check for abnormal conditions, such as pressure transition zones, which signal the approach of an abnormally pressured interval.

Plotting the maximum recorded temperature obtained at each logging depth usually yields a plot similar to that of **Fig. 4.32.** The plot shows two linear segments of markedly different slopes—i.e., geothermal gradients. This change in slope or "elbow" is also observed in wells drilled along the U.S. gulf coast and in Oklahoma.[23]

Several explanations exist for the change in slope. A drastic change in the range of formations' thermal conductivities from low to high can produce such a change in geothermal gradient. This explanation, however, does not apply in this case because shallow formations exhibit lower thermal conductivities owing to reduced consolidations and higher porosities.[23] Transition from a normal to an abnormal pressure environment cannot always explain such a change in slope. The top of the geopressured zone does not always correlate with the depth of the elbow on the temperature/depth curve.[31,32] Another possible explanation is that deep wells usually have a large diameter near the surface and a small diameter in the deeper section. Large-diameter boreholes are expected to display lower temperatures than small holes because greater volumes of mud must be heated after circulation stops. But again, a weak correlation between hole sizes and temperature trend slopes is observed.[24]

It is believed that the assumption used to derive Eq. 4.13, which resulted in a linear temperature distribution, is not valid for deep wells. The temperature distribution is actually a curve, as illustrated in **Fig. 4.33.** The curve can be approximated by linear Segment A in the upper section of the well. This linear segment was observed in shallow wells drilled earlier and promoted confidence in the assumption of linear distribution. In deeper wells, however, a departure from the linear distribution is observed. The lower part of the curve can also be approximated by a linear Segment B. Segments A and B meet, forming an elbow as illustrated in Fig. 4.32.

Assuming linear distribution in deep wells can result in misrepresentation of formation temperature. As Fig. 4.33 shows, defining a geothermal gradient based on Temperature T_1 obtained at intermediate depth and using it for the deeper portion of the hole results in underestimation of the formation temperature. Defining a gradient based on Temperature T_2 measured at TD results in overestimation of formation temperature. Representing temperature distribution with two linear segments as in Fig. 4.32 is valid. It requires, however, that data be available at depths above the elbow, which is not always the case.

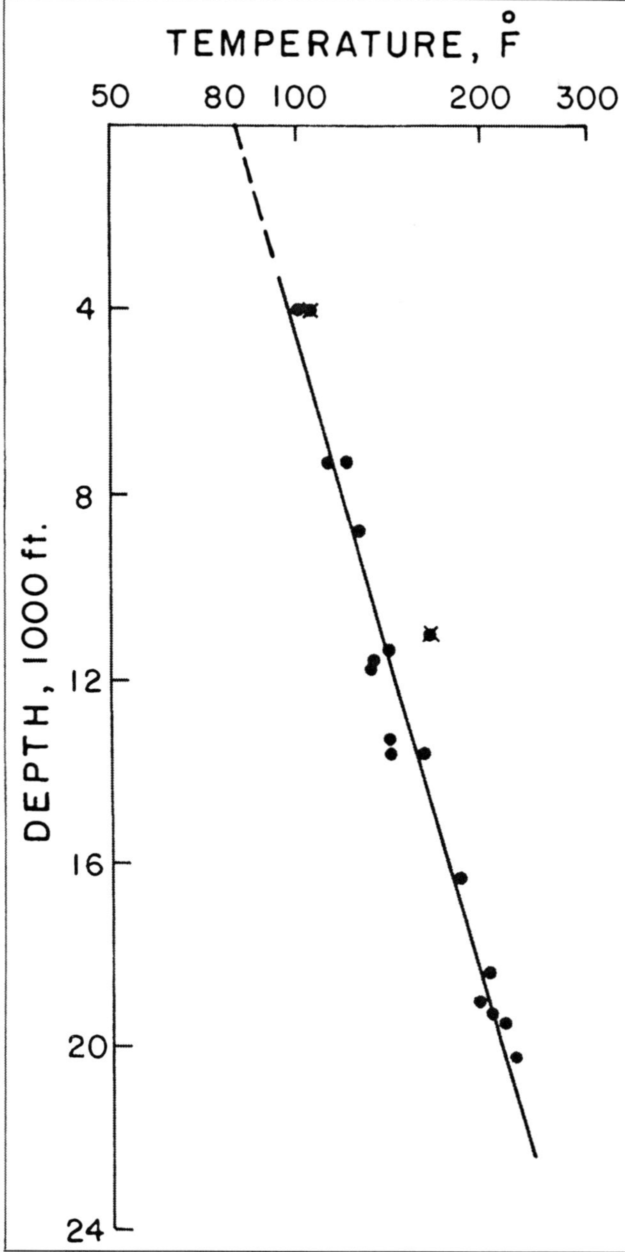

Fig. 4.34—Semilog plot of temperature data obtained in wells drilled in Pecos County, TX.

Plotting the temperature data of Fig. 4.32 on semilog paper results in a linear trend, as shown in **Fig. 4.34.** Similar plots prepared for the U.S. gulf coast area also displayed linear trends.[32] This observation suggested that the temperature/depth relationship can be expressed empirically as[33]

$$T_f = T_s e^{m_T D}, \quad (4.18)$$

where m_T is a constant related to the slope of the linear trend on the semilog plot. m_T varies from region to region and possibly from well to well. The main advantage of the semilog plot of temperature vs. depth is that one curve characterized by one constant m_T expresses temperature distribution from top to bottom of the borehole. With one parameter, m_T, contour maps similar to that of Fig. 4.26 can be constructed to give temperature variation regardless of well depth.

4.6 Record of Measurement Environment

Data pertaining to measurement environment and other relevant data are recorded on a log heading, which is the front page of the folded log. The API recommends the standard heading shown in **Fig. 4.35.** Using a uniform log heading and presentation provides space for all data pertinent to the log interpretation.

The heading is subdivided into several sections, with each section listing several data items. The items identified by numbers in Fig. 4.35 are explained below. The explanation is based on the API recommendation.[5]

1. Logging company name and log name.
2. Well identification, including the operating company, well name and number, and field, county, and state where the well is located.
3. Well identification typed or printed perpendicular to the heading's main orientation for filing purposes.
4. Exact well location according to the system used in the state. The most common designation uses the section/township/range system.
5. Other logs and services run in the well (for completeness and quick reference).
6. A permanent depth datum and its elevation. The permanent datum is selected so that rig removal does not result in a loss of depth reference. The API recommends that the top surface of the surface casing's upper flange, called the bradenhead flange (BHF), be used as the permanent reference plane. Some operators use ground level (GL) or mean water level (MWL) as the permanent datum.
7. The datum from which the log depth is measured and its elevation above the permanent datum. The API recommends that all logs be measured from the rotary kelly bushing (KB).
8. The datum from which the driller's measurements are made. The kelly bushing is also recommended for these measurements.
9. The exact elevation from sea level of the ground level, derrick floor (DF), and kelly bushing. These elevations are helpful in constructing subsurface structural maps and cross sections.
10. The run number. Each time the logging tool is run in a well, the run is assigned a number. The run date and other measurement environment data pertaining to the run are listed. Space is provided for four runs. The lower portion of the heading is duplicated to record additional runs.
11. The depth of the bottom of the hole at the time of logging according to driller's and logger's measurements, respectively. The driller's and logger's measurements of depth usually differ by a few feet, mainly because they use different means of determining depth. The driller uses the length of the rigid drillstring, while the logger relies on the length of the flexible logging cable.
12. The depth of the top and bottom of the interval logged.
13. The casing size and the depth according to driller's and logger's measurements. The section of the well above the logged open hole is cased.
14. The bit size used to drill the borehole being logged.
15. The type of drilling mud in the hole at the time of logging. Also listed is the mud density in pounds per gallon, mud viscosity in Marsh funnel seconds, pH, and fluid loss in cubic centimeters per 30 minutes.
16. The source of the mud sample used to determine the properties listed in Item 15. The API recommends that a circulated mud sample be used. This sample is collected from the circulated flow stream immediately before circulation is stopped and the drillstring is retrieved to prepare for logging. The API also recommends that samples collected from mud tanks ("pits") not be used because they may be nonrepresentative.
17. Mud, mud filtrate, and mudcake resistivities measured according to the procedure outlined in Sec. 4.3.6. The resistivity values are given in ohm meters and the measurement temperature in degrees Fahrenheit.
18. Mud filtrate and mudcake resistivities either measured or calculated with the correlation of Sec. 4.3.7. The sources of R_{mf} and R_{mc} values have to be indicated. "M" is usually used to show that the value is measured on a sample; "C" is usually used to show that the value is calculated or determined from a chart.
19. The value of mud resistivity at BHT, usually estimated with the chart in Fig. 4.15 or a similar chart developed specifically for the mud type in the hole.

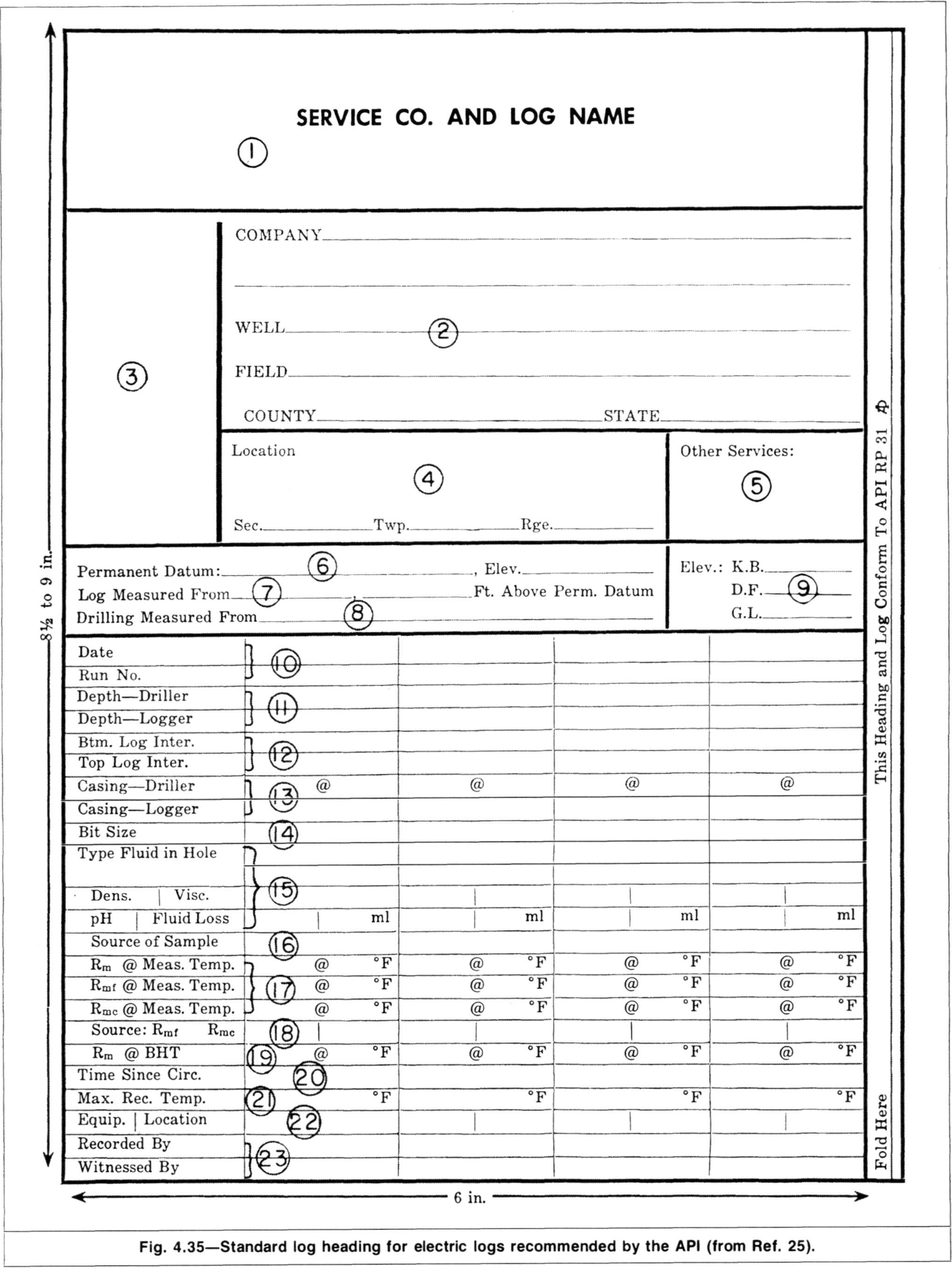

SERVICE CO. AND LOG NAME ①

③

COMPANY

WELL ②

FIELD

COUNTY STATE

Location ④

Sec. Twp. Rge.

Other Services: ⑤

Permanent Datum: ⑥ , Elev.

Log Measured From ⑦ Ft. Above Perm. Datum

Drilling Measured From ⑧

Elev.: K.B. D.F. ⑨ G.L.

Date	⑩				
Run No.					
Depth—Driller	⑪				
Depth—Logger					
Btm. Log Inter.	⑫				
Top Log Inter.					
Casing—Driller	⑬	@	@	@	@
Casing—Logger					
Bit Size	⑭				
Type Fluid in Hole	⑮				
Dens. | Visc.					
pH | Fluid Loss		ml	ml	ml	ml
Source of Sample	⑯				
R_m @ Meas. Temp.	⑰	@ °F	@ °F	@ °F	@ °F
R_{mf} @ Meas. Temp.		@ °F	@ °F	@ °F	@ °F
R_{mc} @ Meas. Temp.		@ °F	@ °F	@ °F	@ °F
Source: R_{mf} R_{mc}	⑱				
R_m @ BHT	⑲	@ °F	@ °F	@ °F	@ °F
Time Since Circ.	⑳				
Max. Rec. Temp.	㉑	°F	°F	°F	°F
Equip. | Location	㉒				
Recorded By	㉓				
Witnessed By					

This Heading and Log Conform To API RP 31

Fold Here

8½ to 9 in.

6 in.

Fig. 4.35—Standard log heading for electric logs recommended by the API (from Ref. 25).

20. Time (in hours) elapsed between the end of mud circulation before logging and when the logging tool reached bottom.

21. The temperature indicated by the maximum-recording thermometer used during the run.

22. The logging truck or equipment number and the truck's base of operation.

23. The names of the logging company and operating company representatives. The logging company representative is in charge of the logging operations. The operating company representative witnesses the logging operation to ensure that the logging program is executed according to company plan and that log quality meets company standards.

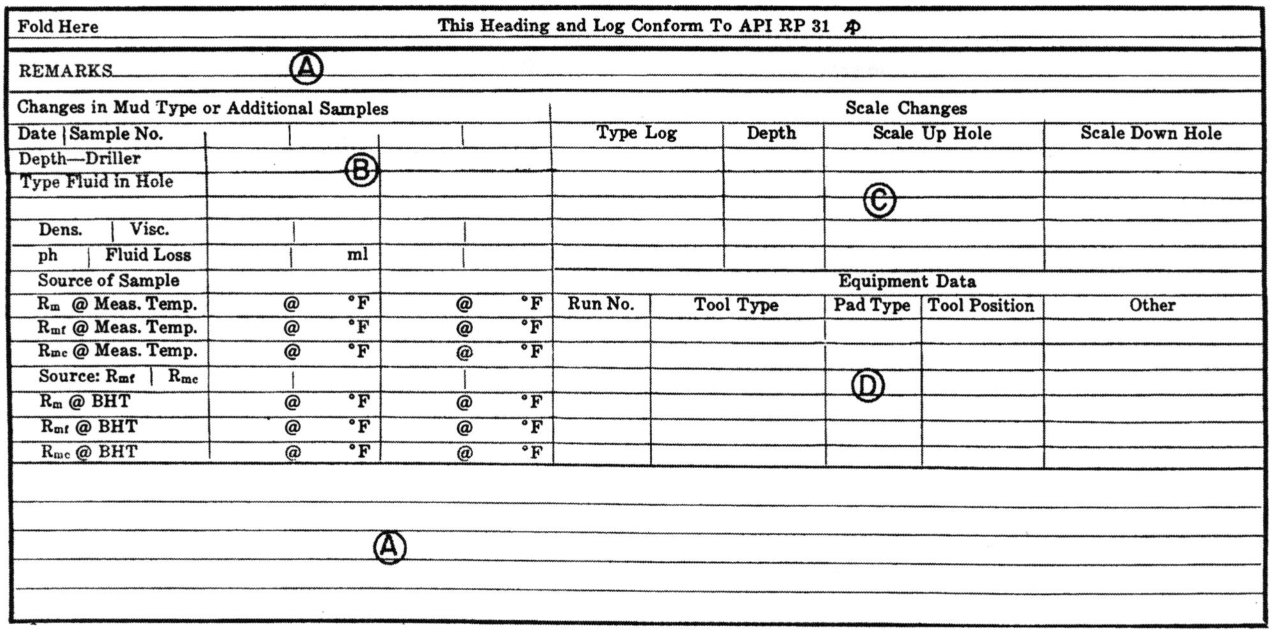

Fold Here — This Heading and Log Conform To API RP 31

REMARKS (A)

Changes in Mud Type or Additional Samples

Date / Sample No.				
Depth—Driller	(B)			
Type Fluid in Hole				
Dens. / Visc.				
ph / Fluid Loss		ml		
Source of Sample				
R_m @ Meas. Temp.	@	°F	@	°F
R_{mf} @ Meas. Temp.	@	°F	@	°F
R_{mc} @ Meas. Temp.	@	°F	@	°F
Source: R_{mf} / R_{mc}				
R_m @ BHT	@	°F	@	°F
R_{mf} @ BHT	@	°F	@	°F
R_{mc} @ BHT	@	°F	@	°F

Scale Changes

Type Log	Depth	Scale Up Hole	Scale Down Hole
		(C)	

Equipment Data

Run No.	Tool Type	Pad Type	Tool Position	Other
		(D)		

(A)

Fig. 4.36—Second fold of the log heading showing the remarks section.

Space is provided for remarks on the second fold of the log. The Remarks section in **Fig. 4.36** includes (a) general remarks; (b) mud data for two additional samples for the particular log run; (c) scale changes if made during logging; (d) and detailed equipment data, including tool type, pad type, and tool eccentricity.

Recent log headings are generated by computer. A computer-generated well heading looks different from that shown in Fig. 4.35 because the grid lines are omitted. The heading, however, lists the same information.

Example 4.6. Figs. 4.37 and 4.38 show the headings of the Dual Induction-Laterolog (DIL[SM]) and Borehole Compensated Log (BHC[SM]) obtained in the Pardee Co. No. 2 well. Using the data recorded on log heading, give the following information.

a. The interval logged.

b. Order of magnitude of the drilling-mud salinity.

c. Depth from sea level of a sand top shown by the log to be at a depth of 6,224 ft.

d. Best estimate of the formation temperature at the well TD (i.e., 9,673 ft). Exact circulation time is not available, but it is estimated to be between 6 and 10 hours.

e. Geothermal gradient.

f. Mud, mud filtrate, and mudcake resistivities at 6,224 ft.

Solution.

a. The log heading indicates that the DIL was run over the entire openhole interval, which extended from the casing shoe at 2,523 ft deep to TD of 9,667 ft. However, the BHC sonic log covered only the interval of 3,300 to 9,668 ft. The recommended practice, however, is to log the entire open borehole.

b. Entering Fig. 1.7 or 4.15 with the resistivity of the mud filtrate, which is 0.94 Ω·m at 70°F, yields a salinity of about 6,500 ppm. This value is based on the assumption that mud filtrate is a predominantly NaCl solution. It gives, however, a representative value called equivalent NaCl salinity.

c. As **Fig. 4.39** illustrates, the depth, D, to the top of the sand measured from sea level can be calculated as $D=6{,}224-201=6{,}023$ ft.

d. Because two tools were run in the borehole, the Horner-type plot can be used to estimate the static formation temperature. Data read from the log heading can be tabulated as shown below.

Tool	T_m (°F)	Time Circulation Stopped	Time Logger on Bottom
DIL	212	4:00	10:30
BHC	229	4:00	19:00

The data are used to calculate δt and $\delta t/(t+\delta t)$. The circulation time is estimated to be between 6 and 10 hours. Both values are used to get a feel for the sensitivity of the estimated formation temperature to the value of t.

T_m (°F)	δt (hours)	$\delta t/(6+\delta t)$	$\delta t/(10+\delta t)$
212	6.5	0.520	0.394
229	15.0	0.714	0.600

The Horner-type plot of **Fig. 4.40** was generated with the above data. Both lines extrapolate to the same temperature of 248°F, which is the best estimate of static formation temperature. Note that the result is not sensitive to the value of circulation time used as long as a reasonable value is chosen.

e. From Eq. 4.15,

$$g_G=\frac{248-70}{9{,}673}\times 100=1.84\text{°F}/100\text{ ft}.$$

f. The log heading lists the following resistivity values: $R_m=1.49$ Ω·m at 64°F, $R_{mf}=0.94$ Ω·m at 70°F, and $R_{mc}=1.24$ Ω·m at 64°. The formation temperature at 6,224 ft is calculated with Eq. 4.16:

$$T_f=70+1.84\frac{6{,}224}{100}=185\text{°F}.$$

From Eq. 1.11,

$$R_m=1.49\frac{64+6.77}{185+6.77}=0.55\ \Omega\cdot\text{m at }185\text{°F},$$

$$R_{mf}=0.94\frac{70+6.77}{185+6.77}=0.38\ \Omega\cdot\text{m at }185\text{°F},$$

and $R_{mc}=1.24\dfrac{64+6.77}{185+6.77}=0.46\ \Omega\cdot\text{m at }185\text{°F}.$

Example 4.7. Refer to the ISF/sonic log heading in **Fig. 4.41** for the following.

Schlumberger

DUAL INDUCTION - LATEROLOG
WITH LINEAR CORRELATION LOG

COUNTY BOSSIER, LA
FIELD IVAN
LOCATION -
WELL THE PARDEE CO. NO.2
COMPANY SUN OIL CO.

COMPANY SUN OIL COMPANY
WELL FILE COPY
WELL THE PARDEE COMPANY NO.2
FIELD IVAN
COUNTY BOSSIER STATE LOUISIANA

LOCATION 1307'FSL & 841'FWL NE SW SW
Other Services: BHC, FDC/GR

API SERIAL NO.	SEC.	TWP	RANGE
17-015-20863	26	21N	11W

Permanent Datum: GROUND LEVEL ; Elev.: 183
Log Measured From KB 18 Ft. Above Perm. Datum
Drilling Measured From KB

Elev.: K.B. 201
D.F. 200
G.L. 183

Date	12-13-76			
Run No.	ONE			
Depth—Driller	9674			
Depth—Logger	9673			
Btm. Log Interval	9667			
Top Log Interval	2523			
Casing—Driller	9 5/8 @ 2525	@	@	@
Casing—Logger	2523			
Bit Size	8 3/4			
Type Fluid in Hole	CAUSTIC-GEL			
Dens. / Visc.	13.5 / 44			
pH / Fluid Loss	8.5 / 5.1 ml	ml	ml	ml
Source of Sample	PIT			
Rm @ Meas. Temp.	1.49 @ 64 °F	@ °F	@ °F	@ °F
Rmf @ Meas. Temp.	.94 @ 70 °F	@ °F	@ °F	@ °F
Rmc @ Meas. Temp.	1.25 @ 64 °F	@ °F	@ °F	@ °F
Source: Rmf / Rmc	M / C			
Rm @ BHT	.05 @ 212 °F	@ °F	@ °F	@ °F
TIME Circulation Stopped	400			
TIME Logger on Bottom	1030			
Max. Rec. Temp.	212 °F	°F	°F	°F
Equip. / Location	7770 / TYLER			
Recorded By	MCCALL			
Witnessed By	GOULD-ALBACK			

Fig. 4.37—DIL heading of Example 4.6 (courtesy Sun Oil Co.).

a. Sketch a cross section of the well where the log was run. Show different boreholes and casings. Mark the borehole sizes, casing sizes, and casing depths. Also mark the mud weight used to drill each section of the well.

b. List the values of the temperatures recorded in the well, the depth at which each value was recorded, and the elapsed time since mud circulation stopped.

c. Give your best estimate of formation temperature at 5,000 and 24,000 ft.

d. Estimate the resistivity of the mud filtrate saturating the flushed zone of a permeable formation situated at 5,000 ft. Comment on the representativity of your estimate.

Solution.

a. This deep well (D=20,556 ft) required the setting of several intermediate casing strings. The data from the five logging runs were used to prepare the well's cross section in **Fig. 4.42.**

b. Data pertaining to temperature measurement in this well are given in **Table 4.3.**

c. Assuming linear distribution yields the following values from Eqs. 4.15 and 4.16:

$$g_G = \frac{360-70}{20{,}556}100 = 1.41\,°\text{F}/100\text{ ft},$$

$$T_{5{,}000} = 70 + 1.41\frac{5{,}000}{100} = 141\,°\text{F},$$

$$\text{and } T_{24{,}000} = 70 + 1.41\frac{24{,}000}{100} = 408\,°\text{F}.$$

However, assuming a linear distribution starting with T_s=70°F is not valid for deep wells. As **Fig. 4.43** illustrates, the line

Schlumberger — **BOREHOLE COMPENSATED SONIC LOG**

COUNTY BOSSIER, LA.
FIELD IVAN
LOCATION -
WELL THE PARDEE CO. NO.2
COMPANY SUN OIL COMPANY

COMPANY SUN OIL COMPANY

WELL FILE COPY

WELL THE PARDEE COMPANY NO.2

FIELD IVAN

COUNTY BOSSIER STATE LOUISIANA

LOCATION 1307'FSL & 841'FWL NE SW SW

Other Services: DIL,FDC/GR

API SERIAL NO	SEC	TWP	RANGE
17-015-20863	26	21N	11W

Permanent Datum: GROUND LEVEL ; Elev.: 183
Log Measured From KB 18 Ft. Above Perm. Datum
Drilling Measured From KB

Elev.: K.B. 201 D.F. 200 G.L. 183

Date	12-13-76			
Run No.	ONE			
Depth—Driller	9674			
Depth—Logger (Schl.)	9673			
Btm. Log Interval	9668			
Top Log Interval	3300			
Casing—Driller	9 5/8@ 2525	@	@	@
Casing—Logger	NOT CHECKED			
Bit Size	8 3/4			
Type Fluid in Hole	CAUSTIC-GEL			
Dens. / Visc.	13.5 / 44			
pH / Fluid Loss	8.5 / 5.1 ml	ml	ml	ml
Source of Sample	PIT			
Rm @ Meas. Temp.	1.49 @ 64 F	@ F	@ °F	@ F
Rmf @ Meas. Temp.	.94 @ 70 F	@ F	@ °F	@ F
Rmc @ Meas. Temp.	1.25 @ 64 F	@ F	@ F	@ F
Source: Rmf / Rmc	M / C			
Rm @ BHT	.05 @212 F	@ F	@ °F	@ F
TIME Circulation Stopped	0400			
TIME Logger on Bottom	1900			
Max. Rec. Temp.	229 F	F	F	F
Equip. / Location	7770 / TYLER			
Recorded By	MCCALL			
Witnessed By Mr.	GOULD-MOYER-ALBACK			

Fig. 4.38—BHC sonic log of Example 4.6 (courtesy Sun Oil Co.).

representing the calculated geothermal gradient of 1.41°F/100 ft does not go through the data points. Moreover, a straight line forced through the data points results in an unrealistic temperatures at the surface ($T_s=-60$°F) and at 5,000 ft ($T_{5,000}=40$°F).

A semilog plot of the data (**Fig. 4.44**) results in a realistic surface temperature of 78°F. The temperature distribution shown in Fig. 4.44 is expressed by Eq. 4.18:

$$T_f=78e^{7.54\times10^{-5}D}.$$

Fig. 4.44 and the above equation yield the best temperature estimates. Then, $T_f=114$°F at 5,000 ft and $T_f=480$°F at 24,000 ft.

d. The formation situated at a depth of 5,000 ft was logged during Run 1. The sampled mud filtrate yielded $R_{mf}=0.51$ Ω·m at 74°F. Assuming that the mud sampled at the time of logging is the same mud present in the borehole at the time of drilling yields

$$R_{mf}=0.51\frac{74+6.77}{114+6.77}=0.34\ \Omega\cdot\text{m at } 114°\text{F}.$$

Review Questions

1. Why does measurement environment have to be known?
2. What are the main parameters that define the measurement environment?
3. What formation types give rise to borehole enlargement? Why?
4. What is mudcake? What are the conditions required for its formation? How does its presence affect borehole diameter?
5. What are the different caliper measurements available? Why do they usually display different readings in the same hole?
6. What are the different uses of caliper information?

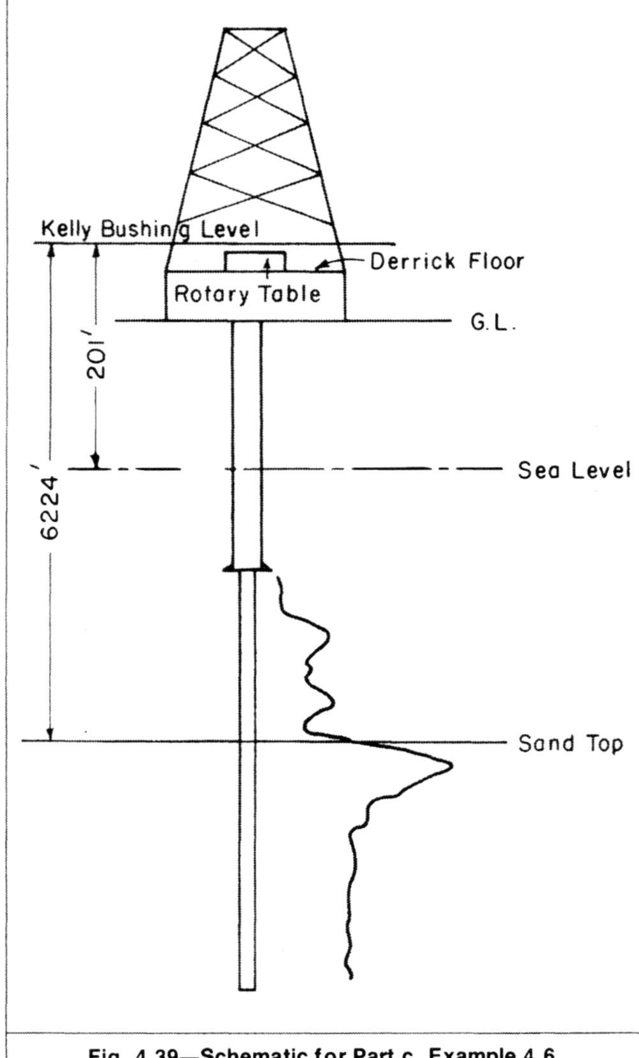

Fig. 4.39—Schematic for Part c, Example 4.6.

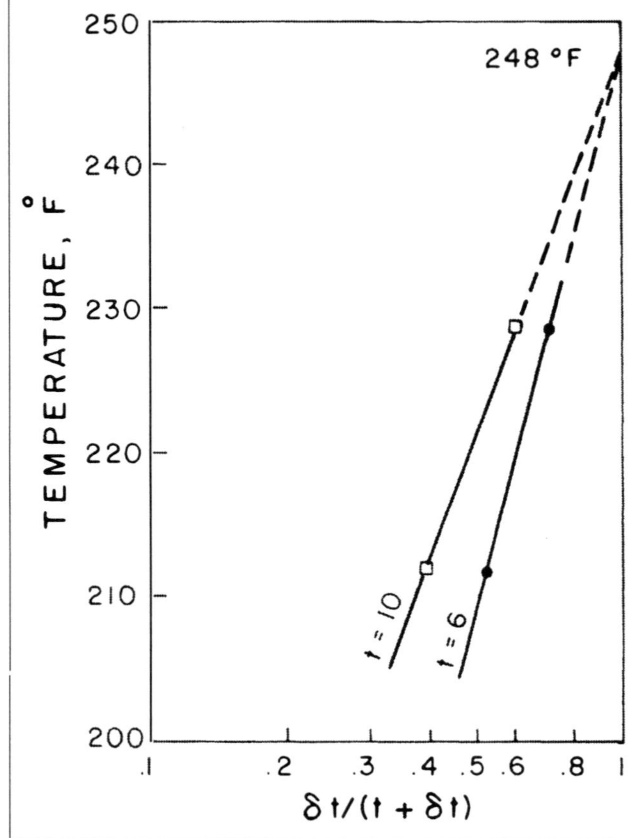

Fig. 4.40—Horner-type temperature buildup plot for Example 4.6.

7. Explain how mudcake buildup can go undetected by caliper logs.
8. What drilling-mud data are pertinent to logging operations?
9. What is the most commonly used mud? What is its makeup?
10. How are the mud, mud filtrate, and mudcake resistivities usually determined?
11. How does temperature affect the mud and mud-filtrate resistivities? How is that effect determined?
12. What is a mud log?
13. What causes variation in drilling-mud resistivity? How does this variation complicate log analysis?
14. Describe the step, transition, and annulus profiles of invasion.
15. What are the parameters that control the mud-filtrate volume invading a permeable formation?
16. Explain why porosity is the essential parameter that determines the depth of invasion? How does the diameter of invasion vary with porosity?
17. Define the geothermal gradient. How is it usually calculated?
18. Why is the BHT indicated by the maximum-recording thermometer usually less than the formation temperature?
19. How does temperature distribution in deep wells differ from that in shallow wells?
20. Why is it recommended to designate a permanent datum and indicate its elevation on the well heading?
21. Why is knowing the elevation of the kelly bushing usually necessary?
22. Why does the well TD indicated by the driller differ from that indicated by the logger?
23. What additional information do you think should be recorded on the log heading?

Problems

4.1 **Fig. 4.45** shows a one-arm device caliper and a three-arm bow-spring device caliper obtained in the same borehole interval. Explain the disagreement between the two logs in both the thick and thin zones.

4.2 **Fig. 4.46** shows an interval of a microcaliper recorded in a Ouachita Parish, LA, well.

TABLE 4.3—TEMPERATURE DATA OF EXAMPLE 4.7

Data Point	D (ft)	Time/Date Circulation Stopped	Time/Date Logger on Bottom	δt (hours)	T_m (°F)
1	13,725	10:30/Oct. 28	22:30/Oct. 28	12	220
2	16,579	N/A*	17:30/Jan. 19	N/A	275
3	20,074	01:00/April 18	11:00/April 18	10	355**
4	20,444	01:30/April 28	00:00/April 29	22.5	360
5	20,553	21:00/May 1	21:00/May 2	24	360

*Missing data.
**Two thermometers were used during this run. The highest reading is retained.

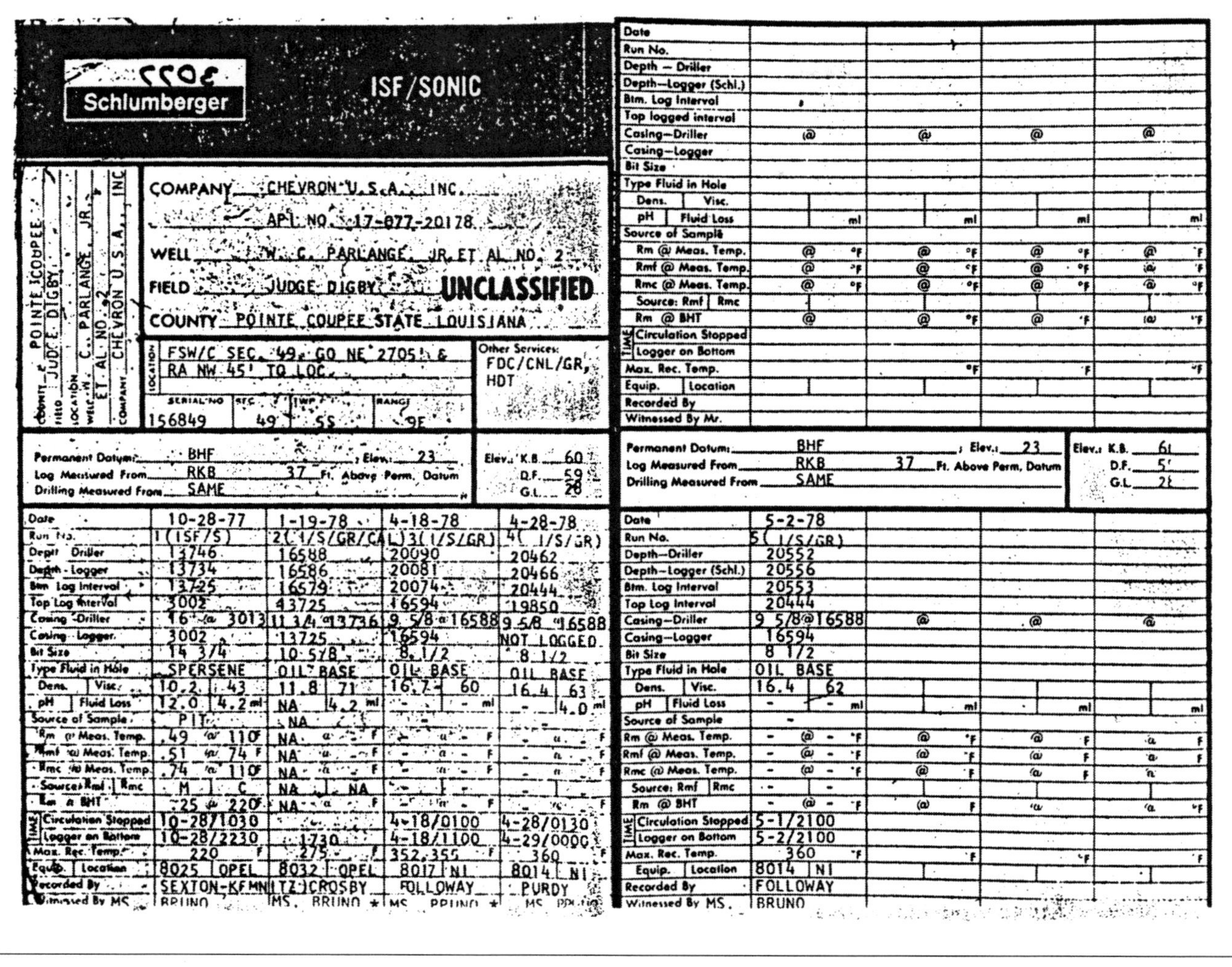

Schlumberger ISF/SONIC

COMPANY CHEVRON U.S.A. INC.
API NO. 17-077-20178
WELL W. C. PARLANGE, JR. ET AL NO. 2
FIELD JUDGE DIGBY UNCLASSIFIED
COUNTY POINTE COUPEE STATE LOUISIANA

LOCATION: FSW/C SEC. 49, GO NE 2705' & RA NW 45' TO LOC.
Other Services: FDC/CNL/GR, HDT
SERIAL NO 156849 SEC 49 TWP 5S RANGE 9E

Permanent Datum: BHF ; Elev.: 23
Log Measured From RKB 37 Ft. Above Perm. Datum
Drilling Measured From SAME
Elev.: K.B. 60 D.F. 59 G.L. 28

Date	10-28-77	1-19-78	4-18-78	4-28-78	5-2-78
Run No.	1 (ISF/S)	2 (I/S/GR/CAL)	3 (I/S/GR)	4 (I/S/GR)	5 (I/S/GR)
Depth—Driller	13746	16588	20090	20462	20552
Depth—Logger	13734	16586	20081	20466	20556
Btm. Log Interval	13725	16579	20074	20444	20553
Top Log Interval	3002	43725	16594	19850	20444
Casing—Driller	16" @ 3013	11 3/4 @ 13736	9 5/8 @ 16588	9 5/8 @ 16588	9 5/8 @ 16588
Casing—Logger	3002	13725	16594	NOT LOGGED	16594
Bit Size	14 3/4	10 5/8	8 1/2	8 1/2	8 1/2
Type Fluid in Hole	SPERSENE	OIL BASE	OIL BASE	OIL BASE	OIL BASE
Dens. / Visc.	10.2 / 43	11.8 / 71	16.7 / 60	16.4 / 63	16.4 / 62
pH / Fluid Loss	12.0 / 4.2 ml	NA / 4.2 ml	- / - ml	- / 4.0 ml	- / - ml
Source of Sample	PIT	NA	-	-	-
Rm @ Meas. Temp.	.49 @ 110°F	NA @ - °F	- @ - °F	- @ - °F	- @ - °F
Rmf @ Meas. Temp.	.51 @ 74°F	NA @ - °F	- @ - °F	- @ - °F	- @ - °F
Rmc @ Meas. Temp.	.74 @ 110°F	NA @ - °F	- @ - °F	- @ - °F	- @ - °F
Source: Rmf / Rmc	M / C	NA / NA	- / -	- / -	- / -
Rm @ BHT	.25 @ 220°F	NA @ - °F	- @ - °F	- @ - °F	- @ - °F
Circulation Stopped	10-28/1030		4-18/0100	4-28/0130	5-1/2100
Logger on Bottom	10-28/2230	1730	4-18/1100	4-29/0000	5-2/2100
Max. Rec. Temp.	220°F	275°F	352, 355°F	360°F	360°F
Equip. / Location	8025 / OPEL	8032 / OPEL	8017 / NI	8014 / NI	8014 / NI
Recorded By	SEXTON-KFMNITZ	CROSBY	FOLLOWAY	PURDY	FOLLOWAY
Witnessed By	MS. BRUNO	MS. BRUNO *	MS. BRUNO *	MS. BRUNO	MS. BRUNO

Permanent Datum: BHF ; Elev.: 23
Log Measured From RKB 37 Ft. Above Perm. Datum
Drilling Measured From SAME
Elev.: K.B. 61 D.F. 5' G.L. 2'

Fig. 4.41—Log heading of Example 4.7 (courtesy Chevron U.S.A. Inc.).

a. What are the minimum and maximum borehole diameters indicated by the microcaliper in this interval?

b. What is the thickest mudcake buildup in this interval?

c. If the zone between 8,891 ft and the bottom of the hole is shale, explain why part of it is washed out and the other is not.

4.3 **Fig. 4.47** shows the heading of a microlog, and **Fig. 4.48** shows the mud log recorded as the tool is lowered into the borehole. Compare the mud-log and mud-sample resistivities at BHT.

4.4 Examine the log heading of **Fig. 4.49.**

a. On linear graph paper, plot the maximum recorded temperature vs. depth.

b. Is the assumption of linear thermal gradient acceptable in this case?

c. Find the mathematical depth/temperature relationship for the subject well.

d. Plot a cross section of the well.

e. Calculate the mud and mud-filtrate resistivities of the drilling fluid sampled for the eight log runs at 75°F. Plot R_{mf} vs. R_m. Does a strong correlation exist between these two parameters? Can this relation be predicted by any of the methods discussed in this chapter?

4.5 At the time of logging, the resistivity of a 10-lbm/gal drilling mud is measured with an instrument on the surface. A value of 1.52 Ω-meter is obtained at a temperature of 90°F.

a. What is the apparent salinity of the mud?

b. What is the estimated mud resistivity at a depth of 8,600 ft in the hole where the formation temperature is 180°F?

c. Estimate the resistivities of the mud filtrate at 90 and 180°F.

d. Can you justify the statistical approximation for predominantly NaCl muds: $R_{mf}=0.75R_m$?

e. Estimate the resistivity of the mudcake at 180°F.

f. Can you justify the statistical approximation for predominately NaCl muds: $R_{mc}=1.5R_m$?

4.6 Determine the values of R_m, R_{mf}, and R_{mc} at 10,000 ft if mud weight=16 lbm/gal, mud resistivity=0.42 Ω·m at 75°F, and geothermal gradient=1.1°F/100 ft.

4.7 **Fig. 4.50** shows a composite plot of mud temperatures measured in wells drilled along the Tuscaloose trend in Louisiana.

a. Explain why data points do not lie on the same straight line.

b. Find the parameter T_s and m_T of Eq. 4.18 for the Tuscaloose trend.

c. What T_{bh} would you expect in a 24,000-ft well drilled along the trend?

4.8 **Figs. 4.51 and 4.52** show the headings of two logs run in the same borehole. Examine the two headings and then provide the following information:

a. Apparent drilling-fluid salinity.

b. Depth from sea level of a formation top shown by the log at 5,603 ft.

c. The best estimate of the static formation temperature at total well depth.

d. Geothermal gradient.

e. R_{mf} and R_m at the 5,603-ft formation.

4.9 Estimate the static formation temperature from the following data (from Ref. 30): depth=7,646 ft, drilling stopped=22:00

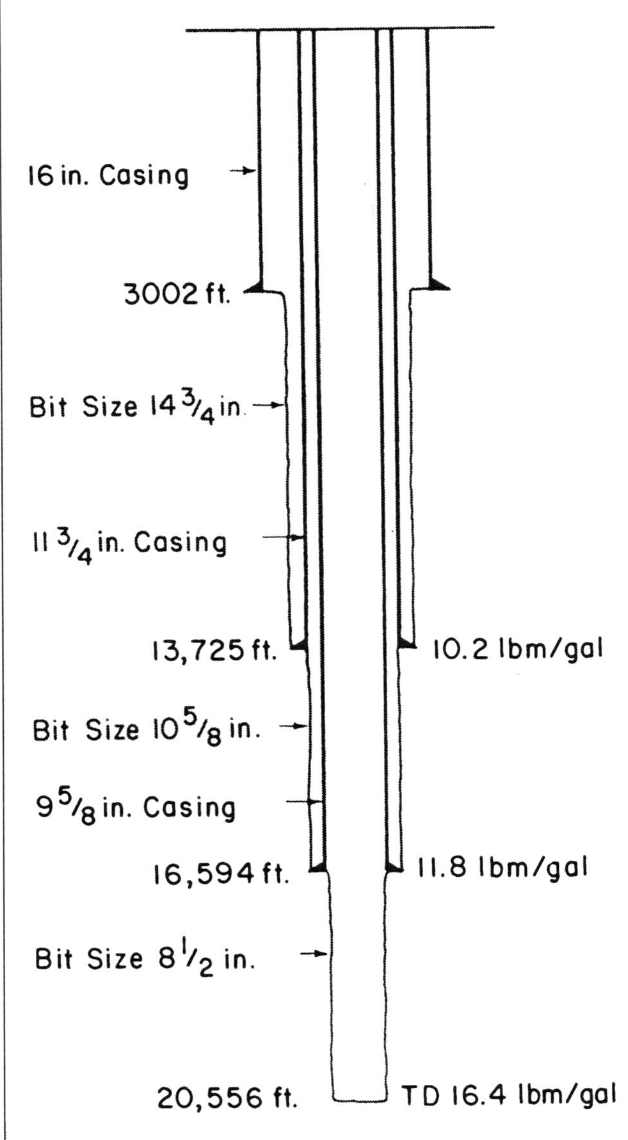

Fig. 4.42—Cross section of the Well in Example 4.7 (ppg-lbm/gal).

on the second day of the month, circulation stopped=02:30 on the third day of the month.

Tool	Thermometer Depth (ft)	Time Off Bottom (time/day)	δt (hours)	T_m (°F)
Sonic	7,608	07:36/third	5:06	99
DIL	7,608	12:48/third	10:18	106
FDC	7,620	14:29/third	14:29	107
SNP	7,620	20:37/third	18:07	110

Nomenclature

A = area, ft^2
C = constant in Eq. 4.17
d = diameter, in.
d_i = diameter of invasion, in.
D = depth, ft
F = formation resistivity factor
g_G = geothermal gradient, °F/100 ft
h = formation thickness, ft
h_{mc} = mudcake thickness, ft
k_h = thermal conductivity, calories/(sec-cm-°C)

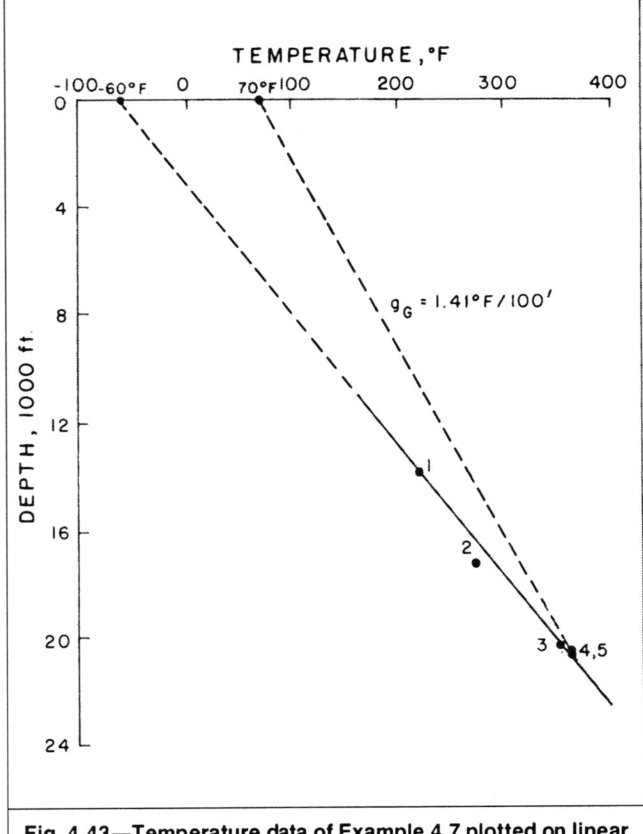

Fig. 4.43—Temperature data of Example 4.7 plotted on linear scales.

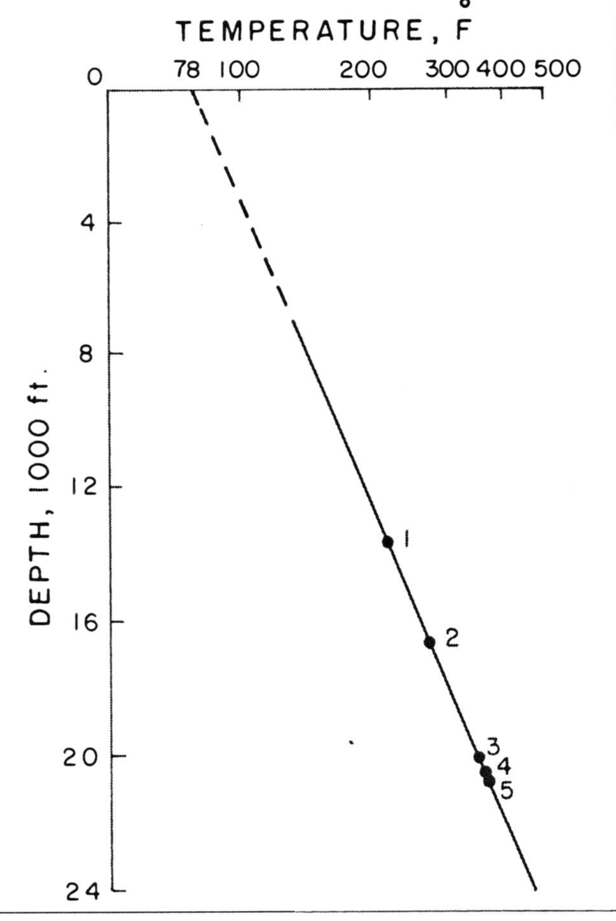

Fig. 4.44—Temperature data of Example 4.7 plotted on semi-log scale.

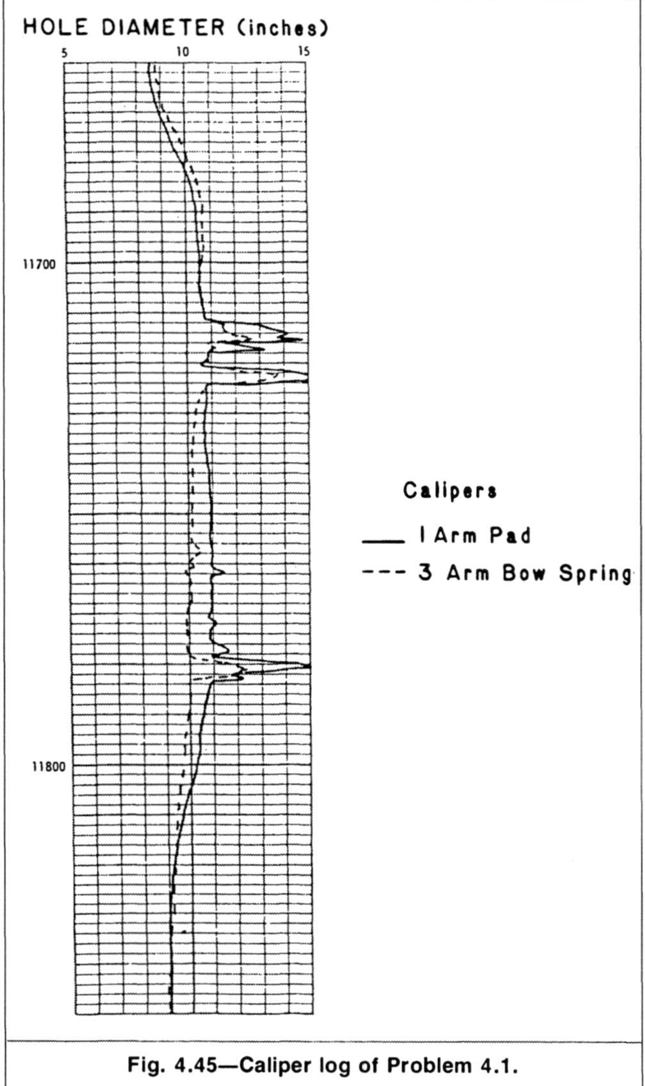

Fig. 4.45—Caliper log of Problem 4.1.

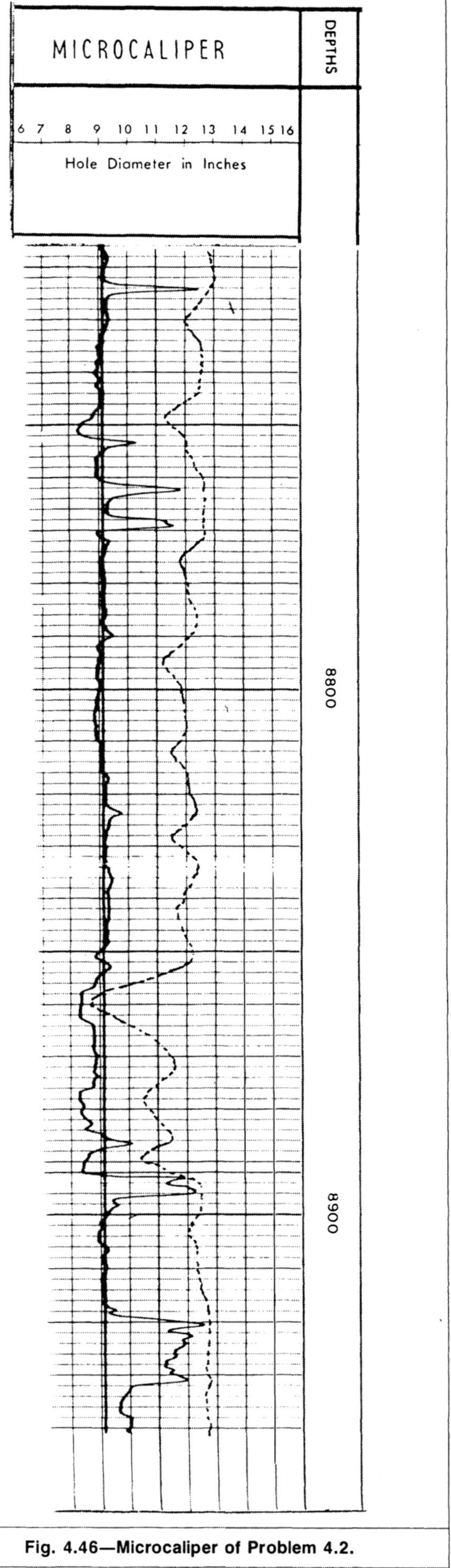

Fig. 4.46—Microcaliper of Problem 4.2.

K_m = coefficient that varies with mud weight
m_T = constant in Eq. 4.18
n = saturation exponent
$\dot{Q}$ = heat flow rate, cal/sec
R_m = mud resistivity, Ω·m
R_{mc} = mudcake resistivity, Ω·m
R_{mf} = mud-filtrate resistivity, Ω·m
R_t = resistivity of uninvaded zone, Ω·m
R_w = formation water resistivity, Ω·m
R_{xo} = resistivity of flushed zone, Ω·m
S_o = oil saturation, fraction
S_{or} = residual oil saturation, fraction
S_w = water saturation, fraction
ΔS_w = variation in water saturation
S_{xo} = saturation of mud filtrate in flushed zone, fraction
t = mud circulation time, hours
δt = time since mud circulation stopped, hours
T = temperature, °F
T_a = mean annual temperature of air, °F
T_f = static formation temperature, °F
T_m = mud temperature, °F
T_s = temperature near surface, °F
V_{mf} = mud filtrate volume, cm^3
ρ = density, g/cm^3
ρ_m = mud density, lbm/gal
ϕ = porosity, fraction

GER

COUNTY LINCOLN, MISSISSIPPI FIELD or BROOKHAVEN LOCATION SEC. 20-3N-7W WELL BROOKHAVEN UNIT #17 COMPANY THE CALIFORNIA COMPANY	COMPANY THE CALIFORNIA COMPANY WELL BROOKHAVEN UNIT # 17 FIELD BROOKHAVEN LOCATION SEC. 20-3N-7W COUNTY LINCOLN STATE MISSISSIPPI	Other Surveys IES-SL-CST Location of Well 660 4' N 662 5' W SE/C SEC Elevation: K.B.: D.F.: 439 or G.L.:

Log Depths Measured From RDB OR 13.57 Ft. above BHF

RUN No.	ONE		
Date	2-18-62		
First Reading	10600		
Last Reading	1353		
Feet Measured	9242		
Depth Reached	10601		
Bottom Driller	10600		
Mud Nat.	CAUST, GEL		
Dens. | Visc.	10.1 | 42	|	|
Mud Resist.	1.39 @ 84 °F	@ °F	@ F
" Res. BHT	0.71 @ 216 °F	@ °F	@ F
" pH	9.5 @ °F	@ °F	@ F
" Wtr. Loss	3.2 CC 30 min	CC 30 min	CC 30 min
" Rmf (M)	1.40 @ 30 °F	@ °F	@ F
" Rmc (C)	0.75 @ 216 °F	@ °F	@ F
Mud Log: Rm	0.34		
Depth	10300		
Bit Size	6 3/4"		
Sonde Type	PMS-A		
Pad Type	HYDRAULIC		
Opr. Rig Time	4 HOURS		
Truck No.	1507-COL		
Recorded By	MARSH		

Fig. 4.47—Log heading of Problem 4.3.

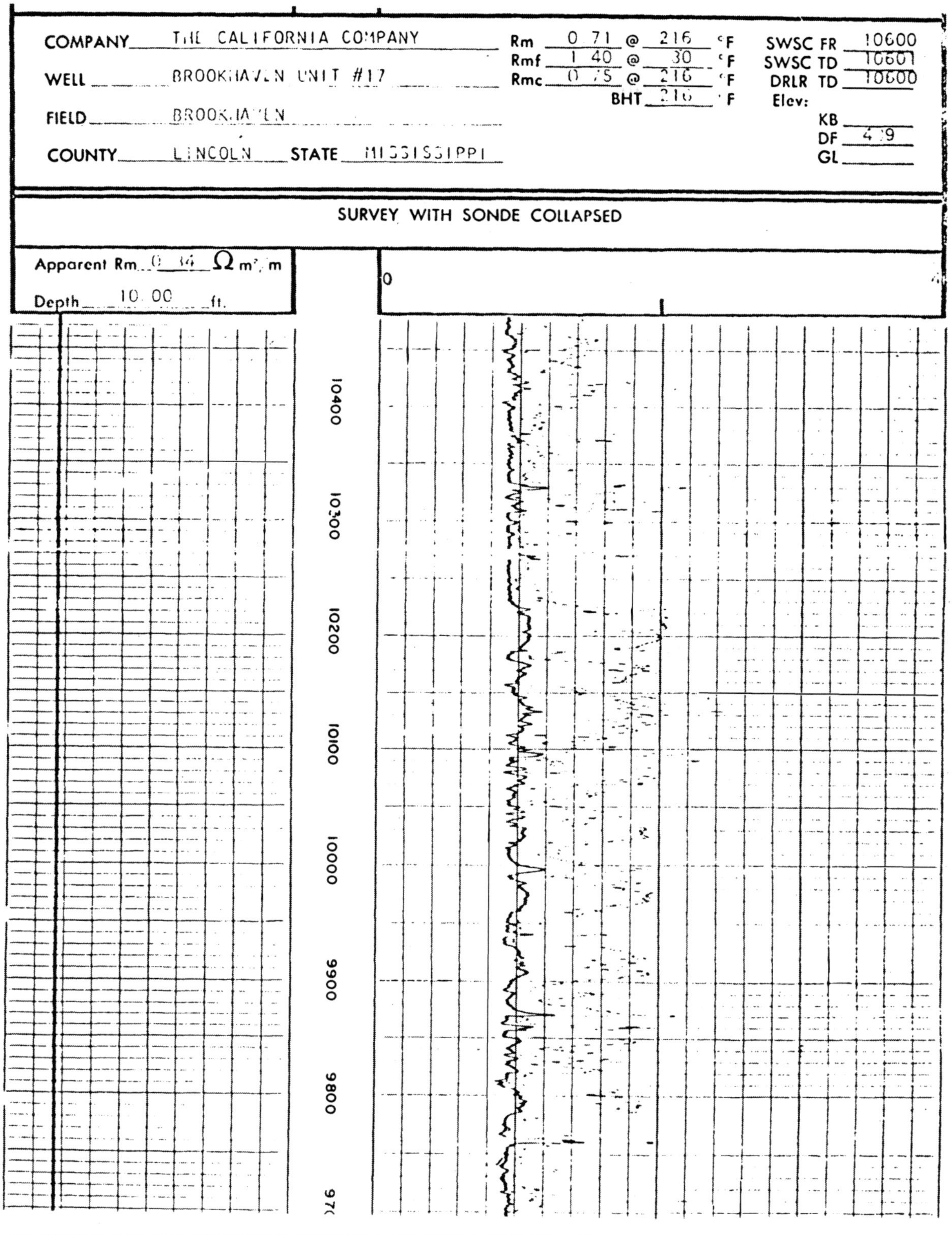

Fig. 4.48—Mud log of Problem 4.3.

Subscripts

bh = bottomhole
h = borehole
i = invasion
m = mud
mc = mudcake
mf = mud filtrate
o = oil
or = residual oil

References

1. Parsons, C.P.: "Caliper Logging," *Trans.*, AIME (1943) **151**, 35–47.
2. Hilchie, D.W.: 'Caliper Logging—Theory and Practice," *Log Analyst* (Jan.-Feb. 1968) 3-12.
3. Boyeldieu, C., Coblentz, A., and Pélissier-Combescure, J.: "Formation Evaluation in Oil Base Mud Wells," paper BB presented at the 1984 SPWLA Symposium, New Orleans, June 10-13.
4. *Principles of Drilling Fluid Control,* 12th edition, Petroleum Extension Service, U. of Texas, Austin (1981).
5. *RP 13B, Standard Procedure for Testing Drilling Fluids,* fifth edition, API, Dallas (1974).

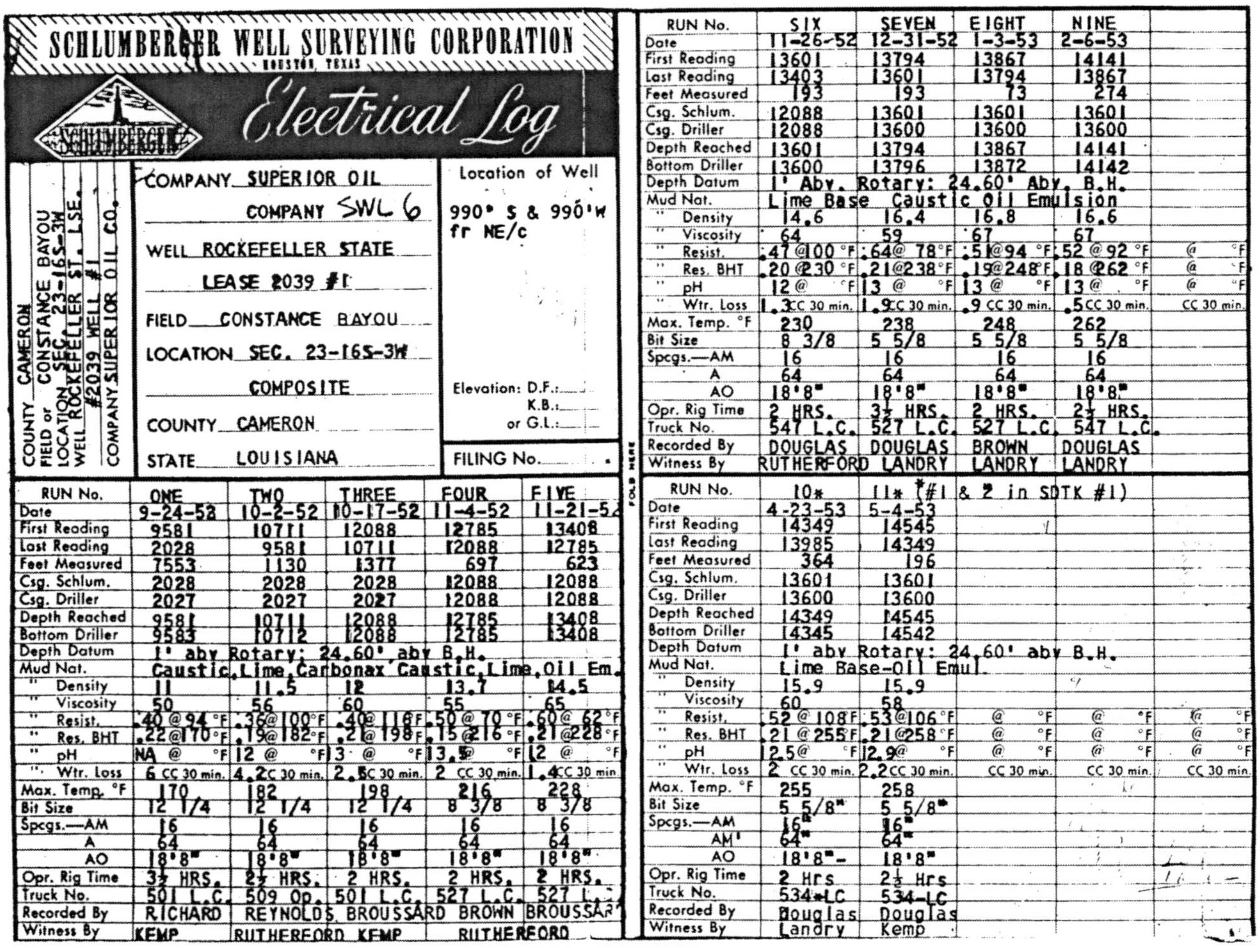

SCHLUMBERGER WELL SURVEYING CORPORATION
HOUSTON, TEXAS

SCHLUMBERGER

Electrical Log

COUNTY CAMERON
FIELD or LOCATION CONSTANCE BAYOU SEC. 23-16S-3W
WELL ROCKEFELLER ST. LSE. #2039 WELL #1
COMPANY SUPERIOR OIL CO.

COMPANY SUPERIOR OIL COMPANY SWL 6

WELL ROCKEFELLER STATE LEASE 2039 #1

FIELD CONSTANCE BAYOU

LOCATION SEC. 23-16S-3W

COMPOSITE

COUNTY CAMERON

STATE LOUISIANA

Location of Well: 990' S & 990'W fr NE/c

Elevation: D.F.: K.B.: or G.L.:

FILING No.

RUN No.	ONE	TWO	THREE	FOUR	FIVE
Date	9-24-52	10-2-52	10-17-52	11-4-52	11-21-52
First Reading	9581	10711	12088	12785	13408
Last Reading	2028	9581	10711	12088	12785
Feet Measured	7553	1130	1377	697	623
Csg. Schlum.	2028	2028	2028	12088	12088
Csg. Driller	2027	2027	2027	12088	12088
Depth Reached	9581	10711	12088	12785	13408
Bottom Driller	9583	10712	12088	12785	13408
Depth Datum	1' abv Rotary: 24.60' abv B.H.				
Mud Nat.	Caustic, Lime, Carbonax Caustic, Lime, Oil Em.				
" Density	11	11.5	12	13.7	14.5
" Viscosity	50	56	60	55	65
" Resist.	.40 @ 94 °F	.36 @ 100 °F	.40 @ 118 °F	.50 @ 70 °F	.60 @ 62 °F
" Res. BHT	.22 @ 170 °F	.19 @ 182 °F	.21 @ 198 °F	.15 @ 216 °F	.21 @ 228 °F
" pH	NA @ °F	12 @ °F	13 @ °F	13.5 @ °F	12 @ °F
" Wtr. Loss	6 CC 30 min.	4.2 CC 30 min.	2.5 CC 30 min.	2 CC 30 min.	1.4 CC 30 min.
Max. Temp. °F	170	182	198	216	228
Bit Size	12 1/4	12 1/4	12 1/4	8 3/8	8 3/8
Spcgs.—AM	16	16	16	16	16
A	64	64	64	64	64
AO	18'8"	18'8"	18'8"	18'8"	18'8"
Opr. Rig Time	3½ HRS.	2½ HRS.	2 HRS.	2 HRS.	2 HRS.
Truck No.	501 L.C.	509 Op.	501 L.C.	527 L.C.	527 L.C.
Recorded By	RICHARD	REYNOLDS	BROUSSARD	BROWN	BROUSSARD
Witness By	KEMP	RUTHERFORD	KEMP	RUTHERFORD	

RUN No.	SIX	SEVEN	EIGHT	NINE	
Date	11-26-52	12-31-52	1-3-53	2-6-53	
First Reading	13601	13794	13867	14141	
Last Reading	13403	13601	13794	13867	
Feet Measured	193	193	73	274	
Csg. Schlum.	12088	13601	13601	13601	
Csg. Driller	12088	13600	13600	13600	
Depth Reached	13601	13794	13867	14141	
Bottom Driller	13600	13796	13872	14142	
Depth Datum	1' Abv. Rotary: 24.60' Abv. B.H.				
Mud Nat.	Lime Base Caustic Oil Emulsion				
" Density	14.6	16.4	16.8	16.6	
" Viscosity	64	59	67	67	
" Resist.	.47 @ 100 °F	.64 @ 78 °F	.51 @ 94 °F	.52 @ 92 °F	@ °F
" Res. BHT	.20 @ 230 °F	.21 @ 238 °F	.19 @ 248 °F	.18 @ 262 °F	@ °F
" pH	12 @ °F	13 @ °F	13 @ °F	13 @ °F	@ °F
" Wtr. Loss	1.3 CC 30 min.	1.9 CC 30 min.	.9 CC 30 min.	.5 CC 30 min.	CC 30 min.
Max. Temp. °F	230	238	248	262	
Bit Size	8 3/8	5 5/8	5 5/8	5 5/8	
Spcgs.—AM	16	16	16	16	
A	64	64	64	64	
AO	18'8"	18'8"	18'8"	18'8"	
Opr. Rig Time	2 HRS.	3½ HRS.	2 HRS.	2½ HRS.	
Truck No.	547 L.C.	527 L.C.	527 L.C.	547 L.C.	
Recorded By	DOUGLAS	DOUGLAS	BROWN	DOUGLAS	
Witness By	RUTHERFORD	LANDRY	LANDRY	LANDRY	

RUN No.	10*	11*			
	*(#1 & 2 in SDTK #1)				
Date	4-23-53	5-4-53			
First Reading	14349	14545			
Last Reading	13985	14349			
Feet Measured	364	196			
Csg. Schlum.	13601	13601			
Csg. Driller	13600	13600			
Depth Reached	14349	14545			
Bottom Driller	14345	14542			
Depth Datum	1' abv Rotary: 24.60' abv B.H.				
Mud Nat.	Lime Base-Oil Emul.				
" Density	15.9	15.9			
" Viscosity	60	58			
" Resist.	.52 @ 108 °F	.53 @ 106 °F	@ °F	@ °F	@ °F
" Res. BHT	.21 @ 255 °F	.21 @ 258 °F	@ °F	@ °F	@ °F
" pH	12.5 @ °F	12.9 @ °F	@ °F	@ °F	@ °F
" Wtr. Loss	2 CC 30 min.	2.2 CC 30 min.	CC 30 min.	CC 30 min.	CC 30 min.
Max. Temp. °F	255	258			
Bit Size	5 5/8"	5 5/8"			
Spcgs.—AM	16"	16"			
AM'	64"	64"			
AO	18'8"-	18'8"			
Opr. Rig Time	2 Hrs	2½ Hrs			
Truck No.	534-LC	534-LC			
Recorded By	Douglas	Douglas			
Witness By	Landry	Kemp			

FOLD HERE

Fig. 4.49—Log heading of Problem 4.4.

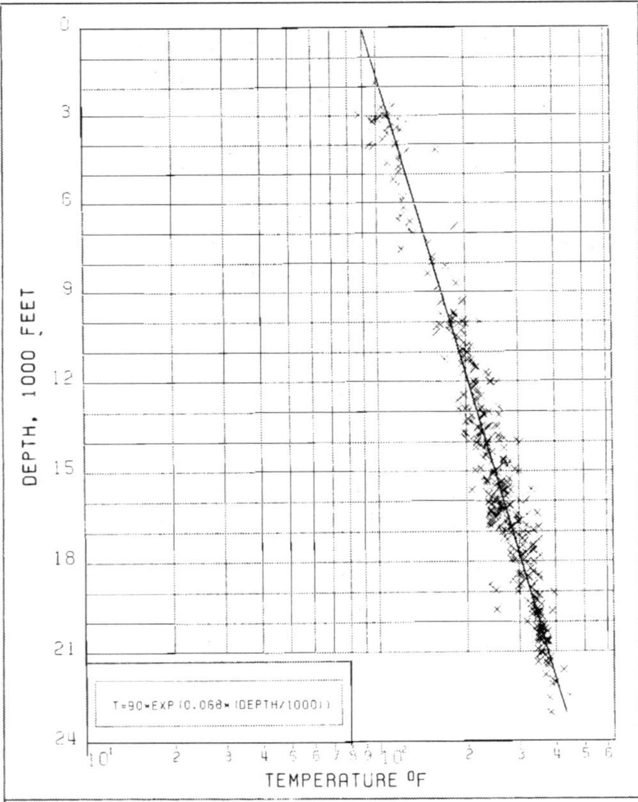

Fig. 4.50—Composite plot of mud temperatures recorded in wells drilled along the Tuscaloose trend, LA (from Ref. 32).

6. Horner, V. *et al.*: "Microbit Dynamic Filtration Studies," *Trans.*, AIME (1957) **210,** 183–95.
7. Lowe, T.A. and Dunlap, H.F.: "Estimation of Mud Filtrate Resistivity in Fresh Water Drilling Muds," *Log Analyst*, (March–April 1986) 77–84.
8. Overton, H.L. and Lipson, L.B.: "A Correlation of the Electrical Properties of Drilling Fluids with Solids Content," *Trans.*, AIME (1958) **213,** 333–36.
9. *Log Interpretation Charts*, Schlumberger, Houston (1972).
10. Wang, G.C., Helander, D.P., Wieland, D.R.: "Laboratory Resistivity Evaluation of Lime, GYP, and Calcium Lignite—Calcium Lignosulfonate Muds," *Trans.*, SPWLA, Third Symposium (1967).
11. Lynn, R.D.: "Effect of Temperature on Drilling Mud Resistivities," paper SPE 1302G presented at the 1959 SPE Annual Meeting, Dallas, Oct. 4–7.
12. *Log Interpretation, Vol. 1: Principles*, Schlumberger, Houston (1972).
13. Ferguson, C.K. and Klotz, J.A.: "Filtration from Mud During Drilling," *Trans.*, AIME (1954) **201,** 29–42.
14. Williams, H. and Dunlap, H.F.: "Short Term Variations in Drilling Fluid Parameters; Their Measurement and Implications," *Log Analyst* (Sept.–Oct. 1984) 3–9.
15. Johnson, H.M.: "The Borehole Environment: Known and Unknown," paper 7051, Canadian Well Logging Soc., 1970.
16. *Basic Concepts of Well Log Interpretation*, Welex, Houston (1978).
17. Phelps, G.D., Stewart, G., and Peden, J.M.: "Analysis of the Invaded Zone Characteristics and Their Influence on Wireline Log and Well-Test Interpretation," paper SPE 13287 presented at the 1989 SPE Annual Technical Conference and Exhibition, Houston, Oct. 16–19.
18. Gondouin, M. and Heim, A.: "Experimentally Determined Resistivity Profiles in Invaded Water and Oil Sands for Linear Flows," *JPT* (March 1964) 337–43; *Trans.*, AIME, **231.**
19. Pirson, S.J.: *Handbook of Well Log Analysis*, Prentice-Hall Inc., Englewood Cliffs, NJ (1963) 59.
20. Guyod, H.: "Temperature Well Logging: Heat Conduction," *Oil Weekly* (Oct. 21, 1946).

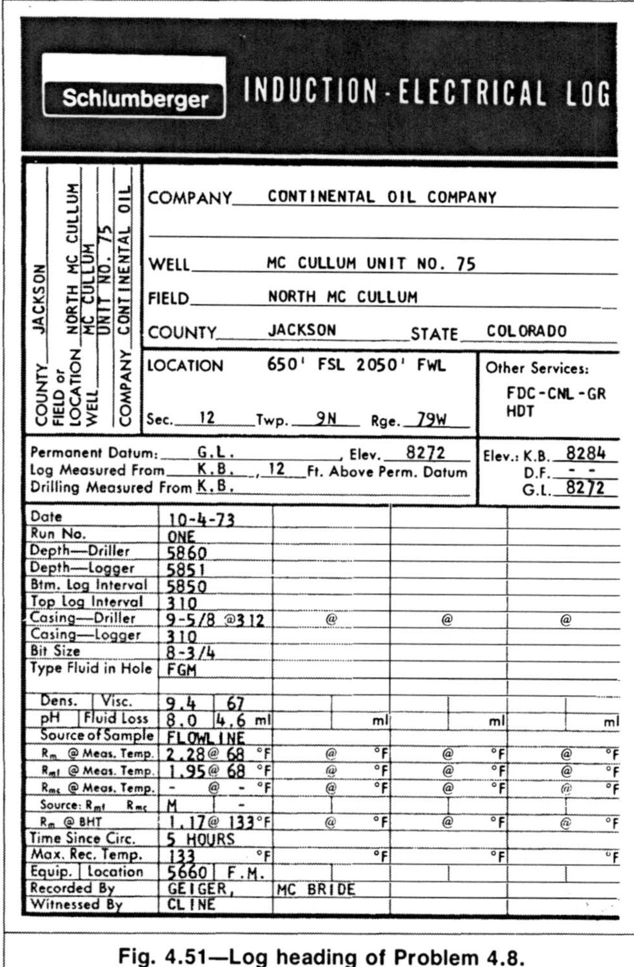

Schlumberger INDUCTION-ELECTRICAL LOG

COUNTY JACKSON
FIELD or LOCATION NORTH MC CULLUM / MC CULLUM
WELL UNIT NO. 75
COMPANY CONTINENTAL OIL

COMPANY CONTINENTAL OIL COMPANY
WELL MC CULLUM UNIT NO. 75
FIELD NORTH MC CULLUM
COUNTY JACKSON STATE COLORADO
LOCATION 650' FSL 2050' FWL
Sec. 12 Twp. 9N Rge. 79W
Other Services: FDC-CNL-GR HDT

Permanent Datum: G.L., Elev. 8272
Log Measured From K.B., 12 Ft. Above Perm. Datum
Drilling Measured From K.B.
Elev.: K.B. 8284 D.F. - - G.L. 8272

Date	10-4-73			
Run No.	ONE			
Depth—Driller	5860			
Depth—Logger	5851			
Btm. Log Interval	5850			
Top Log Interval	310			
Casing—Driller	9-5/8 @312	@	@	@
Casing—Logger	310			
Bit Size	8-3/4			
Type Fluid in Hole	FGM			
Dens. | Visc.	9.4 | 67			
pH | Fluid Loss	8.0 | 4.6 ml	ml	ml	ml
Source of Sample	FLOWLINE			
R_m @ Meas. Temp.	2.28 @ 68 °F	@ °F	@ °F	@ °F
R_{mf} @ Meas. Temp.	1.95 @ 68 °F	@ °F	@ °F	@ °F
R_{mc} @ Meas. Temp.	- @ - °F	@ °F	@ °F	@ °F
Source: R_{mf} | R_{mc}	M | -			
R_m @ BHT	1.17 @ 133 °F	@ °F	@ °F	@ °F
Time Since Circ.	5 HOURS			
Max. Rec. Temp.	133 °F	°F	°F	°F
Equip. | Location	5660 | F.M.			
Recorded By	GEIGER,	MC BRIDE		
Witnessed By	CLINE			

Fig. 4.51—Log heading of Problem 4.8.

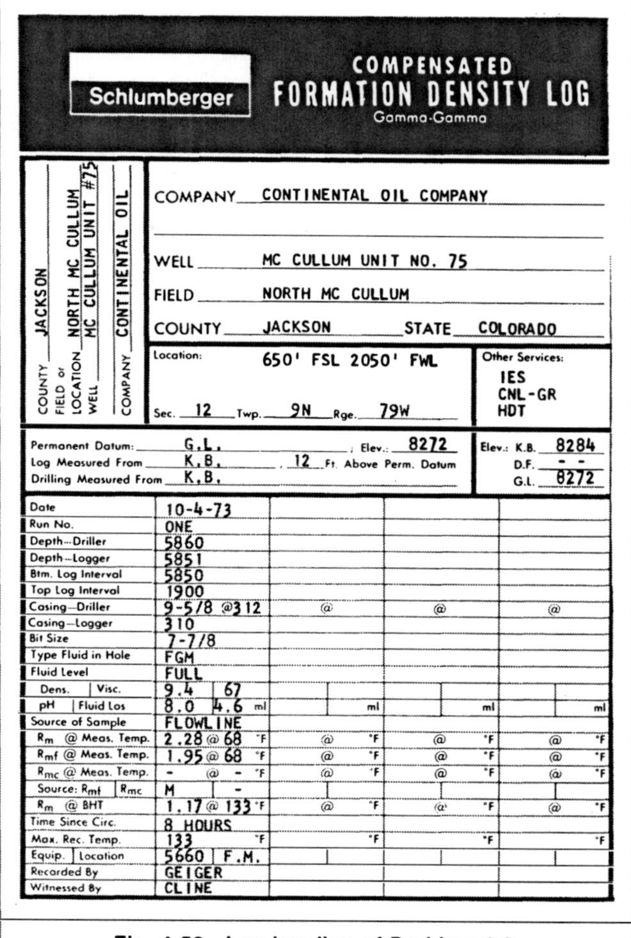

Schlumberger COMPENSATED FORMATION DENSITY LOG Gamma-Gamma

COUNTY JACKSON
FIELD or LOCATION NORTH MC CULLUM
WELL MC CULLUM UNIT #75
COMPANY CONTINENTAL OIL

COMPANY CONTINENTAL OIL COMPANY
WELL MC CULLUM UNIT NO. 75
FIELD NORTH MC CULLUM
COUNTY JACKSON STATE COLORADO
Location: 650' FSL 2050' FWL
Sec. 12 Twp. 9N Rge. 79W
Other Services: IES CNL-GR HDT

Permanent Datum: G.L.; Elev.: 8272
Log Measured From K.B., 12 Ft. Above Perm. Datum
Drilling Measured From K.B.
Elev.: K.B. 8284 D.F. - - G.L. 8272

Date	10-4-73			
Run No.	ONE			
Depth—Driller	5860			
Depth—Logger	5851			
Btm. Log Interval	5850			
Top Log Interval	1900			
Casing—Driller	9-5/8 @312	@	@	@
Casing—Logger	310			
Bit Size	7-7/8			
Type Fluid in Hole	FGM			
Fluid Level	FULL			
Dens. | Visc.	9.4 | 67			
pH | Fluid Los	8.0 | 4.6 ml	ml	ml	ml
Source of Sample	FLOWLINE			
R_m @ Meas. Temp.	2.28 @ 68 °F	@ °F	@ °F	@ °F
R_{mf} @ Meas. Temp.	1.95 @ 68 °F	@ °F	@ °F	@ °F
R_{mc} @ Meas. Temp.	- @ - °F	@ °F	@ °F	@ °F
Source: R_{mf} | R_{mc}	M | -			
R_m @ BHT	1.17 @ 133 °F	@ °F	@ °F	@ °F
Time Since Circ.	8 HOURS			
Max. Rec. Temp.	133 °F	°F	°F	°F
Equip. | Location	5660 | F.M.			
Recorded By	GEIGER			
Witnessed By	CLINE			

Fig. 4.52—Log heading of Problem 4.8.

21. Guyod, H.: "Temperature Well Logging: Temperature Distribution in the Ground," *Oil Weekly* (Nov. 4, 1946).
22. *Temperature Log Interpretation,* Welex, Houston (1979).
23. Hilchie, D.W.: "Maximum Temperatures Recorded in Wellbores," *Log Analyst* (Sept.-Oct. 1968) 21-24.
24. Millikan, C.V.: "Temperature Surveys in Oil Wells," *Trans.,* AIME (1941) **142,** 15-23.
25. *RP31 Recommended Practice and Standard Form for Electrical Logs,* third edition, API, Dallas (1976).
26. Nichols, E.A.: "Geothermal Gradients in Mid-Continent and Gulf Coast Oil Fields," *Trans.,* AIME (1947) **170,** 44-47.
27. Moses, P.L.: "Geothermal Gradients Now Known in Greater Detail," *World Oil* (May 1961) 79-82.
28. Guyod, H.: "Temperature Well Logging: Wells Not in Thermal Equilibrium, A: Rotary Holes," *Oil Weekly* (Dec. 2, 1946).
29. Timko, D.J. and Fertl, W.H.: "How Downhole Temperatures, Pressures Affect Drilling," *World Oil* (Oct. 1972) 73-88.
30. Dowdle, W.L. and Cobb, W.M.: "Static Formation Temperature From Well Logs—An Empirical Method," *JPT* (Nov. 1975) 1326-30.
31. Bebout, D.G. and Gutierrez, D.R.: "Geopressured Geothermal Resource in Texas and Louisiana—Geological Constraints," *Trans.,* Fifth Conference on Geopressure-Geothermal Energy, Louisiana State U., Baton Rouge (1981).
32. Bassiouni, Z.: "Evaluation of Potential Geopressure Geothermal Test Sites in Southern Louisiana," research report, Louisiana State U., Baton Rouge (1980).
33. Bassiouni, Z.: "Temperature Distribution in Deep Wells," research report, Louisiana State U., Baton Rouge (1986).

Chapter 5
Resistivity Logs

5.1 Introduction

The discussion of the nature of electrical resistivity of rocks in Chap. 1 revealed that a relationship exists between rock resistivity and both porosity and water saturation. It also revealed that if the formation resistivity is measured, the presence of hydrocarbon and its quantity can be deduced.

Resistivity measurement was first used in mining exploration. This geophysical exploration method consists of inducing a low-frequency current into the subsurface with metallic electrodes placed at the surface. The potential difference generated by the induced current is measured at the surface. The quadripole ABMN in **Fig. 5.1** is the most commonly used electrode array in surface exploration. A and B are the current electrodes, and M and N are the potential electrodes. A resistivity value, usually called apparent or effective resistivity, is derived from the induced current and measured potential at a surface station. The measurement and calculation of apparent resistivity are repeated at several stations covering the area prospected. Resistivity profiles similar to that in **Fig. 5.2** and resistivity maps similar to that in **Fig. 5.3** are constructed. These profiles and maps are used to obtain quantitative and qualitative information pertaining to type and structure of subsurface formations. The depth of investigation of this technique is controlled mainly by the spacing between the electrodes.

Electric well logging is a spin-off of geophysical prospecting. In March 1921, Marcel Schlumberger and several colleagues took advantage of a 2,500-ft-deep reconnaissance borehole and conducted downhole resistivity measurements.[1] The purpose of these measurements was to enhance the interpretation of surface data. The resistivity measurement did reflect the variation in the nature of subsurface formations penetrated by the wellbore. In early 1927, Conrad Schlumberger outlined the principle of a new exploration method called electrical coring. The first resistivity log was obtained on Sept. 5, 1927, in Pechelbronn, France.[1]

The first resistivity log was obtained by lowering three electrodes, a current Electrode A and two potential Electrodes M and N, into the wellbore at the end of insulated cables (**Fig. 5.4**). The surface casing was used as the other current Electrode B. The value of the current that circulated between Electrodes A and B and the voltage that resulted between Electrodes M and N were used to calculate an apparent resistivity value. The measurement was repeated at 3-ft intervals. A plot of resistivity vs. depth resulted in the first resistivity log, which is shown in **Fig. 5.5**. This successful logging operation was followed by the progressive development of the many resistivity tools discussed in this chapter. These tools may be classified as either conventional electrode, focused current, or induction type.

5.2 Apparent Resistivity

The theory for resistivity measurements is complex. Accurate treatment calls for involved electromagnetic theory. The true complex geometry of a borehole that penetrates several formations also has to be considered.[2] The concept of apparent resistivity, however, can be introduced by using the simplified case of a DC, point-power electrode in a homogeneous, isotropic, and infinitely extended medium.[3]

Consider a DC source located at Point A in a homogeneous, isotropic medium with resistivity $\boldsymbol{R}$ (**Fig. 5.6**). The current return electrode is placed so far from the electrode at Point A that its presence may be neglected during consideration of the current flow around Point A. Because the medium is completely homogeneous, the current density around the source depends only on the distance, r, from Point A. All points equidistant from the power electrode are then at the same potential. The prevailing flow system is spherical with equipotential spheres and radial lines of current flow.

The resistance, $d\rho$, of the spherical shell between the radii r and $r+dr$ is given by

$$d\rho = R\frac{dL}{dA} = R\frac{dr}{4\pi r^2}. \quad (5.1)$$

The resistance between two measuring points, Points 1 and 2, situated on two equipotential spheres with radii r_1 and r_2, respectively, is

$$\rho_{12} = \int_1^2 d\rho = \frac{R}{4\pi}\int_1^2 \frac{dr}{r^2} = \frac{R}{4\pi}\left(\frac{1}{r_1} - \frac{1}{r_2}\right). \quad (5.2)$$

If I is the intensity of the current leaving Electrode A and ΔV is the potential difference between the two spheres, the application of Ohm's law to the conductor between these spheres gives

$$\rho_{12} = \Delta V_{12}/I. \quad (5.3)$$

Combining Eqs. 5.2 and 5.3 results in

$$\Delta V_{12} = \frac{IR}{4\pi}\left(\frac{1}{r_1} - \frac{1}{r_2}\right). \quad (5.4)$$

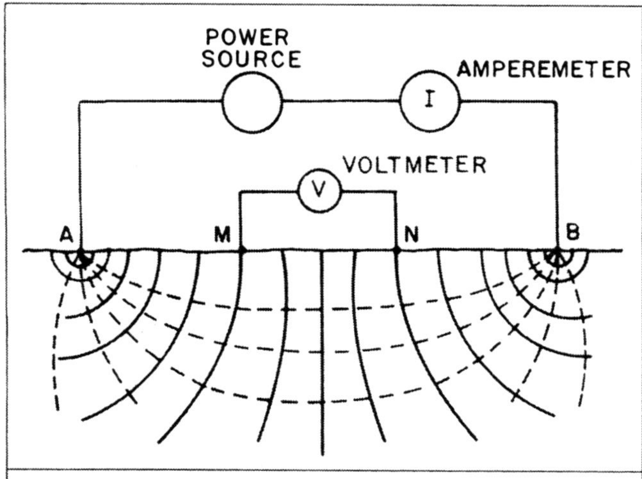

Fig. 5.1—Current flowlines and equipotentials for a typical ABMN quadrupole.

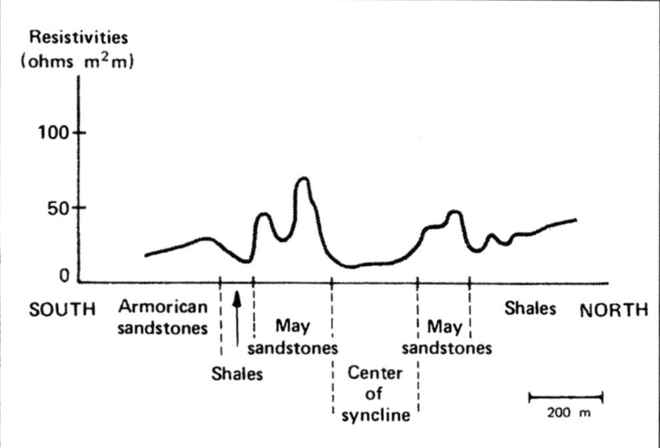

Fig. 5.2—Example of surface resistivity profile illustrated by the first one obtained in 1920 at May St. André, Normandy (from Ref. 1).

Eq. 5.4, giving the potential difference ΔV_{12} between any two points in a homogeneous medium as a result of the flow of a current I, may be solved for R as follows:

$$R=G_T(\Delta V_{12}/I), \qquad (5.5)$$

where $G_T=4\pi[r_1r_2/(r_2-r_1)]$. $\qquad$ (5.6)

G_T is a geometric coefficient for this particular electrode array. It depends only on the spacing between the electrodes on a sonde. A similar expression for R can be obtained for other electrode arrangements. Only the geometric coefficient G_T will be different.

Example 5.1. A constant 1-A current flows from the current Electrode A situated in a 100-Ω·m homogeneous, isotropic medium. What is the voltage measured by two electrodes situated at 17 ft, 4 in. and 20 ft, respectively? Note that this electrode arrangement corresponds to the 18 ft, 8 in. lateral tool described later in Sec. 5.3.2.

Solution. With the assumption of spherical flow and uniform resistivity about the current electrode, Eq. 5.4 can be used to calculate the voltage ΔV_{12}. The distances r_1 and r_2 to the two potential electrodes are r_1=17 ft, 4 in.=208 in.=5.2832 m and r_2=20 ft=240 in.=6.0960 m.

Using Eq. 5.6 gives

$$G_T=\frac{4\pi(5.2832)(6.0960)}{6.0960-5.2832}=497.93 \text{ m}.$$

Because I=1 A, and R=100 Ω·m,

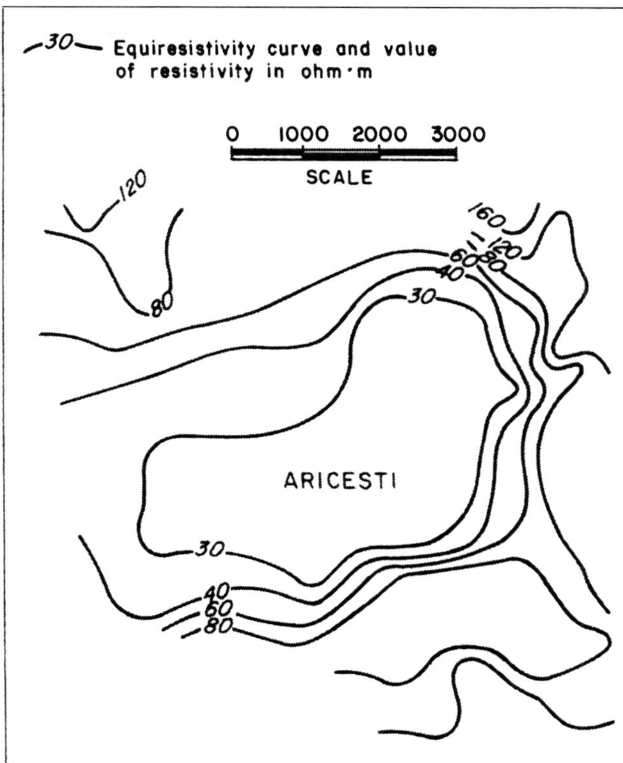

Fig. 5.3—Example of surface resistivity map obtained above Aricesti Salt Dome, Rumania, in 1923 (after Ref. 1).

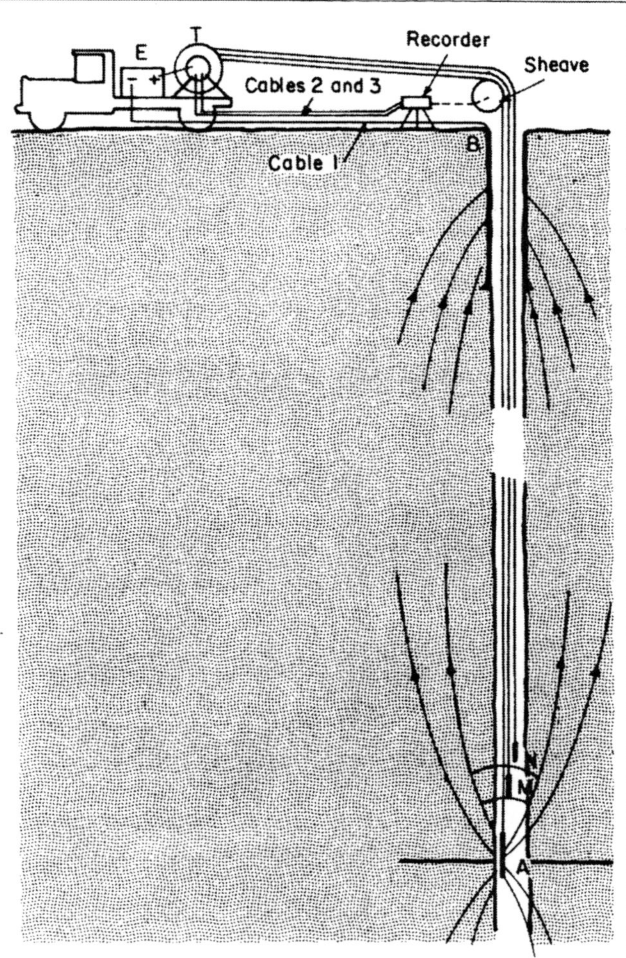

Fig. 5.4—Schematic of the electrodes array used in recording the first resistivity log (from Ref. 1).

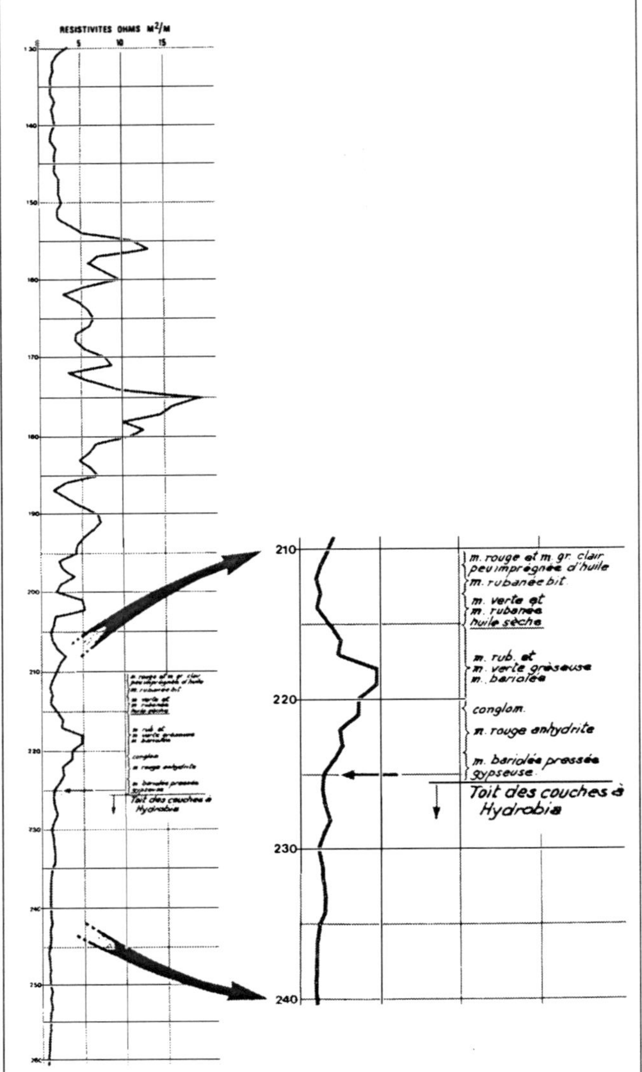

Fig. 5.5—Reproduction of the first resistivity log recorded in Pechelbronn, France, in 1927 (from Ref. 1).

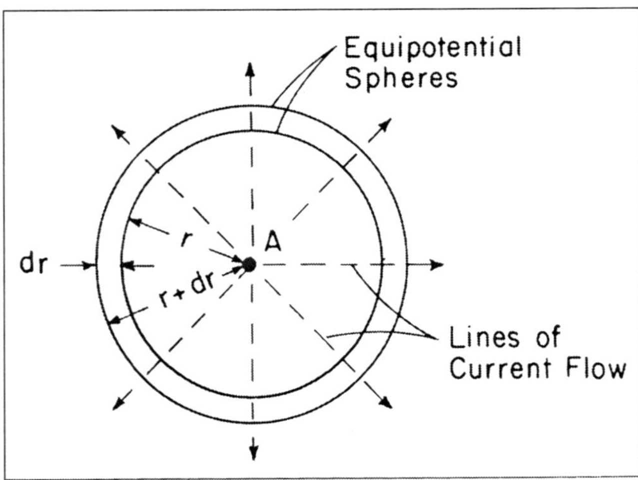

Fig. 5.6—Point power electrode in a homogeneous, isotropic, and infinitely extended medium.

$$\Delta V_{12} = \frac{(1)(100)}{497.93} = 0.200 \text{ V}$$

$$= 200 \text{ mV}.$$

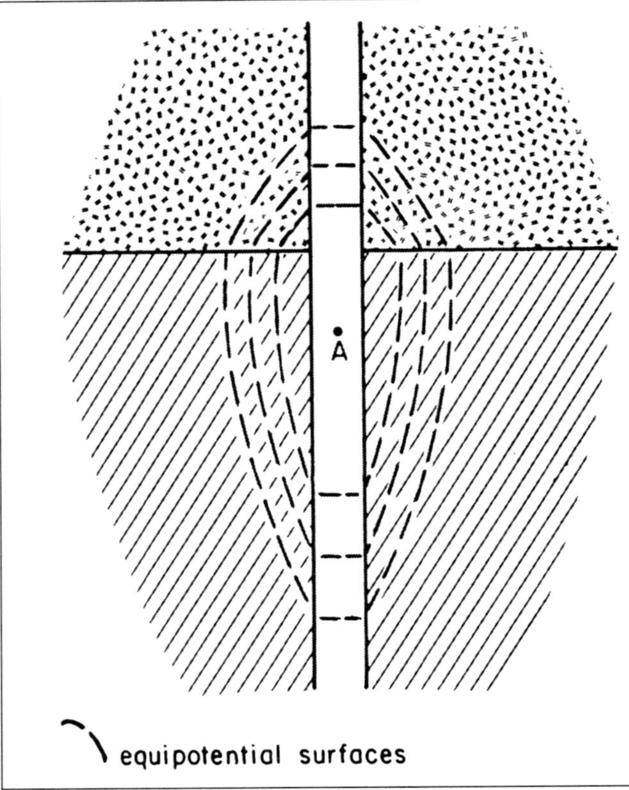

Fig. 5.7—Schematic of equipotential surfaces around a power electrode in a borehole.

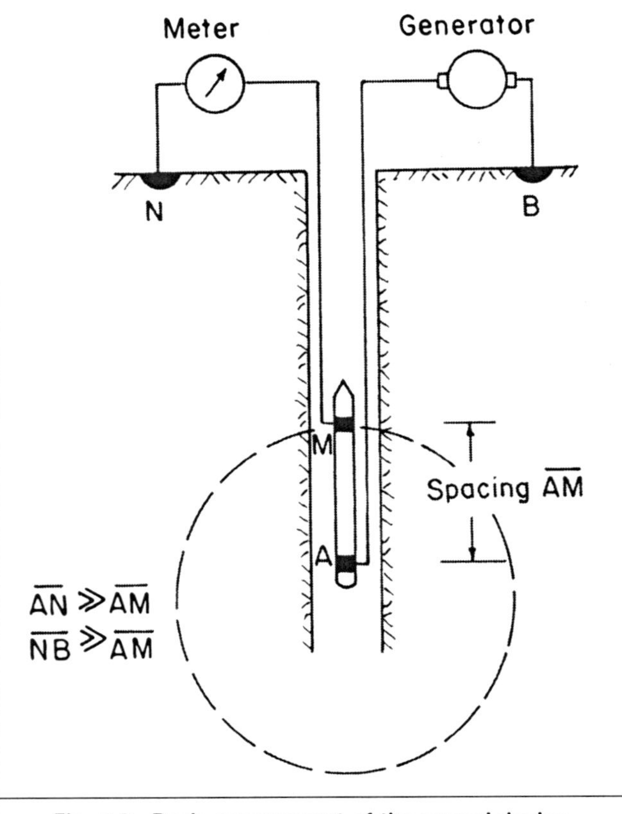

Fig. 5.8—Basic arrangement of the normal device.

Up to this point, a homogeneous, infinite medium has been assumed. An actual medium is less homogeneous in several ways. The existence of a hole filled with drilling mud, the presence of several formations, and the inevitable nonhomogeneities in the formations affect the configuration of the equipotential surfaces and the resulting potential differences observed at the measuring electrodes. **Fig. 5.7** is a schematic of the equipotential surfaces around power Electrode A in a borehole near the boundary between two formations of different resistivities, each more resistive than the mud filling the hole.

In an actual drillhole, a voltage ΔV is still indicated by the measuring electrodes. If ΔV is substituted into Eq. 5.5 with the proper values of I and G_T, a resistivity value can be calculated and plotted vs. depth as a log. This calculated resistivity is called apparent resistivity, R_a. The apparent resistivity can be regarded as the resistivity of a homogeneous, isotropic, and infinitely extended medium that is electrically equivalent to the true medium surrounding the tool.

This concept of apparent resistivity uses the simplified case of spherical flow of DC and electrode-type devices. It holds true in more complex cases of focused current and induction devices. In those cases, the signal recorded by the tool is used to calculate an apparent resistivity, which is the resistivity of a simplified medium electromagnetically equivalent to the true medium surrounding the tool. The simplified medium is usually one where (1) all the media surrounding the measuring device are assumed to be homogeneous and isotropic, (2) the fluid column filling the drillhole has the shape of a circular and infinitely long cylinder, and (3) the electrical logging tool is located on the drillhole axis.

Regardless of the tool type and design and the simplifying assumptions used, an apparent medium resistivity is calculated and presented as a resistivity log. When sufficient information is available regarding the conditions of the measurement, it is often possible to use the log values of apparent resistivity to arrive at the true formation resistivity by means described in Secs. 5.6 and 5.7.

Example 5.2. Three electrodes, A, M, and N, are lowered on a logging sonde into a 3,000-ft wellbore drilled through an unknown medium and filled with drilling fluid. The distances AM and AN are 17 ft, 4 in. and 20 ft, respectively. When a 2-A current is circulated between Electrodes A and B (B is situated at the surface), a signal of 10 mV is measured between Electrodes M and N. Calculate the apparent resistivity of the medium that surrounds the sonde.

Solution. Because Electrode B is so far from Electrode A, Eq. 5.5 can be used:

$$R_a = G_T(\Delta V/I), \qquad (5.7)$$

where R_a is the apparent resistivity of the medium surrounding the tool.

Eq. 5.6 yields G_T=497.93 m, and because I=2 A and V=0.010 V,

$$R_a = 497.93(0.01)/2$$
$$= 2.5 \Omega \cdot \text{m}.$$

Note that none of the zones surrounding the tool would necessarily have a resistivity of 2.5 Ω·m.

5.3 Conventional Electrode Tools

As with the geophysical exploration method, the electrode-type logging devices use an array of four electrodes: A, B, M, and N. Two or three of these electrodes are mounted on a sonde and lowered into the borehole. The remaining electrodes are either grounded at the surface or positioned in the borehole far from the sonde. The electrode arrangements common in well logging are the normal and lateral arrangements. In the normal arrangement, two electrodes are mounted on the sonde; in the lateral arrangement, three electrodes are on the logging tool.

Resistivity logs obtained with the normal and lateral electrode arrangements are not currently in use because of inherent limitations (discussed later). Despite this fact, familiarity with these logs is warranted because of several reasons. First, their concept is the simplest of all developed resistivity tools, and grasping it makes understanding the workings of other devices easy. Second, the normal and lateral devices were used extensively from the late 1930's to the early 1960's and make up almost 45% of all existing logs.[4] Finally, the new technology of log measurement while drilling (MWD) uses a normal device to measure formation resistivity.

5.3.1 Normal Device. **Fig. 5.8** shows the basic electrode arrangement of the normal device. The current Electrode A and potential Electrode M are mounted on the sonde and lowered into the borehole. Electrodes B and N are located at the surface far from Electrodes A and M. In practice, Electrode B is also put in the borehole. The generator induces a low-frequency, constant current I between Electrodes A and B. Because Electrode N is remote from the current electrodes, its potential is practically negligible. The voltmeter measures the potential of Electrode M. This potential can be expressed with Eq. 5.4, where r_1 is replaced by $\overline{AM}$ and r_2 is infinite, resulting in

$$\Delta V = IR/4\pi\overline{AM}. \qquad (5.8)$$

The apparent resistivity calculated from the measured ΔV and I is expressed with equations similar to Eqs. 5.5 through 5.7:

$$R_a = G_N(\Delta V/I), \qquad (5.9)$$

where $G_N = 4\pi\overline{AM}$. (5.10)

G_N is the geometric coefficient of the normal sonde and $\overline{AM}$ is the sonde spacing. The value of R_a is plotted at a depth that corresponds to the midpoint between Electrodes A and M. This point is usually called the inscription point.

As demonstrated later, 50% of the signal measured by a normal device placed in an infinite, homogeneous medium comes from a spherical volume centered at Electrode A, with a radius equal to twice the spacing $\overline{AM}$. Consequently, the normal device is said to have a radius or depth of investigation that is twice its spacing.

Most of the electric tools contained at least two normal devices of two different spacings. The most commonly used normal spacings are the 16-in. short normal and the 64-in. long normal. An intermediate 36-in. spacing was also available on the U.S. gulf coast.[5] The long normal device, which has a radius of investigation of about 10 ft, was used to overcome borehole and invaded-zone effects and to provide a representative value of true formation resistivity, R_t. The resolution of the long normal, however, is poor in thin beds. The short normal resistivity log is used for correlation, for the location of bed boundaries, and for the evaluation of thin beds.

5.3.2 Lateral Device. Electrodes A, M, and N are mounted on the lateral sonde (**Fig. 5.9**). The current I induced between Electrodes A and B creates a voltage difference, ΔV, measured between Electrodes M and N. This voltage and the apparent resistivity derived from it are expressed by Eqs. 5.4 and 5.7. The geometric coefficient G_L of the lateral curve is expressed by Eq. 5.6 and can be written as

$$G_L = 4\pi(\overline{AM}\cdot\overline{AN}/\overline{MN}). \qquad (5.11)$$

The calculated apparent resistivity is transcribed on the log at a depth corresponding to the midpoint, O, of Electrodes M and N. The distance $\overline{AO}$ is called the spacing of the lateral device.

In a homogeneous, infinite medium, the lateral log measures the resistivity of the imaginary spherical shell between potential Electrodes M and N. The most common lateral spacings, $\overline{AO}$ and $\overline{MN}$, are 18 ft, 8 in. and 32 in., respectively. The radius of investigation of this tool is about 19 ft, which exceeds the 10-ft radius of investigation of the long normal. The lateral device largely overcomes the effect of the invaded zone and yields a good R_t value. This is true, however, only for beds 40 ft or more thick. In thinner beds, the lateral loses most of its vertical resolution.

5.3.3 Response of the Normal and Lateral Devices. Insight into the shape of the apparent resistivity curves was gained through both theoretical and experimental work. The theory of electrical images and the principle of superposition[6-8] yield a relatively simple solution for the case of the half-space of **Fig. 5.10**. The half-space is made up of two homogeneous beds of resistivities R_1 and R_2 where $R_1 > R_2$. The logging device consists of point electrodes

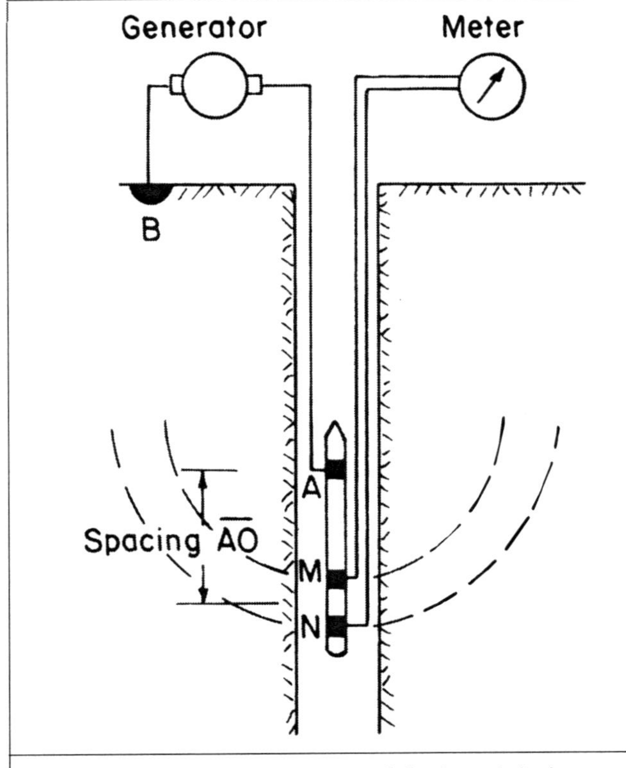

Fig. 5.9—Basic arrangement of the lateral device.

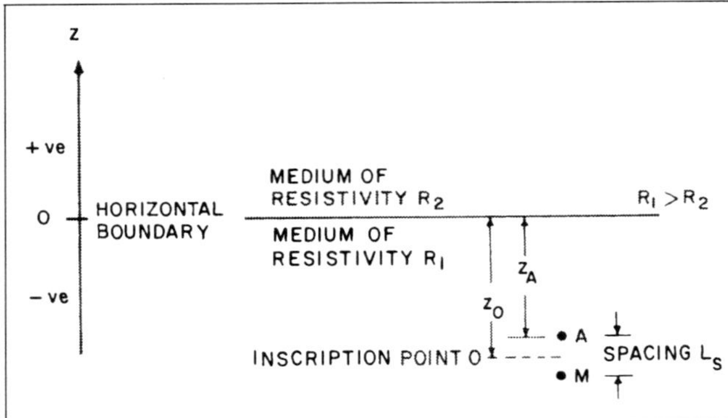

Fig. 5.10—Schematic illustrating the symbols and parameters of Eqs. 5.12 through 5.15.

traveling vertically, perpendicular to the boundary that separates the two beds. For the current Electrode A situated above the potential Electrode M where both electrodes are situated in Medium 1, the apparent resistivity can be calculated from[2]

$$R_a=\left[1+\frac{C_R}{2(z_A/L_s)-1}\right]R_1. \quad (5.12)$$

If only Electrode M is in Medium 1, then

$$R_a=2R_1R_2/(R_1+R_2), \quad (5.13)$$

and when both electrodes are in Medium 2,

$$R_a=\left[1+\frac{C_R}{2(z_A/L_s)+1}\right]R_2 \quad (5.14)$$

where C is usually called the coefficient of reflection defined by

$$C_R=(R_1-R_2)/(R_1+R_2), \quad (5.15)$$

z_A is the coordinate of the current Electrode A measured from the boundary and L_s is the device spacing $\overline{AM}$.

TABLE 5.1—R_a VALUES AS A FUNCTION OF DIMENSIONLESS DEPTH FOR EXAMPLE 5.3

z_A/L_s	z_O/L_s	R_a (Ω·m)	z_O, 16-in. Short Normal (ft)	z_O, 64-in. Long Normal (ft.)
0	−0.5	18.0	0.67	2.67
−0.5	−0.1	54.0	1.33	5.33
−1	−1.5	66.0	2.00	8.00
−2	−2.5	75.7	3.33	13.33
−3	−3.5	79.7	4.67	18.67
−4	−4.5	82.0	6.00	24.00
−5	−5.5	83.5	7.33	29.33
−8	−8.5	85.76	11.33	45.33
−10	−10.5	86.6	14.00	56.00
−20	−20.5	88.2	27.33	109.33
−40	−40.5	89.1	54.00	216.00
−100	−100.5	89.6	134.0	536.00

TABLE 5.2—R_a VALUES AS A FUNCTION OF DEPTH FOR EXAMPLE 5.3

z_A/L_s	z_O/L_s	R_a (Ω·m)	z_O, 16-in. Short Normal (ft)	z_O, 64-in. Long Normal (ft)
0	0.5	18.0	0.67	2.67
1	1.5	12.7	2.00	8.00
2	2.5	11.6	3.33	13.33
3	3.5	11.1	4.67	18.67
4	4.5	10.9	6.00	24.00
5	5.5	10.7	7.33	29.33
10	10.5	10.4	14.00	56.00
20	20.5	10.2	27.33	109.33
40	40.5	10.1	54.00	216.00
100	100.5	10.0	134.00	536.00

Example 5.3. Plot the apparent resistivity curve obtained by a 16-in. short normal and a 64-in. long normal device in a half-space made of 90- and 10-Ω·m resistivity beds.

Solution. C_R given by Eq. 5.15 is

$$C_R=\frac{R_1-R_2}{R_1+R_2}=\frac{(R_1/R_2)-1}{(R_1/R_2)+1}$$

$$=(9-1)/(9+1)$$

$$=0.8.$$

R_a, when both electrodes are in the 90-Ω·m medium, is given by Eq. 5.12:

$$R_a=\left[1+\frac{0.8}{2(z_A/L_s)-1}\right]90.$$

R_a values are given in **Table 5.1** as a function of dimensionless depths (z_A/L_s) and (z_O/L_s). z_O is the depth of Inscription Point O. R_a, when both electrodes are in the 10-Ω·m bed, is given by Eq. 5.14:

$$R_a=\left[1+\frac{0.8}{2(z_A/L_s)+1}\right]10.$$

Table 5.2 gives the values of R_a as a function of depth. Eq. 5.13 gives the value of R_a when only Electrode M is situated in Medium 1.

$$R_a=[2(90)(10)/(90+10)]=18.$$

This occurs as the sonde straddles the boundary over an interval between $z_O/L_s=-0.5$ and $z_O/L_s=+0.5$.

Fig. 5.11 shows R_a vs. depth of the Inscription Point O for both the 16-in. short normal and the 64-in. long normal. The difference in the response of the two devices shows the effect of tool spacing on response.

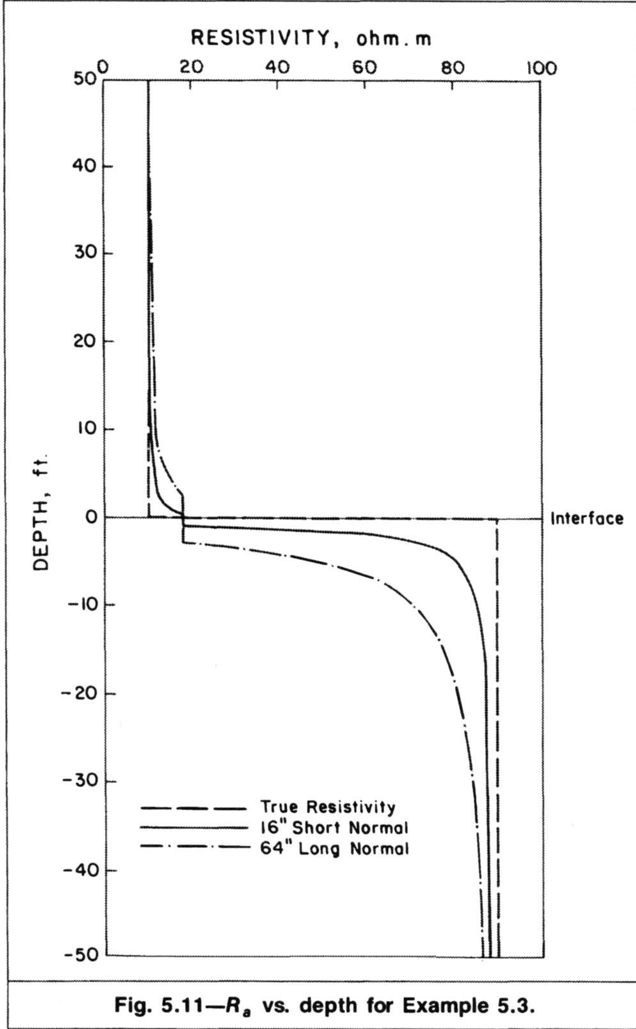

Fig. 5.11—R_a vs. depth for Example 5.3.

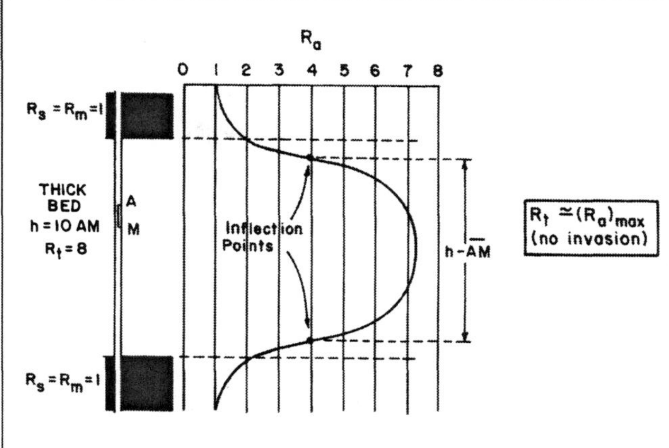

Fig. 5.12—Response of the normal device in a relatively thick bed ($h>>\overline{AM}$) more resistive than adjacent formation ($R_t>R_s$), after Ref. 9.

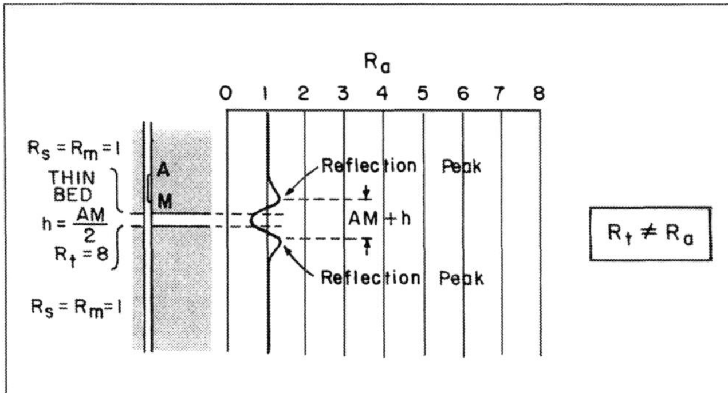

Fig. 5.13—Response of the normal device in a relatively thin bed ($h<\overline{AM}$), more resistive than adjacent formation ($R_t>R_s$) (after Ref. 9).

Figs. 5.12 through 5.20 show typical apparent resistivity curves obtained experimentally for the simple case of a single homogeneous bed of resistivity R_t and thickness h that is bounded by very thick homogeneous formations of resistivity R_s. In these laboratory experiments, the resistivity of the drilling mud that fills the borehole, R_m, is equal to the resistivity of the adjacent formation, R_s.

Fig. 5.12 shows the response of the normal device for a bed that is more resistive than the adjacent beds and where the spacing $\overline{AM}$ is less than the formation thickness. This case corresponds, for example, to a thick oil sand surrounded by conductive shales. The curve is symmetrical with respect to the bed centerline. Moving upward, the apparent resistivity begins increasing at some point above the formation's lower boundary, attaining a maximum where the midpoint O of $\overline{AM}$ coincides with the center of the bed. The apparent resistivity decreases from this maximum value and ultimately reaches the surrounding bed resistivity at a point located several times the spacing $\overline{AM}$ above the top boundary. The two segments of the curves marking the transition from R_s to a maximum apparent resistivity value, $(R_a)_{max}$, display two inflection points situated within the bed boundaries. The vertical distance between these two inflection points is equal to the bed thickness minus one spacing length. Therefore, the shorter the spacing, the easier it is to determine the bed boundaries and thickness.

The maximum value reached by the apparent resistivity depends on the bed thickness, h. It approaches R_t as h increases. When h exceeds four times the spacing $\overline{AM}$, the apparent resistivity value approximates the true formation resistivity, R_t. The 64-in. long normal reads the true formation resistivity in beds 20 ft or more thick. This is true, of course, only in the absence of invasion.

When $\overline{AM}>h$, a slight hump in the curve shows above and below the bed boundaries, as illustrated in Fig. 5.13. The distance between these humps, called reflection peaks, is equal to the bed thickness plus one spacing length. Opposite the resistive bed, the curve is depressed below the R_s value, and it erroneously appears to be conductive. This response, called a reversed signal, is a distortion caused by geometry. It appears on the 64-in. long normal curve opposite beds 5 ft or less thick. In this case, R_t is not reflected by R_a readings.

Figs. 5.14 and 5.15 show the case of a bed that is less resistive than the surrounding formations. This case corresponds, for example, to a saltwater-bearing, porous sandstone bounded by relatively tight shales or to a shale streak within a hydrocarbon-bearing formation. Fig. 5.14 illustrates the case of a relatively thick bed ($h>\overline{AM}$), and Fig. 5.15 illustrates the case of a relatively thin bed ($h<\overline{AM}$). In both cases, the curve is symmetrical with inflection points located outside the bed. The distance between these points is equal to the thickness of the bed plus one spacing length. The minimum apparent resistivity value, $(R_a)_{min}$, displayed by the curve differs considerably from R_t for thin beds. For thick beds, this minimum value approaches the true resistivity value. R_t is approximated by $(R_a)_{min}$ for beds four times thicker than the tool spacing.

In general, lateral curves are asymetrical about the center of the bed. Bed boundaries are harder to recognize from these curves. Fig. 5.16 shows a resistive bed much thicker than the $\overline{AO}$ spacing. The slight depression of R_a above the layer can be explained qualitatively. When Electrodes A, M, and N are situated above the resistive bed boundary, the current is forced upward because this is the path of least resistance. The current intensity at Point O is less than it would be in a homogeneous, infinite medium. The apparent resistivity calculated from Eq. 5.7 with the nominal current value is less than R_s. When Electrodes M and N are situated below the top boundary but Electrode A is still in the upper conductive bed, most of the current is forced upward and low apparent resistivity is calculated over an interval equal to $\overline{AO}$, the "delay" interval. When Electrodes A, M, and N are in the resistive bed, R_a increases rapidly and reaches a plateau. The resistivity of the plateau approximates the true resistivity. A bed thickness greater than 2½ times the spacing $\overline{AO}$ is needed for a plateau to develop. In this condition, the bed appears as infinite to the tool situated in its middle. When Electrode A is near the lower boundary, the current is forced downward toward the lower conductive bed. The current intensity at Point O is much higher than it would be in a homogeneous, infinite medium. The calculated apparent resistivity overshoots R_t. The "overshot" interval ends when Electrodes M and N cross the lower boundary of the resistive bed. Reaching the resistivity R_s is delayed over an interval equal to $\overline{AO}$ as the current electrode and Point O straddle the boundary.

Fig. 5.17 illustrates the case of a resistive bed of a thickness slightly higher than $\overline{AO}$ ($\overline{AO} \leq h \leq 2½\overline{AO}$). The curve shows a depression above the top boundary, a delay below both the top and lower boundaries, and an overshot above the lower boundary. A plateau does not develop, making the determination of R_t more difficult. Empirical relations have been developed to relate R_a to R_t. If the bed is $1.3 \times \overline{AO}$ in thickness, then

$$R_t \simeq (R_a)_{max}, \quad (5.16)$$

and if h is about $1.5 \times \overline{AO}$, then

$$R_t \cong R_{AO} + ⅔[(R_a)_{max} - R_{AO}], \quad (5.17)$$

where R_{AO} is the apparent resistivity displayed by the lateral curve at a level situated a distance $\overline{AO}$ below the top boundary. Eq. 5.17 is known as the two-thirds rule.

When $h < \overline{AO}$, the lateral tool response is characterized by a sharp peak with an apparent resistivity less than R_t (Fig. 5.18). Other characteristic features are a zone of low resistivity below the lower boundary followed by a second peak. This peak, the "reflection peak," is located at a distance $\overline{AO}$ from the lower boundary. The presence of the peak can be explained qualitatively by the higher current intensity at Electrodes M and N where Electrode A is situated close to the resistive boundary, causing the current to deflect downward. The zone of low resistivity, called the "blind zone," occurs as Electrodes A and O straddle the bed and Electrodes M and N are shielded from the current source, Electrode A, by the resistive bed. The blind zone distorts the lateral curve, especially in sequences of thin beds at short distances from one another. R_t is related to R_a readings by[4]

$$R_t \geq [(R_a)_{max}][R_s/(R_a)_{min}], \quad (5.18)$$

where $(R_a)_{max}$ and $(R_a)_{min}$ are the minimum and maximum responses of the tool, respectively.

Figs. 5.19 and 5.20 illustrate the case of a conductive thick bed and a conductive thin bed, respectively. The features of the curves are similar to those of the resistive bed.

5.3.4 Electric Log. The electric log is used to record, simultaneously or alternately, two normal curves of different spacings and a lateral curve. Several apparent resistivity curves of different radii of investigation are required for the evaluation of thin and thick beds. Several readings are also required to detect invasion. **Fig. 5.21** shows an example array of four electrodes that can be used to record short normal, long normal, and lateral curves. While a current is induced between Electrodes 1 and 4, potential is measured between Electrode 2 and a surface electrode to generate a short normal and between Electrode 3 and the same surface electrode to generate a long normal. For the lateral curve, a current is induced between Electrodes 2 and 3, and the potential is measured between Electrode 4 and the surface electrode. This lateral circuit, M-AB, is the reciprocal of A-MN in Fig. 5.9. Both circuits give exactly the same result according to the principle of reciprocity.[9] The current is sent between Electrodes 2 and 3 instead of between Electrode 4 and the surface electrode because Electrode 4 is used to record the self-potential (SP) curve at the same time that the lateral is recorded (see Chap. 6).

The number of electrodes used depends on the number of conductors in the logging cable. The array of Fig. 5.21 requires four conductor cables. Cables with more conductors permit the use of more electrodes.

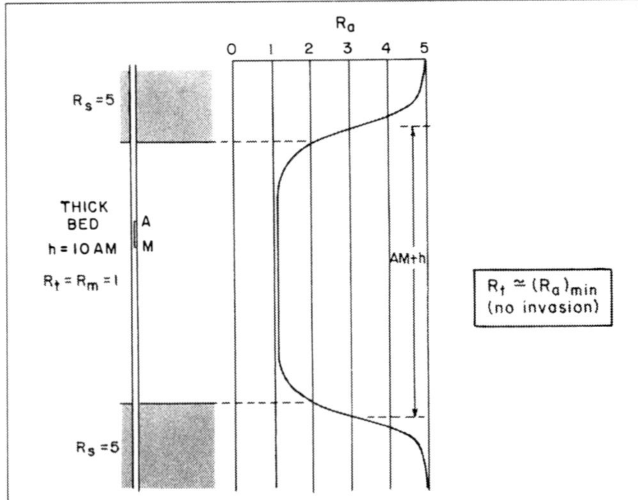

Fig. 5.14—Response of the normal device in a relatively thick bed ($h >> \overline{AM}$) less resistive than the adjacent formation ($R_t < R_s$) (after Ref. 9).

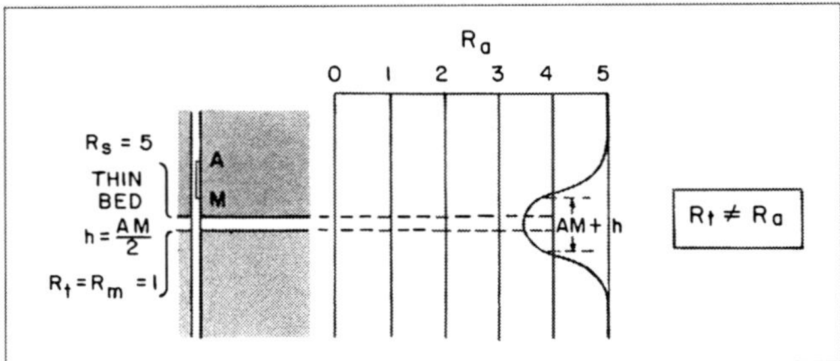

Fig. 5.15—Response of the normal device in a relatively thin bed ($h < \overline{AM}$) less resistive than the adjacent formation ($R_t < R_s$) (after Ref. 9).

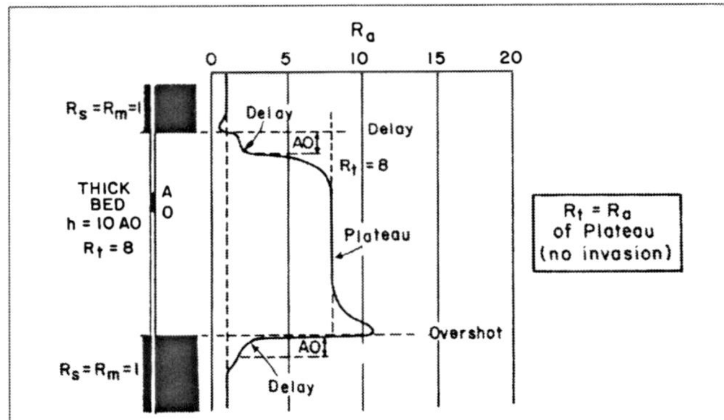

Fig. 5.16—Response of lateral device in relatively thick bed ($h > 2.5\ \overline{AO}$) more resistive than adjacent formations ($R_t > R_s$) (after Ref. 9).

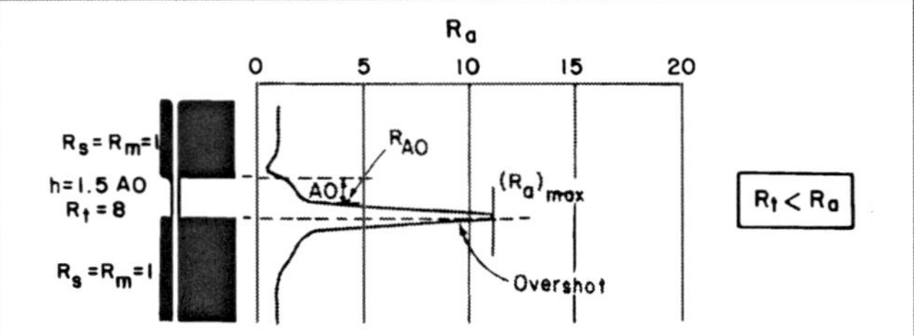

Fig. 5.17—Response of lateral device in a bed more resistive than the surrounding ($R_t > R_s$) and of thickness slightly higher than spacing $\overline{AO}$ (after Ref. 9).

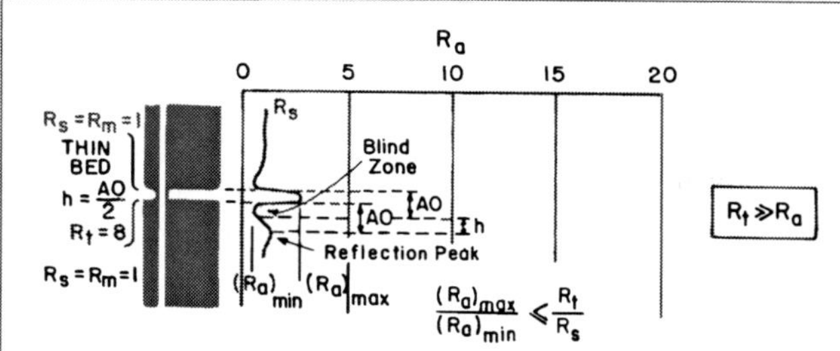

Fig. 5.18—Response of lateral device in a relatively thin bed ($h < \overline{AO}$) more resistive than the adjacent formation ($R_t > R_s$).

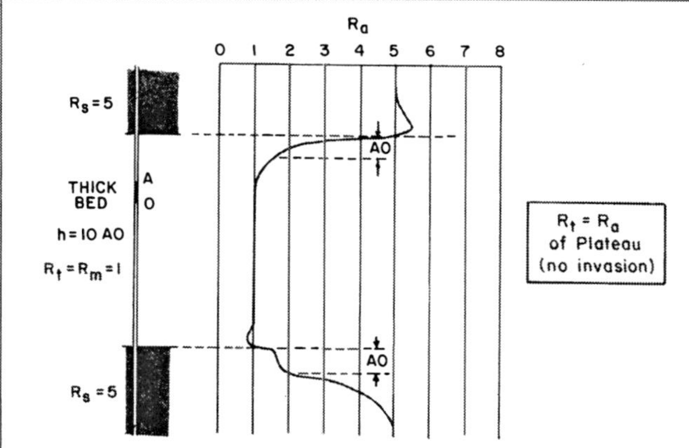

Fig. 5.19—Response of the lateral device in a relatively thick bed ($h > 50$ ft) less resistive than the adjacent beds ($R_t < R_s$) (after Ref. 9).

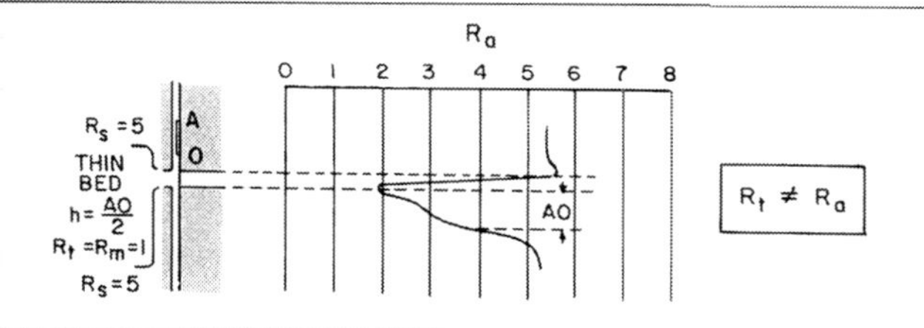

Fig. 5.20—Response of the lateral device in a relatively thin bed ($h < \overline{AO}$) less resistive than the adjacent beds ($R_t < R_s$) (after Ref. 9).

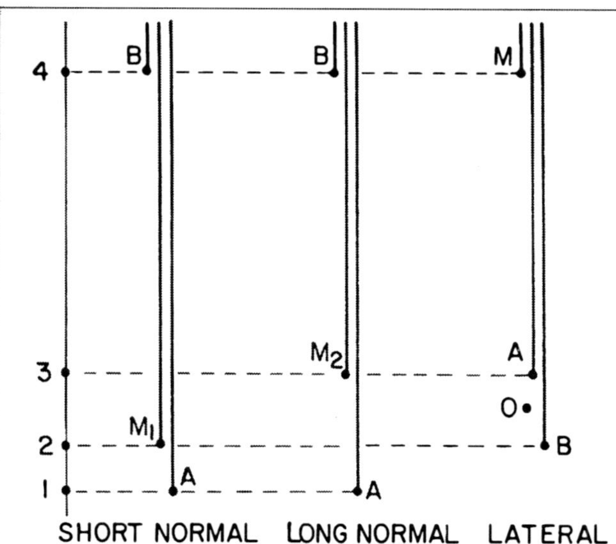

Fig. 5.21—Electrode array used to record the different curves of the electric log (after Ref. 9).

The electric log is recorded on the grid shown in **Fig. 5.22**. The log is divided into three recording tracks. The depth of the inscription point is printed at 100-ft intervals in the vertical blank column that separates Track 1 from Tracks 2 and 3. Logs used for quantitative evaluation are recorded on a detailed depth scale of 5 in. = 100 ft. In this case, the space between two horizontal lines represents 2 ft of hole. Intermediate 10-ft-depth lines are heavier than the 2-ft lines. Also, much heavier lines are printed at 50-ft intervals to facilitate curve reading. Logs used for qualitative evaluation (e.g., geologic correlations) are recorded on a scale of either 1 in. = 100 ft or 2 in. = 100 ft.

Each of the three recording tracks is divided by vertical lines into 10 divisions that are scaled linearly. The range of the scale is selected to accommodate a certain range of the parameter recorded on the track. When the parameter exceeds the upper limit of the scale, its value is divided by a constant, usually 2, 5, or 10, and recorded on a less detailed scale, usually referred to as an "off-scale" or a "backup scale." If a more detailed recording is needed, the parameter is multiplied by a constant, usually 2, 5, or 10, and recorded on a more detailed scale, usually referred to as the "amplified scale."

Fig. 5.23 is an example of an actual electric log. The SP curve discussed in Chap. 6 is recorded on Track 1 on the left side of the log. Tracks 2 and 3 show four apparent resistivity curves. Each resistivity curve shows a sequence of responses similar to those illustrated by Figs. 5.12 through 5.20. A single lateral curve obtained with an 18-ft, 8-in. spacing device is recorded on Track 3 on a 0- to 20-Ω·m scale. Track 2 shows three normal curves—two solid curves and one dashed curve. By convention, when more than one curve is recorded on the same track, the dashed curve represents the response of the tool with the deeper radius of investigation. In this case, it represents the 64-in. normal, which is recorded on a 0- to 20-Ω·m scale. The two solid curves represent the 16-in. short normal. One is recorded on a 0- to 20-Ω·m scale and the second on an amplified 0- to 4-Ω·m scale. Although the amplified scale range is not shown, it can be deduced from the fact that amplified and standard short normal curves should indicate the same readings at any specific depth.

Salient features of this log example include the following.

1. A reversed signal is indicated by the 64-in. long normal in thin resistive Beds B, D, E, and F of thicknesses less than the tool spacing. R_t cannot be determined from this curve in these thin beds.
2. Thin resistive Beds B, D, E, and F are indicated on the lateral curve by a distinctive peak. None of these peaks reach the true resistivity value because of limited bed thickness.
3. Resistive thin Beds B, D, E, and F are also indicated by the short normal curve. The 16-in. normal readings are affected by invasion, making the determination of R_t impractical.
4. Resistive Bed C does not appear on the lateral curve because it is hidden by the blind zone created by Bed B. The reflection peak of Zone B appears about 19 ft below the lower boundary of Bed B. The boundary of Bed B is determined from the 16-in. normal, which displays the shortest spacing.
5. The short normal and long normal curves indicate apparent resistivities of 19 and 5 Ω·m, respectively, in Bed C. This clearly indicates that Bed C is invaded (i.e., permeable).
6. The thickness of Bed C is about 12 ft, or about twice the long normal spacing. R_t of Bed C is greater than 5 Ω·m. An empirical relationship[4,11] indicates that for this thickness $R_t = 2R_a$. Hence, R_t can be approximated to be 10 Ω·m. This determination of R_t, however, carries a degree of uncertainty.
7. Both short and long normals indicate the same R_a in Zone A. This signals a lack of invasion (i.e., an impermeable zone). Zone A is shale, as also indicated by the SP log (see Chap. 6).

Example 5.4. Fig. 5.24 shows an electric log. The log displays several normal and lateral curves obtained for a resistive, 22-ft sand surrounded by thick shales. Give your best estimate of the true resistivity of the sand.

Solution. The following readings are obtained from the different curves.

16-in. normal: $(R_a)_{max}=8\ \Omega\cdot m$.
36-in. normal: $(R_a)_{max}=7\ \Omega\cdot m$.
64-in. normal: $(R_a)_{max}=5.5\ \Omega\cdot m$.
18-ft, 8-in. lateral: not a meaningful reading because bed thickness is close to sonde spacing. Also, the curve shows considerable distortion.
8-ft lateral: the bed thickness is almost three times the tool spacing. A short plateau is reached at $R_a=5.5\ \Omega\cdot m$. Lack of a better-defined plateau results from the heterogeneity of the sand.

The readings of the normal devices differ because of invasion. The relative magnitudes of the readings indicate relatively shallow invasion, such as the reading of the 64-in. normal, which is affected not at all or only a little by invasion. Given this reasoning and that the bed is more than four times the spacing in thickness, then $R_t=R_a=5.5\ \Omega\cdot m$. This conclusion is supported by the reading of the 8-ft lateral.

Quantitative consideration of invasion, borehole, and bed thickness effects is possible through the use of the resistivity-departure-curve method.[12] Elaborate departure curves were developed to give apparent resistivity normalized with respect to mud resistivity, R_a/R_m, as a function of normalized true resistivity, R_t/R_m, and tool spacing normalized with respect to borehole diameter, $\overline{AM}/d_h$. These curves were generated for several combinations of h/d_h, d_i/d_h, and R_i/R_m, which are the normalized values of bed thickness, diameter of invasion, and invaded-zone resistivity. Both h and d_i are normalized by borehole diameter and the invaded-zone resistivity, is normalized by mud resistivity. True resistivity and diameter of invasion are determined from this method by a curve-matching technique.

For this example, $h=22$ ft, $d_h=8$ in., $R_m=0.8\ \Omega\cdot m$, and $R_i \geq R_{16\ in.} \geq 8\ \Omega\cdot m$. These values result in

$$h/d_h=22/(8/12)=33$$

and $R_i/R_m \geq 8/0.8 \geq 10$.

The tool readings give the following values.

$\overline{AM}$ (in.)	R_a ($\Omega\cdot m$)	R_a/R_m ($\Omega\cdot m$)	$\overline{AM}/d_h$ (in.)
16	8	10.0	2
36	7	8.75	4.5
64	5.5	6.875	8

R_a/R_m is plotted vs. $\overline{AM}/d$ and matched to different departure curves constructed for parameters that approximate the parameters of the example. The departure curve in **Fig. 5.25** provides the best match. From this match, $R_t/R_m=6.5$, which yields $R_t=5.2\ \Omega\cdot m$, and $d_i/d_h=2$, which yields $d_i=16$ in.

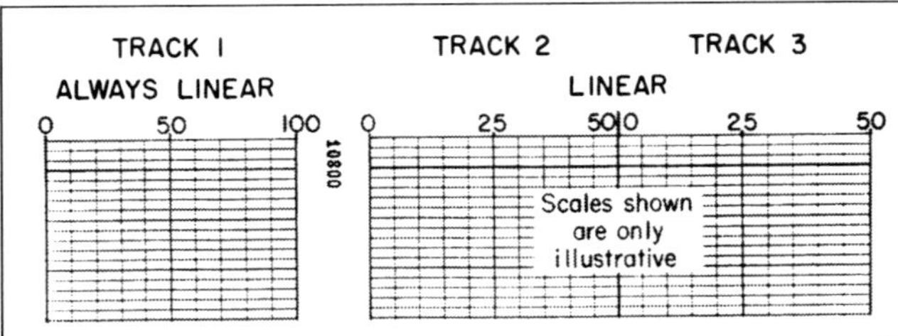

Fig. 5.22—Linear log grid used to record the electric log (after Ref. 10).

The invasion is shallow, extending only 4 in. from the wall of the hole. Interpretation with departure curves confirms that the 64-in. long normal reads the true resistivity.

The discussion of normal and lateral logs indicates that measurement and interpretation have several limitations. These logs, however, have laid the foundation for the modern resistivity tools discussed in Secs. 5.4 and 5.5.

5.3.5 Microlog. Clear delineation of the boundaries of the permeable beds is possible with normal and lateral devices that have spacings of a few inches. To overcome the borehole effect, these devices are pressed against the side of the wall. The Microlog™ (ML) in **Fig. 5.26** consists essentially of a rubber pad with three electrodes (Electrodes A, M_1, and M_2) mounted 1 in. apart in its face. The pad is pressed against the side of the wall by a mechanical and hydraulic system. A current is induced between Electrode A and a distant Electrode B. A potential is measured between Electrode M_2 and another potential electrode far from M_2. This arrangement constitutes a 2-in. short normal device called a Micro-normal. The apparent resistivity calculated and recorded on the log is called the $R_{2\ in.}$. The radius of investigation of this tool is twice its spacing (i.e., 4 in.).

Also recorded is the potential between Electrodes M_1 and M_2. This arrangement corresponds to a lateral device of spacing $\overline{AO}=1½$ in. The apparent resistivity derived from this device's response is referred to as $R_{1\ in.\times 1\ in.}$. This device, called the Micro-inverse, has a radius of investigation of 1½ in.

The mechanical design of the tool permits the recording of a caliper log that represents the distance between the pad that carries the electrodes and a back pad (Fig. 5.26). The pressure applied to the pads is controlled from the surface. With the arms collapsed, the tool is lowered into the borehole. The curves recorded this way give a curve called mud log (see Chap. 4). The 1½-in.-spacing Micro-inverse investigates the mud zone while the tool is collapsed.

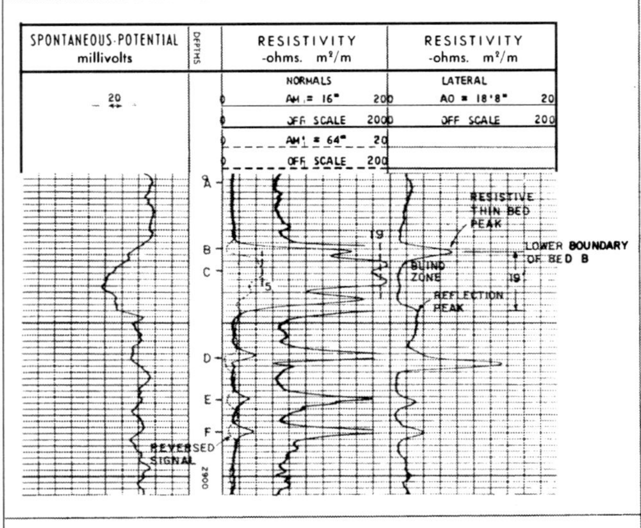

Fig. 5.23—Example of electric log (courtesy Schlumberger).

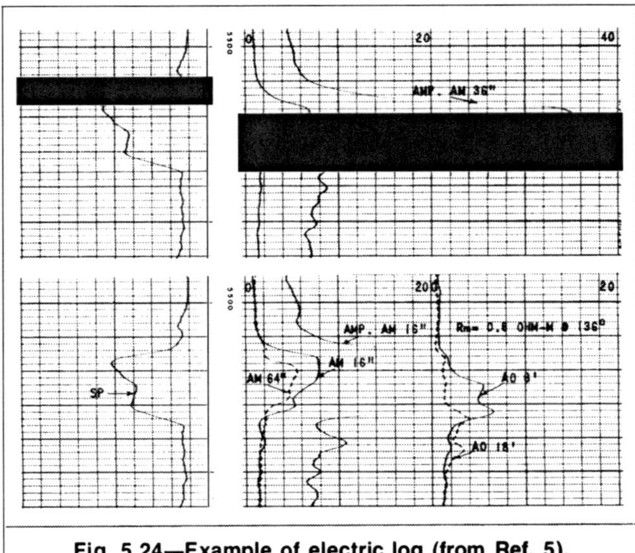

Fig. 5.24—Example of electric log (from Ref. 5).

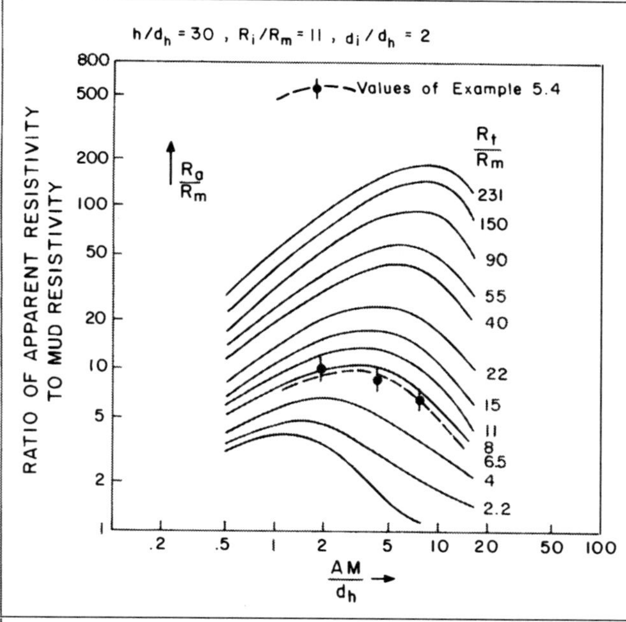

Fig. 5.25—Example resistivity departure curve showing data of Example 5.4 (after Ref. 12).

The tool is opened (i.e., the arms are extended) at the bottom of the hole. The tool is then pulled upward with the pads pressed against the hole wall, and $R_{1\ in.\times 1\ in.}$ and $R_{2\ in.}$ are recorded.

Fig. 5.27 shows an example of a Microlog. The log is recorded on a linear grid like that in Fig. 5.22. The caliper is recorded on the first track on a linear scale that ranges from 5 to 15 in.; each division represents 1 in. The Micro-inverse $R_{1\ in.\times 1\ in.}$ and Micro-normal $R_{2\ in.}$ are recorded on Tracks 2 and 3. The field of both tracks is used and scaled from 0 to 10 Ω·m (i.e., 1 division=0.5 Ω·m). Because the Micro-normal has the deeper radius of investigation, it is traced in a dashed line. The Micro-inverse is traced in a solid line.

Following are some of the salient features of this log.

1. The borehole is enlarged at Level A because of the lithology type (shale). Because shale is impermeable, no invasion or mudcake buildup occurs. The Micro-normal and Micro-inverse are pressed directly against the formation. Because of the relatively limited radius of investigation of both tools, this relatively thick bed appears as infinite. Both tools record the same apparent resistivity, which is close to the true resistivity.

2. Zone B is a permeable formation, as indicated by the caliper log's mudcake buildup. At the level of permeable formations, the tool pad is pressed against the mudcake buildup. The apparent resistivity recorded by both logs is a weighted average of the mudcake resistivity, R_{mc}, and the invaded-zone resistivity, R_{xo}. For extremely shallow invasion, the true resistivity, R_t, is also included in the average.

The Micro-inverse with a 1½-in. spacing is affected by R_{mc} more than the 2-in.-spacing Micro-normal is. Because R_{xo} is several times greater than R_{mc}, the Micro-normal reading is higher than that of the Micro-inverse, and the two curves separate as shown. This separation where $R_{2\ in.} > R_{1\ in.\times 1\ in.}$ is called the "positive separation." Positive separation is a strong qualitative indication of porosity and permeability. As explained later, R_{xo} can be determined from the $R_{1\ in.\times 1\ in.}$ and $R_{2\ in.}$ values.

3. The vertical resolution of the Microlog is excellent. It can detect extremely thin beds, such as Zones C and D. These beds could certainly remain undetected by tools with poorer vertical resolution, such as the long normal, the lateral, and even the short normal.

The Microlog can be extremely useful in detecting permeable zones. It also can be used to derive an R_{xo} value with the method discussed in Sec. 5.7.

5.4 Focused Current Devices

The conventional electrode-type devices introduced in Sec. 5.3 have two main limitations.

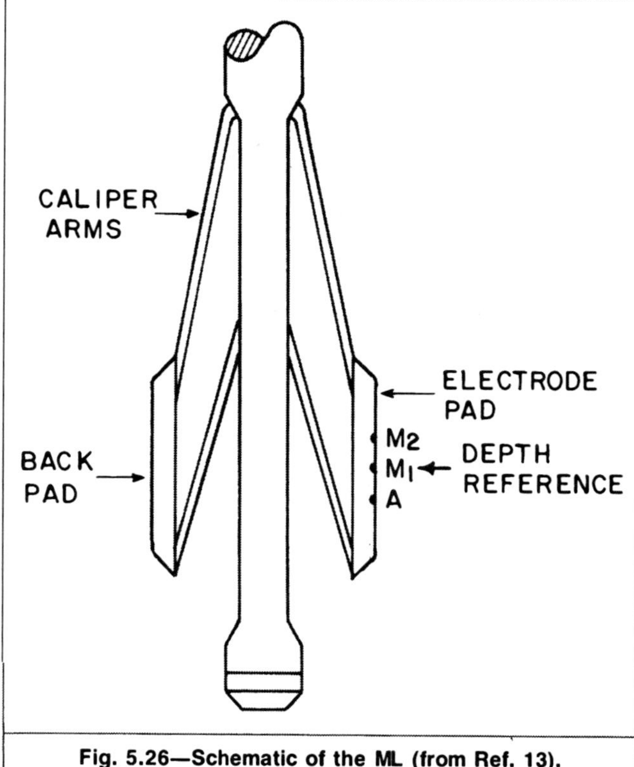

Fig. 5.26—Schematic of the ML (from Ref. 13).

1. In relatively thin beds, the tool response is severely distorted. The recorded apparent resistivity is considerably different from the true resistivity, rendering meaningful quantitative and even qualitative interpretations difficult.

2. In a saltwater-based mud, the current emitted by the tool is confined within the borehole mud column or within the mudcake layer in the case of the Microlog. Recorded apparent resistivity approaches R_m or R_{mc} and is of little or no practical value.

To overcome these limitations, a family of tools that uses different current focusing schemes has been developed.

5.4.1 Laterolog-3. The Laterolog-3 (LL3) tool consists of a long cylinder divided into three isolated electrodes. A middle electrode, Electrode A_0, usually is 1 ft long and is called the measuring electrode. Upper and lower long electrodes, known as guard electrodes, are designated Electrodes G_1 and G_2 in **Fig. 5.28**. Guard electrodes are usually 5 ft long. A constant current I_0 is induced between Electrode A_0 and a remote return electrode. An automatically adjustable current is induced between the guard and remote electrodes. Because the current emitted by electrode A_0 is kept constant, its potential, V_0, with reference to a remote potential electrode varies with the resistance of the medium offered to the flow of current. An apparent resistivity of the medium is calculated by an equation similar to Eq. 5.7:

$$R_a = G_T V_0 / I_0 . \quad (5.19)$$

The parameter G is dependent on tool design. G_T is called the calibration coefficient because it usually is determined experimentally.[14]

The tool vertical resolution is controlled by focusing the current induced by Electrode A_0. This is done by maintaining Electrodes A_0, G_1, and G_2 at the same potential by adjusting the current emitted by the guard electrodes. The current of the middle electrodes is thus forced to flow in a plane perpendicular to the axis of the logging tool. The thickness of the current beam is approximately equal to the length of electrode A_0.[14] The LL3 then has an excellent vertical resolution. It detects thin beds and, in certain cases, even those that are thinner than the focused current beam.

5.4.2 Laterolog-7. An inconvenience of the LL3 is the considerable mass of metal in the sonde. Such mass will, for example, disturb the flow of naturally emitted currents into the borehole. Such a disturbance affects the quality of the SP log unless the electrode

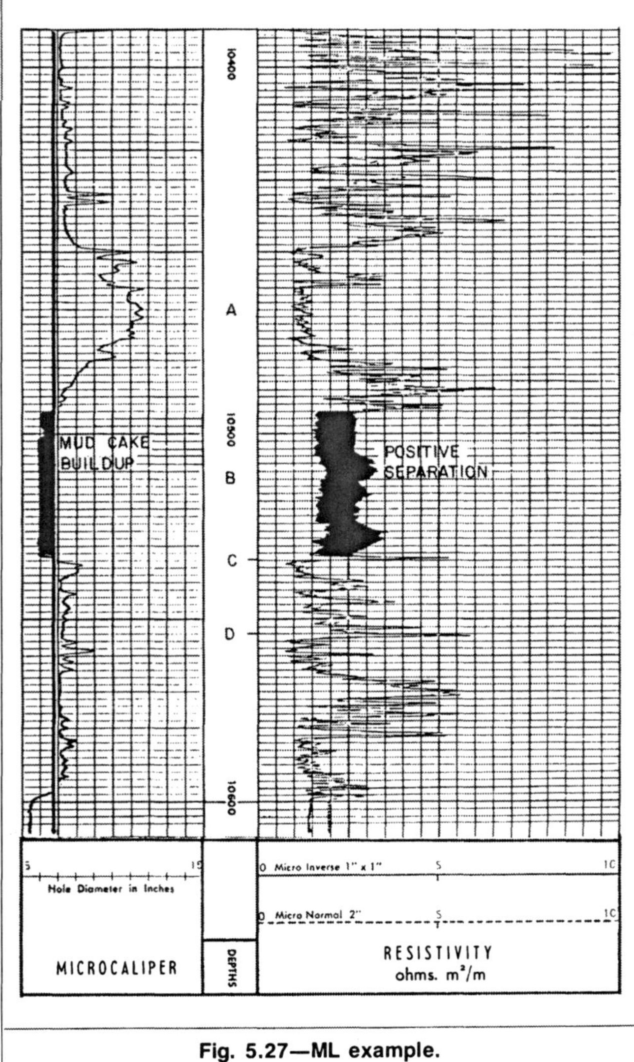

Fig. 5.27—ML example.

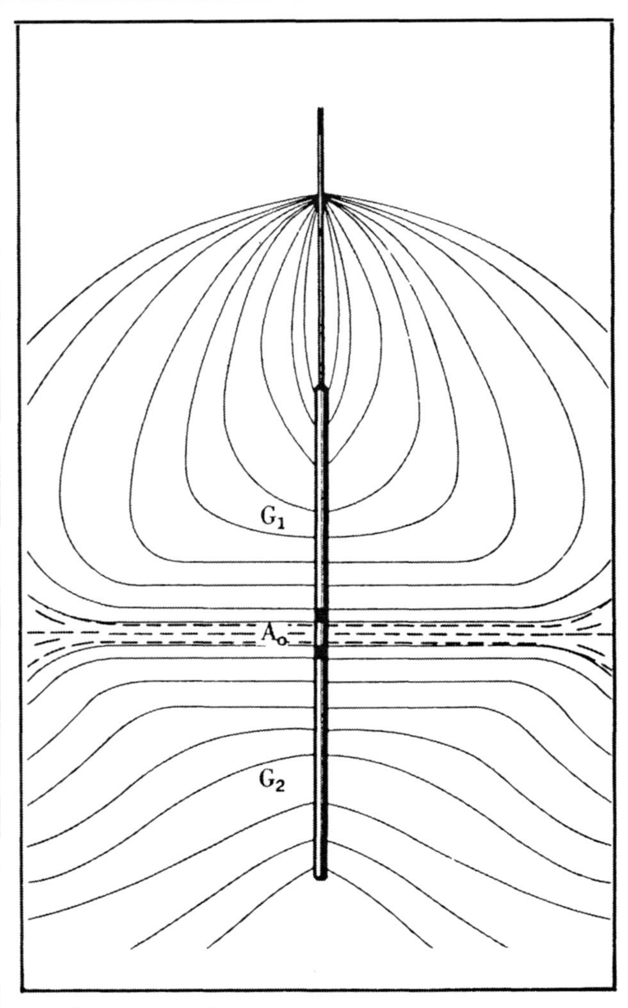

Fig. 5.28—Schematic of a three-electrode focused system (guard log, LL3) showing current flowlines (from Ref. 14).

measuring the SP is placed far from the LL3 sonde, usually 25 ft away.

To overcome this inconvenience, the Laterolog-7 (LL7) focuses the current by using multiple small electrodes arranged as shown in **Fig. 5.29**. The thickness of the current beam emitted by this tool is the distance O_1O_2 in Fig. 5.29, usually 32 in. The length A_1A_2 of the sonde is 80 in. Because of the reduced amount of metal in the sonde, an SP curve may be recorded on depth simultaneously with this laterolog.[15]

The LL3, LL7, and similar devices, such as the Deep Laterolog (LLd) discussed later, overcome the two aforementioned limitations of the conventional electrode-type devices. For example, **Fig. 5.30** shows the responses of the LL7 and conventional logs to a thin bed drilled with saltwater-based mud. As a result of the thinness of the bed, the short and long normal tool responses are distorted and give the characteristic reversed signal discussed in Sec. 5.3.3. The lateral tool shows the bed as more resistive than the surrounding formations. However, the displayed maximum apparent resistivity of 13 Ω·m is only a small fraction of the 100-Ω·m true resistivity. Despite the adverse conditions of the thin bed and saltwater-based mud, the LL7 clearly indicates the resistive bed and displays an average apparent resistivity of the same order of magnitude as the true resistivity.

Laterologs are usually used to investigate formations that display a wide range of resistivity. Linear scales are not suitable. A logarithmic grid (**Fig. 5.31**) usually is used to record laterologs. The scale encompasses four cycles and covers a range of 0.2 to 2000 Ω·m. The linear scale of the left track usually is used to display the gamma ray log. **Fig. 5.32** shows an LL7 recorded on such a grid. The resistivity in the illustrated interval spans a wide range of 5 to 500 Ω·m. The vertical resolution of the device is excellent; it defines zones as thin as 1 ft.

5.4.3 Laterolog Devices of Intermediate Depth of Investigation. To define the resistivity distribution around the wellbore, several apparent resistivity readings of tools of different radii of investigation are required. This need prompted the design of the shallow-investigation Laterolog-8 (LL8) device. Subsequently, the spherically focused log (SFL) was developed as an improvement over the LL8. As **Fig. 5.33** shows, in these devices, the current-return electrode is located a relatively short distance from Electrode A_0. With this configuration, the equipotential surfaces assume a spherical shape. The tool reading is influenced most by the invaded zone. The LL8 and the SFL are usually recorded in conjunction with an LLd or a deep induction log.

5.4.4 Dual Laterolog. The Dual Laterolog is the most advanced Laterolog device. The tool provides a simultaneous measurement of an LLd and a shallow Laterolog (LLs). The tool provides two different current beams of different configurations and frequencies. The current patterns of the two cases are shown in **Fig. 5.34**. The LLd provides the deepest investigation of all available laterologs. The LLd and LLs have essentially the same vertical resolution.

5.4.5 Focused-Current Microdevices. When the resistivity contrast between the invaded zone and the mudcake is high (i.e., high R_{xo}/R_{mc}), the current induced by the Micrmusical log tends to escape through the mudcake (**Fig. 5.35**). To eliminate this problem, focused-current devices—Microlaterolog™ (MLL), the Proximity Log™ (PL), and the Microspherically Focused Log™ (MicroSFL)—were developed.

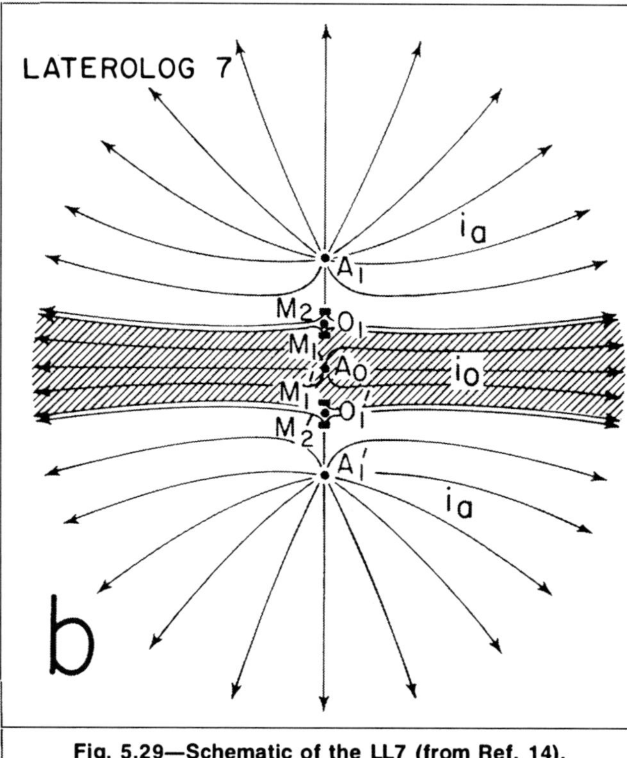

Fig. 5.29—Schematic of the LL7 (from Ref. 14).

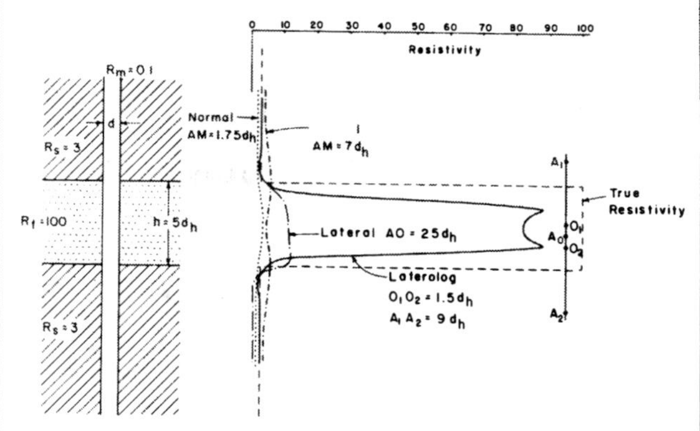

Fig. 5.30—Response of LL7 and conventional logs opposite a thin, resistive bed, uninvaded, with salt mud in the well (after Ref. 15).

The MLL comprises one central electrode (A_0) and three ring electrodes (M_1, M_2, and A_1). These electrodes are imbedded in a pad, as shown in **Fig. 5.36**. The tool functions identically to the LL7. Because of the focusing effect, the MLL tool response is affected more by the invaded-zone resistivity and less by the mudcake resistivity. When invasion is sufficiently deep and mudcake is < ⅜ in. thick, the MLL displays an apparent resistivity value that approximates R_{xo}.

The PL is similar in principle to the MLL. The pad and electrode designs are such that isotropic mudcakes up to ¾ in. have very little effect on the log measurement.[15] However, a diameter of invasion equal to or greater than about 40 in. is needed for the PL to provide direct approximation of R_{xo}. If the invasion is shallow, the reading is influenced by R_t. **Fig. 5.37** is an example PL run simultaneously with an ML. The PL, Micro-normal, and Micro-inverse log readings for Zone Q are 100, 13, and 8 Ω·m, respectively. These values clearly indicate the dramatic effect of mudcake on the Microlog readings and demonstrate the advantage of focused-current devices in obtaining representative resistivity values. Moreover, the log illustrates the importance of the Microlog as a permeability and porosity detection tool. Porous and permeable zones are clearly indicated by the positive separation of the $R_{2\text{ in.}}$ and $R_{1\text{ in.}\times 1\text{ in.}}$ curves.

The MicroSFL is similar in principle to the SFL described in Sec. 5.4.3. The MicroSFL electrodes are smaller and pad-mounted. **Fig. 5.38** shows the electrode arrangement and current distribution of the MicroSFL. This design minimizes mudcake effect without requiring deep invasion, as with the PL. The MicroSFL provides good values of R_{xo} in a wider range of conditions than either the PL or MLL. Another distinct advantage of the MicroSFL is its compatibility with other logging tools, such as the Dual Laterolog and the induction SFL. This eliminates the need for a separate logging run to obtain R_{xo} data.[15] **Fig. 5.39** is a schematic of the Dual Laterolog/MicroSFL combination tool (also known as the Dual Laterolog/R_{xo} tool). **Fig. 5.40** shows a log measured with this combination tool. The MicroSFL, LLs, and LLd are traced on the logarithmic scale in solid, dotted, and dashed lines, respectively. The thick formation located in the depth interval of 5,687 to 5,717 ft is characterized by a separation of the three curves. This separation indicates that the zone is invaded by mud filtrate; hence, it is permeable. This interval was drilled with a saltwater-based mud (R_{mf}=0.056 at 68°F) that has a salinity exceeding that of formation water. The resistivity displayed by the logs increases with the tool's radius of investigation because of the salinity contrast between the mud filtrate and the formation water.

5.5 Induction Devices

Electrode devices, either conventional or focused current, require the presence of conductive fluid in the borehole. The conductive borehole fluid establishes electrical contact between the electrodes and the formations. Current cannot be induced without such a contact. This contact was not possible in early holes drilled with cable tools, which generally are dry. This contact is also impossible in holes drilled with nonconductive oil-based mud. To overcome this adverse condition, conventional electric logging used scratcher electrodes forced by springs to make direct contact with the formation. In general, measurements performed with such electrodes were not reliable.[18]

Induction tools were introduced in the mid-1940's to log empty boreholes and boreholes filled with oil-based mud. The induction tool operates on an entirely different principle from electric tools. It does not require conductive fluid in the borehole or direct physical contact with the formation.

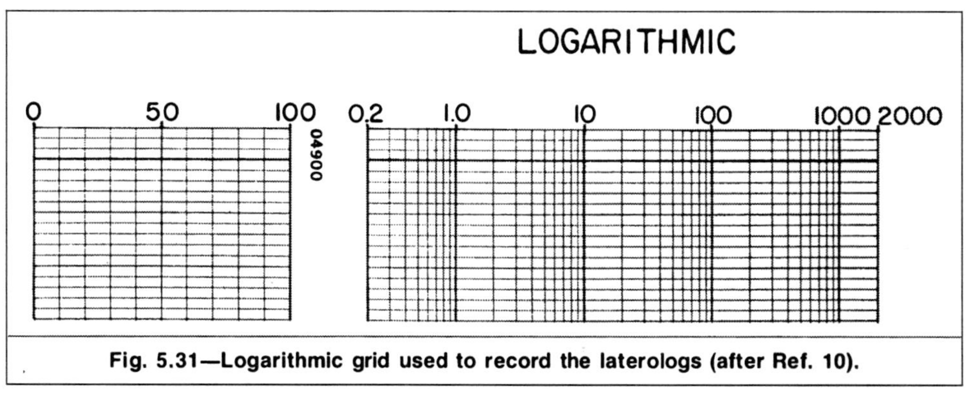

Fig. 5.31—Logarithmic grid used to record the laterologs (after Ref. 10).

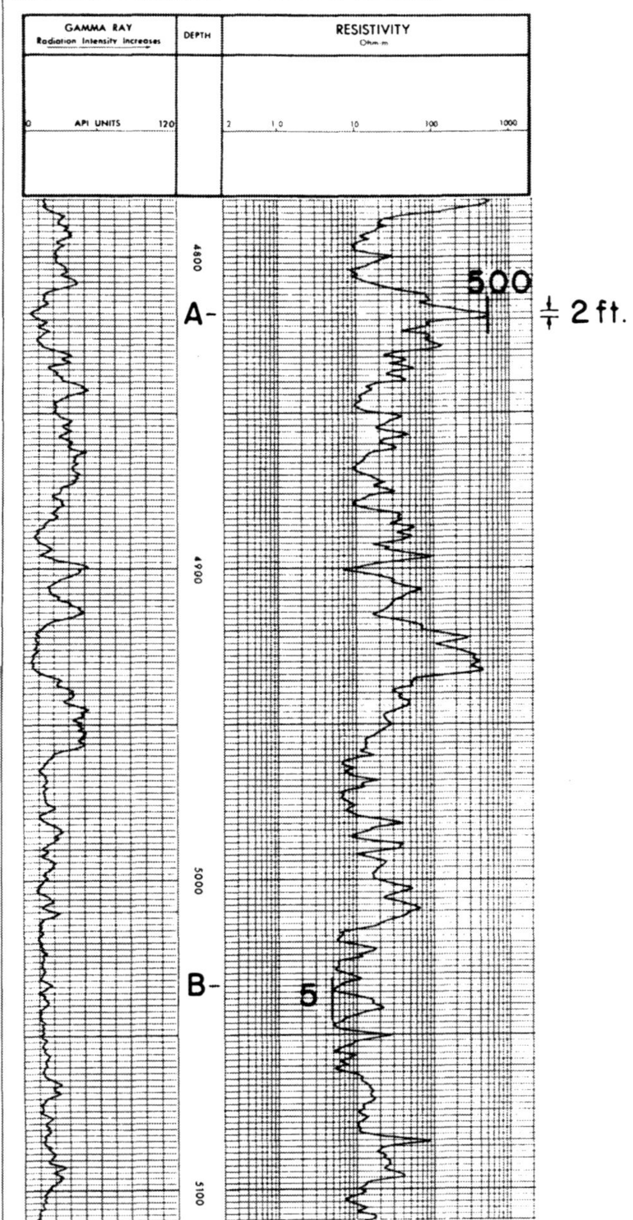

Fig. 5.32—Example of LL7 recorded on a logarithmic grid (after Ref. 14).

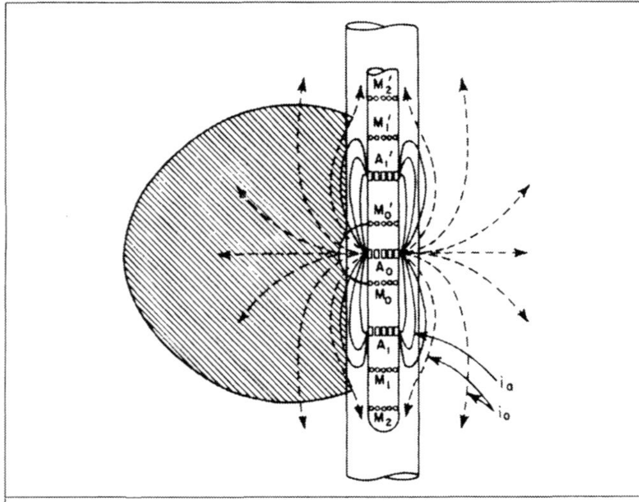

Fig. 5.33—Schematic of current pattern for the SFL (after Ref. 15).

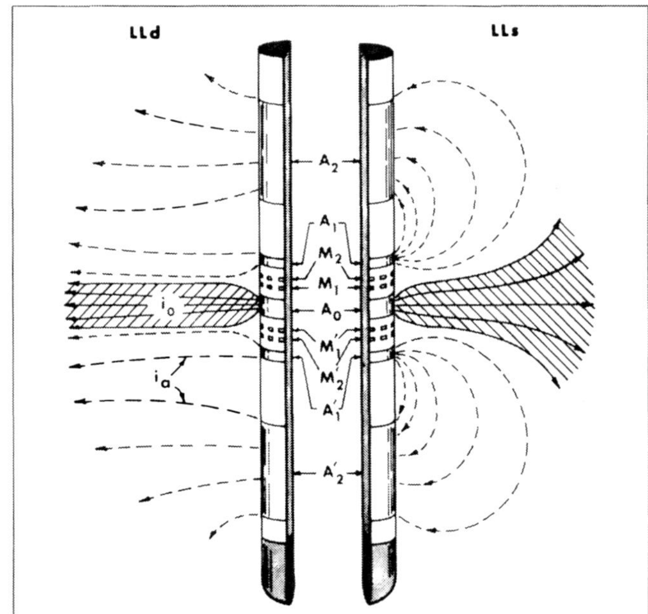

Fig. 5.34—Representative current patterns for the LLd and LLs (from Ref. 16).

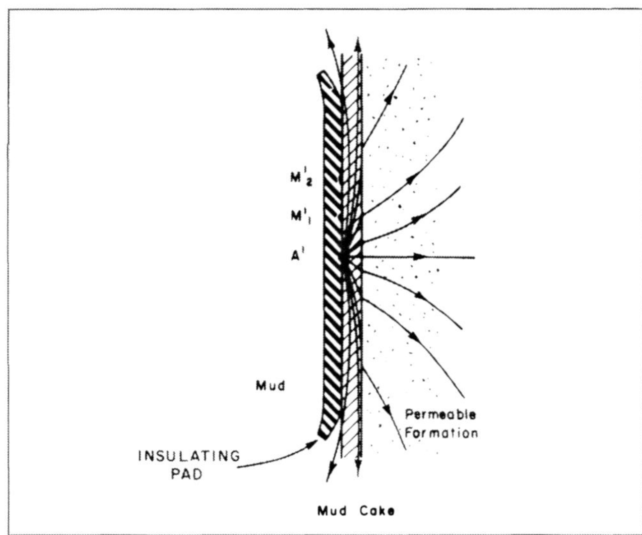

Fig. 5.35—ML current configuration of a permeable formation more resistive than the mudcake (from Ref. 17).

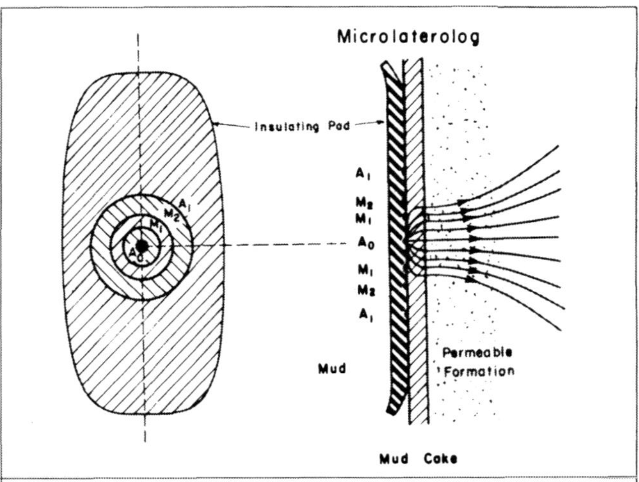

Fig. 5.36—MLL pad and current distribution next to a permeable formation (after Ref. 17).

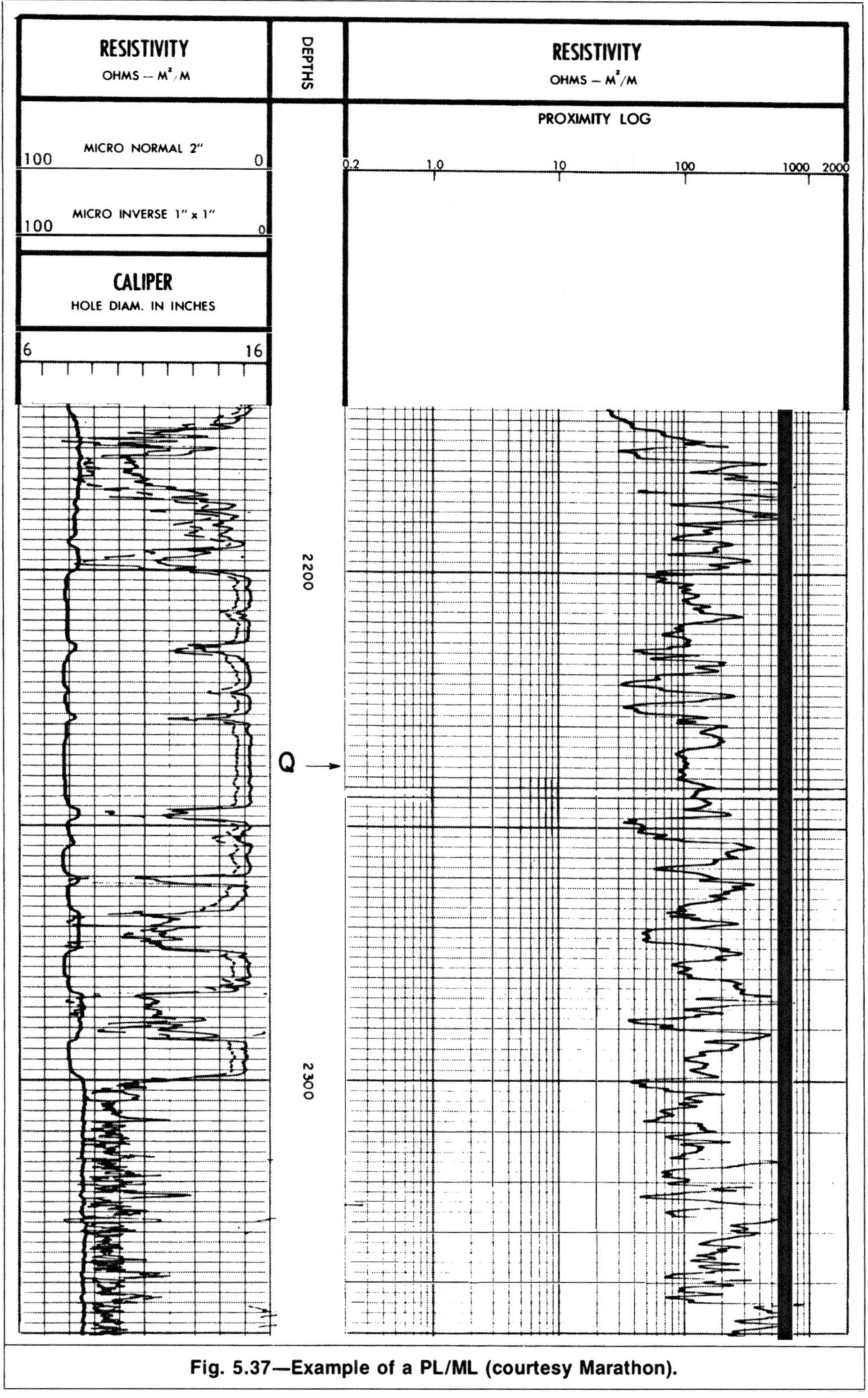

Fig. 5.37—Example of a PL/ML (courtesy Marathon).

Induction tools are focused to minimize the contribution of the borehole, invaded zone, and surrounding formations to the measurement. Focusing is realized by means of a system of several transmitter and receiver coils. The principle of the tool, however, can be simplified with the two-coil system in **Fig. 5.41**. A transmitter coil is excited by an alternating current, I, of medium frequency. This current induces a primary magnetic field, B_p, in the formation that surrounds the borehole. The magnitude and frequency of B_p depends on the transmitter current. The vertical component of this induced magnetic field, $(B_p)_z$, in turn generates an electric field, E, which curls around the vertical axis. This electric field causes a current to flow in the formation in circles concentric with the borehole. The current density, J, is proportional to the electric field, and the formation conductivity, C. The current that flows in a formation ring behaves as a transmitter coil and generates its own secondary magnetic field, B_S. B_S, which is proportional to the formation conductivity, induces an electric signal, V, in the receiver coil. This signal is easily calibrated in terms of formation conductivity. The conductivity is usually converted to resistivity and recorded vs. depth.

The induced magnetic field is a component of an electromagnetic wave that undergoes attenuation and phase shift as it propagates through the formation. This propagation effect, also known as skin effect, depends on the transmitter frequency and the formation's electromagnetic properties. Induction logs are automatically corrected for the skin effect during recording.[20]

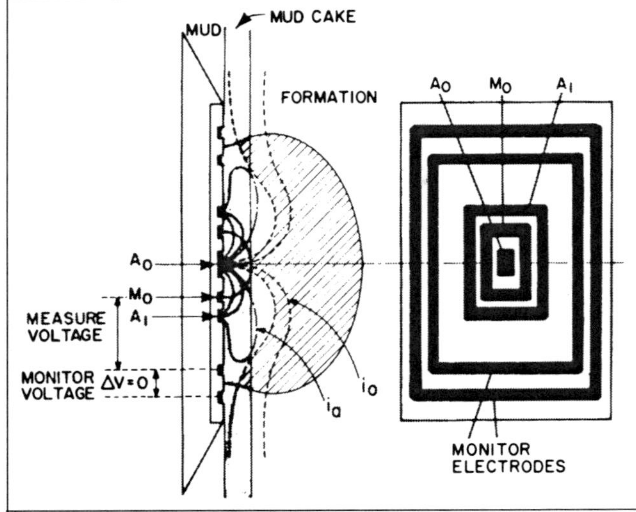

Fig. 5.38—Electrode arrangement and current distribution of the MicroSFL (from Ref. 16).

Fig. 5.42 illustrates an example of an induction log recorded in a borehole filled with an oil-based mud. The gamma ray log, discussed in Chap. 7, is recorded on the first track. The SP log usually displayed on this track cannot be recorded in oil-based mud because it requires a conductive drilling fluid. The resistivity curve is shown on the second track. It is recorded on a linear scale, 0 to 10 Ω·m, increasing from left to right. The induction conductivity is shown on the third track. It is recorded on a linear scale, 0 to 4000 m℧/m, increasing from right to left. The scales are set up in this fashion so that the resistivity and conductivity curves move together in the same direction as the curve passes through zones of high and low resistivities.

Although induction tools were designed for a nonconductive borehole environment, they were found to yield excellent measurements in water-based mud, provided that the mud is not too salty, the formation is not too resistive, and the borehole diameter is not too large.[15] **Fig. 5.43** shows an induction log recorded in freshwater-based mud. In addition to the induction resistivity and conductivity curves, an SP and 16-in. short normal are shown. This combination is known as the induction/electric survey (IES). The short normal is recorded to reflect the near-wellbore resistivity.

In recent applications, the SFL replaced the short normal. Also, a dual induction tool (DIL) was designed. The DIL consists of a regular deep-induction device (ILd) and a medium device (ILm). The ILm has the same vertical resolution as the ILd but about half the depth of investigation.[15] The DIL curves are usually recorded with a shallow Laterolog, such as the LL8 or the SFL. **Fig. 5.44** shows a DIL/LL8 combination log.

Induction logs and Laterologs differ, however, in the invaded-zone effect on the tool response. To a first approximation, the invaded zone and the true resistivity zone are investigated in parallel by the induction tool and in series by the Laterolog tools.[15] **Fig. 5.45** gives the preferred ranges of application of induction logs and Laterologs for R_t determination. Recently introduced induction devices include the Phasor Induction Tool[21] and the Electromagnetic Propagation Tool[22] (EPT). The Phasor Induction Tool uses a DIL/SFL array to record resistivity at three depths of investigation. The tool measures the quadrature signal in addition to usual in-phase induction measurements. Its main advantage is that it gives more accurate values in all resistivity ranges, and it has the capability to log in large boreholes.

The EPT uses a very short-spaced array of transmitters and receivers to record the propagation and attenuation of a 1.1-GHz electromagnetic wave sent through the invaded zone. The magnetic propagation in a medium of relatively low conductivity is related mainly to its dielectric constant. This constant is much larger for water than that of other formation constituents, including hydrocarbons. The measurement, which is predominantly sensitive to the water content, can be used to estimate the mud-filtrate saturation in the invaded zone, S_{xo}.

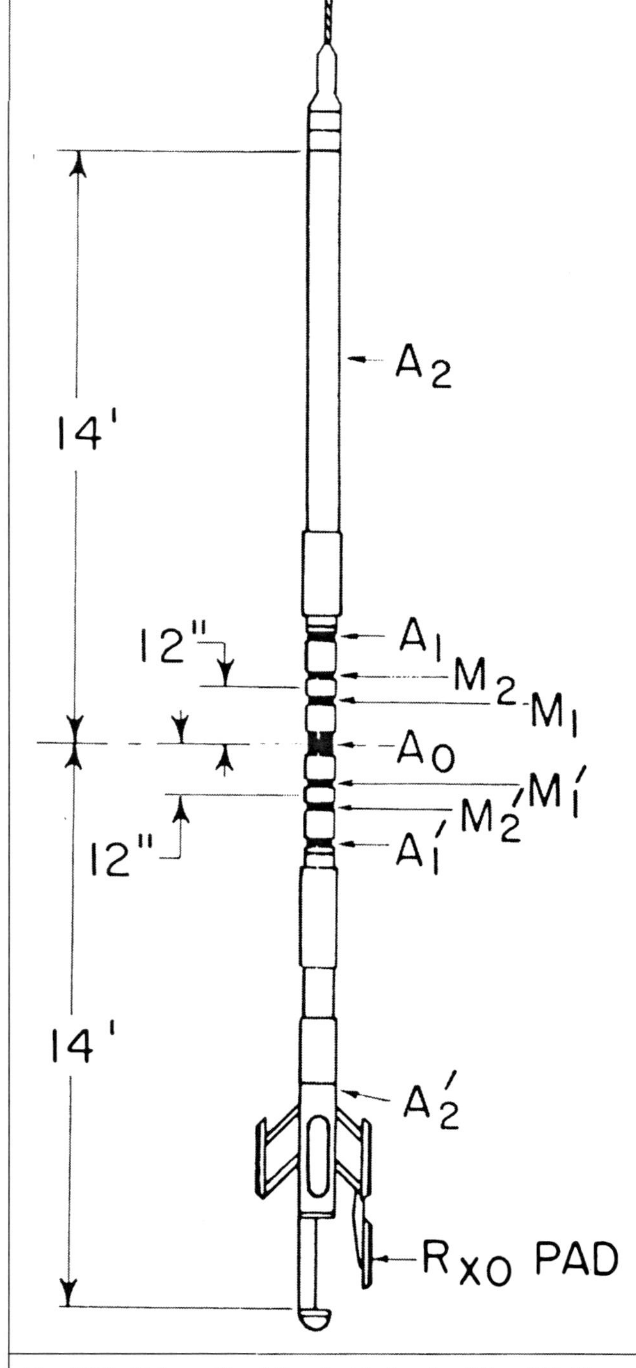

Fig. 5.39—Schematic of the dual Laterolog/MicroSFL (from Ref. 16).

5.6 True Resistivity Determination

Several tools of different designs are used to measure the electric resistivity of formations penetrated by a borehole. These include conventional electrode, focused-current, or induction tools. The response of each of these tools, recorded when they are positioned next to a formation of interest, is a weighted average that reflects the different zones surrounding the tool. These zones are (1) the drilling fluid and mudcake zone, usually called the borehole zone; (2) the first few feet of the formation invaded by the mud filtrate, usually called the invaded or flushed zone; (3) the formation zone beyond the invaded zone, usually known as the uninvaded zone (of primary interest to the log analyst); and (4) the beds adjacent to the formation of interest.

The tool response is used to calculate a resistivity parameter known as apparent resistivity. The contribution of each of the four zones to the calculated value of the apparent resistivity is deter-

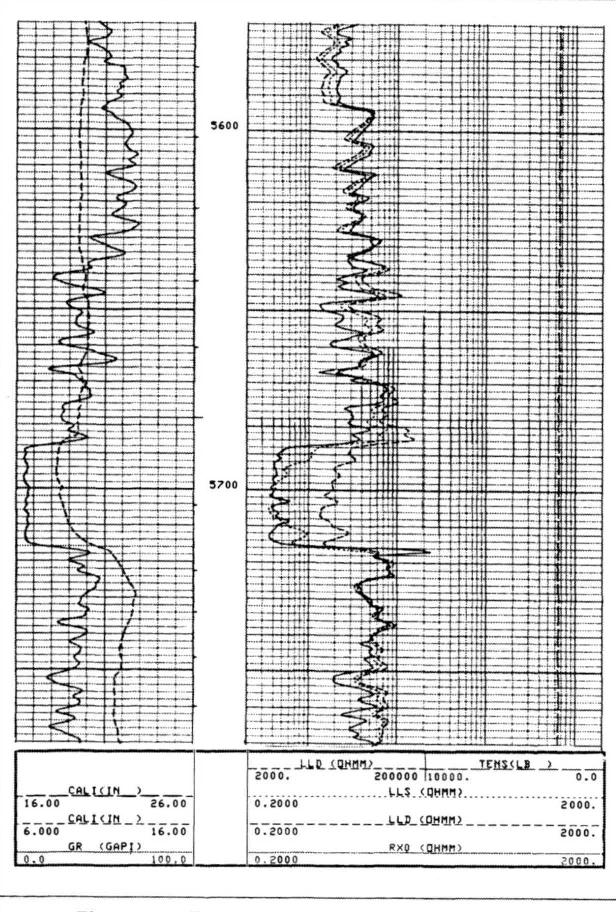

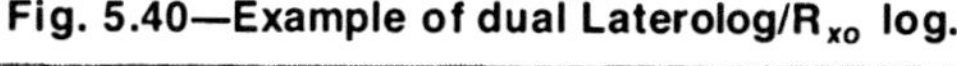

Fig. 5.40—Example of dual Laterolog/R_{xo} log.

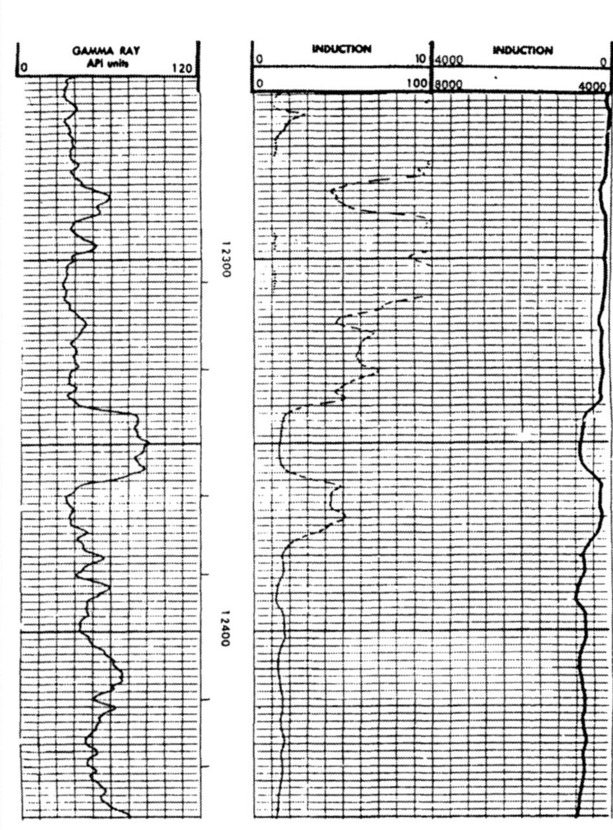

Fig. 5.42—Example of an induction log recorded in a borehole filled with oil-based mud (courtesy Amoco).

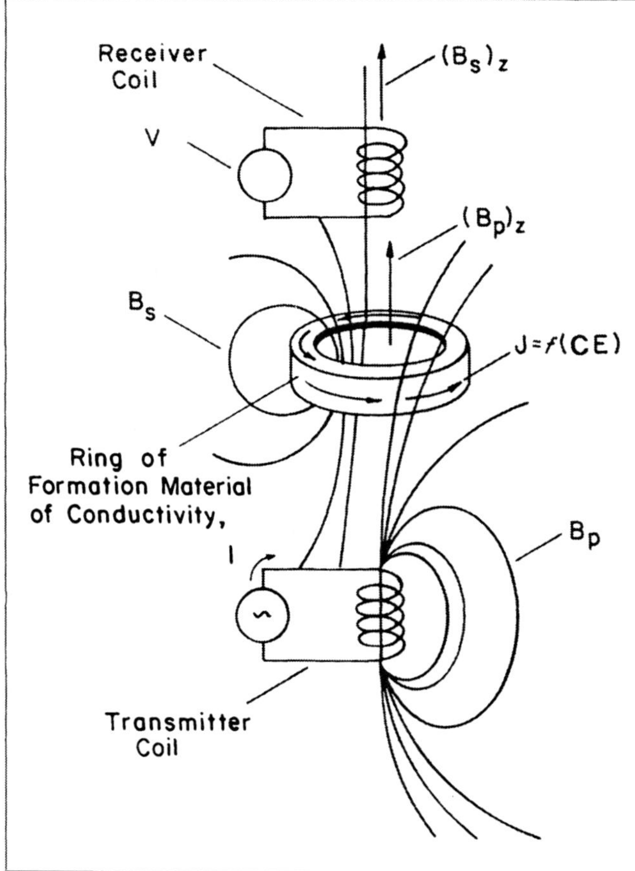

Fig. 5.41—Schematic of a two-coil induction system (after Ref. 19).

mined by the zone resistivities and geometries and the tool type and design.

The value of the apparent resistivity, R_a, generally is different from the formation true resistivity, R_t. It is usually a value between R_t and the resistivity of the flushed zone, R_{xo}. For qualitative interpretation, the apparent resistivity log values can be used as is. However, their use in Eq. 1.42 for quantitative interpretations could yield inaccurate and even misleading results. The determination of R_t and R_{xo} from the R_a value is of great importance in the quantitative interpretation process.

5.6.1 Geometric Factors. The manner in which the various media surrounding the electrodes affect the R_a value obtained with a given device is sometimes indicated by reference to its radius of investigation. This characteristic is dependent on the spacing of the device. When very short spacing is used, the mud in the hole may have a dominant effect on the resistivity value obtained. If the spacing is increased, the volume of the media appreciably influencing the measurement increases, and the effect of adjacent formations becomes dominant. For the determination of true formation resistivities, it is useful to record curves that have large radii of investigation. This is particularly true for permeable formations that have been deeply invaded by the mud filtrate.

The radius of investigation is commonly defined for simplicity on the basis of the potential distribution in the vicinity of the borehole. For example, the radius of investigation of a normal device placed in a homogeneous medium is taken to be equal to twice its spacing. At this distance, the potential has dropped by one-half its original value. A more precise idea regarding the radius of investigation of a device under various conditions may be arrived at by the study of geometric factors.

Fig. 5.46 shows a homogeneous, isotropic, and infinite medium of resistivity R. A current Electrode A and a potential measuring Electrode M are also shown. The current-return Electrode B and the second measuring Electrode N are placed at infinity. The medi-

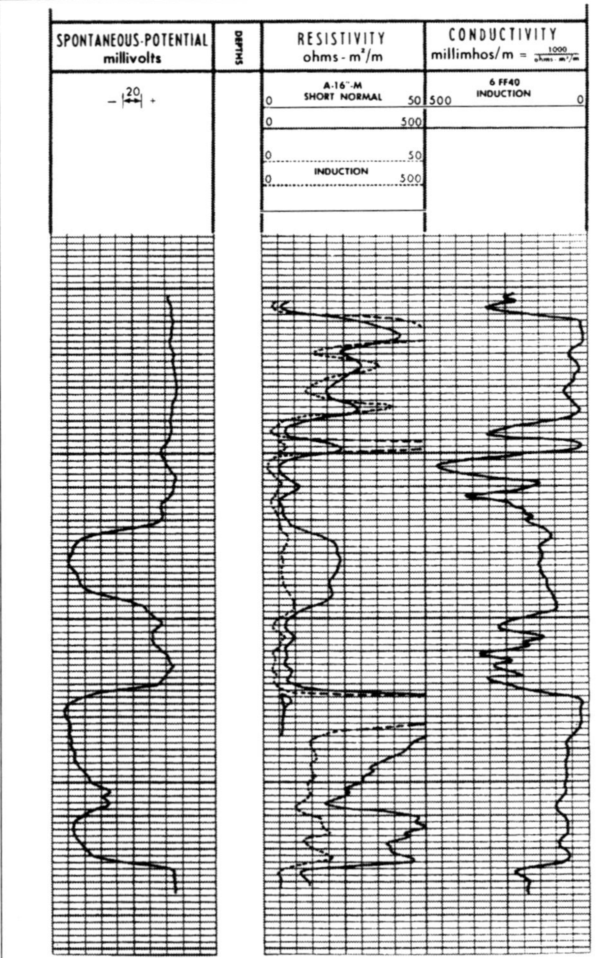

Fig. 5.43—Example of induction electrical log (from Ref. 15).

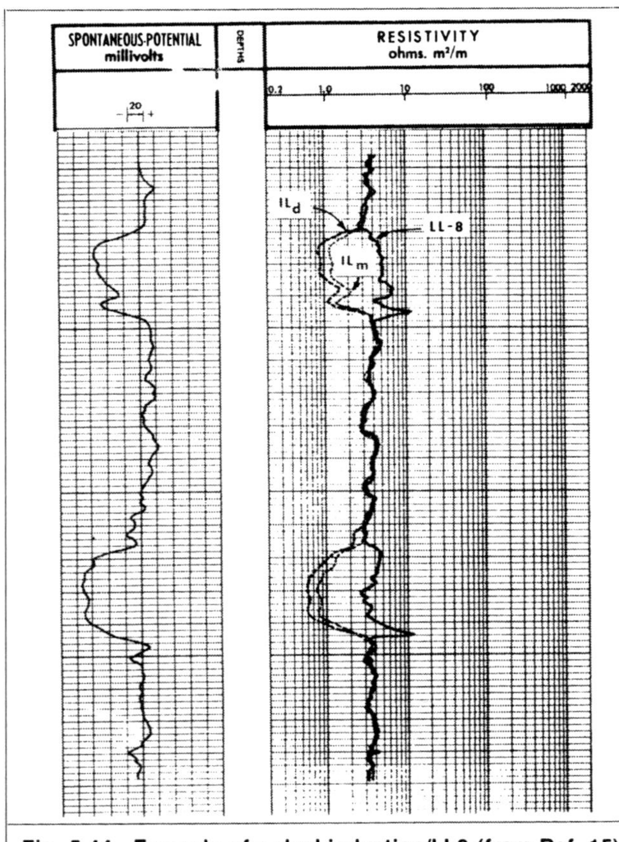

Fig. 5.44—Example of a dual induction/LL8 (from Ref. 15).

um is subdivided into concentric spheres. The center of these spheres is Electrode A, and their radii are a multiple of the spacing, L_s.

The potential difference between the measuring Electrodes M and N, $\Delta V_{M\infty}$, is the sum of the individual potential drops across each spherical shell:

$$\Delta V_{M\infty}=\Delta V_{12}+\Delta V_{23}+\Delta V_{34}+\ldots+\Delta V_{(n-1)n}+\ldots \quad (5.20)$$

Because the current I remains the same, Eq. 5.20 can be written as

$$\Delta V_{M\infty}=I\rho_{M\infty}=I[\rho_{12}+\rho_{23}+\rho_{34}\ldots+\rho_{(n-1)n}+\ldots], \quad (5.21)$$

where ρ_{12}, ρ_{23}, etc., are the resistances of the individual shells. $\rho_{M\infty}$ is the total resistance between the measuring Electrodes M and N.

The contribution of each of these shells to the total signal is measured by the differential geometric factor, G:

$$G=\frac{\Delta V_{(n-1)n}}{\Delta V_{M\infty}}=\frac{\rho_{(n-1)n}}{\rho_{M\infty}}. \quad (5.22)$$

An integrated geometric factor, G_I, is also defined. G_I is a measure of the contribution to the total signal of all shells between the first and nth spherical surface:

$$G_I=\frac{\Delta V_{Mn}}{\Delta V_{M\infty}}=\frac{\rho_{Mn}}{\rho_{M\infty}}. \quad (5.23)$$

With Eq. 5.2, the resistances $\rho_{M\infty}$, ρ_{Mn}, and $\rho_{(n-1)n}$ can be expressed as

$$\rho_{M\infty}=R/4\pi L_s, \quad (5.24)$$

$$\rho_{Mn}=\frac{R}{4\pi}\left(\frac{1}{L_s}-\frac{1}{nL_s}\right)=\frac{R}{4\pi L_s}\frac{n-1}{n}, \quad (5.25)$$

and $$\rho_{(n-1)n}=\frac{R}{4\pi}\left[\frac{1}{(n-1)L_s}-\frac{1}{nL_s}\right]=\frac{R}{4\pi L_s}\frac{1}{(n-1)(n)}. \quad (5.26)$$

Replacing $\rho_{M\infty}$, ρ_{Mn}, and $\rho_{(n-1)n}$ in Eqs. 5.22 and 5.23 with their expressions given in Eqs. 5.24 through 5.26 results in

$$G=1/(n-1)n \quad (5.27)$$

and $$G_I=n-1/n. \quad (5.28)$$

Table 5.3 gives the values of G and G_I for the first nine shells of Fig. 5.46. **Fig. 5.47** shows these values graphically. The shells close to the current electrode contribute more than those far away.

Now, let the concentric spherical shells making up the medium be of different resistivities R_1, $R_2\ldots R_n$, respectively. Eqs. 5.21 and 5.25 can still be used to express the voltage $\Delta V_{M\infty}$ indicated by the tool:

$$\Delta V_{M\infty}=I\left[\frac{R_1}{4\pi L_s}\frac{1}{1(2)}+\frac{R_2}{4\pi L_s}\frac{1}{2(3)}+\ldots\frac{R_n}{4\pi L_s}\frac{1}{n(n+1)}+\ldots\right]. \quad (5.29)$$

Considering the definition of G, Eq. 5.29 becomes

$$\Delta V_{M\infty}=\frac{I}{4\pi L_s}(G_1R_1+G_2R_2+\ldots+G_nR_n+\ldots), \quad (5.30)$$

where G_1, $G_2\ldots G_n$ are the differential geometric factors of the consecutive spherical shells.

Because R_a is calculated by

$$R_a=4\pi L_s(\Delta V_{M\infty}/I), \quad (5.31)$$

Eq. 5.30 simplifies to

$$R_a=G_1R_1+G_2R_2+\ldots+G_nR_n+\ldots, \quad (5.32)$$

which can be expressed in the form

$$R_a=\sum_{i=1}^{\infty}G_iR_i. \quad (5.33)$$

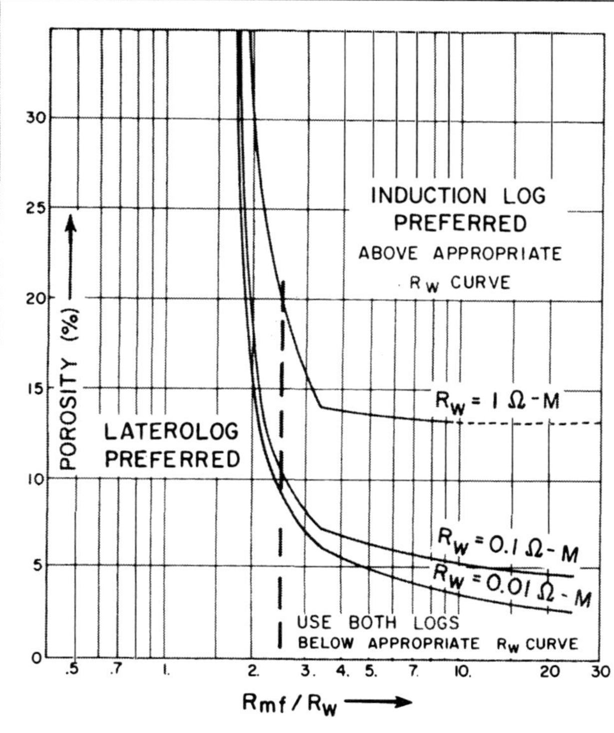

Fig. 5.45—Preferred ranges of application of induction logs and Laterologs (from Ref. 15).

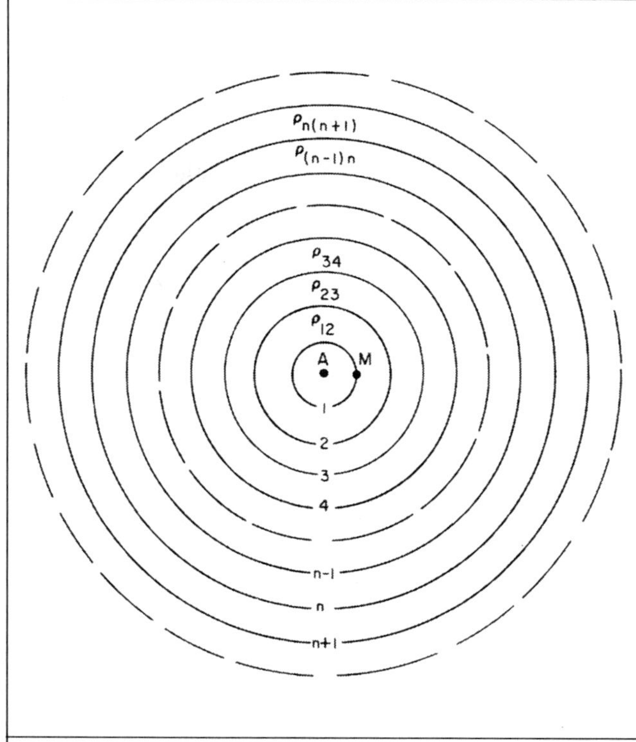

Fig. 5.46—A homogeneous, isotropic, and infinite medium subdivided into concentric spheres.

TABLE 5.3—DIFFERENTIAL AND INTEGRATED GEOMETRIC FACTORS FOR FIG. 5.46

n	$G = 1/(n-1)n$	$G_I = (n-1)/n$
1		0
	0.500	
2		0.500
	0.167	
3		0.667
	0.083	
4		0.750
	0.050	
5		0.800
	0.033	
6		0.833
	0.024	
7		0.857
	0.018	
8		0.875
	0.014	
9		0.889
	0.011	
10		0.900

Note that

$$\sum_{i=1}^{\infty} G_i = 1. \quad (5.34)$$

Example 5.5. A normal electrode array is placed in a medium composed of three homogeneous and isotropic zones centered around Electrode A (**Fig. 5.48**). The first zone extends spherically to a radius of 32 in. and has a resistivity of 20 Ω·m. The second zone extends spherically to a radius of 64 in. and has a resistivity of 5 Ω·m. The third zone extends infinitely beyond the second zone and has a resistivity of 1.0 Ω·m.

a. What voltage is measured between Electrode M and a distant Electrode N for a 1-A current and electrode spacing $\overline{AM}$=16 in.?

b. What is the value of the apparent resistivity calculated with the measured voltage?

Solution. $\Delta V_{M\infty}$ and R_a can be calculated with two approaches.

1. Fig. 5.48 is a schematic of the measurement environment and shows the equivalent electric circuit. The equivalent circuit is composed of three resistances in series, ρ_1, ρ_2, and ρ_3, which represent the three zones of resistivities, R_1, R_2, and R_3, respectively. Eq. 5.20 can be rewritten in this case as

$$\Delta V_{M\infty} = \Delta V_1 + \Delta V_2 + \Delta V_3,$$

where ΔV_1, ΔV_2, and ΔV_3 are the potential drops in the three zones, respectively. According to Eq. 5.4,

$$\Delta V_1 = \frac{(1)(20)}{4\pi}\left[\frac{1}{16(0.0254)} - \frac{1}{32(0.0254)}\right] = 1.958 \text{ V},$$

$$\Delta V_2 = \frac{(1)(5)}{4\pi}\left[\frac{1}{32(0.0254)} - \frac{1}{64(0.0254)}\right] = 0.245 \text{ V},$$

$$\Delta V_3 = \frac{(1)(1)}{4\pi}\left[\frac{1}{64(0.0254)} - \frac{1}{\infty}\right] = 0.049 \text{ V},$$

and $\Delta V_{M\infty} = 1.958 + 0.245 + 0.049 = 2.252$ V.

From Eq. 5.31,

$$R_a = 4\pi(16)(0.0254)\frac{2.252}{1} = 11.5 \ \Omega\cdot\text{m}.$$

2. Eq. 5.33 can be rewritten for this case as

$$R_a = G_1R_1 + G_2R_2 + G_3R_3.$$

The thickness of the first spherical shell is 16 in., which is equal to the spacing L_s. From Table 5.3, $G_1 = 0.500$. The thickness of Zone 2 is 32 in. and is two concentric spherical shells, each 16 in. thick. Again, from Table 5.3, $G_2 = 0.167 + 0.083 = 0.250$. Because $G_1 + G_2 + G_3 = 1$, $G_3 = 0.250$ and

$$R_a = 0.500(20) + 0.250(5) + 0.250(1) = 11.5 \ \Omega\cdot\text{m}.$$

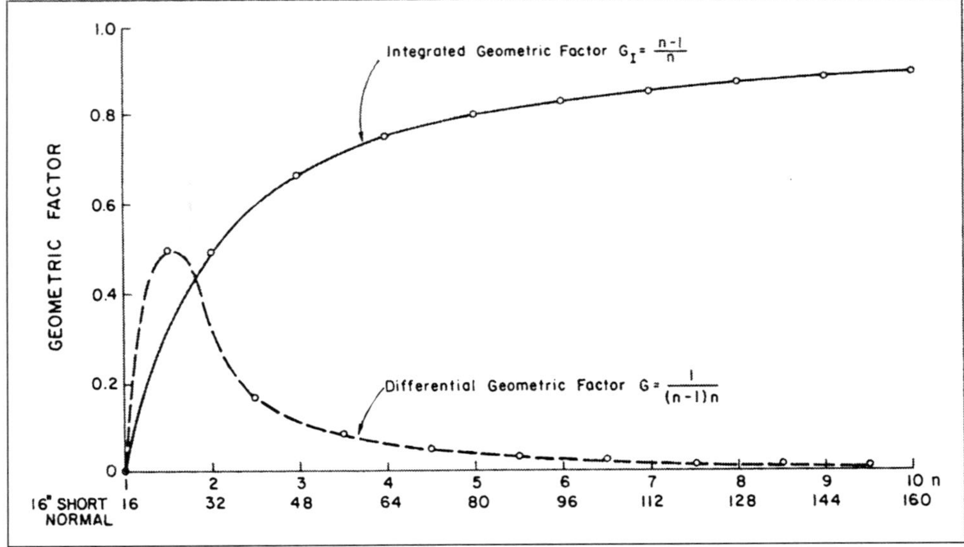

Fig. 5.47—Differential and integrated geometric factors for normal electrode devices in infinite medium.

From Eq. 5.5,

$$\Delta V_{M\infty} = \frac{(1)(11.5)}{4\pi(16)(0.0254)} = 2.252 \text{ V}.$$

The preceding discussion of geometric factors considers the simple case of spherical flow of DC and conventional electrode devices. The concept also is true for other potential distributions, geometries, and tools. Derivations of geometric factors of these more complex cases are available in the literature. Doll[18] treated the geometry of induction logging and gave the relationships to calculate the respective influence of different regions of the medium surrounding the coil system. In the treatment of induction logging, it is more appropriate to use conductivities than resistivities. The value of the apparent conductivity, C_a, can be written as

$$C_a = G_1C_1 + G_2C_2 + \ldots G_nC_n + \ldots, \quad \ldots\ldots\ldots\ldots\ldots (5.35)$$

where $C_1, C_2 \ldots C_n$ are the conductivities of the different regions whose geometric factors are $G_1, G_2 \ldots G_n$, respectively. **Fig. 5.49** is an example of the relative values of geometrical factors. Region A corresponds to the uninvaded zone of a permeable bed, Region B corresponds to the zone of that bed invaded by the mud filtrate, Regions D and F correspond to two very thick beds that constitute the upper and lower strata, and Region M corresponds to the mud column. For the particular dimensions represented, the numerical values of the geometric factors are 0.448, 0.239, 0.078, 0.078, and 0.157, respectively.

It should be stressed again that the geometrical factor, G, for a given region of the medium indicates the fraction of the total signal contributed by such a region when the conductivity is the same for the whole space. Actually, to determine what proportion of the whole signal is contributed by that region, it is necessary to take into account the respective conductivities of the different regions, as is evident from Eq. 5.35. A region having only a small geometric factor can, in fact, greatly influence the measurement simply

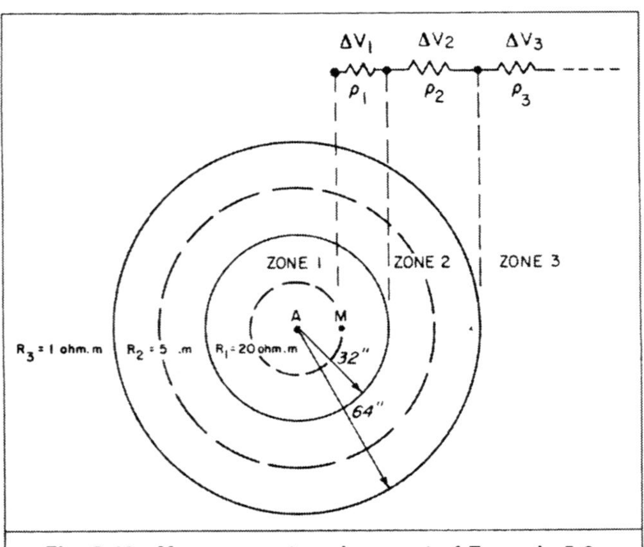

Fig. 5.48—Measurement environment of Example 5.3.

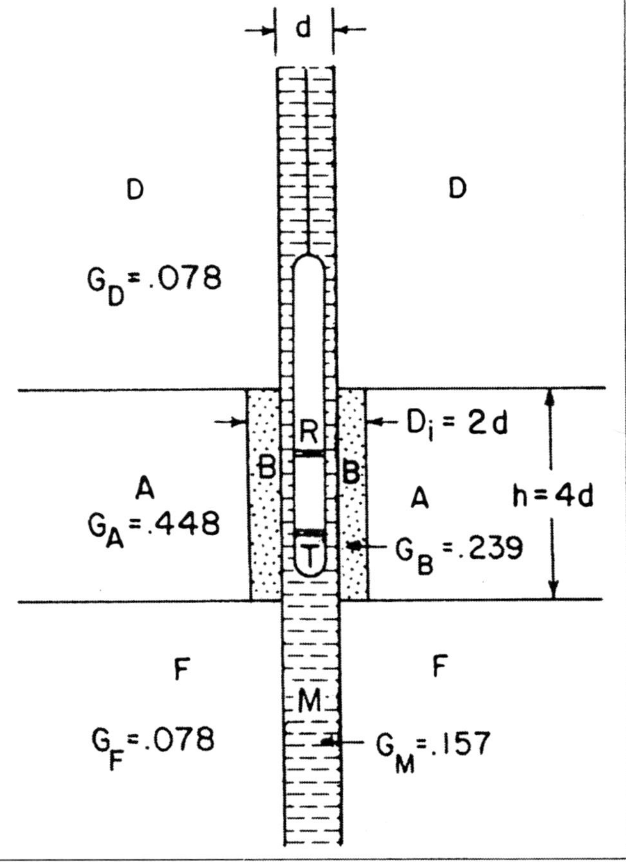

Fig. 5.49—Example of relative values of geometric factor, two-coil sonde; spacing = 1.25 × hole diameter (after Ref. 18).

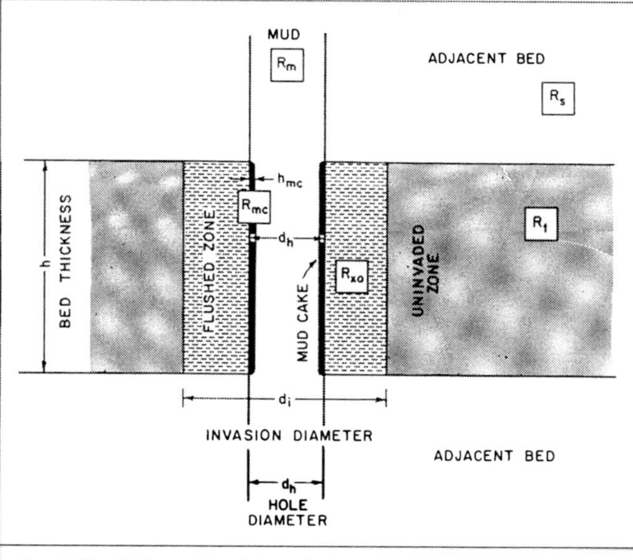

Fig. 5.50—Schematic of downhole measurement environment.

because it has a proportionally large conductivity. This would be the case, for instance, for Regions D and F in Fig. 5.49 if their conductivities are much higher than that of Region A, which the system is supposed to measure.

Example 5.6. If the conductivities of Regions A, B, D, F, and M in Fig. 5.49 are 100, 10, 500, 500, and 1,000 m℧/m, respectively, calculate the apparent resistivity indicated by a two-coil sonde whose spacing is 1.25 h, where h is the bed thickness.

Solution. Eq. 5.34 can be rewritten for this example as

$$C_a = G_A C_A + G_B C_B + G_D C_D + G_F C_F + G_M C_M.$$

Using the values of G indicated in Fig. 5.49 gives

$$C_a = (0.448)(100) + (0.239)(10) + 2(0.078)(500) + (0.157)(1{,}000)$$
$$= 282 \text{ m℧/m}$$

and $R_a = \dfrac{1{,}000}{C_a} = \dfrac{1{,}000}{282} = 3.5\ \Omega\cdot\text{m}$

5.6.2 Parameters Determining R_a. Fig. 5.50 shows the measurement environment considered in the determination of R_t and R_{xo} from R_a. This is a simplified radial model with no transition zone between the flushed and the uninvaded zones and with adjacent beds that extend infinitely. The radial resistivity distribution is hence simplified to a stair-step function, as shown in **Fig. 5.51**. These assumptions substantially simplify the mathematical work and lead to practical results. However, significant errors can result if the true resistivity profile differs considerably from the assumed step profile.

Figs. 5.50 and 5.51 indicate that 10 parameters contribute to the determination of the apparent resistivity value:

1. The true resistivity of the formation, R_t.
2. Resistivity of the invaded (flushed) zone, R_{xo}.
3. The diameter of the invaded zone, d_i.
4. Resistivity of the mud, R_m.
5. Resistivity of the mudcake, R_{mc}.
6. The measuring tool vertical resolution.
7. Mudcake thickness, h_{mc}.
8. Hole diameter, d_h.
9. Bed thickness, h.
10. The resistivity of the adjacent beds, R_s.

Fortunately, some of these factors do not play a significant role in all the cases where resistivity is to be determined. It is possible, in fact, to set up rather broad limits wherein the log reading may be used directly for R_t or with only a simple correction. In the

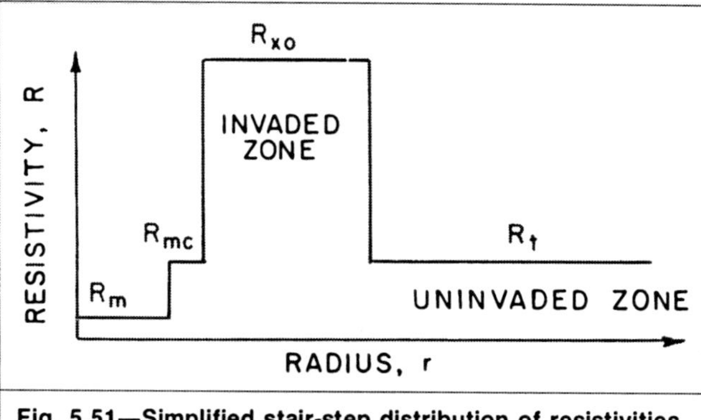

Fig. 5.51—Simplified stair-step distribution of resistivities.

more difficult cases, it is necessary to resort to lengthy computations to make the proper interpretation.

The above parameters can be classified into four groups. Each group determines the effect of one of the four distinct zones, shown in Fig. 5.50, on the apparent resistivity. These four zones and the corresponding parameters are (1) the borehole (R_m, R_{mc}, d_h, h_{mc}), (2) surrounding beds (R_s, h, tool vertical resolution), (3) the invaded zone (R_{xo}, d_i), and (4) the uninvaded zone or true resistivity zone (R_t, d_i).

A relationship similar to Eq. 5.32 can be written for a specific tool of a certain vertical resolution:

$$R_a = G_m(d_h, h_{mc})R_m + G_s(h)R_s + G_{xo}(d_i)R_{xo} + G_t(d_i)R_t, \quad (5.36)$$

where G_m, G_s, G_{xo}, and G_t are the geometric factors of the borehole zone, surrounding beds, invaded zone, and true resistivity zone, respectively.

For induction tools, Eq. 5.36 assumes the form

$$C_a = G_m(d_h, h_{mc})C_m + G_s(h)C_s + G_{xo}(d_i)C_{xo} + G_t(d_i)C_t. \quad (5.37)$$

The calculation of R_t from Eq. 5.36 or 5.37 requires the measured value of R_a and knowledge of the different geometric factors and R_m, R_s, and R_{xo}.

Example 5.7. For the particular case represented in Fig. 5.49, calculate the true conductivity of the uninvaded Zone A of the permeable bed if the tool indicates an apparent conductivity of 2,000 m℧/m; the mud conductivity, C_m, is 1,000 m℧/m; the uninvaded-zone conductivity is 10 times that of the invaded zone; and the adjacent bed conductivity, C_s, is 500 m℧/m.

Solution.

Step 1—Borehole effect.

Borehole signal $= G_m C_m = 0.157(1{,}000) = 157$ m℧/m.

Signal free from borehole effect $= 2{,}000 - 157 = 1{,}843$ m℧/m.

Eq. 5.37 reduces to

$1{,}843 = G_s C_s + G_{xo} C_{xo} + G_t C_t$.

Step 2—Surrounding-bed effect.

Surrounding-bed effect $= G_s C_s = 2(0.078)(500) = 78$ m℧/m.

Signal free from borehole and surrounding beds effects $= 1{,}843 - 78 = 1{,}765$ m℧/m.

Eq. 5.37 reduces further to

$1{,}765 = G_{xo} C_{xo} + G_t C_t$.

Step 3—Invasion effect.

Invasion effect $= G_{xo} C_{xo}$.

Knowing that $C_{xo} = 0.1\ C_t$, the invasion effect becomes

invasion effect $= (0.239)(0.1)C_t = 0.0239 C_t$.

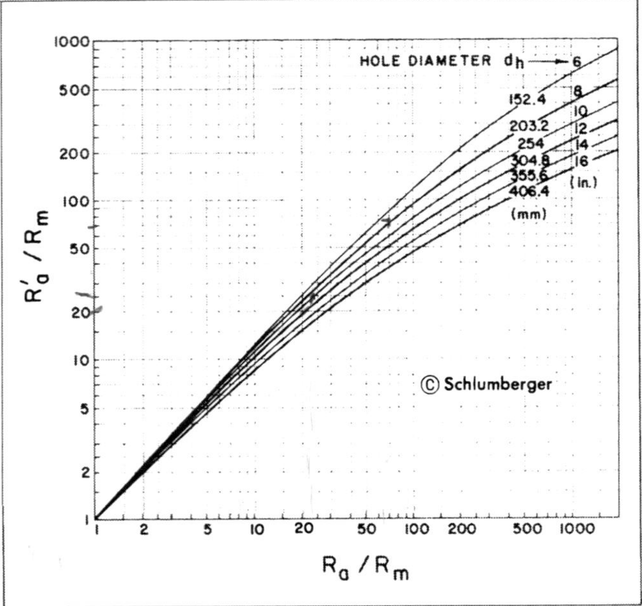

Fig. 5.52—Borehole-effect correction chart for 16-in. short normal (after Ref. 23).

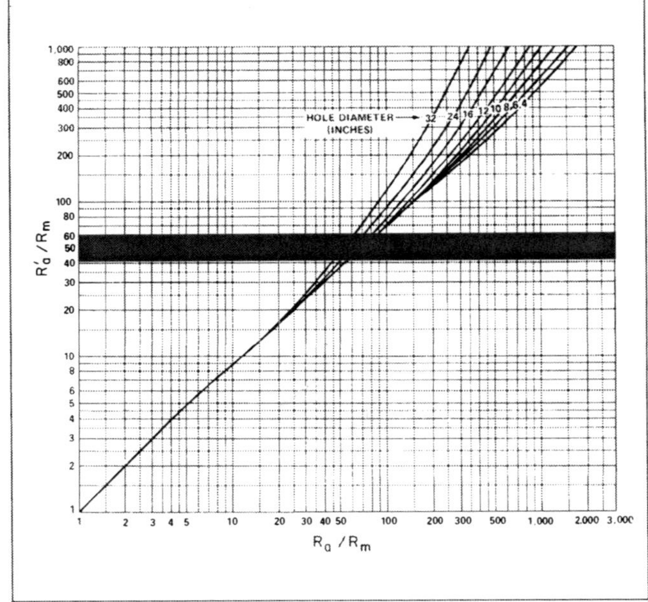

Fig. 5.53—Borehole-effect correction chart for 64-in. long normal (from Ref. 22).

Hence,

$1,765 = 0.0239C_t + 0.448C_t$

and $C_t = 3,740$ m℧/m.

The calculation of R_t or C_t from Eq. 5.36 or 5.37 is performed in several steps in a manner similar to that illustrated by Example 5.7. However, the effects of different zones are obtained directly from charts published periodically by well-logging service companies. The concept and use of these charts are discussed in the following sections.

5.6.3 Borehole Effect. The borehole contribution to the overall response is determined by R_m, R_{mc}, d_h, and h_{mc}. For resistivity

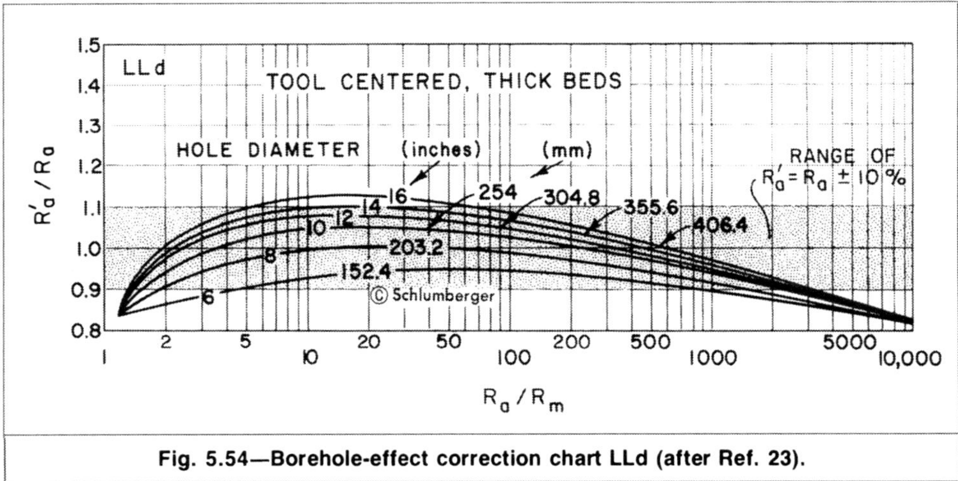

Fig. 5.54—Borehole-effect correction chart LLd (after Ref. 23).

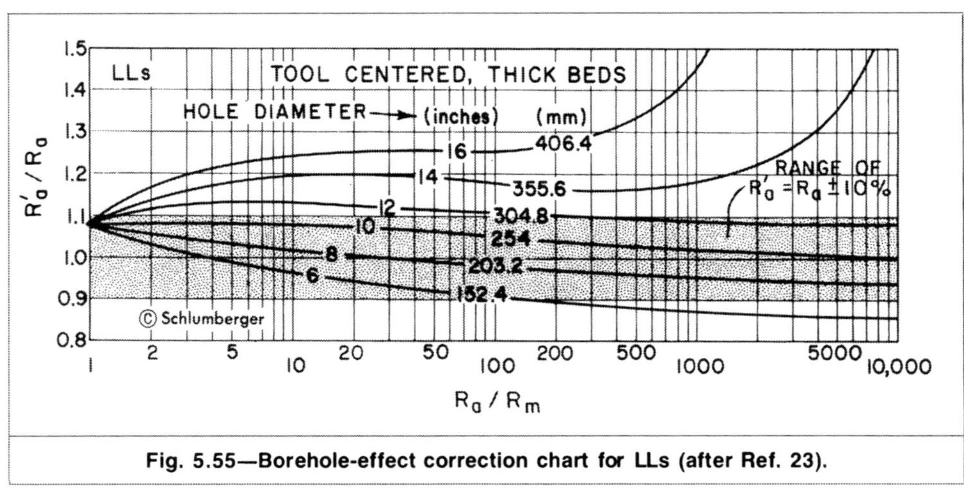

Fig. 5.55—Borehole-effect correction chart for LLs (after Ref. 23).

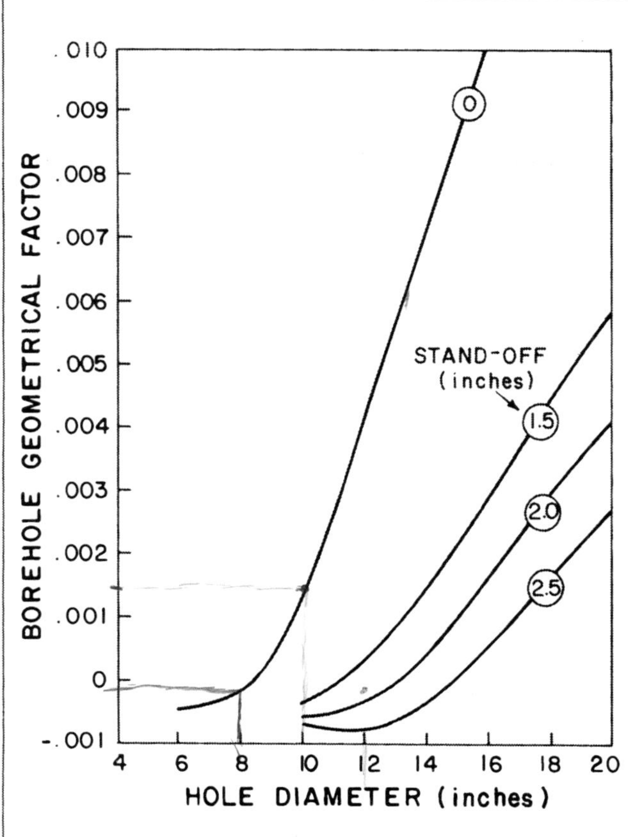

Fig. 5.56—Borehole-effect chart for deep induction tools (after Ref. 23).

tools with a relatively deep radius of investigation, such as focused-current and induction devices, the effect of the mudcake zone (R_{mc}, h_{mc}) is negligible, and the borehole effect is determined by only d_h and R_m. On the other hand, for pad-type tools (microresistivity devices) the effect of d_h and R_m is eliminated because the electrodes are pressed against the formation face. Only R_{mc} and h_{mc} determine the borehole zone effect.

The borehole effect is annulled with correction charts. In general, the input to the chart is the dimensionless ratio (R_a/R_m). The output is another dimensionless ratio, (R_a'/R_m) or (R_a'/R_a). R_a' is the value of the apparent resistivity corrected for the borehole effect. R_a' is called the "borehole-effect-free" apparent resistivity.

Fig. 5.52 is the borehole correction chart for the 16-in. normal resistivity recorded with Schlumberger's Induction Electric Log[23] (6FF40-16"N). **Fig. 5.53** is an example of a borehole correction chart for 64-in. normal readings.[24] For conventional electrode devices, the borehole contribution increases as borehole diameter increases and mud resistivity decreases.

These charts, like all other correction charts, were based on theoretical and experimental work. Assumptions and simplifications had to be made to perform the calculations and experiments. The major assumptions are that (1) the tool is centered in the borehole, (2) the bed is thick, and (3) the borehole is a uniform circle. These conditions are not usually fulfilled in real applications. The correction given by these charts is then an approximation. The higher the correction, the higher is the uncertainty in the output value.

The correction charts, however, can be very useful in determining the range of resistivity and geometric parameters where the tool will provide an apparent resistivity value with negligible borehole correction. As Figs. 5.52 and 5.53 reflect, for conventional electrode tools, the borehole correction is extremely small (less than 10%) or nonexistent for reasonable borehole diameters (8 to 10 in.) and for $R_a/R_m < 10$.

Figs. 5.54 and 5.55 are borehole correction charts for LLd and LLs readings for a centered sonde. Eccentricity has little effect on the LLd but can be detrimental to the LLs when R_a/R_m is high. Fig. 5.54 shows that the LLd in the most common measurement environment will read within 10% of the correct value, but that value still includes surrounding-bed and invasion effects. Fig. 5.55 shows that the LLs also will read within about 10% of its correct value in holes up to 11 in. in diameter. For larger holes, the corrections can be quite important.

Fig. 5.56 gives the borehole geometric factor for the deep induction tools. The borehole signal is obtained as the product of the mud conductivity and the geometric factor. The hole-conductivity signal is to be subtracted from the apparent conductivity, C_a, to obtain a "borehole-effect-free" apparent conductivity, C_a'.

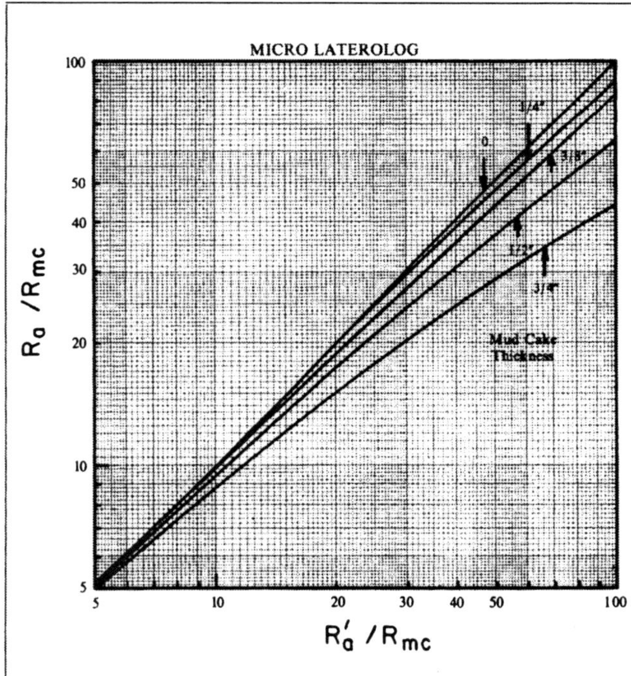

Fig. 5.57—Borehole-effect correction chart for the MLL (after Ref. 25).

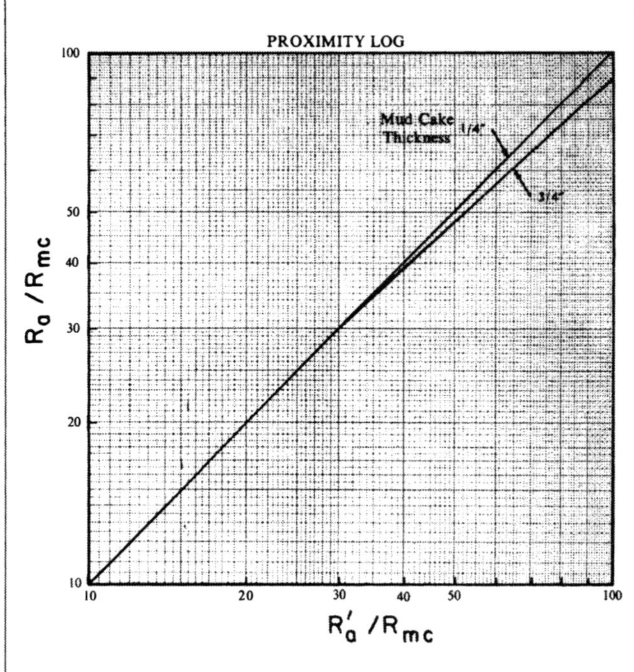

Fig. 5.58—Borehole-effect correction chart for the PL (after Ref. 25).

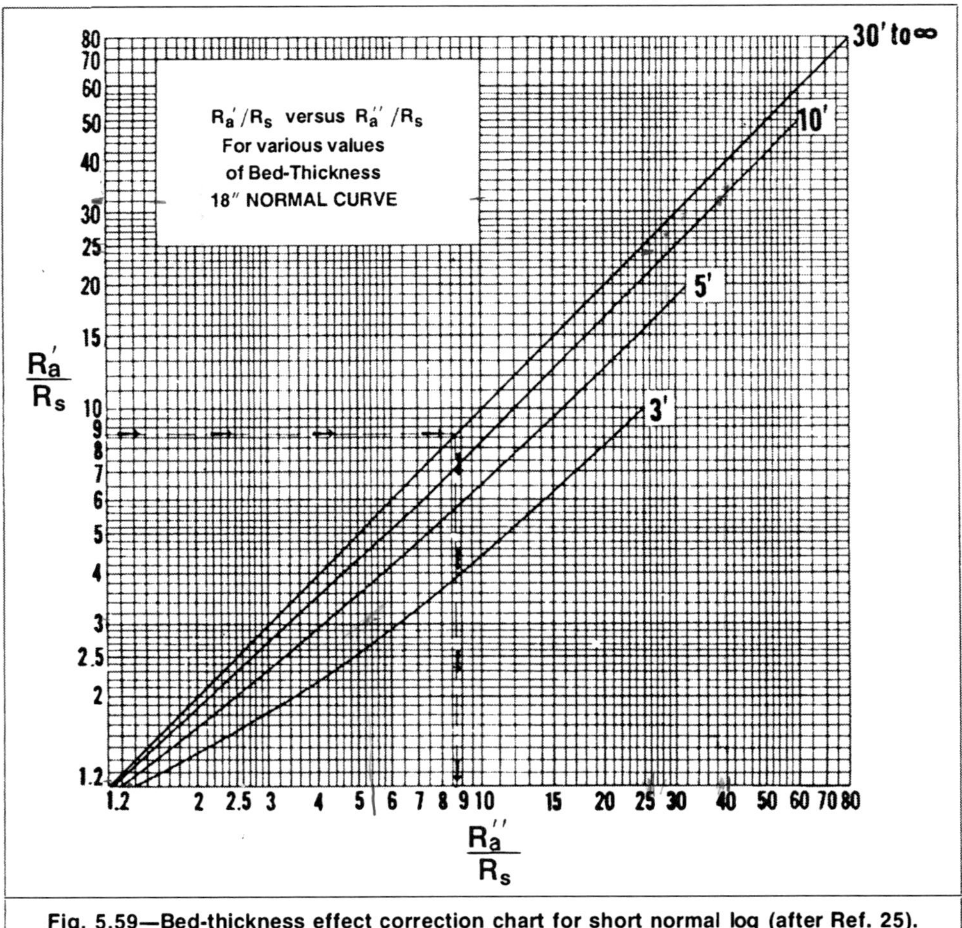

Fig. 5.59—Bed-thickness effect correction chart for short normal log (after Ref. 25).

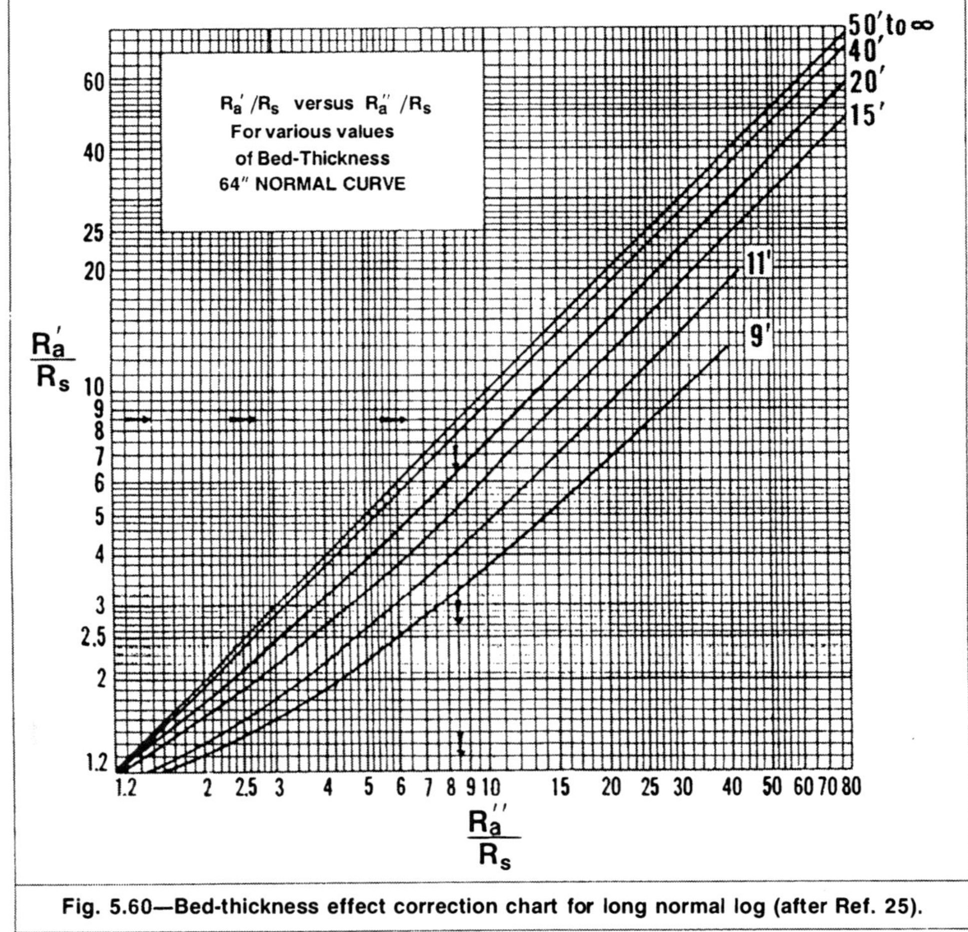

Fig. 5.60—Bed-thickness effect correction chart for long normal log (after Ref. 25).

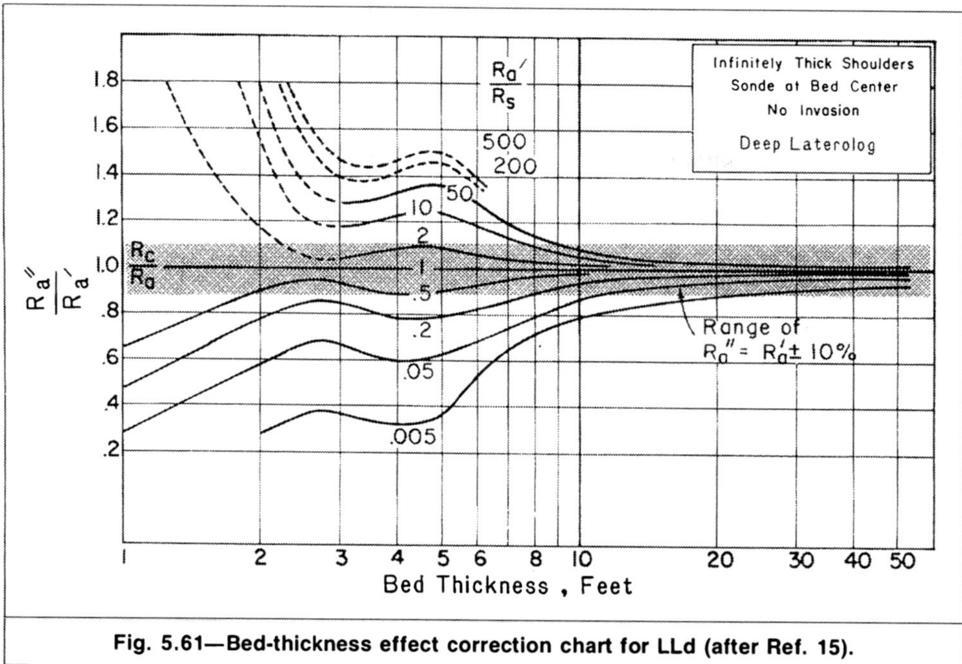

Fig. 5.61—Bed-thickness effect correction chart for LLd (after Ref. 15).

The response of the induction tool to borehole fluids is small when the tool is centered. For that reason, the induction tool is usually fitted with rubber standoffs that hold the tool away from the borehole wall. Fig. 5.56 gives geometrical factors for various standoff values.

The borehole contribution to the overall response of the induction tool is small in high-conductivity formations and usually can be neglected. However, when formation conductivity is low, the borehole signal contribution can become significant. The nominal borehole signal, based on bit size, is often removed from the recorded log. Log headings should always be consulted to ascertain whether this correction was made. This precaution applies primarily to medium and deep induction tools.

Figs. 5.57 and 5.58 give borehole correction (mudcake correction) for the MLL and PL. The effect of mudcake on the MLL is negligible up to a mudcake thickness of ¼ in. The effect increases rapidly with greater mudcake thicknesses. The proximity tool pad and electrode design are such that mud cakes up to ¾ in. thick have very little effect on the measurement (see Fig. 5.58). Accordingly, the PL is preferred over the MLL when deep invasion and mudcakes thicker than ⅜ in. are expected.[20]

Example 5.8. An apparent resistivity value of 10 Ω·m is indicated by an LLd recorded in a 10-in. hole. Calculate the borehole-effect-free apparent resistivity if R_m=0.1 Ω·m at measurement conditions.

Solution. Because $R_a=10$ Ω·m and $R_m=0.1$ Ω·m, then, $R_a/R_m=100$. From Fig. 5.54, for a 10-in. hole and $R_a/R_m=100$, $R_a'/R_a=1.025$ and $R_a'=10(1.025)=10.25$ Ω·m.

Note, that in this case, if the borehole-effect correction is neglected, only a small error of 2.5% would be introduced.

Example 5.9. A DIL was used to log an interval of interest of a 12-in. borehole. The tool was equipped with a 1.5-in. standoff, and the average mud resistivity over the interval of interest was 0.15 Ω·m. If the maximum formation resistivity measured in that interval is 10 Ω·m and a maximum of 5% borehole effect is considered negligible, show that the recorded log values will for practical purposes, be essentially free of borehole effect.

Solution. From Fig. 5.56, the borehole geometric factor is G_m=0.0003 and the hole signal is

$$C_m G_m = G_m/R_m = \frac{0.0003}{0.15/1{,}000} = 2 \text{ m℧/m}.$$

The maximum conductivity measured = 1,000/10 = 100 m ℧/m. The maximum borehole effect expected = (2/100)100 = 2%. The expected maximum borehole effect is much less than the tolerable 5%. The recorded log values could, for practical purposes, be considered essentially free of borehole effect.

As Examples 5.8 and 5.9 show, borehole effects can be negligible under certain measurement conditions. In fact, the different resistivity tools are designed to be free of borehole effect for the optimum range of borehole size and mud resistivity. Nonnegligible borehole effect should be expected, however, if the tool is used under conditions outside the optimum range. An inexperienced log analyst should routinely calculate the borehole effect to understand the magnitude of the effect in different conditions. After some experience, the analyst will be able to recognize cases where the correction for borehole effect is needed.

5.6.4 Bed-Thickness Effect. The effect of surrounding beds, usually called the bed-thickness effect, is determined by the bed thickness, the vertical resolution of the tool, and the resistivity contrast R_t/R_s between the main bed of interest and the adjacent bed. The thicker the bed is (relative to the tool vertical resolution), the smaller the effect is. For a resistive main bed (i.e., $R_t>R_s$), adjacent bed effects will lower the apparent resistivity value. For thin conductive beds (i.e., $R_t<R_s$), the apparent resistivity value will increase because of the adjacent bed effect.

The reading of the microresistivity tools is free from the effect of adjacent beds because of the small spacing, which is usually only a few inches. Readings from other tools will carry a bed-thickness effect. This effect is corrected for with charts. Borehole corrections should be made before these charts are used. Bed-thickness-effect correction charts are usually entered with the borehole-effect-free apparent resistivity, R_a', or the dimensionless resistivities ratio R_a'/R_s. The correction chart will yield an apparent resistivity value, R_a'', corrected for both borehole and bed-thickness effects. R_a'' will be called the "bed-thickness-effect-free" apparent resistivity.

Figs. 5.59 and 5.60 are examples of correction charts for the normal devices' readings for thin beds. For the short normal (Fig. 5.59), bed-thickness effect decreases progressively as the bed thickness increases. Once the bed thickness reaches 30 ft, the adjacent bed effect becomes nil. For the long normal device, the bed thickness has to reach 50 ft before the adjacent bed-effect disappears (Fig. 5.60).

Fig. 5.61 shows the bed-thickness correction chart for a LLd situated at the center of a homogeneous bed surrounded by infinitely

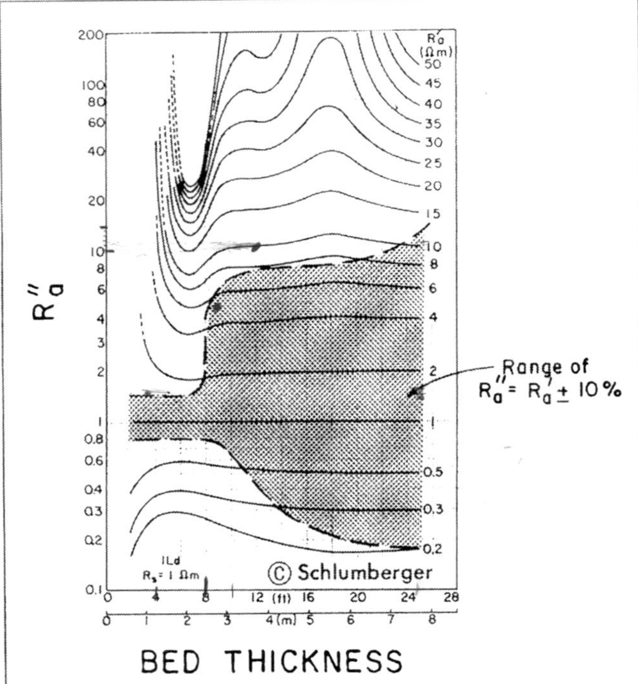

Fig. 5.62—Bed-thickness effect correction chart for ILd (after Ref. 23).

thick shoulders. The correction factor R_a''/R_a' is considerable for thin beds where $h<10$ ft with high R_a'/R_s or R_s/R_a' resistivity contrast. The shaded area designates the region where $0.9<R_a''/R_a'<1.1$. If the 10% error introduced by the surrounding beds can be considered negligible for practical purposes, then log readings falling in the shaded area can be considered free of bed-thickness effect. Moreover, if the reading is also free of borehole effect, R_a'' values can be read directly off the log.

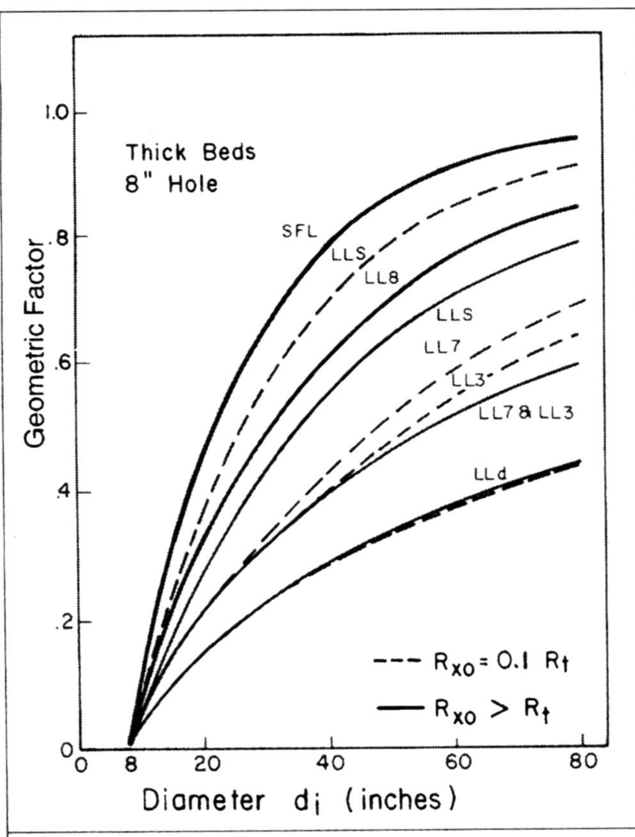

Fig. 5.64—Integrated geometric factors for focused-current tools (from Ref. 15).

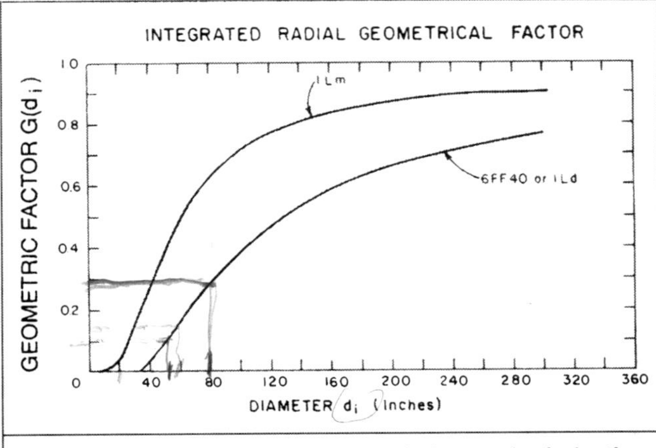

Fig. 5.63—Integrated radial geometric factors for induction tools (after Ref. 15).

Corrections for induction tools are provided by charts similar to **Fig. 5.62**. This chart is for a deep-induction reading recorded with a shoulder-bed resistivity setting of 1 Ω·m. This setting controls the vertical focusing of the sonde and can be adjusted by the logging operator. The shaded area represents the range of formation resistivity and bed thickness in which the bed-thickness correction is zero or negligible.

Example 5.10. A 64-in. long normal device reads 10 Ω·m in a 9-ft bed. The borehole diameter is 10 in. and is full of 0.1-Ω·m mud. If the bed is surrounded by thick shales of 1-Ω·m resistivity, determine the bed-thickness-effect-free apparent resistivity, R_a''.

Solution. The apparent resistivity value $R_a=10$ Ω·m first has to be corrected for borehole effect with Fig. 5.53: $R_a/R_m=10/0.1=100$. From Fig. 5.53, $R_a'/R_m=70$ and $R_a'=7$ Ω·m.

The borehole-effect-free value $R_a'=7$ Ω·m can now be corrected for bed-thickness effect with Fig. 5.60: $R_a'/R_s=7/1=7$ and, from Fig. 5.60, $R_a''/R_s=21$ and $R_a''=21$ Ω·m.

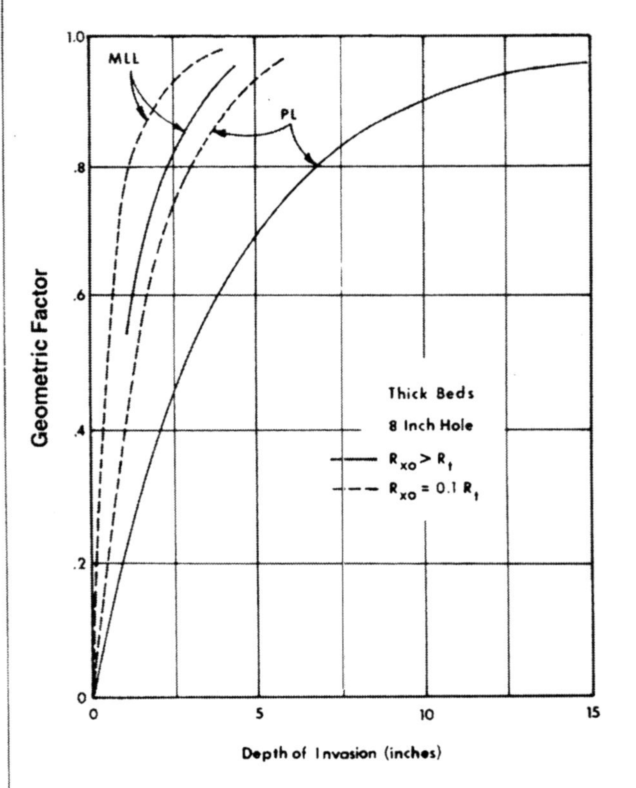

Fig. 5.65—Integrated geometric factors for microresistivity tools (from Ref. 15).

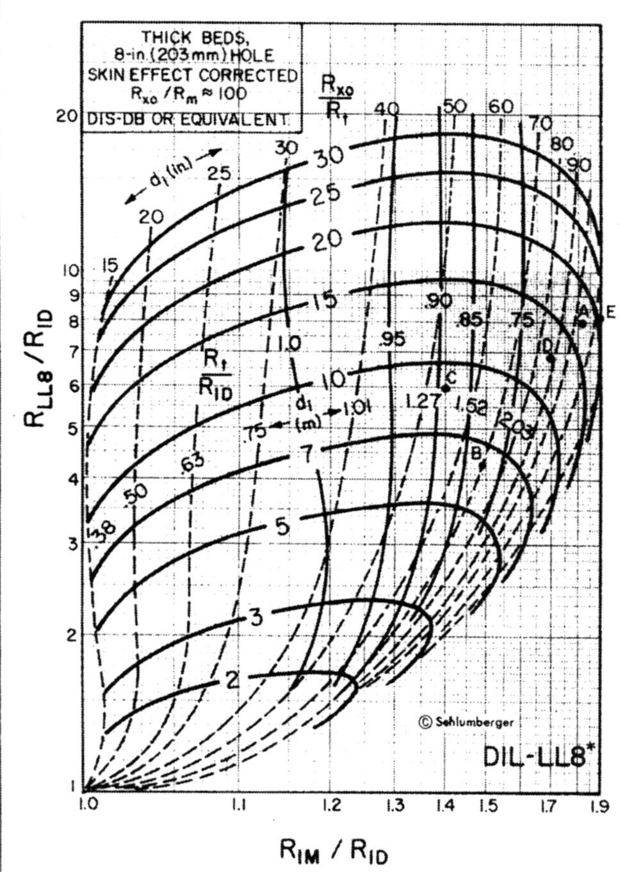

Fig. 5.66—Interpretation chart for DIL/Laterolog tool (from Ref. 23).

The value of R_a'' is more than double the value of R_a read by the tool. This represents a combined 110% bed thickness and borehole effects. Such a large correction suggests that the measurement environment (bed thickness, mud resistivity, etc.) was not optimum for that particular tool. Correction charts were constructed from several assumptions and should not be considered error-free in nonoptimum environments. The value of $R_a''=21$ Ω·m can only be considered an order of magnitude. The parameters calculated with this value will bear some degree of uncertainty.

Example 5.11. An apparent resistivity value of 10 Ω·m is indicated by an LLd recorded in a 10-in. hole full of 0.1-Ω·m mud. This reading was obtained with the tool centered in a 20-ft bed surrounded by thick shales of 1-Ω·m resistivity. Determine the value of R_a''.

Solution. Experience gained from Example 5.8 indicates that the borehole effect is very small and can be ignored. Then, $R_a'=R_a=$ 10 Ω·m. Fig. 5.61 can be used to find the correction factor R_a''/R_a'. For a bed thickness of 20 ft and a resistivity contrast $R_a'/R_s=10$, the correction factor is almost unity. Hence, $R_a''=R_a'=R_a=10$ Ω·m.

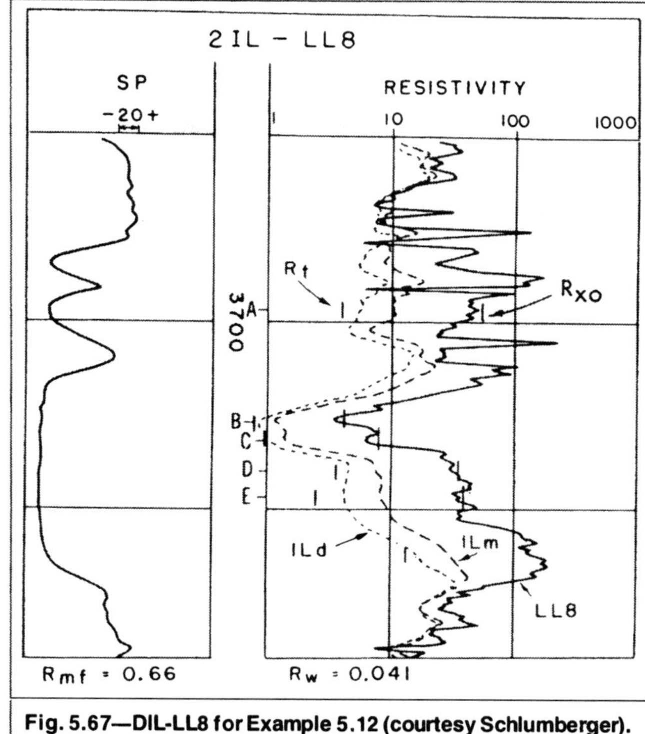

Fig. 5.67—DIL-LL8 for Example 5.12 (courtesy Schlumberger).

In this case, the measurement environment is extremely favorable because the log reads a value that is free from both borehole and bed-thickness effects.

5.6.5 Invasion Effect. The value R_a'' represents the hypothetical response of the tool in an infinite medium. No borehole or adjacent beds are present. However, a flushed zone of resistivity R_{xo} and invasion diameter d_i is surrounding the tool, separating it from the uninvaded zone of resistivity R_t. The effect of this invaded zone can be seen from the concept of geometric factors. The R_a'' value is affected by only two zones: the flushed zone and the uninvaded zone with geometric factors of G_{xo} and G_t, respectively. Because $G_{xo}+G_t=1$, Eq. 5.36 reduces to

$$R_a''=G_{xo}(d_i)R_{xo}+[1-G_{xo}(d_i)]R_t. \quad\quad (5.37)$$

G_{xo} is the differential radial geometric factor of the flushed zone. Because of the position assumed by the zone next to the tool, G_{xo} is also the integrated radial geometric factor of the zone that extends to a diameter d_i. It is the only geometric factor in the equation, so the subscript *xo* is usually dropped for simplicity. Three unknown parameters appear in Eq. 5.37: d_i, R_{xo}, and R_t. Note that, for induction tools, conductivities should replace resistivities in Eq. 5.37.

When d_i, R_{xo}, and R_t are unknown, three resistivity curves are required for determination of R_t. When three resistivity tools are run simultaneously or successively, Eq. 5.37 can be written for each of the three apparent resistivity values $(R_a'')_1$, $(R_a'')_2$, and $(R_a'')_3$ indicated for the bed of interest:

$$(R_a'')_1=G_1(d_i)R_{xo}+[1-G_1(d_i)]R_t, \quad\quad (5.38)$$

TABLE 5.4—RESISTIVITY VALUES FOR EXAMPLE 5.12

	Read From Logs					From Fig. 5.66				
Level	R_{ILd}	R_{ILM}	R_{LL8}	R_{ILM}/R_{ILd}	R_{LL8}/R_{ILd}	R_t/R_{ILd}	R_{xo}/R_t	d_i	R_{xo}	R_t
A	5.5	10.0	44.0	1.82	8.00	0.62*	17.0	102	58.0	3.4
B	0.8	1.2	3.5	1.50	4.38	0.80	6.3	70	4.0	0.6
C	1.0	1.4	6.0	1.4	6.00	0.89	9.0	53	8.0	0.9
D	4.4	7.5	30.0	1.7	6.82	0.69*	12.5	86	37.5	3.0
E	4.2	8.0	35.0	1.9	8.33	0.59*	20.0	110	50.0	2.5

*Extrapolated values.

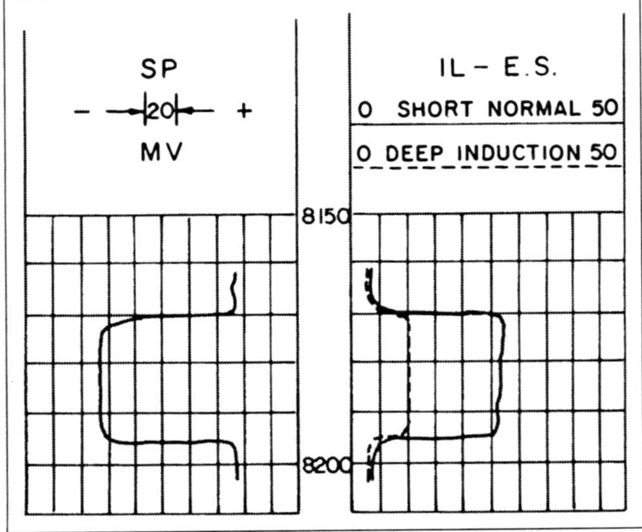

Fig. 5.68—Schematic of induction/electric survey, Example 5.13.

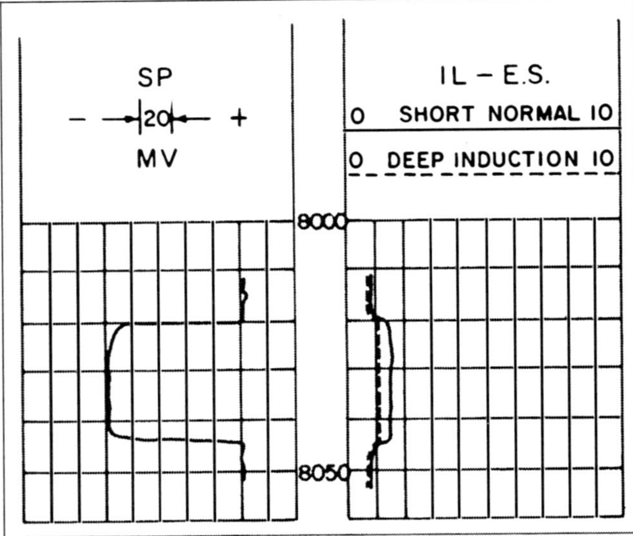

Fig. 5.69—Schematic of induction/electric survey, Example 5.12.

$$(R_a'')_2 = G_2(d_i)R_{xo} + [1 - G_2(d_i)]R_t, \quad \text{(5.39)}$$

and $(R_a'')_3 = G_3(d_i)R_{xo} + [1 - G_3(d_i)]R_t,$ ....(5.40)

where G_1, G_2, and G_3 are the geometric factors of the three resistivity tools.

This system of equations in three unknowns (d_i, R_{xo}, and R_t) can be solved analytically. The values of $G(d_i)$ can be obtained from **Figs. 5.63 through 5.65**. The analytical solution to this system of equations is tedious. The system can be solved graphically

TABLE 5.5—APPROXIMATE VALUES OF d_i AS A FUNCTION OF POROSITY (from Ref. 20)

Porosity Range (%)	d_i
5 to 10	$10d_h$
10 to 15	$5d_h$
15 to 20	$2.5d_h$

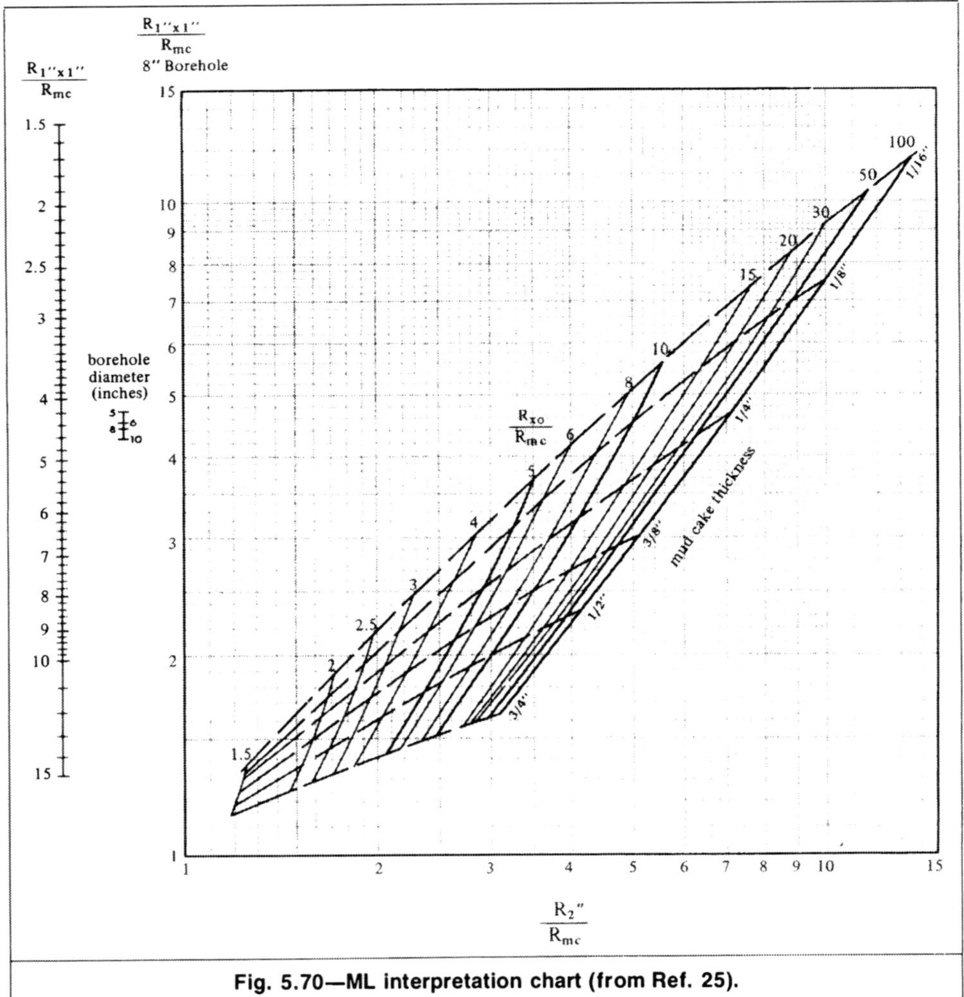

Fig. 5.70—ML interpretation chart (from Ref. 25).

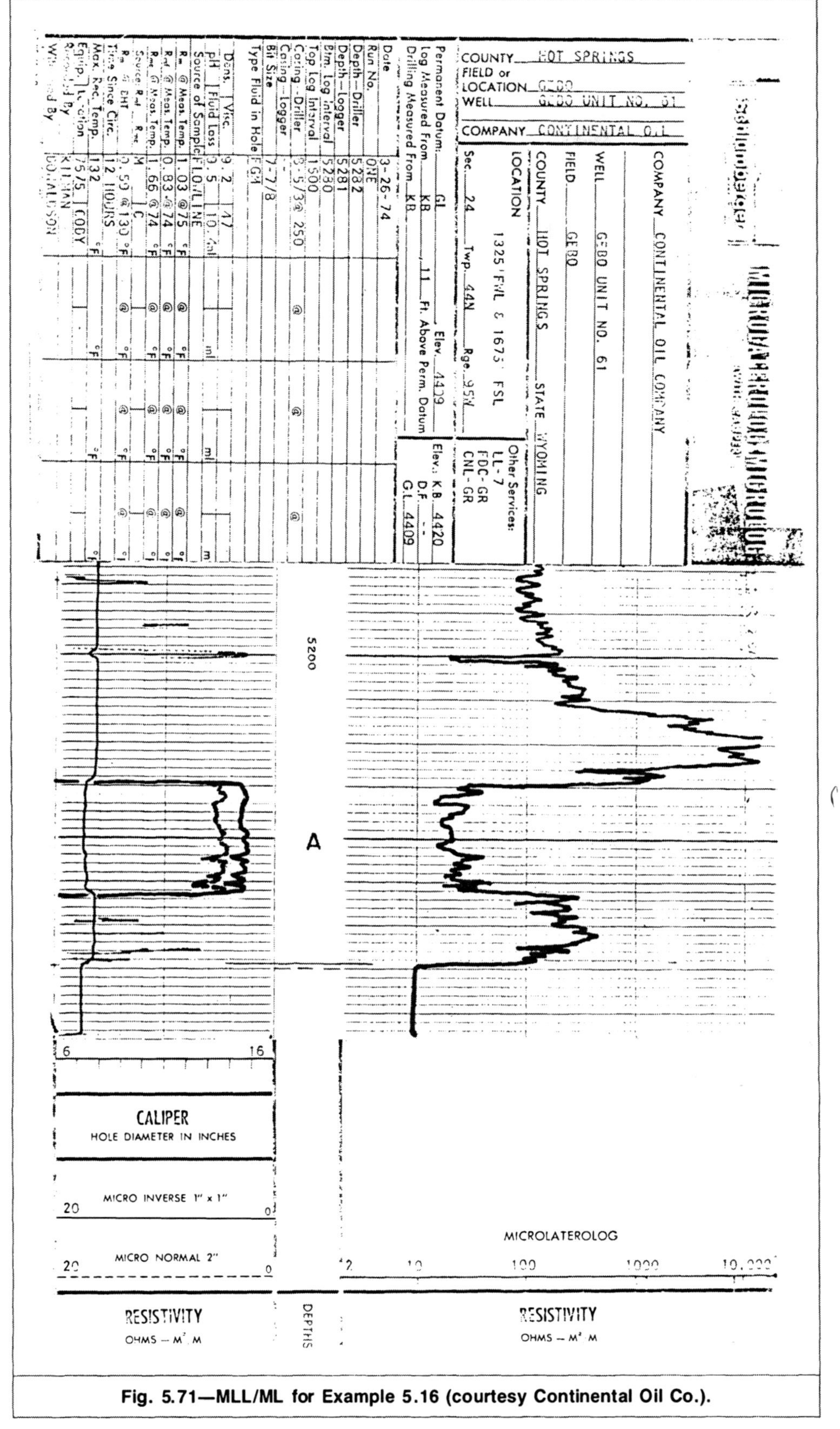

Fig. 5.71—MLL/ML for Example 5.16 (courtesy Continental Oil Co.).

with charts provided by logging companies. **Fig. 5.66** is an example of such a chart. These charts are usually called "tornado charts" because of their distinctive shape.

Fig. 5.66 was constructed for Schlumberger's Dual Induction-Laterolog-8 tool. This tool records three resistivity curves: deep induction, R_{ILd}, medium induction, R_{ILm}, and relatively shallow LL8, R_{LL8}. It was assumed in the construction of this chart that a step profile of invasion exists. The chart also assumes that the induction readings are automatically corrected for the skin effect. To use the chart, the ratio R_{LL8}/R_{ILd} is plotted in ordinate vs. R_{ILm}/R_{ILd} in abscissa. The resulting plot yields values of R_t/R_{ILd}, R_{xo}/R_t, and d_i. Because the value of R_{ILd} is known, R_t and R_{xo} are readily computed.

Fig. 5.66 is based on a hole diameter of 8 in. and is constructed for thick beds. If borehole and bed-thickness corrections are required, they should be made before the data are entered in the chart.

Fig. 5.66 assumes freshwater-based mud and $R_{xo}/R_m \cong 100$. The ratio R_{xo}/R_m reflects the porosity of the formation because $R_{xo}/R_{mf} = F$ in water-bearing formations.

Well-logging service companies provide tornado charts for different tool combinations. These charts, however, are for representative values of R_{xo}/R_m. Jenson and Gartner[26] found that significant (up to 21%) errors in interpreted values could arise when a chart with an inappropriate R_{xo}/R_m ratio is used. They also determined, however, that, if the actual situation has an R_{xo}/R_m value within a factor of 2.5 of the chart value, the average error in either d_i or R_{xo}/R_t is less than 5%.

Example 5.12. The Dual Induction-Laterolog survey in **Fig. 5.67** was recorded in an 8-in. hole. Determine R_t, R_{xo}, and d_i values at the five levels marked on the log by the letters A through E. Experience in this type of environment indicates negligible borehole and bed-thickness effects.

Solution. Because the borehole and bed-thickness effects are negligible, R_a'' values can be read directly from the three resistivity curves. These values are listed in **Table 5.4**.

The ratios $R_{\text{ILm}}/R_{\text{ILd}}$ and $R_{\text{LL8}}/R_{\text{ILd}}$ are calculated for each level and plotted in Fig. 5.66. Values of R_t/R_{ILd}, R_{xo}/R_t, and d_i were read and are listed in Table 5.4, together with the calculated values of R_{xo} and R_t. The calculated values of R_{xo} and R_t were also marked on the log.

Although the d_i values calculated in Example 5.12 are subject to the assumption of a step profile, they show that invasion can vary greatly in a formation. Such variations are not apparent from observation of the resistivity curves alone. This fact may be demonstrated in Fig. 5.66, where, for example, for $R_{xo}/R_t = 10$, $R_{\text{LL8}}/R_{\text{ILd}} = 5.0$ could correspond to a $d_i = 25$ or 105 in. At $d_i = 25$ in., $R_{\text{ILd}} = R_t$, but for $d_i = 105$ in., R_{ILd} requires a correction factor of about 0.65 to give R_t. Addition of the ILm data permits an estimate of d_i.

When a significant invasion effect is expected, it is recommended that three resistivity tools be run. If only two resistivity logs are available, the invasion effect is essentially indeterminate. Approximate values of R_t and R_{xo} are obtained if the third unknown, d_i, can be estimated independently. An empirical relationship between porosity and diameter of invasion may be established. Such a correlation is given in **Table 5.5**, where d_i is given as a function of borehole diameter and porosity range.

Empirical correlations yield d_i values that are, of course, rough estimates. They were found to give reasonable results when the porosity range is unequivocally recognized and when the correlation has been tested and proved valid in the type of formation of interest. These values, however, may lead to erroneous results in particular cases because d_i is also controlled by mud filtration properties, mud hydrostatic pressure, and other drilling variables.

With the $(R_a'')_1$ and $(R_a'')_2$ resistivity values obtained from the two available resistivity curves for the formation of interest, Eq. 5.37 can be written twice:

$$(R_a'')_1 = G_1(d_i)R_{xo} + [1 - G_1(d_i)]R_t \quad \text{.................} \quad (5.41)$$

and $$(R_a'')_2 = G_2(d_i)R_{xo} + [1 - G_2(d_i)]R_t. \quad \text{...............} \quad (5.42)$$

The values of $G_1(d_i)$ and $G_2(d_i)$ can be obtained for the estimated d_i from Fig. 5.64, 5.65, or 5.66. Only two unknowns remain in the system of Eqs. 5.41 and 5.42 and can readily be solved for R_{xo} and R_t. Tornado charts provide an easier solution. The appropriate chart is entered by the value of $(R_a'')_1/(R_a'')_2$. The point whose coordinates are the ratio $(R_a'')_1/(R_a'')_2$ and the estimated d_i yields values of the ratios $R_t/(R_a'')_2$ and R_{xo}/R_t. Because $(R_a'')_2$ is known, R_t and R_{xo} are readily computed.

Example 5.13. Fig. 5.68 schematically shows the response of an electrical survey of a relatively low-porosity formation. If the porosity is expected to range from 10% to 15%, estimate R_t and R_{xo}. The mud resistivity and borehole diameter are 1.2 Ω·m and 12 in., respectively.

Solution. The two resistivity curves shown on the logs are shallow (16-in. short normal) and deep (6 FF 40 induction) investigation tools. They indicate the following apparent resistivity values: $(R_a)_{\text{SN}} = 27$ Ω·m and $(R_a)_{\text{IL}} = 10$ Ω·m (100 m℧/m).

Borehole Effect. The borehole is relatively large. Borehole-effect calculations must be performed. Fig. 5.52 can be used to calculate $(R_a')_{\text{SN}}$ as follows: $(R_a)_{\text{SN}}/R_m = 27/1.2 = 22.5$, and from Fig. 5.52 $(R_a')_{\text{SN}}/R_m = 25$ and $(R_a')_{\text{SN}} = 1.2(25) = 30$ Ω·m.

Whether the induction tool was fitted with standoffs is unknown. Assume the worst case, from the borehole-effect viewpoint, that standoffs were not used. Then, from Fig. 5.56 $(G_m)_{\max} = 0.0042$ and $(G_m C_m) = (0.0042/1.2) \times 1{,}000 = 3.5$ m℧/m.

The maximum possible borehole signal is only 3.5 m℧/m, which is negligible compared with the measured apparent conductivity of 100 m℧/m. So, $(R_a')_{\text{IL}} = (R_a)_{\text{IL}} = 10$ Ω·m. Had the assumption of no standoffs resulted in a nonnegligible hole signal, better information concerning the use and size of standoffs would have to be obtained.

Bed-Thickness Effect. The bed of interest is 25 ft thick and can be considered thick. However, the adjacent bed resistivity is only 2.5 Ω·m, which yields a resistivity contrast $(R_a')_{\text{SN}}/R_s$ of 12. This relatively high contrast could result in a significant adjacent-bed effect. The bed-thickness effect must be investigated.

Fig. 5.59 gives $R_a''/R_s = 13$ for $R_a'/R_s = 12$ and $h = 20$ ft. Then, $(R_a'')_{\text{SN}} = 2.5(13) = 32.5$ Ω·m.

For the induction reading, Fig. 5.62, which is based on $R_s = 1$ Ω·m, indicates that the adjacent-bed effect is negligible for a 25-ft bed (R_a'' value lies within the shaded area). If the effect is negligible for $R_s = 1$ Ω·m, then it also must be negligible for the case when $R_s = 2.5$, which represents a smaller resistivity contrast. Fig. 5.62 was used for convenience only. Charts for R_s values other than 1 Ω·m are available in service companies' chart books. For the induction log,

$$(R_a'')_{\text{IL}} = (R_a')_{\text{IL}} = (R_a)_{\text{IL}} = 10 \ \Omega\cdot\text{m}.$$

Invasion Effect. Table 5.5 indicates that $d_i \cong 5d_h$ for the estimated porosity range of 10% to 15%. Therefore, $d_i = 5(12) = 60$ in.

For this diameter of invasion, the geometric factors are $G_{\text{SN}} = 0.73$ from Fig. 5.47 and $G_{\text{IL}} = 0.16$ from Fig. 5.63. Note that the geometric factors G_{SN} given by Fig. 5.47 were derived for spherical zones. Their use in this example to calculate R_t and R_{xo} of cylindrical zones is a simplification.

Substituting into Eqs. 5.41 and 5.42, keeping in mind that conductivities should be used for the induction tool, yields

$$32.5 = 0.73 \ R_{xo} + 0.27 \ R_t$$

for the short normal tool and

$$\frac{1}{10.0} = \frac{0.16}{R_{xo}} + \frac{0.84}{R_t}$$

for the Dual Induction-Laterolog.

Solving for R_t and R_{xo} yields two sets of values because a quadratic equation is involved: $R_{xo} = 41.3$ Ω·m and $R_t = 8.7$ Ω·m, or $R_{xo} = 1.7$ Ω·m and $R_t = 115.7$ Ω·m. Only the first set of values is likely.

Determining R_t from one resistivity reading is possible only when the invasion effect is negligible. The invasion effect could be negligible in the case of shallow invasion (i.e., small d_i), such as in high-porosity formations, and when a deep investigation resistivity tool, such as the deep induction, is used.

Consider the case of an 8-in. borehole and a formation with a porosity exceeding 20%. Table 5.5 shows that d_i will be less than 20 in. For this value, Fig. 5.63 shows that the geometric factor of the invaded zone is negligible. In fact, for these tools, d_i must be greater than 50 in. for the geometric factor to exceed 0.1. When optimum conditions prevail, the reading of deep induction tools is free from invasion effect and $R_t = R_a''$.

Sec. 5.6.3 showed that, under optimum conditions of small boreholes filled with high-resistivity mud, the borehole effect is negligible and $R_a' = R_a$. Sec. 5.6.4 showed that $R_a'' = R_a'$ for thick beds.

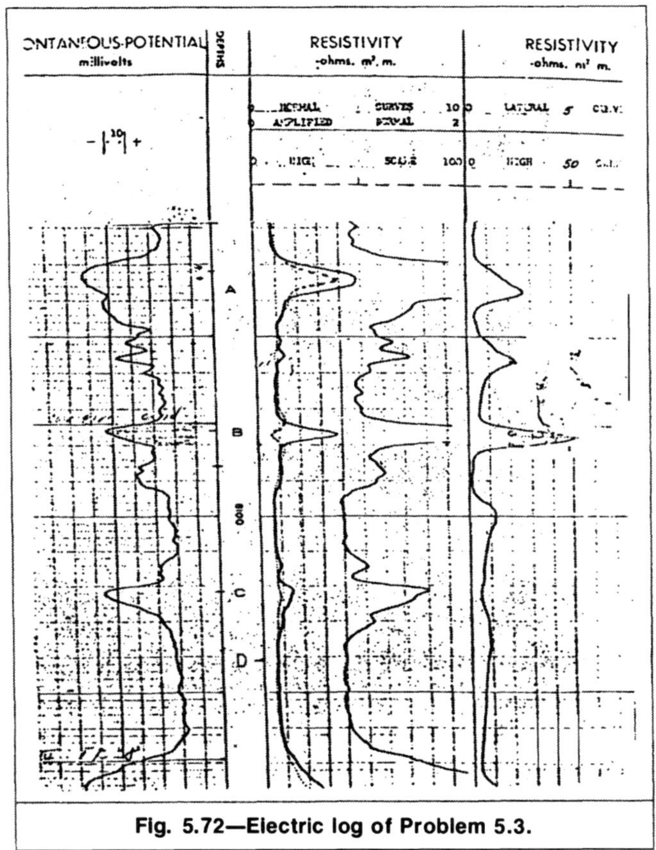

Fig. 5.72—Electric log of Problem 5.3.

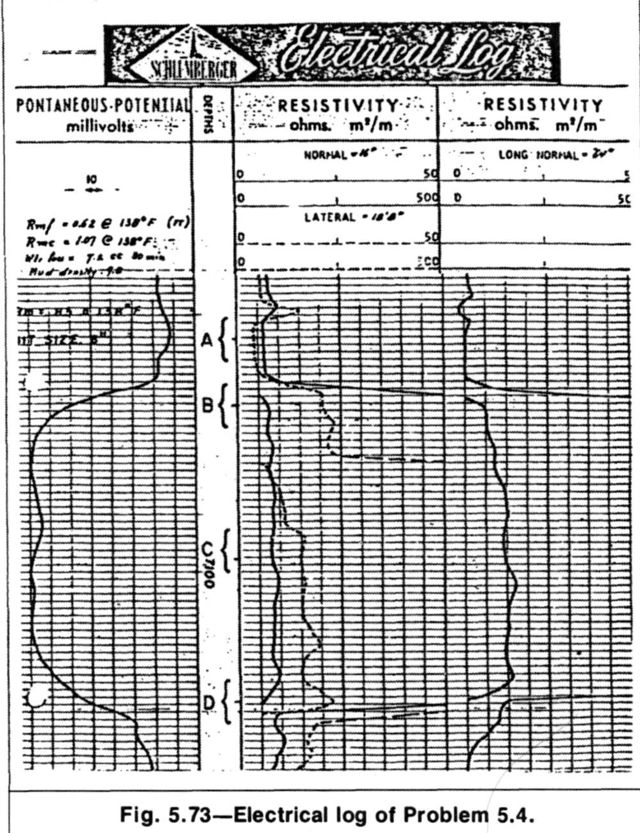

Fig. 5.73—Electrical log of Problem 5.4.

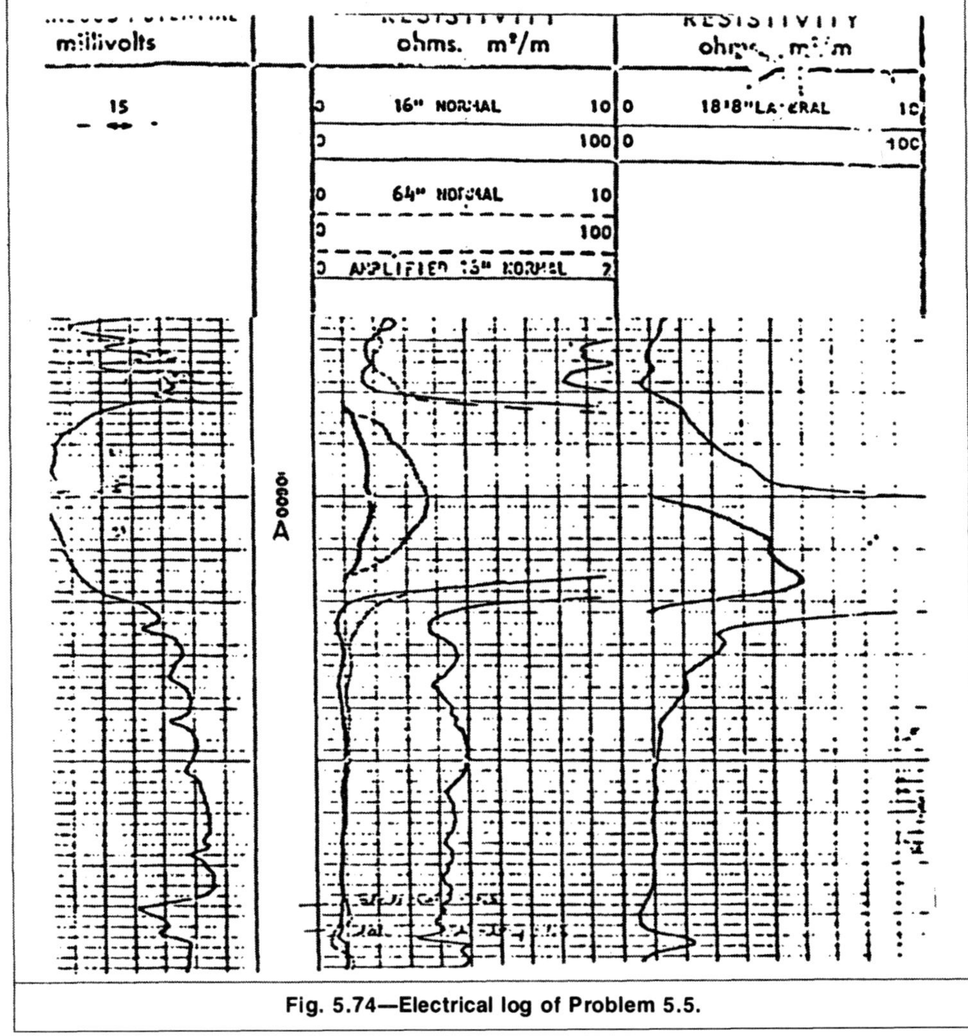

Fig. 5.74—Electrical log of Problem 5.5.

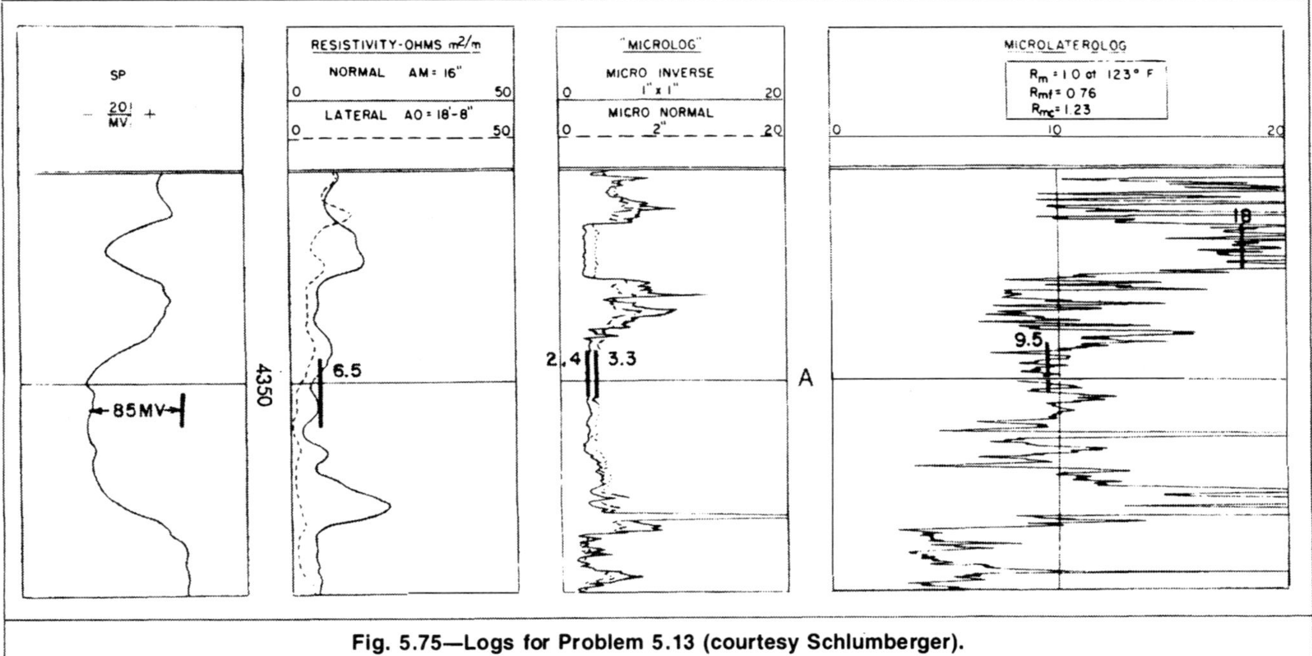

Fig. 5.75—Logs for Problem 5.13 (courtesy Schlumberger).

In high-porosity, thick sands drilled with freshwater-based mud, the apparent resistivities recorded by deep induction tools in small boreholes generally are very close to the true resistivity values. The true resistivity can be read directly from the log. These conditions are often found in the U.S. gulf coast region.

Example 5.14. **Fig. 5.69** schematically shows the response of an electrical survey of a relatively high-porosity formation. If the porosity is expected to be higher than 20%, estimate R_t and R_{xo}. The mud resistivity and borehole diameter are 1.2 Ω·m and 12 in., respectively.

Solution. The induction curve shown on the log reads $(R_a)_{IL} = 1.0$ Ω·m. The log also indicates that $h=25$ ft and $R_s=0.7$

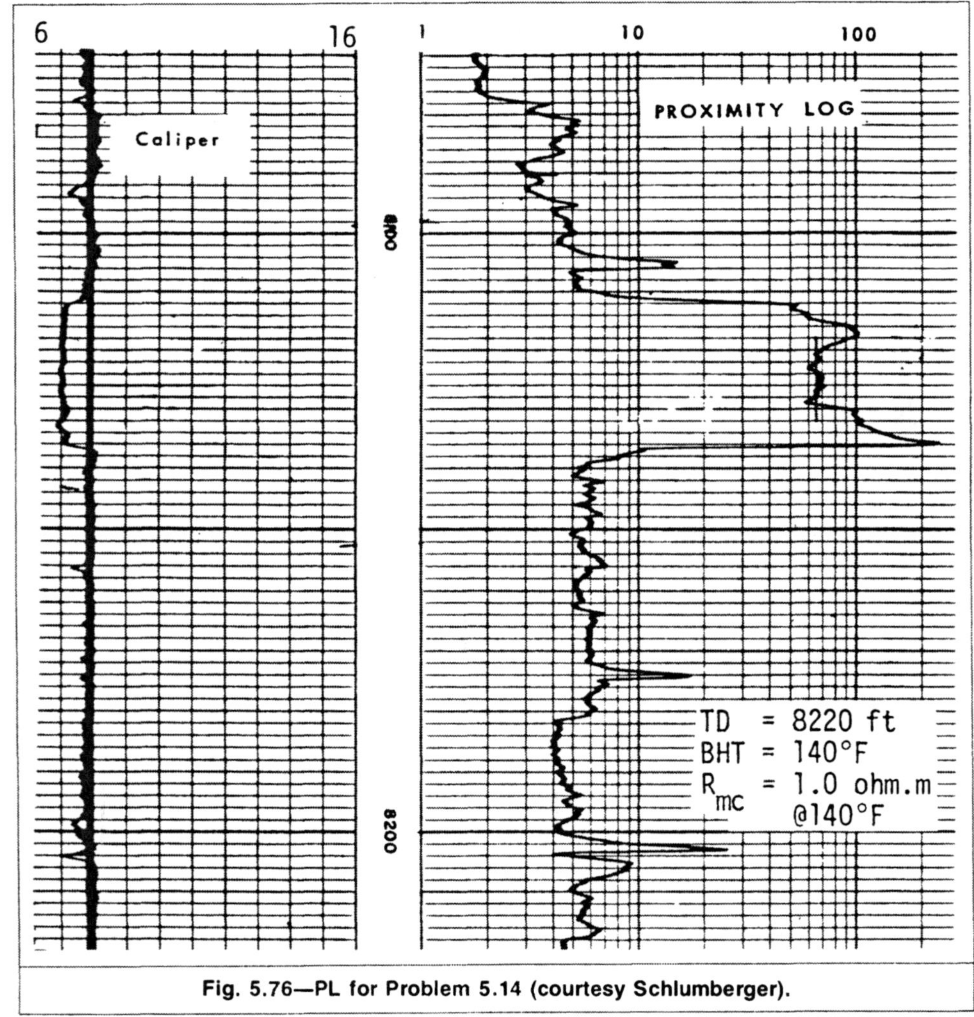

Fig. 5.76—PL for Problem 5.14 (courtesy Schlumberger).

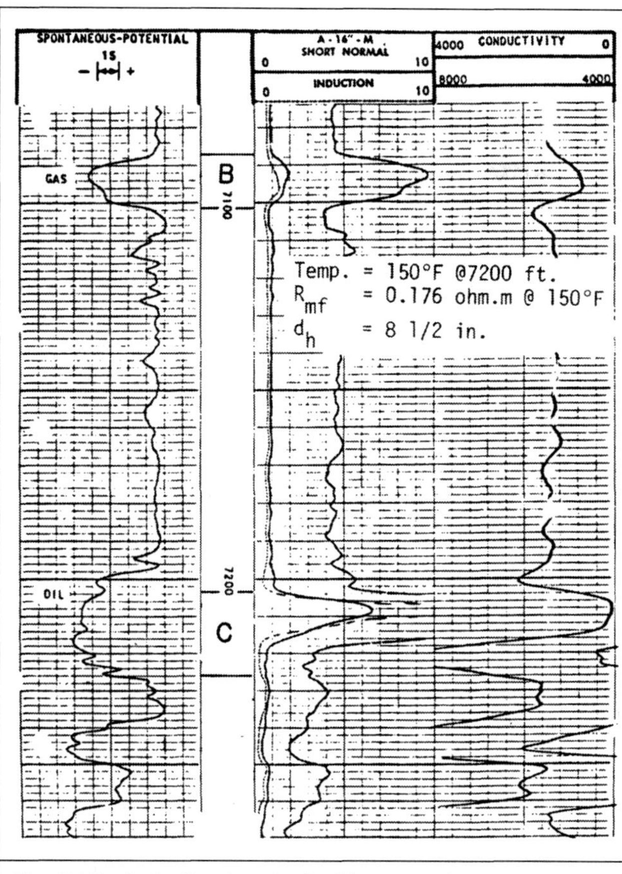

Fig. 5.77—Induction log for Problem 5.15 (courtesy Amoco Oil Co.).

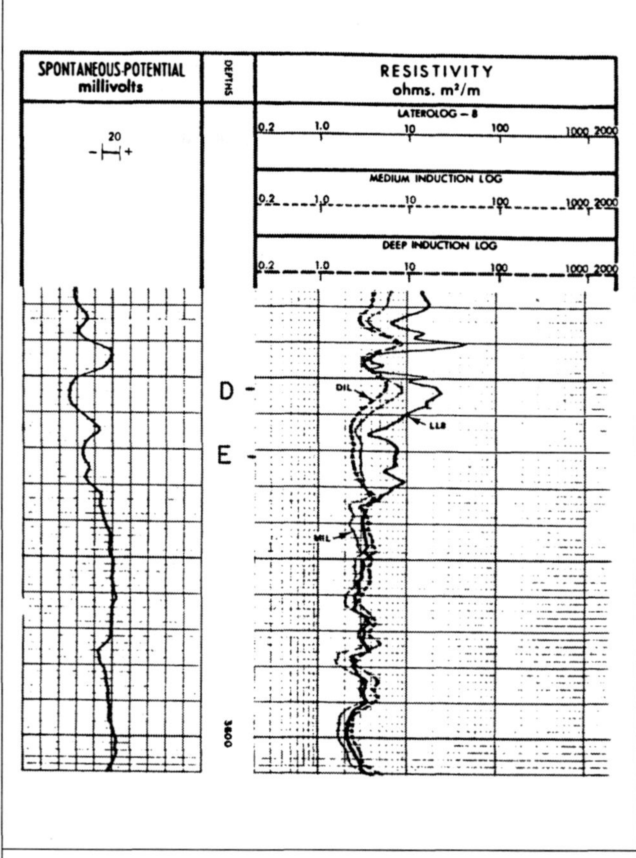

Fig. 5.78—DIL for Problem 5.16 (courtesy Continental Oil Co.).

Ω·m. Previous experience (Example 5.13) indicates that borehole and bed-thickness effects were insignificant in much more severe conditions. They can be neglected in this case, resulting in $(R_a'')_{IL}=R_a=1.0$ Ω·m.

We know that $\phi>20\%$. Then from Table 5.5, $d_i<2.5d_h<30$ in., and from Fig. 5.63, $G_{xo}=0$.

Eq. 5.37 becomes $(R_a'')_{IL}=(0)(R_{xo})+(1)(R_t)=R_t=1.0$ Ω·m.

Because of shallow invasion, $(R_a)_{SN}$ is greatly affected by the uninvaded zone. A reliable R_{xo} value cannot be determined from the furnished data.

5.7 Determination of R_{xo} From Microresistivity Tool Readings

Microresistivity tools can be considered free from adjacent-bed effect because of their small spacings. They are, however, affected by the mudcake and invaded and uninvaded zones:

$$R_a=G_{mc}(h_{mc})R_{mc}+G_{xo}(d_i)R_{xo}+G_t(d_i)R_t. \quad (5.43)$$

When the first microresistivity tools were introduced, the effect of the uninvaded zone was assumed to be negligible. The tool is then affected only by the mudcake zone and the flushed zone, whose geometric factors are G_{mc} and G_{xo}, respectively. Because $G_{mc}+G_{xo}=1$, Eq. 5.43 simplifies to

$$R_a=G_{mc}(h_{mc})R_{mc}+[1-G_{mc}(h_{mc})]R_{xo}. \quad (5.44)$$

Two equations of this form can be written for the Microlog readings, one with the microinverse reading, $(R_a)_1$, and the geometric factor G_1,

$$(R_a)_1=G_1R_{mc}+(1-G_1)R_{xo}, \quad (5.45)$$

and another using the micronormal reading, $(R_a)_2$, and geometric factor G_2,

$$(R_a)_2=G_2R_{mc}+(1-G_2)R_{xo}. \quad (5.46)$$

Eqs. 5.45 and 5.46 can be solved simultaneously for the two unknowns R_{xo} and h_{mc}. **Fig. 5.70** is a graphical solution of this system of equations constructed for the ML tool.

Example 5.15. The ML gives microinverse and micronormal readings of 2.7 and 3.0 Ω·m, respectively. If the borehole diameter is 10 in. and the mudcake resistivity is 0.81 Ω·m at formation temperature, determine the flushed-zone resistivity.

Solution. $(R_a)_1/R_{mc}=2.7/0.81=3.3$ and $(R_a)_2/R_{mc}=3/0.81=3.7$. Entering the chart of Fig. 5.70 with these two ratios gives $R_{xo}/R_{mc}=8$ and $h_{mc}=½$ in., from which $R_{xo}=8(0.81)=6.5$ Ω·m.

A microresistivity tool other than the Microlog carries only one electrode arrangement and displays one resistivity curve. The MLL, PL and MicroSFL readings are first corrected for mudcake effect with the appropriate chart. The resulting R_a'' value is expressed by Eq. 5.37. The determination of R_{xo} from one microresistivity reading alone is possible only when the effect of the uninvaded zone is negligible. This effect can be negligible when deep invasion prevails, as in low-porosity formations.

Consider the case of an 8-in. hole and a formation with a porosity $<20\%$. Table 5.5 indicates that d_i will be greater than 20 in. For this value, Fig. 5.65 shows that the geometric factor of microresistivity tools is equal or very close to unity. In fact, for the MLL, G_{xo} reaches a value of unity once d_i exceeds a mere 6 in. When optimum conditions prevail, the reading of microresistivity tools is free from the uninvaded-zone effect and $R_{xo}=R_a''$. Sec. 5.6.3 showed that, under optimum conditions of thin, resistive mudcakes, the borehole effect is negligible and $R_a''=R_a$. In low- to moderate-porosity formations, the apparent resistivity recorded by the microdevices generally is very close to the flushed-zone resistivity value. R_{xo} can be read directly from the log.

For shallow invasion, R_{xo} can be determined only if R_t and d_i values are available. Eq. 5.37 can then be solved for R_{xo}. $G_{xo}(d_i)$ values are read from Fig. 5.65.

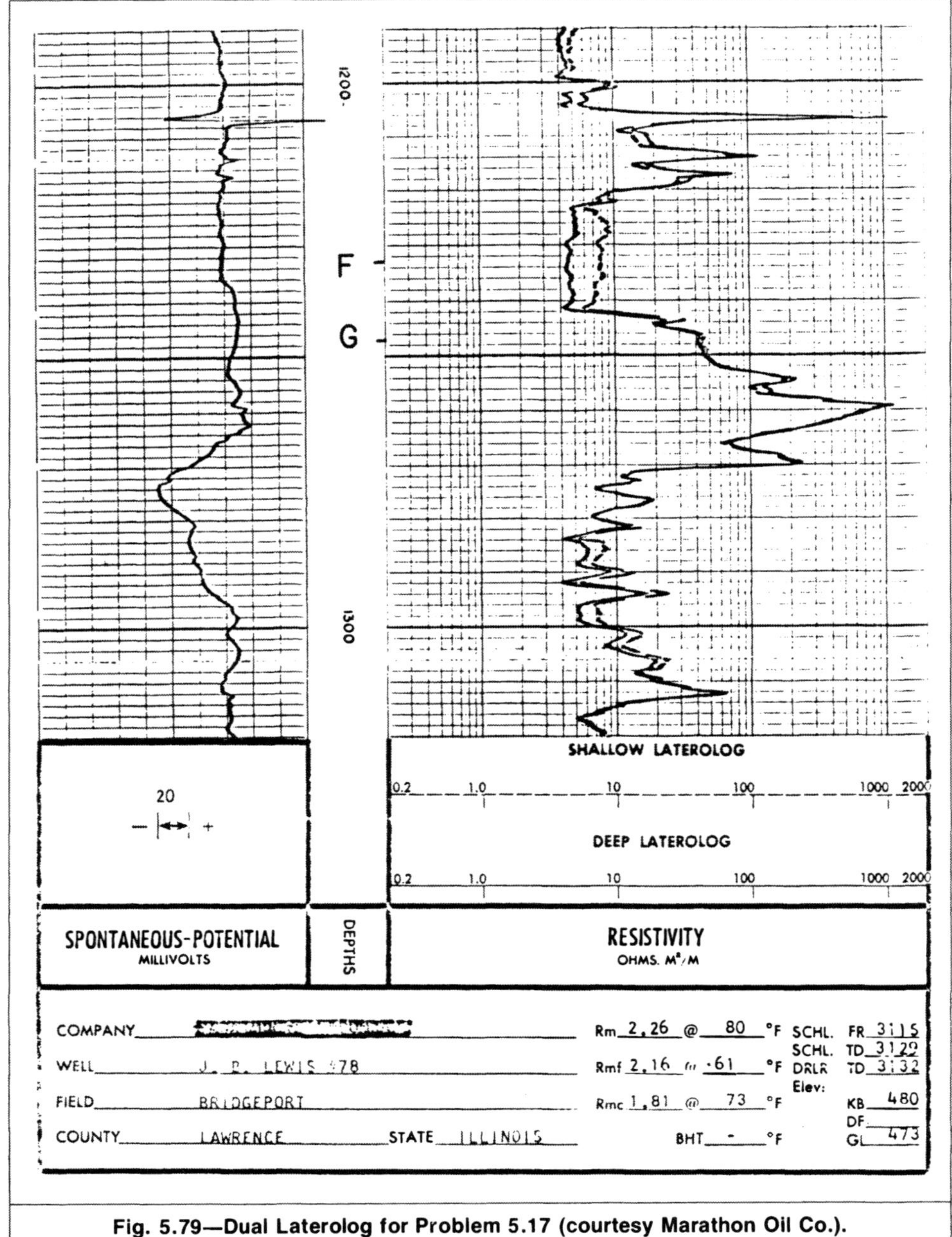

Fig. 5.79—Dual Laterolog for Problem 5.17 (courtesy Marathon Oil Co.).

Example 5.16. Fig. 5.71 shows a section of an MLL/ML run. Determine the resistivity of the flushed zone of Interval A.

Solution. The following data are obtained for the log.

$(R_a)_{MLL}=21\ \Omega\cdot m$ from the MLL curve.

$(R_a)_1=2.7\ \Omega\cdot m$ from the ML curve.

$(R_a)_2=4.4\ \Omega\cdot m$ from the ML curve.

Bit size $=7\frac{7}{8}$ in.

Hole size $=7\frac{3}{8}$ in. from the caliper curve.

Depth interval $=5{,}248$ to $5{,}250$ ft.

Temperature at $5{,}281$ ft $=132°F$.

$R_{mc}=1.66\ \Omega\cdot m$ at 74°F.

$R_{mf}=0.83\ \Omega\cdot m$ at 74°F.

R_{mc} and R_{mf} at Formation Temperature. The interval of interest is only a few feet above the bottom of the borehole. The temperature at this level is very close to the maximum recorded temperature of 132°F. R_{mc} and R_{mf} can be calculated from Eq. 1.11:

$$R_{mf} \text{ at } 132°F=0.83\frac{74+6.77}{132+6.77}=0.483\ \Omega\cdot m.$$

$$R_{mc} \text{ at } 132°F=1.66\frac{74+6.77}{132+6.77}=0.966\ \Omega\cdot m.$$

R_{xo} and h_{mc} from the ML.

$$\frac{(R_a)_1}{R_{mc}}=\frac{2.7}{0.966}=2.80$$

$$\frac{(R_a)_2}{R_{mc}}=\frac{4.4}{0.966}=4.55.$$

Entering the chart of Fig. 5.70 with the above two values, 2.80 and 4.55, yields $R_{xo}/R_{mc}=40$ and $R_{xo}=40\ (0.966)=38.6\ \Omega\cdot m$. R_{xo}/R_{mc} is greater than 15, which is the limit above which Fig. 5.70 yields erroneous results.[10] The value of $R_{xo}=38.6$ has to be disregarded.

Borehole Correction of the MLL Reading. The mudcake thickness, h_{mc}, can be approximated from the caliper reading:

$$h_{mc}=\tfrac{1}{2}(\text{bit size}-\text{hole size})$$

$$=\tfrac{1}{2}(7\tfrac{7}{8}-7\tfrac{3}{8})=\tfrac{1}{4} \text{ in.}$$

Entering the chart of Fig. 5.57 with h_{mc} and $(R_a)_{MLL}/R_{mc}=21/0.966=21.7$ gives $(R_a')_{MLL}/R_{mc}=22.5$, from which

$(R_a')_{MLL}=21.7\ \Omega\cdot m$. Because the MLL reading is free from adjacent bed effects,

$$(R_a'')_{MLL}=(R_a')_{MLL}=21.7\ \Omega\cdot m.$$

Order of Magnitude of d_i. The drilling fluid in the borehole is freshwater-based mud. Therefore, $R_{mf}>R_w$ and, most likely, $R_{xo}>R_t$.

If the invasion is such that the $(R_a'')_{MLL}$ value is affected by the uninvaded zone, then $R_{xo}>(R_a'')_{MLL}$. F can be expressed as

$$F=\frac{R_{xo}}{R_{mf}}>\frac{(R_a'')_{MLL}}{R_{mf}}.$$

Incorporating this value into Eq. 1.16 yields

$$\phi^{2.15}=\frac{0.62}{F}<\frac{0.62\ R_{mf}}{(R_a'')_{MLL}}<\frac{0.62(0.483)}{21.7}.$$

$\phi<14\%$.

According to Table 5.5,

$$d_i>5d_h$$

$$>5(7\tfrac{7}{8})>39\text{ in.}$$

R_{xo} From the MLL Reading.

The geometric factor G_{xo} for $d_i>39$ in. is unity. The MLL reading is free from the R_t zone effect; hence,

$$R_{xo}=(R_a'')_{MLL}=21.7\ \Omega\cdot m.$$

Note that R_{xo} differs from $(R_a)_{MLL}$ by only 0.7 Ω·m, or about 3%. Thus, when such a measuring environment is encountered, it would be practical to read R_{xo} directly off the log.

5.8 Summary

Resistivity devices display apparent resistivity values. The apparent resistivity, R_a, measured at a depth of interest is affected by the resistivity and geometry of four zones that surround the tool: the borehole, adjacent beds, and the invaded and uninvaded zones of the bed of interest. The apparent resistivity value should be corrected for the borehole and adjacent-bed (also called bed-thickness) effects. The corrected value, R_a'', bears the influence of the invaded and uninvaded zones. Depending on the type and number of resistivity logs available, R_a'' is used to calculate R_t, R_{xo}, or both. Departure curves (charts) are usually used to perform these corrections and calculations.

Determining R_t and R_{xo} requires three steps.

1. The apparent resistivity, R_a, is first corrected for borehole effects by use of R_m and d_h. Correction of readings of microresistivity devices requires R_{mc} and h_{mc} instead. For each resistivity tool, optimum measurement conditions exists in which the borehole effect is nil or negligible.
2. The value obtained from Step 1, R_a', is then corrected for bed-thickness effect with the resistivity of the adjacent bed, R_s, and the thickness of the bed in question. No adjacent-bed corrections are necessary when the bed thickness exceeds a certain value. This value depends on the tool's vertical resolution and the resistivity contrast R_t/R_s. Readings of microresistivity tools are free from this effect.
3. The last step is to use the value R_a'' obtained in Step 2 to calculate R_t and R_{xo}. The calculation calls on the geometric factors. Departure curves (tornado charts) or a system of equations is used. If the diameter of invasion is small, the effect on invasion is negligible and $R_t=R_a''$. On the other hand, for microresistivity devices, when invasion is deep, the effect of the uninvaded zone becomes negligible and $R_{xo}=R_a''$.

In certain measurement environments, the three effects (borehole, bed thickness, and invasion) are negligible. In these cases, R_t or R_{xo} can be read directly from the log.

In some instances, the data available lead to unrealistic results, such as negative values, or the data fall off the tornado chart. In most instances, such an occurrence is the result of values improperly read or improperly corrected for borehole or bed-thickness effects. However, in cases of very deep invasion with an invasion profile that cannot be approximated by a step profile, such a case may result.[27,28] Treatment of other invasion profiles, such as the transition and annulus profiles, requires extensive data; four resistivity readings are usually needed.

Review Questions

1. What is the resistivity value calculated from a resistivity tool response called? How is it defined?
2. What is the geometric coefficient of an electrode array?
3. Does the apparent resistivity recorded by a resistivity tool equal the true formation resistivity? Explain.
4. Describe the spacing and inscription point of a logging tool.
5. What are the similarities and differences between the normal and lateral resistivity devices?
6. What is the relationship between a tool spacing and its depth of investigation?
7. Describe the response of a normal curve in a thick, uninvaded, resistive bed.
8. Describe the response of a normal curve in a thin ($h<AM$), uninvaded, resistive bed.
9. Describe the response of a lateral curve recorded opposite a thick, uninvaded, resistive bed.
10. Under what conditions does a blind zone appear on the response of the lateral device?
11. How thick should a resistive bed be to result in a plateau response on the lateral curve?
12. What curves are usually displayed on a conventional electrical survey log? What grid and scales are commonly used in their recording?
13. Why does the recording of electrode-type devices require the presence of a conductive drilling fluid in the borehole? What happens to the quality of the recording as the drilling-fluid conductivity increases?
14. Why are the ML electrodes placed on a pad pressed against the borehole wall?
15. What are the curves recorded by the ML?
16. What is meant by a positive separation of the ML resistivity curves? Why is such a separation a strong indication of formation permeability?
17. In a porous and permeable formation, what is the range of resistivities, expressed in terms of mud resistivity, displayed by the ML?
18. How does the focused-current device overcome the limitations of conventional electrode devices?
19. Why are different focused-current tools having different current configurations needed for optimum interpretation?
20. What is the major advantage of the focused microresistivity devices over the ML?
21. How does the MLL differ from the PL?
22. What is the major advantage of the MicroSFL over the other microdevices?
23. What is the concept of the induction tool?
24. What is the "skin effect?"
25. Name the different zones of the medium surrounding a resistivity tool that contribute to the tool's response.
26. The contribution of each zone around the tool can be represented by a mathematical product. What is this product?
27. What does the geometrical factor G for a given zone indicate?
28. How is the borehole usually represented in the mathematical models used in resistivity calculations? Which position does the tool assume in the borehole?
29. What is the invasion profile considered in the construction of tornado charts? Why was this profile selected?
30. Can R_t of a permeable bed be determined when only one resistivity log is available?
31. Can R_{xo} of a permeable bed be determined when only one resistivity log is available?
32. How many resistivity curves are usually needed for true resistivity determination in low-porosity formations? Explain.
33. Can the true resistivity of shale beds, R_{sh}, be read directly off the short normal? Explain.

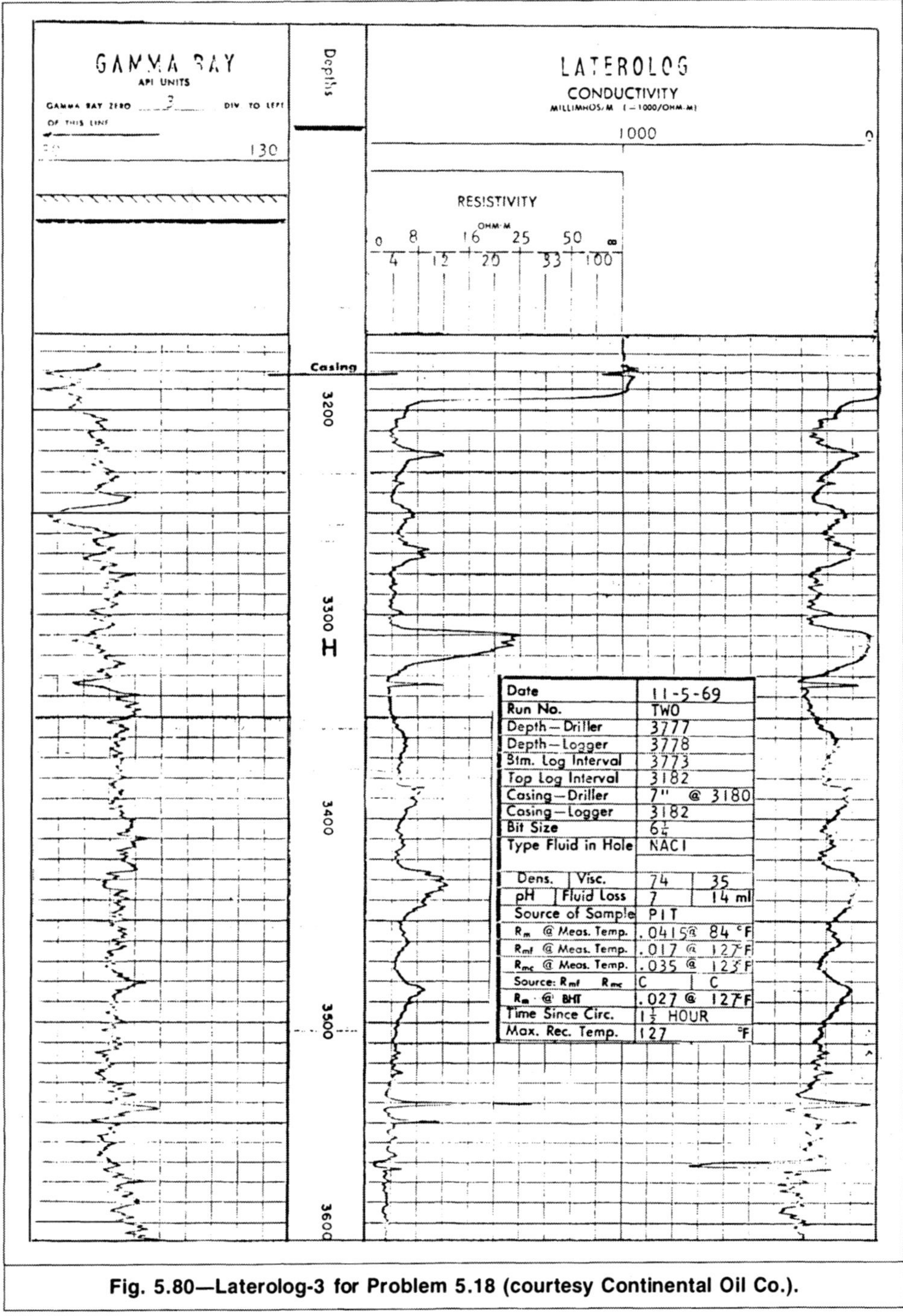

Fig. 5.80—Laterolog-3 for Problem 5.18 (courtesy Continental Oil Co.).

Problems

5.1 For $r_2 = 1{,}000\ r_1$, what error is introduced in the calculated value of the resistance by neglecting the term $1/r_2$ in Eq. 5.2?

5.2 Using the results of Example 5.3, approximate the response of a 16-in. short normal and a 64-in. long normal in a 90-$\Omega\cdot$m, 1,600-in.-thick bed bounded by two 10-$\Omega\cdot$m, very thick formations.

5.3. Examine the electric log of **Fig. 5.72**; then:

a. On the log, circle and label log responses caused by geometric distortion.

b. Mark the upper and lower boundaries of permeable Sands A through C marked on the log. What is the gross thickness of each sand?

c. Give the apparent resistivities of Zone A indicated by the short and long normal tools. Explain why the two tools indicate different values.

d. Explain why the short and long normal tools read the same in Zone D, which is an impermeable shale. What is this reading?

e. Give your best evaluation of the apparent resistivities of Zone B as indicated by the short normal and the lateral. If the drilling fluid is freshwater-based mud, explain why $R_{18'18''} > R_{16''}$.

5.4. Refer to the electrical log of **Fig. 5.73** in providing information on the following.

a. The log shows a thick sand. Give your best estimate of the formation thickness.

b. Discuss the responses of the lateral curve at Levels A through D marked on the log.

c. Explain why the lateral and long normal tools display the same resistivity reading at Level C.

The log headings indicate that:
Bit size = 8 in.
Mud density = 9.8 lbm/gal.
Temperature at 7,100 ft = 138°F.
R_m at 138°F = 0.85 $\Omega\cdot$m.
R_{mf} at 138°F = 0.62 $\Omega\cdot$m.
R_{mc} at 138°F = 1.07 $\Omega\cdot$m.

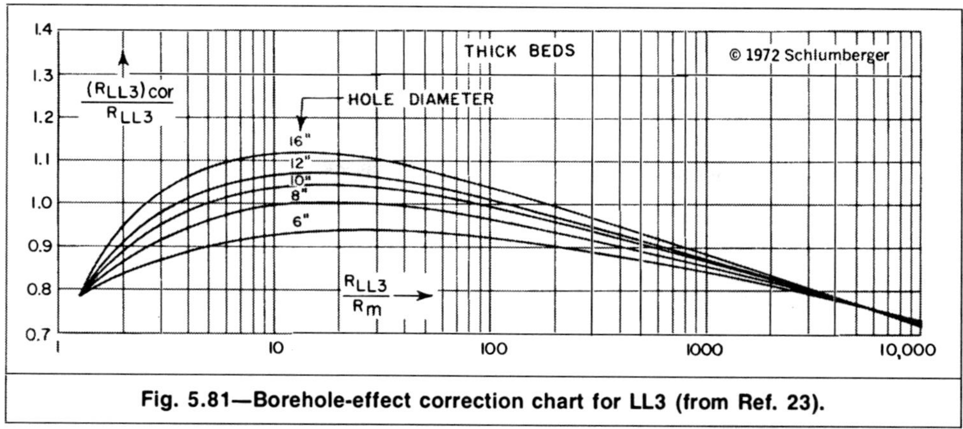

Fig. 5.81—Borehole-effect correction chart for LL3 (from Ref. 23).

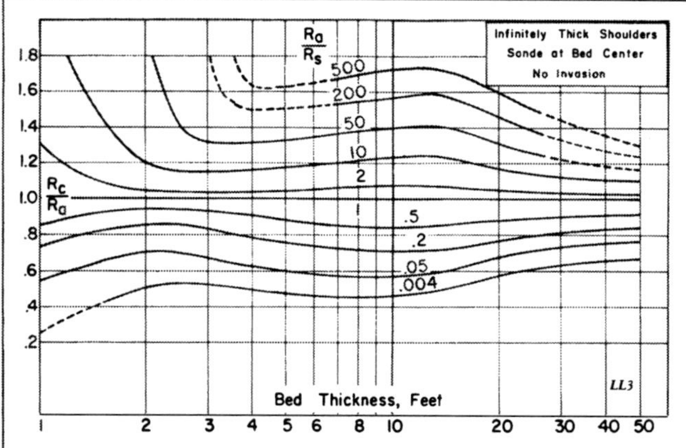

Fig. 5.82—Bed-thickness effect correction chart for LL3 (from Ref. 15).

5.5. Give your best estimate of the true resistivity of Sand A in **Fig. 5.74.**

5.6. Consider a current source located at Point A in a certain medium. The current-return Electrode B is so far from the electrode at Point A that its presence may be neglected when considering the current flow around Point A.

Suppose that Point A is the center of a zone of resistivity $R_1=50\ \Omega\cdot m$, which extends spherically to a radius of 32 in. A second zone of resistivity $R_2=10\ \Omega\cdot m$ extends spherically beyond the first zone to infinity.

a. What voltage is measured between an Electrode M and a distant Electrode N for 0.5-A current and electrode spacing AM=16 in.?
b. Using Eq. 5.5, calculate the apparent resistivity of the medium.
c. Using the geometric factors shown by Fig. 5.47, calculate the apparent resistivity of the medium.
d. Repeat Steps a through c for a 64-in. electrode spacing.
e. Do the results of Step d explain why longer spacing is less affected by the invaded zone?

5.7. A section of an 8-in. borehole is filled with saltwater-based mud of 0.1-$\Omega\cdot m$ resistivity. If a maximum of 5% borehole effect can be considered negligible, indicate the range of formation resistivity within which the readings of the LLd and LLs can be considered free from borehole effect.

5.8. A deep induction log is used to log a section of a 16-in. borehole where the mud resistivity is 0.2 $\Omega\cdot m$. If a maximum hole signal of 10 m℧/m can be tolerated, what minimum standoff size is recommended?

5.9. A short normal resistivity log has been run in a borehole that has the following characteristics.

Bit size=9⅞ in.
Mud density=9.8 lbm/gal.
$R_m=1.47\ \Omega\cdot m$ at 77°F.
Bottomhole temperature=156°F.
Bottomhole depth=10,463 ft.

The short normal readings for the 8,048- to 8236-ft interval are listed below.

Interval (ft)	Lithology	Short normal Reading (Ω·m)
8,048 to 8,079	Shale	2
8,079 to 8,111	Sand	55
8,111 to 8,195	Shale	2
8,195 to 8,200	Sand	27
8,200 to 8,236	Shale	2

Determine the short normal resistivity value corrected for borehole diameter and bed-thickness effects.

5.10 Opposite a 10-ft bed, the LLd reads a center bed value of 50 $\Omega\cdot m$. The adjacent beds have a resistivity of 1.0 $\Omega\cdot m$. What is the resistivity value corrected for adjacent bed effect? What is the corrected value if the bed is only 5 ft thick? Assume negligible borehole effect.

5.11 The R_t and R_{xo} values of a very thick, permeable bed are 10 and 40 $\Omega\cdot m$, respectively. Assume that the borehole effect can be neglected and that $d_i=20$ in.

a. Determine the reading of the SFL, the LLs, and the LLd.
b. Repeat Part a for $d_i=80$ in.

5.12 In a very thick, permeable zone, $d_i=30$ in., $R_t=2.0\ \Omega\cdot m$, and $R_{xo}=20.0\ \Omega\cdot m$. Ignoring borehole signal, compute the theoretical responses for ILm and ILd induction curves.

5.13 Consider Zone A in the composite log of **Fig. 5.75.**

a. Calculate R_{xo} of Zone A, knowing that R_m, R_{mf}, and R_{mc} are 1.0, 0.76, and 1.23 $\Omega\cdot m$, respectively, at formation temperature (123°F).
b. Using the R_{xo} value from Part a, estimate the porosity of Zone A. Assume the zone to be 100% water-saturated.
c. For the same Zone A, the MLL reads an average value of 9.5 $\Omega\cdot m$. Calculate R_{xo}.
d. Why are the R_{xo} values that are calculated in Parts a and c different? Give your best estimate of R_{xo}.
e. Why is the R_{xo} value obtained above greater than the reading of the 16-in. short normal?

5.14 **Fig. 5.76** shows a PL. The zone of interest is from 8,120 to 8,128 ft.

a. Give h_{mc} and $(R_a)_{PL}$ of the zone of interest.
b. Determine $(R_a'')_{PL}$.
c. If $R_t=30\ \Omega\cdot m$ and $d_i=5$ in., estimate R_{xo}.
d. If $R_t=30\ \Omega\cdot m$ and $d_i=15$ in., estimate R_{xo}.

5.15 The IES of **Fig. 5.77** shows a gas Formation B and an oil Formation C.

a. If the change in resistivity within Formation C is from a change in fluid type, determine the apparent resistivity in the water and oil zones.
b. Is the borehole correction negligible?
c. What is the resistivity of adjacent shales?
d. What is the thickness of the oil zone?
e. Use Fig. 5.62 to correct the resistivity values obtained in Part a for bed-thickness effect.

5.16 a. Using the chart of Fig. 5.66, determine d_i, R_{xo}, and R_t of Zones D and E on the DIL of **Fig. 5.78**. Borehole and adjacent-bed effects can be considered negligible.
b. Sketch the radial resistivity profile for Zones D and E.

5.17 **Fig. 5.79** illustrates a Dual Laterolog obtained in a 7⅞-in. borehole drilled through carbonate formations.
a. What are the apparent resistivities at Level G indicated by the two resistivity curves?
b. Determine d_i, R_{xo}, and R_t at Level G.
c. What are the apparent resistivities at Level F indicated by the LLd and LLs?
d. Explain the difference between the two readings in Part c.
e. Show that the borehole and adjacent-bed corrections are so small that they can be neglected.
f. Assuming that $5\% < \phi < 10\%$, estimate R_{xo} and R_t of Zone F.
g. Assuming that $10\% < \phi < 15\%$, estimate d_i, R_{xo}, and R_t of Zone F.

5.18 The LL3 in **Fig. 5.80** shows a resistive Formation H.
a. What is the apparent resistivity of Zone H?
b. Correct the above value for borehole effect with **Fig. 5.81**.
c. What is the average resistivity of beds adjacent to Formation H?
d. Using the chart of **Fig. 5.82**, determine the R_a'' value for Zone H.
e. Calculate R_{xo} for Zone H using the following data: core samples indicated a residual oil saturation of 25%; a porosity of 12% was calculated from the density log; and the formation water resistivity obtained from water samples is 0.05 Ω·m at 75°F.
f. Estimate the geometric factor of the flushed zone of Formation H.
g. Determine the true resistivity of Formation H.
h. Calculate the oil saturation.
i. Calculate the oil saturation using the uncorrected apparent resistivity value indicated by the log.

Nomenclature

C = conductivity, m℧/m
C_a' = borehole-effect-free apparent conductivity, m℧/m
C_R = coefficient of reflection
C_s = adjacent-bed conductivity, m℧/m
C_t = conductivity of uninvaded zone, m℧/m
d = diameter, in.
d_i = invasion diameter, in.
G = differential geometric factor
G_I = integrated geometric factor
G_L = geometric coefficient of lateral tool
G_m = mud-zone geometric factor
G_N = geometric factor of normal sonde
G_s = adjacent-bed geometric factor
G_t = geometric factor of uninvaded zone
G_T = tool geometric factor
G_{xo} = geometric factor of flushed zone
h = thickness, ft
I = current, A
L_s = device spacing, in.
r = radius, ft or m
R = resistivity, Ω·m
R_a = apparent resistivity, Ω·m
R_a' = borehole-effect-free apparent resistivity, Ω·m
R_a'' = bed-thickness-effect-free apparent resistivity, Ω·m
R_{AO} = apparent resistivity displayed by lateral curve at a level situated a distance $\overline{AO}$ below top boundary Ω·m
R_{mc} = mudcake resistivity, Ω·m
R_s = resistivity of adjacent formation, Ω·m
R_t = true formation resistivity, Ω·m
R_{xo} = invaded-zone resistivity, Ω·m
S = saturation
V = voltage, V
ΔV = potential difference, V
z_A = vertical coordinate of current Electrode A
z_O = depth of tool Inscription Point O
ρ = resistance, Ω
ϕ = porosity

Subscripts

a = apparent
h = borehole
i = invasion, invaded zone
IL = induction log tool
m = mud
max = maximum
mc = mudcake
mf = mud filtrate
min = minimum
SN = short normal tool
xo = flushed zone

References

1. Allaud, L. and Martin, M.: *Schlumberger, The History of a Technique*, John Wiley & Sons Inc., New York City (1977).
2. Dakhnov, V.N.: "Geophysical Well Logging," *Quarterly of the Colorado School of Mines* (April 1962) **57**, No. 2, Chap. 3.
3. "Interpretation Hand-Book for Resistivity Logs," Schlumberger Well Surveying Corp., Document No. 4 (July 1950).
4. Frank, R.W.: "Prospecting with Old E-Logs," Schlumberger Educational Services, Houston (1986).
5. McCray, D.L.: "An Analysis of Electrical Logging Devices: Their Advantages and Limitations," Perforating Guns Atlas Corp. publication (1957).
6. Guyod, H.: *Guyod's Electrical Well Logging*, Welex, Houston (1944) Part 5.
7. Pirson, S.J.: "Factors Affecting Measured Apparent Formation Resistivities," *Oil & Gas J.* (Nov. 22, 1947) 63-87.
8. Pirson, S.J.: "The Resistivity Curves," *Oil & Gas J.* (Oct. 25, 1947) 92-115.
9. *Interpretation Hand-Book for Resistivity Logs*, Schlumberger Well Surveying Corp., Houston (July 1950).
10. *Log Interpretation/Application*, Schlumberger, Houston (1974).
11. *Log Interpretation Charts*, Chart B-8, Schlumberger, Houston (1958).
12. *Resistivity Departure Curves*, Schlumberger, Houston (1949).
13. Doll, H.G.: "The Micro Log—A New Electrical Logging Method for Detailed Determination of Permeable Beds," *Trans.*, AIME (1950) **189**, 129-42.
14. Hammack, G.W.: *Laterolog*, technical bulletin, Dresser Atlas, Houston (1971).
15. Log Interpretation/Principles, Schlumberger, Houston (1972).
16. Suau, J. *et al.*: "The Dual Laterolog R_{xo} Tool," paper SPE 4018 presented at the 1972 SPE Annual Meeting, San Antonio, Oct. 8-11.
17. Doll, H.G.: "The Microlaterolog," *Trans.*, AIME (1953) **198**, 17-32.
18. Doll, H.G.: "Introduction to Induction Logging and Application to Logging of Wells Drilled with Oil Base Mud," *Trans.*, AIME (1949) **186**, 148-62.
19. Ellis, D.V.: *Well Logging for Earth Scientists*, Elsevier, New York City (1987).
20. *Log Review 1*, Dresser Atlas, Houston (1971).
21. *Phasor Induction Tool*, Schlumberger, Houston (1986).
22. *Electromagnetic Propagation Tool*, Schlumberger, Houston (1984).
23. *Log Interpretation Charts*, Schlumberger Well Surveying Corp., Houston (1972).
24. *Formation Evaluation Data Handbook*, Go Wireline Services (1974).
25. *Log Interpretation Charts*, Dresser Atlas, Houston (1983).
26. Jensen, J.L., and Gartner, M.L.: "Tornado Chart Sensitivity Analysis," *Trans.*, SPWLA (1983) paper N.
27. Doll, H.G., Dumanoir, J.L., and Martin, M.: "Suggestions for Better Electric Log Combinations and Improved Interpretations," *Geophysics* (Aug. 1960) 854-82.
28. Tixier, M.P. *et al.*: "Determination of Formation Resistivity and Water Saturation With the Dual Induction-Laterolog—A New Technique," paper SPE 713 presented at the 1963 SPE Annual Meeting, New Orleans, Oct. 6-9.

Chapter 6
The Spontaneous Potential Log

6.1 Naturally Occurring Electrical Potentials

Naturally occurring electrical potentials are observed at the earth's surface and in its subsurface. These potentials, usually called spontaneous potentials, have been used in mineral exploration. They are associated with weathering of mineral bodies, variation of rock properties at geologic contacts, bioelectric activity of organic material, thermal and pressure gradients in underground fluids, and other phenomena.[1] During early resistivity measurements at Pechelbronn, France, a potential was observed between Electrodes MN of Fig. 5.4 when no current was emitted.[2] The value measured, which is a potential gradient, varied with depth, as shown by the gradient curve in **Fig. 6.1.** This curve was integrated and resulted in the potential curve, which represents the voltage that would have been measured between an electrode in the borehole and a surface electrode. This potential originates from the contact of the drilling fluid filling the borehole and the formation. Although not actually a naturally occurring potential, it is also called a spontaneous potential, self-potential, or simply SP because it originates without an artificial source of current.

Despite the complex geology at Pechelbronn, a bed of conglomerate generated an SP response distinct from surrounding formations, as shown clearly in Fig. 6.1. This early encouraging result prompted the simultaneous recording of the SP curve and resistivity logs. **Fig. 6.2** is a schematic of the circuit used to measure the SP. The main circuit components are a downhole mobil electrode, a fixed surface electrode, a voltmeter, and a bucking circuit composed of batteries and a variable resistor. The absolute voltage measured between the two electrodes could be several hundred millivolts. This voltage consists of two components: a major component reflecting the naturally occurring potentials associated with the major geologic subsurface structure around the wellbore, and a minor component reflecting geologic changes in the immediate wellbore vicinity. The latter is the component of interest in well logging, so the major component is offset with the bucking circuit, which introduces a voltage of similar magnitude but different polarity.

6.2 The SP Log

Fig. 6.3 is a schematic of an SP curve recorded in a shale/sand formation sequence with the circuit of Fig. 6.2. Such a curve, run simultaneously with other logs in almost all boreholes, is traced on the first track on a linear scale. Because a bucking circuit is used, the SP measurement emphasizes the variation of potential rather than its absolute value. The sensitivity of the linear scale is chosen so that the variations remain on the track and can be read easily. The sensitivity is usually 10 or 20 mV per scale division.

Correlation of the SP log with stratigraphic data indicated that the log response in shale formations fall on a straight line called the "shale baseline." Tops and bottoms of the sands are marked by sharp negative deflections of several tens of millivolts. The inflection points on these deflections correspond to bed boundaries.

The SP log can be used to distinguish between impermeable shales and permeable and porous sands. The boundaries of each permeable bed can be defined and its thickness calculated. In thick beds, the deflection reaches a maximum and stabilizes as a plateau. The response in thin beds falls short of the plateau and exhibits a round shape. Maximum SP deflection is also affected by shaliness. A shale-content index can be calculated by comparing the deflection in the shaly sand to that of a clean sand. The shape of the SP deflection is also used to correlate between wells and in sedimentological studies.[3]

Example 6.1. Fig. 6.4 shows an electric log obtained in the 8,900- to 9,040-ft interval of a well drilled through sands and shales.

1. Determine the SP deflection displayed by the four sands marked on the log by the letters A through D.
2. The SP deflection indicates permeable and porous beds. Do other data displayed by the log corroborate the SP indication?
3. What is the gross thickness of each bed?
4. What is the net thickness of Bed A?
5. Calculate the shale-content index of Zone D.

Solution.

1. The shale baseline is first traced through shale beds. SP deflections off this line are −98, −74, −82, and −30 mV for Beds A through D, respectively.

2. Yes. The separation between the short normal and deep induction displayed by Beds A through C indicates invasion, which is associated with porous and permeable formations. Note the strong correlation between the separation of the resistivity log and the SP deflection. The lack of clear separation in Bed D is the result of shaliness.

Although the separation is used in this case to corroborate the presence of porosity and permeability, its absence does not necessarily indicate the opposite.

3. From the inflection points on the SP curve and the separation between the resistivity curves, bed boundaries are traced on the log. Gross thicknesses of Beds A through D are 48, 4, 8, and 6 ft, respectively.

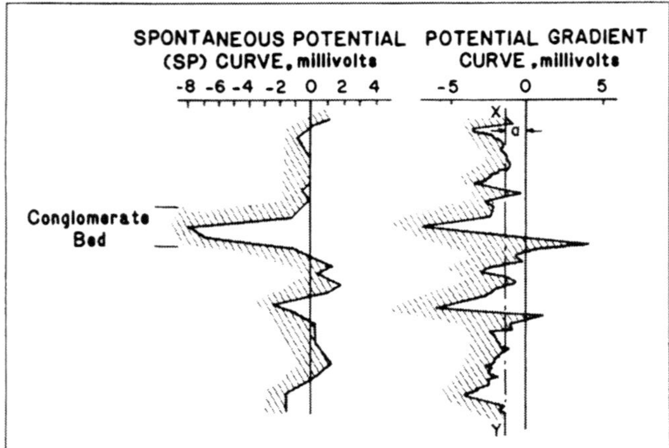

Fig. 6.1—The first SP measurement obtained in Pechelbronn, France (after Ref. 2).

4. The gross thickness of Bed A, which corresponds to the distance between the top and the bottom of the sand unit, was determined to be 48 ft. Sands are usually interbedded with impermeable shale streaks. These streaks are usually indicated by depression in the SP deflection. The total thickness of these streaks is subtracted from the gross thickness to obtain the net thickness or the effective thickness of the permeable and porous zone.

Two depressions of the SP deflection appear within Bed A at Levels X and Y. The severity of the depression and the disappearance of the separation between the two resistivity curves at Level Y indicate a shale streak. For opposite reasoning, Interval X is slightly shaly but is invaded, indicating permeability. Only the thickness of Zone Y (3 ft) is subtracted from the gross thickness. The net thickness of Sand A is then 45 ft.

5. The SP deflection in Zone D is only −30 mV, compared with −98 mV full deflection in clean Sand A. The shale-content index of Sand D is (98−30)/98=0.69 (69%).

The SP deflection is partially decreased as a result of the bed-thickness effect (see Sec. 6.4). The 69% shale-content index is an overestimate.

6.3 Origin of the SP

The SP observed in boreholes is of electrokinetic and electrochemical origin. Electrochemical potential, believed to be the major contributor to the SP, consists of two components: the diffusion potential and the membrane potential.

6.3.1 Diffusion Potential. Diffusion potential, also known as liquid junction potential, arises when two electrolytes of different chemical activities are separated by a clean porous medium. Activity is related to both number and type of ions present. **Fig. 6.5** illustrates a configuration that will result in the generation of a diffusion potential, E_d, when concentrated and dilute solutions of sodium chloride are used. In this case, both positive sodium ions (Na^+) and negative chloride ions (Cl^-) will diffuse across the porous medium from the concentrated to the dilute electrolyte. The Cl^- ions will diffuse more rapidly than the Na^+ ions because of their smaller size and lesser affinity for water. The Cl^- ions are said to have a higher mobility (the speed at which an ion moves under a fixed potential gradient). Because of this diffusion phenomenon, the dilute solution becomes negatively charged. The rate of diffusion slows as the negative charge repels further migration of negative ions. When the cell reaches equilibrium, a steady E_d is established. When the two solutions are connected by an electric conductor, a current flows from the positively charged concentrated solution, through the conductor to the dilute solution, then through the porous medium and back to the concentrated solution.

E_d can be expressed by the Nernst equation[3,4]:

$$E_d = -(t_{Cl} - t_{Na})(RT_a/F)\ln(a_1/a_2) \quad \text{(6.1)}$$

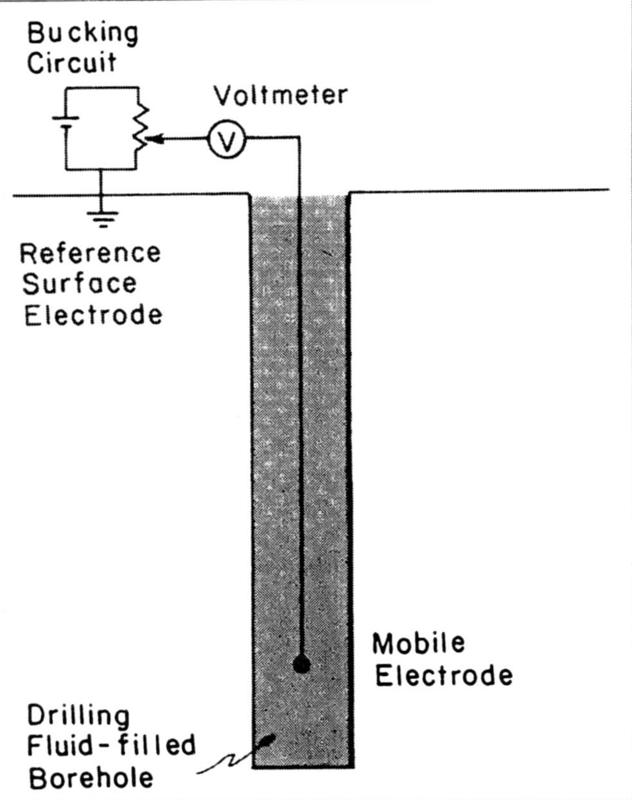

Fig. 6.2—Schematic of the circuit used to measure the SP Log.

where E_d=diffusion potential, V; t_{Cl}=chlorine anion transference number; t_{Na}=sodium cation transference number; R=gas constant, 8.314 J/°C; T_a=absolute temperature, °K; F=Faraday constant, 96,516 C; and a_1 and a_2=the activities of the two electrolytes.

The transference number is the fraction of current carried by the designated ion. In a sodium chloride solution, t_{Cl} and t_{Na} are related to the ion mobilities by

$$t_{Cl} = U_{Cl}/(U_{Cl} + U_{Na}) \quad \text{(6.2)}$$

and $$t_{Na} = U_{Na}/(U_{Cl} + U_{Na}), \quad \text{(6.3)}$$

where U_{Cl} and U_{Na} are the mobilities of the Cl^- and Na^+ ions, respectively. By definition of transference numbers, $t_{Cl} + t_{Na} = 1$ and Eq. 6.1 becomes

$$E_d = -(2t_{Cl} - 1)(RT_a/F)\ln(a_1/a_2). \quad \text{(6.4)}$$

6.3.2 Membrane Potential. Membrane potential, also known as shale potential, arises when two electrolytes of different concentrations are separated by a porous medium where the pore walls carry an electrical double layer. As discussed in Chap. 1, clay minerals have such a layer, and because shale is rich in clays, it constitutes such a membrane. **Fig. 6.6** shows a membrane-potential generating cell where two electrolytes of concentrated and dilute NaCl solutions are separated by a shale membrane. In this case, both Cl^- ions and Na^+ ions try to diffuse from the concentrated solution to the dilute solution. However, the pore walls of the shale membrane are negatively charged because of the existence of the electric double layer, so the passage of Cl^- anions is greatly restricted. The Na^+ cations, however, can enter the pores freely. Consequently, the concentrated solution will be negatively charged with respect to the dilute solution, and a potential, E_m, is created across the membrane. When the two solutions are connected by an electric conductor, a current flows from the positively charged dilute solution, through the conductor to the concentrated solution, through the shale, then back to the dilute solution.

This membrane potential, E_m, can be expressed by Eq. 6.1. Certain shale membranes can completely block the passage of Cl^-

ions, and charge transport is entirely the result of the Na^+ ion. Such a membrane is called a perfect membrane. The maximum possible shale potential occurs across a perfect membrane because $t_{Cl}=0$ and $t_{Na}=1$, and according to the Nernst equation,

$$E_m=(RT_a/F)\ln(a_1/a_2). \quad (6.5)$$

When a shale membrane is imperfect, a Cl^- anion leak occurs and the membrane potential decreases. In this case, a correction factor should be introduced into Eq. 6.5.

6.3.3 Electrochemical Component of the SP. The two situations illustrated in Figs. 6.5 and 6.6 can be combined into one cylindrical cell. **Fig. 6.7** shows a cylindrical trough that contains two solutions of sodium chloride of activities a_1 and a_2. The two solutions are separated at the upper part of the trough by a perfect shale membrane and at the lower part of the trough by a clean porous medium. As discussed, an electric current will flow across the shale from the concentrated solution to the dilute solution, creating the membrane potential, E_m. The electric current will flow across the clean porous medium from the dilute solution to the concentrated solution, creating a diffusion potential, E_d. This configuration, first suggested by Mounce and Rust,[5] constitutes a closed electric circuit in which E_d and E_m are created electrochemically. These two potentials are additive, and their sum is known as the electrochemical potential, E_c:

$$E_c=E_d+E_m. \quad (6.6)$$

Using the expression of E_d and E_m given by Eqs. 6.4 and 6.5 yields

$$E_c = -2t_{Cl}(RT_a/F)\ln(a_1/a_2). \quad (6.7)$$

Fig. 6.7 shows that the components forming the closed electric circuit are found at the boundary between a shale bed and a saltwater-bearing permeable formation traversed by a borehole filled with drilling mud. The saltwater-bearing formation plays the roles of both a concentrated NaCl solution of activity a_w and a clean porous membrane. The shale is, of course, the ion-selective membrane. The drilling mud in the borehole and the mud filtrate in the permeable invaded zone constitute the dilute solution of activity, a_{mf}.

Upon substitution of a_w and a_{mf} for a_1 and a_2, respectively, and conversion of the natural logarithm to the logarithm of the base 10, Eq. 6.7 becomes

$$E_c = -K\log(a_w/a_{mf}), \quad (6.8)$$

where $K=2(2.303)t_{Cl}RT/F=4.606\,t_{Cl}RT_a/F$. (6.9)

At 25°C, the chloride and sodium mobilities are 7.91×10^{-4} and 5.19×10^{-4} cm/s·V, respectively. According to Eq. 6.2,

$$t_{Cl}=\frac{7.91\times10^{-4}}{(7.91+5.19)\times10^{-4}}=0.604.$$

The transference number of the chloride ion, t_{Cl}, is relatively independent of concentration and temperature over the ranges normally experienced in well-logging applications.[4] Substituting 0.604 for t_{Cl} in Eq. 6.9, together with the values of the gas and Faraday constants, results in

$$K=4.606(0.604)(8.314/96{,}516)T_a=0.00024T_a.$$

Expressing K in millivolts and temperature in degrees Celsius gives

$$K=0.24(273.2+T)$$

or $K=65.5+0.24T$. (6.10)

If temperature, T, is expressed in degrees Fahrenheit,

$$K=61.3+0.133T. \quad (6.11)$$

6.3.4 Electrokinetic Component of the SP. Mud filtrate may be forced into the drilled formations as a result of the difference between formation pressure and the hydrostatic pressure of the mud column. This filtration process results in an electrokinetic potential,

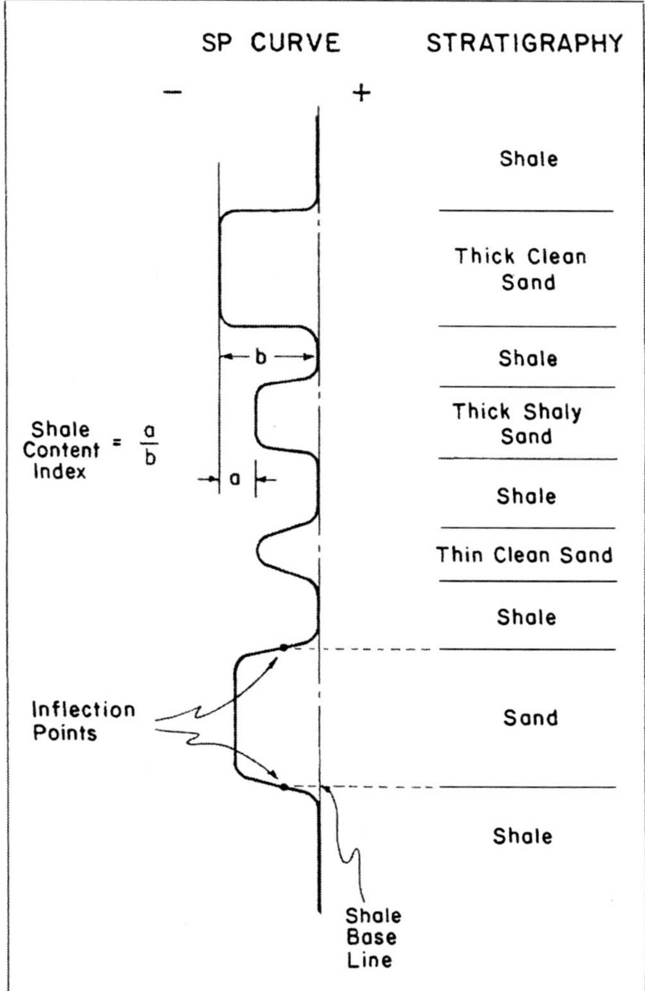

Fig. 6.3—Schematic of an SP curve recorded in a shale/sand formation sequence.

also known as the streaming potential, E_k. A mudcake is usually formed across permeable formations. The mudcake contains clay particles that have an electric double layer. Because of the difference in ionic concentration between the bound clay water and the free water, a potential difference exists. This difference between the free water and the boundary of the bound-water layer is known as the zeta potential, ξ. If pressure is applied to the solution, the solution will flow past the solid surface, carrying with it charges that have a potential ξ. The movement of these charges creates the electrokinetic potential. Wyllie's[6] experimental investigations of the electrokinetic potential developed across a mudcake established a relationship of the form

$$E_k=xp^y, \quad (6.12)$$

where E_k=electrokinetic potential, p=differential pressure, and x and y=constants related to mud composition and resistivity.

The greater the mud resistivity and the pressure differential are, the greater the electrokinetic potential is. Experimental data for 1,000-psi differential pressure are represented by the upper curve in **Fig. 6.8.**

Shale is similar, in kind and properties, to a well-formed mudcake. Gondouin and Scala[7] experimentally proved the existence of electrokinetic potential across shales. Their data are represented by the lower curve in Fig. 6.8. Because the SP is a relative measurement with respect to shale, the combined contribution of potentials of electrokinetic origin would be the difference between the electrokinetic potential across the shale and that across the mudcake. At low mud resistivity and low differential pressure, this difference is generally small, as shown by Fig. 6.8. These differences may vary considerably in practice, however, depending on the relative electrokinetic properties of a particular mudcake and shale system.

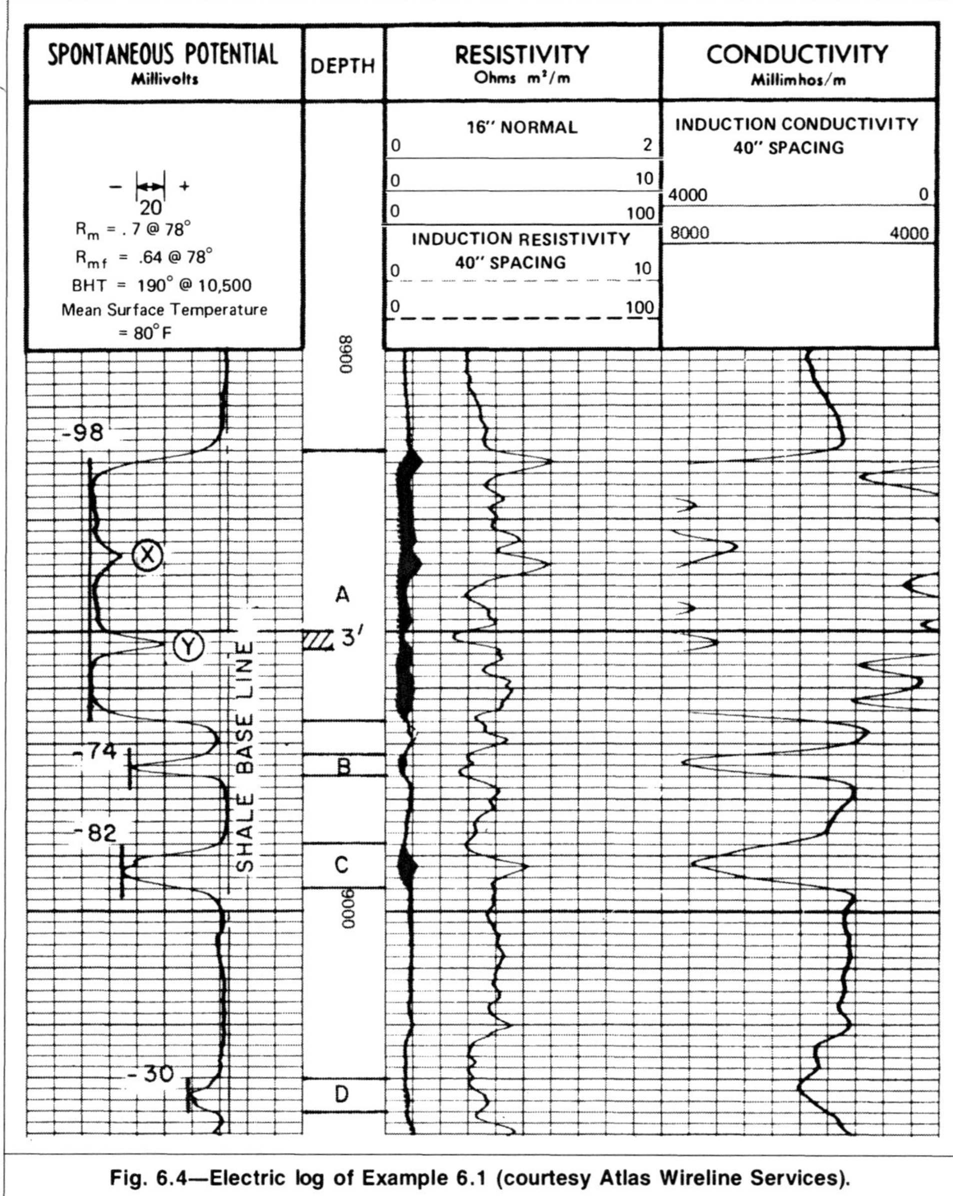

Fig. 6.4—Electric log of Example 6.1 (courtesy Atlas Wireline Services).

Experience indicates that electrokinetic potentials are generally negligible for low mud resistivities and 500-psi or less differential pressure.[8]

Electrokinetic potentials are usually ignored in SP calculations. The total naturally occurring electric potential, called the static self-potential, E_{SSP}, is assumed to equal E_c and is expressed with Eqs. 6.6 and 6.8:

$$E_{SSP}=E_c=E_d+E_m=-K\log(a_w/a_{mf}). \quad (6.13)$$

The negative sign is added because, by convention, a negative SP is measured relative to the shale when $a_w>a_{mf}$.

6.4 Theoretical E_{SSP} vs. Measured SP

The theoretical potential, E_{SSP}, expressed by Eq. 6.13 is determined by the formation temperature, which controls the value of K, and the chemical activities. The formation temperature can be determined as discussed in Chap. 4. The mud filtrate is sampled and its properties are measured. Consequently, if E_{SSP} can be deduced from the measured SP, Eq. 6.13 can be solved for formation water properties. As the equivalent circuit of **Fig. 6.9** shows, the current, I, circulating at the sand/shale boundary is generated by the total potential:

$$E_{SSP}=I(r_f+r_{sh}+r_m), \quad (6.14)$$

where r_f, r_{sh}, and r_m are the resistances encountered by the electric current while flowing through the sand formation, shale formation, and mud column, respectively. The measured SP is the ohmic potential drop as the electrode passes from shale to sand. It is expressed by

$$E_{SP}=Ir_m. \quad (6.15)$$

Combining Eqs. 6.14 and 6.15 yields

$$E_{SP}=\frac{r_m}{r_f+r_{sh}+r_m}E_{SSP}. \quad (6.16)$$

Eq. 6.16 indicates that E_{SSP} is greater than E_{SP}. The E_{SSP}/E_{SP} ratio depends on the borehole's resistivities and geometry, as well as the shale and sand formations. Formation resistivity is typically greater than mud resistivity. However, the current has a much wider cross-sectional path in the formation, resulting in a low resistance compared with that of the restricted path in the borehole. r_m can be expressed as

$$r_m \propto (R_m/d_h^2)h, \quad (6.17)$$

where R_m=mud resistivity, h=bed thickness, and d_h=borehole diameter.

We can see from Eq. 6.17 that, for relatively low-resistivity formations, freshwater-based mud, thick beds, and small boreholes, it is practical to assume that $E_{SP}=E_{SSP}$ because $r_m \gg (r_f+r_{sh})$.

To account for the difference between E_{SP} and E_{SSP}, an empirical correction chart (**Fig. 6.10**) was constructed from data taken

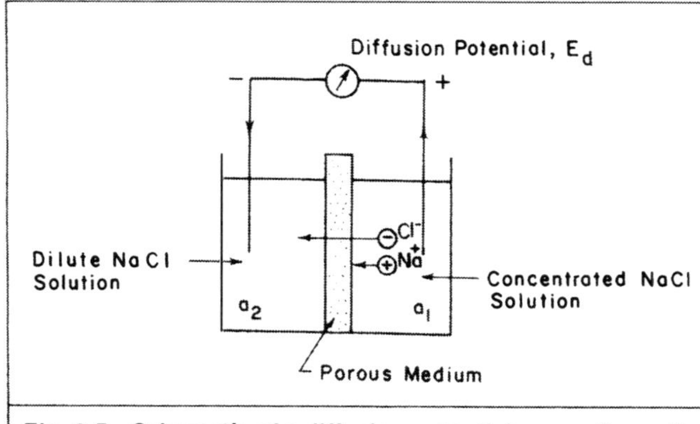

Fig. 6.5—Schematic of a diffusion-potential generating cell.

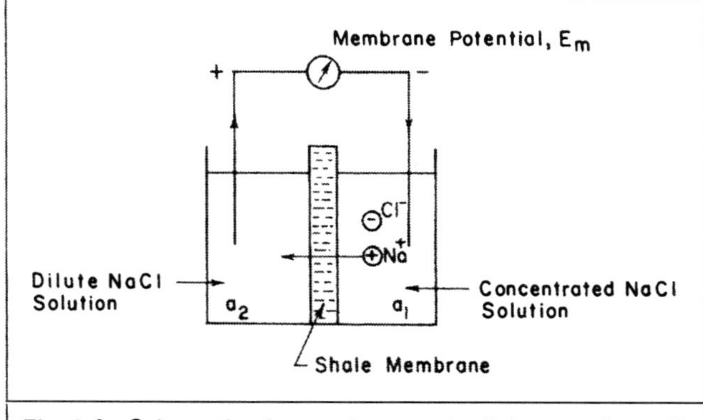

Fig. 6.6—Schematic of a membrane-potential generating cell.

on a resistor network analog.[9] In this chart, R_i is the resistivity of the invaded zone, which can be approximated by the reading of the shallow resistivity tool. The value of the diameter of invasion, d_i, can be approximated with data from Table 5.5. The chart shows that, for average R_i/R_m and d_i values, $(E_{SP}/E_{SSP})>0.9$ for beds ≥ 10 ft thick. Only thin beds warrant a correction.

6.5 Determination of Formation Water Resistivity

A representative value of formation water resistivity, R_w, is essential to quantitative log calculation of water saturation, S_w. Models relating E_{SSP} to R_w exist and are frequently used to derive an R_w value.

6.5.1 Relationship Between E_{SSP} and R_w for Predominantly NaCl Waters. The use of Eq. 6.13, which is expressed in terms of a_w and a_{mf}, is possible; however, in well-logging interpretation, it is more practical to use the resistivities. **Fig. 6.11** shows the relation between activity and resistivity of pure NaCl solutions. To express the relation analytically in a form similar to that of Eq. 6.13, Gondouin *et al.*[11] introduced the concept of "equivalent resistivity" of the formation water. By definition, the equivalent resistivity is proportional to the reciprocal of the activities:

$$(R_w)_{eq}=A/a_w, \quad \text{(6.18)}$$

where $(R_w)_{eq}$=the equivalent resistivity of formation water and A=proportionality factor.

If the mud filtrate is essentially a pure sodium chloride solution, then

$$(R_{mf})_{eq}/(R_w)_{eq}=a_w/a_{mf}, \quad \text{(6.19)}$$

and the E_{SSP} can be expressed by

$$E_{SSP}=-K[\log(R_{mf})_{eq}/(R_w)_{eq}]. \quad \text{(6.20)}$$

As Fig. 6.11 shows, $(R)_{eq}=R$ for pure NaCl solutions of resistivity larger than 0.1 Ω·m at 75°F. This generally is the case for freshwater-based drilling mud. Eq. 6.20 can then be written when $R_{mf}>0.1$ Ω·m:

$$E_{SSP}=-K\log R_{mf}/(R_w)_{eq}. \quad \text{(6.21)}$$

Because R_w, not $(R_w)_{eq}$, is the parameter needed in well-logging interpretation, Gondouin *et al.*[11] used actual resistivity measurements and literature values of activities to construct **Fig. 6.12,** which relates R_w and $(R_w)_{eq}$. If the E_{SSP} and R_{mf} values are known, $(R_w)_{eq}$ is first calculated from Eq. 6.21 and the corresponding R_w is obtained graphically from **Fig. 6.13.**

Fig. 6.14 shows a more convenient chart for SP log interpretation. This chart takes into account the effect of temperature and concentration on the Na^+ transport numbers.[12] R_w is easily determined by taking the following steps (**Fig. 6.15**).

1. Determine the magnitude of the SP, E_{SP}, from the log for the zone of interest.
2. If warranted, correct the SP value for bed-thickness and invasion effects.
3. Determine R_{mf} at formation temperature.
4. Enter the chart with the value of R_{mf}. The intercept with the appropriate temperature line defines the value E_{c1}.
5. Determine the value E_{c2} by subtracting the negative E_{SSP} value from E_{c1}. Add the positive E_{SSP} value to E_{c1} to determine E_{c2}.
6. Determine the value of R_w at the intercept of E_{c2} with the line representing the formation temperature, Point B.

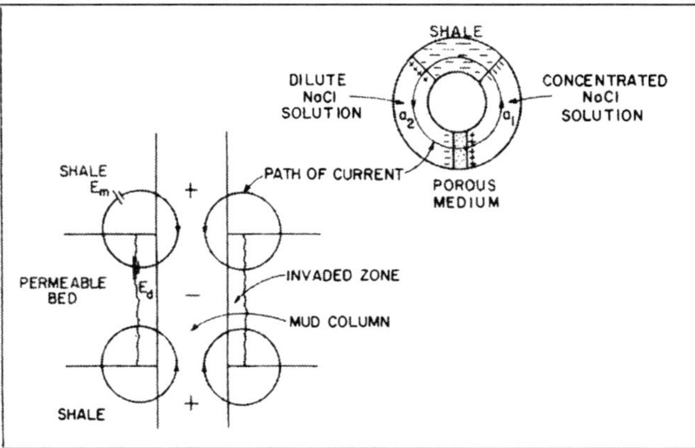

Fig. 6.7—Schematic of the current flow caused by the electrochemical potential (after Refs. 4 and 5).

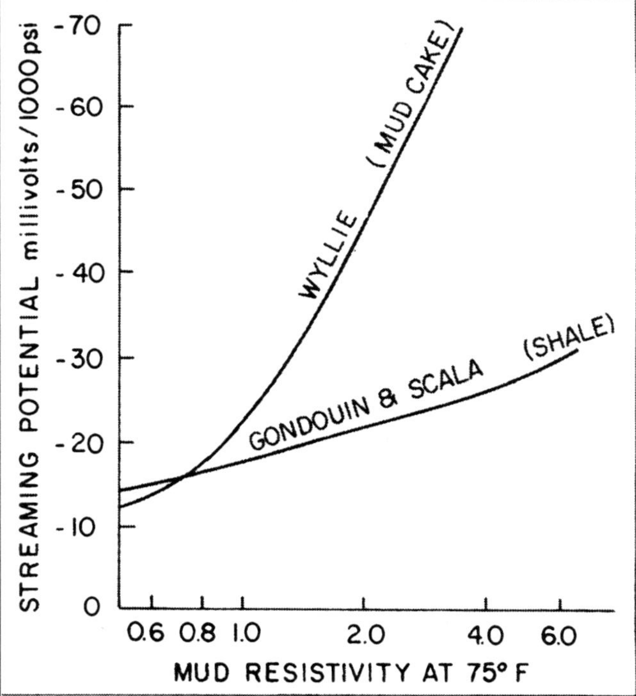

Fig. 6.8—Streaming potential across mudcake and shale vs. mud resistivity at 75°F (after Ref. 4).

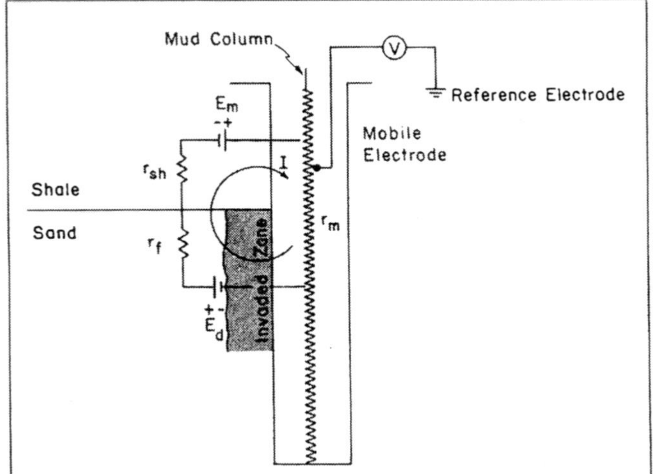

Fig. 6.9—Equivalent circuit representing SP measurement.

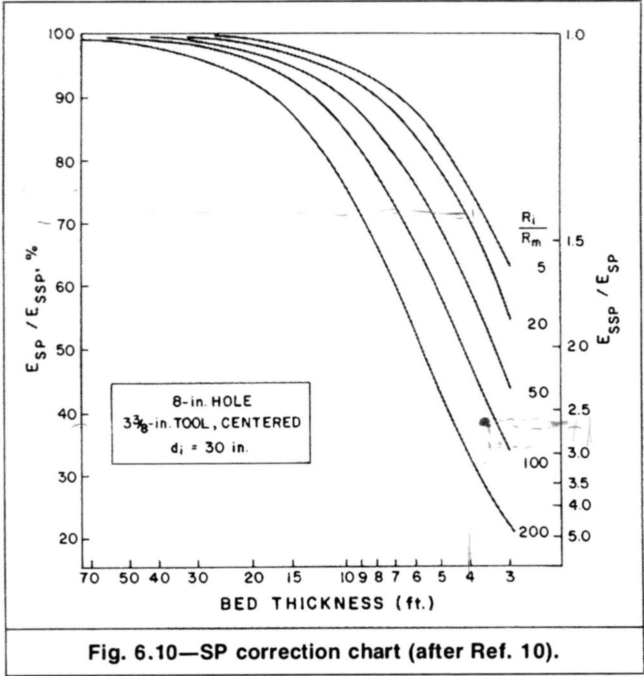

Fig. 6.10—SP correction chart (after Ref. 10).

Example 6.2. Calculate the E_{SSP} for a clean, predominantly NaCl water-bearing sand drilled with a freshwater-based mud (also predominantly NaCl). The formation temperature is 180°F, and R_{mf} and R_w measured at this temperature are 0.54 and 0.37 Ω · m, respectively.

Solution. Because the mud filtrate is essentially a NaCl solution with resistivity greater than 0.1 Ω·m at 75°F (R_{mf}>0.1 Ω·m at 180°F), $(R_{mf})_{eq}=R_{mf}$ and Eq. 6.21 applies:

K = 61.3 + 0.133(180) = 85.2 mV

from Eq. 6.11,

$(R_w)_{eq}$ = 0.03 Ω·m

from Fig. 6.13,

and E_{SSP} = − 85 log (0.54/0.03)

= − 107 mV.

Note that a similar value is obtained from Fig. 6.14.

Example 6.3. Determine R_w and porosity for Zone A of Fig. 6.4, which is a water-bearing sand.

Solution. Because Sand A is relatively thick, $E_{SSP}=E_{SP}=-98$ mV. Using Eq. 1.11 knowing that R_{mf}=0.64 at 78°F yields

R_{mf}=0.64(78+6.77)/(180+6.77)

=0.29 Ω·m at 180°F.

From the one-step chart, Fig. 6.14, R_w=0.028 Ω·m at 180°F. From the conductivity curve, C_o=5,200 m℧/m, so R_o=1,000/5,200=0.19 Ω·m and $F=R_o/R_w$=0.19/0.028=6.87. Using Eq. 1.19 relating F and porosity yields ϕ=35%.

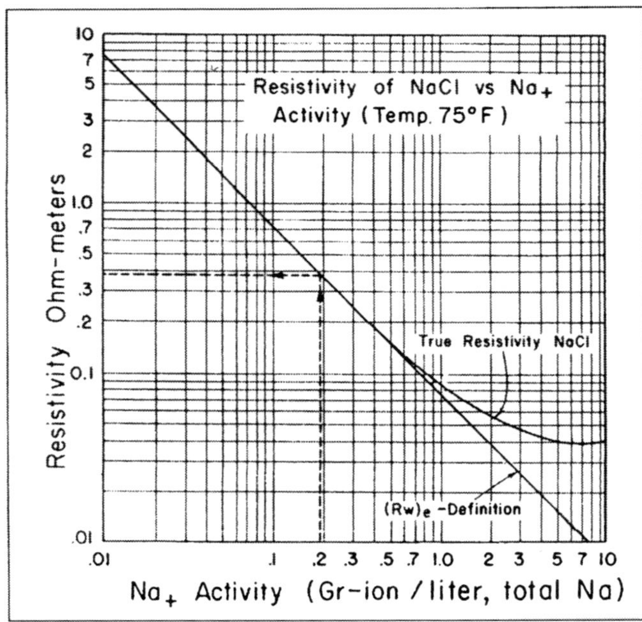

Fig. 6.11—Activity of Na⁺ ions vs. NaCl resistivity (from Ref. 11).

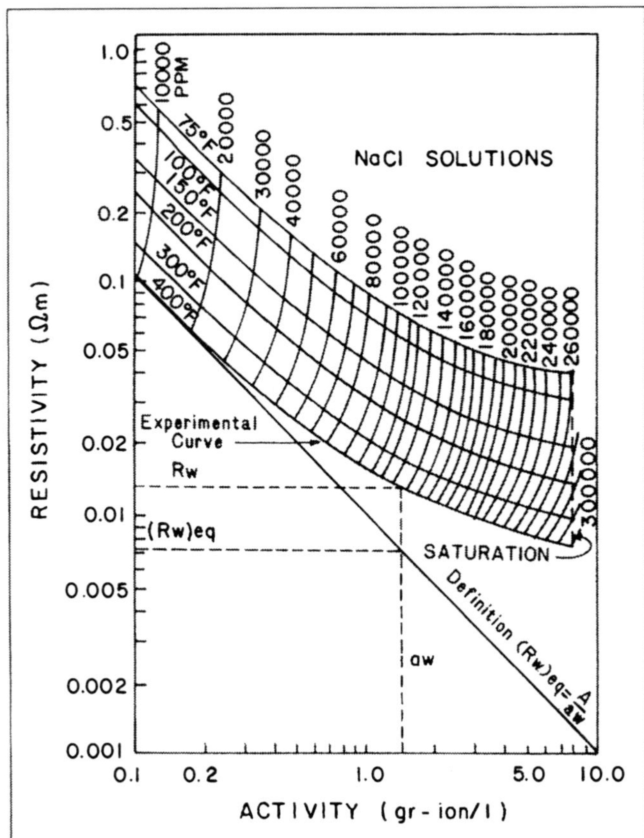

Fig. 6.12—Activity vs. resistivity for concentrated NaCl solutions at different temperatures (from Ref. 11).

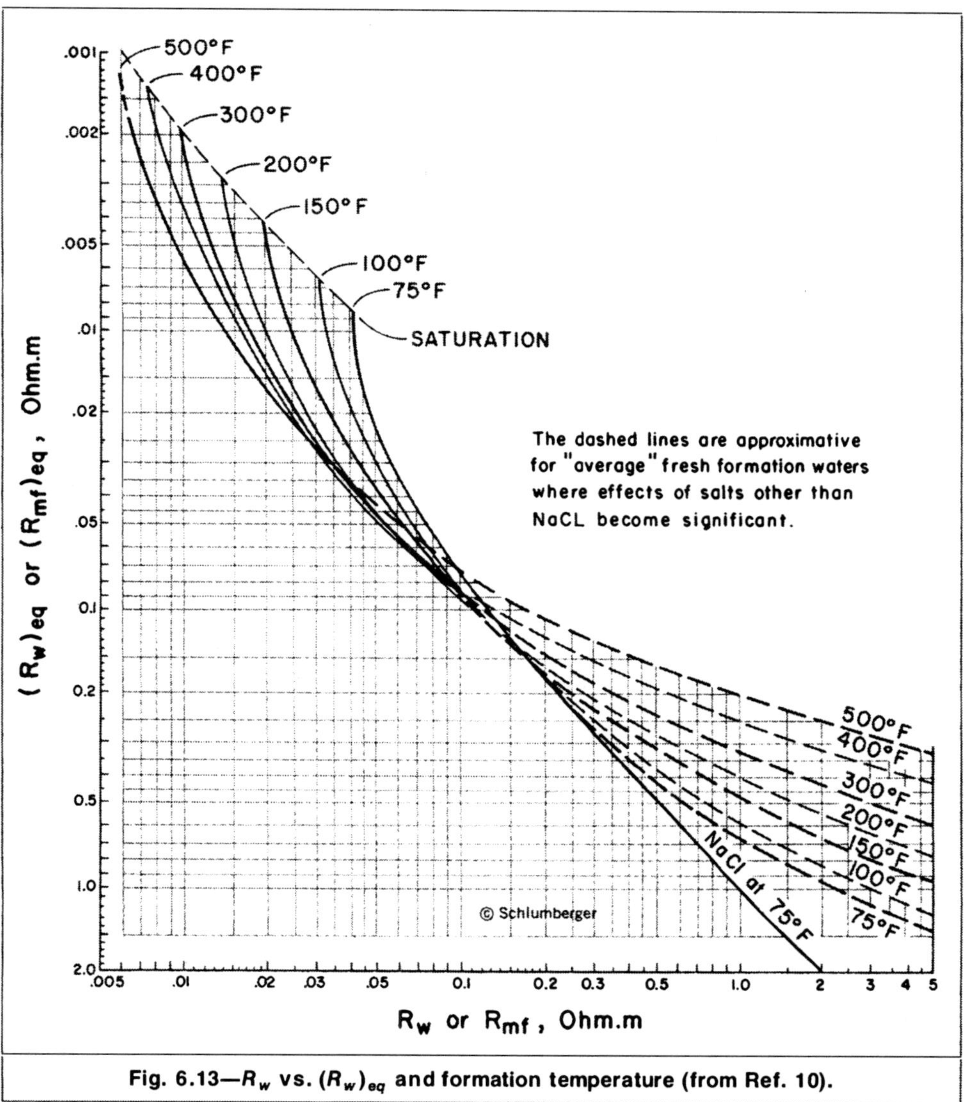

Fig. 6.13—R_w vs. $(R_w)_{eq}$ and formation temperature (from Ref. 10).

6.5.2 E_{SSP}/R_w Relationship for Water Containing Salts Other Than NaCl. The presence of divalent ions (like calcium, Ca^{++}, and magnetism, Mg^{++}) in either the mud filtrate or the formation water considerably affects the magnitude of E_{SSP} when the salinity of either fluid is low. Gondouin *et al.*[11] used an empirical approach to find that E_{SSP} in such cases can be expressed by

$$E_{SSP}=-K\log\frac{[a_{Na}+\sqrt{(a_{Ca}+a_{Mg})}\,]_w}{[a_{Na}+\sqrt{(a_{Ca}+a_{Mg})}\,]_{mf}}, \qquad (6.22)$$

where a_{Na}, a_{Ca}, and a_{Mg} are the activities of the Na^+, Ca^{++}, and Mg^{++} ions, respectively. **Fig. 6.16** shows the relation between cation concentration and activity.

Example 6.4. A typical relatively freshwater analysis (in ppm) from a well in Moffat County, CO,[11] with a Ca^{++} concentration about 25% of the Na^+ concentration is given below.

Ion	Concentration (ppm)
Sodium	1,315
Calcium	310
Magnesium	52
Bicarbonate	427
Chloride	768
Sulfate	2,320
Total	5,192

Measured water resistivity = 1.45 Ω·m at 75°F

Calculate E_{SSP} in a sand saturated with the above formation water at 75°F. The borehole is assumed to be drilled with a pure NaCl solution whose measured resistivity is 4.66 Ω·m at 75°F.

Solution. Because the divalent cation concentration is about 25% of the monovalent sodium cation concentration, the use of Eq. 6.22 is recommended. From Fig. 6.16,

$(a_{Na})_w=0.05$ g-ion/L.

$[(\text{ppm})_{Ca}+(\text{ppm})_{Mg}]_w=310+52=362$ ppm.

Again, from Fig. 6.16,

$(\sqrt{a_{Ca}+a_{Mg}}\,)_w=0.06$ g-ion/L.

$(a_{Na}+\sqrt{a_{Ca}+a_{Mg}}\,)_w=0.05+0.06=0.11$ g-ion/L.

From Fig. 6.11,

$(a_{Na})_{mf}=0.015$ g-ion/L.

From Eq. 6.10,

$K=61.3+0.133(75)=71.3$ mV.

Substituting in Eq. 6.22 yields

$E_{SSP}=-71.3\log(0.11/0.015)=-62$ mV.

This value is equal to the value actually measured in the borehole.

Assuming that the water is predominantly NaCl, Eq. 6.20 results in $E_{SSP}=-71.3\log(4.66/1.45)=-36$ mV, which differs considerably from the measured value.

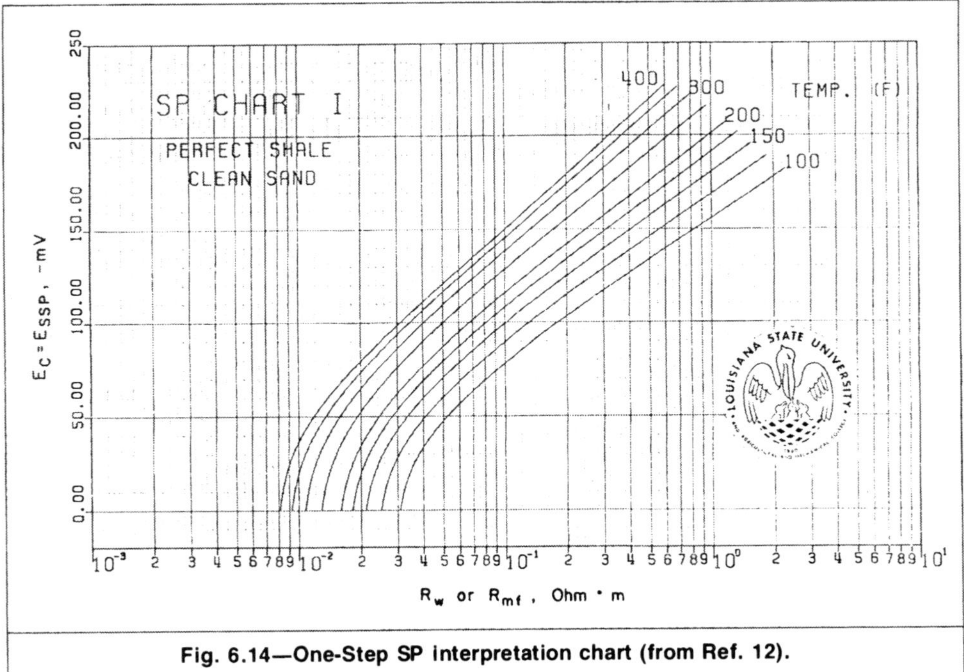

Fig. 6.14—One-Step SP interpretation chart (from Ref. 12).

Example 6.4 illustrates how the value of E_{SSP} can be predicted when water analysis is known. The real interpretation problem, however, is determining the electrical resistivity of the formation water, R_w, from the SP log. An empirical relationship between R_w and $(R_w)_{eq}$ may be developed for a geologic environment known for a constant ion assemblage.[14]

Evers and Iyer[15] applied this principle in Wyoming's Big Horn and Wind River basins. They used published water analyses to predict the E_{SSP} from Eq. 6.22, which was then used to calculate $(R_w)_{eq}$ from Eq. 6.21. Mud filtrates were considered NaCl solutions. R_w values were obtained from published data and plotted vs. $(R_w)_{eq}$. **Fig. 6.17** presents their results. For water with only NaCl in solution, the $R_w = (R_w)_{eq}$ relationship is given by the straight line. Curves of R_w vs. $(R_w)_{eq}$ of other waters are displaced upward so that, for the same value of $(R_w)_{eq}$, the R_w value is greater than for the pure NaCl water. The curves of Fig. 6.17 are applicable only to Wyoming's Big Horn and Wind River basins. They emphasize the importance of collecting and using local data to develop empirical relations to be used in well-log interpretation.

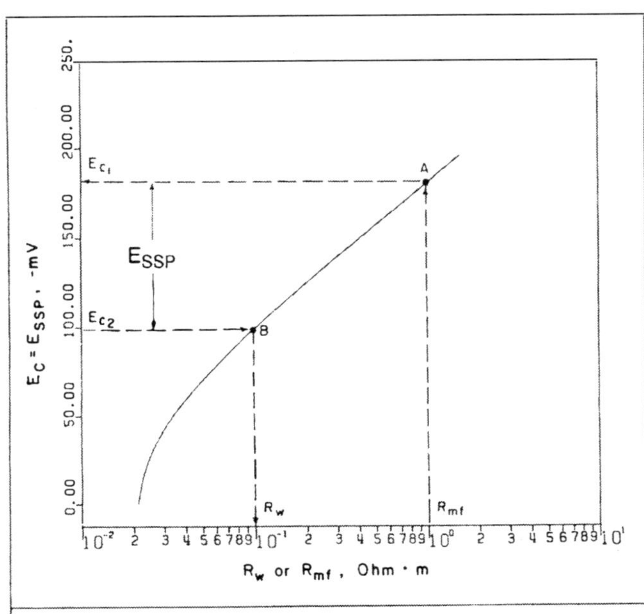

Fig. 6.15—R_w determination using Fig. 6.14 (from Ref. 12).

6.5.3 E_{SSP}-R_w Relationship for Nonideal Shale Membrane. Eqs. 6.21 and 6.22 assume that shale formations behave as ideal cationic selective membranes. Because the clay contents and shale porosities may vary appreciably, shale membrane potentials also vary. Laboratory measurements made on shale membranes cut from cores commonly show a potential that is different from that calculated for an ideal membrane.[11] Assuming perfect membrane behavior in such a case will not yield an accurate value of formation water resistivity. Eq. 6.21 tends to overestimate R_w, which in turn results in underestimation of the potential of hydrocarbon zones.[16,17] This fact is evident in **Fig. 6.18**, which shows the correlation between R_w values inferred from water samples and those calculated from Eq. 6.21 for 32 selected cases in Grand Isle field, LA.[17]

Smits[18] developed a theoretical equation that describes the membrane potential in shales and shaly sands as a function of the cation exchange capacity per unit PV of the rock, Q_V. Consequently, changes in membrane potentials caused by changes in clay content can be accounted for. However, because of the difficulty associated with obtaining reliable Q_V values, log analysts have been reluctant to accept this theory.

Silva and Bassiouni[16,17] developed an empirical graphical relationship between E_{SSP} and R_w for nonideal shale membranes. The readily available shale electrical resistivity, R_{sh}, is used to reflect the changes in salinity, clay content, shale porosity, and hence membrane efficiency. **Fig. 6.19** gives the relationship between E_{SSP}

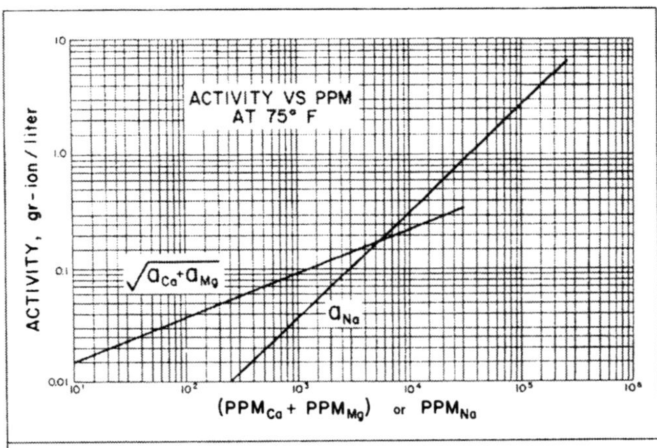

Fig. 6.16—Activity vs. concentration, Na^+ and Ca^{++} plus Mg^{++} (from Ref. 13).

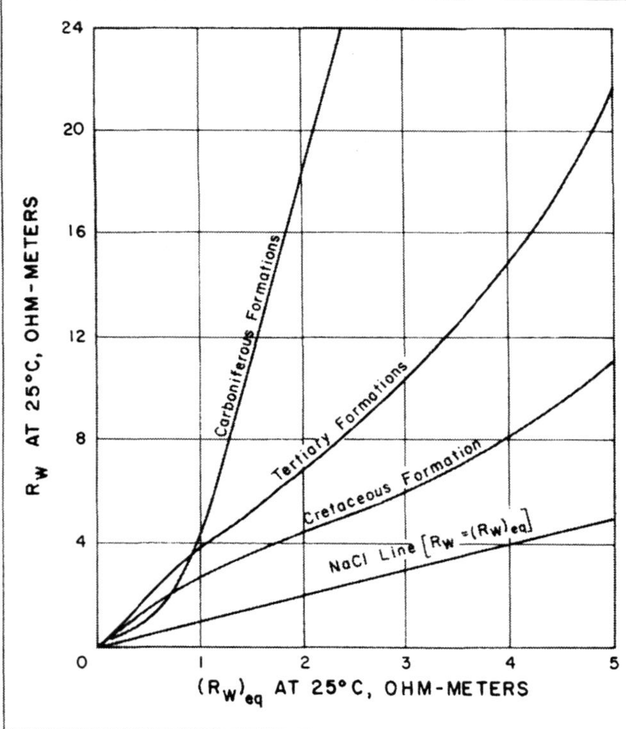

Fig. 6.17—R_w vs. $(R_w)_{eq}$ for relatively fresh waters of Wyoming's Big Horn and Wind River basins (from Ref. 15).

values and R_{sh}, R_{mf}, and R_w. Use of this chart, which was developed with U.S. gulf coast data, is restricted to cases where NaCl constitutes more than 95 wt% of the total dissolved solids (TDS). To determine R_w, the E_{SSP} value, R_{mf} at formation temperature, and average resistivity of the purest adjacent sha es, R_{sh}, are required.

To use Fig. 6.19 to estimate R_w, the abscissa is entered with the E_{SSP} value. The intersection with the curve of the proper R_{sh}/R_{mf} ratio determines the corresponding R_{mf}/R_w ratio and, in turn, R_w.

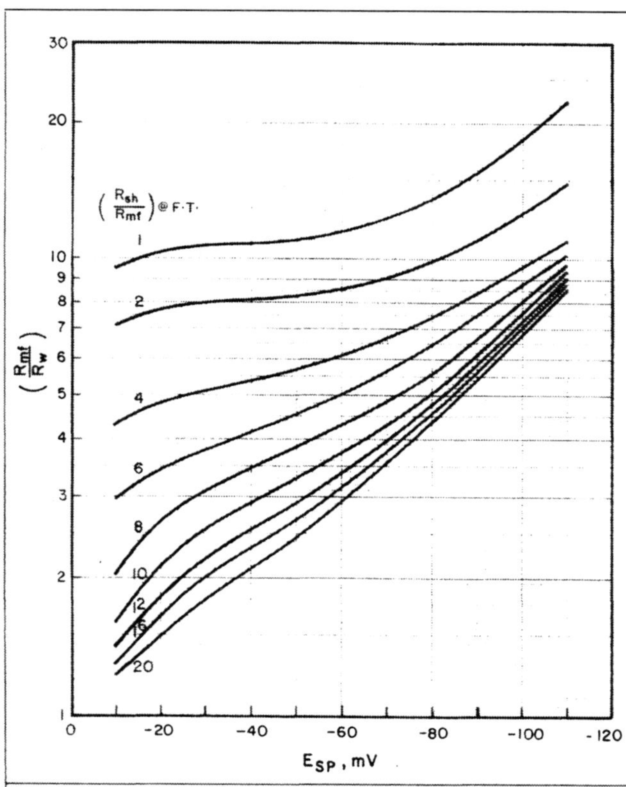

Fig. 6.19—Chart relating the SP log reading to R_{sh}, R_{mf}, and R_w (from Ref. 17).

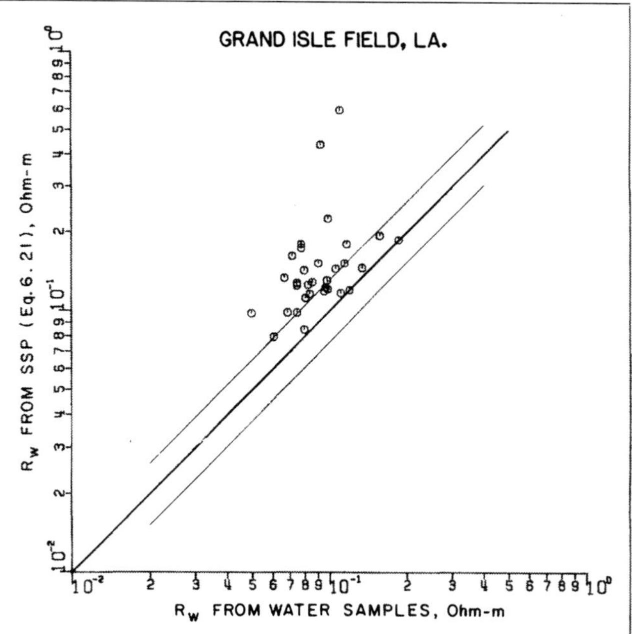

Fig. 6.18—R_w from SP log (Eq. 6.21) vs. R_w from water samples from Grand Isle field, LA (from Ref. 17).

Fig. 6.20 shows the improvement in R_w estimation that results from the use of Fig. 6.19. Fig. 6.20 shows the correlation between R_w values inferred from chemical analysis and those derived from the chart for the 32 data points used in Fig. 6.18 representing Grand Isle field, LA.

Lau and Bassiouni[19] introduced an SP model that incorporates the concept of water-transport phenomenon and shale-membrane efficiency, m_{eff}. This term measures the nonideal behavior of a shale membrane. An ideal membrane displays 100% efficiency. The membrane efficiency relates empirically to the shale resistivity, R_{sh}, by[20]

$$m_{eff} = 0.47 + 0.3R_{sh} \leq 1. \quad (6.23)$$

The value of m_{eff} calculated from Eq. 6.23 should not exceed unity.

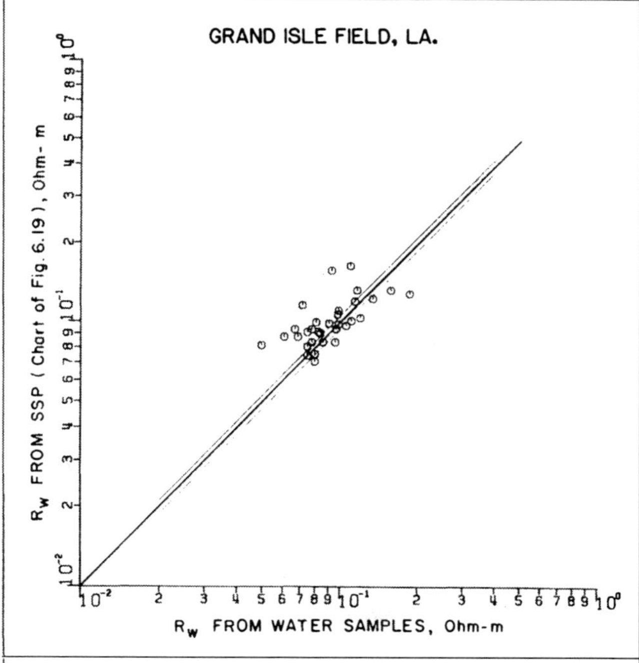

Fig. 6.20—R_w from SP log (Fig. 6.19) vs. R_w from chemical analysis, Grand Isle field, LA (from Ref. 17).

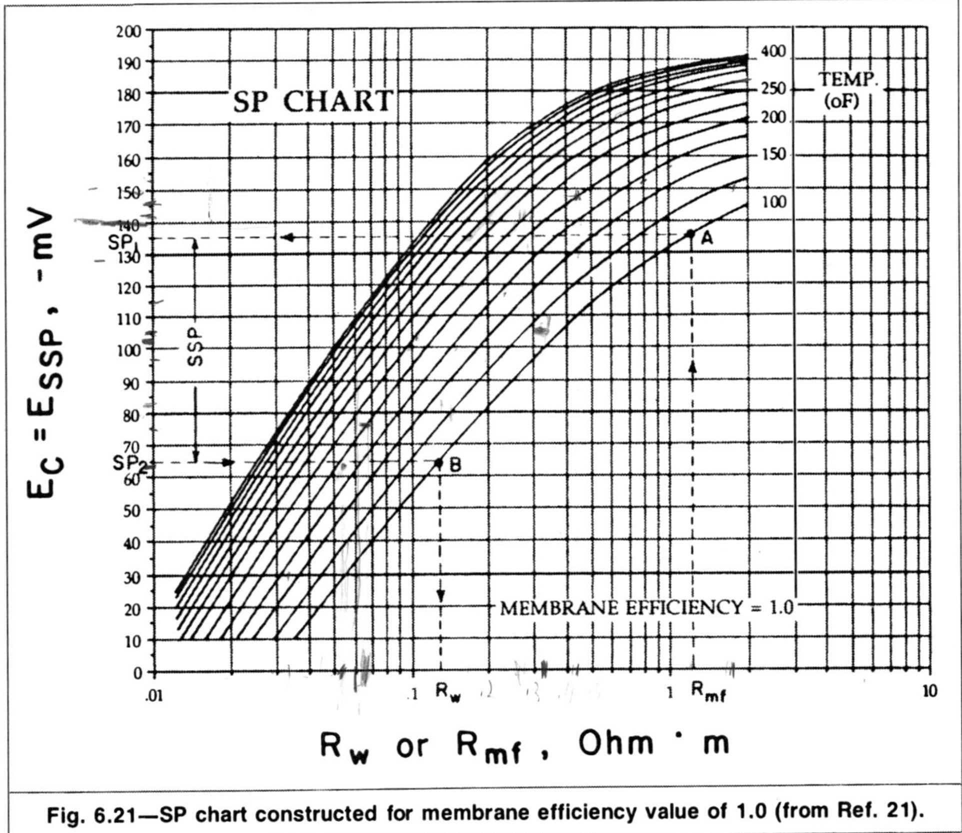

Fig. 6.21—SP chart constructed for membrane efficiency value of 1.0 (from Ref. 21).

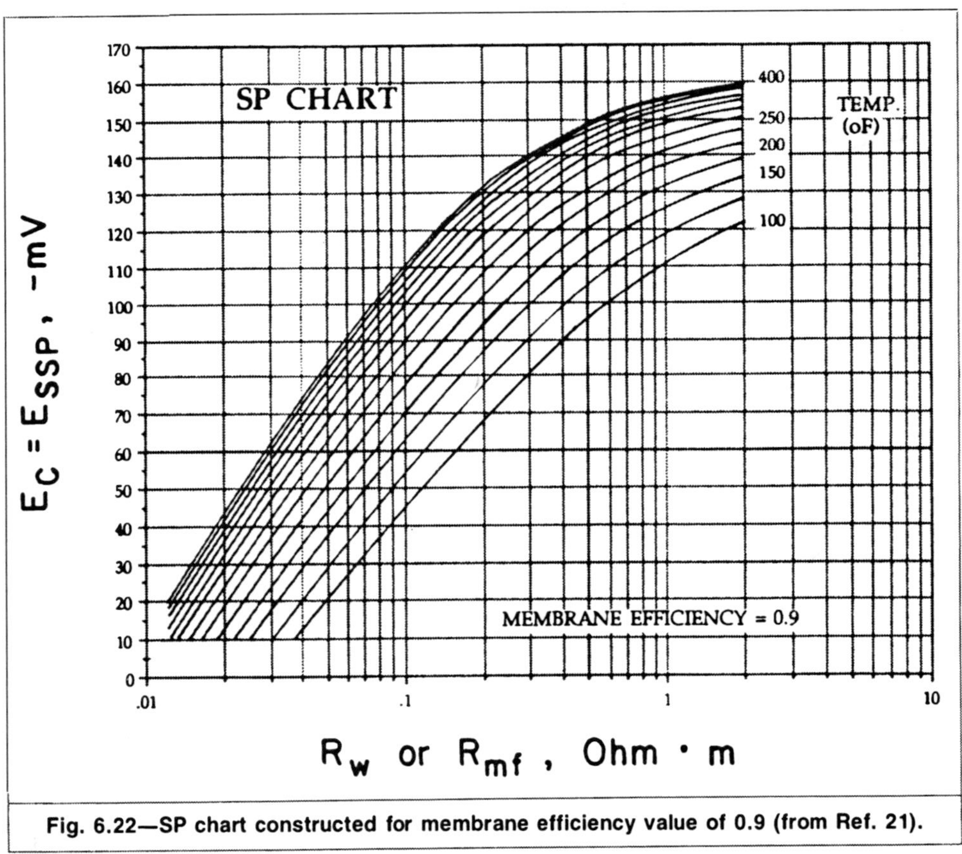

Fig. 6.22—SP chart constructed for membrane efficiency value of 0.9 (from Ref. 21).

Figs. 6.21 through 6.24 illustrate interpretation charts for different membrane efficiencies.

Example 6.5. The following data pertain to a Miocene sand in the 2,812- to 2,878-ft interval of a St. Charles Parish, LA, well (see **Fig. 6.25**).

E_{SSP}, mV	−55
Formation temperature, °F	108
R_{mf} at 108°F, Ω·m	0.89
R_{sh} at 108°F, Ω·m	0.9
R_w from reliable water samples, Ω·m at 108°F	0.059

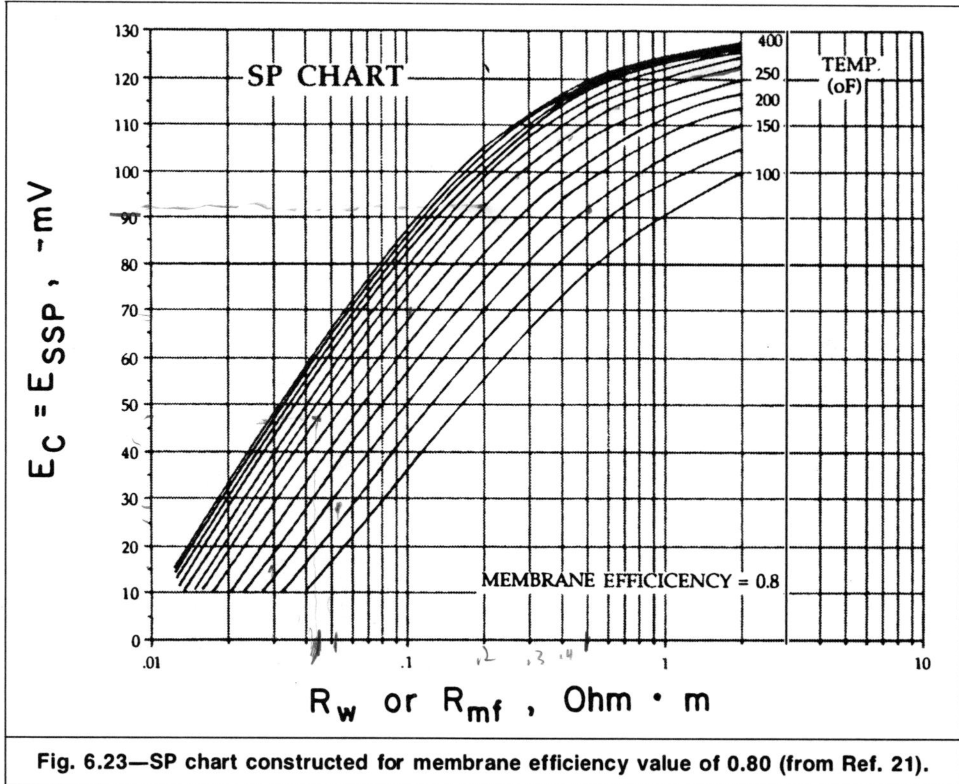

Fig. 6.23—SP chart constructed for membrane efficiency value of 0.80 (from Ref. 21).

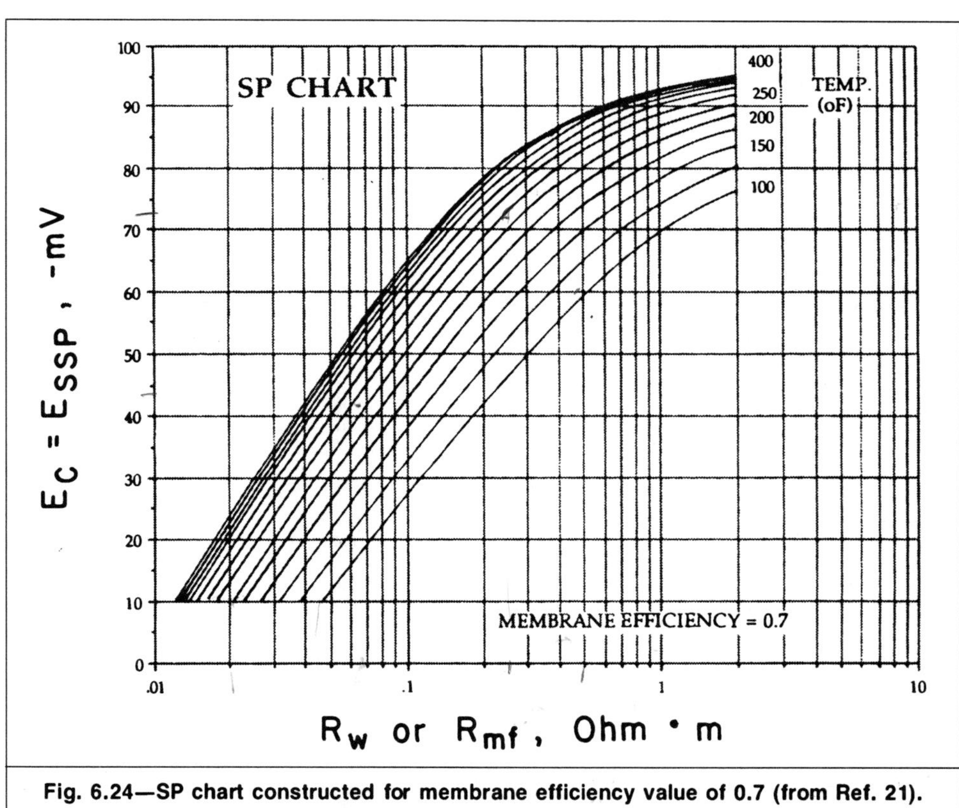

Fig. 6.24—SP chart constructed for membrane efficiency value of 0.7 (from Ref. 21).

Estimate R_w using different approaches. Compare the estimated values to the value obtained from water samples.

Solution. Eq. 6.21 assumes that shale behaves as a perfect membrane. Its use yields

$$K=61.3+0.133(108)=75.7 \text{ mV}$$

and $(R_w)_{eq}=\dfrac{R_{mf}}{10^{-55/-75.5}}=0.167\ \Omega\cdot\text{m}.$

TABLE 6.1—R_w VALUES OF EXAMPLE 6.5

Method	R_w at 108°F	Deviation (%)	Error in S_w Calculation (%)
Water analysis	0.059	—	—
Eq. 6.21	0.167	183	+68
Fig. 6.19	0.082	39	+18
Membrane-efficiency concept	0.067	14	+7

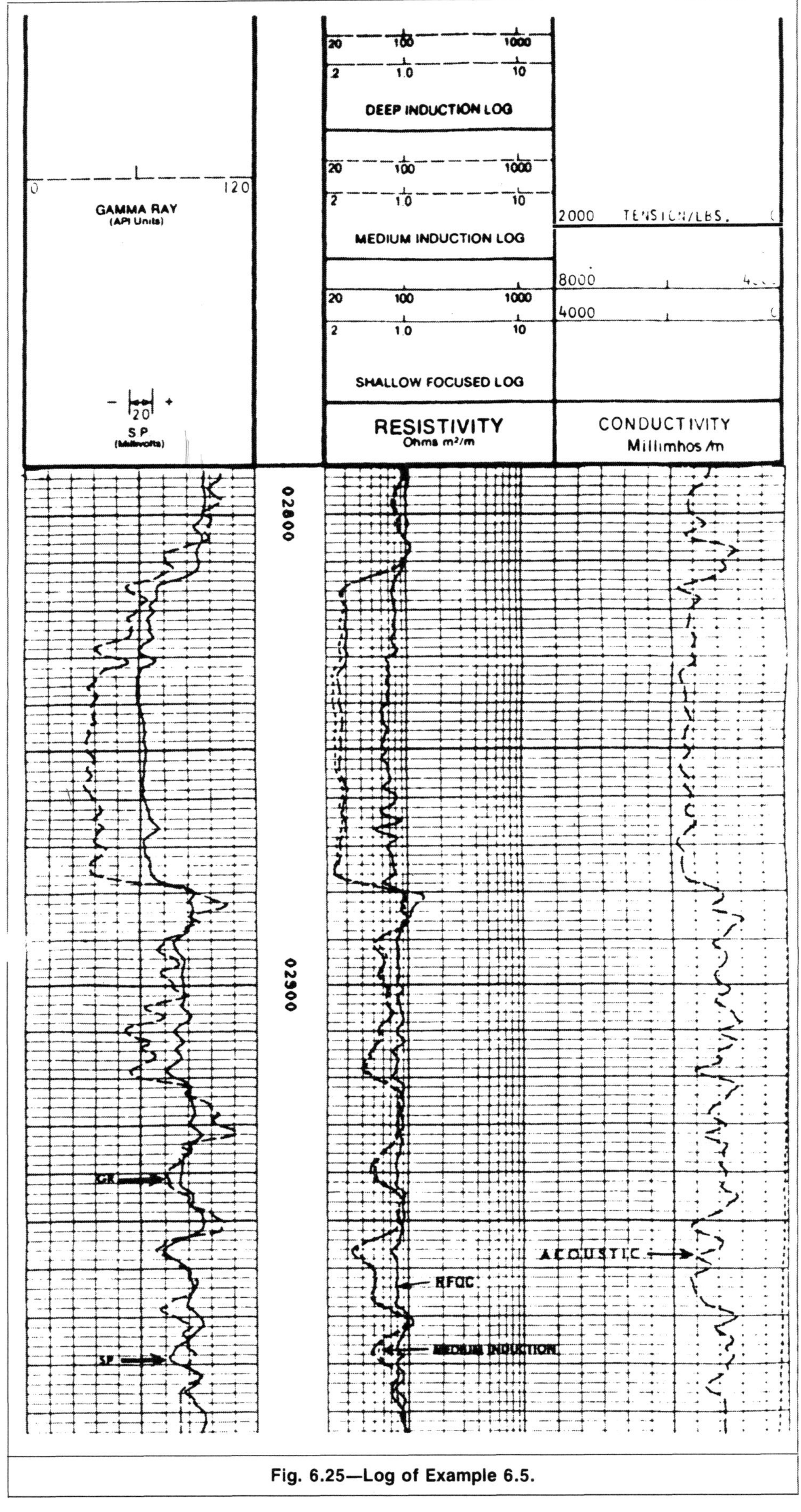

Fig. 6.25—Log of Example 6.5.

Because $(R_w)_{eq}>0.1$,

$R_w=(R_w)_{eq}=0.167\ \Omega\cdot\text{m}$ at 108°F.

Fig. 6.19 attempts to take the nonideal shale-membrane behavior into consideration:

$R_{sh}/R_{mf}=0.9/0.89\cong 1\ \Omega\cdot\text{m}$.

$R_{mf}/R_w=10.8$.

$R_w=0.89/10.8=0.082\ \Omega\cdot\text{m}$ at 108°F.

The concept of the shale-membrane efficiency results in

$$m_{eff}=0.47+0.3(0.9)$$
$$=0.74. \quad (6.23)$$

From **Figs. 6.23 and 6.24**, $R_w=0.088\ \Omega\cdot\text{m}$ for $m_{eff}=0.8$ and $R_w=0.053\ \Omega\cdot\text{m}$ for $m_{eff}=0.7$. By interpolation, $R_w=0.067\ \Omega\cdot\text{m}$ for $m_{eff}=0.74$.

The R_w values obtained by these different methods are listed in **Table 6.1**, together with deviation and possible error in S_w determination. As Table 6.1 shows, the values obtained when the nonideal shale behavior is taken into consideration are in closer agreement with that obtained from chemical analysis and result in a lower error in S_w value.

6.6 Character and Shape of the SP Deflection

In clean formations, the deflection of the SP curve off the shale baseline is expressed by Eq. 6.21. Considering a relatively short interval of a few hundred feet of a wellbore, the temperature and, subsequently, K and R_{mf} remain practically unchanged. $(R_w)_{eq}$ is directly proportional to R_w. Therefore, the higher R_w is, the lower the SP deflection is; i.e., the higher the formation water salinity is, the higher the SP deflection is.

In addition to the absolute value of R_w, the contrast R_{mf}/R_w or precisely $R_{mf}/(R_w)_{eq}$ determines the magnitude of SP deflection. The contrast between mud filtrate and formation water salinity results in one of three typical cases (**Fig. 6.26**).

1. If the mud is relatively fresher than the formation water, such as $R_{mf}>R_w$, a negative SP deflection occurs.
2. If the salinities of the mud filtrate and formation water are about the same, such as $R_{mf}\cong(R_w)_{eq}$, then little or no SP deflection occurs because $\log R_{mf}/(R_w)_{eq}\cong 0$.
3. If formation water is fresher than the mud filtrate, such as $R_{mf}<(R_w)_{eq}$, positive SP deflection occurs.

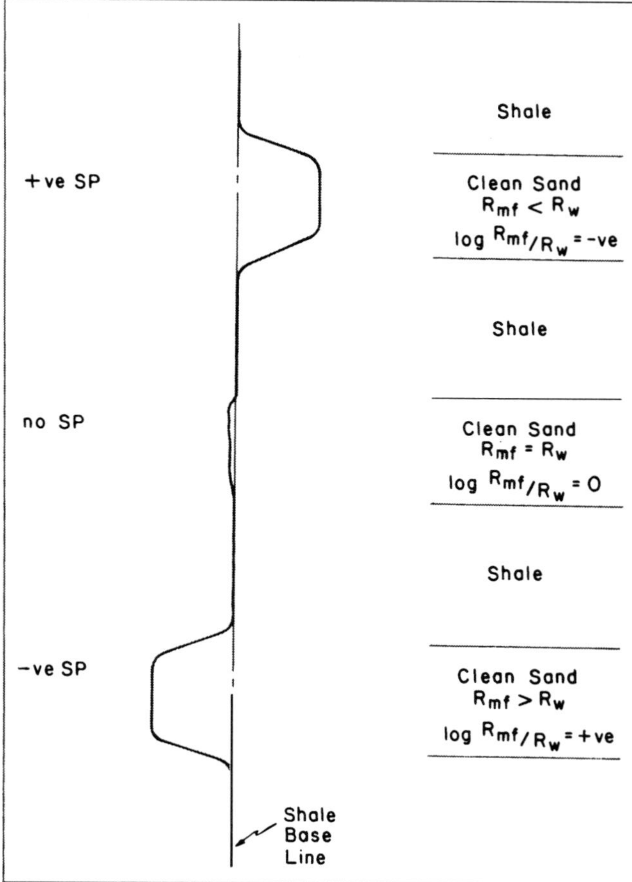

Fig. 6.26—Schematic of the effect of contrast between R_{mf} to R_w on SP deflection.

Example 6.6. Determine the formation water salinity for Sands A through C in **Fig. 6.27**. The log heading lists the following information: $R_m=0.30$ at 66°F, $R_{mf}=0.23$ at 68°F, and $T=148$°F at 7,665 ft.

Solution. The geothermal gradient is

$$g_G=[(148-70)/7,665]100=1\text{°F}/100\text{ ft}.$$

Formation temperature for all three sands $=70+1(6,650/100)=137$°F.

Mud filtrate resistivity is

$$R_{mf}=0.23(68+6.77)/(137+6.77)=0.12\ \Omega\cdot\text{m}.$$

Sand A. Considering maximum deflection off the shale baseline, $E_{SSP}\cong E_{SP}=+33$ mV. From Eq. 6.11,

$$K=61.3+0.133(137)=79.5\text{ mV}.$$

Solving Eq. 6.21 for $(R_w)_{eq}$ gives

$$\log[R_{mf}/(R_w)_{eq}]=E_{SSP}/-K$$
$$(R_w)_{eq}=R_{mf}/10^{E_{SSP}/-K}$$
$$=0.12/10^{+33/-79.5}$$
$$=0.31\ \Omega\cdot\text{m}.$$

Because $(R_w)_{eq}>0.1$, $R_w=(R_w)_{eq}=0.31\ \Omega\cdot\text{m}$ at 137°F.

Assuming the solution is predominantly NaCl, Fig. 1.7 indicates that the formation water salinity is 11,000 ppm.

Fig. 6.28 shows how a similar value of R_w is obtained from Fig. 6.14.

Sand B. Because $E_{SP}=0$, $R_w\cong R_{mf}=0.12\ \Omega\cdot\text{m}$, and from Fig. 1.7, water salinity is 30,000 ppm.

Mud filtrate salinity is also about 30,000 ppm. It is saltier than the formation water of Sand A, which explains the positive SP deflection.

Sand C. Here, $E_{SSP}=E_{SP}=-10$ mV. Using Eq. 6.21 or Fig. 6.14 results in $R_w=0.09\ \Omega\cdot\text{m}$, which in turn indicates a salinity of 38,000 ppm.

Example 6.7. Explain why Sands W through Z in **Fig. 6.29** display different E_{SP} values. Sand X is known to be clean and $R_m=0.25\ \Omega\cdot\text{m}$ at formation temperature.

Solution. The difference in the SP deflection displayed by the different zones can be explained by one or a combination of the following: a change in formation water salinity, different bed-thickness and formation-resistivity effects, and the fact that Zones W through Z are shaly sands.

The probable reason(s) is deduced by elimination. First, major changes in water salinity are improbable given the closeness of the four sands. Second, the bed-thickness and formation-resistivity effects are investigated. Data aiding in the investigation are listed below.

	Sands			
	X	W	Y	Z
Thickness, ft	12	6	7	6
E_{SP}, mV	46	27	39	24
R_i (short normal reading) $\Omega\cdot\text{m}$	2.2	5.3	2.3	2
R_i/R_m	8.8	21.2	9.2	8
E_{SP}/E_{SSP} (Fig. 6.10)	0.96	0.84	0.89	0.87
E_{SSP}	48	32	44	28

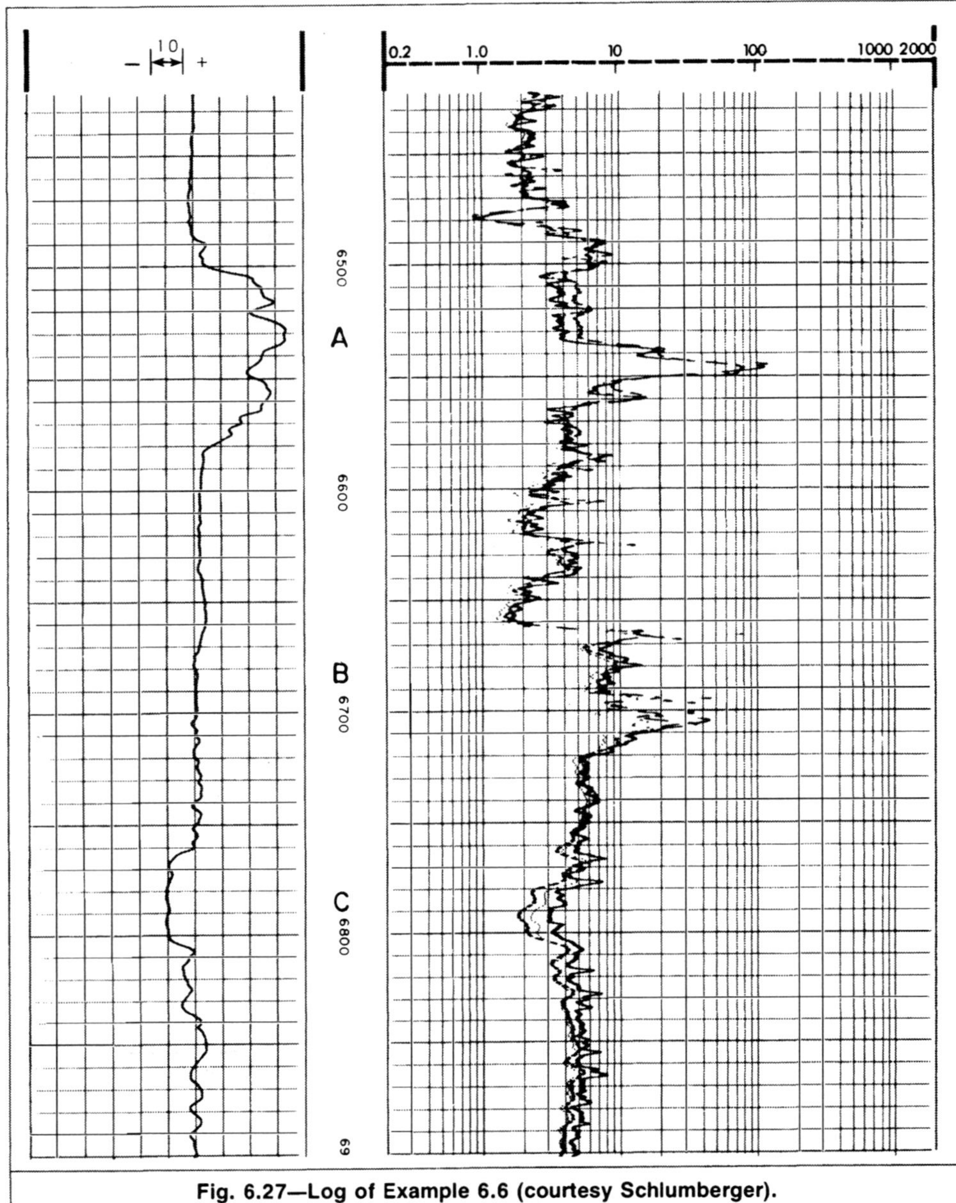

Fig. 6.27—Log of Example 6.6 (courtesy Schlumberger).

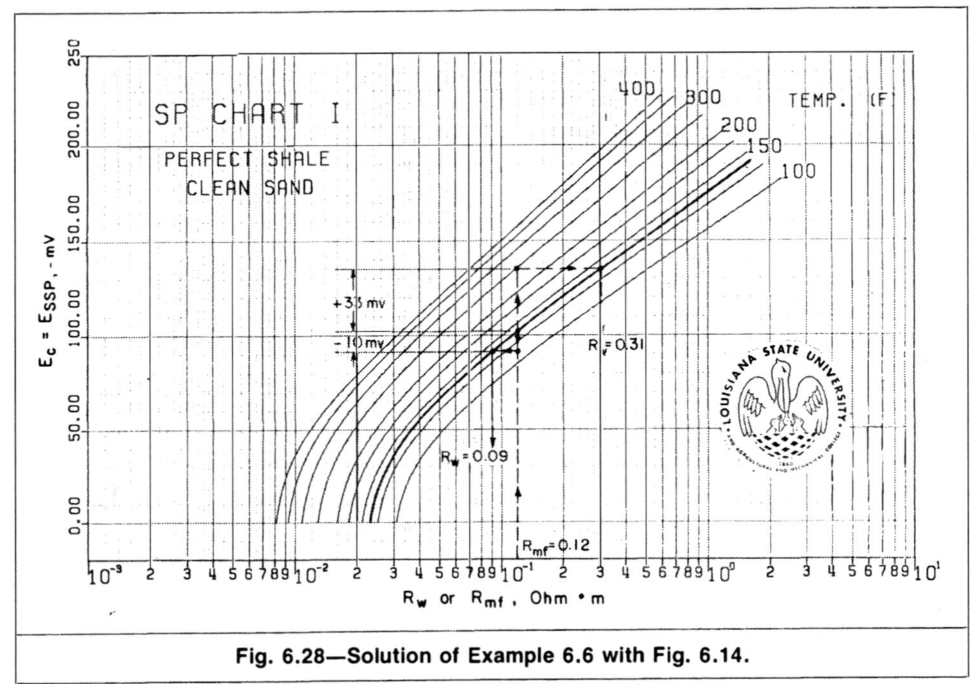

Fig. 6.28—Solution of Example 6.6 with Fig. 6.14.

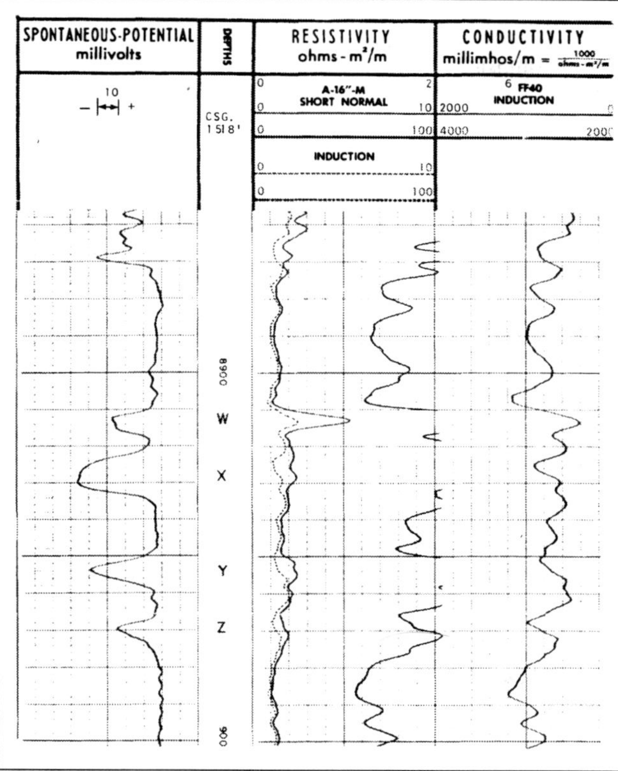

Fig. 6.29—Electric log of Example 6.7 (courtesy Schlumberger).

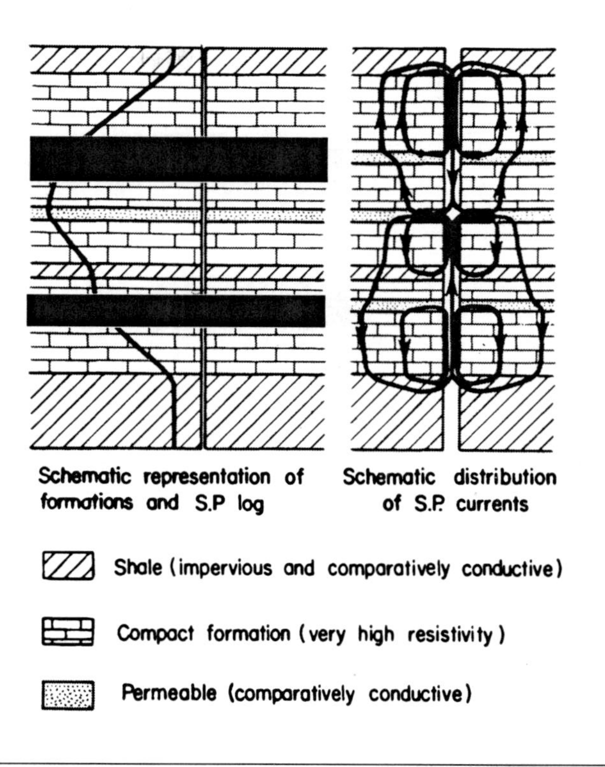

Fig. 6.30—Schematic of the SP curve in highly resistive formations (from Ref. 13).

Because the effect of salinity changes is considered minor, the E_{SSP} value should be the same for all clean zones. However, Fig. 6.10 is empirical. The E_{SSP} value is approximate and should be treated only as an order of magnitude. It follows that Sand Y is relatively clean. It displayed an SP value less than that of Sand X, mainly because of the bed-thickness effect. The E_{SSP} values of Zones W and Z are considerably different from that of clean Zone X because these sands are shaly. Their SP deflection is different from that of Zone X owing to both bed-thickness and shaliness effects. The shale index, I_{sh}, for these zones can be calculated as

$(I_{sh})_W=(48-32)/48=0.33$

and $(I_{sh})_Z=(48-28)/48=0.42$.

The formation resistivity, as discussed in Sec. 6.4, affects the value of the SP deflection. In a highly resistive formation, however, the shape of the SP curve itself is affected. As **Fig. 6.30** illustrates, the flow of the current next to highly resistive beds is largely confined to the wellbore because it presents the path of least resistance. The intensity of the current remains constant, which in

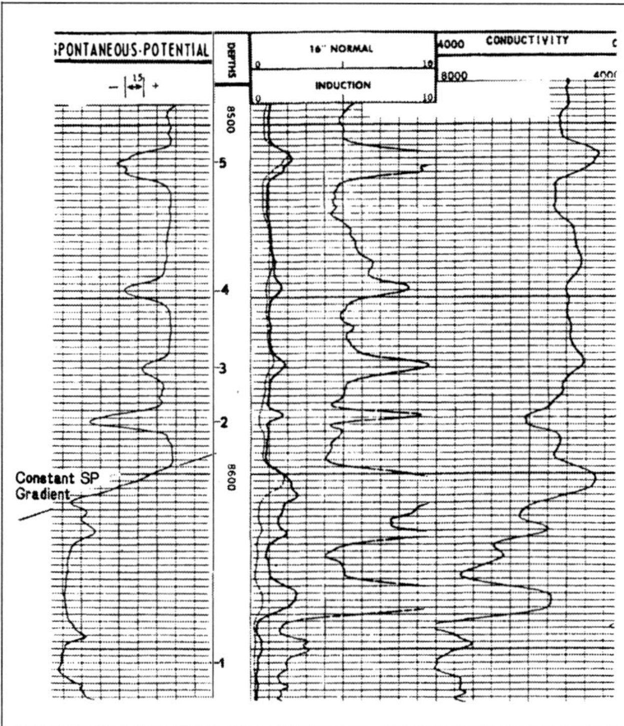

Fig. 6.31—Log showing a constant SP gradient next to a highly resistive formation (courtesy Amoco Production Co.).

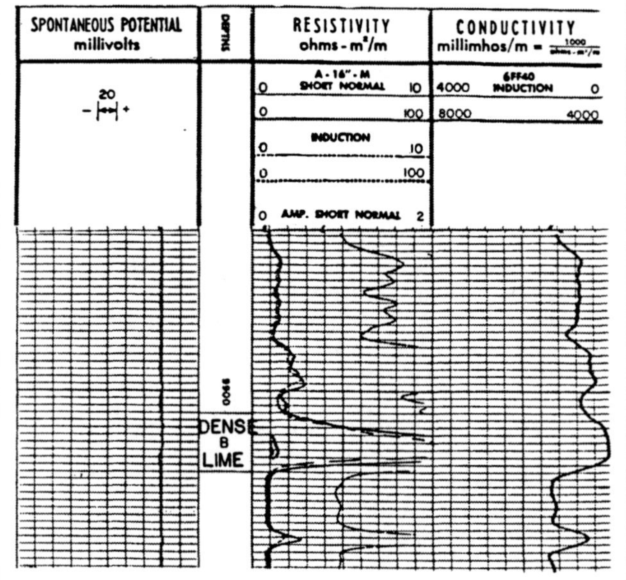

Fig. 6.32—Log showing the changes in SP next to a highly resistive bed between two shales (courtesy Marathon Oil Co.).

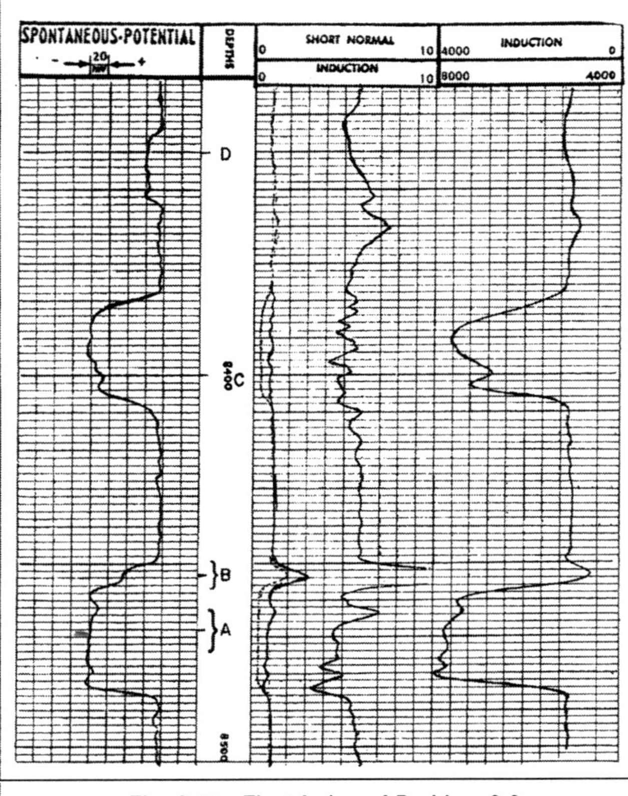

Fig. 6.33—Electric log of Problem 6.6

turn results in a constant potential gradient. The SP curve then assumes a straight-line shape next to highly impermeable formations. The slope and/or current change next to permeable beds and shales where the current enters and leaves the wellbore. **Fig. 6.31** shows a resistive bed in the 8,597- to 8,608-ft interval on top of a thick permeable sand. The SP clearly displays a constant gradient over this interval, indicating that the current flow remained largely confined to the wellbore. When the highly resistive bed is between two shales, the potential does not change. This is illustrated by **Fig. 6.32**, which shows a highly resistive lime bed between two thick shales.

Review Questions

1. What are the sources of naturally occurring electrical potentials?
2. Strictly speaking, is the variation of the potential observed in wellbores of natural origin?
3. Explain how and why the SP measurement is relative rather than absolute.
4. Why does an SP deflection occur at the boundary between a shale bed and a permeable formation?
5. Explain how the diffusion, membrane, and streaming potentials are generated. What is the relative magnitude of each?
6. Why is the electrokinetic component of the SP usually ignored in SP quantitative interpretation?
7. How is the E_{SSP} value defined?
8. Why does the measured SP differ from E_{SSP}?
9. Under what conditions are E_{SP} and E_{SSP} close in value?
10. Under what conditions are the resistivities of pure NaCl solutions inversely proportional to their activities?
11. What is meant by the formation water equivalent resistivity, $(R_w)_{eq}$?
12. How would you relate E_{SSP} to R_w in relatively freshwater formations that contain salts other than NaCl?
13. In SP quantitative interpretation, shale is usually assumed to be a perfect cationic membrane. How valid is this assumption, and what effect will it have on the calculated R_w and S_w values?
14. What are the factors affecting the polarity and order of magnitude of the SP deflection?

Problems

6.1 Calculate the relative magnitude of the membrane potential compared with the diffusion potential for clean sands at 80°F.

6.2 a. Estimate E_{SSP}, assuming that the shale membrane is perfect, if formation temperature=200°F, R_{mf} at 200°F=0.5 Ω·m, and R_w at 200°F=0.1 Ω·m.
b. Taking into consideration the nonideality of the shale membrane, estimate E_{SSP} if R_{sh}=2 Ω·m.

6.3 Estimate E_{SSP} for a formation with the following characteristics.

Formation temperature, °F	90
Formation thickness, ft	100
R_{mf}, Ω·m at 90°F	2
R_w, Ω·m at 90°F	2.5

a. Assume that the formation water is a pure NaCl solution.
b. Take into account the fact that water is fresh, so the effect of salts other than NaCl has to be considered.

6.4 The electric log in Fig. 6.31 was obtained in a well drilled with freshwater-based mud where R_{mf}=0.65 at 85°F. Maximum temperature recorded in the well was 165°F at 8,700 ft. If Zone 1 is a clean water-bearing sand and formation water salinity is practically the same in all sands, determine R_w and I_{sh} of the different permeable zones.

6.5 Determine the formation water salinity of the formations discussed in Example 6.7.

6.6 The heading of the electric log in **Fig. 6.33** lists the following information.

Total depth, ft	9,500
Maximum recorded temperature, °F	168
R_m, Ω·m at 168°F	0.56
Mud weight, lbm/gal	11

a. Estimate the formation water resistivity and salinity for the bottom permeable Zone A.
b. Explain the reduction of SP deflection displayed at Level B.
c. Explain the reduction of SP deflection displayed at Level C.
d. Explain the reduction of the SP deflection at Level D.

Nomenclature

a = chemical activity of an electrolyte
A = proportionality factor in Eq. 6.18
C = conductivity, m℧/m
d = diameter, in.
E_c = electrochemical potential, V
E_d = diffusion potential, V
E_k = electrokinetic potential, V
E_m = membrane potential, V
E_{SP} = measured self-potential, mV
E_{SSP} = static self-potential, mV
F = Faraday constant (96,516 Coulombs)
g_G = geothermal gradient, °F/100 ft
h = bed thickness, ft
I = electric current, A
I_{sh} = shale index
K = temperature-dependent coefficient in SP equation, mV
m_{eff} = shale-membrane efficiency
p = pressure, psia
Q_V = cation exchange capacity, meq/cm³
r = resistance, Ω
R = gas constant, (8.314 J/°C)
R = resistivity, Ω·m
S = saturation
t_{Cl} = chlorine anion transference number
t_{Na} = sodium cation transference number
T_a = absolute temperature, °K
T = temperature, °F
U = mobility, cm/s·V
ϕ = porosity

Subscripts

Ca = calcium
Cl = chlorine
eq = equivalent
f = sand formation
h = borehole
i = invaded zone
m = mud
mf = mud filtrate
Mg = magnesium
Na = sodium
o = oil
sh = shale formation
w = water

References

1. Telford, W.M. *et al.*: *Applied Geophysics,* Cambridge U. Press, Cambridge, MA (1976) 458–68.
2. Allaud, L. and Martin, M.: *Schlumberger, The History of a Technique,* John Wiley and Sons, New York City (1977) 116–23.
3. Pirson, S.F.: *Geological Well Log Analysis,* Gulf Publishing Co., Houston (1970) 36–58.
4. Lynch, E.J.: *Formation Evaluation,* Harper and Row Publishers, New York City (1962) 99–109.
5. Mounce, W.D. and Rust, W.M. Jr.: "Natural Potentials in Well Logging," *Trans.*, AIME (1944) **155**, 49–55.
6. Wyllie, M.R.J.: "An Investigation of the Electrokinetic Component of the Self Potential Curve," *Trans.*, AIME (1951) **192**, 1–18.
7. Gondouin, M. and Scala, C.: "Streaming Potential and the SP Log," *Trans.*, AIME (1958) **213**, 170–79.
8. Althaus, V.E.: "Electrokinetic Potentials in South Louisiana Tertiary Sediment," *The Log Analyst* (May–July 1967) 29–34.
9. Segesman, F.F.: "New SP Correction Charts," *Geophysics* (Dec. 1962) **27**, No. 6, Part I, 815–28.
10. *Log Interpretation Charts,* Schlumberger, Houston (1979).
11. Gondouin, M., Tixier, M.P., and Simard, G.L.: "An Experimental Study on the Influence of the Chemical Composition of Electrolytes on the SP Curve," *Trans.*, AIME (1957) **210**.
12. Silva, P.L. and Bassiouni, Z.: "One Step Chart for SP Log Interpretation," paper presented at the Canadian Well Logging Soc. 10th Formation Evaluation Symposium, Sept. 1985.
13. *Log Interpretation Principles,* Schlumberger, Houston (1972).
14. Alger, R.P.: "Interpretation of Electric Logs in Water Wells in Unconsolidated Formations," *Proc.*, SPWLA, Tulsa (1966) paper CC.
15. Evers, J.F. and Iyer, B.G.: "A Statistical Study of the SP Log in Fresh Water Formations of Wyoming's Big Horn and Wind River Basins," *Trans.*, Canadian Well Logging Soc. (1975) **5**.
16. Silva, P.L. and Bassiouni, Z.: "A New Approach to the Determination of Formation Water Resistivity from the SP Log," *Proc.*, SPWLA, Mexico City (1981) paper G.
17. Silva, P.L. and Bassiouni, Z.: "Applications of New SP Interpretation Charts to Gulf Coast Louisiana Fields," *The Log Analyst* (March 1983) **25**, No. 2, 12–15.
18. Smits, L.J.M.: "SP Log Interpretation in Shaly Sands," *SPEJ* (July 1968) 123–36; *Trans.*, AIME, **243**.
19. Lau, M.N. and Bassiouni, Z.: "Development and Field Applications of Shaly Sand Petrophysical Models: Part 2—The Spontaneous Potential Model," paper SPE 20387 available at SPE headquarters, Richardson, TX.
20. Combariza, G.: "By-Passed Oil Due to Misinterpretation of Well Logs," MS thesis, Louisiana State U., Baton Rouge (Dec. 1990).
21. Combariza, G., Lau, M.N., and Bassiouni, Z.: "Formation Water Resistivity Determination from SP Log in Case of Imperfect Shale Membrane," *Proc.*, 13th European Formation Evaluation Symposium, Budapest (Oct. 1990) paper E.

Chapter 7
Gamma Ray Log

7.1 Introduction

The gamma ray log is a continuous recording of the intensity of the natural gamma radiations emanating from the formations penetrated by the borehole vs. depth. All rocks have some radioactivity. The most abundant source of natural radioactivity is the radioactive isotope of potassium, K^{40}, and the radioactive elements of the uranium and thorium series. The radioactive material originally occurred in igneous rocks. It was subsequently distributed unequally throughout sedimentary formations during erosion, transport, and deposition. In sedimentary formations, radioactive elements tend to concentrate in clay minerals, which, in turn, concentrate in shales. **Table 7.1** gives the average gamma ray activity of sedimentary rocks, and **Fig. 7.1** shows the relative degree of radioactivity of the most common sedimentary rocks. Natural radioactivity is a function of the type of formation, its age, and the method of deposition. In general, sandstones, limestones, and dolomites have very little radioactive content. Black shales and marine shales exhibit the highest levels of radioactivity. Radioactivity is related to lithology, but not directly or rigorously. It can be used to distinguish between shale and nonshale formations and to estimate the shale content of shaly formations. A high level of radioactivity is not always associated with the presence of clay minerals. Such anomalous cases include potash salts, which have a high potassium content, and sandstones that contain uranium or thorium salts. Use of natural radioactivity in lithology differentiation requires good knowledge of the local lithology.

The gamma ray log is usually recorded with porosity-type logs—i.e., density, neutron, and sonic. As **Fig. 7.2** illustrates, the gamma ray curve is recorded on the first track of the log with a linear scale. All recordings are positive, with the radioactivity level increasing to the right. Because shales normally display the highest level of natural radioactivity, the gamma ray curve generally appears similar to the self-potential (SP) curve of the electric logs.

In empty boreholes or boreholes drilled with oil-based mud, an SP curve cannot be recorded. The gamma ray curve replaces the SP curve on the first track of the induction log (see Fig. 5.42).

7.2 Detection and Measurement of Nuclear Radiation

Four different detectors—the ionization chamber, the proportional counter, the Geiger-Mueller counter, and the scintillation counter—have been used in radiation well logging. The proportional counter currently is used only in neutron logging. The other three devices have been used in both gamma ray and neutron logging. The concept of the ionization chamber, proportional counter, and Geiger-Mueller counter is based on the ability of gamma rays to cause ionization upon passing through a medium, such as gas. The concept of the scintillation detector is based on the ability of gamma rays to produce tiny flashes of light as they decay in certain crystals.

The ionization chamber, shown schematically in **Fig. 7.3**, consists of a metallic housing that encloses a high-pressure (1,000 to 1,500-psia) gas. The steel case functions as an electrode. The second electrode is a central thin wire. A voltage, typically 100 V, is applied across these two electrodes to create an electric field. Ions and electrons produced by ionization resulting from gamma radiation entering the chamber are collected at these two electrodes. A flux of gamma rays produces a minute electric current that requires considerable amplification before it can be recorded.[2] The log deflection is proportional to the amount of current, which, in turn, is a measure of the intensity of the radiation. The ionization chamber is of simple construction and low voltage. However, it presents several disadvantages. It uses a high-pressure gas, and the amplification required could result in drift and instability. The ionization chamber also has a poor counting efficiency that ranges from 5% to 10%, depending on gas pressure and chamber length. Use of the ionization chamber has been mostly discontinued.[3]

The Geiger-Mueller counter and the proportional counter (**Fig. 7.4**) are designed like the ionization chamber. The voltage differential between the electrodes, however, is much higher in magnitude than in the ionization chamber. The gas enclosed in the counter is rarefied rather than pressurized, which causes the chamber to act differently. The introduction of an ionizing particle into the gas starts a chain reaction of multiple ionization. Because of the high potential difference, the initial ionization products are accelerated at a high speed toward the electrode of opposite polarity. They collide with other atoms and cause ionization. The process is repeated, and the result is an "avalanche" of ions.

In the Geiger-Mueller counter, this avalanche creates a uniform voltage pulse that is independent of the number of ions formed by the initial ionizing event. These pulses are counted, and the log deflection is proportional to the number of pulses per unit time. The signal from a Geiger-Mueller counter is much greater than that from an ionization chamber. The counter is sensitive to very weak radiation. The device is also quite stable because the number of pulses rather than magnitude is measured.

The major disadvantage of the Geiger-Mueller counter is its relatively slow reaction and poor measuring efficiency. It cannot distinguish between different radiation energies. With proper circuitry,

TABLE 7.1—GAMMA RAY ACTIVITY OF SEDIMENTARY ROCKS (after Ref. 1)

Lithology Type	Average Radioactivity in Radium Equivalent per gram × 10^{-12}
Black and grayish-black shale	26.1
Shale	20.3
Sandy shale	11.0
Siltstone	10.3
Calcareous shale	8.5
Shaly and silty sand	7.1
Granite wash	6.9
Sand	4.1
Limestone	3.8
Dolomite	3.1

the Geiger-Mueller counter is modified into a proportional counter that produces a signal proportional to the number of ions formed. Thus, it can discriminate between the various types of radiation. Moreover, it operates 100 times faster than the Geiger-Mueller counter.[2]

The scintillation counter (**Fig. 7.5**) is composed of a fluorescent crystal and a photomultiplier tube. An example of a scintillator is a sodium iodide crystal deliberately contaminated with a small amount of thallium iodide. When the crystal is struck by gamma radiation, it emits ultraviolet or blue-light photons. These photons strike a photosensitive surface (photocathode), which causes the emission of photoelectrons into an electric field. The photoelectrons strike anodes placed at successively higher potentials. At each anode, the electrons multiply as a result of secondary electron emissions. A large number of electrons are collected at the last anode. The electric charge is amplified and recorded.

The scintillation counter is the most suitable instrument for radiation detection. It is characterized by fast reaction and high detection efficiency (50% to 60%). These characteristics allow it to detect thin beds. It also senses gamma energies, making spectrometry possible. The scintillation counter, however, is temperature-sensitive. The crystal scintillating efficiency varies with temperature. The gain of the photomultiplier is also a function of temperature.[5]

7.3 Unit of Measurement

When gamma ray logging was first introduced, comparisons of logs run by different service companies were virtually impossible because they used different units of measurement (e.g., counts per minute, counts per second, radiation units, micrograms of radium-equivalent per ton of formation, and microroentgens per hour).[6]

This lack of standardization prompted the American Petroleum Inst. (API) to appoint a subcommittee to develop a standard practice that would create uniformity to allow direct comparison of radioactivity logs. The subcommittee designed a standard log heading and form and established a standard API unit of measurement for both gamma ray and neutron logs. A calibration facility for nuclear logs was designed and promoted. A standard procedure for presenting calibration data also was developed.[6,7]

The calibration facility, located at the U. of Houston and illustrated in **Fig. 7.6**, is primarily a pit 4 ft in diameter and 25 ft deep that is filled with three 8-ft-thick zones of radioactive concrete. A 5½-in. casing extends through the concrete sections 15 ft below the bottom of the pit. Radioactivity was obtained by adding radioactive material to a mixture of cement and Ottawa sand. The top and bottom concrete zones are low in activity. The center section, containing about 12 ppm uranium, 24 ppm thorium, and 4% potassium, has approximately twice the radioactivity of an average shale.

The industry adopted the "API gamma ray unit" as the standard for gamma ray log measurements. One API gamma ray unit is defined as 1/200 of the difference in log deflection between the two lower concrete zones of low and high radiation in the calibration pit. All gamma ray tools calibrated to API standards record gamma radiation in the same units of measurements. A tool can be calibrated in the pit. In practice, however, a secondary portable standard is calibrated with the pit. This portable standard, in turn, is used for

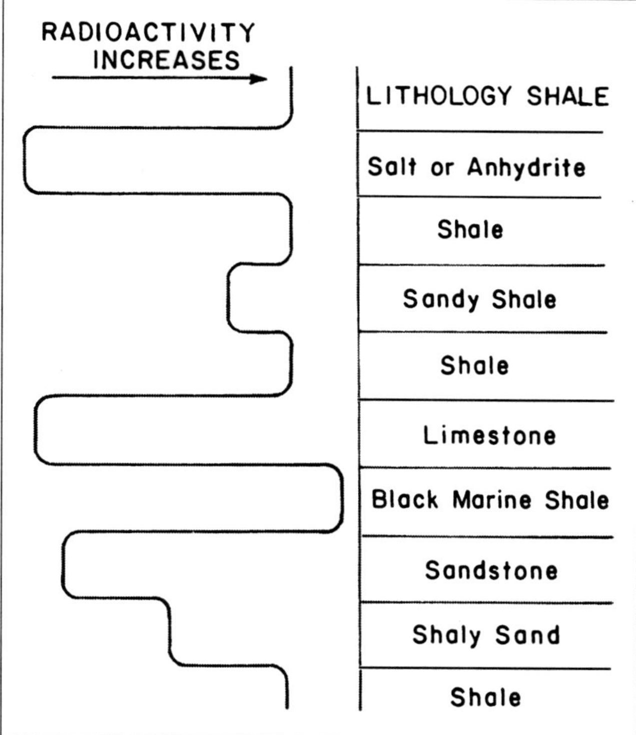

Fig. 7.1—Relative degree of radioactivity of the most common sedimentary rocks.

field calibration. To calibrate a secondary standard, which is usually a jig containing a radioactive pill, a specific tool first is calibrated in the pit. Then, the jig is positioned at a fixed, standardized distance from the tool, and its strength in API units is determined. This calibration is specific to each tool size, source spacing, and detector type.

7.4 Statistical Variations

Even with the gamma ray sonde stationary in the borehole, the number of gamma rays that reaches the detector varies with time. This is the result of the random nature of radioactive disintegrations. These fluctuations follow mathematical laws of probability and are known as statistical variations. To obtain a representative and reproducible value of radiation intensity for a specific formation, the number of gamma rays should be observed and averaged over a period of time. For good readings, a few seconds are usually required.

Statistical variations have been averaged by using a resistor-condenser (rc) smoothing electric circuit similar to that shown in **Fig. 7.7.** The electric pulses generated by the detector are passed through a circuit that impresses a charge, Q, on the condenser at each pulse. The current entering the capacitor, I_{in}, is

$$I_{in}=JQ, \quad (7.1)$$

where J is the pulse rate. The capacitor discharges through the resistor, r. The current leaving the capacitor, I_{out}, is

$$I_{out}=V/r, \quad (7.2)$$

where V is the potential across the condenser. At equilibrium, where $I_{in}=I_{out}$,

$$V=JrQ. \quad (7.3)$$

Then, the recorded voltage, V, is proportional to the pulse rate—i.e., the radioactivity of the medium generating the pulses.

If the pulse rate varies from J_1 to J_2, the recording device does not react instantaneously to the change. The voltage lags the change. It builds up or draws down vs. time according to[9,10]

$$V(t)=rQ[J_1+(J_2-J_1)(1-e^{-t/rc})]. \quad (7.4)$$

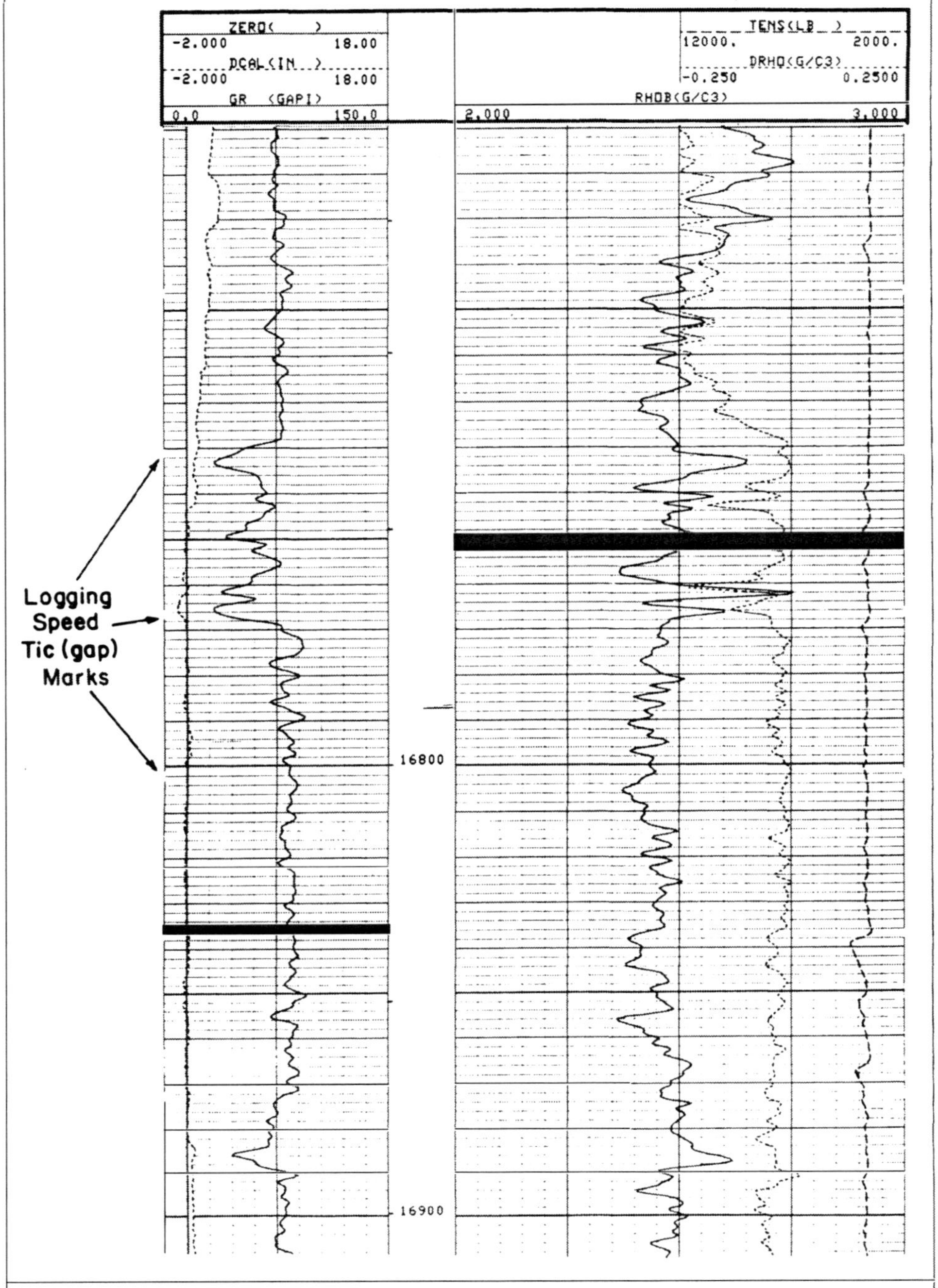

Fig. 7.2—Gamma ray curve recorded with the compensated-neutron/formation-density log.

Eq. 7.4 also can be written as

$$J_a(t)=[V(t)/rQ]=J_1+(J_2-J_1)(1-e^{-t/rc}), \quad \ldots\ldots\ldots\ldots (7.5)$$

where $J_a(t)$ is the apparent intensity reflected by the recorder. The meter then provides a running average of the pulse rate. This results in the smoothing out of statistical variations. The product rc is called the time constant of the electric circuit. When $t=rc$, then $(1-e^{-t/rc})=0.63$. One time constant is then the time necessary for the equipment to record 63% of the change in counts from the previous level. This constant determines the time over which the

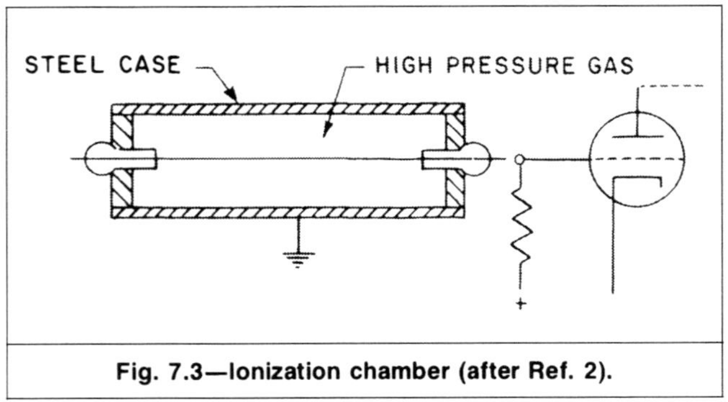

Fig. 7.3—Ionization chamber (after Ref. 2).

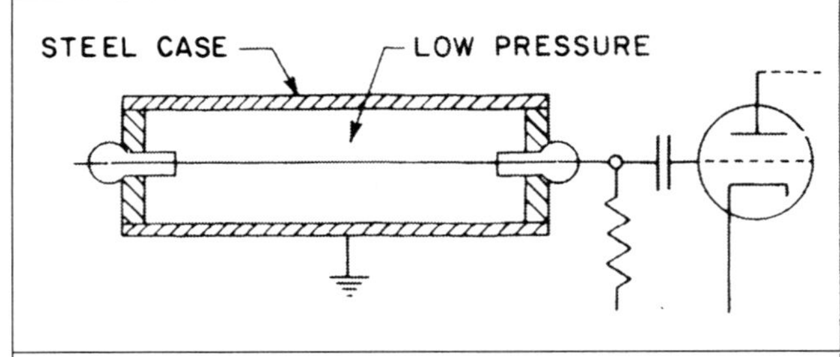

Fig. 7.4—Proportional counter, or Geiger-Mueller Counter (after Ref. 2).

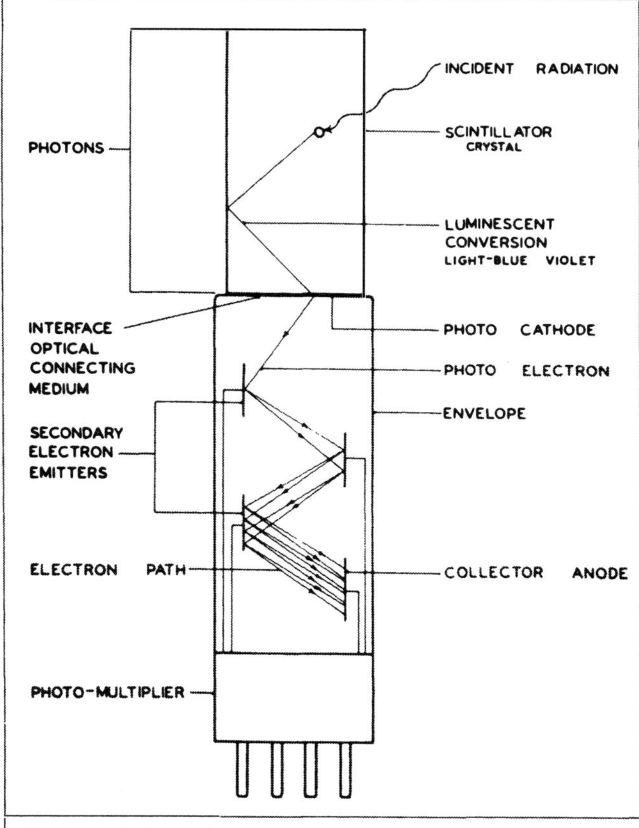

Fig. 7.5—Schematic of scintillator and photomultiplier tube (from Ref. 4).

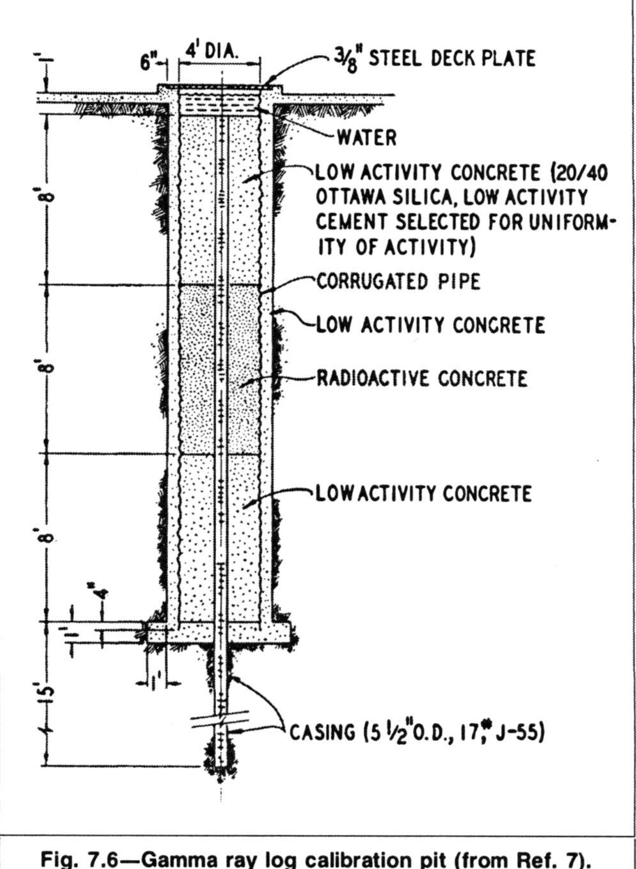

Fig. 7.6—Gamma ray log calibration pit (from Ref. 7).

radioactive counts are averaged. The time constant of a measuring device can be changed by a slight modification of the circuitry. Statistical variations are more pronounced for lower count rates, so a larger time constant is required for adequate averaging of the variations. In most cases, however, a time constant of 2 seconds is sufficient.[11] To check the appropriateness of the selected time constant, radioactivity of a relatively active formation is recorded for several minutes. This test is called a statistical check. For a log of acceptable quality, the residual fluctuations should not exceed a certain percentage, usually 5% to 15%, of the total change in radioactivity between shales and clean formations.

Example 7.1. Fig. 7.8 shows, hypothetically, the fluctuation of pulse rate vs. time. Show the smoothed-out response generated by the rc circuit having a time constant of 1 second. Repeat for a time constant of 2 seconds. Let the pulse rate immediately before these events be 100 counts/sec.

Solution. Using Eq. 7.5 for the first second gives

$$J_a(t)=100+10(1-e^{-t/\mathrm{rc}}).$$

Values of J_a for rc=1 and rc=2 calculated from this expression vs. time are listed below.

t (seconds)	J_a, rc=1 (counts/sec)	J_a, rc=2 (counts/sec)
0	100	100
0.25	102.2	101.2
0.50	103.9	102.2
0.75	105.3	103.1
1.0	106.3	103.9

Calculations for the other time intervals are performed similarly. For example, $J_a(t)$ for rc=1 is expressed for the following second as

$$J_a(t)=106.3+20[1-e^{-(t-1)/\mathrm{rc}}].$$

The smoothed-out response (**Fig. 7.9**) indicates that the larger the time constant, the smoother the profile.

7.5 Logging Speed

The number of pulses averaged by the detector depends on the radiation intensity, the counter's efficiency, the time constant, and the logging speed. An increase in logging speed is equivalent to an apparent delay of equipment reactions to a change in radiation intensity: the higher the speed, the smoother the tool response and vice versa. **Fig. 7.10** shows the effect of speed on log quality. The same section of a well is logged at speeds of 720 and 2,700 ft/hr. The quality of these logs is different. Beds are not defined as well on the 2,700-ft/hr log. But bed resolution also depends on the time constant; a better resolution calls for a smaller time constant. A good-quality log should be run with an optimum combination of logging speed and time constant.

Example 7.2.

a. A 10-ft-thick hypothetical bed with an average radioactive intensity of 100 counts/sec lies between two infinite beds of zero radioactivity. Show the response of a point detector having a time constant of 4 seconds run at a speed of 3,600 ft/hr.

b. Repeat for a speed of 900 ft/hr.

c. If a logging speed of 1,800 ft/hr is used, what time constant, if any, reproduces the response calculated in Part b?

Solution.

a. From Eq. 7.5,

$$J_a(t)=100(1-e^{-0.25t}),$$

where t is the time elapsed since the tool left the lower bed boundary.

If z is the tool location above the same boundary, then for a speed of 3,600 ft/hr, the following data are obtained.

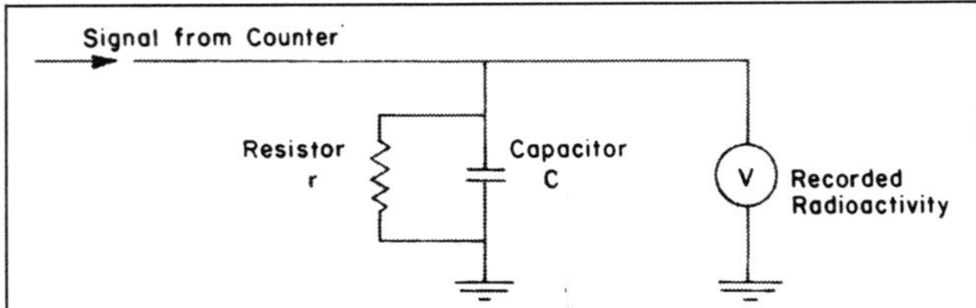

Fig. 7.7—Circuit for averaging statistical variations (from Ref. 8).

t (seconds)	z (ft)	$J_a(t)$ (counts/sec)
0	0	0
1	1	22
2	2	39
3	3	53
4	4	63
5	5	71
6	6	78
7	7	82
8	8	87
9	9	90
10	10	92

b. For a speed of 900 ft/hr, the values given below result.

t (seconds)	z (ft)	$J_a(t)$ (counts/sec)
0	0	0
2	0.5	39
4	1	63
6	1.5	78
8	2	87
10	2.5	92
20	5	99
40	10	100

Fig. 7.11 graphically presents these two cases. The intensity above the top bed boundary for the 3,600- and 900-ft/hr cases is calculated from the following two equations, respectively:

$J_a(t)=92e^{-0.25(t-10)}$

and $J_a(t)=100e^{-0.25(t-40)}$.

c. J_a given by Eq. 7.5 can be expressed as a function of z:

$$J_a(z)=J(0)\{1-e^{-[z/v(rc)]}\}, \quad (7.6)$$

where v is the logging speed in feet per second. Eq. 7.6 indicates that identical responses are obtained vs. depth for different speeds, provided that the product $v(rc)$ remains the same. The profile of Part b, obtained with a speed of 900 ft/hr and a time constant of 4 seconds, can be reproduced at a logging speed of 1,800 ft/hr but at a time constant of 2 seconds.

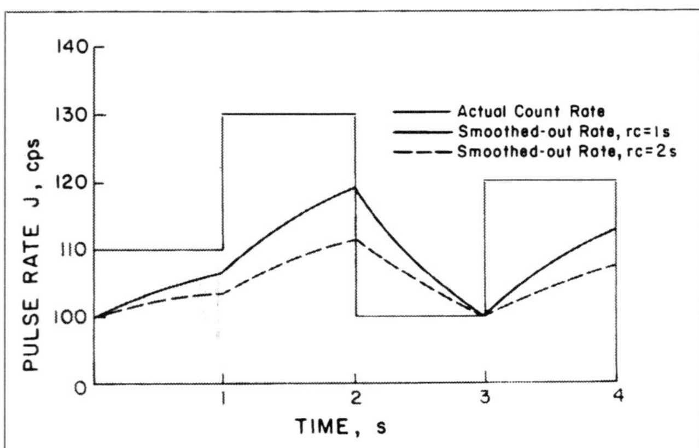

Fig. 7.9—Smoothed-out fluctuation using rc circuit, Example 7.1.

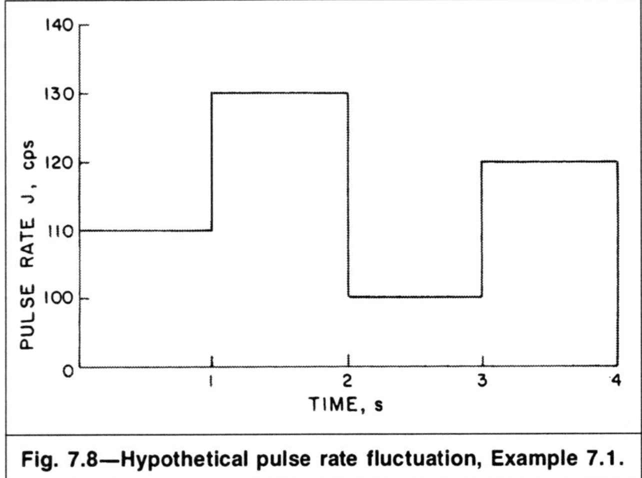

Fig. 7.8—Hypothetical pulse rate fluctuation, Example 7.1.

The logging-speed/time-constant combination results in two effects.

1. The log response is not representative of a bed whose thickness is less than the critical thickness—i.e., the distance traveled by the sonde in one time constant.

2. An anomaly is shifted in the direction in which the tool is moving. This lag, as well as the critical thickness, is expressed by[10]

$$h_c=vt_c \quad (7.7)$$

where h_c=lag or critical thickness, ft; v=logging speed, ft/sec; and t_c=time constant, seconds. To avoid excessive distortion of the gamma ray curve, v and t_c are chosen so that the lag h_c is about 1 ft. Most common logging speeds and corresponding optimum time constants are as shown below.

v (ft/hr)	t_c (seconds)
3,600	1
1,800	2
1,200	3
900	4

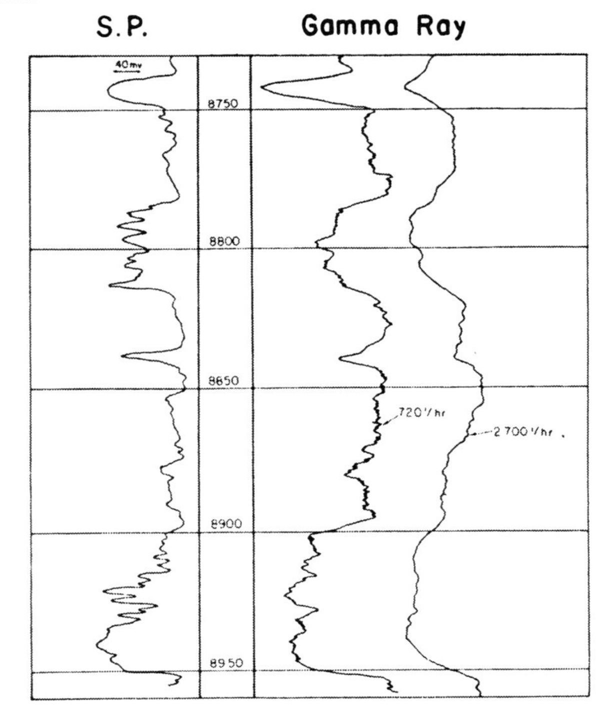

Fig. 7.10—Effect of speed. Log of the same section recorded at a speed of 720 and 2,700 ft/hr (from Ref. 9).

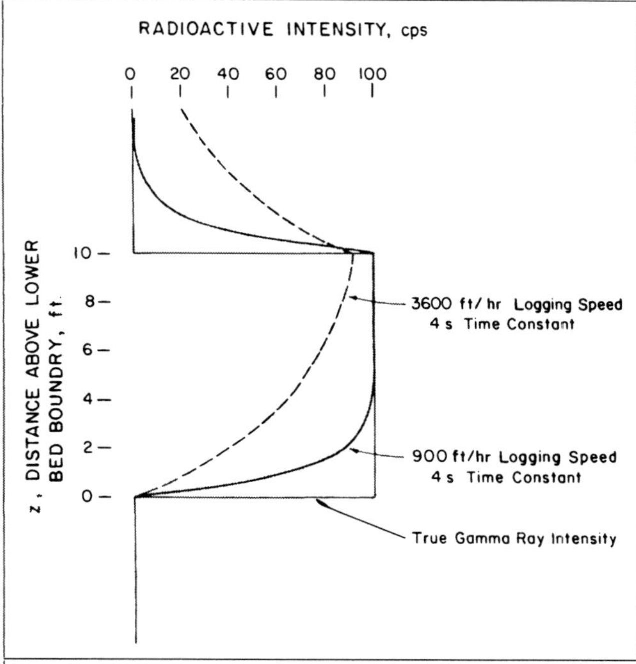

Fig. 7.11—Effect of logging speed on log response, Example 7.2.

The logging speed is usually indicated by tic marks or gaps in the far left vertical line of the first track. These tics are spaced 1 minute apart. For example, the log of Fig. 7.2 shows gaps spaced 35 ft apart; this indicates that the logging speed was 35 ft/min or 2,100 ft/hr.

The theoretical tool responses shown so far assume a point detector. A detector of appreciable length is used in actual logging. This results in additional smoothing of the curve, as illustrated by **Fig. 7.12.**

New equipment uses an averaging technique other than the rc circuit. One method consists of averaging the reading obtained at one depth with the readings that immediately precede and follow.

7.6 Tool Response

The gamma ray tool response, recorded with an optimum speed and time constant with the tool situated opposite a given formation, depends on several factors: specific formation radioactivity, a—i.e., gamma rays/sec-g; formation bulk density, ρ_b; specific activities of the borehole fluid; density of the borehole fluid; borehole diameter; characteristics of the detector and the counting system; and position of the detector in the borehole—i.e., eccentricity. In a cased hole, the tool response also depends on the specific activities of the casing and cement and on the thicknesses and densities of the casing and cement.

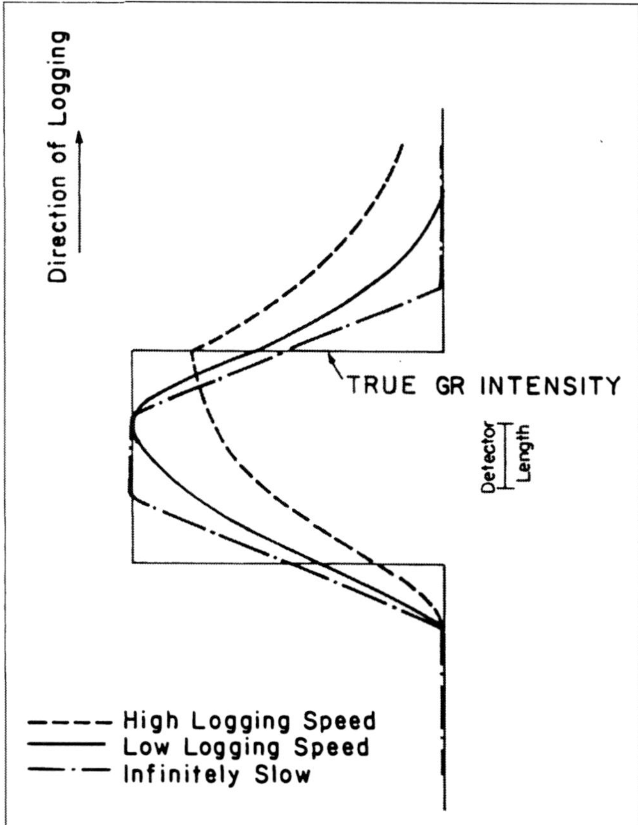

Fig. 7.12—Effect of detector length on gamma ray curve shape (after Ref. 9).

For qualitative interpretation, the borehole-environment effect (i.e., borehole fluid, cement, and casing) can be neglected because the interpretation relies on relative variation from one zone to another and because the borehole conditions affecting the gamma ray remain practically unchanged over long borehole sections. For quantitative interpretation, the effect should be taken into consideration, especially in large and cased holes.

Service company chart books contain charts to correct for borehole effect. **Fig. 7.13** is an example. This chart, which is based largely on laboratory experiments, gives the ratio of corrected gamma ray, γ_{cor}, to the gamma ray log response, γ_{log}, as a function of hole size and mud weight for centered and eccentered tools. γ_{cor} is defined as the response of a 3⅝-in. tool eccentric to an 8-in. hole filled with 10-lbm/gal mud. These are considered normal measurement conditions. In most cases, boreholes deviate from the vertical and the sonde can be assumed eccentric.

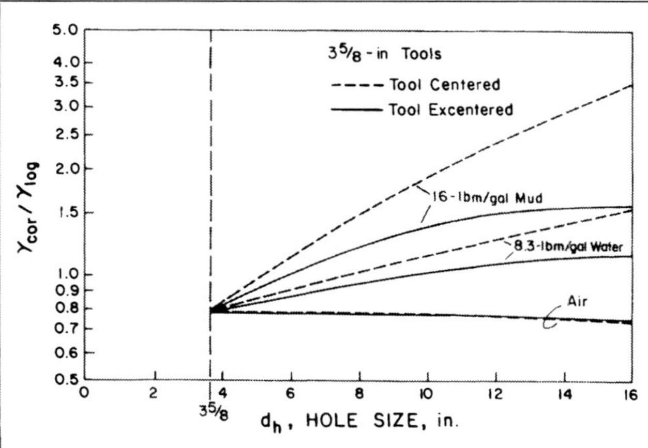

Fig. 7.13—Gamma Ray Correction for hole size and mud weight, after Ref. 12.

Example 7.3. A gamma ray log was recorded in an empty, open hole drilled with a 7⅞-in. bit. The gamma ray device is 3⅝ in.

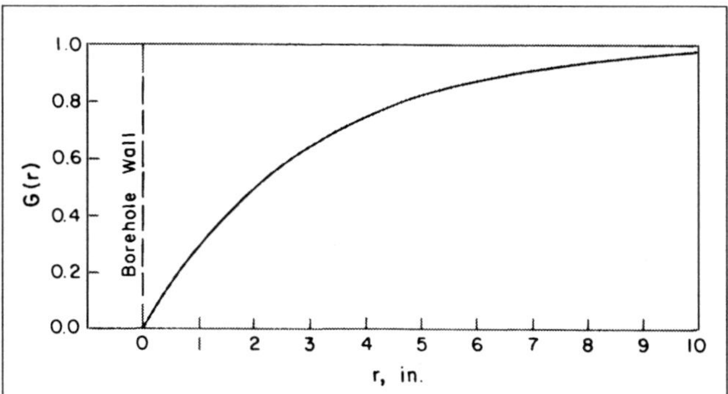

Fig. 7.14—Integrated radial geometric factor of the gamma ray log under the conditions stated in Example 7.4.

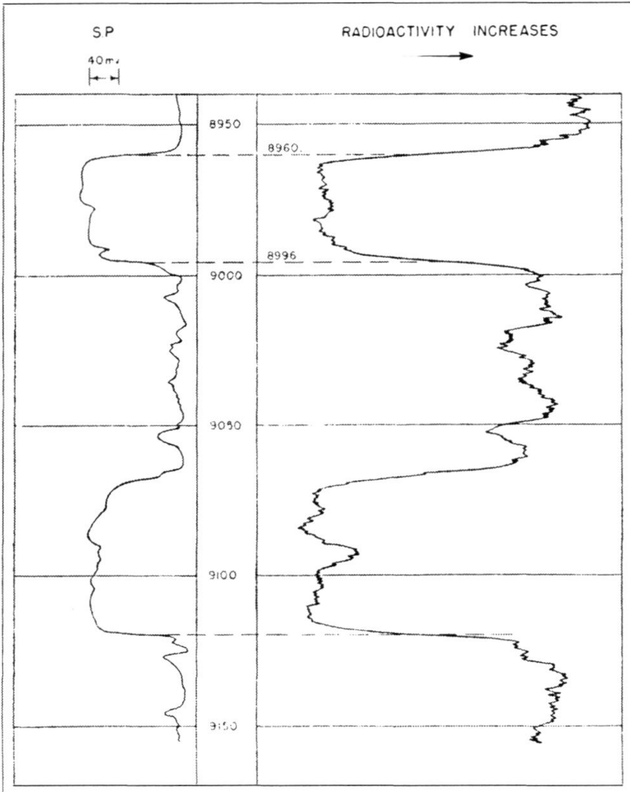

Fig. 7.15—SP and gamma ray logs recorded simultaneously in an open hole (from Ref. 9).

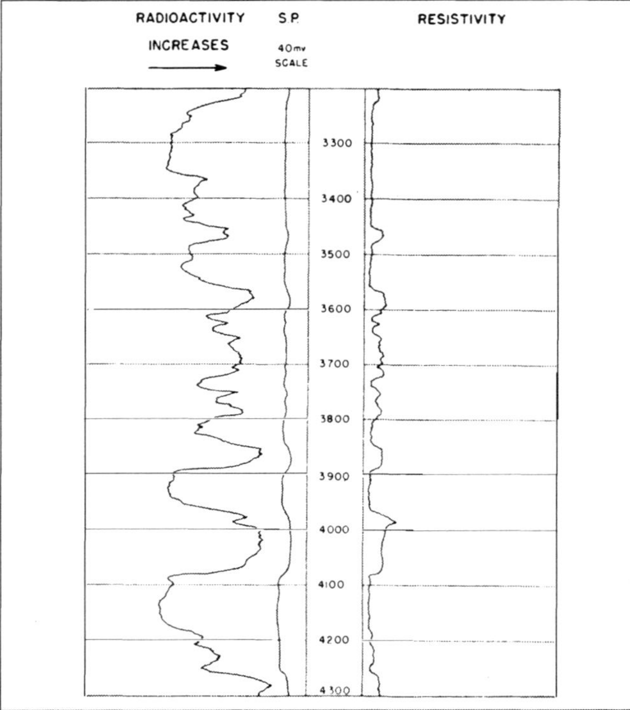

Fig. 7.16—Gamma ray and SP log recorded in high-salinity mud, $R_{mf} \cong R_w$ (from Ref. 9).

in diameter and runs simultaneously with a Formation Density Compensated (FDC^{SM}) tool.

a. Determine the corrected radioactivity, γ_{cor}, of a zone that registered 30 API units on the log.

b. On a subsequent run, with the same equipment, the hole has been filled with 10-lbm/gal mud. What level of radioactivity would now be displayed by the log for the same zone?

c. Repeat Part b for a 16-lbm/gal mud.

d. Repeat Parts b and c for the case where the gamma ray device is run simultaneously with a Borehole Compensated Sonic (BHC^{SM}) tool.

Solution.

a. From Fig. 7.13,

$$\gamma_{cor}/\gamma_{log}=0.79$$

and $\gamma_{cor}=0.79(30)=23.7 \cong 24$ API units.

b. Because the FDC is run in an eccentric fashion, the borehole environment closely matches that used to define γ_{cor}: 8-in. borehole, 10-lbm/gal mud, and tool eccentricity. Therefore, $\gamma_{log}=\gamma_{cor}=24$ API units.

c. For prevailing conditions and 16-lbm/gal mud, the chart in Fig. 7.13 gives

$$\gamma_{cor}/\gamma_{log}=1.2$$

and $\gamma_{log}=24/1.2=20$ API units.

d. Gamma ray devices are run simultaneously with the BHC centered. This affects the log reading. For a 10-lbm/gal mud,

$$\gamma_{cor}/\gamma_{log} \cong 1.1$$

(by interpolation) and

$$\gamma_{log}=24/1.1 \cong 22 \text{ API units.}$$

For the 16-lbm/gal mud,

$$\gamma_{cor}/\gamma_{log}=1.46$$

and $\gamma_{log}=24/1.46 \cong 16$ API units.

After the borehole-environment effect is normalized, the corrected reponse is proportional to the weight concentration of radioactive materials in the formation. It can be expressed for a formation containing n specific radioactive minerals as[11]

$$\gamma=\frac{1}{\rho_b}\sum_{i=1}^{n}\rho_i V_i a_i \quad (7.8)$$

$$=\frac{1}{\rho_b}\sum_{i=1}^{n}B_i V_i, \quad (7.9)$$

where ρ_i=density of the radioactive mineral, V_i=bulk volume fraction of the mineral, a_i=proportionality factor that corresponds to the radioactivity of the mineral and depends on the detector used and the sonde design, and $B_i=\rho_i a_i$=a constant for each specific radioactive mineral.

For Schlumberger devices, the relation between the concentration of K, Th, or U and the total gamma ray signal recorded in common sedimentary rocks can be approximated by[13]

$$\gamma=4C_{Th}+8C_U+C_K, \quad (7.10)$$

where γ=total gamma ray, API units; C_K=potassium concentration, wt%, and C_{Th} and C_U=thorium and uranium concentrations, respectively, in ppm. If the effect of formation density needs to be taken into account, the gamma ray log may be normalized by multiplying it by the bulk density value obtained from the density log.

The depth of investigation of the gamma ray tool (i.e., the volume of the formation contributing the major portion of the tool response) is difficult to determine by experimentation. An analytical treatment using Monte Carlo simulation shows that, in general, 90% of the signal comes from a shell 6 in. thick.[14] A less sophisticated calculation shows that the integrated radial geometric factor, $G(r)$, can be expressed as[15]

$$G(r)=1-e^{-r/\bar{h}}, \quad (7.11)$$

where r is the radial distance measured from the borehole wall and $\bar{h}$ is the mean free path. The mean free path, discussed in Sec. 2.7.3, is the average distance traveled by a photon in a medium between

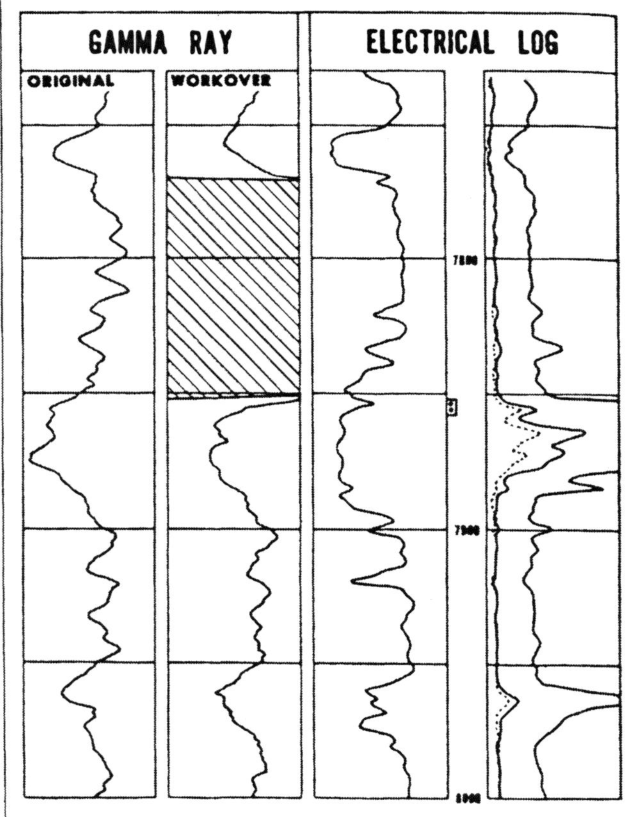

Fig. 7.17—Original and workover gamma ray log showing water channeling behind the casing (from Ref. 15).

successive interactions. $\bar{h}$ depends on photon energy and the medium density.

Example 7.4. Plot the integrated geometric factor vs. radius for an 8-in. borehole with a formation density of 2.65 g/cm^3, where the K^{40} radiation is detected by the tool.

Solution. From Eqs. 2.38 and 2.32,

$$\bar{h}=1/\alpha_\ell=1/\rho\alpha_m,$$

where ρ is the medium bulk density and α_ℓ and α_m are the linear and mass absorption coefficients, respectively. K^{40} emits gamma rays of 1.46-MeV energy. From Table 2.5, $\alpha_m=0.05127$ cm^2/g. Then,

$$\bar{h}=1/(2.65\times 0.05127)=7.36 \text{ cm}=2.9 \text{ in.}$$

Eq. 7.11 reduces to

$$G(r)=1-e^{-0.345r},$$

where r is in inches. **Fig. 7.14** is a plot of this function. Fig. 7.14 indicates that the first 7 in. of the formation generates 90% of the signal under the stated conditions.

7.7 Applications of the Gamma Ray Log

With few exceptions, the gamma ray log correlates very well with the SP log, as illustrated in **Fig. 7.15.** Like the SP log, the gamma ray log can be used to delineate shale beds and to correlate between wells. The gamma ray log substitutes for the SP log when the SP is flat as a result of low contrast between R_{mf} and R_w **(Fig. 7.16)** and when the SP log cannot be recorded, as in the case of oil-based mud, empty holes, and cased holes.

When potassium is the only or the major contributor to shale radioactivity, the gamma ray log response is used to estimate the shale content. A shale index, I_{sh}, is calculated from

$$I_{sh}=(\gamma_{log}-\gamma_{min})/(\gamma_{sh}-\gamma_{min}), \quad (7.12)$$

where γ_{log} is the log response in the zone analyzed, and γ_{sh} and γ_{min} are the log responses in shales and in zones of minimum radioactivity, respectively. I_{sh} is converted to shale content, V_{sh}, with the method detailed in Sec. 15.3.

In mineral exploration, gamma ray logging is used to detect radioactive minerals, such as potash or uranium, and nonradioactive minerals, such as coal.

The gamma ray log is particularly useful in cased-hole applications in both completion and production operations. The simultaneous recording of a casing-collar locator and the gamma ray is indispensible for accurate positioning of perforating guns and other downhole devices. Openhole information is tied to cased-hole conditions by correlation of an openhole SP or gamma ray log to the cased-hole gamma ray log, which, in turn, is tied to the casing-collar locations. Instead of measured depth, which usually differs from true depth by a few feet, the gun or any other tool is positioned with respect to the collars.

Application of gamma ray logging in production operations abounds. For example, prolonged fluid migration behind casing or through perforations causes radioactive anomalies that are easily recognizable on gamma ray logs.[15] **Fig. 7.17** shows openhole electrical and gamma ray logs recorded in a well completed in the 7,852 to 7,858-ft interval. It also shows a workover gamma ray run 2 years later to identify the source of a sudden increase in water production. Because the workover log showed an abnormally high radioactivity throughout the interval between the top of the perforation and the base of the next-higher saltwater sand, it was concluded that water production was caused by water channeling from this sand through the cement sheath. With this information, appropriate measures were taken, and the well was restored to water-free production.

The gamma ray log is also at the heart of radioactive-tracer logging. In this process, the fluid pumped into a borehole or a formation carries gamma-ray-emitting isotopes, either in solution or in suspension. A gamma-ray log run later will display an increase in radioactivity opposite zones that experienced fluid intake. Radioactive-tracer logging is used to determine waterflood injection profiles, zone response to fracture treatment, loss of circulation, casing leaks, perforated zones, and cement location.[16] **Fig. 7.18** shows an example of a gamma ray used to locate cement carrying a radioactive tracer. Gamma ray Run 1 is a base log obtained before cementing. Gamma ray Run 2 was made after 125 sacks of radioactive cement was pumped through perforations located at 3,678 to 3,692 ft.

7.8 Gamma Ray Spectrometry Log

The gamma ray log provides a measure of the total natural radioactivity of a formation, regardless of its energy level or energy spectrum. The spectral gamma ray log, or gamma ray spectrometry tool, also detects the naturally occurring gamma rays and defines the energy spectrum of the radiations. Because potassium, thorium, and uranium are responsible for the energy spectrum observed by the tool, their respective elemental concentrations can be calculated.[17]

Fig. 2.6 shows the gamma-ray-emission spectrum of the potassium, uranium, and thorium series. The observed spectrum is of a continuous rather than a discrete form. This results mainly from a detector-type depth of investigation and logging speed. **Fig. 7.19** illustrates the continuous spectrum obtained with a sodium iodide crystal scintillation detector. The spectrum shows three distinctive peaks characteristic of the three sources of natural radioactivity. These peaks are at energy levels of 2.62, 1.76, and 1.46 MeV. They correspond to gamma ray emissions associated with the decay of thalium (Tl^{208}) bismuth (Bi^{214}) and potassium (K^{40}), respectively. The three peaks are used to distinguish the thorium, uranium, and potassium because they are sharp and of relatively high magnitude.

One method for analyzing the pulse height is to divide the energy spectrum into several energy ranges known as windows. Fig. 7.19 shows the five window systems Schlumberger uses. The pulses that correspond to each window are recorded with a specific de-

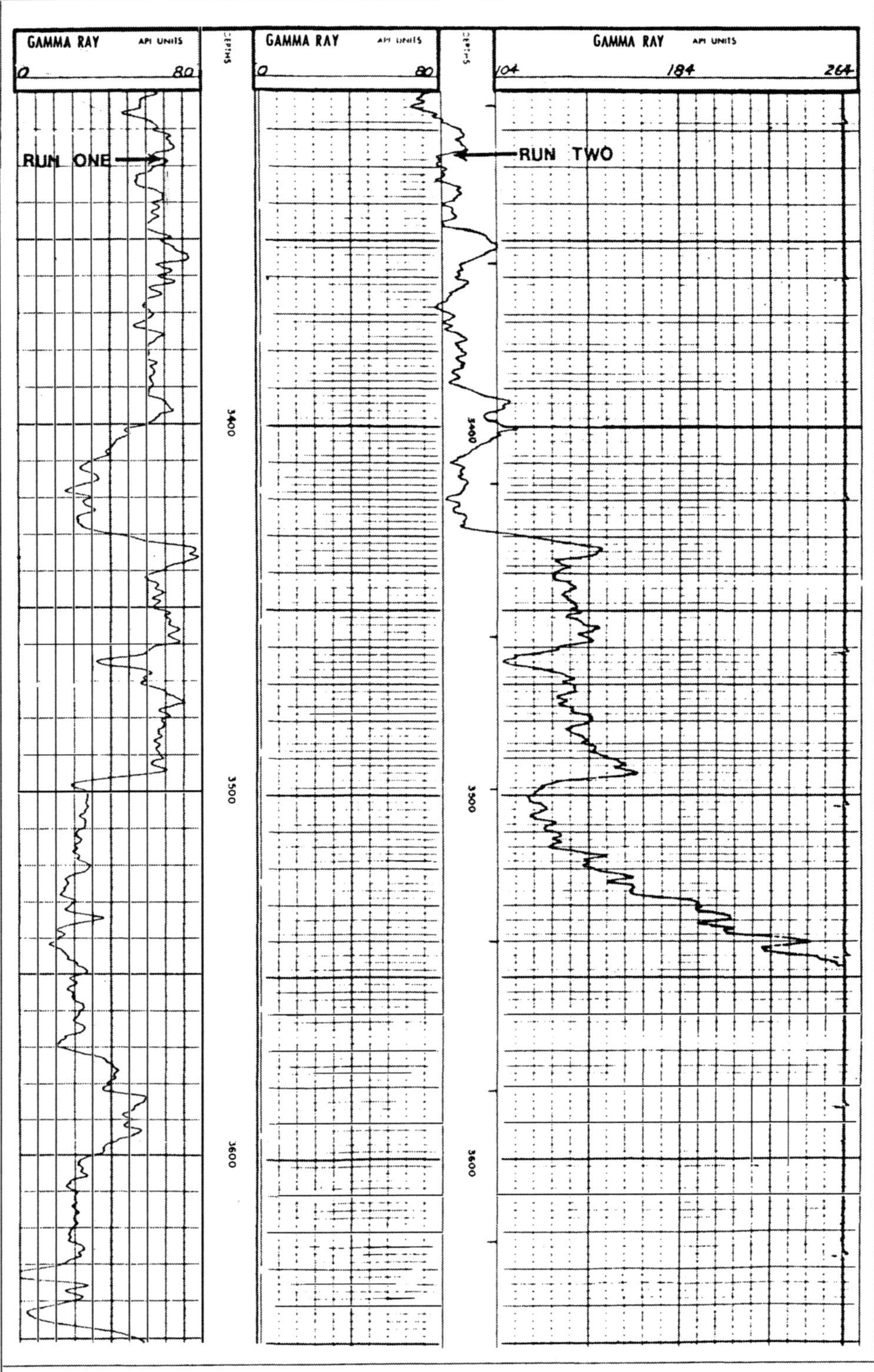

Fig. 7.18—Example of gamma ray logs used to locate radioactive cement (courtesy Schlumberger).

tector. The elemental concentrations are estimated by the matrix relationship[17]

$$[\boldsymbol{\Theta}]=[\mathbf{M}][\mathbf{W}], \quad (7.13)$$

where $[\boldsymbol{\Theta}]$ is the vector of Th, U, and K concentrations,

$$[\boldsymbol{\Theta}]=\begin{bmatrix} C_{\mathrm{Th}} \\ C_{\mathrm{U}} \\ C_{\mathrm{K}} \end{bmatrix}; \quad (7.14)$$

$[\mathbf{W}]$ is the vector of five window count rates, R_i,

$$[\mathbf{W}]=\begin{bmatrix} R_1 \\ R_2 \\ R_3 \\ R_4 \\ R_5 \end{bmatrix}; \quad (7.15)$$

and $[\mathbf{M}]$ is the 3×5 matrix of estimation coefficients, B_{ij}, for Window i and Element j.

$$[\mathbf{M}]=\begin{bmatrix} B_{1,\mathrm{Th}} & B_{1,\mathrm{U}} & B_{1,\mathrm{K}} \\ B_{2,\mathrm{Th}} & B_{2,\mathrm{U}} & B_{2,\mathrm{K}} \\ \vdots & \vdots & \vdots \\ B_{5,\mathrm{Th}} & B_{5,\mathrm{U}} & B_{5,\mathrm{K}} \end{bmatrix} \quad (7.16)$$

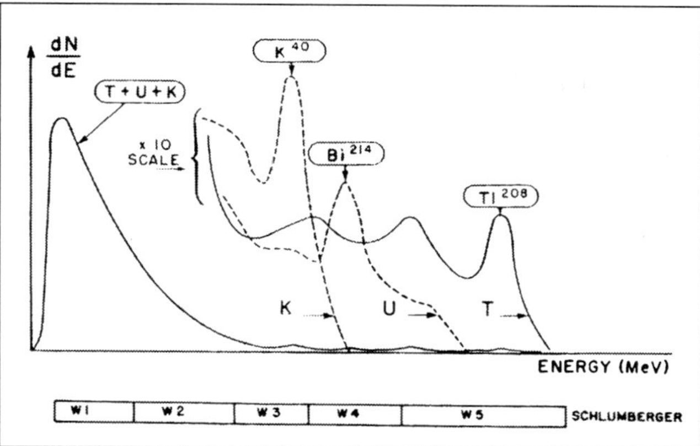

Fig. 7.19—Natural gamma ray energy spectrum determined with sodium iodide crystal detector (from Ref. 17).

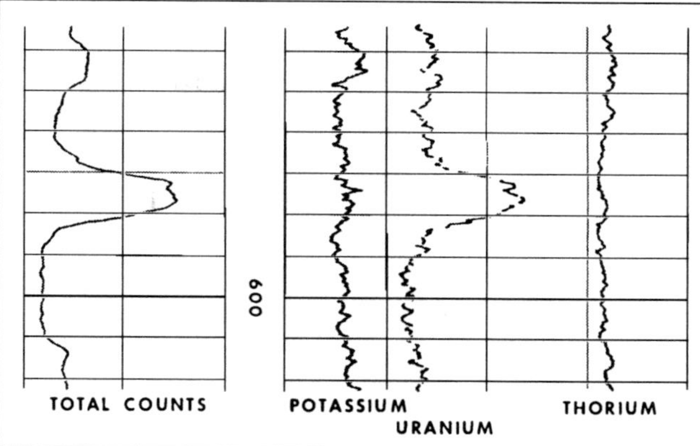

Fig. 7.21—Apparent shale response caused by high-uranium streak in a northern California well (from Ref. 18).

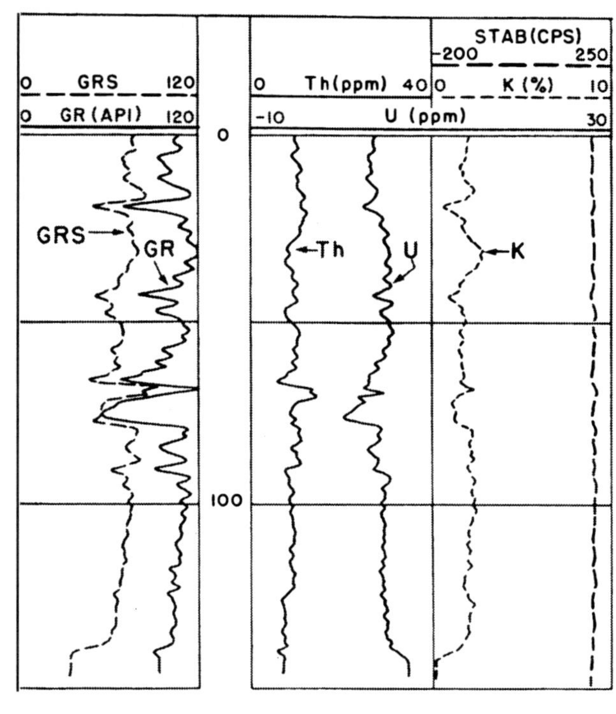

Fig. 7.20—Typical presentation of the gamma ray spectrometry tool (from Ref. 17).

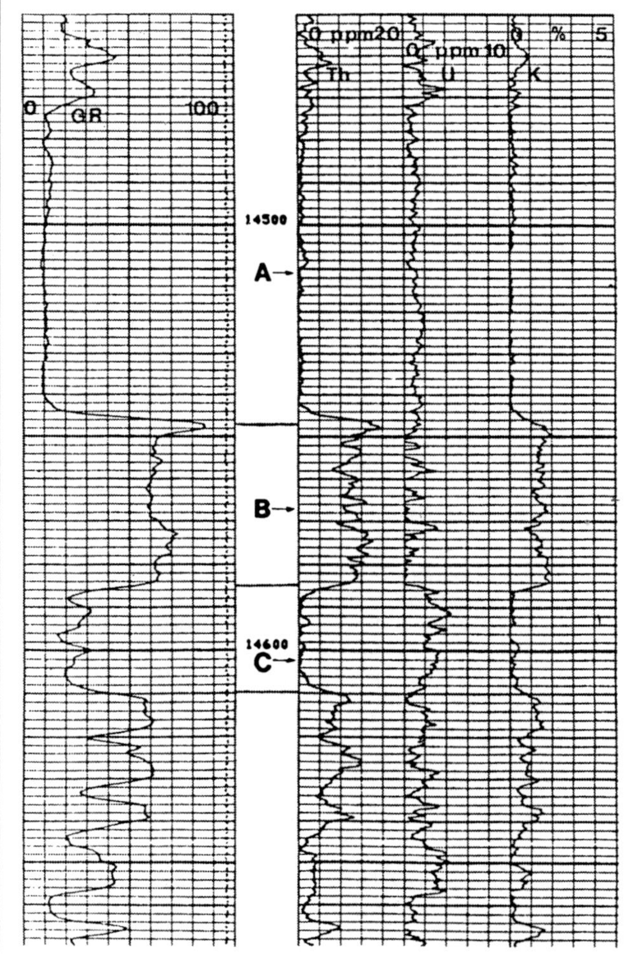

Fig. 7.22—Clean Zone C appears as shaly on the gamma ray log because of the presence of uranium (courtesy Atlas Wireline Services).

The matrix coefficients B_{ij} are determined for each tool in calibration pits built of formations that contain known quantities of the three elements.

Eq. 7.13 is solved for each level of the borehole, and the results are represented in a form similar to that shown in **Fig. 7.20.** The log shows thorium and uranium curves scaled in parts per million and a potassium curve scaled in percent. The conventional total gamma ray log is also presented. It is obtained by linear combination of the three elements' individual responses and by use of a relationship similar to that expressed in Eq. 7.10. A "uranium free" gamma ray curve (GRS in Fig. 7.20) obtained by combining only the potassium and thorium can also be presented. The log quality is indicated by a stabilization curve recorded in some cases.

The gamma ray spectrometry log has several potential applications in geological and engineering studies. The amounts and types of elements present in a formation are determined by the way the formation is deposited and what has happened to it since deposition. The concentration curves calculated show a correlation to depositional environment, diagenetic processes, clay type, and clay volume.[17]

One major application is the estimation of shale content. On the conventional gamma ray log, high-radioactivity zones were considered to be shale and were not analyzed; or if they were analyzed, a shale correction was applied that could have resulted in a misleading interpretation. An example (**Fig. 7.21**) was obtained through a sandstone encountered in a northern Oklahoma well. The high-API value appearing on the gamma ray log at 570 to 580 ft could be interpreted as a shale streak. With the benefit of the spectrometry log, it was determined that the high radioactivity was actually caused by high uranium concentration in a sandstone streak. This uranium was deposited from solutions migrating through the permeable streaks of the formation. This zone, in fact, was found to be a productive hydrocarbon zone.[18]

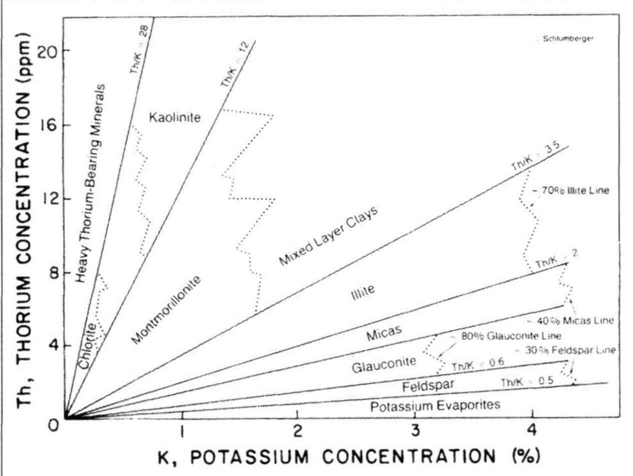

Fig. 7.23—Mineral identification from natural gamma ray spectrometry log (from Ref. 12).

A shale index, I_{sh}, can be calculated from the gamma spectrometry log information:

$$(I_{sh})_{Th}=[(C_{Th})_{log}-(C_{Th})_{min}]/[(C_{Th})_{sh}-(C_{Th})_{min}], \quad \ldots (7.17)$$

$$(I_{sh})_{K}=[(C_{K})_{log}-(C_{K})_{min}]/[(C_{K})_{sh}-(C_{K})_{min}], \quad \ldots (7.18)$$

and/or $(I_{sh})_{uf}=[(\gamma_{uf})_{log}-(\gamma_{uf})_{min}]/[(\gamma_{uf})_{sh}-(\gamma_{uf})_{min}]$. . .(7.19)

C_{Th}, C_K, and γ_{uf} are the log responses indicated by the thorium-, potassium-, and uranium-free curves, respectively. The subscripts *sh* and min refer to the logs' responses in shales and in zones that indicate a minimum level of radioactivity. The I_{sh} values calculated from Eqs. 7.17 through 7.19 are more representative than those obtained from the gamma ray. This results from the exclusion of uranium, which is associated with radioactive minerals other than those found in shale—i.e., organic materials. For example, Zone C in **Fig. 7.22** appears shaly on the gamma ray log. However, it contains almost no thorium or potassium, which indicates that it is clean.

Clay minerals can be identified from the crossplot in **Fig. 7.23** from potassium and thorium concentrations. Because the compositions of many clay minerals vary somewhat, a mineral location on the plot is not a unique point but a general area. Zone B in Fig. 7.22 plots as mixed-layer clays.

Example 7.5. Calculate the uranium-free gamma ray (γ_{uf}) for Levels A through C of Fig. 7.22. Estimate the I_{sh} of each zone using the total and uranium-free gamma ray levels. Which of the two is the better estimate?

Solution. Log readings at the three levels are as given below.

Level	γ (API units)	C_{Th} (ppm)	C_U (ppm)	C_K (%)
A	10	0	1.25	0
B	60	8.5	0	1.7
C	24	0	3	0

γ_{uf} is estimated with Eq. 7.10:

$$\gamma_{uf}=\gamma-8C_U,$$

where γ_{uf} and γ are in API units and C_U is in ppm.

The shale indices $(I_{sh})_1$ and $(I_{sh})_2$ are derived from total gamma ray, γ, and uranium-free, γ_{uf}, responses with Eqs. 7.12 and 7.19:

$$(I_{sh})_1=(\gamma\text{-}10)/(60\text{-}10)$$

and $(I_{sh})_2=(\gamma_{uf}\text{-}0)/(60\text{-}0)$.

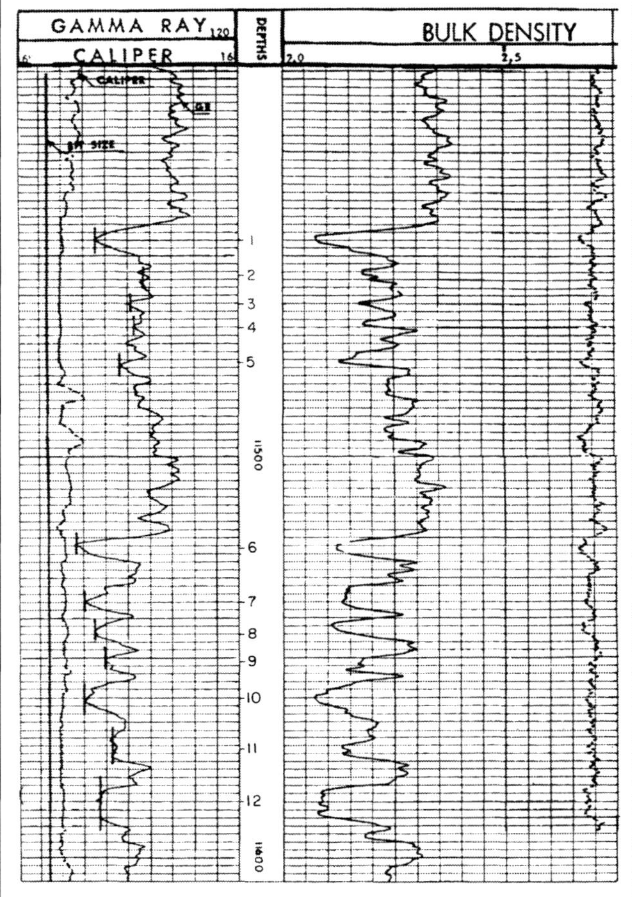

Fig. 7.24—IES log of Problem 7.3 (courtesy Amoco Production Co.).

The values of γ_{uf}, $(I_{sh})_1$, and $(I_{sh})_2$ are listed below.

Level	γ_{uf} (API units)	$(I_{sh})_1$ (%)	$(I_{sh})_2$ (%)
A	0	0	0
B	60	100	100
C	0	28	0

The value of $(I_{sh})_2$ derived from the uranium-free value is more representative.

Review Questions

1. What formation property does the gamma ray log reflect?
2. Is the gamma ray response a rigorous lithology indicator? Explain.
3. Why do the SP and gamma ray curves correlate in shale and sand sequences of sediments?
4. Discuss the concepts, advantages, and disadvantages of the different detectors used in radiation logging.
5. What is meant by the efficiency of a radiation detector? What determines such efficiency?
6. What is the standard unit of gamma ray measurement? What prompts the use of this unit? How is it defined?
7. How are gamma ray logging devices calibrated to this standard unit?
8. Even with the tool stationary in the borehole, the amount of radiation passing through the detector fluctuates with time. What is this phenomenon called? Describe two methods used to smooth out these fluctuations.
9. What is meant by the "time constant" of a gamma ray logging tool? How does it control measurement quality?
10. How does logging speed affect log quality? How is the optimum speed selected?

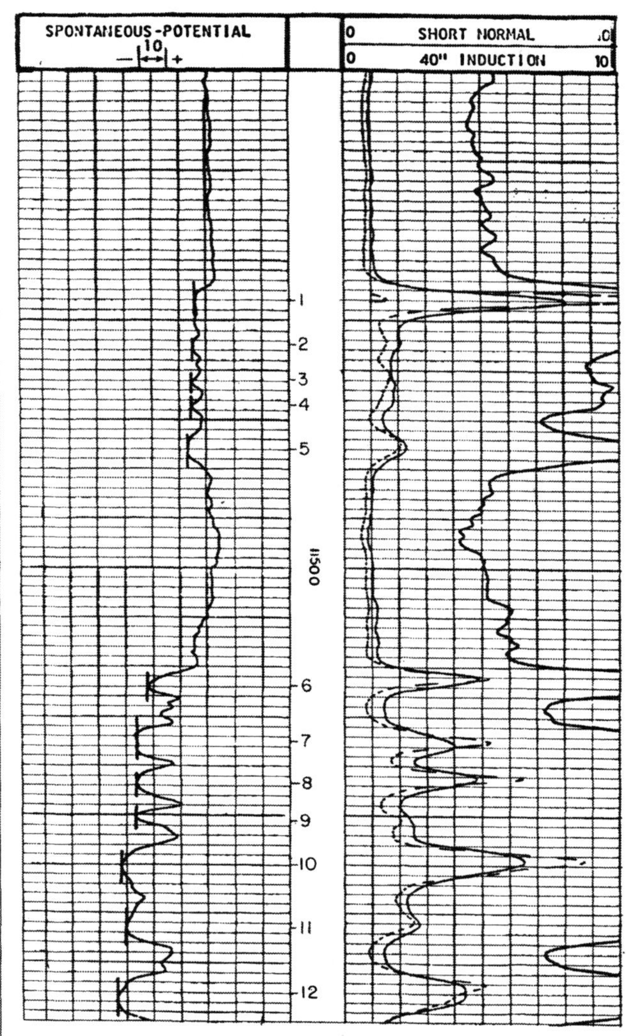

Fig. 7.25—Density log of Problem 7.3 (courtesy Amoco Production Co.).

11. What parameters determine the gamma ray tool response when it is stationary in the borehole and opposite a given formation?

12. How is the borehole-environment effect treated in both qualitative and quantitative interpretations?

13. On average, what is the depth of investigation of the gamma ray tool? What factors control this parameter?

14. Discuss the different applications of the gamma ray log.

15. Is the formation shale index derived from the gamma ray log always representative? Explain.

16. Discuss the concept and application of the gamma ray spectrometry log.

Problems

7.1 Using the rc smoothing circuit, how long must a tool stay next to a formation to receive 95% of the counts available to it?

7.2 Prepare a plot of gamma ray vs. bulk density using the response of zones within the interval shown in Fig. 7.2. Does a correlation exist between these two parameters? If yes, provide an explanation.

7.3 **Figs. 7.24 and 7.25** show a 200-ft interval on the Induction Electric Survey™ and density logs run in a southern Louisiana well. Twelve zones were selected for analysis. The zones were marked on both the SP and gamma ray curves. Prepare a gamma ray vs. SP plot.

a. Mark and label the pattern that represents relatively clean sands.

b. Mark and label the pattern that represents shaly sands.

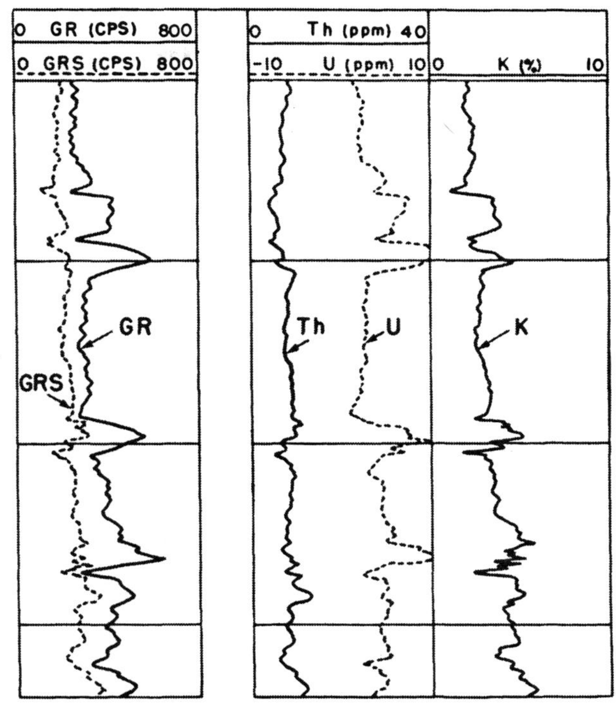

Fig. 7.26—Gamma ray spectrometry log of Problem 7.5 (from Ref. 17).

c. One of the twelve zones considered is abnormal—i.e., it does not fall into either of the two previous patterns. Which zone is it? How does it differ from the other zones? What is the nature of its abnormality?

7.4 A typical average shale contains 12 ppm, 3.7 ppm, and 2.7% of thorium, uranium, and potassium, respectively. Calculate its response on a gamma ray log calibrated to API standards.

7.5 The gamma ray spectrometry log of **Fig. 7.26** was recorded opposite a thick, Permian Basin carbonate reservoir. If thorium is known to be the best shale indicator in these sediments, can this log be used to detect permeable zones? If yes, mark these zones on the log. Explain the basis for your interpretation.

7.6 **Fig. 7.27** is a composite log of the openhole SP and microlog and a workover gamma spectrometry log obtained in a sandstone reservoir made up of several permeable zones. Can these data be used to identify the zone(s) responsible for the high produced-water/oil ratio?

7.7 Zone A marked in **Fig. 7.28** exhibits high levels of radioactivity. Explain the source of this radioactivity and estimate the shale index. Note that this zone tested at 447 Mcf/D.

Nomenclature

a = specific radioactivity, counts/sec-g
c = capacitance, F
C_K = potassium concentration, wt%
C_{Th} = thorium concentration, ppm
C_U = uranium concentration, ppm
G = integrated radial geometric factor
h = thickness, ft
$\bar{h}$ = mean free path, cm
h_c = lag or critical thickness, ft
I = electric current entering the capacitor, A
I_{sh} = shale index, fraction
J = pulse rate, counts/s
n = number of radioactive minerals
Q = electric charge, C
r = radial distance measured from borehole wall, in.
r = resistance, Ω
R_{mf} = mud-filtrate resistivity, Ω·m

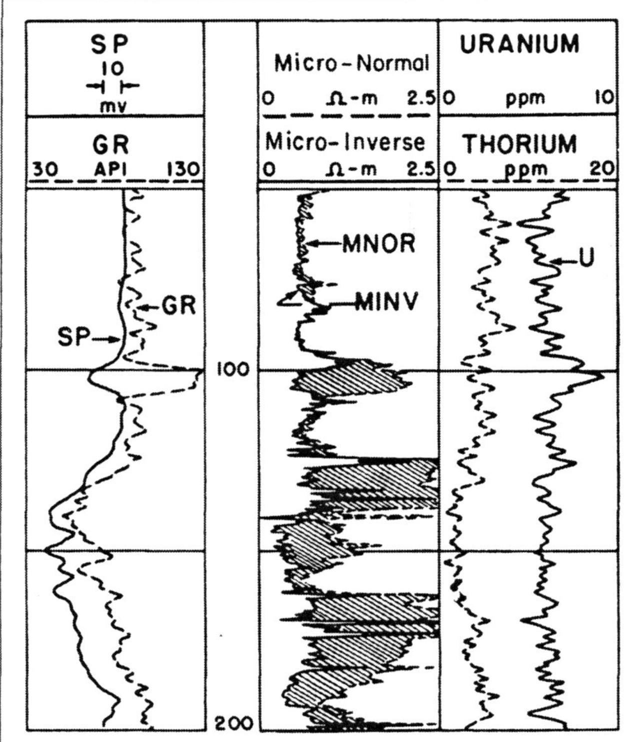

Fig. 7.27—Gamma ray spectrometry log of Problem 7.6 (from Ref. 17).

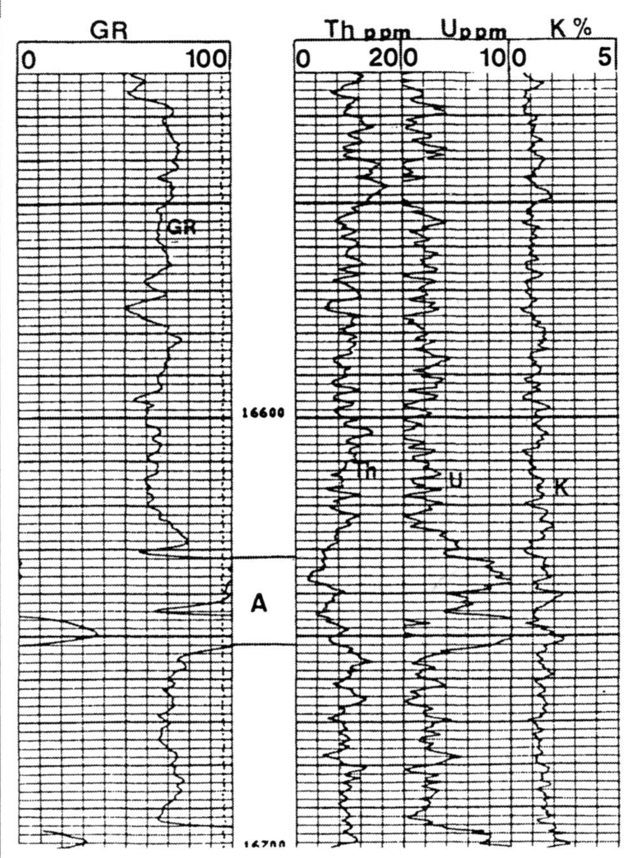

Fig. 7.28—Gamma spectrometry log of Problem 7.7 (courtesy Atlas Wireline Services).

R_w = formation-water resistivity, Ω·m
t = time, seconds
t_c = time constant, seconds
v = logging speed, ft/hr
V = potential, V
V = volume, fraction
V_{sh} = shale content, fraction
z = vertical distance, ft
α_ℓ = linear absorption coefficient, cm^{-1}
α_m = mass absorption coefficient, cm^2/g
γ = gamma ray activity, API units
ρ = density, g/cm^3
ρ_b = formation bulk density, g/cm^3

Subscripts

a = apparent
c = critical
cor = corrected
K = potassium
log = log response
min = minimum
sh = shale
Th = thorium
U = uranium
uf = uranium-free

References

1. Russell, W.L.: "The Total Gamma Ray Activity of Sedimentary Rocks as Indicated by Geiger Counter Determinations," *Geophysics* (April 1944) **IX,** No. 2, 180-216.
2. Wakefield, E.H.: "Nuclear Radiation—Its Detection and Measurement," *ASTM Bulletin* (FP39) (April 1953) 33-38.
3. *Log Review 1,* Dresser Atlas, Houston (1974) Sec. 8.
4. "Scintillation Spectrometer Well Logging," McCullough Tool Co., presented at the 1955 Joint Meeting of Rocky Mountain Petroleum Section/AIME, Denver, May 26-27.
5. Youmans, A. and Monaghan, R.: "Stability Requirements for Scintillation Counters Used in Radioactivity Logging," *JPT* (March 1964) 319-28; *Trans.*, AIME, **231.**
6. Belknap, W.B.: "Standardization and Calibration of Nuclear Logs," *Pet. Eng.* (Dec. 1959) B24-27.
7. *RP 33, Recommended Practice for Standard Calibration and Form for Nuclear Logs,* API, Dallas (Sept. 1959).
8. Lynch, E.J.: *Formation Evaluation,* Harper & Row Publishers, New York City (1962) 227-67.
9. Kokesh, F.P.: "Gamma Ray Logging," *Oil & Gas J.* (July 26, 1951).
10. Hearst, J.R. and Nelson, P.H.: *Well Logging for Physical Properties,* McGraw-Hill Book Co. Inc., New York City (1985) 191-278.
11. *Log Interpretation Principles,* Schlumberger, Houston (1972).
12. *Log Interpretation Charts,* Schlumberger, Houston (1979 and 1985).
13. Ellis, D.V.: *Well Logging for Earch Scientists,* Elsevier Scientific Publishers, Amsterdam (1987) 161-200.
14. Wahl, J.S.: "Gamma Ray Logging," *Geophysics* (1983) **48,** No. 11.
15. Killion, H.W.: "Fluid Migration Behind Casing Revealed by Gamma Ray Logs," *The Log Analyst* (Jan.-March 1966) **6,** No. 5, 46-49.
16. Flagg, A.H. *et al.*: "Radioactive Tracers in Oil Production Problems," *Trans.*, AIME (1955) **204**, 1-6.
17. Serra, O. *et al.*: "Theory and Practical Application of Natural Gamma Ray Spectroscopy," *Trans.*, 21st SPWLA Symposium, Lafayette (1980) paper Q.
18. Kowalski, J.J. and Asekun: "It May Not Be Shale," *Proc.*, 20th SPWLA Symposium, Tulsa (1979) paper p.

Chapter 8
Gamma Ray Absorption Logs

8.1 Introduction

The first commercial tool to use the physical phenomena of gamma ray scattering and absorption was introduced in the early 1950's.[1] The tool, known as the gamma-gamma, was developed initially to measure bulk density in situ as an aid to geophysicists in gravity-meter data interpretation. The tool consisted of a gamma ray source and a detector.[2,3] The sources used are cesium 137 and cobalt 60. The cesium, preferred because of its stability, decays with a half-life of about 30 years, emitting gamma rays of about 0.66 MeV. The cobalt yields gamma rays of 1.17 and 1.33 MeV and has a half-life of 5.2 years. The gamma rays emitted by the source, held in a skid in contact with the borehole wall, are directed toward the formation. Some of these gamma rays are absorbed and some are scattered away from the detector, but others are scattered into the detector and counted (**Fig. 8.1**). Gamma rays detected have energies ranging from 0.2 to 0.6 MeV. As explained in Chap. 2, the energy lost by scattering is dependent on the number of electrons in the formation and hence is proportional to the bulk density of the formation: the higher the density of the formation, the lower the response at the detector.

By the early 1960's, the formation density log provided by this early tool was accepted as a source of porosity because porosity can be related rigorously to bulk density by Eq. 2.50.

The response of the density tool is used alone to determine porosity when matrix and fluid densities are known. Porosity determination from in-situ density measurement did not become reliable, however, until a two-detector tool was developed in the late 1960's. The tool, called the Formation-Density Compensated (FDC^{SM}) tool corrects for certain borehole effects.[4]

The Litho-Density Tool^{SM} was introduced in the early 1980's.[5] It is designed to measure a photoelectric index, P_e, in addition to bulk density, ρ_b. P_e is highly dependent on lithology and can be used to determine matrix density, ρ_{ma}, which, in turn, is used with the corresponding bulk density to calculate porosity. Actually, P_e and ρ_b are used to solve for matrix density and porosity simultaneously.[6]

8.2 Single-Detector Formation Density Tool

Fig. 2.7 shows that the gamma rays emitted by the cesium 137 or cobalt 60 (energy range of 0.66 to 1.33 MeV) are attenuated predominantly by Compton scattering. Combining Eqs. 2.45, 2.47, and 2.48 shows that the detector count rate, R, can be expressed by

$$R=\alpha e^{-\beta\rho_b}, \qquad (8.1)$$

where α and β are coefficients. α depends mainly on the source strength and the fluid in the borehole. β depends mainly on detector/source spacing.

Eq. 8.1 shows that, if the count rate is measured, the tool can be calibrated to obtain directly the ρ_b of the medium traversed by the gamma rays. The early tools were calibrated in test pits containing blocks of materials with known bulk densities and boreholes of different diameters filled with muds of different densities. **Fig. 8.2** shows an example calibration curve.

Except for aluminum windows at the source and detector, the tool is insulated by a lead shield. Because the source and detector are kept in contact with the borehole wall, most of the gamma rays reaching the detector pass through the formation. In permeable formations, however, the gamma rays reaching the detector also pass through a mudcake with a density considerably different from that of the formation. The detector count rate is affected by the two media and results in a bulk density value between the formation and mudcake densities. The mudcake contribution can be corrected with charts such as that in **Fig. 8.3**. This correction is not always reliable because of uncertainties in mudcake thickness and composition. The interpretation of a single-detector tool was further complicated by the necessity for manual conversion of log counting rates to density.[7] This shortcoming prompted the development of the dual-detector formation density tool.

8.3 Dual-Detector Density Tool

A dual-detector tool was developed to compensate for mudcake and minor borehole irregularities automatically.[4] A schematic of this tool, called the compensated device, is shown in **Fig. 8.4**. It is equipped with two detectors situated at different spacings from the source. The long-spaced detector is at the same spacing as in the single-detector, uncompensated device and yields the same responses. The short-spaced detector, because of its closeness to the radiation source, is particularly sensitive to the mudcake and hole irregularities; thus, it yields a response different from that of the long-spaced detector. The difference in the detectors' count rates reflects the spacings; the formation bulk density, ρ_b; and the mudcake density, ρ_{mc}, and thickness, h_{mc}. **Fig. 8.5** shows the responses of the two detectors for a case without mudcake. The line of Fig. 8.5, usually called a "spine," reflects the effect of change in formation bulk density and is scaled in ρ_b values.

Fig. 8.6 illustrates the effect of mudcake thickness for formation and mudcake densities of 2.5 and 1.5 g/cm^3, respectively. The curved line that leaves the spine, called the "rib," indicates the effect of increase in h_{mc}. As h_{mc} becomes quite large, both detec-

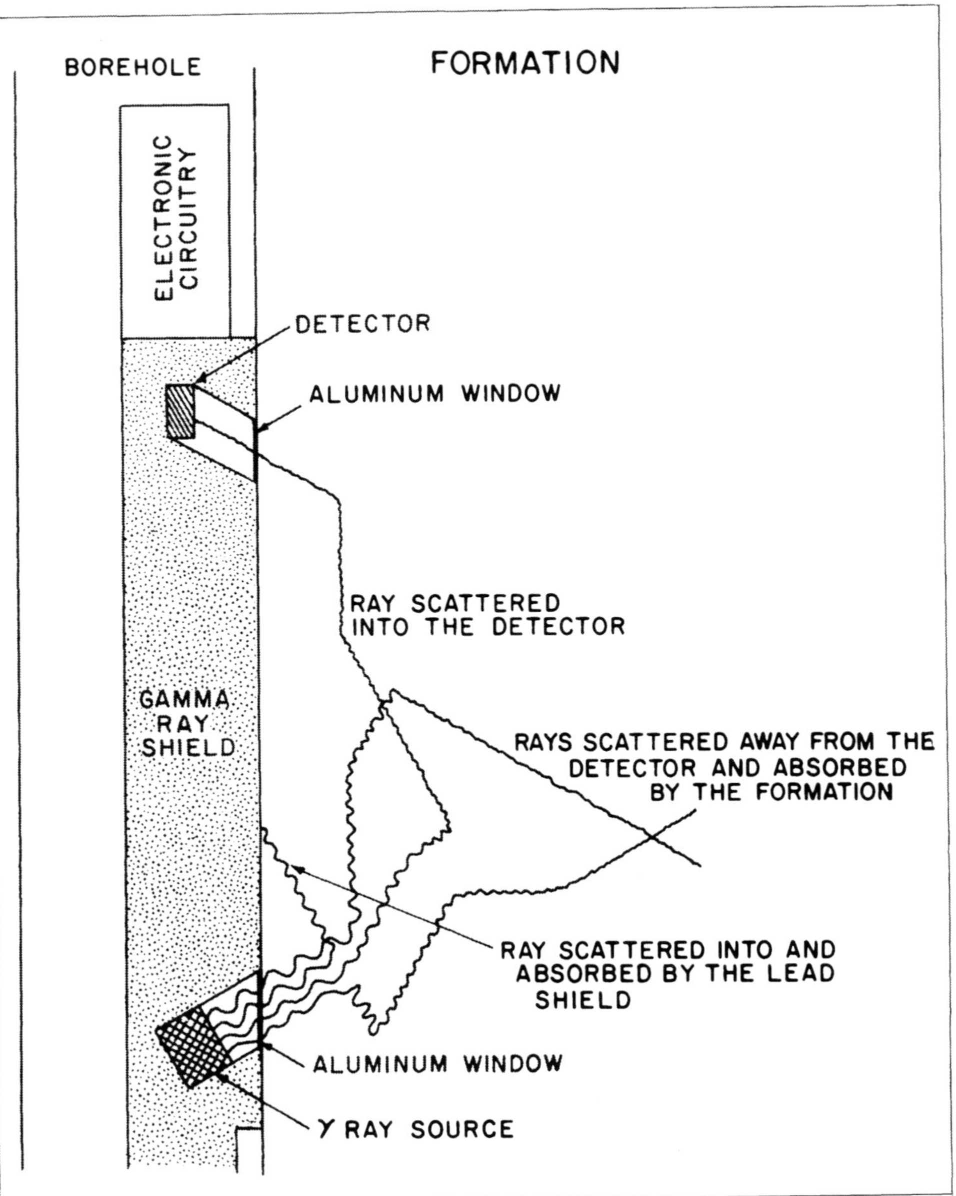

Fig. 8.1—Schematic of the early single-detector gamma-gamma density tool.

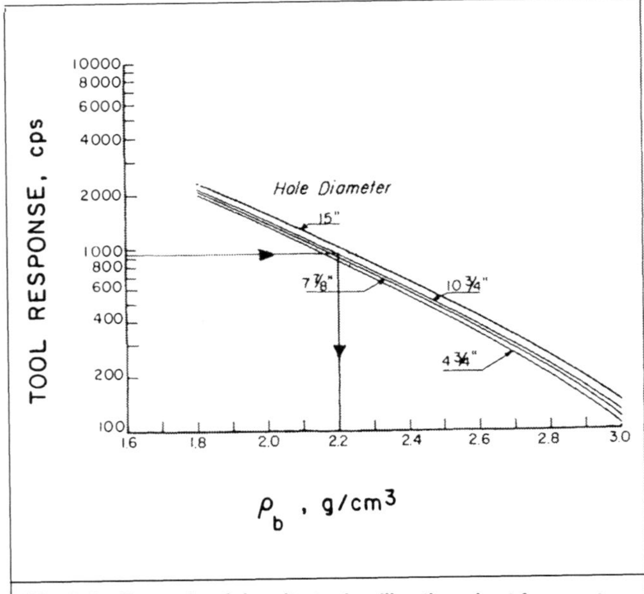

Fig. 8.2—Example of density tool calibration chart for a water-filled hole and no mudcake (from Ref. 7).

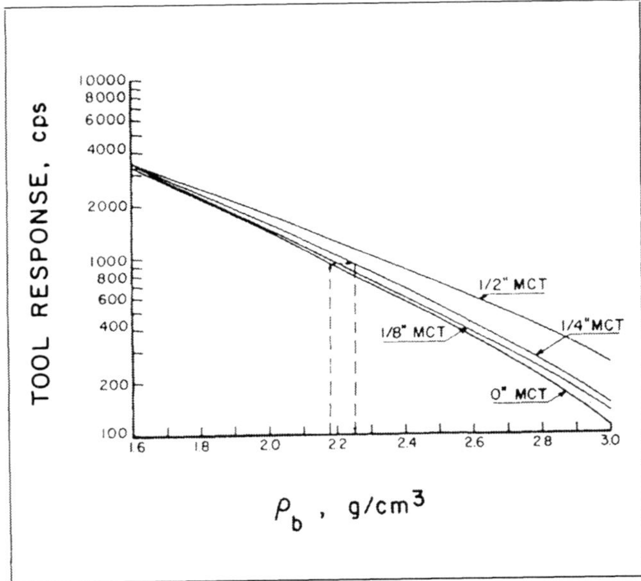

Fig. 8.3—Chart showing corrections for mudcakes having a density of 1.5 g/cm^3 to be applied to Fig. 8.2 (from Ref. 7).

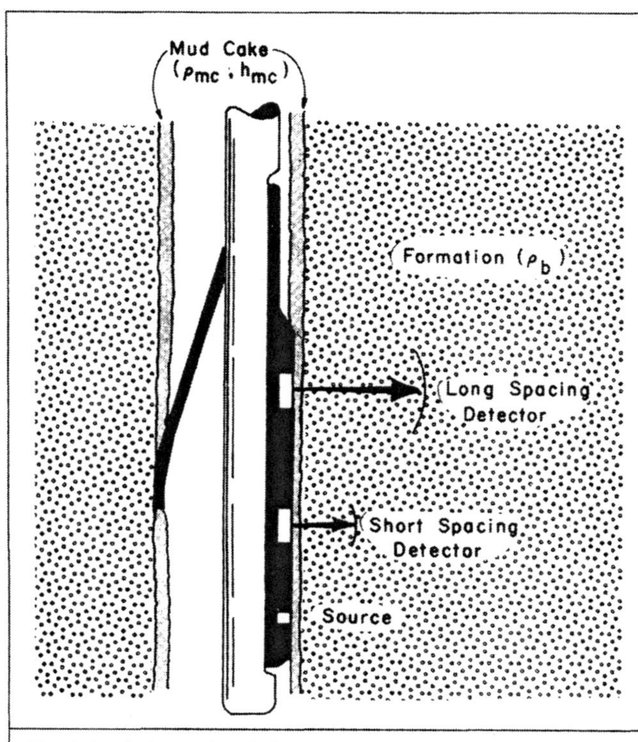

Fig. 8.4—Schematic of the dual-detector density tool (from Ref. 4).

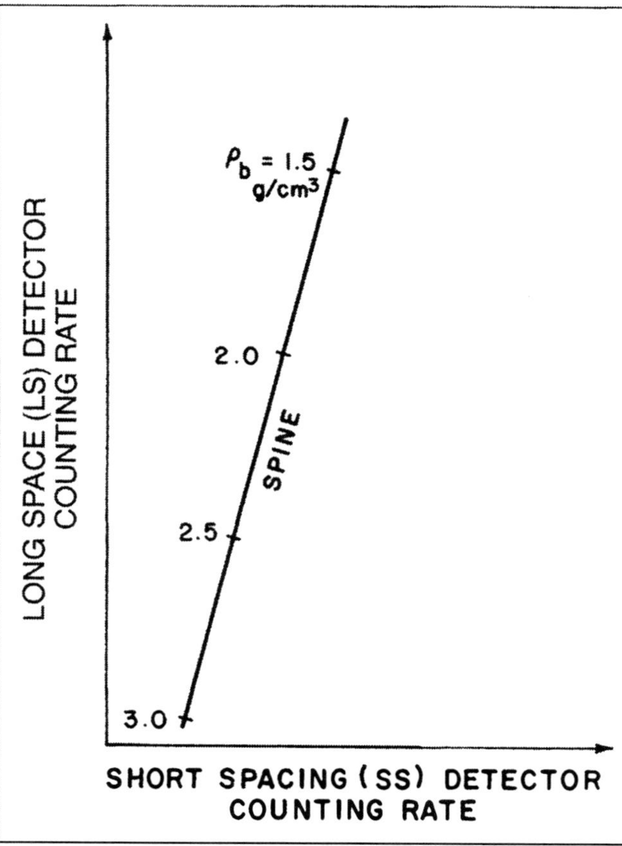

Fig. 8.5—Effect of formation bulk density on long- and short-spaced detector responses in the absence of mudcake (after Ref. 4).

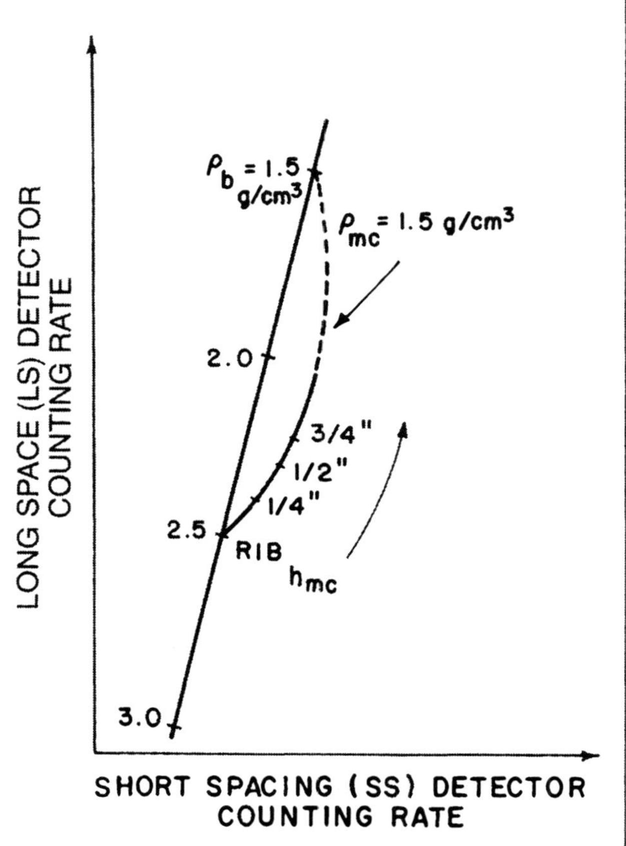

Fig. 8.6—Effect of mudcake thickness on long- and short-spaced detector responses (after Ref. 4).

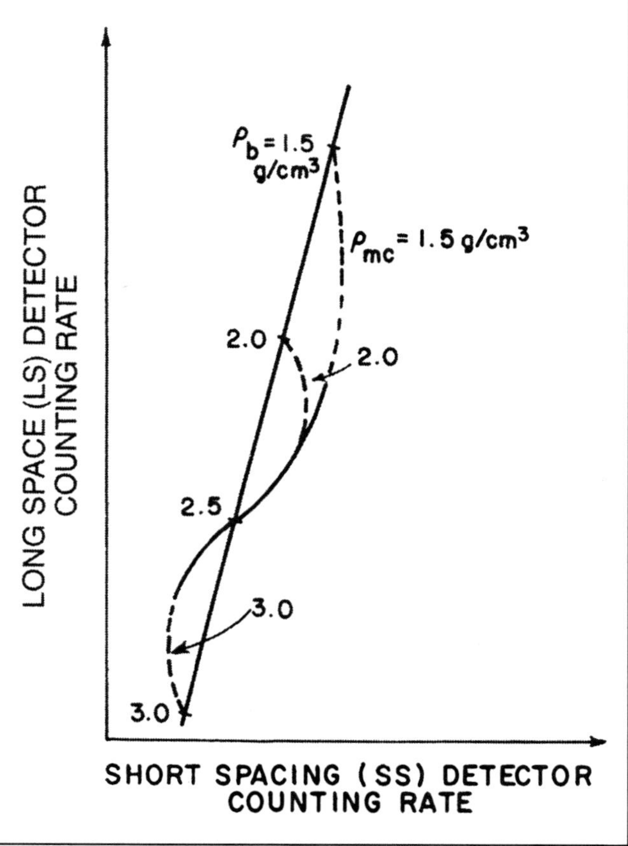

Fig. 8.7—Effect of mudcake density on short- and long-spaced detector responses (after Ref. 4).

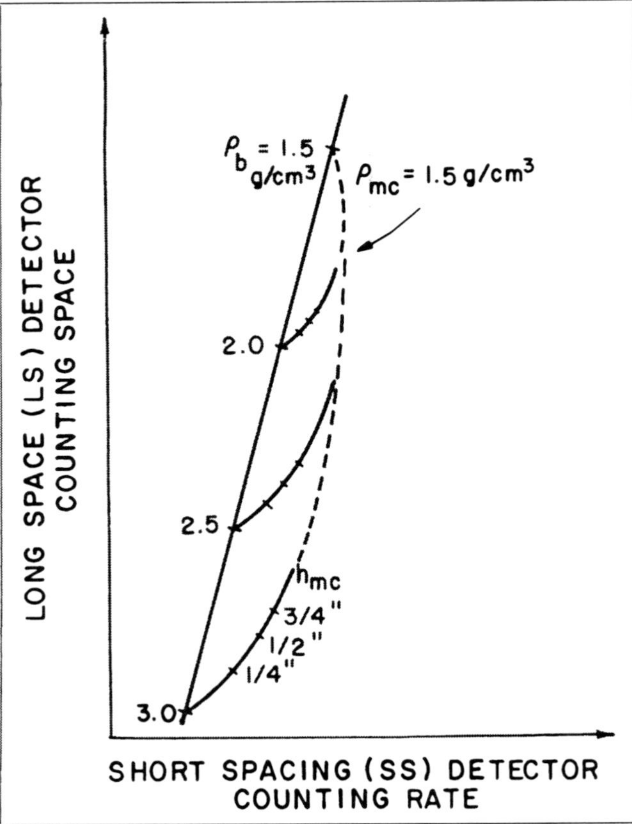

Fig. 8.8—Effect of formation density on the short- and long-spaced detector responses in the presence of a mudcake (after Ref. 4).

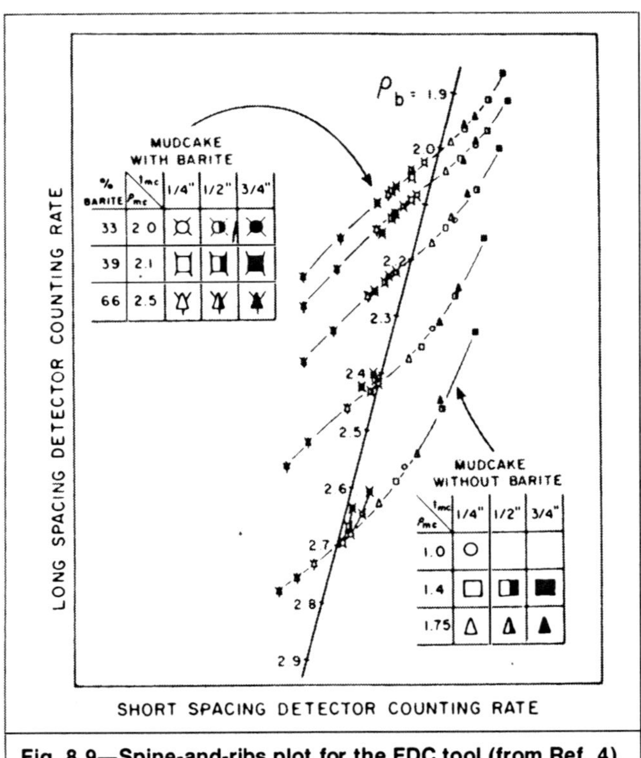

Fig. 8.9—Spine-and-ribs plot for the FDC tool (from Ref. 4).

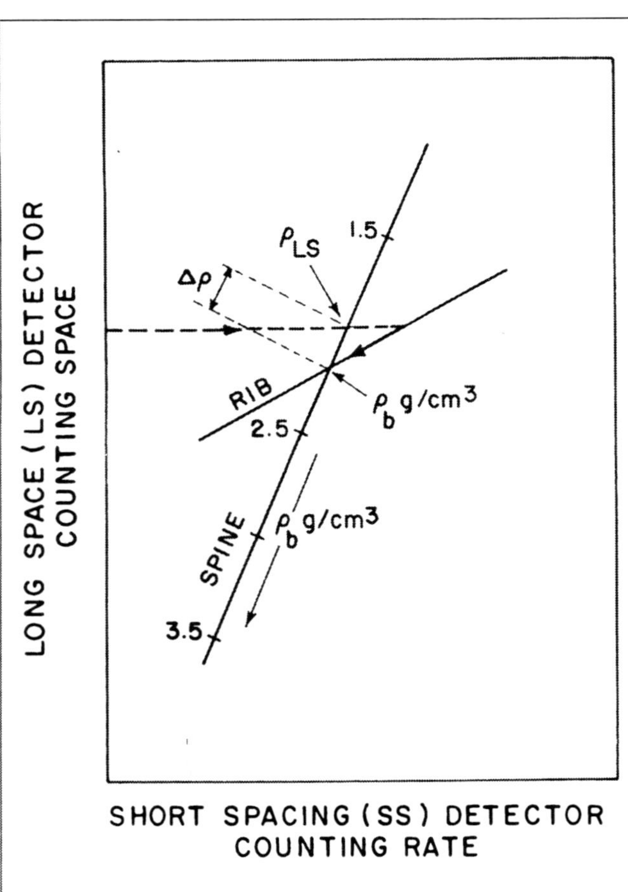

Fig. 8.10—Graphical presentation of the compensation method.

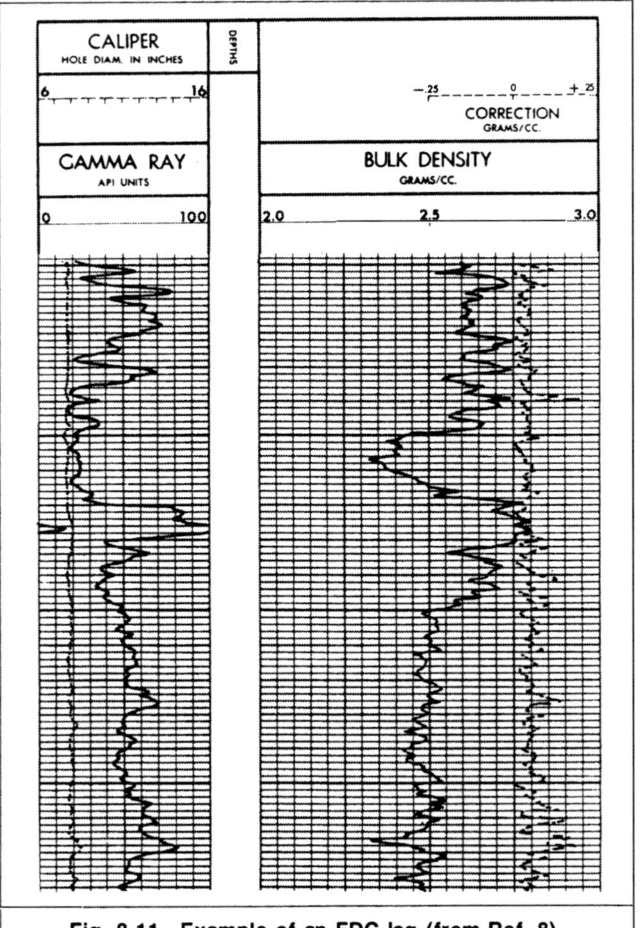

Fig. 8.11—Example of an FDC log (from Ref. 8).

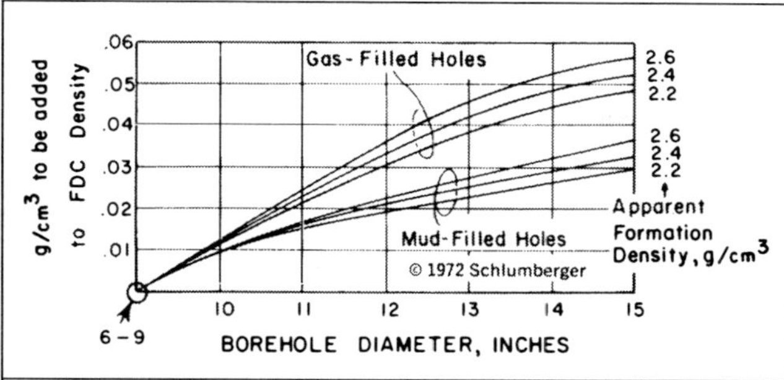

Fig. 8.12—FDC borehole correction for gas- and mud-filled boreholes (from Ref. 9).

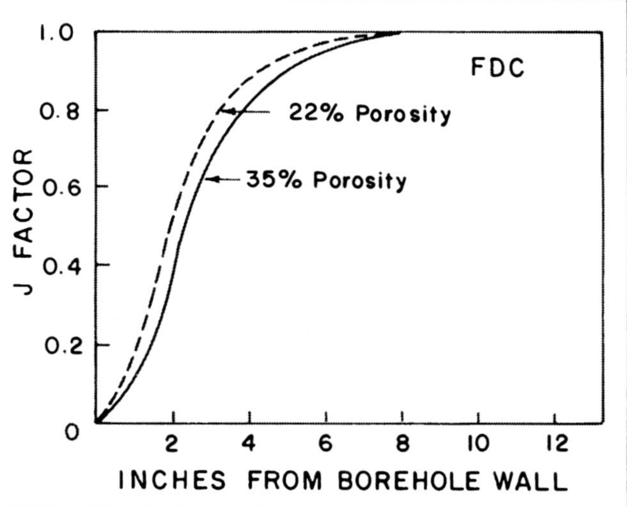

Fig. 8.13—Integrated geometric factor curves for FDC tool (after Ref. 10).

tors are affected only by the mudcake, and the rib returns to the spine at a point where $\rho_b = \rho_{mc} = 1.5$ g/cm³.

Fig. 8.7 illustrates the effect of varying mudcake density. When mudcake density exceeds formation density, the rib extends to the left side of the spine because both detectors count less than when mudcake is absent. Finally, **Fig. 8.8** shows the effect of varying formation density in the presence of mudcake.

8.3.1 Tool Compensation Method. **Fig. 8.9** shows the actual spine-and-ribs plot obtained experimentally for the FDC tool. Ribs were developed for five limestone formation density values and for a variety of synthetic mudcakes. A single rib that is practically independent of mudcake thickness, density, or composition corresponds to each formation density.[4] These ribs can be scaled in terms of true formation density.

Conceptually, a compensated density value is obtained by entering such a graph with the counts of the two detectors. In practice, the two counts are processed by a surface computer. The computer derives a value of uncompensated density using the long-spaced detector response. It also computes a density correction, $\Delta\rho$, using both counts and an algorithm base on the spine-and-rib plot. The correction is added to the uncompensated values to obtain the compensated bulk density, ρ_b:

$$\rho_b = \rho_{LS} + \Delta\rho, \quad (8.2)$$

where ρ_{LS} is the long-spaced detector uncompensated density. A relatively light mudcake shows a negative correction; a relatively heavy mudcake shows a positive correction.

The compensation method is depicted graphically in **Fig. 8.10**. The accuracy of this technique is limited. For an extremely thick mudcake or rugose borehole (see Fig. 4.3), both detectors will see mainly mudcake or a mud pocket. The tool interprets this as a formation with a bulk density very close to the mud or mudcake density.

Fig. 8.11 is an example of an FDC log that shows the compensated bulk density and the $\Delta\rho$ curves. The bulk density curve is shown in Tracks 2 and 3 as a solid line. The tool response is in grams per cubic centimeter. The conventional scale is from 2.0 to 3.0 g/cm³ across both tracks, or 0.05 g/cm³ per division. The correction curve is traced in Track 3 as a dotted curve. It is normally scaled from −0.25 to +0.25 g/cm³. The $\Delta\rho$ curve is given for quality control purposes. Bulk density values derived with a correction that exceeds 0.06 g/cm³ should be used only qualitatively.

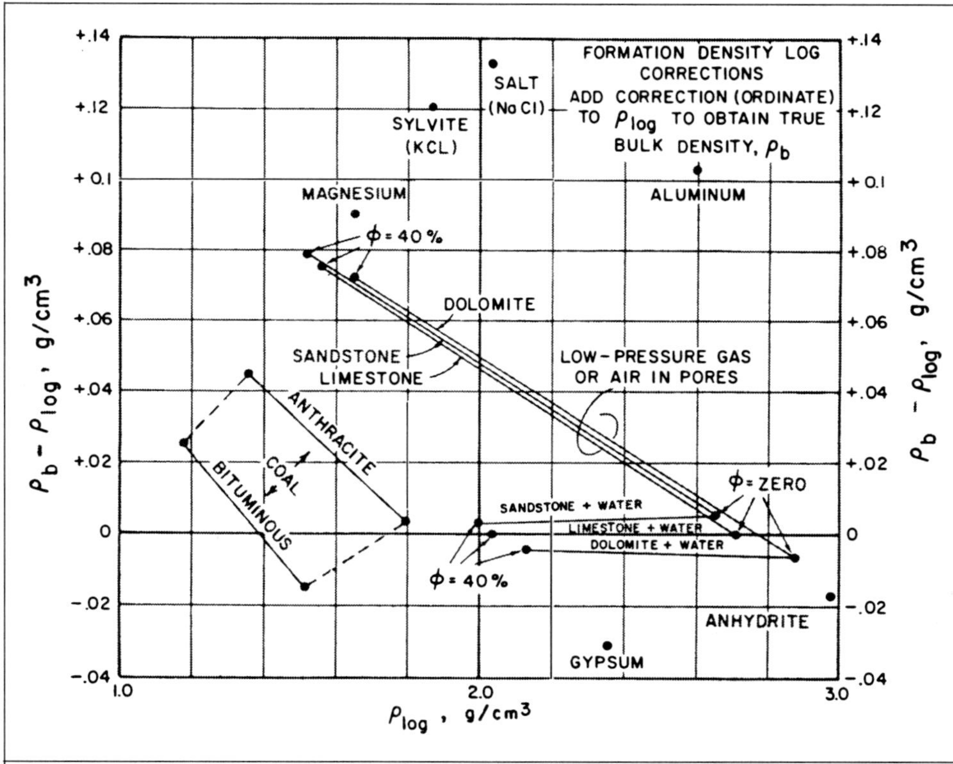

Fig. 8.14—Correction needed to obtain true bulk density from apparent log density (from Ref. 11).

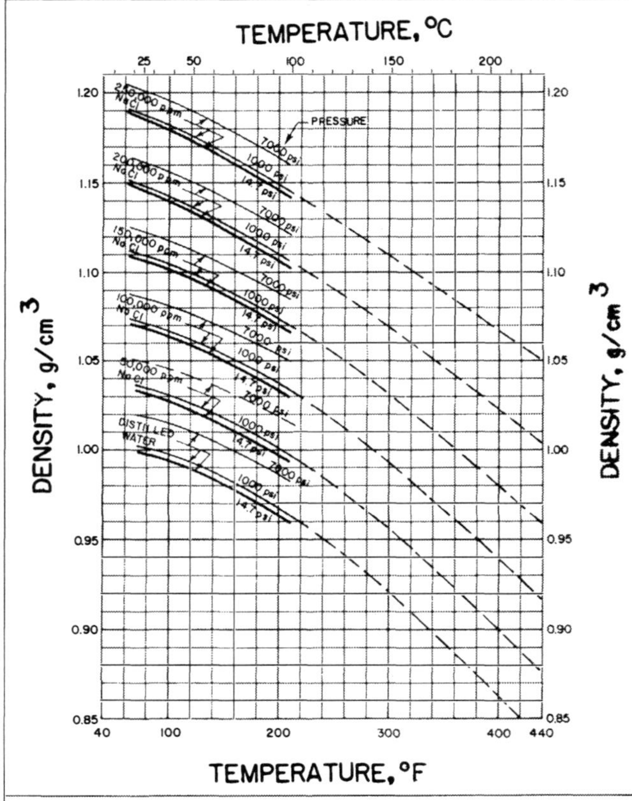

Fig. 8.15—Densities of water and NaCl solutions at various temperatures and pressures (from Ref. 8).

In addition to the compensation for mudcake effect, the density curve is corrected automatically for borehole size. **Fig. 8.12** shows the order of magnitude of this correction.

8.3.2 Depth of Investigation. The integrated geometric factor curves obtained experimentally for the FDC tool[10] are shown in **Fig. 8.13**. These experimental results indicate that the tool investigates only the first few inches of the region next to the tool. Half of the tool response reflects the region within about 2 in., while 90% reflects the region within 5 in. of the borehole wall. Consequently, the density tool investigates the invaded zone of permeable formations.

8.4 Tool Calibration

The Compton scattering effect, on which density logging is based, is proportional to the number of electrons per unit volume, N_e, defined by Eq. 2.2. N_e is, in turn, proportional to the electron density index, ρ_e. ρ_e is arbitrarily defined by Eqs. 2.45 and 2.46 as

$$\rho_e = C\rho_b, \quad (8.3)$$

where $C = 2(Z/A)$. (8.4)

Z and A are the atomic number and weight, respectively.

For most elements and compounds found in sedimentary rock and listed in Tables 2.6 and 2.7, the quantity $2(Z/A)$, to a high degree of approximation, can be considered equal to unity. To consider the effect of variations in the value of Z/A, the tool is calibrated in pits containing a sequence of identical formations with very accurately known grain densities and variable freshwater-filled porosities.[11] The tool is calibrated using limestone as the "standard" with freshwater-filled porosities ranging from 0% to 40%. For freshwater-bearing limestone, the standard formula given by Eq. 2.50 can be written in terms of both the true density and electron density index:

$$\phi = (\rho_{ls} - \rho_b)/(\rho_{ls} - \rho_w) \quad (8.5)$$

and $\phi = [(\rho_{ls})_e - \rho_e]/[(\rho_{ls})_e - (\rho_w)_e]$, (8.6)

where ρ_{ls} = limestone bulk density = 2.71 g/cm^3, ρ_w = freshwater bulk density = 1 g/cm^3, $(\rho_{ls})_e$ = limestone electron density index, and $(\rho_w)_e$ = freshwater electron density index.

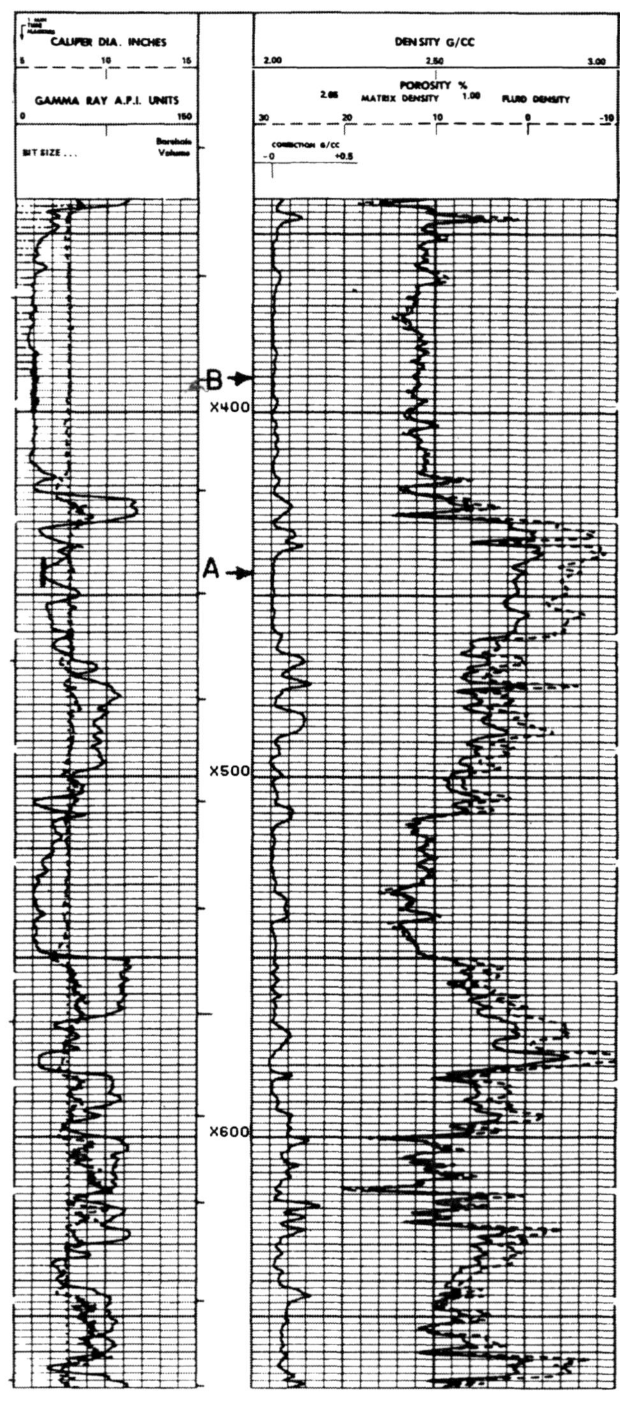

Fig. 8.16—Example of a density log displaying the density porosity curve.

Using Eq. 8.3 and (Z/A) values from Table 2.7 gives

$$(\rho_{ls})_e = 0.9991\ \rho_{ls} \quad (8.7)$$

and $(\rho_w)_e = 1.1101\ \rho_w$. (8.8)

Eliminating ϕ between Eqs. 8.5 and 8.6 and then replacing ρ_{ls}, ρ_w, $(\rho_{ls})_e$, and $(\rho_w)_e$ by their numerical values results in

$$\frac{2.71 - \rho_b}{2.71 - 1.0} = \frac{0.9991(2.71) - \rho_e}{0.9991(2.71) - 1.1101},$$

and finally in $\rho_b = 1.0704\ \rho_e - 0.1883$. (8.9)

Eq. 8.9 is the density tool calibration relationship used to transform the measured electron density index to bulk density.

The bulk density displayed by the log is rigorously correct only in freshwater-bearing limestone formations. In an environment different from the standard, the log displays an apparent value ρ_{log}. **Fig. 8.14** shows the deviation between the true and apparent densities ($\rho_b - \rho_{log}$). We can see from Fig. 8.14 that for oilwell-logging applications, the difference between ρ_b and ρ_{log} is negligible. This difference is significant, however, in formations like gypsum and salt and in low-pressure gas-bearing formations.

The calibration pits representing the "primary" calibration standard cannot be transported. A set of secondary standards is available for tool calibration at logging company bases. The secondary calibrators consist of blocks of aluminum, magnesium, or sulfur of accurately known densities and geometries. These blocks, weighing about 400 lbm, also are not easily transportable, so a field calibrator containing two small gamma ray sources is used at the wellsite to reproduce the same count rates as the secondary calibration blocks.[12]

Example 8.1. Calculate the response of the density tool next to a pure sulfur formation with an actual bulk density of 2.07 g/cm^3.

Solution. From Table 2.6, sulfur atomic weight, A, is 32.07 and its atomic number, Z, is 16. From Eq. 8.3,

$$\rho_e = 2(16/32.07)(2.07) = 2.0655,$$

and from Eq. 8.9,

$$\rho_{log} = 1.07(2.0655) - 0.188 = 2.022 \text{ g/cm}^3.$$

The log response deviates from the true density value by 0.048 g/cm^3 because of the Z/A effect.

8.5 Porosity From Density Log Response

The bulk density is the overall gross or weight-average density of a unit of the formation. It can be expressed by

$$\rho_b = \phi\rho_f + (1-\phi)\rho_{ma}. \quad (8.10)$$

Solving for porosity yields

$$\phi = (\rho_{ma} - \rho_b)/(\rho_{ma} - \rho_f), \quad (8.11)$$

where ρ_f is the average density of the fluids in pore spaces. Common values of ρ_{ma} are given below.

Rock Type	Matrix Density (g/cm^3)
Sand or sandstone	2.65
Limestone	2.71
Dolomite	2.87
Anhydrite	2.98

Because the tool's depth of investigation is shallow, it investigates the invaded zone, and ρ_f is expressed by

$$\rho_f = S_{xo}\rho_{mf} + (1-S_{xo})\rho_h, \quad (8.12)$$

where ρ_{mf} = mud-filtrate density, S_{xo} = mud-filtrate saturation in the invaded zone, and ρ_h = invaded-zone hydrocarbon density.

In water-bearing zones where $S_{xo} = 1$,

$$\rho_f = \rho_{mf}. \quad (8.13)$$

Assuming that the mud filtrate is predominantly sodium chloride, ρ_{mf} can be obtained from **Fig. 8.15** for different salinities, temperatures, and pressures.

In practice, ρ_{mf} can be approximated according to the mud type.

Mud Base	ρ_{mf} (g/cm^3)
Oil	0.9
Fresh water	1.0
Saturated salt water	1.1

These values are also used to approximate ρ_{mf} in oil zones. This approximation is supported by the low value of residual oil saturation, S_{or}, and the small difference between ρ_h and ρ_{mf}. These assumptions usually make small changes to calculated porosities. The case of gas zones is treated in Chap. 16.

Example 8.2. The density porosity is usually calculated assuming $\rho_f = 1$ g/cm^3. Compare this "apparent" porosity, ϕ_a, to the true porosity, ϕ_t, that corresponds to a bulk density of 2.1 g/cm^3 in the following environments: (1) a water-bearing sandstone invaded by a mud filtrate of 1.05-g/cm^3 density, (2) a 0.8-g/cm^3 oil-bearing sandstone characterized by S_{or} = 30%, and (3) a low-pressure gas-bearing sandstone characterized by 30% residual gas saturation.

Solution. The apparent porosity calculated with $\rho_{mf} = 1$ g/cm^3 is

$$\phi_a = (2.65-2.1)/(2.65-1) = 0.33 \text{ or } 33\%.$$

For the salty-mud-filtrate case,

$$\phi_t = (2.65-2.1)/(2.65-1.05) = 0.34 \text{ or } 34\%.$$

Then, $\phi_t \simeq \phi_a$.

For the oil-bearing zone, from Eq. 8.12,

$$\rho_{mf} = 0.7(1) + 0.3(0.8) = 0.94 \text{ g/cm}^3$$

and $\phi_t = (2.65-2.1)/(2.65-0.94) = 0.32$ or 32%.

Again $\phi_t \simeq \phi_a$.

For the low-pressure gas-bearing zone, assume $\rho_g = 0$. Then, from Eq. 8.12, $\rho_f = 0.7$,

$$\phi_t = (2.65-2.1)/(2.65-0.7) = 0.28 \text{ or } 28\%,$$

and $\phi_t < \phi_a$.

Thus, except for gas-bearing formations, it is reasonable to assume that $\rho_{mf} = 1$.

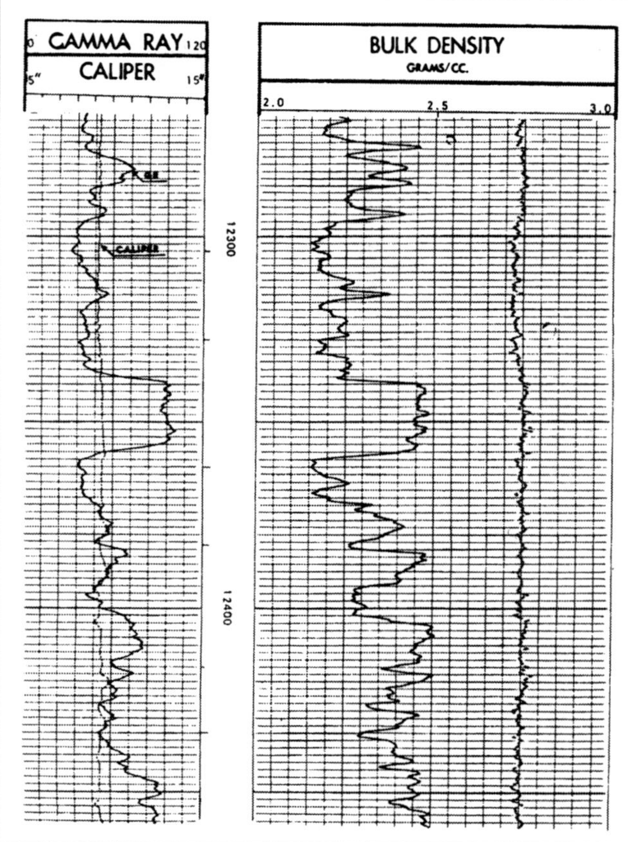

Fig. 8.17—Density log of Example 8.5 (courtesy Amoco Production Co.).

Example 8.3. A water-bearing sandstone displays a 2.1-g/cm^3 bulk density. Calculate the density porosity using a matrix density of 2.68 g/cm^3. Compare this value to that calculated with the normally assumed matrix density of 2.65 g/cm^3.

Solution. From Example 8.2,

$$\phi_a = 33\%.$$

$$\phi_t = (2.68-2.1)/(2.68-1) = 0.345 \text{ or } 34.5\%.$$

This represents a deviation of −1.5 porosity units and a relative error of about −4%.

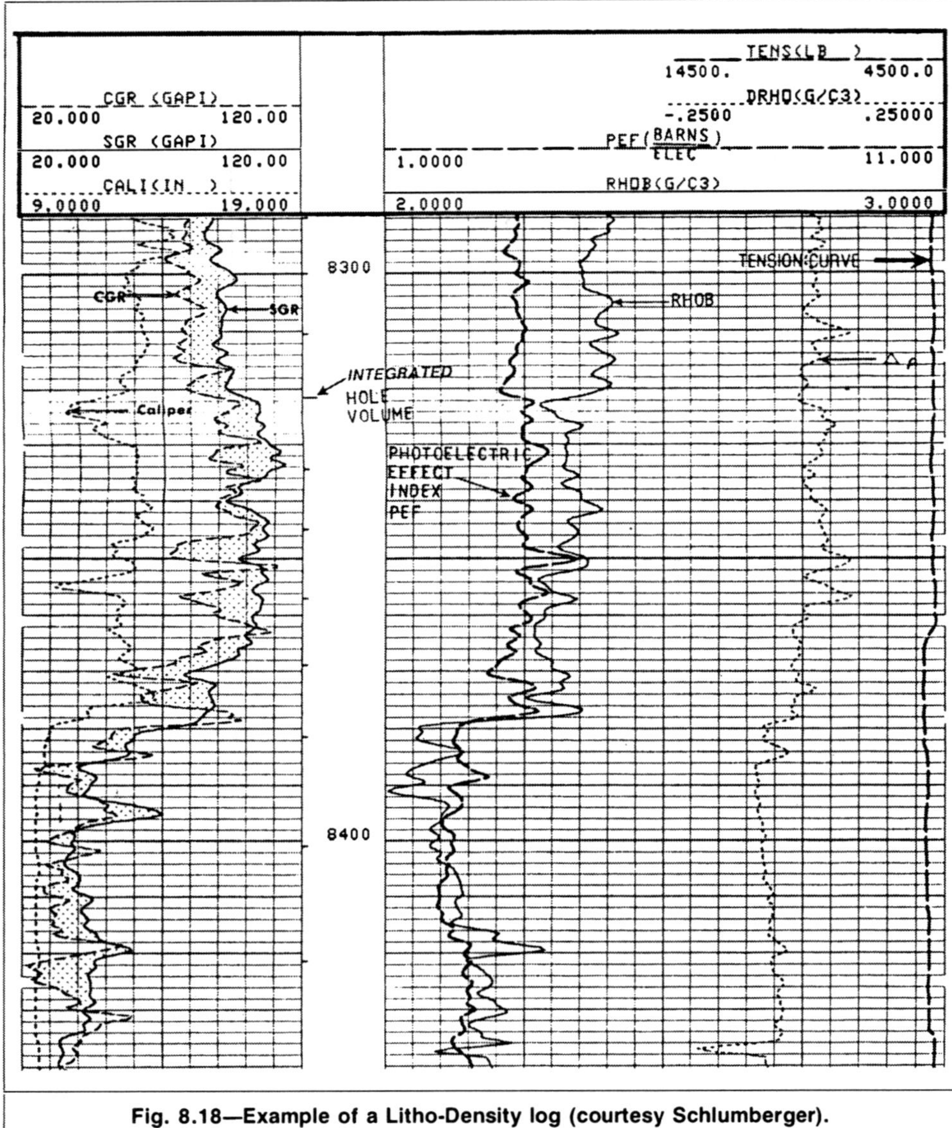

Fig. 8.18—Example of a Litho-Density log (courtesy Schlumberger).

The determination of porosity from the density log response discussed in this chapter applies only to relatively simple environments. In complex environments, such as shaly sands, gas-bearing formations, and complex lithology, the density log is combined with other logs for better evaluation. These cases are treated in later chapters.

The density log response also can be displayed in terms of porosity calculated with Eq. 8.11. Matrix and fluid densities are selected and entered into the computer. **Fig. 8.16** is an example of the generated porosity curve. This curve, called the density porosity, ϕ_D, displays an apparent porosity value. The log displays a true porosity value only if the selected matrix and fluid density values correspond to those of the formation of interest.

Example 8.4. In answering the following questions, refer to the log of Fig. 8.16 obtained in a borehole drilled with freshwater-based mud.

a. What are the values of ρ_b, $\Delta\rho$, gamma ray, caliper, and ϕ_D in Zones A and B?

b. What is the true porosity value of Zones A and B?

Solution.

a. The following values are displayed by the log.

	Zone A	Zone B
ρ_b, g/cm^3	2.75	2.45
ϕ_D, %	−6	12
$\Delta\rho$, g/cm^3	0	0
Gamma ray, API	22	15
Caliper, in.	8	8

b. The porosity curve of Zone A was calculated with freshwater-filled sandstone parameters (see log heading). The log displays a negative apparent porosity of −6% next to Zone A. Eq. 8.11 indicates that a negative value results from the fact that the measured ρ_b is greater than the assumed ρ_{ma}. It follows that the true matrix density is greater than the true sandstone density. It is also greater than the density of limestone (ρ_b=2.75>2.71). Knowledge of lithology is required for accurate determination of the true porosity of this zone.

Although the porosity of Zone B is positive, it can still be an apparent porosity value because the lithology can be something other than sandstone. The logs indicate (1) that the borehole is regular and drilled to gauge (bit size is most certainly 7⅞ in.); (2) that gamma ray response is low; (3) that $\Delta\rho$=0, indicating possible absence of mudcake, which, in turn, suggests a low-permeability formation; and (4) that lithologies other than sandstone are present—e.g., Zone A.

These facts indicate that Zone B is probably not sandstone. Its true porosity can be determined only after accurate lithologic identification.

Example 8.5. Fig. 8.17 shows the density tool's response obtained over a sand/shale series.

a. Calculate the average porosity of the clean sand interval.

b. Using a matrix density of 2.65 g/cm^3, calculate the average density porosity of shales. How does the value compare to the true porosity? Explain the difference, if any.

TABLE 8.1—MATRIX AND FLUID VALUES OF INTEREST[14]

	P_e	Specific gravity	$\rho_{b(log)}$	U
Quartz	1.81	2.65	2.64	4.78
Calcite	5.08	2.71	2.71	13.8
Dolomite	3.14	2.87	2.88	9.00
Anhydrite	5.05	2.96	2.98	14.9
Halite	4.65	2.17	2.04	9.68
Siderite	14.7	3.94	3.89	55.9
Pyrite	17.0	5.00	4.99	82.1
Barite	267	4.48	4.09	1,065
Water (fresh)	0.358	1.00	1.00	0.398
Water (100,000 ppm NaCl)	0.734	1.06	1.05	0.850
Water (200,000 ppm NaCl)	1.12	1.12	1.11	1.36
Oil [n(CH_2)]	0.119	ρ_o	$1.22\rho_o - 0.188$	$0.136\rho_o$
Gas (CH_4)	0.095	ρ_g	$1.33\rho_g - 0.188$	$0.119\rho_g$

TABLE 8.2—POROSITY EFFECT ON P_e (from Ref. 6)

Matrix	ϕ_t	100% H_2O (ρ_w = 1 g/cm³)	100% CH_4 (ρ_g = 0.01 g/cm³)
Quartz	0.00	1.81	1.81
	0.35	1.54	1.76
Calcite	0.00	5.08	5.08
	0.35	4.23	4.96
Dolomite	0.00	3.14	3.14
	0.35	2.66	3.07

Solution.

a. The clean sand intervals identified by the lowest natural gamma ray display an average density of 2.16 g/cm³. Using matrix and fluid densities of 2.65 and 1 g/cm³, respectively, in Eq. 8.11 yields

$$\phi_D = (2.65 - 2.16)/(2.65 - 1) = 0.297 \text{ or } 30\%.$$

This value is representative of the true porosity.

b. The shale zone located in the interval of 12,339 to 12,359 ft displays an average bulk density of 2.46 g/cm³. Using 2.65 and 1 g/cm³ for the matrix and fluid densities gives

$$\phi_D = (2.65 - 2.46)/(2.65 - 1) = 0.115 \text{ or } 12\%.$$

This value is very small compared with shale's relatively high porosity. This difference is probably the result of tool calibration performed in limestone formations. This calculation indicates that the porosity curve, such as that of Fig. 8.16, displays an apparent porosity value in shales. The apparent density porosity in shales differs considerably from the true value.

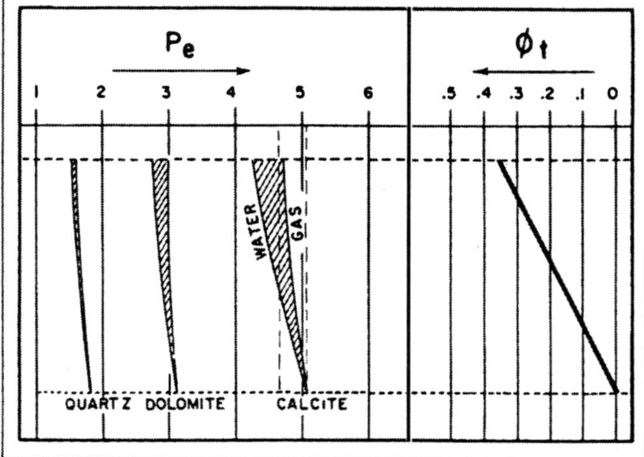

Fig. 8.19—Effect of porosity and fluid type on P_e (from Ref. 6).

8.6 Litho-Density Tool

The Litho-Density Tool is a relatively new density measurement device. Because of changes in design and its ability to provide additional measurement, it is a considerable improvement over the compensated tool. In the newer design, the spacing of the detectors has been shortened.[13] Thus, the counting rates have been increased and the statistical uncertainty of the measurements has decreased. This design also makes the density measurement less sensitive to the presence of mudcakes, especially those containing high Z-value additives, such as barite.

Spectral analyses of the detected gamma ray make possible the measurement of the effective photoelectric absorption cross-section index for the formation, P_e. P_e, which has the unit of barns per electron, is defined as[13]

$$P_e = (Z/10)^{3.6}. \quad (8.14)$$

A volumetric photoelectric absorption cross-section, U, that has the unit of barns per volume can be defined as

$$U = P_e \rho_e. \quad (8.15)$$

Fig. 8.18 shows an example of a Litho-Density log. In addition to the bulk density curve (solid) and the correction curve (dotted), the P_e curve (dashed) is presented on Tracks 2 and 3 on a scale of 1 to 11 barns/electron.

Table 8.1 lists values of P_e, ρ_b, ρ_{log}, and U for matrices and fluids of interest. The P_e of freshwater mud filtrate, oil, and gas are small compared with those of matrices. The measured P_e is then affected slightly by porosity. **Table 8.2 and Fig. 8.19** show the effect of porosity on the P_e value for sandstones, limestones, and dolomites. As can be seen, the three ranges of P_e remain quite distinct, and P_e is affected very little by the fluid type. P_e, then, is a good lithology indicator, especially when simple lithology (i.e., one dominant matrix) exists.

The P_e value is also used in combination with the density value to analyze two-mineral matrices and to determine porosity. It is combined with density and neutron porosity to analyze more complex lithologies. These uses are discussed in Chap. 14.

Example 8.6.

a. What is the predominant lithology present in the interval shown by the Litho-Density log of **Fig. 8.20**?

b. What is the range of porosity of this predominant lithology?

c. Describe the lithology of the thin zones marked by O's.

d. Describe the lithology of the thin zones marked by X's.

Solution.

a. The P_e of the predominant lithology is about 3 barns/electron. As Fig. 8.19 indicates, this is dolomite. This determination is supported by the fact that $\rho_b > 2.8$ g/cm³.

b. The bulk density value ranges from 2.8 to 2.85 g/cm³. Using $\rho_{ma} = 2.87$ and $\rho_f = 1$ in Eq. 8.11 results in a porosity value ranging from 1% to 4%.

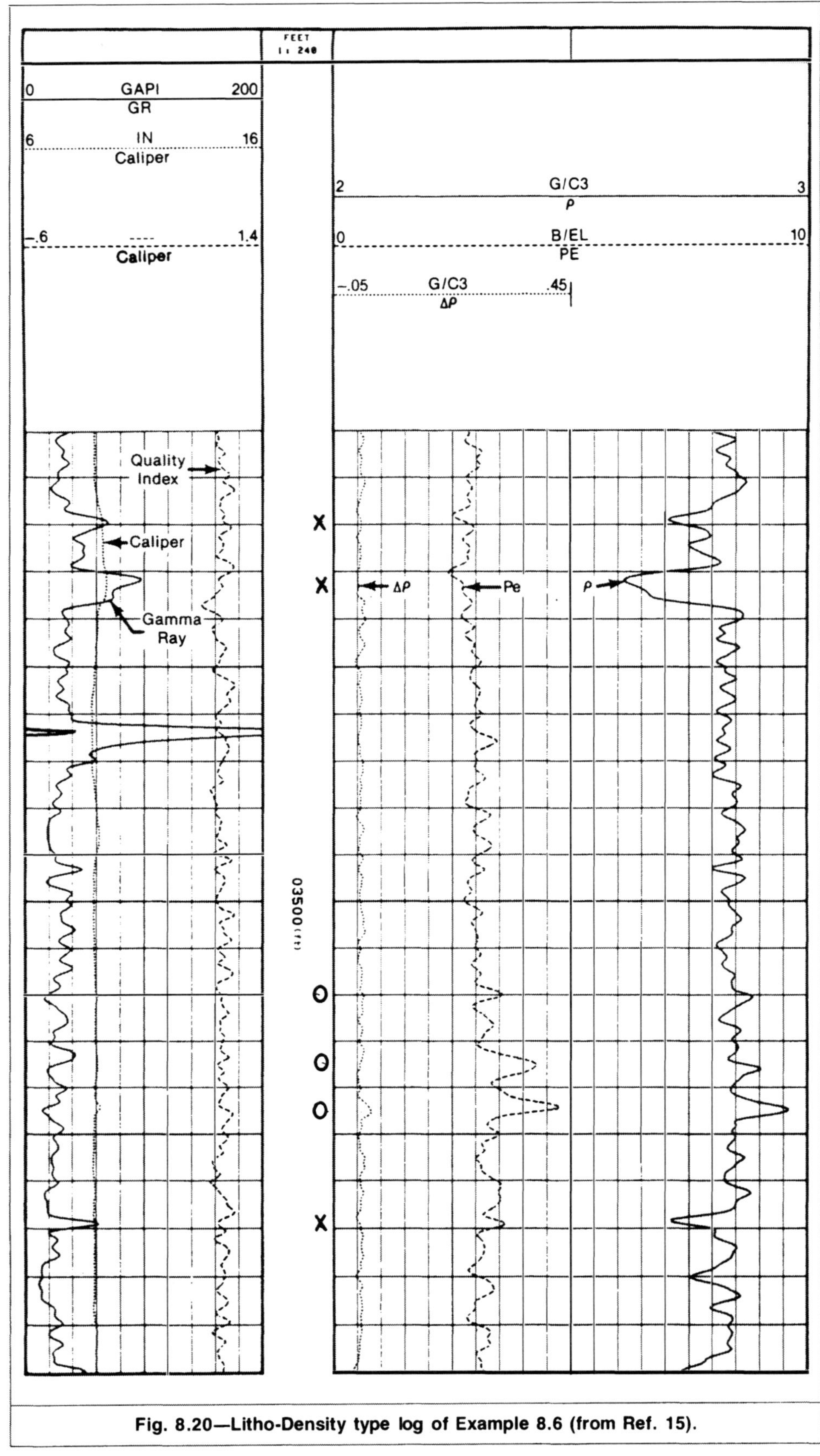

Fig. 8.20—Litho-Density type log of Example 8.6 (from Ref. 15).

c. The P_e values of the zones marked by O's exceed that of dolomite and approach that of anhydrite, which is about 5.05 barns/electron. The lithology is probably a mixture of dolomite and anhydrite.

d. The P_e curve is not helpful here. However, the gamma ray increases, which tends to indicate shale streaks.

Review Questions

1. Explain why the source and detectors of the density tool are mounted in a shielded skid pressed against the borehole wall.

2. Why is cesium 137 usually preferred over cobalt 60 as a source of gamma rays in density logging?

3. Using the concepts presented in Chap. 2, show that the relationship between the density tool detector count and bulk density

can be expressed by Eq. 8.1. Find an expression for the coefficients α and β.

4. The random occurrence of radioactive events introduces a degree of uncertainty into the measurement. Explain why, in density logging, the degree of uncertainty increases as porosity decreases.

5. Why is the dual-detector density tool superior to the single-detector density tool?

6. What is the use of the spine-and-ribs plot? How was it generated?

7. Discuss the Z/A effect.

8. What conditions should prevail for the tool to yield a true bulk density reading?

9. What conditions should prevail for Eq. 8.11 to yield a true porosity value?

10. Why is the density shale porosity considerably different from the true porosity?

11. Why is the Litho-Density Tool better than its predecessor, the FDC tool?

12. What is the parameter P_e? What formation property does it measure?

13. Why is the P_e curve affected only slightly by the formation porosity and fluid type?

Problems

8.1 The compensated density log of **Fig. 8.21** was recorded in a 9⅞-in. borehole drilled with freshwater-based mud in a sand/shale series.
   a. Select zones representative of the sections that show borehole enlargement and of the regular borehole sections.
   b. Read and tabulate the caliper, bulk density, and $\Delta\rho$ values for each of these zones.
   c. Plot $\Delta\rho$ vs. caliper reading. Does a correlation between these two parameters emerge? Explain why or why not.

8.2 Calculate the response of the density tool next to pure aluminum and magnesium blocks of 2.713- and 1.777-g/cm³ bulk density, respectively.

8.3 A density tool calibrated in terms of freshwater-filled limestone generated a bulk density log. This log, in turn, was used to calculate a density porosity log using limestone matrix density and unit fluid density. Prepare a chart that can be used to convert the porosity values displayed by the log to true porosity values for sandstone, limestone, dolomite, and anhydrite formations.

8.4 Construct a chart that represents the transform between the electron density and the density tool bulk density. Show the limestone and water points on the transform.

8.5 Calculate the parameter C of Eq. 8.3 for shale made up entirely of kaolinite. The composition of kaolinite is $Al_4(Si_4O_{10})(OH)_8$.

8.6 Give your best estimate of the lithology and porosity of Zones X and Y in **Fig. 8.22**.

8.7 **Fig. 8.23** shows two density logs run through sandstone Formation A in a gas injection well. Run 1 was made when the formation was fully saturated with water of 50,000 ppm salinity before gas injection. Run 2 was made several months after gas injection started. Estimate the gas saturation in the vicinity of the wellbore. The injected gas is mostly methane. Formation temperature and pressure are 100°F and 2,000 psia, respectively.

8.8 Refer to the density log of **Fig. 8.24** in answering the following questions.
   a. How representative is the log response at the levels marked with X's? Explain.
   b. If the clean sands of the interval illustrated by the log are known to display a more-or-less constant porosity, give your best estimate of that porosity value.
   c. Calculate a density porosity at Level Y. Explain why this value is different from that calculated in Part b.

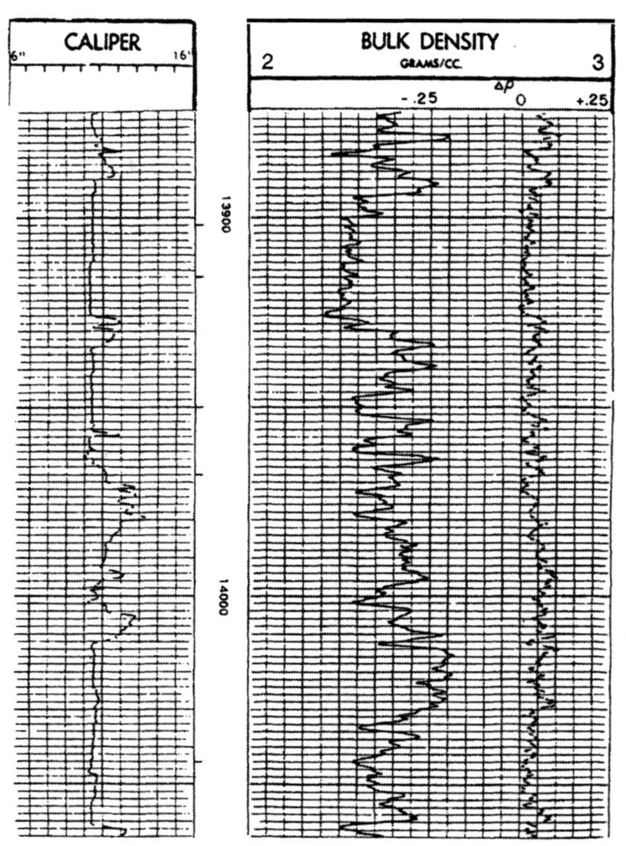

Fig. 8.21—Compensated density log of Problem 8.1 (courtesy Amoco Production Co.).

Nomenclature

A = atomic weight
h = thickness, ft
P_e = photoelectric absorption cross-section index, barns/electron
R = detector count rate, counts/sec
S = saturation, fraction
U = volumetric photoelectric cross section, barns/cm³
Z = atomic number
α,β = coefficients in Eq. 8.1
ρ = density, g/cm³
ϕ = porosity, fraction

Subscripts

a = apparent
b = bulk
D = density tool
e = electron
f = fluid
g = gas
h = hydrocarbon
log = logging tool response
ls = limestone
LS = long-spaced detector
ma = matrix
mc = mudcake
mf = mud filtrate
o = oil
or = residual oil
t = true
w = water
xo = flushed zone

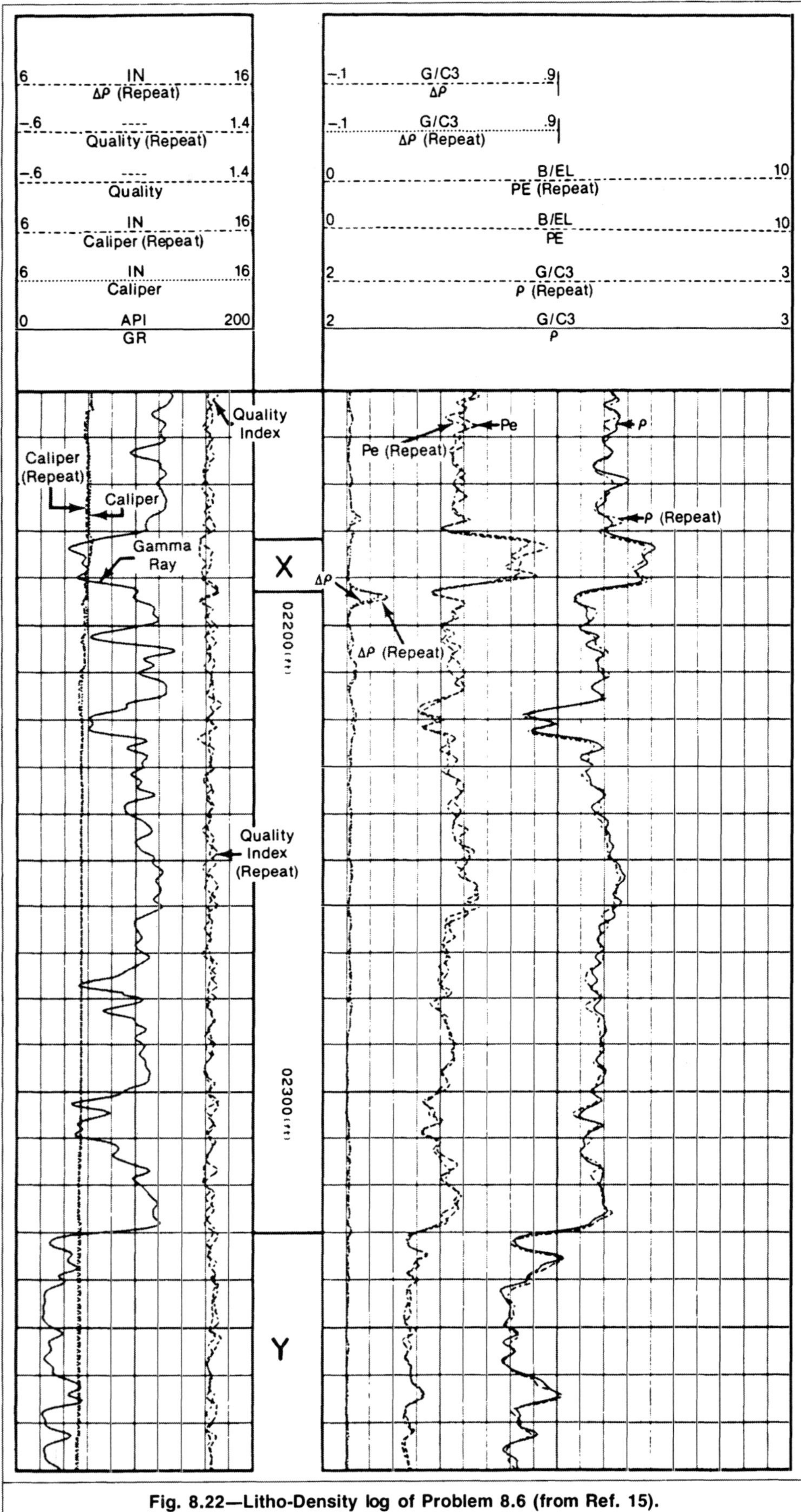

Fig. 8.22—Litho-Density log of Problem 8.6 (from Ref. 15).

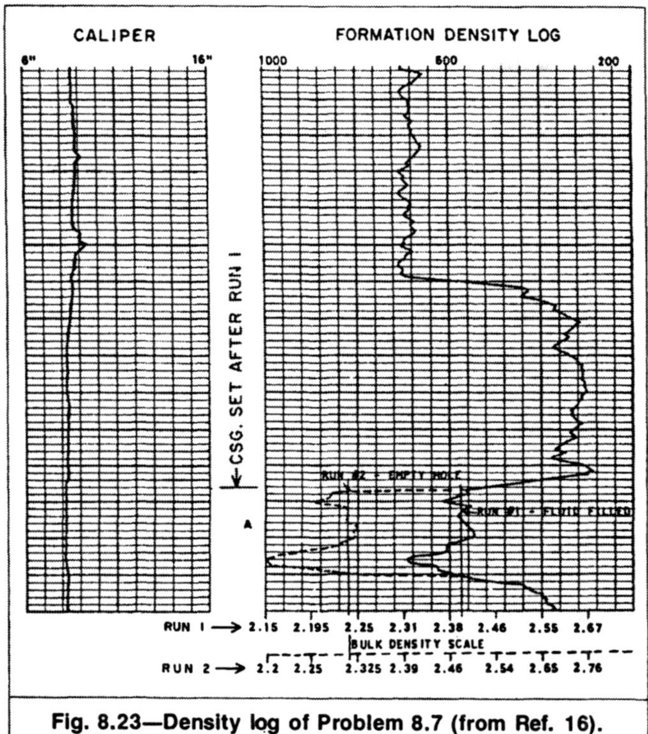

Fig. 8.23—Density log of Problem 8.7 (from Ref. 16).

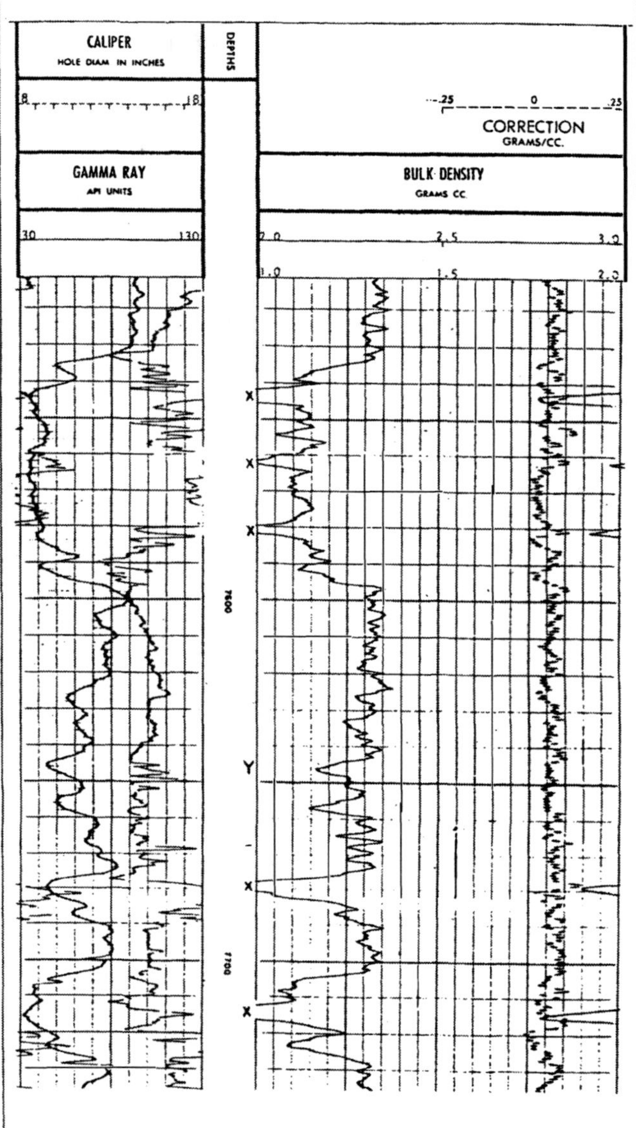

Fig. 8.24—Density log of Problem 8.8 (courtesy Amoco Production Co.).

References

1. Newton, G.R. *et al.*: "Subsurface Formation Density Logging," *Geophysics* (1954) **19**, No. 3, 636.
2. Baker, P.E.: "Density Logging with Gamma Rays," *Trans.*, AIME (1957) **210**, 289–94.
3. Pickell, J.J. and Heacock, J.G.: "Density Logging," *Geophysics* (Aug. 1960) **25**, No. 4, 891–904.
4. Wahl, J.S. *et al.*: "The Dual Spacing Formation Density Log," *JPT* (Dec. 1964) 1411–16; *Trans.*, AIME, **231**.
5. Bertozzi, W., Ellis, D.V., and Wahl, J.S.: "The Physical Foundations of Formation Lithology Logging with Gamma Rays," *Geophysics* (Oct. 1981) **46**, No. 10, 1439.
6. Gardner, J.S. and Dumanoir, J.L.: "Litho-Density Log Interpretation," *Proc.*, SPWLA Symposium, Lafayette, LA (1980) paper N.
7. Alger, R.P. *et al.*: "Formation Density Log Applications in Liquid-Filled Holes," *JPT* (March 1963) 321–32; *Trans.*, AIME, **228**.
8. *Log Interpretation Principles*, Schlumberger, Houston (1972).
9. *Log Interpretation Charts*, Schlumberger, Houston (1972).
10. Sherman, H. and Locke, S.: "Effect of Porosity on Depth of Investigation of Neutron and Density Sondes," paper SPE 5510 presented at the 1975 SPE Annual Technical Conference and Exhibition, Dallas, Sept. 28–Oct. 1.
11. Tittman, J. and Wahl, J.S.: "The Physical Foundations of Formation Density Logging (Gamma-Gamma)," *Geophysics* (April 1965) **30**, No. 2, 284–94.
12. Bateman, R.M.: *Log Quality Control*, IHRDC, Boston (1985) 157–67.
13. Ellis, D.V. *et al.*: "Litho-Density Tool Calibration," *SPEJ* (Aug. 1985) 515–20.
14. Edmundson, H. and Raymer, L.L.: "Radioactive Logging Parameters for Common Minerals," *Log Analyst* (Sept.-Oct. 1979) **XX**, No. 5, 38–47.
15. Gearhart, D.A. and Mathis, G.L.: "Development of a Spectral Litho-Density Logging Tool by Use of Empirical Methods," *Log Analyst* (Sept.-Oct. 1987) **28**, No. 5, 470–87.
16. Lovan, T.E. Jr. and McCluskey, L.D.: "Logging Observation Wells in Gas Storage," *JPT* (July 1964) 745–50.

Chapter 9
Neutron Logs

9.1 Introduction

Inelastic scattering, elastic scattering, and absorption are the basic phenomena that occur after a fast neutron is introduced into a formation. The fast neutrons are slowed first by inelastic scattering (which takes place at a high neutron energy level) and then by elastic scattering. The neutrons eventually slow to a level of energy at which they coexist with the formation nuclei in thermal equilibrium. Neutrons in this state are called thermal neutrons. Thermal neutrons continue to scatter off of formation nuclei elastically and diffuse through the formation. Each thermal neutron eventually is captured by one of the nuclei. The nucleus instantaneously emits gamma rays, called capture gamma rays. Neutrons that have slowed almost to the thermal energy level yet are still energetic enough to avoid capture are known as epithermal neutrons.

Several logging tools are based on these phenomena. The neutron porosity log is based on the elastic scattering of neutrons as they collide with the nuclei in the formation. Each neutron scatters off a nucleus with less kinetic energy. Energy and momentum conservation in elastic collisions dictates that the presence of hydrogen in the formation dominates the slowing process. The reason for this is that the mass of the hydrogen nucleus is approximately equal to that of the incident neutron. Consequently, at a point sufficiently removed from the neutron source, formations with high hydrogen content display low concentrations of epithermal and thermal neutrons and capture gamma rays. Inversely, formations with low hydrogen content display high concentrations of epithermal and thermal neutrons and capture gamma rays. **Fig. 9.1** illustrates this relation between hydrogen content and count rate at the detector. Because most of the hydrogen is part of the fluids located in the pore space, this concentration is inversely related to porosity. The concentration, indicated by a detector, also varies with tool design and borehole environment. The representativity of the derived porosity is affected by the variation of the hydrogen content of pore-space fluid and by the eventual presence of hydrogen within the formation matrix itself.

9.2 Types of Detectors

Most neutron porosity tools use chemical sources containing about 16 curies of americium and produce roughly 4×10^7 neutrons/sec. The various neutron tools differ in design by the type and number of detectors used and the source-to-detector spacing.

9.2.1 Gamma Ray Detectors. Capture gamma rays can be detected by a Geiger counter or a scintillation tube (see Sec. 7.2). A tool containing this type of detector is known as a neutron-gamma (N-G) tool. An N-G tool has the advantage of a relatively larger depth of investigation because it counts the product of the decay of neutrons throughout a volume extending into the formation. Neutron detectors, in contrast, count only those neutrons that reach the tool. A gamma ray detector also provides a high count with moderate source activity. Depending on the source type, an N-G tool could have the disadvantage of also detecting gamma rays that originate in the source itself in association with neutron generation. In general, these gamma rays are of lower energy than the capture gamma rays. Their effect can be reduced by properly shielding the detector and by filtering out energy pulses below a certain level. The intensity of natural gamma rays, which may also reach the detector, is usually small compared with that of induced radiation.

Because capture gamma rays that originate in the formation have to travel to the tool to be detected, formation and borehole materials substantially affect the detector count. For example, the presence of chlorine in the drilling fluid or formation water causes the counting rate to increase because chlorine is an efficient thermal neutron absorber that emits several gamma rays of higher energies upon capture of a thermal neutron.[1]

Some of the gamma rays reaching the detector are generated by a thermal neutron that has reached the tool and been absorbed by the material of the detector itself. The resulting measurement contains a thermal neutron component and is called a "hybrid measurement." In practice, it is difficult to have a pure N-G tool.[2]

9.2.2 Thermal Neutron Detectors. Slow-neutron detectors are also available. They are predominantly sensitive to thermal neutrons but also have varying sensitivities to capture gamma rays, thus giving a hybrid measurement. A tool using this type of detection is called a neutron-thermal neutron tool, or simply a thermal neutron tool.

The spatial distribution of thermal neutrons is approximated by Eq. 2.65. This complex relationship includes epithermal and thermal neutron diffusion parameters and slowdown parameters. Most of these parameters are related to properties of all elements in the formations that are not necessarily porosity-related. Elements with high capture cross section, such as chlorine in salt water and boron in shale, will cause a significant decrease in the count rate. Therefore, thermal neutron tool response is often difficult to interpret quantitatively in terms of porosity in highly saline or shaly environments.[3]

Because the spatial distribution of capture gamma rays in the formation is essentially the same as thermal neutron distribution, thermal neutron and N-G tools are similar. However, a significant

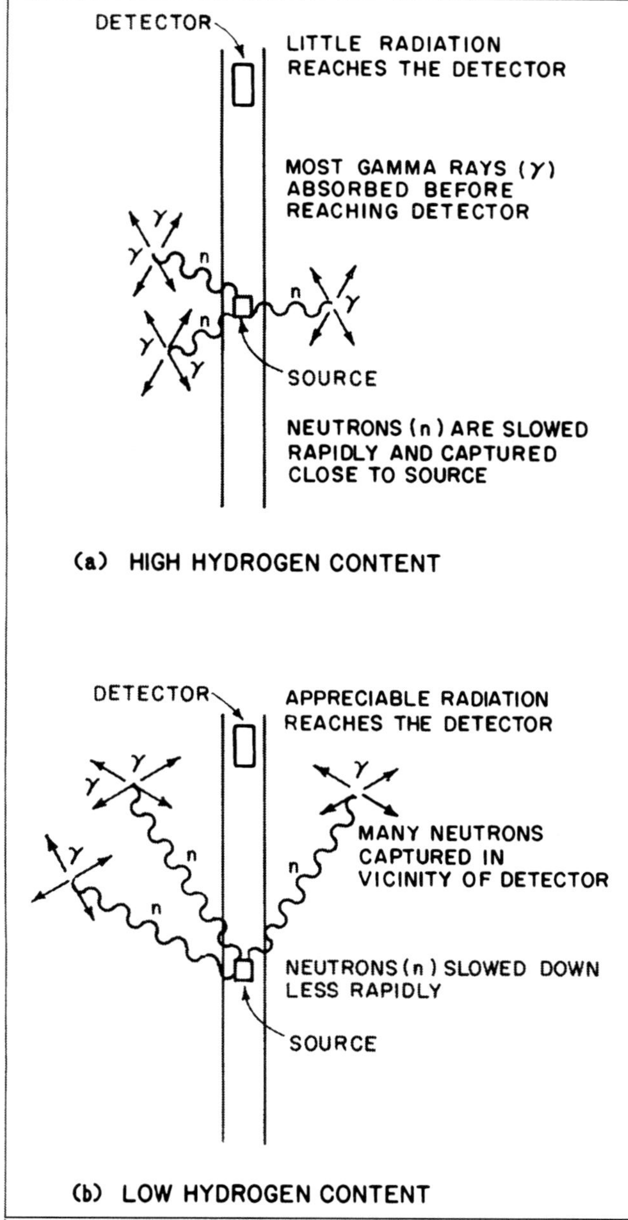

Fig. 9.1—Schematic of the relation between hydrogen content and count rate at the detector.

difference in their responses occurs in zones rich in efficient thermal neutron absorbers, such as chlorine and boron. The count rate in these zones is reduced because a substantial number of thermal neutrons is absorbed by the chlorine and boron. This absorption, however, tends to increase the gamma ray count of N-G tools.

9.2.3 Epithermal Neutron Detectors. Some detectors respond to epithermal neutrons and ignore both thermal and gamma rays. Eq. 2.63, which approximates the spatial distribution of epithermal neutrons, shows that this spatial distribution is primarily a function of the slowdown length and not of capture characteristics. Therefore, it is not affected by elements that are efficient thermal neutron absorbers. Epithermal neutron detectors' count rates are less sensitive to lithology and formation-water salinity than the thermal neutron detectors' rates. Tools using this type of detection, epithermal neutron tools, yield a reponse that accurately reflects the formation hydrogen index and subsequently derived porosity value.

Most epithermal neutron detectors are, in fact, thermal neutron detectors wrapped in a highly efficient thermal neutron absorber, such as cadmium. The wrap prevents low-energy neutrons from reaching the detector, which results in a low count overall. To improve the detector's efficiency, epithermal neutron tools are characterized by a short source-to-detector spacing and, in turn, a more shallow depth of investigation.[4]

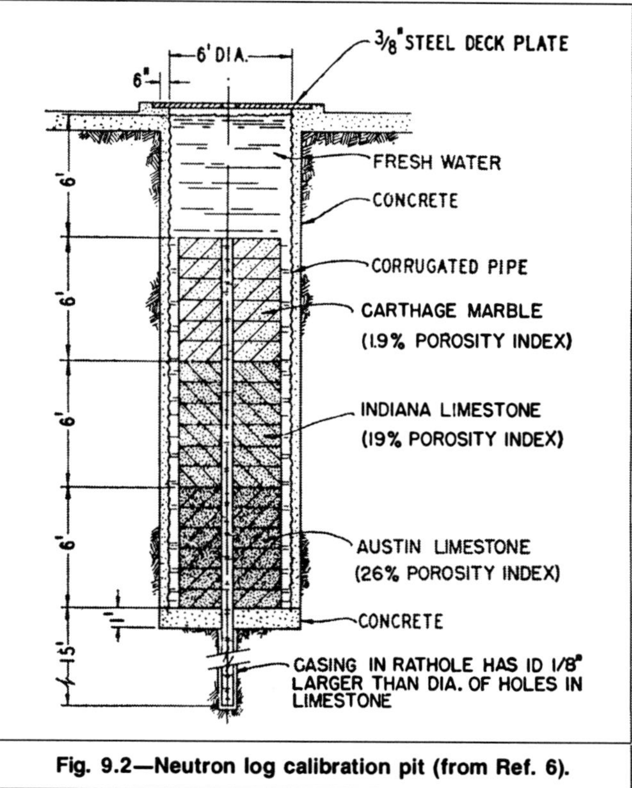

Fig. 9.2—Neutron log calibration pit (from Ref. 6).

9.3 N-G Tool

Early N-G tools used a hybrid measurement detector sensitive to both high-energy capture gamma rays and thermal neutrons. Schlumberger, for example, provided a series of N-G tools called the GNT. Several source-to-detector spacings were available so that the best spacing could be selected for a specific borehole condition and porosity range. The responses of the early tools were expressed in standard counts per second. This unit of measurement was determined arbitrarily and varied from one service company to another. A strong gamma ray source usually is placed at a fixed distance from the detector, and the resulting deflection is arbitrarily divided into a set number, usually 200, of standard neutron counts.[5] To provide a standard unit for neutron log measurements, the American Petroleum Inst. (API) adopted the "API neutron unit." One API neutron unit is defined as 1/1,000 of the difference between instrument zero (i.e., tool response to zero radiation) and log deflection opposite a 6-ft zone of Indiana limestone in a neutron calibration pit at the U. of Houston (**Fig. 9.2**).[6] The pit is 24 ft deep, with a 15-ft rathole. The pit contains three 6-ft-thick limestone zones. Each zone is made up of six 1-ft-thick octagonal blocks. The zones consist of Carthage, Indiana, and Austin limestones displaying average porosities of 1.9%, 19%, and 25%, respectively. The rock is saturated with fresh water. A 6-ft layer of fresh water atop the rock blocks provides a 100% porosity reference point. A 7⅞-in. borehole extends through the pit center.

Fig. 9.3 is an example GNT log. The neutron tool response is plotted on a linear scale (in API units) that covers Tracks 2 and 3. A natural gamma ray log is always recorded in conjunction with the neutron log and presented on Track 1.

To relate the tool response to porosity, tools are run in test pits that consist of blocks of various lithologies and known porosities. The tool response is observed in different borehole sizes and fluids. The API neutron test pit serves as the basis for normalizing the data obtained in test pits with other borehole sizes and fluids. Calibration charts, usually called departure curves, are available for the different tools (**Fig. 9.4**). The chart gives the response in limestone porosity index, which is equal to the true porosity only if the formation is limestone. If the formation is sandstone or dolomite, the limestone porosity index is corrected to the proper matrix with **Fig. 9.5**.

Example 9.1. The N-G log of Fig. 9.3 was obtained with a GNT-F tool with a 19½-in. source-to-detector spacing in an 8-in. open hole filled with 13-lbm/gal freshwater-based mud. The average temperature over the interval shown by the log is 200°F.

a. Estimate the porosity of the most porous interval within Sand A.

b. Estimate the average porosity of limestone Formation B.

c. Estimate and comment on the value of the apparent porosity of Zone C.

Solution.

a. The zone with the highest porosity displays the lowest response, which is 1,280 API neutron units. Borehole and tool design correspond to the chart in Fig. 9.4. In general, the appropriate chart to be used is selected from a chart book according to the tool spacing and mud salinity.

Entering the lower section of the chart with the mud type and borehole diameter results, as illustrated, in a calibration line for these specific borehole conditions. The upper section of the chart is entered with the log response. The reading is corrected, however, for temperature effects before a limestone porosity index of 16% is read from the calibration line. The formation is not limestone; therefore, a value of 19% corrected for the actual lithology, which is sandstone, is determined from Fig. 9.5.

b. The average tool response in Zone B is 2,560 API neutron units. Use of the same calibration line as in Part a yields a 3% porosity. Note that no lithology correction is needed because the actual lithology is limestone.

c. The log reading in Zone C is 480 API units, and this yields a porosity value that exceeds 50%. This is, of course, an apparent porosity and does not reflect the true porosity. The low log response results from the high content of thermal neutron absorbers present in shales. The shale environment is not represented by the calibration charts.

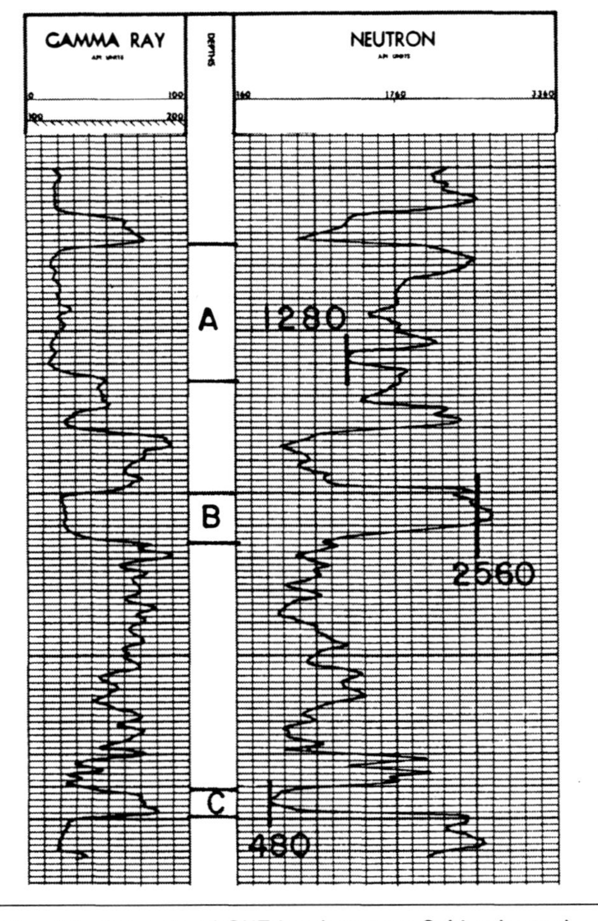

Fig. 9.3—Example of GNT log (courtesy Schlumberger).

In the absence of appropriate calibration charts, an empirical relationship between the tool response and porosity can be established by use of core porosities for a specific well, field, or region. Once established, this relationship can be applied to other wells with similar conditions. As illustrated by Example 2.11, Part f, and Fig. 2.21, for porosities below 40%, the porosity-neutron flux relationship can be approximated by an equation of the form

$$\phi=\alpha-\beta \log N, \quad (9.1)$$

where N is the neutron tool response and α and β are complex constants that involve both formation properties and tool design. Eq. 9.1 expresses a linear relationship between porosity and the logarithm of the tool response. Another more widely accepted empirical method is the logarithmic method. This method, which has no firm theoretical foundation, states that log ϕ plotted vs. tool response provides a linear relationship of the form[7]

$$\log \phi=a-bN, \quad (9.2)$$

where a and b are constants.

In the absence of core data, the linear relationship can be generated with the tool response in two zones for which porosity values can be surmised. A dense zone displaying the highest neutron curve deflection is selected as one point and assigned a porosity value usually ranging from 1% to 3%. The average shale response is selected as the second point and is assigned a porosity value usually between 35% and 40%. Experience in the area defines the values assigned to these two points.[8]

Although the empirical technique based on Eq. 9.1 was defined during the early stages of neutron log technology development, it can be useful in current major field studies that attempt to use both old and modern neutron logs. An example of this application is the SACROC Unit reservoir description project.[9] This unit is the major part of the giant Kelly-Snyder field. **Fig. 9.6** shows porosity vs. logarithm of the deflection of a specific tool used in the field. The data represent several wells but were normalized to remove the effects of different borehole conditions and tool calibration.

Example 9.2. Data listed in **Table 9.1** were obtained from a cored section of the Reef limestone in the North Snyder field. The neutron log was recorded in a 6¼-in. open hole. Show that the log reading can be related to porosity by the empirical relationship expressed in Eq. 9.2, the logarithmic method.

Solution. **Fig. 9.7** plots log ϕ vs. neutron log reading. This figure displays an excellent correlation between the neutron reading and porosity. This correlation is limited, however, to similar tool design, lithology, and borehole conditions.

9.4 Sidewall Neutron Log

Sidewall-neutron-type tools were developed to use epithermal neutron detectors. Epithermal neutron detection minimizes the perturbational influences of the thermal neutron absorption properties of rock matrices and formation water. Log response in various environments can then be predicted more easily and accurately. Because of the relatively low efficiency of these detectors, the source-to-detector spacing is short. To enhance the depth of investigation, the source and detector are mounted on a directionally sensitive pad that is pressed firmly against the borehole wall with a backup shoe. This design minimizes borehole effects and provides a means of obtaining a caliper log.

Fig. 9.8 is an example of a sidewall neutron porosity (SNP) log. The neutron log in Tracks 2 and 3 is scaled directly in porosity. Detector count rate is converted to porosity by the surface panel according to calibration curves similar to those in **Fig. 9.9**. The tool was calibrated by measuring its response in a variety of laboratory formations of accurately known matrix compositions and porosities, including sand, limestone, and dolomite of different

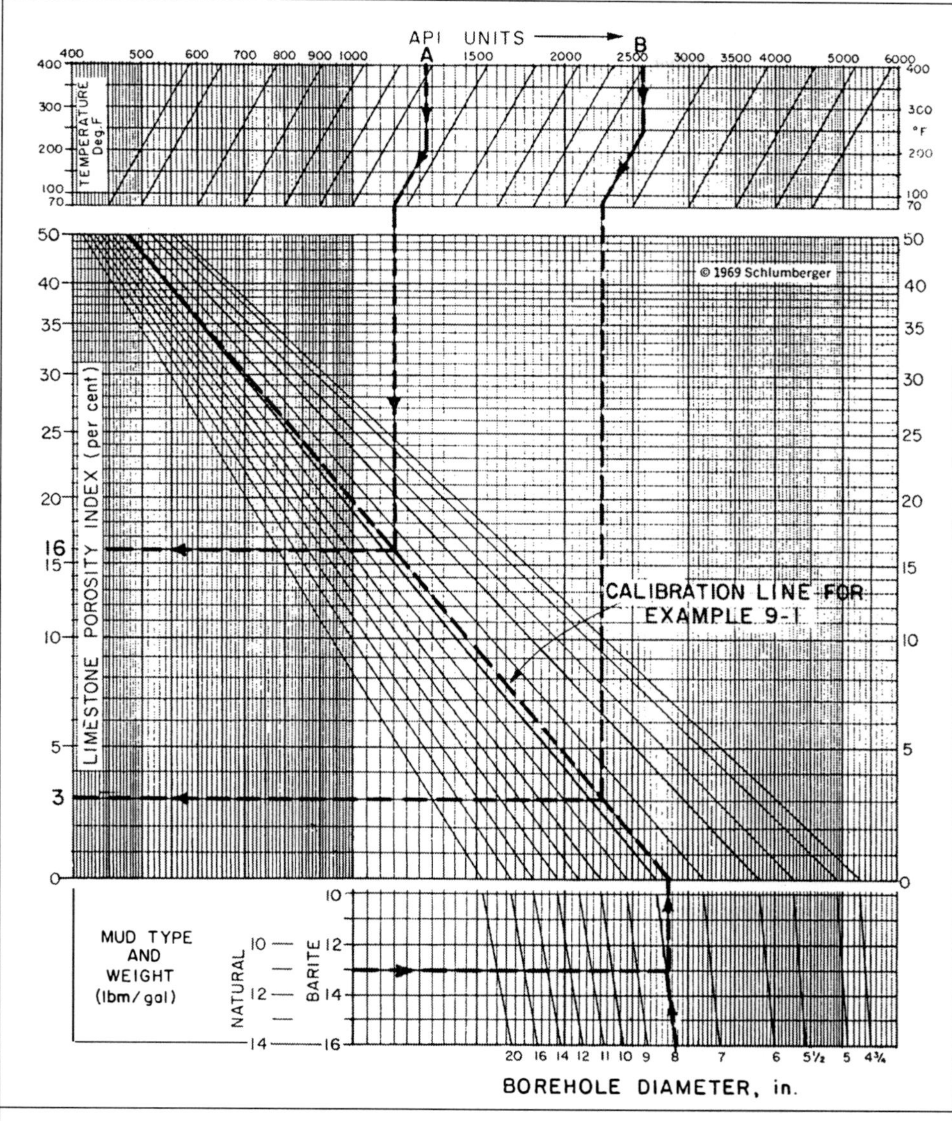

Fig. 9.4—Departure curve for GNT tool run in an uncased hole filled with fresh mud. Source-to-detector spacing is 19½-in. (courtesy Schlumberger).

porosities.[4] The anticipated matrix, which may be sandstone, limestone, or dolomite, can be set on the panel and is usually indicated on the log scale.

The calibration curves of Fig. 9.9 were obtained under the following standard conditions: fresh water in the formation and borehole (for gas-filled holes, the formations contained fresh water), borehole diameter of 7⅞ in., temperature of 75°F, atmospheric pressure, and no mudcake. Corrections are needed when actual logging conditions depart from these standard conditions. These corrections are usually small; most are made automatically by the surface panel.

The borehole-size effect is very small because the source-to-detector pad is pressed against the wall of the borehole. When liquid-filled holes are logged, the panel incorporates a borehole correction based on the reading of the simultaneously recorded caliper. **Fig. 9.10** shows the magnitudes of the correction made by the panel for 6- and 10-in.-diameter boreholes. The shaded areas indicate residual uncertainty.

The panel also applies corrections for formation-water salinity automatically. These corrections assume that the formation-water salinity is equal to that of the drilling fluid. This assumption is supported by the fact that the tool essentially investigates only the invaded zone because of its shallow radius of investigation. **Fig. 9.11** presents experimentally derived geometric factors that help define the tool's depth of investigation. These curves indicate that most of the tool signal is affected by the first 10 to 12 in. from the borehole wall. These first inches are usually invaded by mud filtrate. To enable an appreciation of the order of magnitude of this correction, the effect of saturated salt water on SNP logs is shown in **Fig. 9.12.**

The panel applies corrections for mud weight. Because the hydrogen contents of natural and barite mud are considerably different, the panel also allows the selection of either of these two mud types. **Fig. 9.13** shows the corrections applied automatically by the panel for 12- and 16-lbm/gal barite mud.

In deep wells, measurement temperature and pressure can deviate considerably from standard conditions. However, the temperature and pressure effects tend to offset each other. The temperature increase with depth results in a decrease in hydrogen concentration caused by material expansion. Pressure increase with depth causes an increase in hydrogen concentration because of compression of the borehole and formation material. The combined correction applied by the panel is small, as indicated by the dotted line of **Fig. 9.14**. To arrive at the formation temperature and pressure needed to calculate the correction, temperature and pressure gradients of 1°F/100 ft and 52 psi/100 ft, respectively, are usually assumed. The shaded band indicates the correction required for a temperature of 175°F at atmospheric pressure.

There is no provision for automatic correction for the presence of mudcake because the mudcake is squeezed away by the pad. Only a residual, very thin mudcake is usually present. If a thick mudcake is suspected, a manual correction can be applied from the nomogram of **Fig. 9.15**.

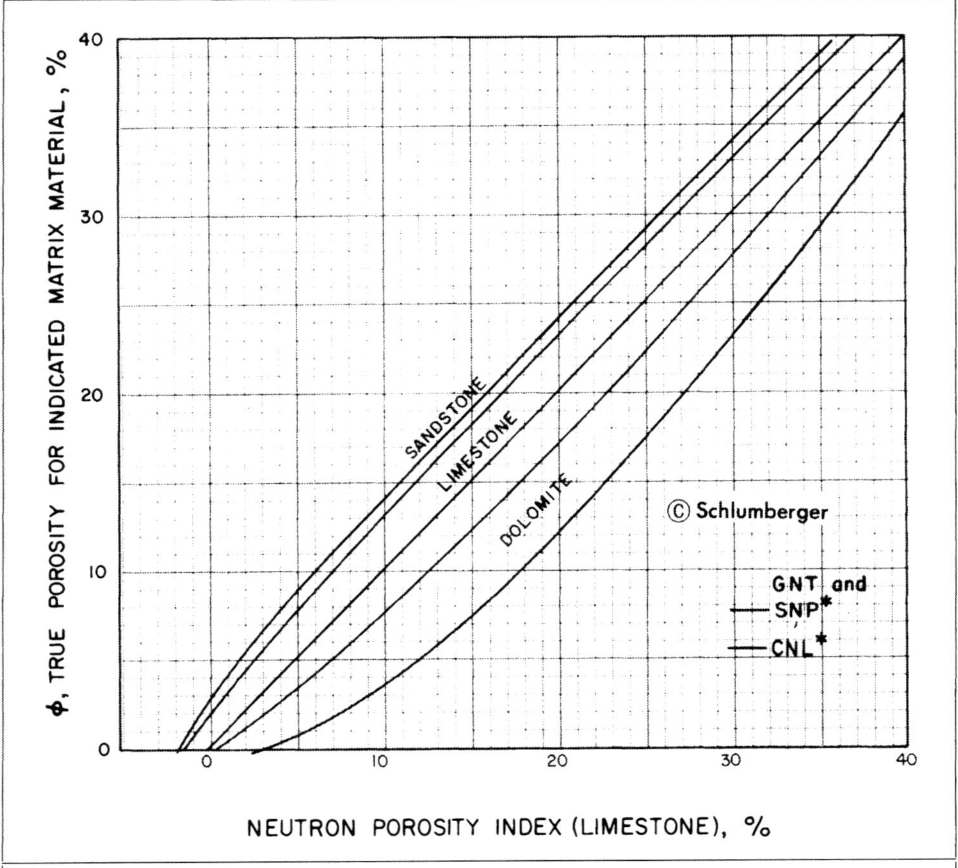

Fig. 9.5—Neutron porosity equivalence curves for GNT, SNP, and CNL tools (courtesy Schlumberger).

Example 9.3. A 20% porosity was determined from the SNP calibration curves. If the actual measurement conditions are a borehole diameter of 10 in., a 16-lbm/gal salt-saturated barite mud as the drilling fluid, a formation temperature of 175°F, and a formation pressure of 5,200 psia, estimate the order of magnitude of the correction applied by the panel.

Solution. Using Figs. 9.10 and 9.12 through 9.14 results in a borehole correction of −1%, a salinity correction of +2.5%, a mud-weight correction of +3%, and a temperature and pressure correction of +0.5%. This represents a total correction of +5%.

9.5 Dual-Detector Neutron Tool

The effect of changes in formation capture properties, caused primarily by salinity and shaliness, on the detection of thermal neutrons can be minimized by the use of two detectors positioned fair-

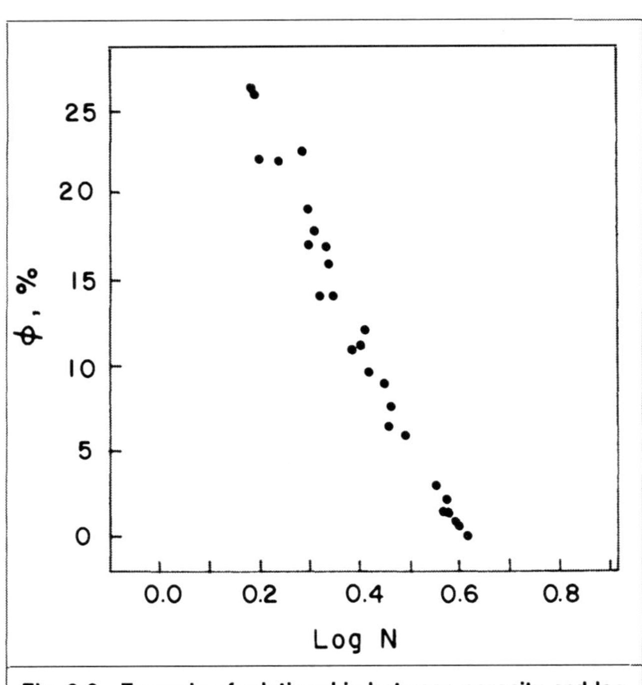

Fig. 9.6—Example of relationship between porosity and log deflection, SACROC Unit, Kelly Snyder field (from Ref. 9).

TABLE 9.1—CORE POROSITY AND NEUTRON LOG DEFLECTION FROM A 6¼-in. OPENHOLE SECTION, REEF LIMESTONE, KELLY-SNYDER FIELD (from Ref. 10)

Core Porosity (%)	Neutron Log Deflection (API standard units)
1	1.160
2	1.010
3	0.920
4	0.852
5	0.801
6	0.762
7	0.728
8	0.699
9	0.672
10	0.647
11	0.627
12	0.607
13	0.590
14	0.575
15	0.560
16	0.542
17	0.532
18	0.520
19	0.505
20	0.495
25	0.442
30	0.402

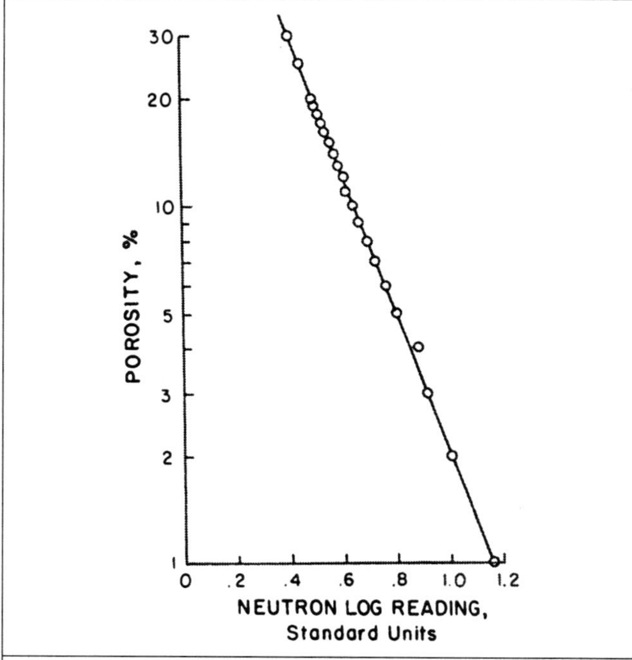

Fig. 9.7—Porosity/neutron-log-reading correlation, Example 9.2.

ly distant from the source. Formations of interest are characterized by an epithermal neutron slowdown length, L_e, that is greater than that of thermal neutron, L_t. Thus, when the source-to-detector radial distance, r, is large enough that $r>L_e$, then $e^{-r/L_t}<e^{-r/L_e}$. Eq. 2.65 expresses the thermal neutron flux, $\Psi_t(r)$, and can be simplified to[12]

$$\Psi_t(r)\cong\frac{N_N L_t^2}{4\pi D_t}\frac{L_t^2}{(L_e^2-L_t^2)}\frac{e^{-r/L_e}}{r}. \qquad (9.3)$$

Eq. 9.3 has the same form as Eq. 2.63, which relates the flux of epithermal neutrons to the distance from the source.

The ratio of thermal neutron flux at two detectors located at sufficient distances of r_1 and r_2 from the source, F, can be approximated by

$$F=(r_2/r_1)e^{-(r_1-r_2)/L_e}. \qquad (9.4)$$

As for epithermal neutron detection (see Sec. 9.2.3), F is primarily a function of the slowdown length and not of capture characteristics. The response F should accurately reflect hydrogen content and, in turn, porosity. The advantage of this method is that thermal neutron detectors are more efficient than epithermal detectors, and count rates are increased by roughly an order of magnitude.[12]

Theory also predicts that F is free of borehole effect. In practice, however, source-to-detector spacing, at least for the near detector, is not large enough to fulfill the condition $r>L_e$. The measurement of F retains, then, residual borehole and thermal absorption effects.

Tools based on this concept of measurement are called compensated neutron logs (CNL's). An example of such a tool is Schlumberger's CNL tool.[13] **Figs. 9.16 and 9.17** are schematics of the CNL tool and its log presentation. The tool is run eccentered in the hole to minimize borehole effects. The eccentricity is accomplished in holes ranging from 6 to 16 in. in diameter by using a bowstring (Fig. 9.16). In holes smaller than 6 in. in diameter, the tool is assumed to follow the low side of the hole without forced eccentralization.[13]

The ratio of the count rates of the near and far detectors is calculated in the surface control panel. The ratio is converted to porosity value. After appropriate borehole environment corrections are applied, the porosity is recorded on the log on a linear scale (Fig. 9.17). The panel is set to record the porosity curve on a scale that corresponds to the predominant formation matrix. The matrix is

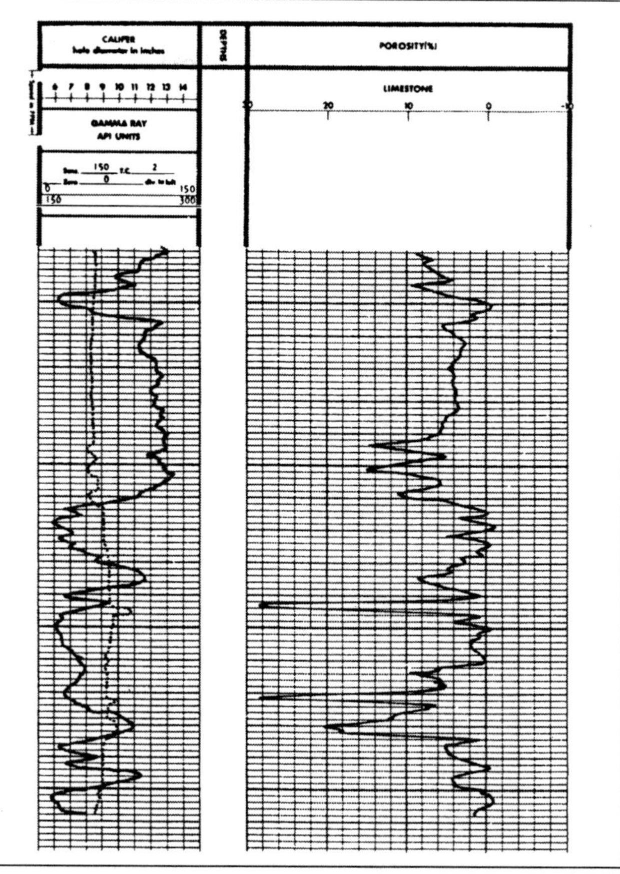

Fig. 9.8—Example SNP log (courtesy Schlumberger).

selected as sandstone, limestone, or dolomite. The compensated neutron tool is usually run in combination with the formation density. Both neutron porosity and density porosity are plotted on the same scale.

As for the SNP log, the conversion of the ratio F to porosity is based on tool responses measured in a large number of laboratory formations of known porosities and matrix compositions. The laboratory data were supplemented by API test pit runs[6] and by field logs where good porosity and lithology data were available. The standard calibration conditions are 7⅞-in. borehole diameter, fresh water in the borehole and formation, no mudcake or tool standoff, 75°F temperature, atmospheric pressure, and tool eccentered in the hole.[13]

In general, departures from these conditions result in a small combined correction. This correction can be obtained by summing the individual corrections of Charts A through H in the nomogram of **Fig. 9.18**. Each correction is determined by drawing a line parallel to the chart trend lines. This line is drawn from the actual formation conditions to the index axis, which is marked by an asterisk.

Chart A is used to correct for borehole size. Actually, a nominal borehole diameter, usually 8 in., is preset in the panel. The reading is corrected only for the difference between the actual borehole size and the nominal diameter. The borehole size correction is built into the surface control panel and is applied automatically when a neutron-density combination, which includes a caliper measurement, is run. When mudcake is present, the caliper senses a reduction in hole diameter, and the panel applies a correction that corresponds to the actual borehole size, which was reduced by the mudcake thickness. This tends to overcorrect for the borehole region, which should include the mudcake. Chart A is used to redress this overcorrection. The "borehole size minus the panel setting" parameter is then a negative amount equal to the mudcake thickness. The panel setting is, in this case, the reading of the caliper used to apply the automatic correction.

The actual effect of mudcake properties on the tool reading is obtained from Chart B. This chart was prepared using properties of a typical mudcake.

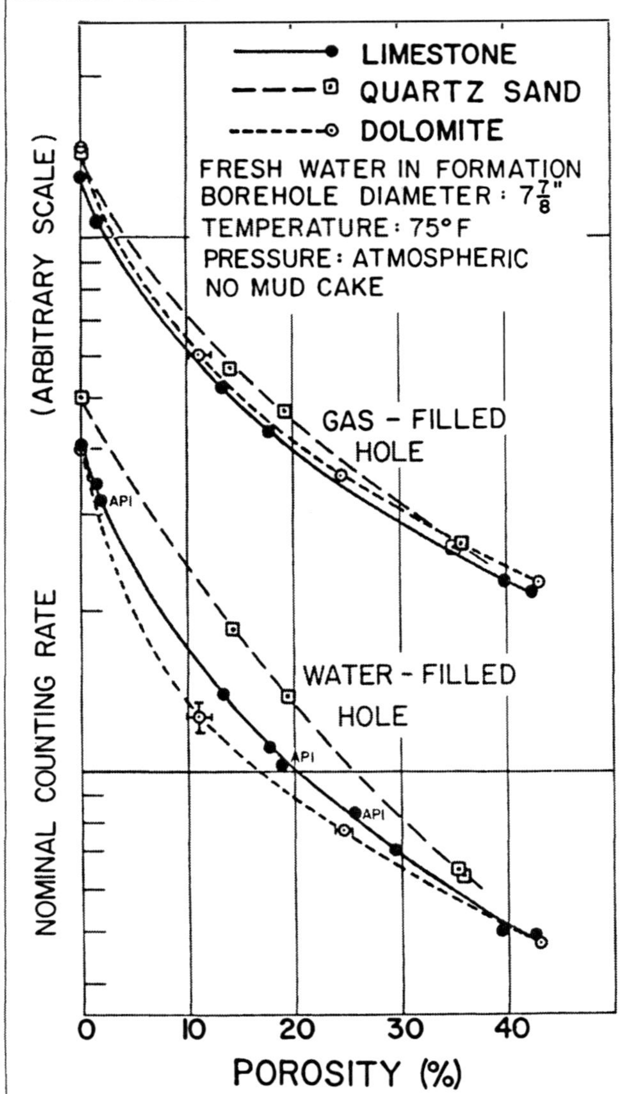

Fig. 9.9—SNP response under standard conditions in limestone, dolomite, and sandstone. Points labeled "API" refer to API test formations at the U. of Houston (from Ref. 4).

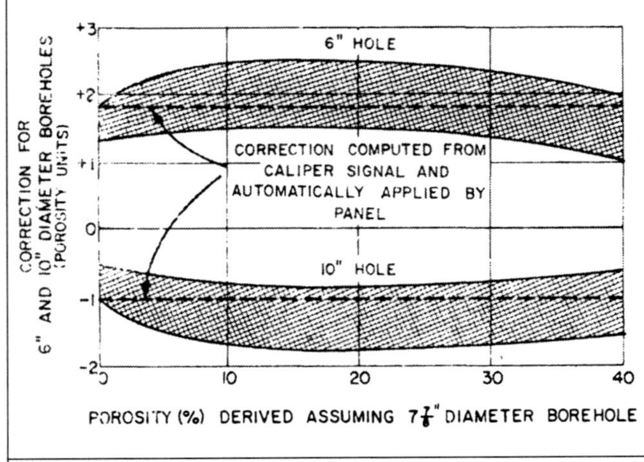

Fig. 9.10—Corrections to be added for borehole-size effect for 6- and 10-in. borehole (from Ref. 4).

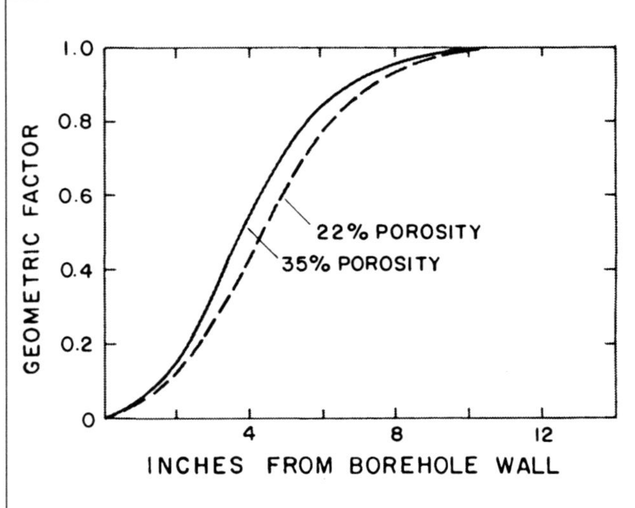

Fig. 9.11—Experimentally derived *J*-factor curves for apparent porosity, from SNP tool for 22% and 35% porosity laboratory formations (from Ref. 11).

The salinity effect is a combination of hydrogen displacement by sodium chloride and thermal neutron absorption by chlorine. Borehole salinity causes a decrease in apparent porosity, whereas formation-water salinity causes an increase in the porosity reading.[13] Charts C and D are used to correct for the salinity effect. Because of invasion, the salinity of the formation water generally can be assumed to equal that of the mud filtrate.

An increase in the mud weight results in a decrease in the hydrogen content in the borehole fluid because the liquid phase is displaced by solids. This, in turn, results in a decrease in apparent porosity. Correction of this effect is obtained from Chart E.

Tool standoff—i.e., the distance that the tool is displaced from the borehole wall—results in an increase in the apparent porosity value indicated by the log. This results because the tool is affected by a zone of 100% equivalent porosity lying between the tool and the borehole wall. Chart F can be used to correct for the tool standoff. Unfortunately, it is seldom known when a tool is off the wall.

Charts G and H correct for the departure of measurement temperature and pressure from standard calibration conditions. The order of magnitude of each of these corrections could be substantial in deep wells. Fortunately, they offset each other, resulting in a small net correction.

The correction charts undergo revisions as more data become available.[14] The log analyst should always look for the most recent chart that corresponds to the design of the tool used in logging the borehole.

Example 9.4. A 1980 CNL run in combination with an FDC indicates a 28% apparent porosity in a zone of interest. Correct this value for borehole conditions that are different from standard calibration conditions. The measurement environment is as follows: mudcake thickness is ½ in., mud weight is 12 lbm/gal, drilling-fluid salinity is 100,000 ppm, formation water salinity is 150,000 ppm, formation temperature is 225°F, formation depth is 21,000 ft, and the tool was run eccentric.

Solution. Enter the nomogram of Fig. 9.18 at 28% porosity value. Proceed vertically to the index line marked with an asterisk on Chart A. Follow the trend lines to −½ in. This value of ½ in. corresponds to the mudcake thickness.

Draw a vertical baseline at this point, which corresponds to an apparent porosity of 27%. In Charts B through H, move parallel to the trend lines from the intersection of the baseline and the actual measurement value to the index line, and read the value of the individual correction.

Chart B—correction of mudcake presence= +2%.
Chart C—borehole-salinity correction= +0.2%.
Chart D—formation-salinity correction= −1.4%.
Chart E—mud-weight correction= +1.7%.
Chart F—standoff correction=0.
Chart G—pressure correction= −3.4%.
Chart H—temperature correction= +4.3%.
Total correction= +3.4%.
Corrected porosity=27+3.4=30.4%.

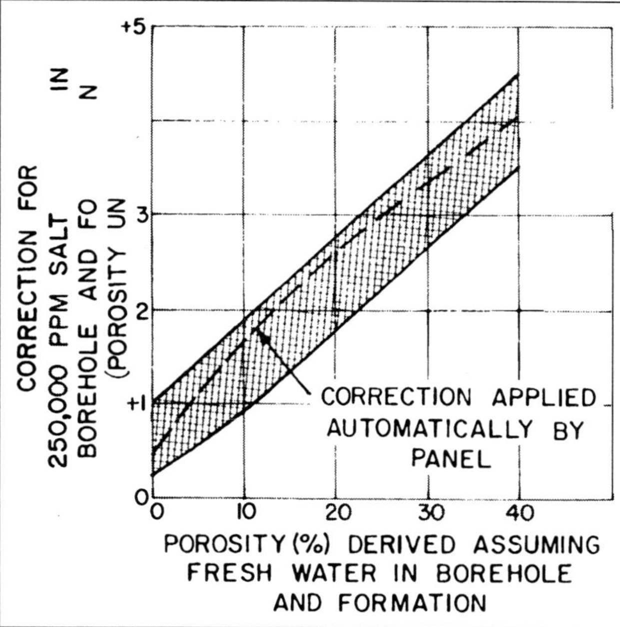

Fig. 9.12—Porosity correction added by the SNP panel to overcome the effect of salt-saturated water in the borehole and formation (from Ref. 4).

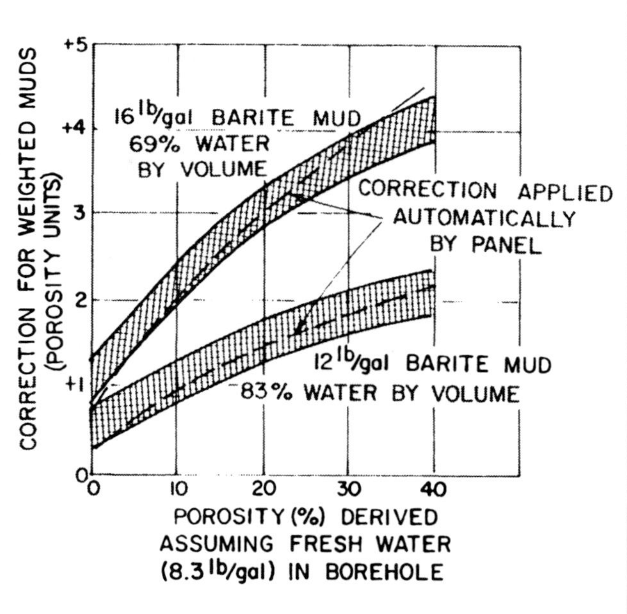

Fig. 9.13—Porosity correction added by the SNP panel to correct for mud weight departure from the standard condition (from Ref. 4).

Assuming that the zone investigated by the tool is invaded with mud filtrate results in a −1% correction instead of −1.4%. The total correction and corrected porosity, in this case, would be +3.8% and 30.8%, respectively.

A measurement based on dual epithermal neutron detectors can also be obtained. One tool that uses such a measurement is Schlumberger's CNT-G. This tool incorporates two thermal and two epithermal detectors for separate porosity measurements.[15] **Figs. 9.19 and 9.20** show a schematic of the tool and a log example.

Fig. 9.20 shows two porosity curves: a thermal neutron porosity curve, $(\phi_N)_t$, derived from the response of the thermal detectors, and an epithermal neutron porosity curve, $(\phi_N)_e$, derived from the response of the epithermal neutron detectors. The two porosity curves differ. The thermal neutron porosity is essentially higher than the epithermal neutron porosity because of the effect of thermal neutron absorbers, such as chlorine and boron. For example, in the clean sand at 5,350 ft, the epithermal neutron porosity, which is more representative of the true porosity, overlies the density porosity. The thermal neutron porosity, however, is 4% higher on average because of the chlorine effect.

The epithermal neutron logs are the most accurate porosity tools because they are not sensitive to capture properties. Another ad-

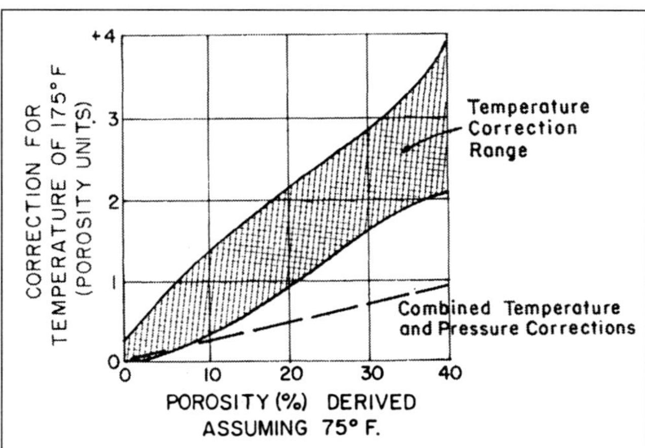

Fig. 9.14—Correction for temperature and pressure effects (from Ref. 4).

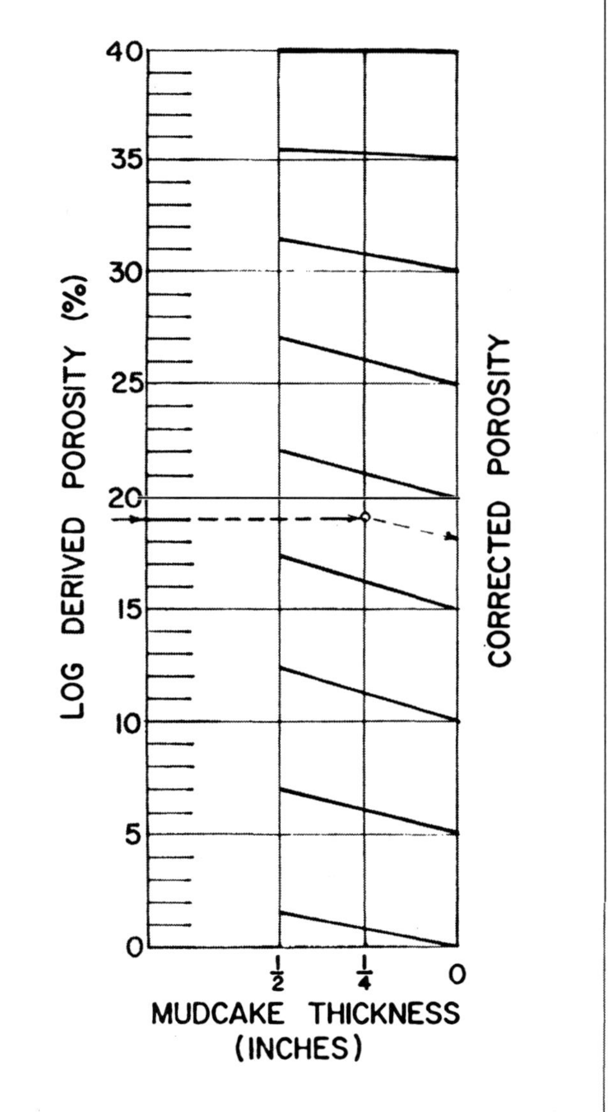

Fig. 9.15—Nomogram for mudcake correction (from Ref. 4).

vantage of using these logs is that the ratio technique, which is no longer applicable because of the short spacing, is replaced by the spine-and-ribs technique for calculating porosity. This method results in a better indication and account of the amount of tool standoff.

The spine-and-ribs technique is similar to that used for the FDC tool (Sec. 8.3.1). **Fig. 9.21** is an example of the spine-and-ribs plot constructed for an 8-in. borehole and limestone formations of various porosities at different values of tool standoff. The heavy solid line, or spine, on the log-log plot of the two epithermal detectors' count rates corresponds to a fully eccentered tool. The thin solid lines, or ribs, are drawn through points of constant porosity with various tool standoffs. The deviation of a data point from the spine is a measure of the effective tool standoff. The standoff effect can be eliminated by bringing the data point to the spine parallel to the ribs.

9.6 Pulsed Neutron Tools

The pulsed neutron tool periodically emits a burst of high-energy (14-MeV) neutrons and then measures the time required for a certain fraction of the neutrons to be absorbed by the formation. Of the common earth elements, chlorine is by far the strongest neutron absorber. In clean formations, the tool response is determined primarily by the chlorine present in the formation water. The tool response can be used qualitatively to differentiate between water-, oil-, and gas-bearing formations. It can also be used quantitatively to estimate water saturation. Because the tool can be run in cased holes, it has become an important device for evaluating old wells and monitoring new wells.

In a homogeneous medium, the decay of thermal neutrons is theoretically an exponential function of time:

$$n = n_0 e^{-t/\tau}, \quad (9.5)$$

where n_0 = number of thermal neutrons at a time t_0 following neutron burst, n = number of thermal neutrons at time t measured from time t_0, and τ = thermal decay time.

With Schlumberger's Thermal Decay Time (TDT^{SM}) tool, τ is defined as the time required for the number of neutrons to diminish to a fraction $1/e$, which is about 37%. The thermal decay time depends on the composite capture cross section of the formation, Σ. Σ and τ are related by[16]

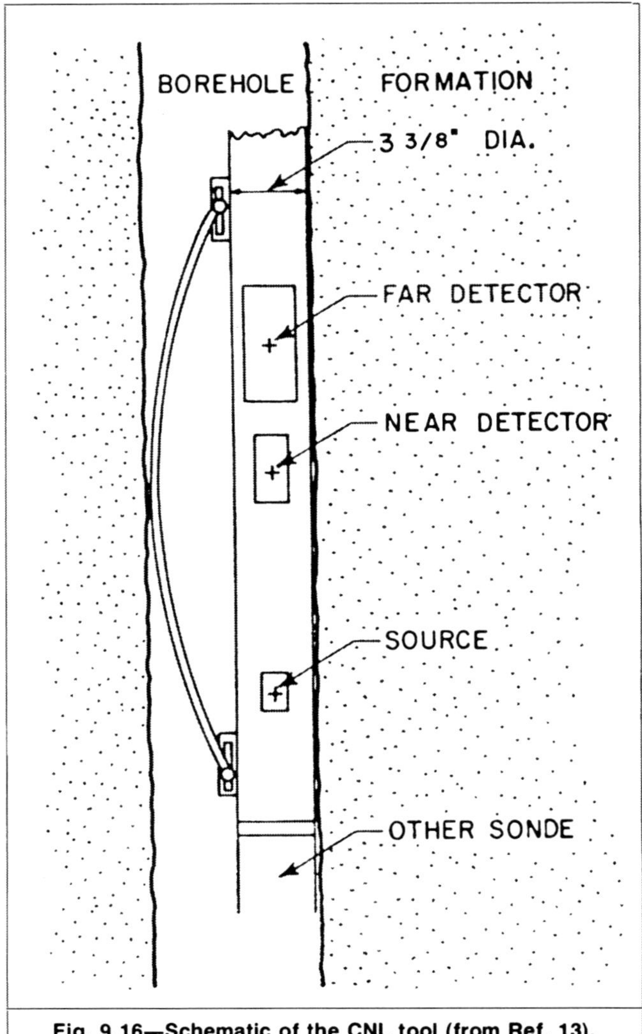

Fig. 9.16—Schematic of the CNL tool (from Ref. 13).

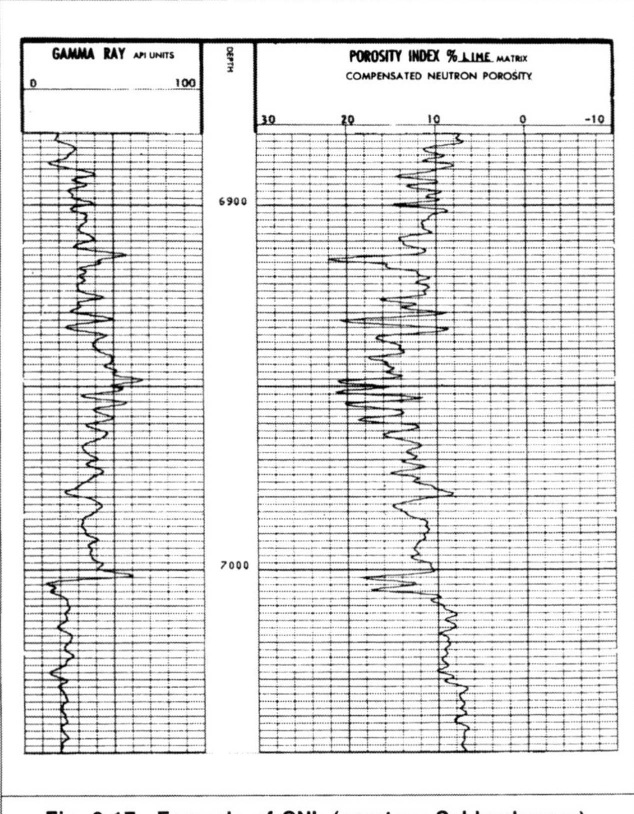

Fig. 9.17—Example of CNL (courtesy Schlumberger).

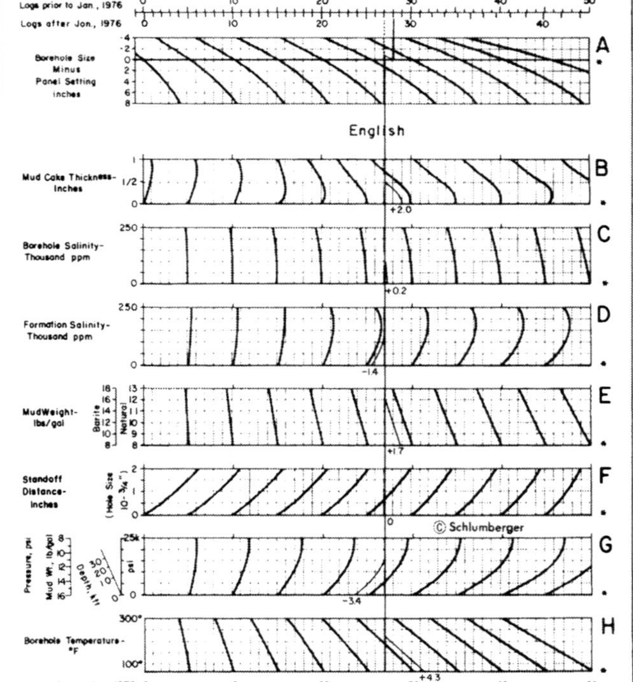

Fig. 9.18—Dual spacing CNL correction nomogram for open hole (courtesy Schlumberger).

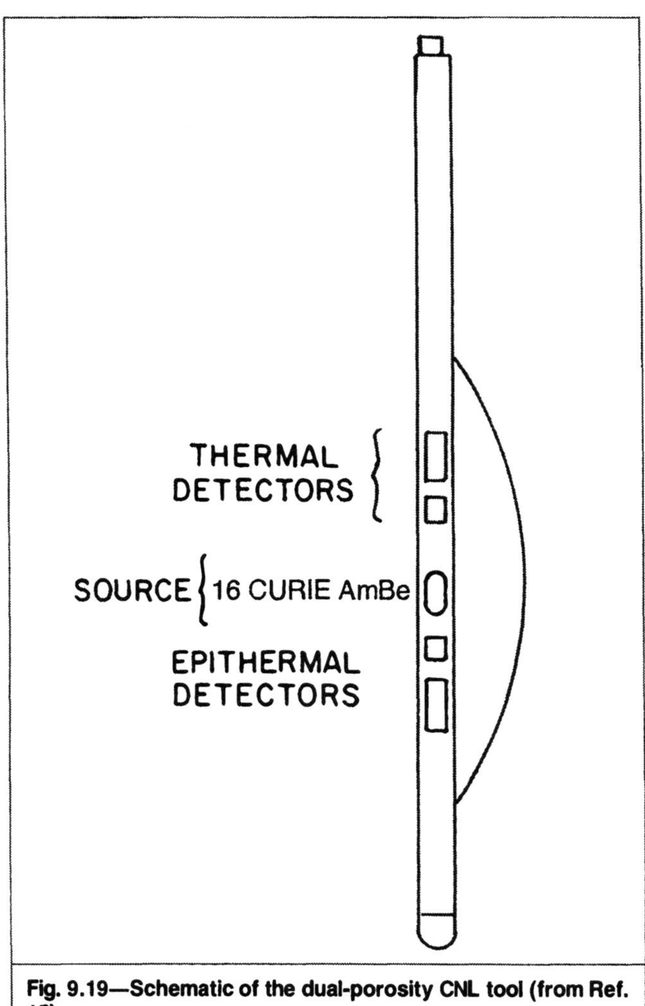

Fig. 9.19—Schematic of the dual-porosity CNL tool (from Ref. 15).

$$\Sigma = 4.55/\tau. \quad (9.6)$$

With Atlas Wireline's Neutron Lifetime Log (NLL^{SM}), τ is defined as the time required for a 50% decay.[17] In this case,

$$\Sigma = 3.15/\tau. \quad (9.7)$$

Fig. 9.22 is a typical curve of neutron decay vs. time for saltwater- and oil-bearing sands. **Fig. 9.23** illustrates the method the TDT tool uses to compute τ. After emission of the neutron bursts and a time sufficiently long for the borehole effect to become negligible, the detector circuit is switched on, and gamma rays are counted during three time intervals or "gates."[16] The background, which consists of natural radioactivity and long-lived induced radioactivity, is measured during the third gate and subtracted from the readings of the first and second gates. The two net readings are combined to compute τ and Σ from Eqs. 9.5 and 9.6. To ensure that the counts of the first two gates are within the region of formation decay, the tool automatically adjusts the timing of the gates so that all three are related to τ, as shown in Fig. 9.23.

Fig. 9.24 shows an example of early TDT logs. The Σ and τ curves are recorded on the combined Tracks 2 and 3. The dotted curve is a quality-control curve. This log was run in 1978 through a clean gas-bearing formation. The log clearly shows the contrast in Σ between the gas and water zones. This contrast reflects the change in chlorine content. The gas/water contact is at 10,084 ft. The initial gas/water contact indicated by the openhole resistivity log recorded in 1972 was at 10,138 ft. This is a case of a strong waterdrive reservoir.

The measured capture cross section, $\Sigma_{\log}$, is related to the constituents of the formation. For a clean, porous formation containing water and hydrocarbon,[18]

$$\Sigma_{\log} = (1-\phi)\Sigma_{ma} + \phi S_w \Sigma_w + \phi(1-S_w)\Sigma_h, \quad (9.8)$$

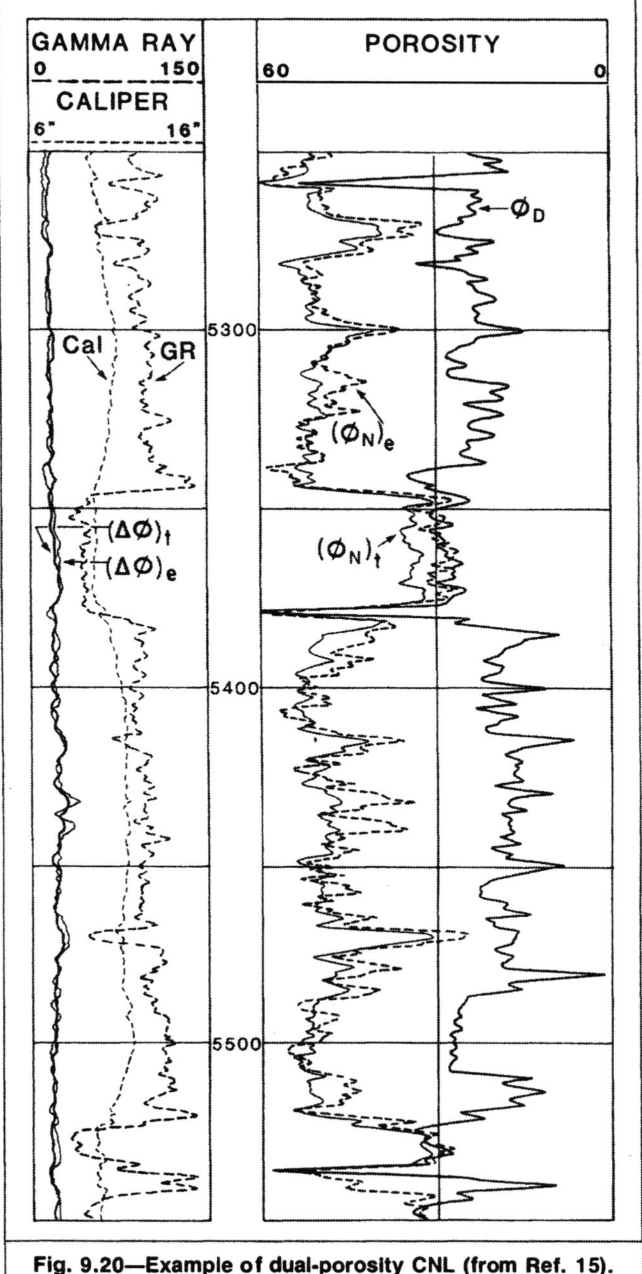

Fig. 9.20—Example of dual-porosity CNL (from Ref. 15).

where Σ_{ma}, Σ_w, and Σ_h are the capture cross sections of rock matrix, water, and hydrocarbon, respectively.

Solving Eq. 9.8 for water saturation gives

$$S_w = \frac{(\Sigma_{\log} - \Sigma_{ma}) - \phi(\Sigma_h - \Sigma_{ma})}{\phi(\Sigma_w - \Sigma_h)}. \quad (9.9)$$

For a clean water-bearing formation, Eq. 9.8 reduces to

$$\Sigma_{\log} = (1-\phi)\Sigma_{ma} + \phi\Sigma_w. \quad (9.10)$$

Water capture cross section, Σ_w, is primarily salinity-dependent. **Fig. 9.25** shows Σ_w as a function of the total sodium chloride content at temperatures of 75 and 200°F. Formation water may contain elements other than chlorine and sodium. Only boron and lithium are important in the interpretation because of their large capture cross sections.[18] Equivalent NaCl salinity may be computed by adding to the parts per million of the chloride ion 80 times the parts per million of boron and 11 times the parts per million of lithium and then multiplying this sum by 1.65 before entering it into Fig. 2.25.

Σ_h for oil can be derived from **Fig. 9.26** when API gravity and the solution GOR are known. A value of 21 c.u. (1 c.u. $=10^{-3}$ cm^2/cm^3) is a good approximation of most field conditions.[18]

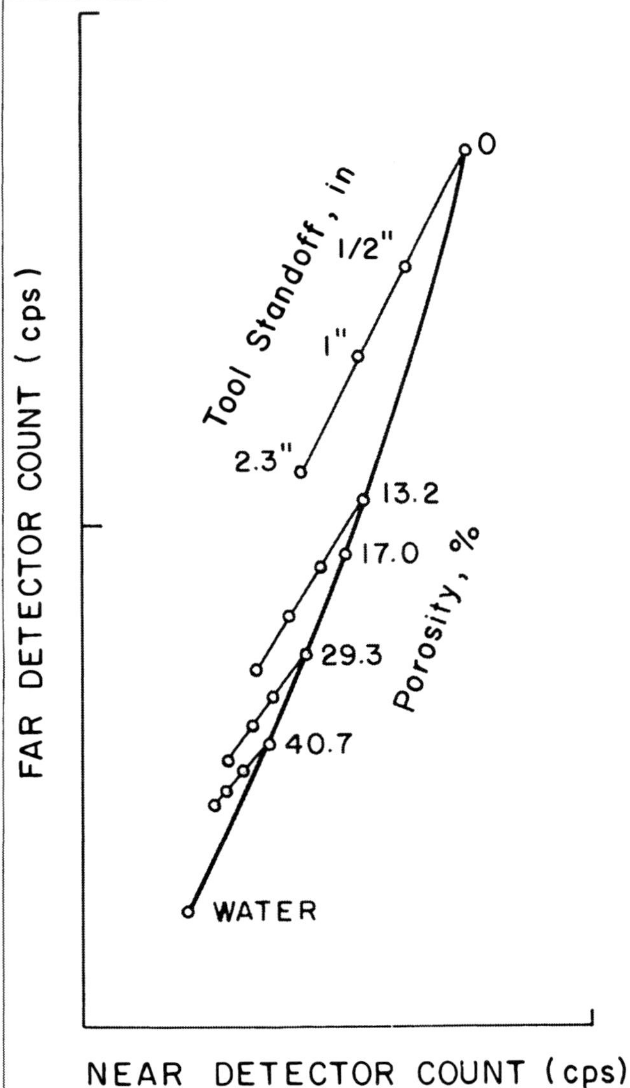

Fig. 9.21—Epithermal neutron spine-and-ribs plot for an 8-in. borehole and limestone formations (from Ref. 15).

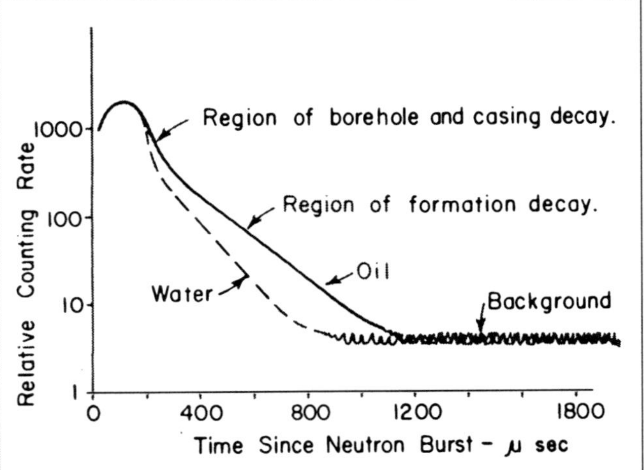

Fig. 9.22—Typical neutron decay curve for oil- and water-bearing formations (from Ref. 16).

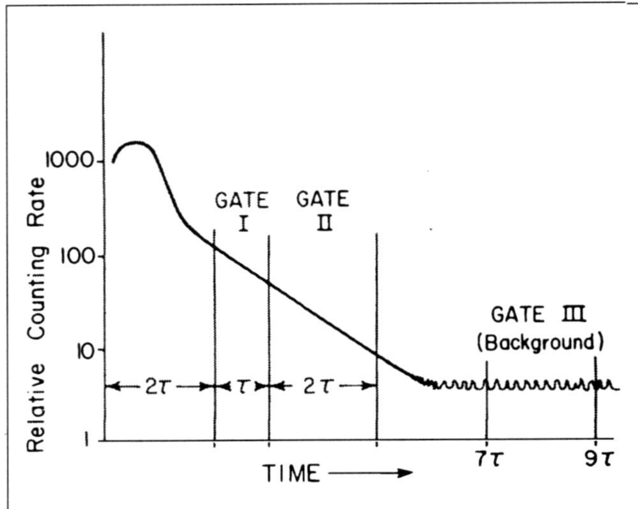

Fig. 9.23—Adjustable-gate system used in TDT tool to compute τ (from Ref. 16).

Gas capture cross section varies with gas composition, pressure, and temperature. **Fig. 9.27** shows the capture cross section for methane, Σ_{CH_4}, as a function of pressure and temperature. Capture cross section of other gases, Σ_g, can be approximated as[18]

$$\Sigma_g = \Sigma_{CH_4}(0.23 + 1.4\gamma_g), \quad (9.11)$$

where γ_g is the gas specific gravity.

Average values of matrix capture cross section recommended for use are 9, 12, and 8 c.u. for sandstones, limestones, and dolomite, respectively.[18]

Example 9.5. Estimate the gas saturation in the unswept and swept Zones G and D of the sand formation shown in Fig. 9.24. The following data are also available.

Formation temperature, °F	200
Reservoir pressure (at the time the TDT log was run), psia	4,500
Water equivalent salinity, ppm	250,000
Gas specific gravity	0.65

Solution. Using Eq. 9.11 and Fig. 9.27 yields

$\Sigma_{CH_4} = 8.1$ c.u.

and $\Sigma_g = 8.1(0.23 + 1.4 \times 0.65)$

$= 9.2$ c.u.

From Fig. 9.26, $\Sigma_w = 120$ c.u.

Zone W is a water zone displaying a Σ_{log} of 40 c.u. Solving Eq. 9.10 for porosity, ϕ, yields

$$\phi = (\Sigma_{log} - \Sigma_{ma})/(\Sigma_w - \Sigma_{ma}). \quad (9.12)$$

Taking $\Sigma_{ma} = 9$ c.u. yields

$\phi = (40-9)/(120-9) = 0.28.$

S_w can now be estimated from Eq. 9.9. For the unswept Zone G where $\Sigma_{log} = 12.5$ c.u.,

$$S_w = \frac{(12.5-9) - 0.28(9.2-9)}{0.28(120-9.2)} = 0.11 \text{ or } 11\%,$$

and for the swept Zone D where $\Sigma_{log} = 28$ c.u.,

$$S_w = \frac{(28-9) - 0.28(9.2-9)}{0.28(120-9.2)} = 0.61 \text{ or } 61\%.$$

New generations of the TDT tool (e.g., TDT-K and TDT-M) have increased neutron output, modified techniques for measuring and recording decay time, and a second detector.[20,21] The second detector provides more information (**Fig. 9.28**). In addition to the Σ curve, the log displays an overlay of the count rates of the far and near detectors (Curves F1 and N1, respectively). Also displayed is a ratio curve derived from the count rate of the two detectors.

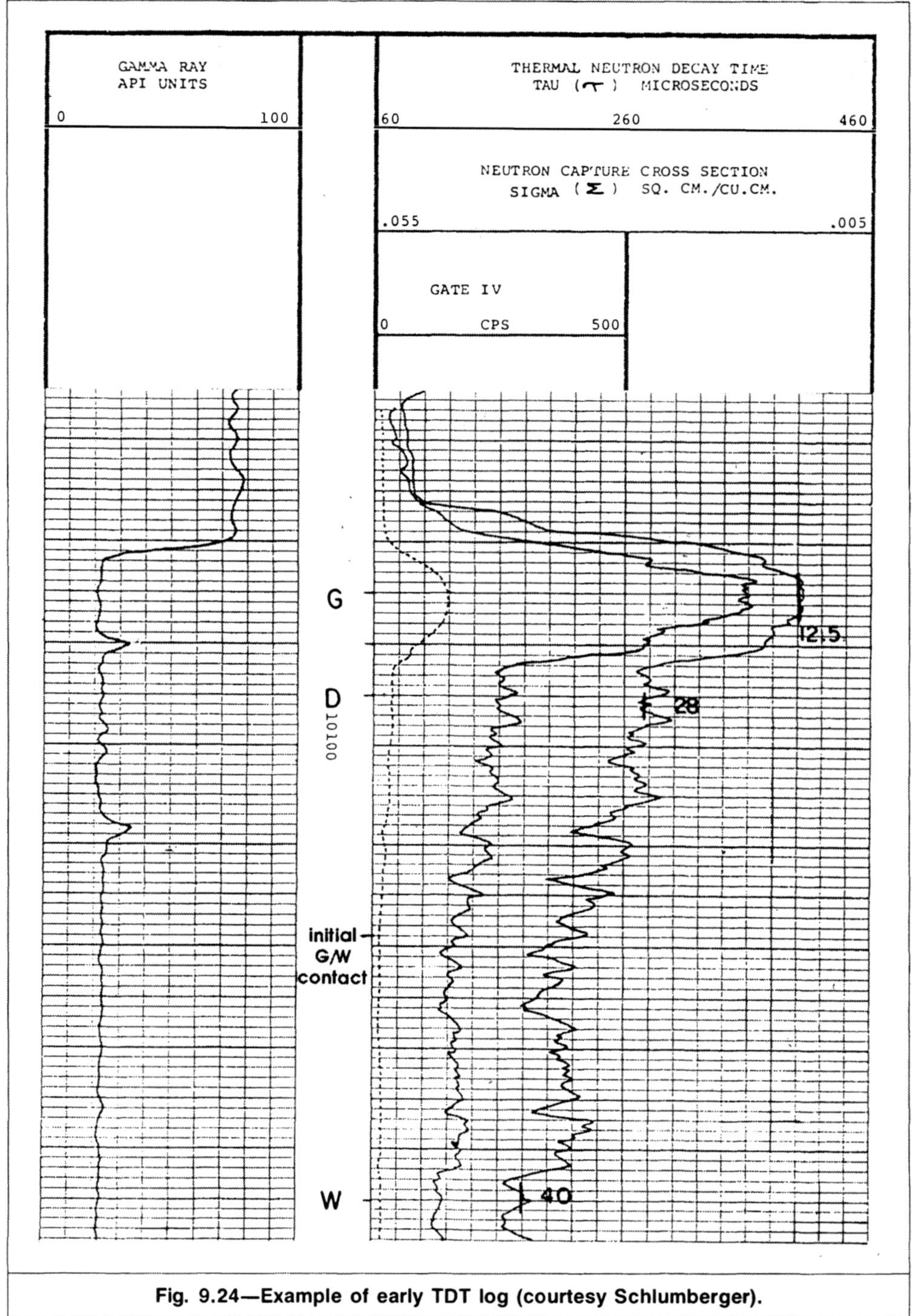

Fig. 9.24—Example of early TDT log (courtesy Schlumberger).

The background curve labeled Curve F3 is recorded for quality-control purposes.

The scales used to record Curves F1 and N1 are selected so that the two curves superimpose in clean water-bearing zones (Zone A). The two curves separate slightly in oil zones (Zone B) but show a large separation in gas zones (Zone C). The separation between these curves and its magnitude result from the difference in hydrogen atom density (number of atoms per unit volume) between water, oil, and gas.

The ratio curve value is derived from the count rates of Curves N1 and F1 corrected for background radiations.[20] The ratio and Σ values permit a quick quantitative evaluation of a formation that does not require the knowledge of Σ_w, Σ_{ma}, and Σ_h. The evaluation method uses experimentally constructed crossplots of Σ vs. ratio similar to those shown in **Fig. 9.29**.

Entering the appropriate crossplot with the zone's Σ and ratio values provides a water salinity, C, and porosity, ϕ, values. These values approximate the true values only if the zone is clean and water-bearing. In hydrocarbon zones, the crossplot yields apparent water salinity, C_a, and apparent porosity, ϕ_a. In oil zones, oil appears like fresh water; this results in an apparent water salinity that is less than the true one. A quick method for estimating the water saturation is to use the water-salinity-index ratio:

$$S_w \simeq C_a/C. \quad (9.13)$$

If the actual value of C is not available, it is obtained from an associated water zone with the ratio/Σ crossplot.

Observations indicate that for gas zones, $C_a \geq C$. To estimate S_w in gas zones, a porosity-index ratio is used to compare the crossplot apparent porosity, ϕ_a, and the true porosity, ϕ:

$$S_w \simeq \phi_a/\phi. \quad (9.14)$$

Example 9.6. The upper sand of the TDT log of **Fig. 9.30** contains gas, oil, and water zones. Here, 5½-in. casing is cemented

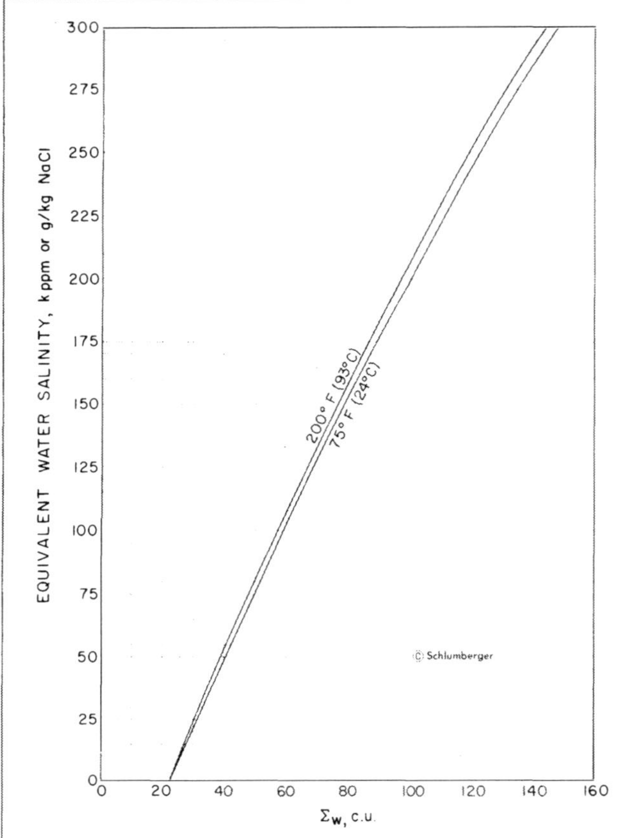

Fig. 9.25—Water capture cross section, Σ_w, as a function of water equivalent salinity (from Ref. 19).

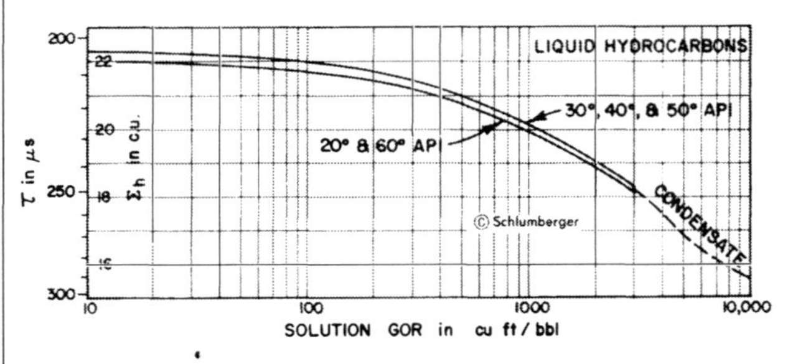

Fig. 9.26—Oil capture cross section, Σ_o, as a function of oil gravity and GOR (from Ref. 19).

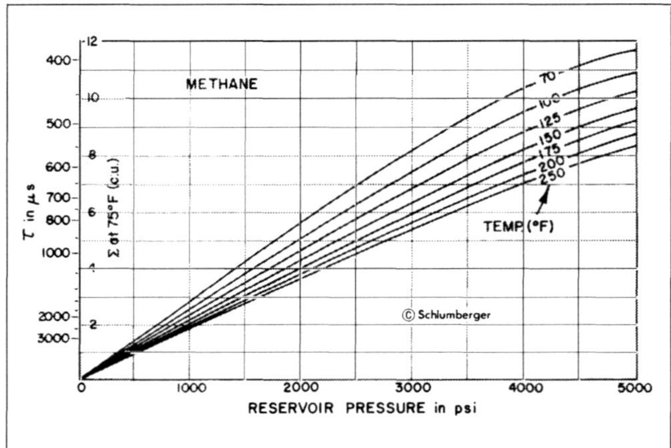

Fig. 9.27—Methane capture cross section, Σ_{CH_4}, as a function of pressure and temperature (from Ref. 19).

in an 8⅝-in. hole, a 2⅜-in. tubing is present, and the annulus is filled with salt water.

a. Determine the depth of the gas/oil and oil/water contacts.

b. For the three zones indicated on the log, estimate the water saturation, assuming that the sand displays a constant porosity value.

Solution.

a. The top zone displays a separation of Curves F1 and N1 that is consistent with the presence of gas. The separation disappears around 6,344 ft, which is the gas/oil contact. In general, oil zones indicate a Σ value higher than that of associated water zones. The oil/water contact can then be placed at a depth of 6,352 ft.

b. The table below lists the log reading and ϕ_a and n_a obtained from Crossplot C of Fig. 9.29.

Zone	Σ (c.u.)	Ratio	ϕ_a (%)	C_a (1,000 ppm)
Gas	18	2.0	13	150
Oil	18.5	2.9	28	60
Water	32	3.7	35	145

The water-saturation values in the gas and oil zones are estimated from Eqs. (9.14) and (9.13), respectively.

For the gas zone, $S_w \cong 13/35 = 0.37$, and for the oil zone, $S_w \cong 60/145 = 0.41$.

Review Questions

1. Explain why logs using epithermal neutron detection depend on a smaller number of formation-characterizing parameters than those using thermal neutron or capture gamma ray detection.
2. Explain why epithermal neutron tools are limited to short source-to-detector spacings and pad-mounted devices.
3. Why is the spatial distribution of capture gamma rays very similar to that of thermal neutrons?
4. Explain what is meant by "hybrid measurement."
5. Why do N-G tools have a relatively larger depth of investigation?
6. Define the API neutron unit.
7. Explain why the count rate at the detector is inversely proportional to porosity.
8. To what extent does natural gamma radiation affect the capture gamma ray detector?
9. What are the advantages and disadvantages of thermal neutron detectors?
10. What are the advantages and disadvantages of epithermal neutron detectors?
11. Describe the API neutron test pit. Where is it located?
12. Describe how the departure curves of the GNT tool and other similar tools were constructed.
13. In the absence of departure curves, how can the log response be calibrated in terms of porosity?
14. Is the shale porosity derived from the neutron logs a true one? Explain.
15. What are the salient features of an SNP tool?
16. How and under what conditions were the SNP tool calibration curves obtained?
17. What are the corrections applied automatically by the SNP surface control panel? Why are they needed?
18. Comment on the order of magnitude and sign of the corrections applied to the SNP tool.
19. What are the advantages of the tool having dual thermal neutron detectors?
20. What are the advantages of the tool having dual epithermal neutron detectors?
21. What are the advantages of pulsed neutron tools?
22. What information is provided by the dual-detector pulsed neutron tool but not by the single-detector tool?

Problems

9.1 **Table 9.2** lists data obtained from a cored section of the reef limestone formation of the Kelly-Snyder field. The neutron log was recorded in a 7-in. cased hole.

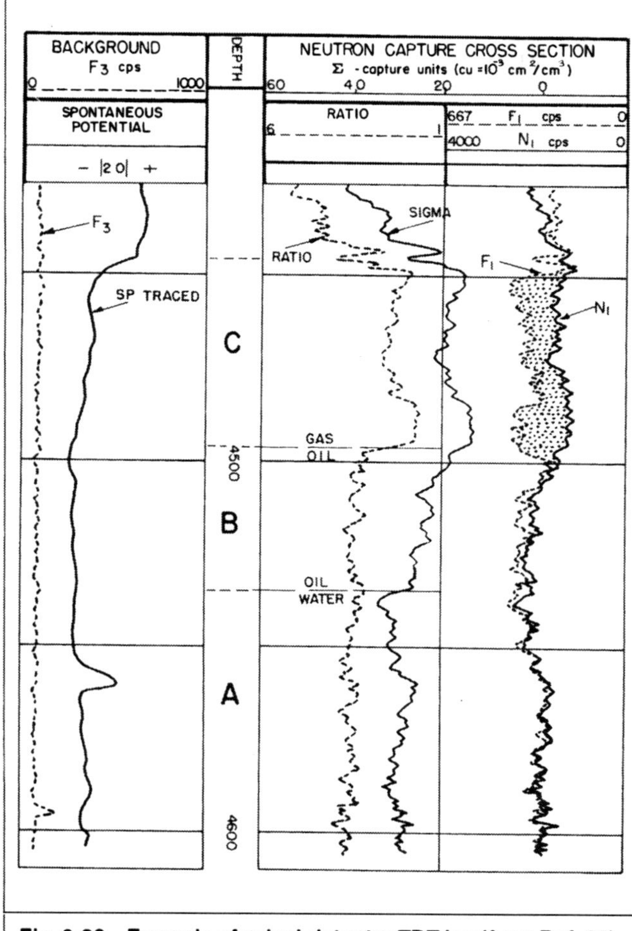

Fig. 9.28—Example of a dual-detector TDT log (from Ref. 22).

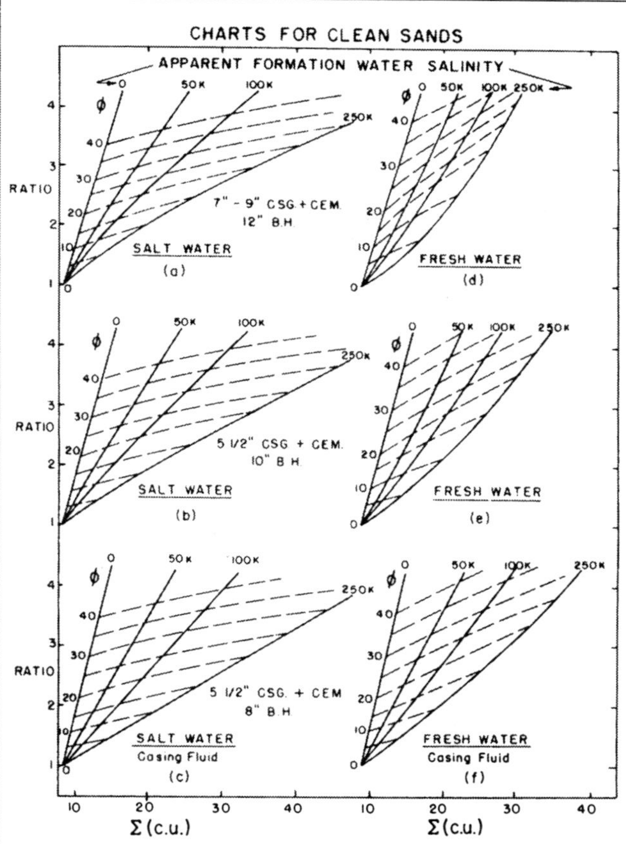

Fig. 9.29—Laboratory crossplots of Σ vs. ratio for several borehole situations in clean sand formations. Apparent porosity and apparent water salinity are derived from these crossplots (from Ref. 20).

a. Show that the log reading can be related to porosity by the empirical relation of Eq. 9.2.
b. Compare this correlation to that of Example 9.2, also prepared from the reef limestone of the Kelly-Snyder field. Comment on the differences and the reason for such differences, if any.

9.2 **Table 9.3** lists core porosities and neutron log deflection data taken from a well in the Short Junction field of Cleveland County, OK. The formation is the Bois d'Arc section of the Hunton limestone. Also listed is the core porosity's running average calculated over a 3-ft interval. Usually, the core porosity log and neutron log are not on-depth. The values listed are already correlated for depth match. In this particular example, there is a 6-ft correction. Examine and comment on the correlation between porosity and neutron deflection.

9.3 Show the borehole-diameter effect on the reading of a 19½-in. GNT tool for the following measurement environment: 12-lbm/gal barite mud, 10% limestone formation true porosity, and 250°F formation temperature. Hint: A picture is worth a thousand words!

9.4 For the same porosity and formation temperature as in Problem 9.3, show the effect of mud weight and type on the tool response in an 8-in. borehole.

9.5 For the same porosity and mud as in Problem 9.3, show the effect of temperature on the tool response in an 8-in. borehole.

9.6 a. Prepare a plot of the apparent limestone porosity index indicated by the SNP tool vs. the true porosity of water-bearing dolomite formations.
b. Repeat for CNL and density tools.
c. Compare and explain the sensitivity of the three different tools to lithology.

9.7 Prepare a plot of gamma ray deflection vs. the porosity indicated by the SNP log of **Fig. 9.31**. Comment on the correlation, or the lack of a correlation, between the gamma ray response and SNP porosity.

9.8 A 20% SNP porosity was determined from the calibration curves. The actual measurement conditions are as follows: borehole diameter is 6 in., drilling fluid is 12-lbm/gal barite freshwater-based mud, formation temperature is 175°F, and formation pressure is 5,200 psia.
a. Estimate the order of magnitude of the correction applied by the SNP panel.
b. Compare this correction to that of Example 9.3. Explain the difference, if any.

9.9 A 1980 CNL log run in combination with the FDC tool indicated a 28% apparent porosity in a zone of interest. Correct this value for borehole conditions that are different from standard calibration conditions. The measurement environment is as follows: mudcake thickness is unavailable; mud weight is 12 lbm/gal; freshwater-based, low-solids-content mud; formation water salinity is unavailable; formation temperature is 225°F; formation depth is 21,000 ft; and the tool was run eccentric. Compare the order of magnitude of this correction to that of Example 9.4.

9.10 Show the borehole diameter effect on the CNL response in a 30% porosity formation. Measurement conditions other than hole size are standard.

9.11 The middle sand displayed by the TDT log of **Fig. 9.32** contains gas, oil, and water zones.
a. Indicate the depth of the gas/oil and oil/water contacts.
b. Knowing that the porosity is the same throughout the sand and that the formation water salinity is 185,000 ppm, estimate the gas and oil saturations.

9.12 The lower sand in Fig. 9.30 contains an oil zone underlaid by water.
a. Indicate the depth of the oil/water contact.
b. Estimate the porosity and oil saturation.

9.13 **Fig. 9.33** shows two neutron logs run through sandstone Formation A in a gas injection well. Run 1 was made when the formation was fully saturated with water of 50,000-ppm sa-

TABLE 9.2—CORE POROSITY AND NEUTRON LOG READING FROM A 7-in. CASED HOLE, REEF LIMESTONE, KELLY-SNYDER FIELD (from Ref. 10)

Core Porosity (%)	Neutron Log Reading (standard units)
1	0.662
2	0.600
3	0.560
4	0.535
5	0.513
6	0.497
7	0.481
8	0.470
9	0.460
10	0.450
11	0.440
12	0.432
13	0.426
14	0.419
15	0.412
16	0.405
17	0.400
18	0.395
19	0.390
20	0.385
25	0.365
30	0.347

linity before gas injection. Run 2 was made several months after gas injection started. Estimate the gas saturation in the wellbore vicinity. The injected gas is mostly methane. Formation temperature and pressure are 60°F and 500 psia, respectively.

Nomenclature

a,b = coefficients in Eq. 9.2
C = salinity, ppm
D_t = diffusion coefficient of thermal neutrons

Fig. 9.30—Dual-detector TDT log of Example 9.6 (from Ref. 20).

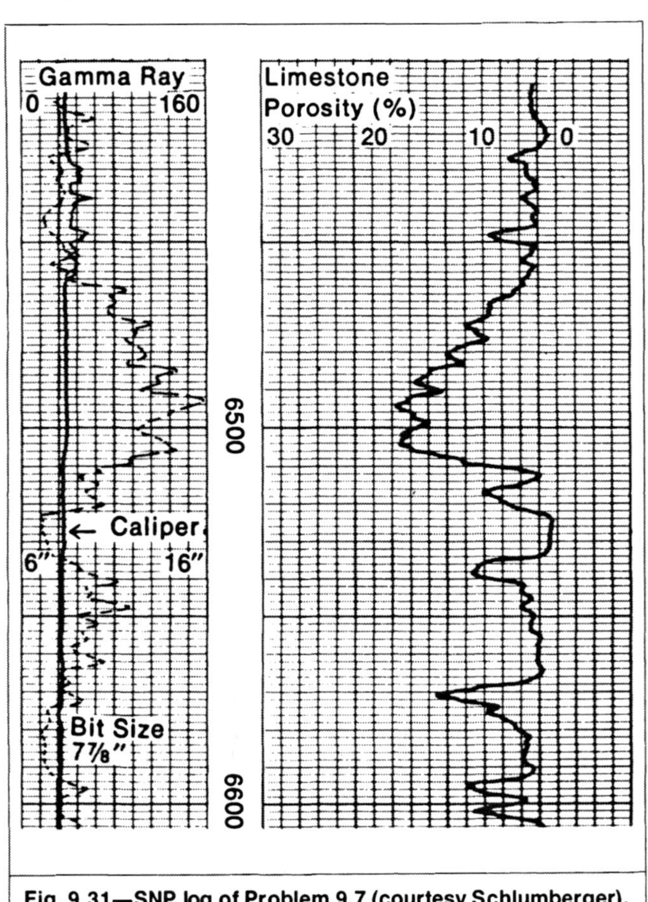

Fig. 9.31—SNP log of Problem 9.7 (courtesy Schlumberger).

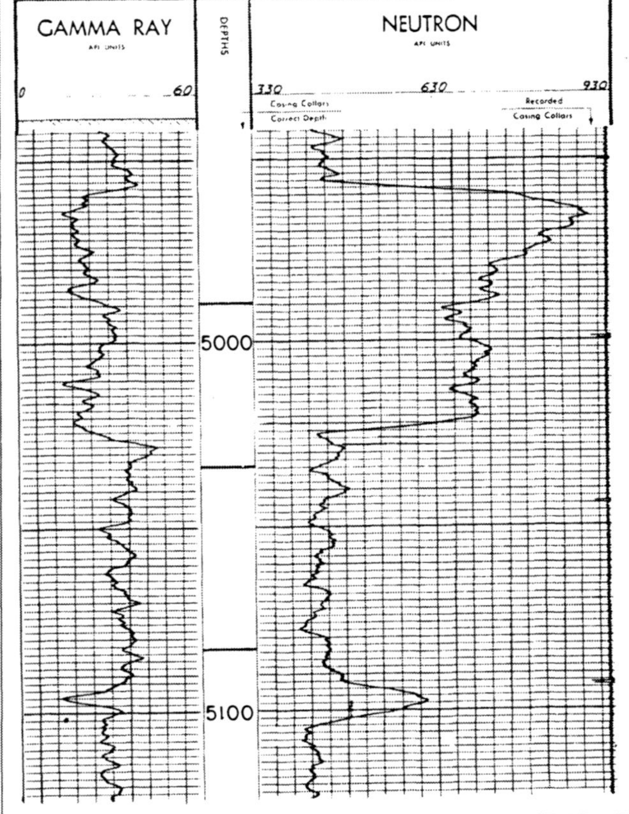

Fig. 9.32—TDT log of Problem 9.11 (courtesy Schlumberger).

TABLE 9.3—CORE AND NEUTRON LOG DATA FROM THE BOIS D'ARC SECTION OF THE HUNTON LIME, SHORT JUNCTION FIELD, CLEVELAND COUNTY, OK (from Ref. 8)

Core Analysis Depth (ft)	Log Depth (ft)	Core Porosity (%)	Average Porosity (%)	Diameter (in.)	Neutron-Derived Porosity (%)
8,149	8,143	2.0	—	4.17	3.2
8,150	8,144	2.7	2.3	4.0	3.2
8,151	8,145	1.8	2.4	3.8	4.6
8,152	8,146	2.7	2.6	3.6	6.0
8,153	8,147	3.5	2.6	3.82	4.8
8,154	8,148	1.8	2.6	3.7	5.7
8,155	8,149	2.7	2.8	3.9	4.0
8,156	8,150	4.0	3.5	3.8	4.3
8,157	8,151	3.8	3.6	3.6	6.0
8,158	8,152	3.1	2.9	3.35	7.5
8,159	8,153	1.7	2.6	3.6	5.9
8,160	8,154	2.9	2.6	3.6	5.8
8,161	8,155	3.1	3.0	3.7	5.2
8,162	8,156	3.0	3.2	3.72	5.0
8,163	8,157	3.4	3.3	3.73	5.2
8,164	8,158	3.5	4.0	3.55	5.9
8,165	8,159	5.1	4.2	3.48	6.4
8,166	8,160	4.0	4.9	3.32	8.0
8,167	8,161	5.6	5.3	3.48	6.5
8,168	8,162	7.0	6.9	3.47	7.0
8,169	8,163	8.1	7.2	3.46	7.8
8,170	8,164	6.6	8.2	3.42	6.8
8,171	8,165	10.0	8.3	3.25	8.9
8,172	8,166	8.4	10.2	3.1	10.0
8,173	8,167	12.2	10.0	3.1	10.0
8,174	8,168	8.4	9.1	3.15	9.1
8,175	8,169	9.0	8.9	3.38	7.7
8,176	81,70	9.2	9.1	3.4	7.3
8,177	8,171	9.1	9.4	3.6	6.0
8,178	8,172	10.1	8.9	3.5	7.0
8,179	8,173	7.7	7.6	3.3	7.6
8,180	8,174	5.0	5.9	3.8	5.2
8,181	8,175	5.0	5.1	4.0	4.5
8,182	8,176	5.3	4.7	4.15	3.5
8,183	8,177	3.9	4.6	4.0	3.5
8,184	8,178	4.5	4.6	3.75	4.9
8,185	8,179	5.4	6.1	3.95	4.8
8,186	8,180	9.3	7.2	3.7	5.4
8,187	8,181	2.0	6.9	3.55	6.4
8,188	8,182	4.4	6.9	3.75	5.2
8,189	8,183	9.4	6.9	3.5	6.9

F = ratio of thermal neutron flux at two detectors
L_e = epithermal slowdown length, cm
L_t = thermal neutron slowdown length, cm
n = number of thermal neutrons at time t measured from time t_0
n_0 = number of thermal neutrons at a time t_0 following neutron burst
N = neutron tool response
N_N = source strength, neutrons/s
r = source-to-detector distance
S = saturation, fraction
t = time, seconds
α, β = coefficients in Eq. 9.1
γ_g = gas specific gravity
Σ = capture cross section, cm^2/cm^3
τ = thermal decay time, seconds
ϕ = porosity, fraction
$\Psi_t(r)$ = thermal neutron flux, $neutron/cm^2 \cdot s$

Subscripts

a = apparent
CH_4 = methane
e = epithermal neutrons
g = gas
h = hydrocarbon
log = log response
ma = matrix
N = neutron
o = oil
t = thermal neutrons
w = water
0 = initial

References

1. Tittman, J.: "Radiation Logging," paper presented at the Petroleum Engineering Conference, U. of Kansas, Lawrence, April 1956.
2. *Log Review 1: Section 8,* Dresser Atlas Wireline Services, Houston (1971).
3. "Neutron Logging," Welex, 1979.
4. Tittman, J. *et al.*: "The Sidewall Epithermal Neutron Porosity Log," *JPT* (Oct. 1966) 1351-62; *Trans.*, AIME, **237**.
5. Pirson, S.J.: *Handbook of Well Log Analysis for Oil and Gas Formation Evaluation*, Prentice-Hall, Englewood Cliffs, NJ (1963) 193-213.
6. *RP 33, Recommended Practice for Standard Calibration and Form for Nuclear Logs*, API, Dallas (Sept. 1959).
7. Brown, A.A. and Bowers, B.: "New Approach Improves Accuracy of Porosity Determinations from Neutron Logs," *Pet. Eng.* (May 1958) B-30-B-34.
8. Storseth, B.J.: "Radioactivity Logging in Carbonate Reservoirs," paper presented at the fifth Biennial Symposium on Subsurface Geology, U. of Oklahoma, Tulsa, March 1957.
9. Swulius, T.M.: "Porosity Calibration of Neutron Deflection Logs, SACROC Unit," *JPT* (April 1986) 468-76; *Trans.*, AIME, **281**.
10. Bush, R.E. and Mardock, E.S.: "Quantitative Application of Radioactivity Logs," *Trans.*, AIME (1951) **192**, 191-98.

TABLE 9.3—CORE AND NEUTRON LOG DATA FROM THE BOIS D'ARC SECTION OF THE HUNTON LIME, SHORT JUNCTION FIELD, CLEVELAND COUNTY, OK (from Ref. 8) (Continued)

Core Analysis Depth (ft)	Log Depth (ft)	Core Porosity (%)	Average Porosity (%)	Diameter (in.)	Neutron-Derived Porosity (%)
8,190	8,184	6.8	8.7	3.3	7.8
8,191	8,185	8.9	7.2	3.6	6.2
8,192	8,186	5.0	5.8	3.65	5.7
8,193	8,187	3.6	3.7	3.9	4.1
8,194	8,188	2.5	2.8	4.2	3.2
8,195	8,189	2.3	2.3	3.98	4.2
8,196	8,190	2.3	2.3	4.08	4.0
8,198	8,192	2.3	2.4	3.95	4.1
8,199	8,193	2.7	2.5	3.9	4.3
8,200	8,194	2.5	2.6	4.3	2.8
8,201	8,195	2.6	2.6	4.0	3.3
8,202	8,196	2.6	2.5	4.3	2.7
8,203	8,197	2.3	2.4	4.4	2.8
8,204	8,198	2.4	2.0	4.4	2.7
8,205	8,199	1.4	2.0	4.28	2.6
8,206	8,200	2.7	2.0	4.28	2.7
8,207	8,201	2.0	2.4	4.45	2.8
8,208	8,202	2.4	2.8	4.1	3.2
8,209	8,203	4.0	3.6	4.0	4.0
8,210	8,204	4.3	5.9	3.85	4.2
8,211	8,205	9.6	6.6	3.6	5.8
8,212	8,206	5.9	6.7	3.8	5.0
8,213	8,207	4.5	4.8	3.75	4.6
8,214	8,208	4.0	3.6	3.9	4.5
8,215	8,209	2.2	2.9	3.65	5.5
8,216	8,210	2.4	2.3	4.0	3.3
8,217	8,211	1.8	2.1	4.05	3.4
8,218	8,212	2.0	2.0	4.42	2.5
8,219	8,213	2.1	2.1	4.6	2.1
8,220	8,214	2.2	2.2	4.7	2.3
8,221	8,215	2.3	2.3	4.4	2.6
8,222	8,216	2.3	2.3	4.4	2.7
8,223	8,217	2.2	2.5	4.28	2.6
8,224	8,218	2.9	2.5	4.52	2.3
8,225	8,219	2.5	3.4	4.5	2.3
8,226	8,220	3.7	3.2	4.7	2.4
8,227	8,221	3.4	3.0	—	2.0
8,228	8,222	2.0	2.6	—	2.4
8,229	8,223	2.6	2.4	—	2.6
8,230	8,224	2.7	—	—	—
8,231	8,225	1.8	—	—	—

11. Sherman, H. and Locke, S.: "Effect of Porosity on Depth of Investigation of Neutron and Density Sondes," paper SPE 5510 presented at the 1975 SPE Annual Technical Conference & Exhibition, Dallas, Sept. 28–Oct. 1.
12. Tittman, J.: *Geophysical Well Logging*, Academic Press Inc., New York City (1986) 90–105.
13. Alger, R.P. *et al.*: "The Dual Spacing Neutron Log—CNL," *JPT* (Sept. 1972) 1073–83.
14. Gilchrist, W.A. Jr. *et al.*: "Improved Environmental Corrections for Compensated Neutron Logs," *SPEFE* (June 1988) 371–76; *Trans.*, AIME, **285**.
15. Davis, R.R. *et al.*: "A Dual Porosity CNL Logging System," paper SPE 10296 presented at the 1981 SPE Annual Technical Conference and Exhibition, San Antonio, Oct. 4–7.
16. *Thermal Decay Time Log*, Schlumberger, Houston (1968).
17. *Neutron Lifetime Log*, Atlas Wireline Services, Houston (1970).
18. Clavier, C., Hoyle, W.R., and Meunier, D.: "Quantitative Interpretation of TDT Logs," *JPT* (June 1971) 743–63.
19. *Log Interpretation Charts*, Schlumberger, Houston (1979).
20. Dewan, J.J. *et al.*: "Thermal Decay Time Logging Using Dual Detection," *Log Analyst* (1973) **14**, No. 5, 13–26.
21. Hall, J.E. *et al.*: "A New Thermal Neutron Decay Logging System—TDT-M," *JPT* (Jan. 1982) 199–207.
22. McGhee, B.F., McGuire, J.A. and Vacca, H.L.: "Examples of Dual Spacing Thermal Neutron Decay Time Logs in Texas Coast Oil and Gas Reservoirs," *Proc.*, 17th SPWLA Annual Logging Symposium, Denver (1976) paper GG.

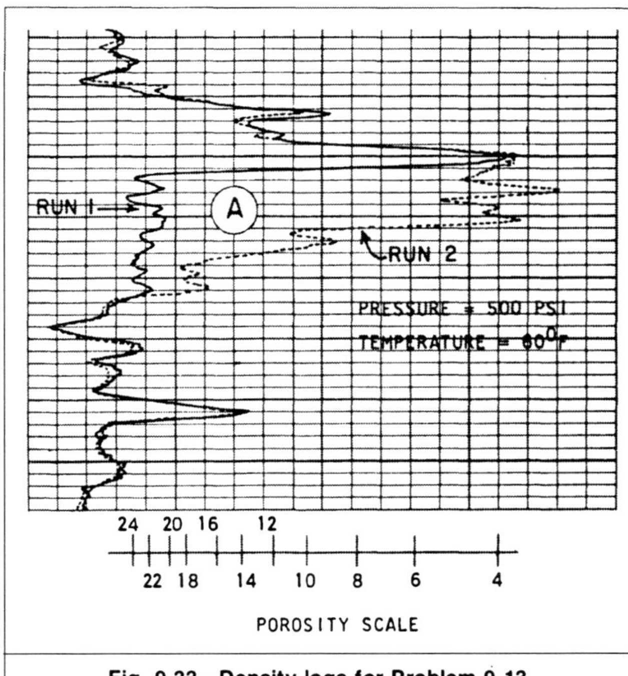

Fig. 9.33—Density logs for Problem 9.13.

Chapter 10
Sonic Porosity Log

10.1 Introduction

Sonic devices were first introduced for seismic velocity determination. These "continuous velocity logs" were widely used in petroleum exploration and development once it was discovered that a reliable formation porosity value could be extracted from the log response. Conventional sonic tools measure the reciprocal of the velocity of the compressional wave. This parameter is called interval travel time, Δt, or slowness, and is expressed in microseconds per foot. Porosity of consolidated formations is related to Δt by Wyllie's equation (see Sec. 3.5):

$$\phi=(\Delta t-\Delta t_{ma})/(\Delta t_f-\Delta t_{ma}), \quad (10.1)$$

where Δt_{ma} and Δt_f are the slownesses of the matrix and pore fluid, respectively. The average values of matrix slowness usually used in Eq. 10.1 are 55.5, 47.5, and 43.5 μsec/ft for sandstone, limestone, and dolomite, respectively. The average fluid slowness used is 189 μsec/ft.

Eq. 10.1, which is derived with a simplified physical model, overestimates porosity in unconsolida formations. In such formations, Eq. 10.1 is modified to

$$\phi=[(\Delta t-\Delta t_{ma})/(\Delta t_f-\Delta t_{ma})]/B_{cp}, \quad (10.2)$$

where B_{cp} is a compaction correction factor related empirically to the average slowness of adjacent shale, Δt_{sh}, by

$$B_{cp}=\Delta t_{sh}/100. \quad (10.3)$$

In general, Δt and ϕ can be related by the following linear equation:

$$\Delta t=A+B\phi. \quad (10.4)$$

This relationship is determined by correlating log response to core porosities. The coefficients A and B do not correspond to well-defined physical parameters but depend on matrix and pore compressibilities. If Wyllie's relationship prevails, then

$$A=\Delta t_{ma} \quad (10.5)$$

and $B=(\Delta t_f-\Delta t_{ma})$. (10.6)

The presence of secondary porosity—i.e., vuggs and fractures—complicates the quantitative evaluation of sonic logs. In such cases, the above models yield an apparent porosity that reflects only the primary (intergranular) porosity. Comparing this apparent value to that of total porosity, obtained from the density or the neutron log, gives an order of magnitude of the secondary porosity.

10.2 Single-Receiver System

Early sonic tools were equipped with a single receiver (**Fig. 10.1**). In this system, a pulse is initiated at the transmitter situated at a distance from the receiver, L_s (also called the spacing). The time measured, $t_{\log}$, is between the initiation of the pulse and the first arrival of acoustic energy at the receiver. **Fig. 10.2** shows the waveform at the receiver and the time measured. The path of the first arrival is also shown in Fig. 10.1. The measured time can be expressed as

$$t_{\log}=\frac{L_s}{v}+\frac{d_h-(d_t+2l_c)}{v_m}\sqrt{1-\left(\frac{v_m}{v}\right)^2} \quad (10.7)$$

where d_h=borehole diameter, d_t=tool diameter, v=formation compressional velocity, and v_m=mud compressional velocity.

In deriving Eq. 10.7, we assume that the tool axis is parallel to the borehole axis but may be displaced from it by a distance l_c. It is also assumed that $v>v_m$.

The slowness of the formation is expressed as

$$\Delta t=(t_{\log}-t_m)/L_s, \quad (10.8)$$

where t_m is a mud-path correction time:

$$t_m=\Delta t_m[d_h-(d_t+2l_c)]\sqrt{1-(\Delta t/\Delta t_m)^2}. \quad (10.9)$$

Fig. 10.3 shows this correction for a specific geometry. The order of magnitude of this correction can be considerable and should be accounted for in quantitative interpretations. In practice, however, it is difficult to estimate the parameters d_h, l_c, and Δt_m. Moreover, the correction is a function of the sought-after parameter Δt itself.

Example 10.1. A compacted sandstone formation of interest, detected by an old single-receiver sonic log, displays a slowness of 90 μsec/ft. Estimate the formation porosity if the bit is 7⅞ in., tool diameter is 3 in., centralizers were used, and the tool spacing is 6 ft.

Solution. Because centralizers were used, l_c=0. From the compressional velocities of Table 3.3, the mud velocity is estimated to be 5,000 ft/sec, which corresponds to 200 μsec/ft. In Eq. 10.9, Δt is taken, as a first approximation, to be the measured value of 90 μsec/ft:

$$t_m=200[(7.875-3)/12]\sqrt{1-(90/200)^2}=72.6 \text{ seconds}.$$

From Eq. 10.8,

$$\Delta t = 90 - (72.6/6) = 77.9\ \mu\text{sec/ft},$$

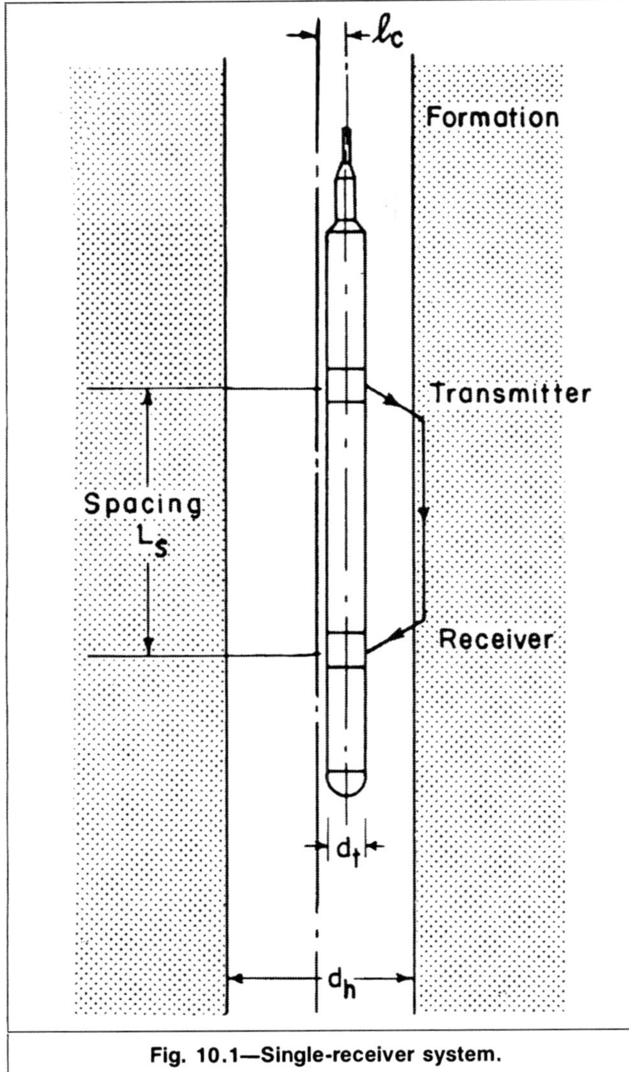

Fig. 10.1—Single-receiver system.

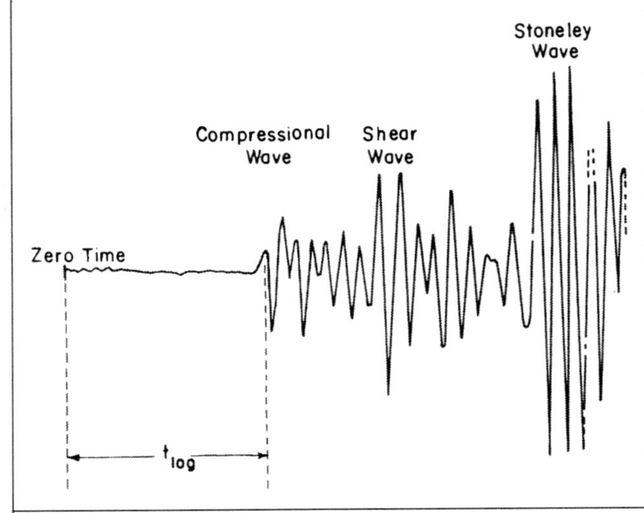

Fig. 10.2—Schematic of the waveform at the receiver showing time measured by the single-receiver system.

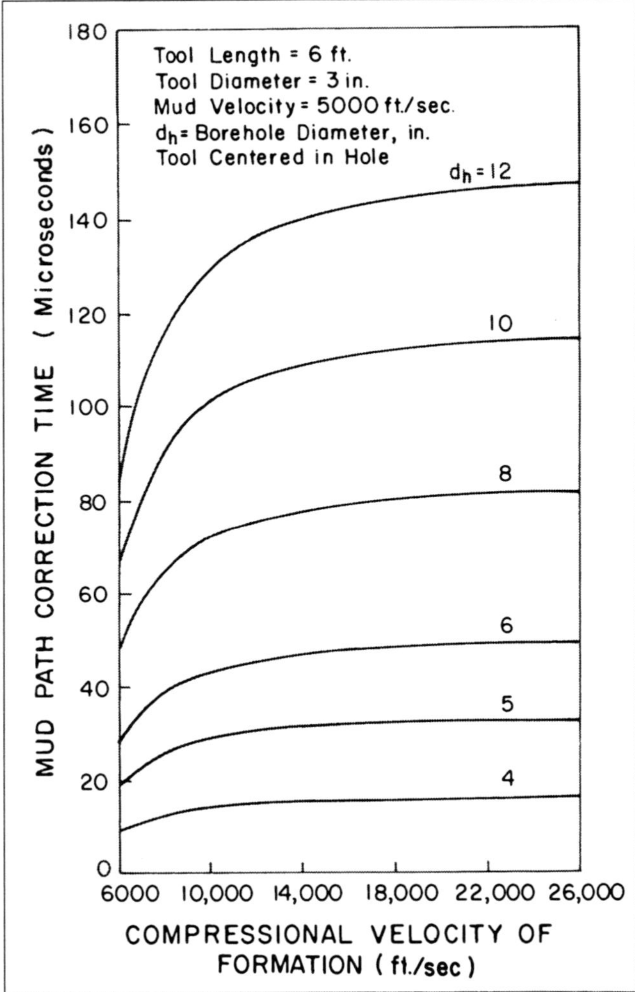

Fig. 10.3—Mud-path correction time as a function of formation compressional velocity (after Ref. 1).

and from Eq. 10.1, with Δt_{ma}=55.5 μsec/ft and Δt_f=189 μsec/ft,

$$\phi=(77.9-55.5)/(189-55.5)=0.167 \text{ or } 17\%.$$

Note that using the uncorrected Δt of 90 μsec/ft results in an apparent porosity of 26%, greatly exceeding the true porosity.

10.3 Dual-Receiver System

The dual-receiver system was introduced to remove the mud-path contribution from the response of sonic tools.[2] **Fig. 10.4** shows a schematic of one of the first tools that incorporated the two-receiver system. The tool consists of a transmitter and three receivers located 3, 4, and 6 ft from the transmitter. The transmitter emits acoustic waves at 10 waves/sec. The first arrival of acoustic energy at each receiver triggers its response system. A two-receiver system can be viewed as a very accurate stopwatch. The stopwatch starts when the acoustic energy arrives at the first receiver and stops when it arrives at the second receiver. The time indicated by the watch is the time required for the sound wave to traverse a length of the formation equal to the spacing between the two receivers. The arrival times t_1 and t_2 at the two receivers can be expressed as

$$t_1=(a/v_m)+(b/v)+(c/v_m) \quad \text{(10.10)}$$

and $$t_2=(a/v_m)+(b/v)+(d/v)+(e/v_m), \quad \text{(10.11)}$$

where a through e are the paths of the acoustic wave (Fig. 10.4).

Assuming that the hole size is constant and that the tool is parallel to the borehole wall yields

$$a=c=e \quad \text{(10.12)}$$

and $$d=L_s. \quad \text{(10.13)}$$

Subtracting Eq. 10.10 from Eq. 10.11 yields

$$t_2-t_1=L_s/v \quad \text{(10.14)}$$

and $$\Delta t=(t_2-t_1)/L_s. \quad \text{(10.15)}$$

The time indicated by Eq. 10.14 is free from mud-path contribution provided that the conditions set by Eqs. 10.12 and 10.13 are met. **Fig. 10.5** shows the responses of single- and dual-receiver

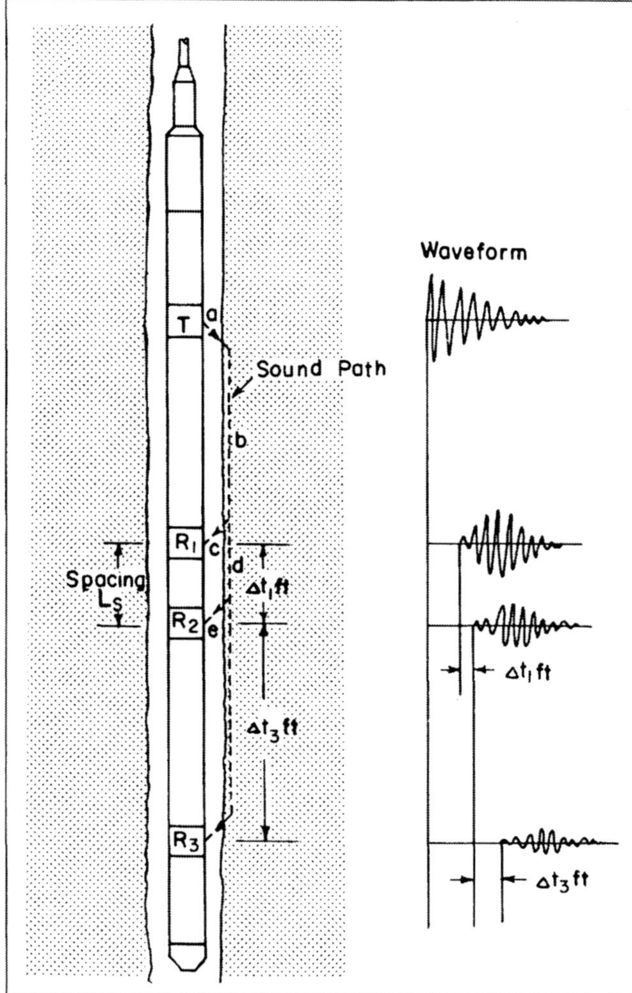

Fig. 10.4—Schematic of the sonic tool incorporating the dual-receiver system concept.

systems recorded in the same borehole. This figure clearly shows the considerable mud signal removed by the two-receiver system.

Usually, Δt is plotted on Tracks 2 and 3 of a linear scale, increasing from right to left (**Fig. 10.6**). A self-potential (SP) or gamma ray log is customarily recorded on Track 1 for depth control. The travel time can be integrated and represented as pips at the edge of the sonic log track. The interval between successive pips represents 1 msec of integrated travel time. This integrated time is useful both for interpretation of seismic records and for determination of average porosity.

Example 10.2. Estimate the average porosity of the thick sand interval shown on the sonic log of Fig. 10.6.

Solution. The average travel time over a thick interval of interest can be determined easily from the integrated travel time. The 134-ft sand spans a time interval made up of 10 pips or 10 msec. Hence,

$$\Delta t = 10(1{,}000)/134 = 74.5 \ \mu\text{sec/ft}.$$

From Eq. 10.1,

$$\phi = (74.5-55.5)/(189-55.5) = 14.2 \text{ or } 14\%.$$

This is, of course, an estimate because the travel time of shale streaks present within the 134-ft interval was also integrated.

10.3.1 Span Between Receivers and Tool Resolution. Several parameters are involved in the design and performance of sonic tools. The distance between the receivers, or span, determines the tool's vertical resolution—i.e., the thinnest bed that can be detected by the measurement. As a rule, tool resolution equals the span between the receivers.

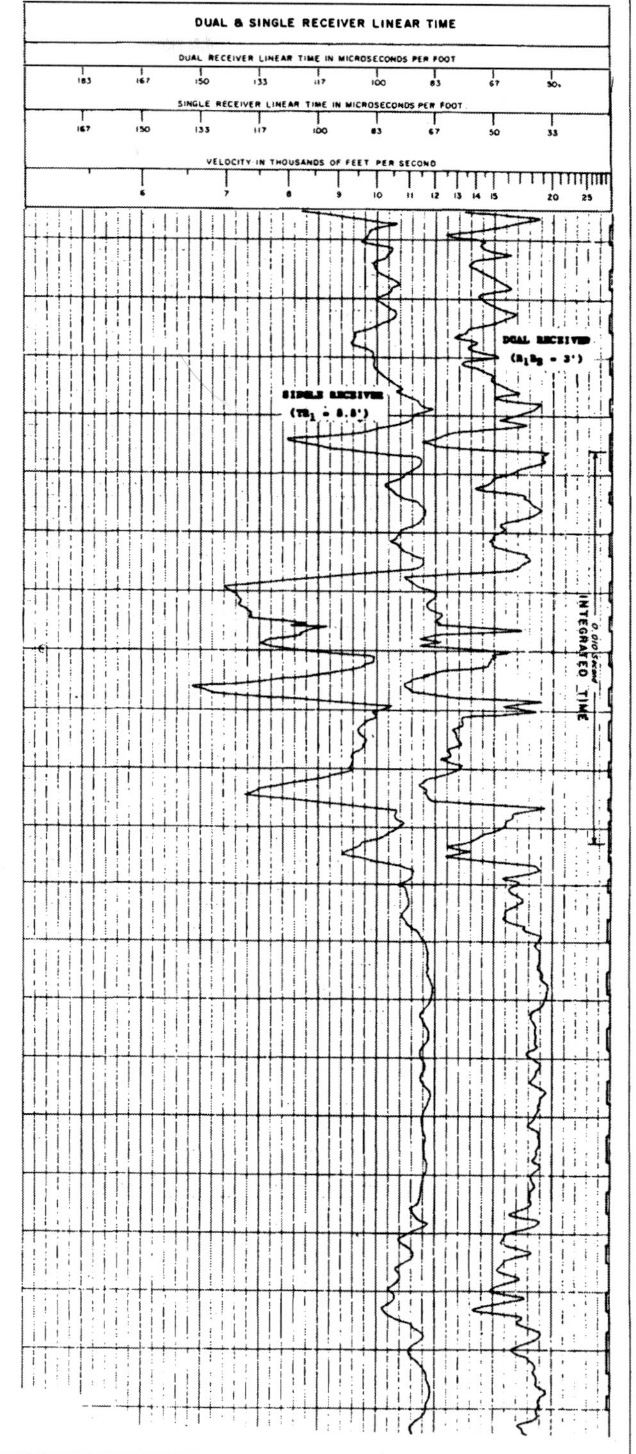

Fig. 10.5—Comparison between single- and dual-receiver system responses (from Ref. 3).

Two sonic curves can be obtained with the tool in Fig. 10.4: a curve based on the travel time between Receivers R1 and R2, which are 1 ft apart, and a second curve based on the travel time between Receivers R2 and R3, which are 3 ft apart. **Fig. 10.7** shows a section of a well in west Texas logged with both 3- and 1-ft receiver spacings. The two curves' overall appearances are similar, but they differ in the details. The smaller the spacing is between the receivers, the greater the detail. It is apparent from this example that the 1-ft log has better vertical resolution than the 3-ft log. Many of the thin beds, in fact, remain undetected by the 3-ft tool.

10.3.2 Spacing Between Transmitter and Receivers. The distance to the farthest receiver is limited by the power of the transmitter

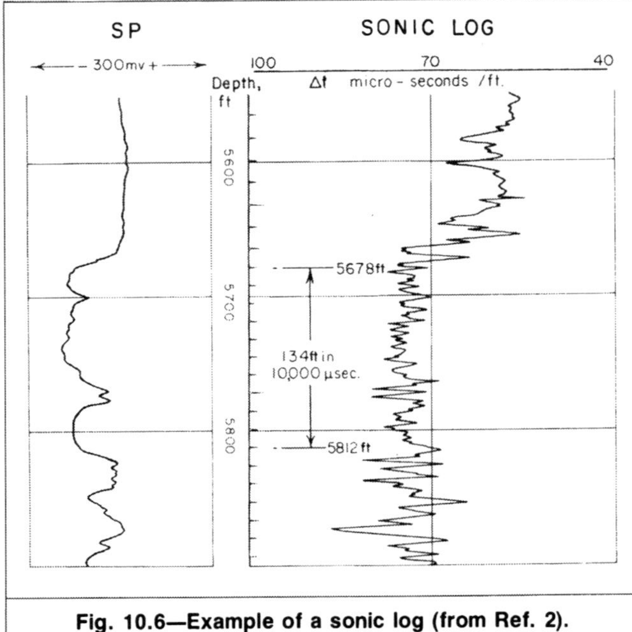

Fig. 10.6—Example of a sonic log (from Ref. 2).

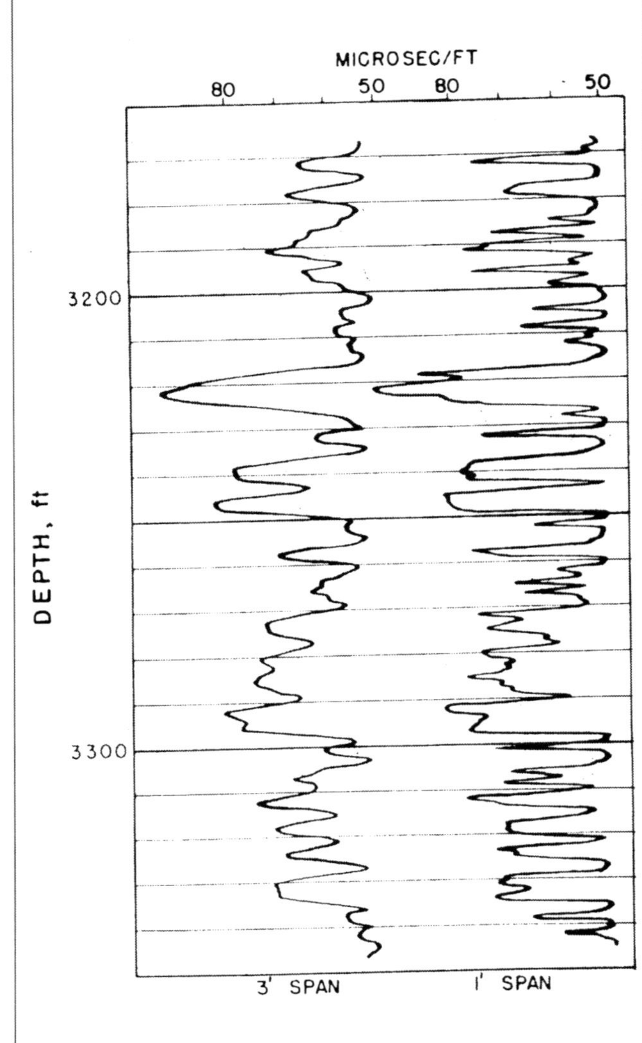

Fig. 10.7—3- and 1-ft-spacing sonic logs recorded in a west Texas well (from Ref. 4).

and the level of noise in the borehole.[4] However, the distance between the transmitter and the first receiver should be large enough that the first arrivals are the refracted compressional waves (see Sec. 3.3) that traveled a considerable distance through the formation.

From the physical concepts discussed in Sec. 3.3, a graph of the transmitter-to-receiver spacing, L_s, vs. transmitter-to-receiver time can be constructed (**Fig. 10.8**). This figure shows that, for a spacing less than the critical spacing, $(L_s)_c$, the fastest path from the transmitter to the receiver is through the mud. At greater spacing, the fastest path is that of the refracted wave passing through the formation. The $(L_s)_c$ needed between the transmitter and the first receiver can be written as[4]

$$(L_s)_c = 2s_{\text{off}}\sqrt{(1+C_{mf})/(1-C_{mf})}\,, \quad \ldots\ldots\ldots\ldots\ldots\ldots (10.16)$$

where s_{off} is the tool standoff—i.e., the separation between the wall and the tool—and C_{mf} is the mud/formation velocity contrast. **Fig. 10.9** plots $(L_s)_c$ vs. C_{mf}. The figure shows clearly that s_{off} should be minimized for quality measurement. It is not necessary for the

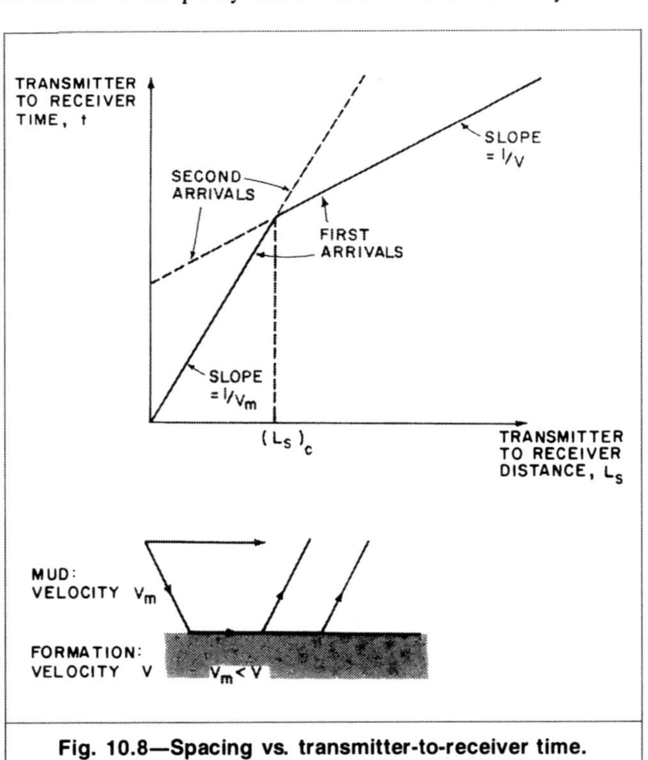

Fig. 10.8—Spacing vs. transmitter-to-receiver time.

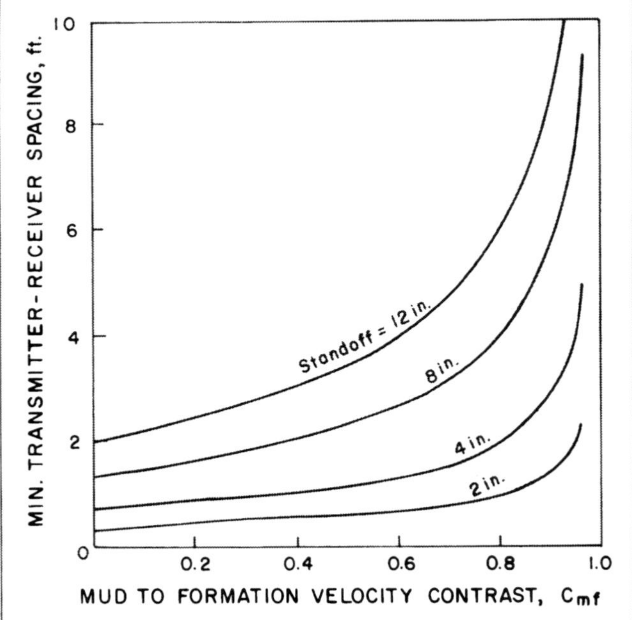

Fig. 10.9—Minimum critical transmitter/receiver spacing, $(L_s)_c$, required to ensure that first arrivals are compressional waves (after Ref. 4).

tool to be centered in the borehole to ensure that first arrivals are compressional waves. On the other hand, centering does increase signal strength because waves generated at all azimuths arrive simultaneously at the receiver.[4]

Example 10.3. Comment on the quality of the sonic-log reading recorded in shallow, soft formations characterized by $\Delta t = 180$ μsec/ft. The mud compressional velocity is 5,000 ft/sec, borehole diameter is 16 in., tool diameter is 3⅜ in., and transmitter-to-first-receiver spacing is 3 ft. Centralizers were used.

Solution.

$s_{off} = (16 - 3⅜)/2 = 6.3125$ in.

From Fig. 10.9,

$(C_{mf})_{max} = 0.8,$

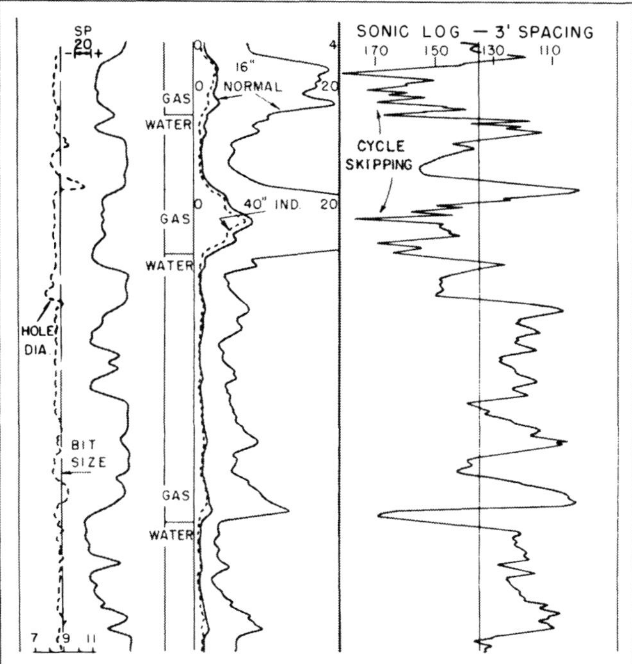

Fig. 10.10—Sonic log showing cycle skipping caused by slow gas formations (from Ref. 2).

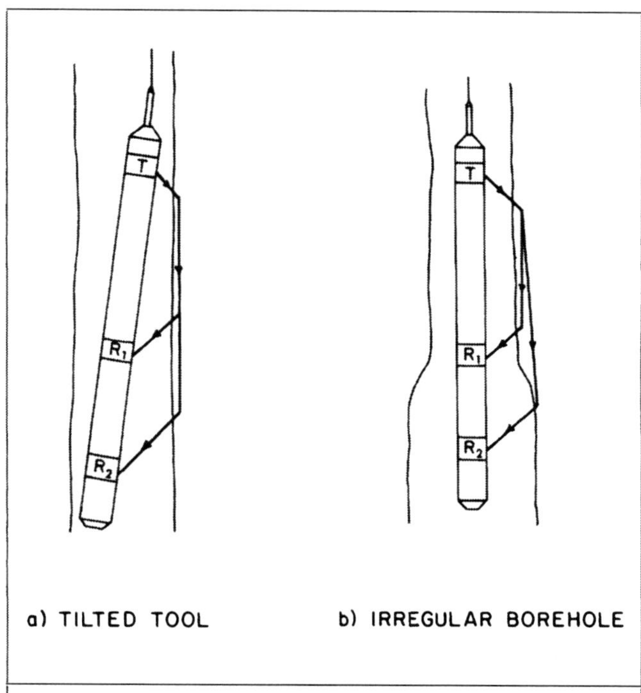

Fig. 10.12—Two-receiver system wave path in case of (a) tilted tool and (b) irregular borehole.

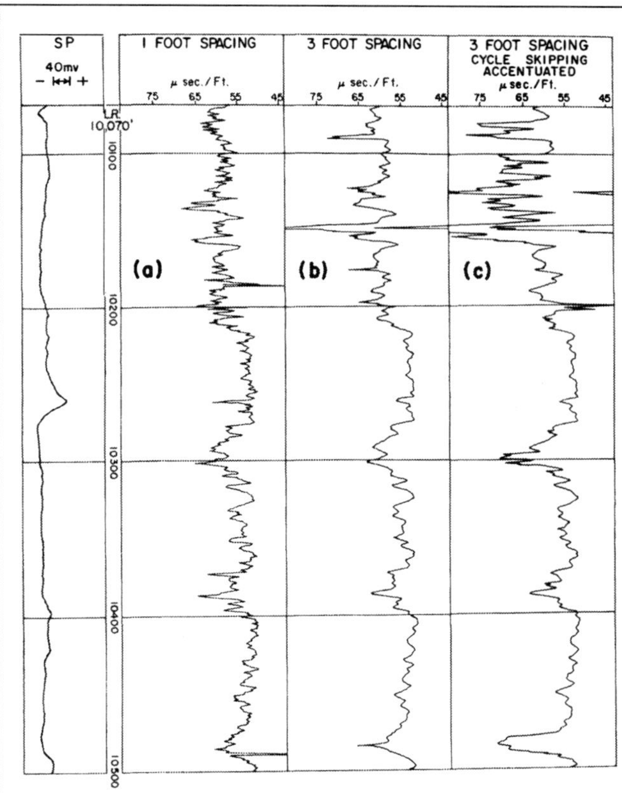

Fig. 10.11—Sonic logs obtained in fractured and fissured Edwards limestone, south Texas (from Ref. 2).

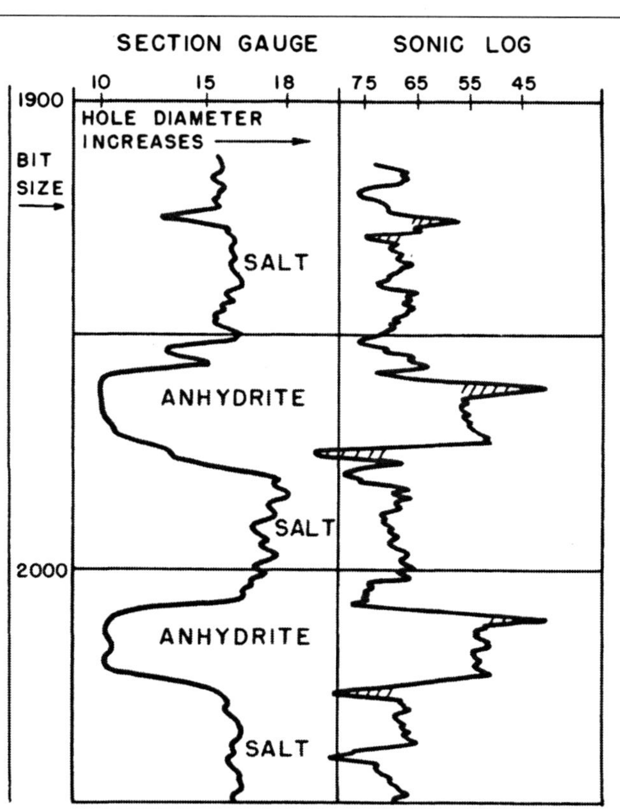

Fig. 10.13—Log example of effect of borehole enlargement in thick and thin beds (from Ref. 2).

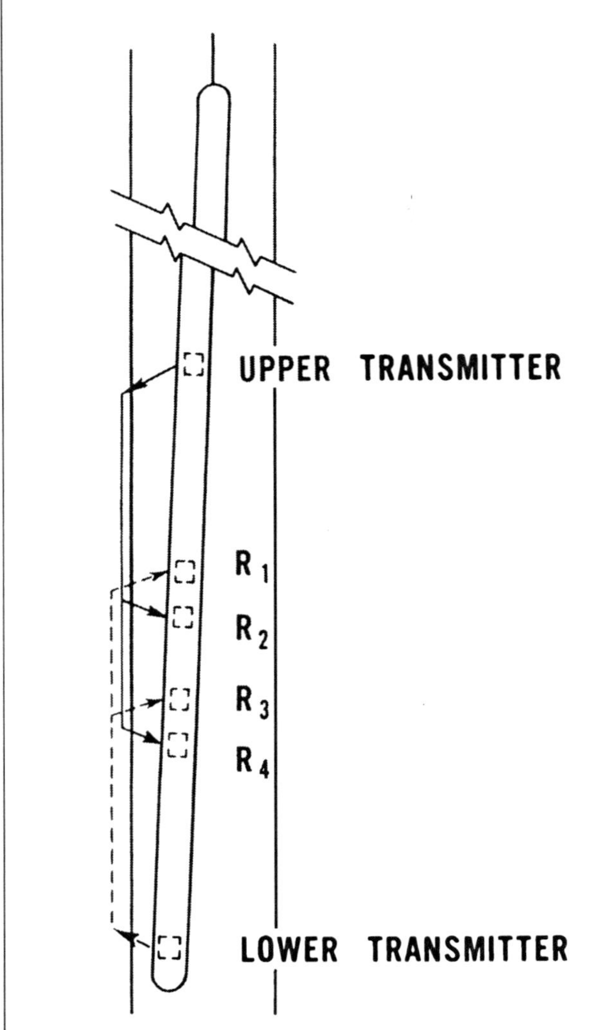

Fig. 10.14—BHC tool in a titled position with respect to hole axis (from Ref. 5).

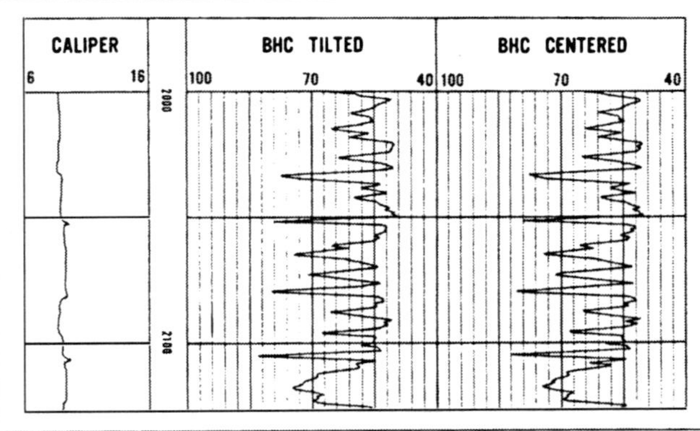

Fig. 10.15—Comparison of a BHC log run with a centered sonde to one run with a deliberately tilted sonde (from Ref. 5).

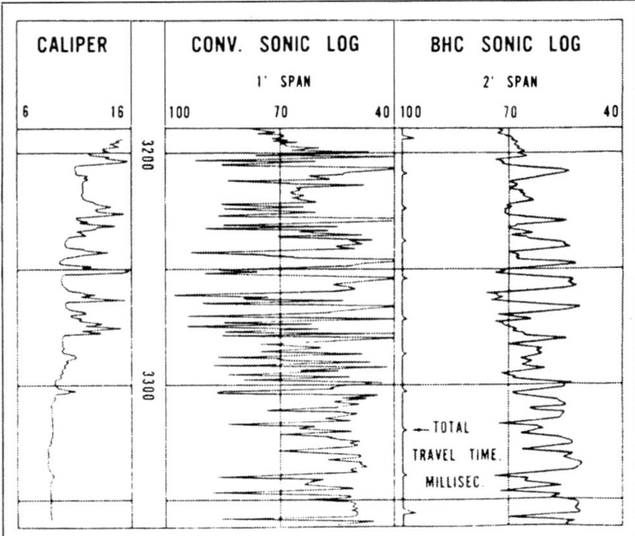

Fig. 10.16—Comparison of 1-ft-span conventional sonic log and 2-ft-span BHC sonic log (from Ref. 5).

$$v_{min}=v_m/0.8$$
$$=5{,}000/0.8=6{,}250 \text{ ft/sec},$$

and $\Delta t_{min}=10^6/6{,}250=160$ μsec/ft.

Because $\Delta t>\Delta t_{min}$, the log quality is poor because the wave reaching the first receiver would have traveled only a short distance through the formation of interest. This distance can be calculated with the concepts of Chap. 3 to be only 1.6 ft.

10.3.3 Cycle Skipping. The acoustic wave attenuates as it propagates through the formation and borehole environment. If the wave is attenuated beyond the threshold of the receiver, the receiver misses the first arrival and detects a later event. If we use the stopwatch analogy, a severe attenuation makes the clock run longer than it should. In this case, the log displays an abnormally longer travel time. Numerically, the travel time indicated by the 1-ft spacing system is increased in multiples of 33 μsec/ft. For the 3-ft span, the increase is in multiples of 11 μsec/ft. **Fig. 10.10** shows an example of this phenomenon, called cycle skipping.

Cycle skipping commonly occurs in series of thin beds of different velocities, gas sands, gas-cut mud, poorly consolidated formations, and fractured formations. Basically, cycle skipping yields an incorrect reading. It can be useful, however, as an indicator for gas-bearing formations and fractured formations. The cycle skipping in Fig. 10.10 is correlatable to gas-bearing formations. **Fig. 10.11** shows an example of cycle skipping associated with fractured formations. Fig. 10.11b shows cycle skipping in the interval of 10,090 to 10,150 ft. This correlates with an Edwards limestone section known, from cores, to be broken by horizontal fissuring. The sonic wave is attenuated by refraction as it crosses the fissures. Cycle skipping is not apparent in Fig. 10.11a, which was recorded by the 1-ft-span system. This results because the attenuation over the 1-ft span is less than that over the 3-ft span. Fig. 10.11c, recorded with a 3-ft span, shows cycle skipping that was accentuated purposely.

10.3.4 Depth of Investigation. Influenced by the rays' presentations of acoustic wave propagation, as in Fig. 10.1, one might assume that the wave path is restricted to the "skin" of the formation. This, in turn, leads one to conclude that the log response is influenced only by the fraction of the formation within a few inches of the borehole wall. Although this assumption is not completely unsound, the acoustic tools' depths of investigation vary with the wavelength, λ, which is related to the formation velocity, v, and signal frequency, f, by Eq. 3.14 rearranged here to

$$\lambda=v/f. \quad (10.17)$$

In sedimentary rocks, v varies from 5,000 to 25,000 ft/sec. For a wave frequency of 20 kHz, λ varies from 0.25 to 1.25 ft. Laboratory experiments indicate that a thickness of at least 3λ is necessary to support a compressional wave over an appreciable distance.[1] Hence, for a 20-kHz wave, the depth of investigation varies from 0.75 ft for soft formations to 3.75 ft for hard formations.

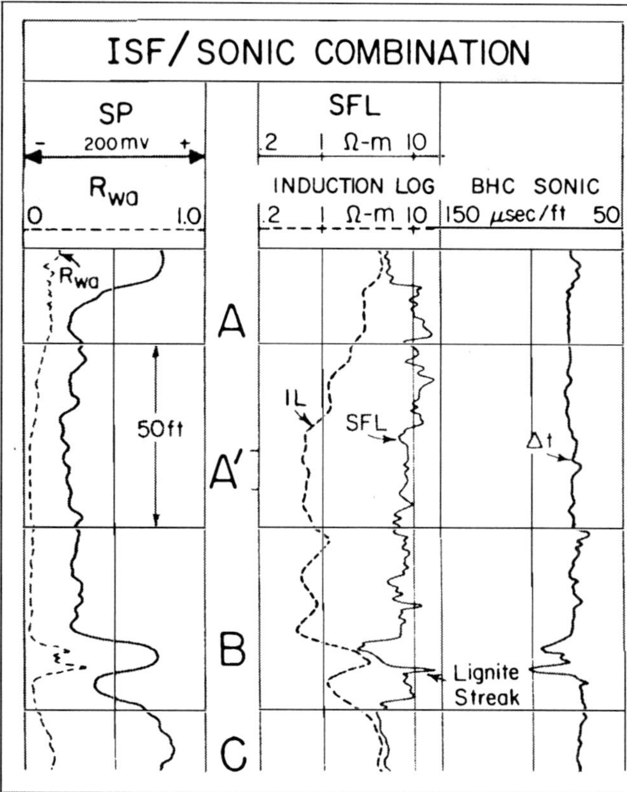

Fig. 10.17—ISF/sonic combination log showing the Wilcox formation of Louisiana (from Ref. 6).

Example 10.4. Estimate the depth of investigation of a 20,000-Hz sonic tool into a consolidated sandstone formation. The formation is water-bearing and has a 20% porosity.

Solution. Rearranging Eq. 10.1 yields

$$\Delta t = \Delta t_m + (t_f - \Delta t_{ma})\phi$$

$$= 55.5 + (189 - 55.5)0.2$$

$$= 82.2\ \mu\text{sec/ft}.$$

Then, $v = 10^6/82.2 = 12{,}165$ ft/sec and $\lambda = 12{,}165/20{,}000 = 0.61$ ft. The depth of investigation is roughly three times the wavelength, approximately 1.8 ft.

10.4 Borehole-Compensated Dual-Transmitter System

As **Fig. 10.12** shows, because of borehole diameter irregularity and eventual tool tilt, the two-receiver system will not remove the mud-path error completely. In an enlarged borehole, the reading of the two-receiver system is distorted only at the top and bottom of the bed. The distortion takes on the shape of spikes, or horns (**Fig. 10.13**). This phenomenon is not crucial unless the bed is thin, as in the top bed. To suppress the possible error resulting from unequal mud travel time to the two receivers, a logging system incorporating two transmitters was developed. An example of these borehole-compensated tools is the BHC^{SM} tool developed by Schlumberger.[5] **Fig. 10.14** shows the BHC tool in a tilted position with respect to the hole axis. The BHC tool incorporates two transmitters and four receivers. It is commonly run with a 3-ft spacing between each transmitter and its near receiver and with a 2-ft receiver span. As the schematic shows, the effect of tool tilt on the upper half of the array (a transmitter and two receivers) is in the opposite direction of the effect on the lower half. An average of the Δt measurement from the two halves cancels the error.[5] **Fig. 10.15** compares a BHC curve run with a centered sonde and one run with a deliberately tilted sonde. The curves are almost identical; Δt differences are only ± 1 μsec/ft.

Similarly, the two-transmitter array also eliminates Δt errors caused by borehole irregularities. **Fig. 10.16** compares a conventional single-transmitter, two-receiver sonic log and a two-transmitter, four-receiver BHC log. The anomalous horns caused by borehole-size erratic changes on the conventional log are effectively suppressed by the BHC tool.

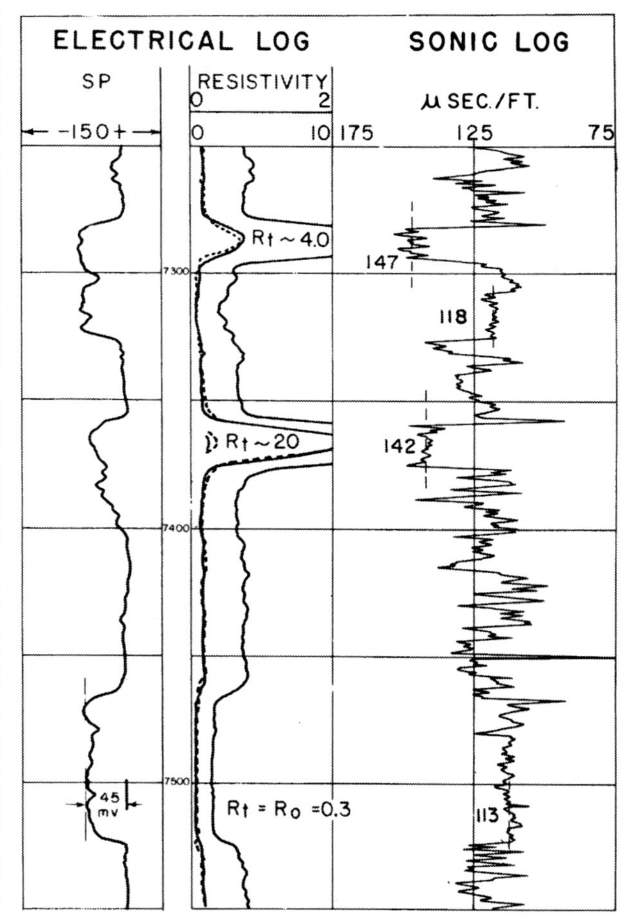

Fig. 10.18—Electrical and sonic logs of Example 10.6 (courtesy Schlumberger).

Example 10.5. Fig. 10.17 shows an Induction Spherically Focused (ISF)/sonic combination log obtained in the Wilcox sandstone formation in Louisiana. Provide a qualitative and quantitative evaluation of the porosity of the thick sand interval.

Solution. The SP curve suggests that the sand is clean. The constant Δt value displayed by the BHC sonic log supports this supposition because it also suggests a well-sorted sand of constant porosity. The log averages $\Delta t = 80$ μsec/ft. Using Eq. 10.1 yields

$$\phi = (80 - 55.5)/(189 - 55.5) = 0.183 \text{ or } 18\%.$$

The calculated porosity value depends on both the accuracy of the log measurement and the representativity of the porosity-Δt transform used. The BHC tool should provide a good-quality log in this environment. Also, Eq. 10.1 has proved to be representative in clean, consolidated sandstone like the Wilcox.

Example 10.6. Fig. 10.18 presents the electrical and sonic logs obtained in shale/sand sequence. The electric log shows three relatively thick, clean sands. The bottom sand is water-bearing. The two upper sands contain a gas zone in the top section. Estimate the average porosity of the gas-bearing zones.

Solution. The top gas interval displays $\Delta t = 147$ μsec/ft. If this value is substituted into Eq. 10.1, together with $\Delta t_{ma} = 55.5$ μsec/ft and $\Delta t_f = 189$ μsec/ft, it yields a porosity value of 69%. This, of course, is an apparent porosity that incorporates two effects: the gas effect (see Chap. 16), which is difficult to correct for, and the

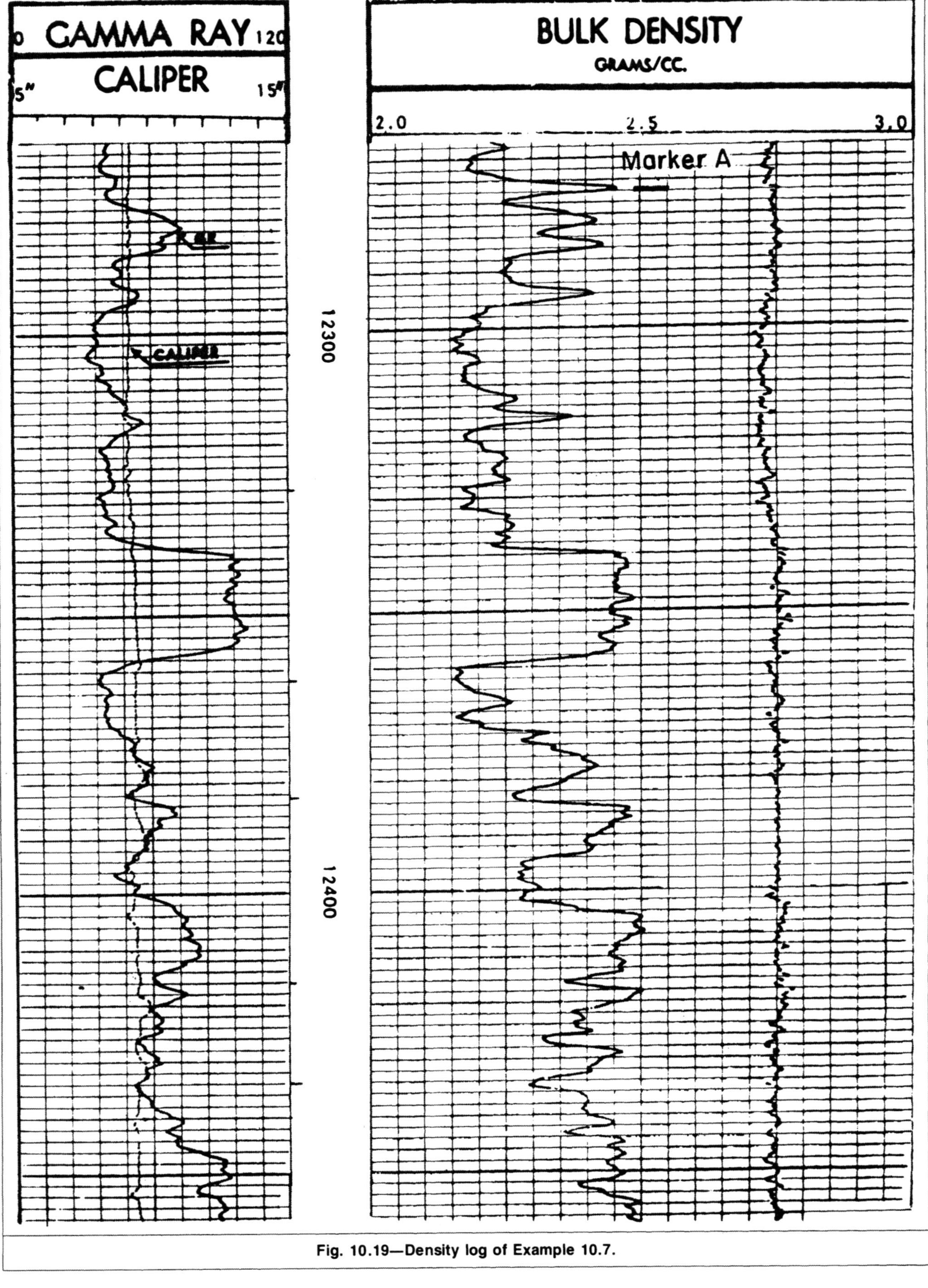

Fig. 10.19—Density log of Example 10.7.

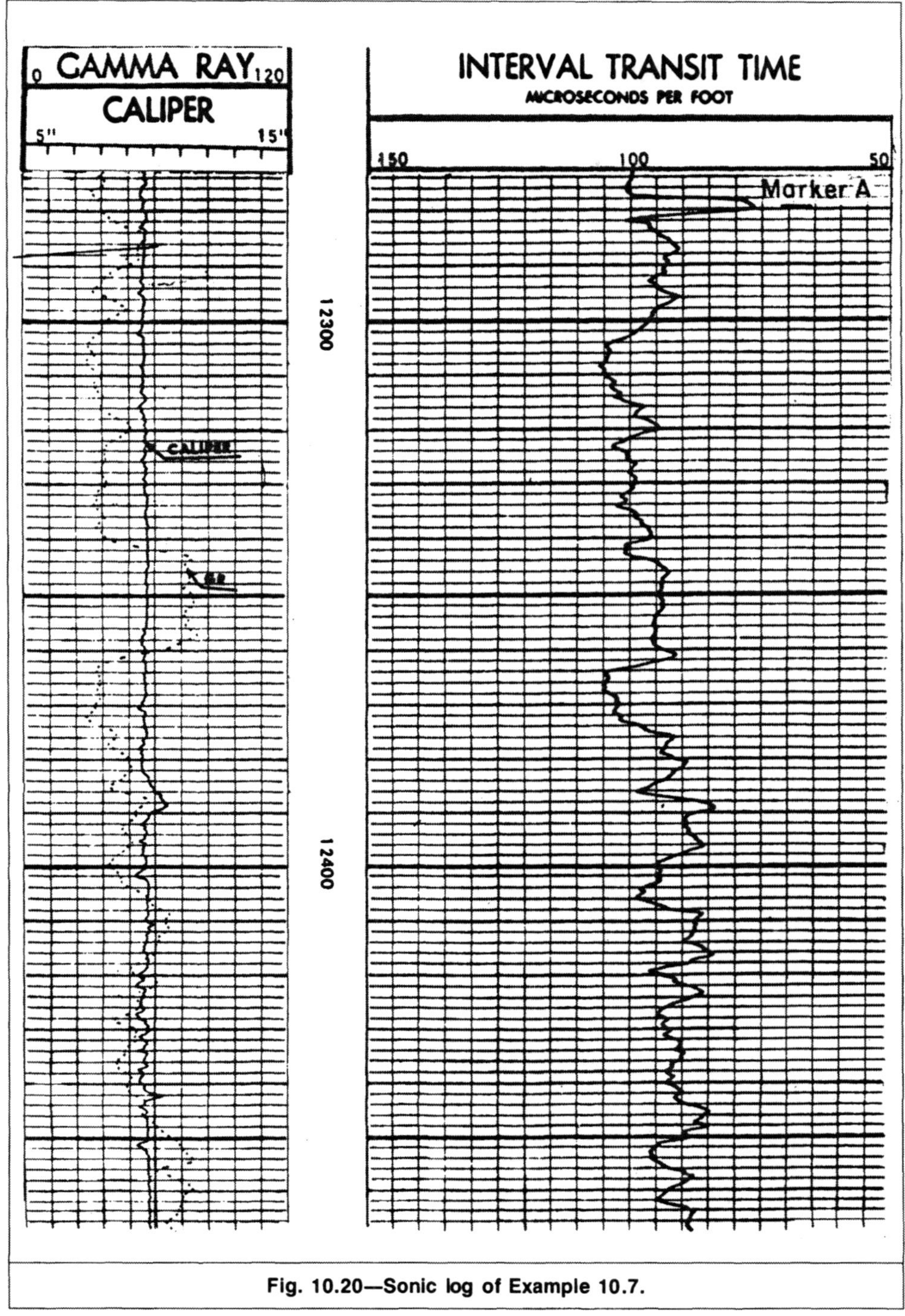

Fig. 10.20—Sonic log of Example 10.7.

TABLE 10.1—BULK DENSITY AND INTERVAL TRAVEL TIME FOR SELECT ZONES, EXAMPLE 10.7

Zone	Depth (Bulk Density Curve) (ft)	Bulk Density (g/cm^3)	Density Porosity (%)	Interval Travel Time (μsec/ft)
1	12,304	2.16	29.7	105
2	12,309	2.17	29.1	103
3	12,325	2.23	25.5	100
4	12,335	2.26	23.6	96
5	12,360	2.16	29.7	105
6	12,369	2.16	29.7	102
7	12,383	2.26	23.8	98
8	12,397	2.28	22.4	95
9	12,416	2.36	17.6	91
10	12,427	2.32	20.0	93
11	12,434	2.30	21.2	92
12	12,443	2.36	17.6	87

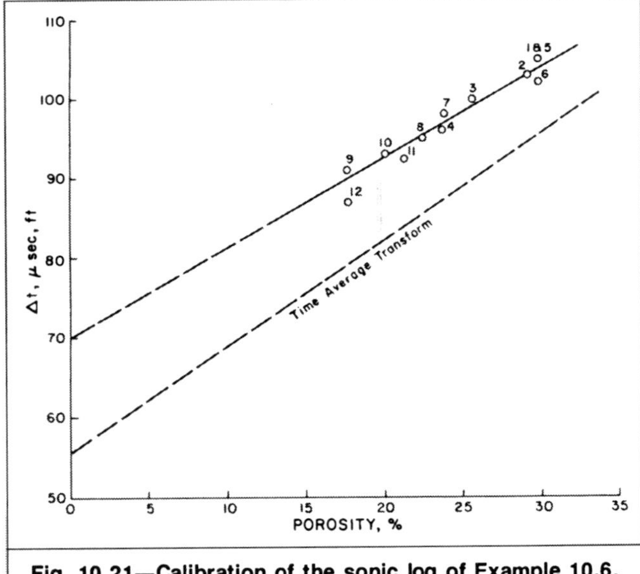

Fig. 10.21—Calibration of the sonic log of Example 10.6.

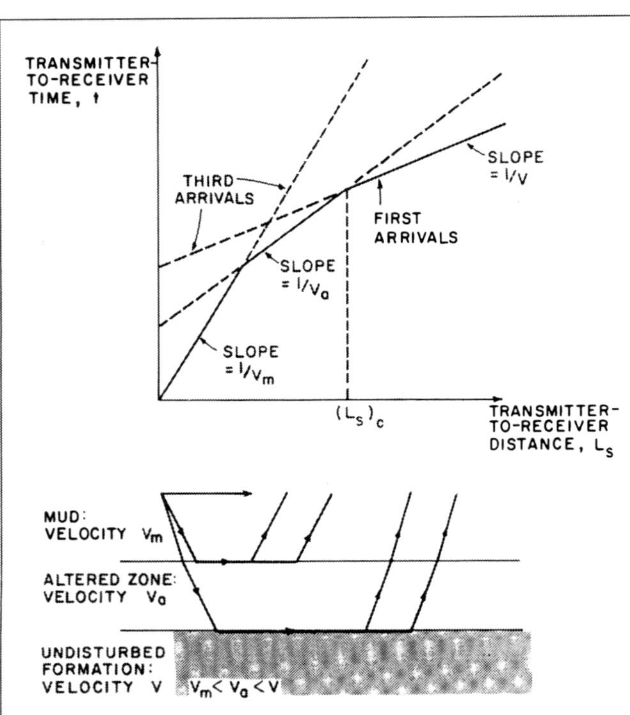

Fig. 10.22—Spacing vs. transmitter/receiver time for altered formations.

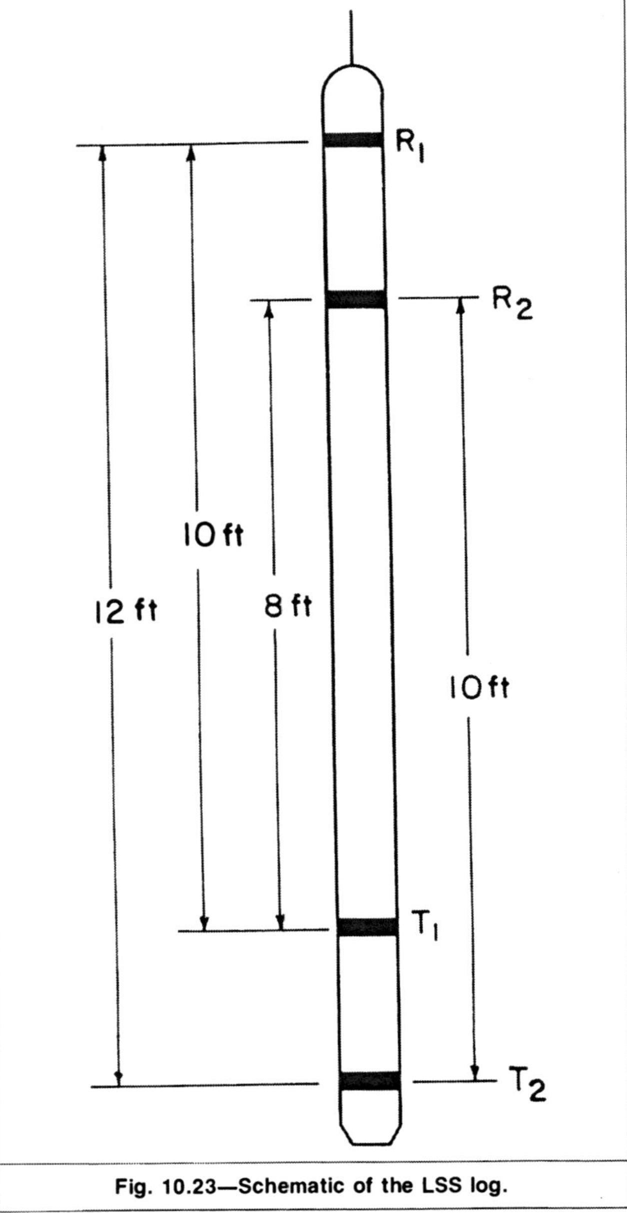

Fig. 10.23—Schematic of the LSS log.

compaction effect, which is indicated by a shale slowness in excess of 100 μsec/ft.

The compaction correction factor is estimated from Eq. 10.3 with the average shale slowness value of 120 μsec/ft displayed by adjacent shales:

$$B_{cp}=120/100=1.2.$$

A rule of thumb states that the porosity of a gas-bearing formation is 0.7 of the value calculated from the time-averaged equation.[2] Using these two empirical corrections for the compaction and gas effects gives a porosity of

$$\phi=(69/1.2)0.7=40\%.$$

To ascertain the representativity of this value, the porosity of the associated water zone is calculated. Because only a compaction effect is present, Eq. 10.2 is used:

$$\phi=[(118-55.5)/(189-55.5)]/1.2$$
$$=0.39 \text{ or } 39\%.$$

This value agrees well with that calculated in the gas-bearing zone.

The porosity of the lower gas-bearing zone is estimated with a similar approach:

$$\phi=0.7[(142-55.5)/(189-55.5)]/1.2$$
$$=0.377 \text{ or } 38\%.$$

Example 10.7. Figs. 10.19 and 10.20 are the density and sonic logs, respectively, obtained over the same series of shales and sands in a south Louisiana well. Using the density tool, calibrate the sonic log in terms of porosity. Comment on the calibration and compare it with that of Eq. 10.1.

Solution. The first step is to ensure that the two logs recorded on separate runs are on depth. Using the top tight zone as a marker (Marker A) indicates that the two logs are 4 ft off depth.

With this shift taken into consideration, several zones were selected; their bulk density and interval transit time values were read and are listed in **Table 10.1.** Clean sandstone zones were preferentially selected to ensure a representative correlation free of shale effects. The two curves also can be digitized mechanically.

The density porosity is calculated with matrix and fluid densities of 2.65 and 1 g/cm³, respectively. The density porosity value, also listed in Table 10.1, is plotted vs. Δt in **Fig. 10.21**. The best fit, shown as a solid line, represents the calibration. In determining the best fit, we give low weight to Point 12, which is thin and sha-

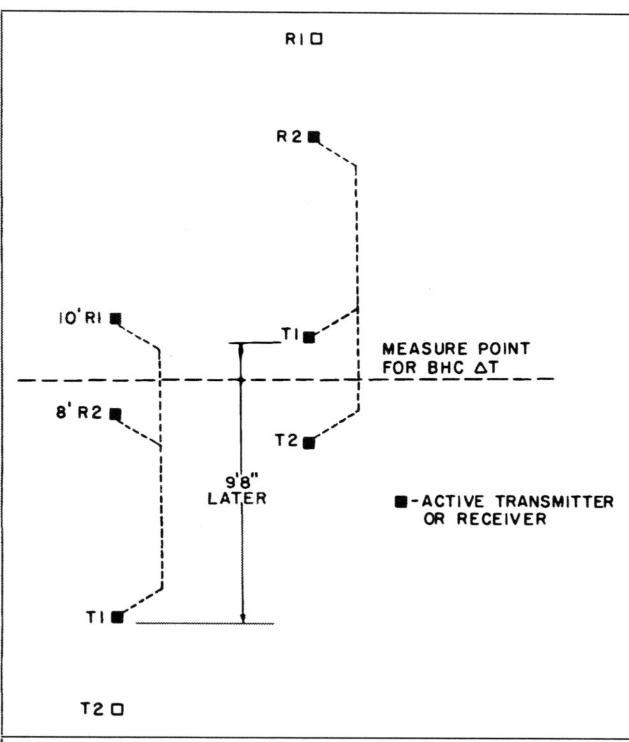

Fig. 10.24—Concept of the long-spaced compensated measurement.

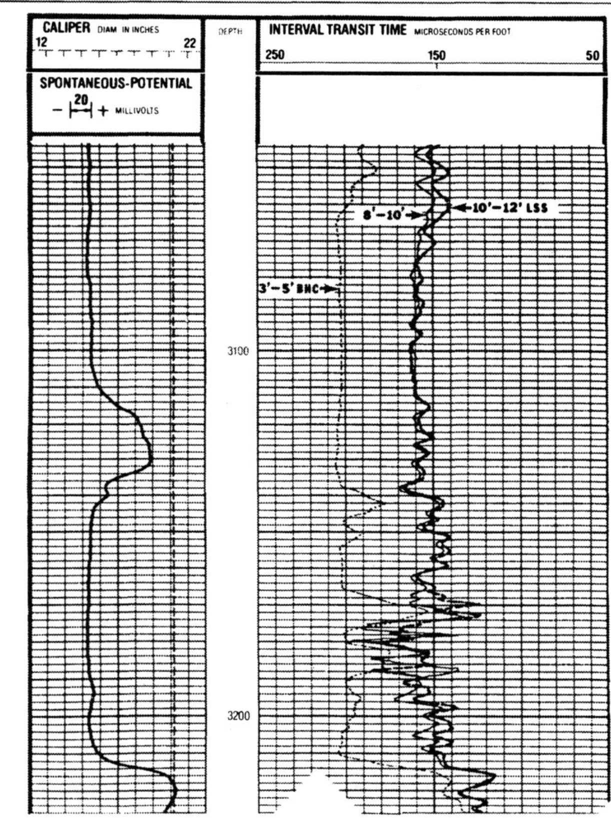

Fig. 10.25—BHC and LSS logs run in a washed-out borehole (courtesy Schlumberger).

ly. The time-averaged equation is represented on the same graph by the dotted line.

The time-averaged transform overestimates the porosity by at least 5%. In this case and in logs run in similar environments, the calibration curve obtained from the density log response is recommended over the time-averaged equation. This recommendation is based

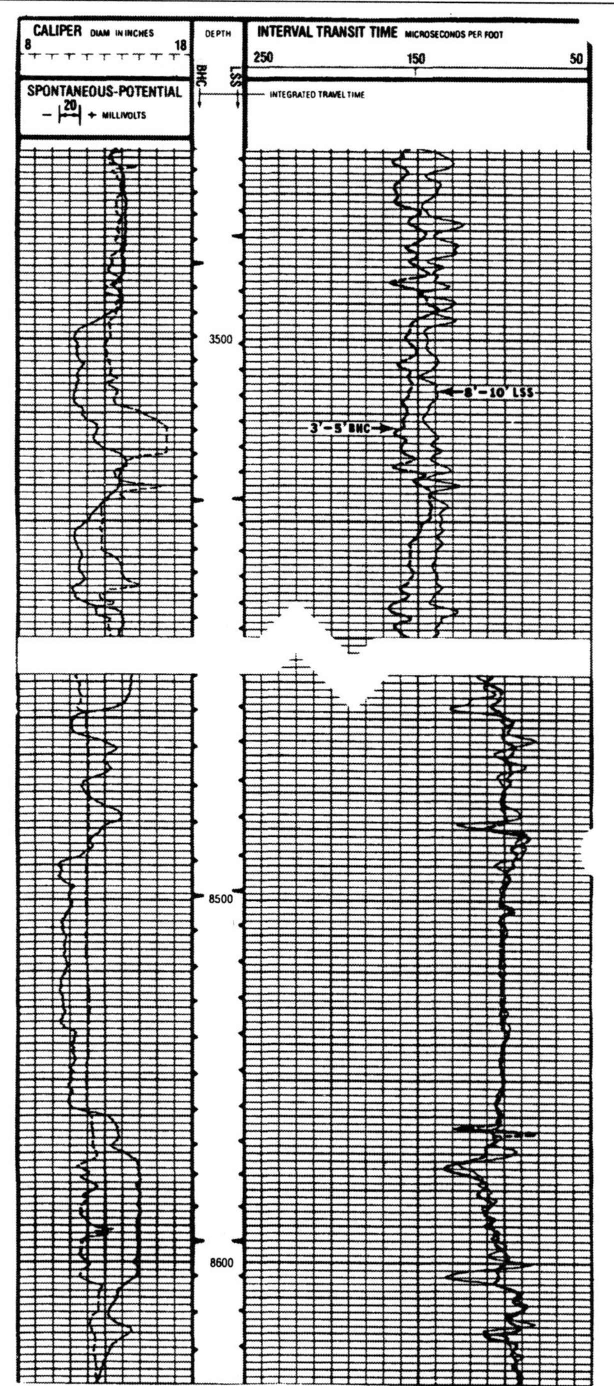

Fig. 10.26—BHC and LSS logs in a Louisiana gulf coast sand/shale sequence (courtesy Schlumberger).

on the fact that the model used in calculating the density porosity is reliable in a clean sand formation, provided that the selected matrix and fluid densities are representative.

10.5 Long-Spacing System

Eq. 10.16 indicates that the critical transmitter/receiver spacing, $(L_s)_c$, increases with increasing borehole diameter, d_h. The standard BHC tool provides a poor-quality log in large-diameter holes drilled in slow formations. This is particularly true in shallow, unconsolidated shale/sand sequences, which are altered by drilling stresses, filtration of the drilling fluids, and/or hydration. The result is an altered zone of low velocity. As **Fig. 10.22** shows, the first arrivals to the receivers of a standard sonic tool are the refracted waves passing through the altered zone. This log displays, then, a transit time that corresponds to the altered zone and not to the

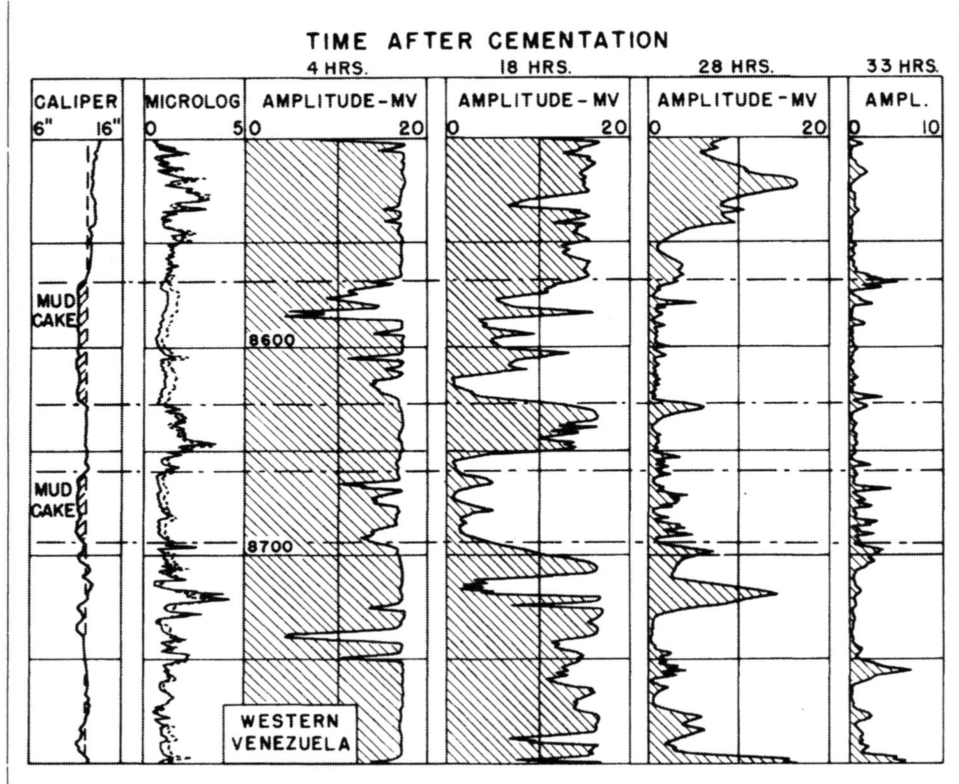

Fig. 10.27—Example of casing wave amplitude logs showing increased cement bonding with time (from Ref. 8).

true formation. Fig. 10.22 also shows that a longer spacing is needed for the waves that traveled through the undisturbed formation to first reach the receivers.

For tools with standard spacing, the limitation caused by a large borehole and the presence of an altered zone is particularly serious in seismic applications of sonic logs.[7] Tools with a longer transmitter/receiver spacing are available for regions where these limitations exist.

Fig. 10.23 is a schematic of Schlumberger's Long-Spaced Sonic (LSS[SM]) log. Two two-receiver arrays are possible: an 8- and 10-ft array, and a 10- and 12-ft array. The 2-ft receiver's span produces the same vertical resolution as the standard sonic tool. The LSS array can also be used to produce a BHC measurement called the depth-derived BHC. As **Fig. 10.24** illustrates, four arrival times are measured. Times t_1 and t_2 are the times of first arrivals from the first transmitter, Transmitter T1, to the 10- and 8-ft receivers, Receivers R1 and R2, respectively. These time values are electronically memorized. When the tool assumes a position 9⅔ ft up the hole, Times t_3 and t_4 are measured. These are the times of first arrivals from the Transmitters T1 and T2 to the second receiver, Receiver R2. The compensated Δt value is calculated as

$$\Delta t=[(t_1-t_2)+(t_4-t_3)]/4. \quad (10.18)$$

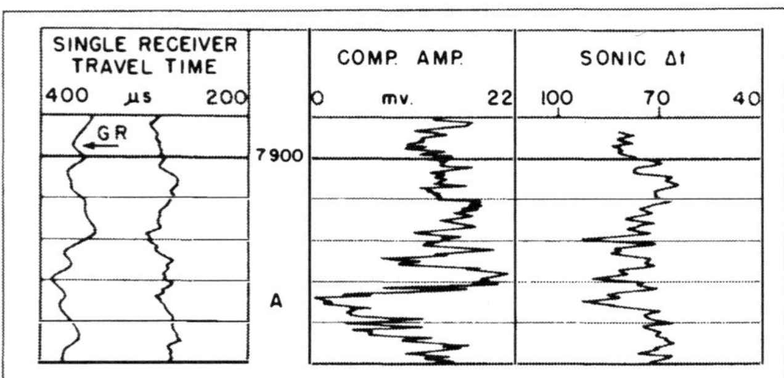

Fig. 10.28—Sonic and sonic amplitude logs in a section of Mississippi limestone (from Ref. 2).

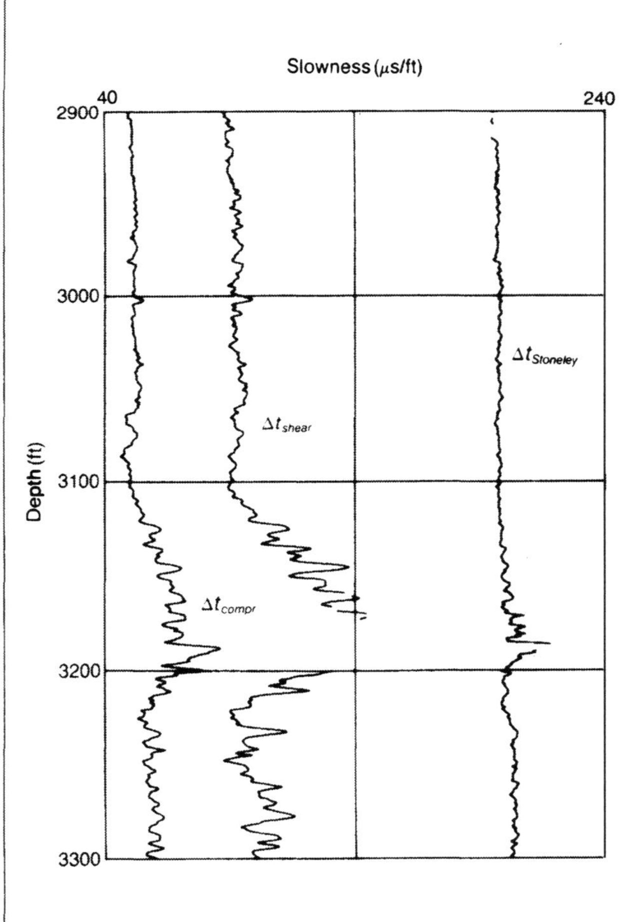

Fig. 10.29—Log of different wave slowness obtained by the array sonic tool (from Ref. 10).

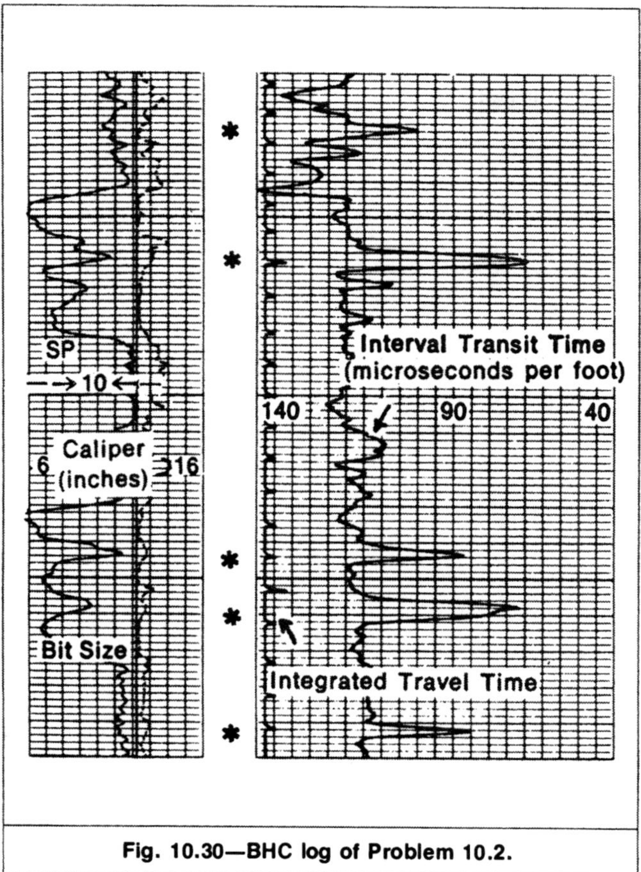

Fig. 10.30—BHC log of Problem 10.2.

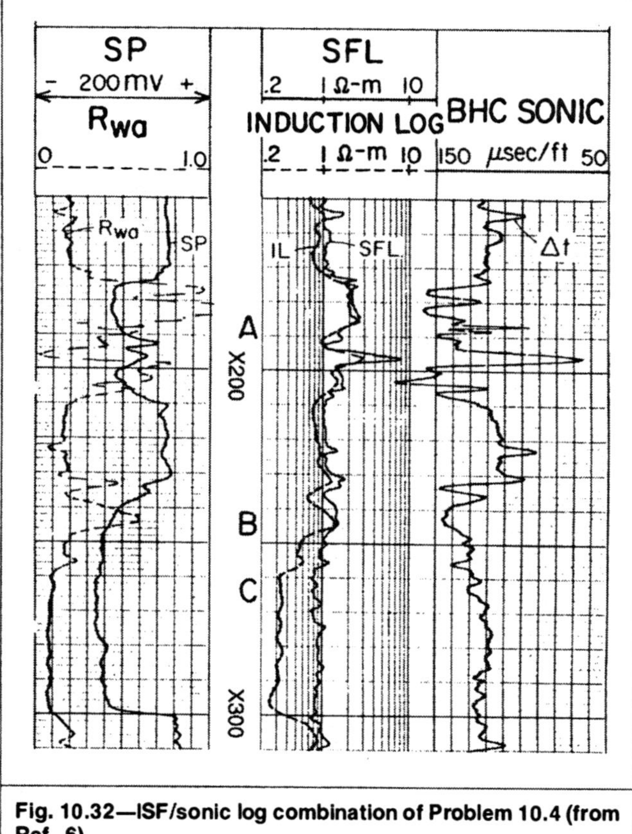

Fig. 10.32—ISF/sonic log combination of Problem 10.4 (from Ref. 6).

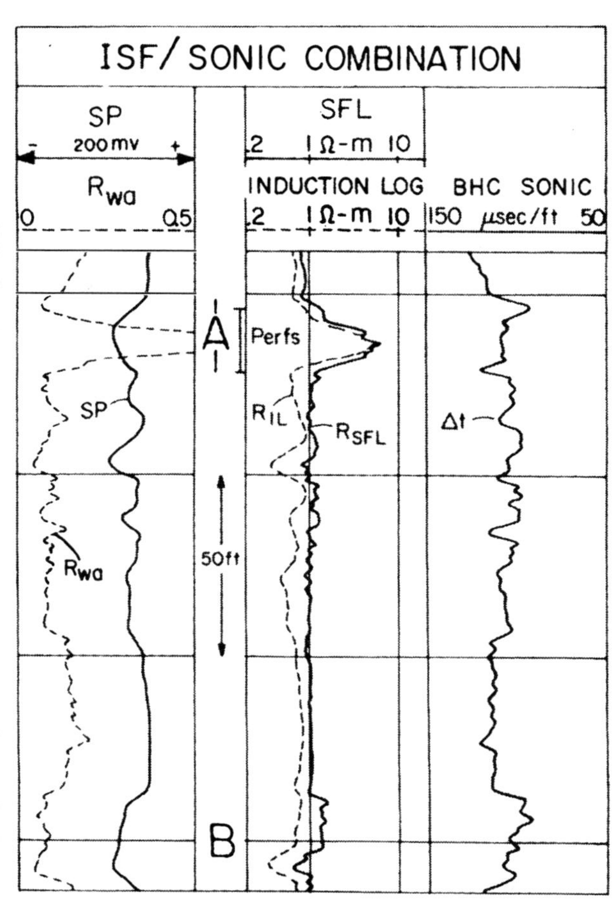

Fig. 10.31—ISF/sonic log combination in Miocene sand formations of Louisiana (from Ref. 6).

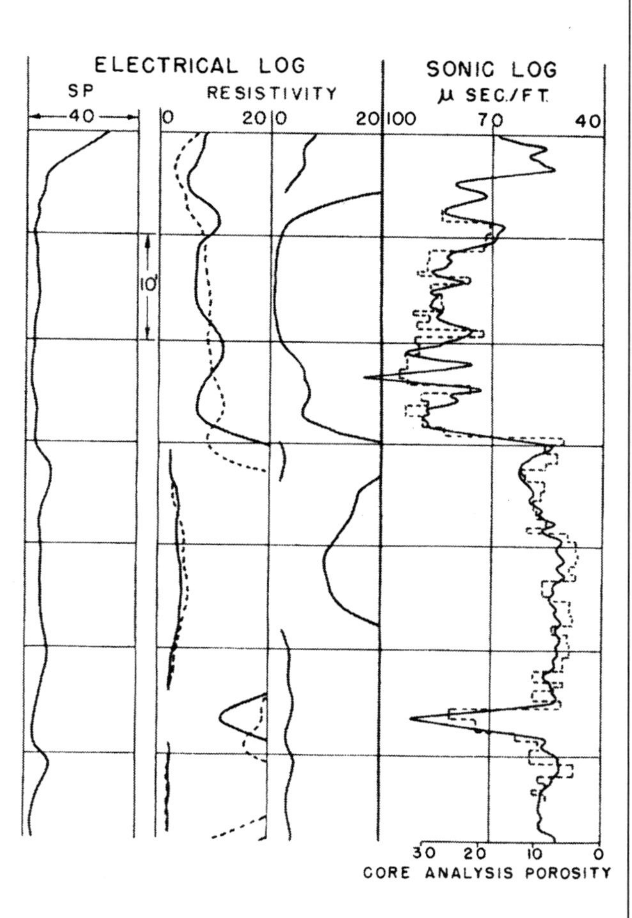

Fig. 10.33—Electrical, sonic, and core porosity logs of Problem 10.5 (from Ref. 2).

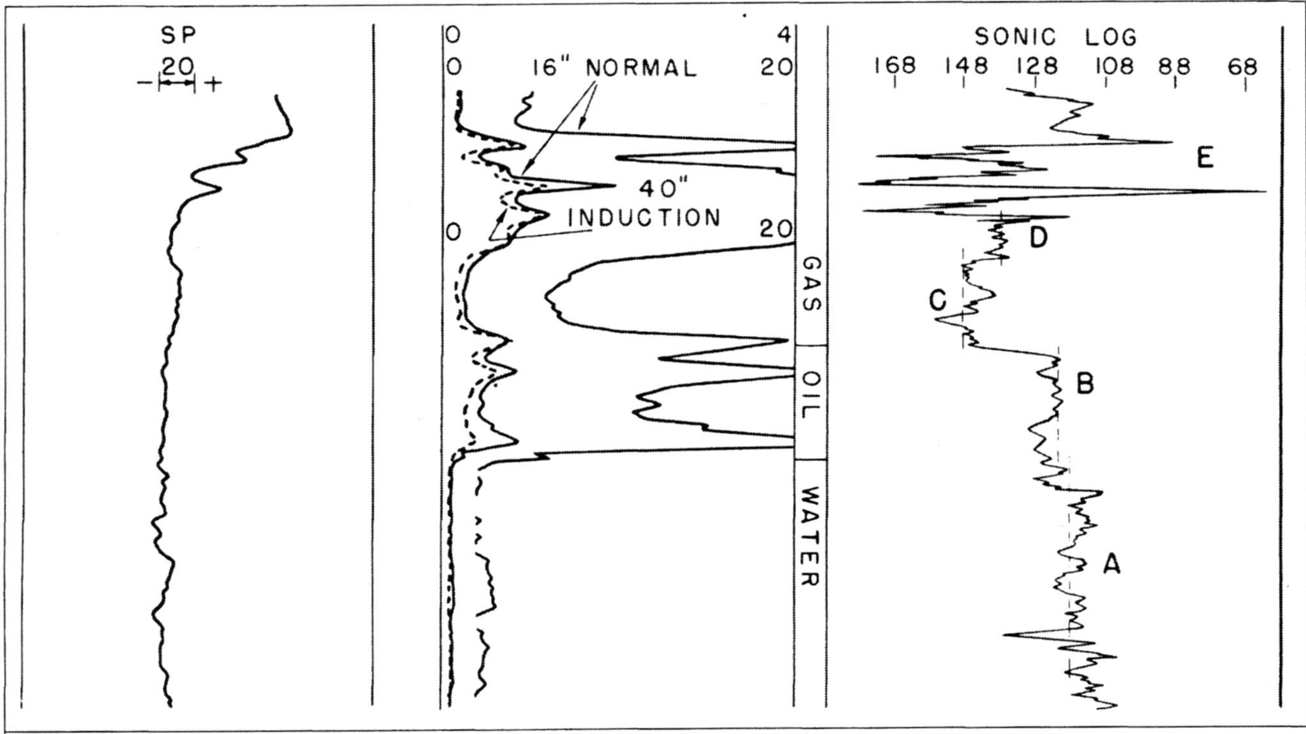

Fig. 10.34—Induction/electrical and sonic logs in a south Texas well with known gas/oil/water contacts (from Ref. 2).

Fig. 10.25 compares the responses of the standard 3- and 5-ft-spacing BHC, the 8- and 10-ft LSS, and the 10- and 12-ft LSS in a washed-out borehole more than 20 in. in diameter. The BHC is reading approximately 45 μsec/ft too high and is probably responding only to the arrival of mud waves.

Fig. 10.26 compares the responses of the standard 3- and 5-ft-spacing BHC and the 8- and 10-ft-spacing LSS log. In the shallow part of the well, both shales and sands seem to be altered by drilling stresses and/or drilling fluids. The BHC, which is affected by these alterations, is reading about 15 μsec/ft higher than the LSS. Deeper in the well, where the formations are more compact and less likely to be altered, the two logs agree reasonably well.

10.6 Amplitude and Waveform Sonic Systems

The sonic logging tools discussed previously use compressional wave propagation. Also available are specialized sonic logging techniques that use other components and characteristics of the sonic waveform shown in Fig. 10.2. These characteristics are the arrival times of the shear waves and the Stoneley waves, the relative magnitude of these waves, and the attenuation of the acoustic energy.

The cement bond log (CBL) is a sonic method for analyzing the quality of casing-string cementation.[8] It is based on the principle that the energy of a sonic pulse transmitted by the casing is greatly attenuated when the casing is bonded to cement, which has a sonic velocity substantially less than that of the casing. **Fig. 10.27** shows logs of the casing wave's amplitude, expressed in millivolts, recorded at different time intervals after cementation. The logs clearly show how the energy is attenuated as the cement setting is advanced. They also show that a good bond is nearly achieved after 33 hours. The technology of CBL's has greatly advanced since the logs of Fig. 10.27 were recorded. Current CBL's present additional information in a more complex manner.

Development of amplitude measurements for the CBL's led to the investigation of amplitude reading in open holes. **Fig. 10.28** displays the sonic and sonic amplitude logs in a section of Mississippi limestone. The amplitude of the compressional wave is greatly

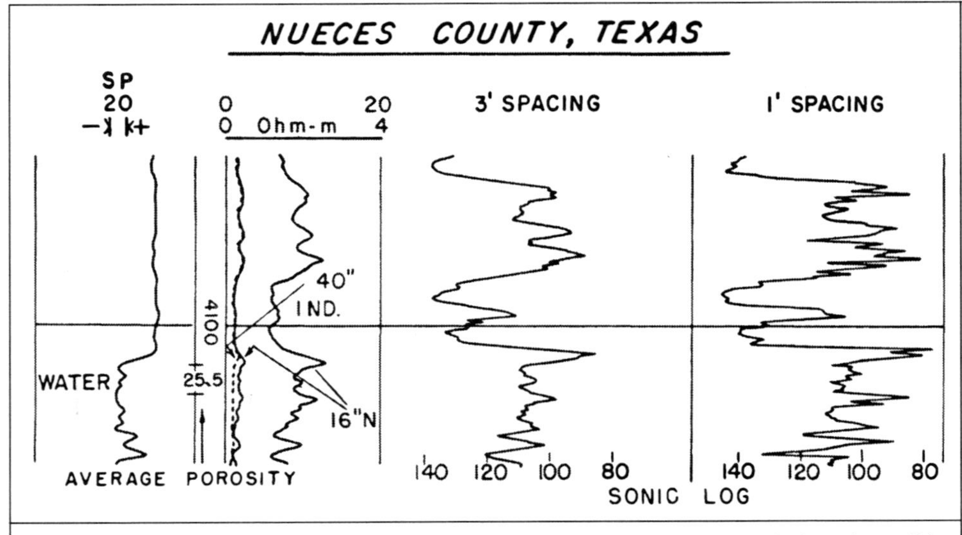

Fig. 10.35—Induction/electrical log, 3-ft-span sonic log, and 1-ft-span sonic log of a well in Texas (from Ref. 2).

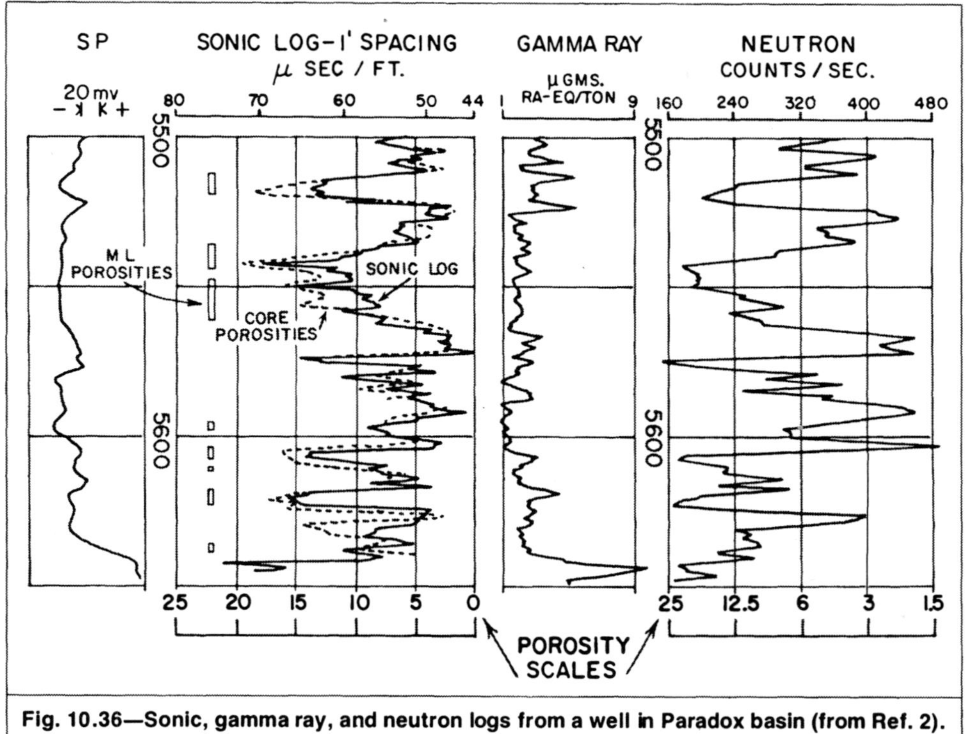

Fig. 10.36—Sonic, gamma ray, and neutron logs from a well in Paradox basin (from Ref. 2).

reduced over Zone A, which is described by cores as extensively fractured. Fracture orientation can also be detected by a comparison of the relative attenuation of the compressional and shear waves. For example, horizontal fractures (those perpendicular to the borehole axis) cause little or no attenuation of the compressional wave. Shear waves, on the other hand, are significantly attenuated by horizontal fractures.[9] Detecting fractures from variations in signal amplitude is rather complex and requires experience with the environment of interest.

Rather than recording just the compressional and shear components, recently developed waveform processing techniques detect and analyze all propagating waves in the full waveform.[10,11]. **Fig.**

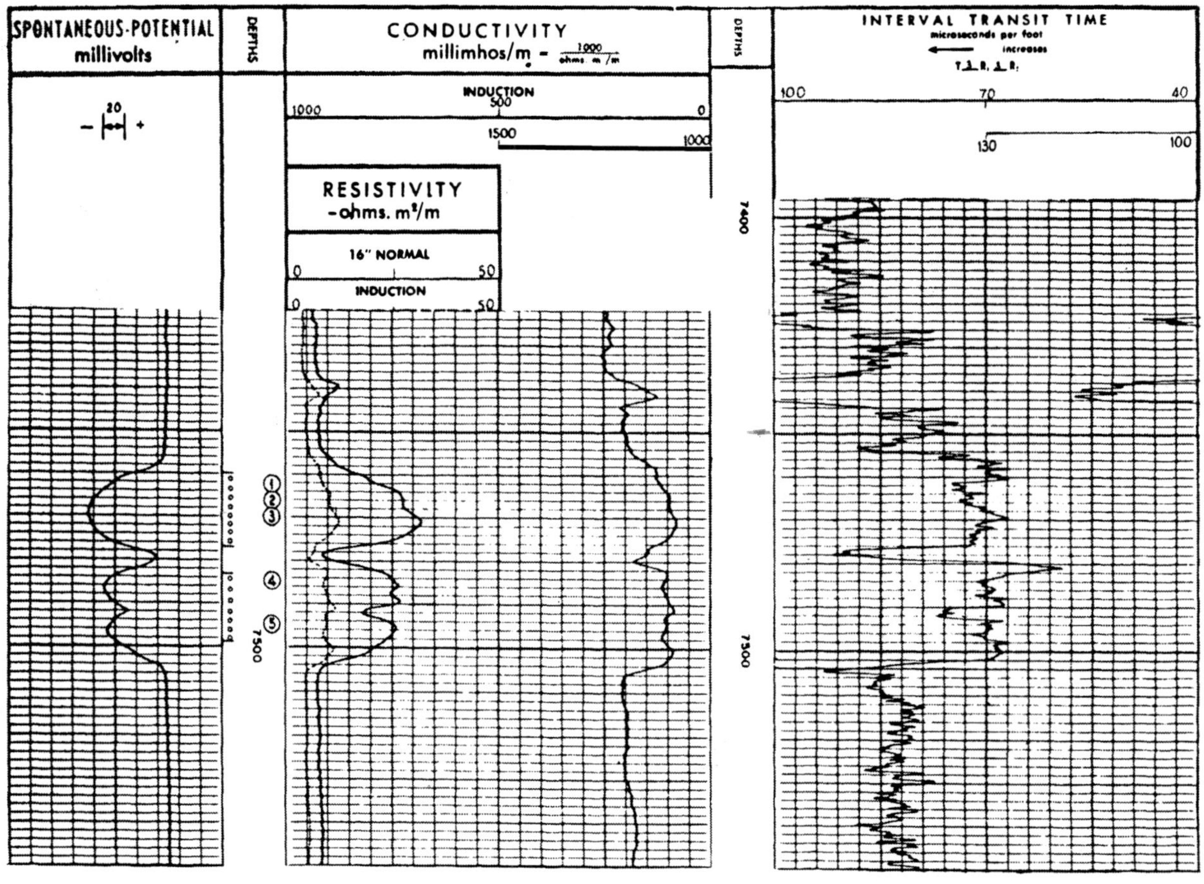

Fig. 10.37—Induction/electric and sonic logs of the Morrow sand, Oklahoma (from Ref. 12).

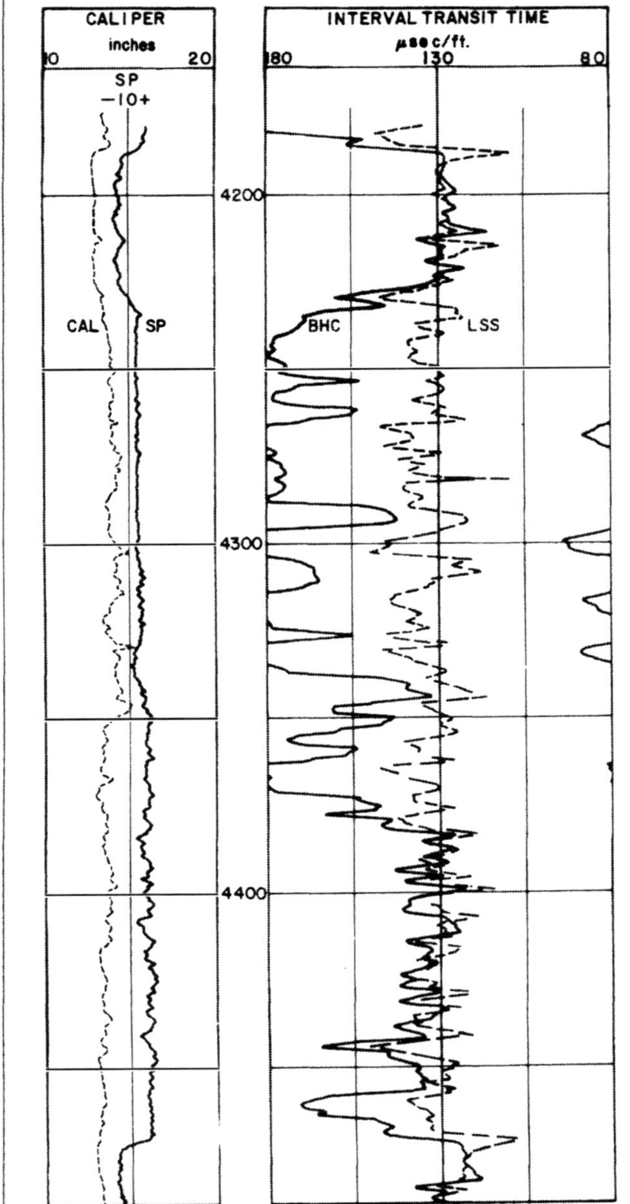

Fig. 10.38—BHC and LSS logs of Problem 10.10 (courtesy Schlumberger).

10.29 compares the slowness of the compressional, shear, and Stoneley waves. These data usually are combined with bulk density values to characterize the stress environment around the borehole.

Review Questions

1. What physical parameter is measured by the sonic porosity tool?
2. How is the formation slowness related to formation porosity?
3. What are the matrix and fluid slowness values usually used in the time-averaged transform?
4. How does an uncompacted environment manifest itself on the sonic log?
5. How is the uncompaction effect considered when porosity is calculated?
6. How does the apparent sonic porosity compare with the true porosity of formations containing vuggs and fractures?
7. What is the main shortcoming of the single-receiver system initially used in acquiring the sonic log?
8. How was the shortcoming of the single-receiver system addressed?
9. What is the main shortcoming of the two-receiver system? How is it overcome?
10. What is meant by the integrated travel time? How is it useful?
11. What are the determining factors for selecting the spacing between the transmitter and receiver and the span between the receivers?
12. What causes cycle skipping? How can it be recognized on the log?
13. How deeply do sonic tools investigate into a formation? What parameters define that depth?
14. Describe how the dual-transmitter system compensates for borehole effect.
15. What are the advantages of the long-spacing sonic system over the standard type?
16. List the logs, other than the sonic porosity tool, that make use of formation acoustic properties. Explain the concept of each.

Problems

10.1 Using the parameters given for Fig. 10.3, prepare a similar plot, this time for l_c=2½ in. From this exercise, and for the purposes of minimizing mud-path correction, would you recommend the centralization of the single-receiver tool? Explain.

What is the correction if the tool is assumed to be against the borehole wall?

10.2 Examine the BHC log of **Fig. 10.30**.
- a. What is the probable lithology of the thin streaks, marked with an asterisk, present within the sands?
- b. Use the integrated travel time to estimate the average porosity of the sands. Comment on the accuracy of your estimate.
- c. List the possible explanations for the spikes of high travel time present within the top 30 ft of the log.

10.3 **Fig. 10.31** presents an ISF/sonic combination log showing Miocene formations in Louisiana. Estimate the porosity of Zone A. This zone initially produced 160 B/D of 45°API oil on an 8/64-in. choke.

10.4 Using two different independent methods, determine the compaction correction factor for the unconsolidated formation of the ISF/sonic log combination of **Fig. 10.32.** Formation water resistivity is 0.1 Ω·m at formation temperature. Comment on the difference in the two values, if any.

10.5 **Fig. 10.33** shows the electrical, sonic, and core porosity logs for a thick limestone formation.
- a. Prepare a calibration chart for the sonic log.
- b. Determine parameters A and B of Eq. 10.4.
- c. Compare the calibration to the time-averaged equation expressed by Eq. 10.1.

10.6 **Fig. 10.34** shows a section of the induction/electrical log and 3-ft-span sonic log of a well in south Texas. Fluid contacts are also shown. The high-resistivity peaks are caused by streaks of limestone.
- a. Explain the spikes displayed in Zone E.
- b. The average core porosities of Zones A through D are 33%, 32%, 34%, and 31%, respectively. Calculate the log porosities of these zones and compare them with those of the cores. Experience with these sediments indicates a compaction correction factor of 1.44.

10.7 Examine the electrical and two sonic logs of **Fig. 10.35.**
- a. Which of the two sonic logs has a vertical resolution similar to that of the 16-in. normal? Explain.
- b. Divide the water-bearing sand into several distinct subzones. List both the 16-in. short normal and sonic log readings for each of these zones.
- c. Plot Δt vs. R_{SN}. Explain the reasons behind the evident correlation between these two parameters.

10.8 Plot the neutron porosity vs. sonic porosity for the formation shown by **Fig. 10.36.** Comment on the quality of the correlation between these two parameters. What are the values of the matrix and fluid travel times used to prepare the sonic porosity scale? Can the correlation be improved if other values are selected?

10.9 **Fig. 10.37** shows induction/electric and sonic logs of the Morrow sand of Oklahoma. Estimate the porosity of the five Zones 1 through 5 marked on the log.

10.10 On the sonic logs of **Fig. 10.38,** indicate the zones of altered shales, if any.

Nomenclature

A,B = coefficients in Eq. 10.4
B_{cp} = compaction correction factor
C = velocity contrast $= v_m/v$
d = diameter, in.
f = frequency, Hz
L_s = tool spacing or span, ft
s_{off} = standoff, in.
t = time, seconds
Δt = interval travel time, μsec/ft
v = velocity, ft/sec
λ = wavelength, cm
ϕ = porosity, fraction

Subscripts

c = critical
f = pore fluid
h = borehole
log = log response
m = mud
ma = matrix
max = maximum
mf = mud filtrate
min = minimum
sh = shale
t = tool

References

1. Stripling, A.A.: "Velocity Log Characteristics, *Trans.*, AIME (1958) **213**, 207-12.
2. Tixier, M.P., Alger, R.P., and Doh, C.A.: "Sonic Logging," *Trans.*, AIME (1959) **216**, 106-14.
3. Thurber, C.H.: "The Geoacoustic Spectrum," *Oil & Gas J.* (Oct. 12, 1959).
4. Kokesh, F.P. and Blizard, R.B.: "Geometrical Factors in Sonic Logging," *Geophysics* (Feb. 1959) **24**, No. 1, 64-76.
5. Kokesh, F.P. *et al.*: "A New Approach to Sonic Logging and Other Acoustic Measurements," *JPT* (March 1965) 282-86.
6. Schuster, M.A., Baden, J.D., and Robbins, E.R.: "Application of the ISF/Sonic Combination Tool to Gulf Coast Formations," *Trans.*, Gulf Coast Assn. of Geologic Scientists (1971).
7. Thomas, D.H.: "Seismic Applications of Sonic Logs," *Log Analyst* (Jan.-Feb. 1978) **19**, No. 1, 23-32.
8. Grosmangin, H., Kokesh, F.P., and Majani, P.: "A Sonic Method for Analyzing the Quality of Cementation of Borehole Casings," *JPT* (Feb. 1961) 165-71; *Trans.*, AIME, **222**.
9. Morris, R.L., Grine, D.R., and Arkfeld, T.E.: "Using Compressional and Shear Acoustic Amplitudes for the Location of Fractures," *JPT* (June 1964) 623-32.
10. Morris, C.F., Little, T.M., and Letton, W. III: "A New Sonic Array Tool for Full Waveform Logging," paper SPE 13285 presented at the 1984 SPE Annual Technical Conference and Exhibition, Houston, Sept. 16-19.
11. Harrison, A.R. *et al.*: "Acquisition and Analysis of Sonic Waveforms From a Borehole Monopole and Dipole Source for the Determination of Compressional and Shear Speeds and Their Relation to Rock Mechanical Properties and Surface Seismic Data," paper SPE 20557 presented at the 1990 SPE Annual Technical Conference and Exhibition, New Orleans, Sept. 23-26.
12. Millard, F.S.: "Resistivity-Velocity Log Evaluation of the Morrow Sand in Western Oklahoma," *Trans.*, First SPWLA Symposium (May 1960).

Chapter 11
Conventional Interpretation Techniques

11.1 Introduction

The most important phase of well-logging operations is interpretation. During this phase, geologists, geophysicists, engineers, and log analysts use well logs to obtain information necessary to perform their tasks. Logs have many uses. Exploration geologists use logs to recognize deposition environments and other significant geologic features. Development geologists use them mostly to correlate and to map potential formations. Logs are valuable tools for geophysicists interpreting seismic data. Drilling engineers use log information to detect overpressured zones and to estimate expected pore pressure and fracture gradient, information that is indispensable for safe and efficient drilling operations. Logs are also used during completion. Log data are extremely valuable in reservoir engineering calculations, especially in reserve estimation. The most critical use of logs, however, is the detection of hydrocarbons and the estimation of the potentials of hydrocarbon-bearing formations. The remainder of this book focuses on the last application—the detection and evaluation of hydrocarbon-bearing formations.

Several interpretation techniques have been used to detect hydrocarbon-bearing zones and to estimate their porosities and fluid saturations. The optimum interpretation technique for analyzing a formation of interest depends on the quantity and the quality of the data available to the log analyst. It also depends on the type of "problem" at hand.

Interpretation problems can be classified into exploration and development problems. Exploration problems are associated with exploratory wells (wildcats) drilled in previously untested geologic formations. In these cases, interpretation is marred by a lack of supporting data such as cores and production tests from other wells, and the log analyst must depend on interpretation techniques previously untested in the subject formation. Development problems, on the other hand, are associated with development wells drilled in previously tested geologic formations. In this situation, the analyst can review data available from other wells. He or she may also benefit from interpretation techniques that were previously tested in the environment of interest.

In general, the additional investment needed to test and complete an exploratory well far exceeds that of a development well. This financial risk renders the exploration problem more complex and involved.

When production from a formation ends because the production rate has reached its economic limit or the well has developed mechanical problems, recompletion in a higher zone is usually contemplated. Well logs are used to evaluate the candidate zones. Interpretation problems pertaining to recompletion can be classified as exploration or development problems, depending on available data.

Log analysts are faced with four main questions.

1. Does a specific formation or zone contain hydrocarbons?
2. Which hydrocarbon is present, oil, gas, or both?
3. Is the hydrocarbon saturation high enough to indicate sufficient effective permeability to hydrocarbons?
4. Is the hydrocarbon accumulation large enough to warrant the completion of the well?

If the log analyst can answer all four questions conclusively and positively, the well is usually completed in the zone of interest. If the answers are conclusively negative, the formation is abandoned. More frequently, especially in exploration, the role of well logs is limited by the complexity of the problem to identifying the relatively high-potential zones. These zones will undergo additional testing, such as sidewall coring, fluid sampling, and drillstem testing, before the final decision to complete or abandon the well is made. Under certain circumstances, however, additional tests cannot be performed or are inconclusive. In such cases, the decision is based on well-log interpretation.

The available interpretation techniques vary from "quick look" techniques, which provide relatively quick answers using interpretation aids likely to be on hand at a wellsite, to sophisticated and comprehensive interpretations using all available data. This chapter is devoted to "conventional" or "traditional" interpretation techniques. The conventional calculations usually use only log responses and are based on generalized petrophysical models developed mainly for clean formations. Some of these models are approximations. Where approximations are not appropriate, a degree of uncertainty is introduced. The application of the conventional interpretation technique is hence limited to development problems.

11.2 Acquiring Raw Data From Logs

11.2.1 Correlation Between Logs. Absolute-depth measurement with wireline tools is very difficult, and depth variation between logs recorded in the same borehole may exist. These variations are caused mainly by borehole irregularities and tool type. Some tools, such as the induction, are cylindrical, so withdrawing them from the wellbore is easy. Other tools, such as the density, sidewall neutron, and sonic log with caliper, have arms that drag on the side of the hole. In an inclined borehole, even the cylindrical tools may drag. Variations in hole size and tool shape cause variations in the

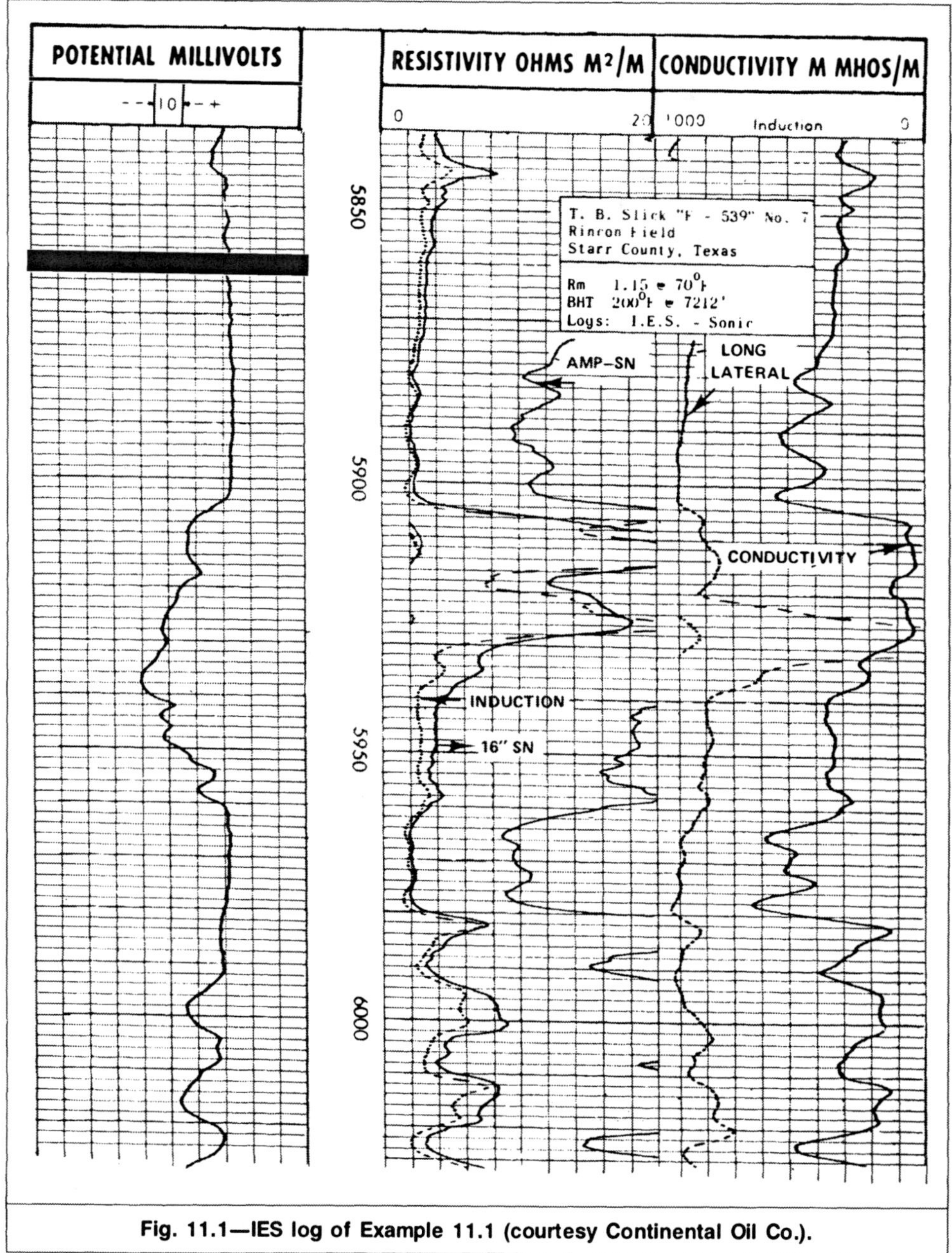

Fig. 11.1—IES log of Example 11.1 (courtesy Continental Oil Co.).

amount of drag, affecting the degree of logging-cable stretch and thus the location of a bed on the log. When this happens, a bed will show at different recorded depths on different logs. These logs are designated "off-depth." The off-depth error can be consistent or inconsistent. The logs may be off a consistent 2 ft, for example, or the error may be 2 ft at one point and 4 ft at another point up the hole. This inconsistency results mainly from a change in borehole conditions. Combining logging devices to produce simultaneous readings of several different logs is possible. Combination log systems have the advantage of providing depth-matched recordings.

The first step in any interpretation technique that uses more than one log is the correlation of the logs to ensure that they are on-depth relative to each other. To correlate between logs, a "marker" has to be selected. A marker is an anomaly or a distinctive response that appears on all logs. Shale and low-porosity stringers are usually good markers. Because of the inconsistency of the error, more than one marker should be used. Two markers, one at the top and one at the bottom of the log section analyzed, are usually recommended. Once the markers are recognized, logs are put on the same depth reference. The resistivity log is usually used as the prime reference. Failure to correlate logs when an appreciable error exists could completely invalidate the interpretation.

Example 11.1. Figs. 11.1 and 11.2 show sections of the Induction Electric Survey (IES) and sonic logs obtained in the 5,850- to 6,030-ft interval of a well drilled in Starr County, TX. Correlate the logs and place them on-depth relative to each other. If they were originally off-depth, what is the magnitude of the error?

Solution. Correlation markers first have to be identified. The sonic log displays two anomalous, low-transit-travel-time zones that correspond to low-porosity stringers. The two markers (Markers A and B) are shown in **Fig. 11.3**. Marker A is at the top of the permeable formation indicated by the self-potential (SP) log and is located at an apparent depth of 5,902 ft. Marker B is located at 5,978 ft in the impermeable zone underlying the relatively thick permeable formation.

Low-porosity or tight zones usually display high resistivity relative to the surrounding formations. Relatively high-resistivity zones are identified on the IES log of **Fig. 11.4** at the same structural positions of Markers A and B. Their apparent depths are 5,906 and 5,982 ft, respectively. The markers' depths from the two logs are listed below.

	Log Depth (ft)		
	IES Log	Sonic Log	Depth Difference (ft)
Marker A	5,906	5,902	4
Marker B	5,982	5,978	4

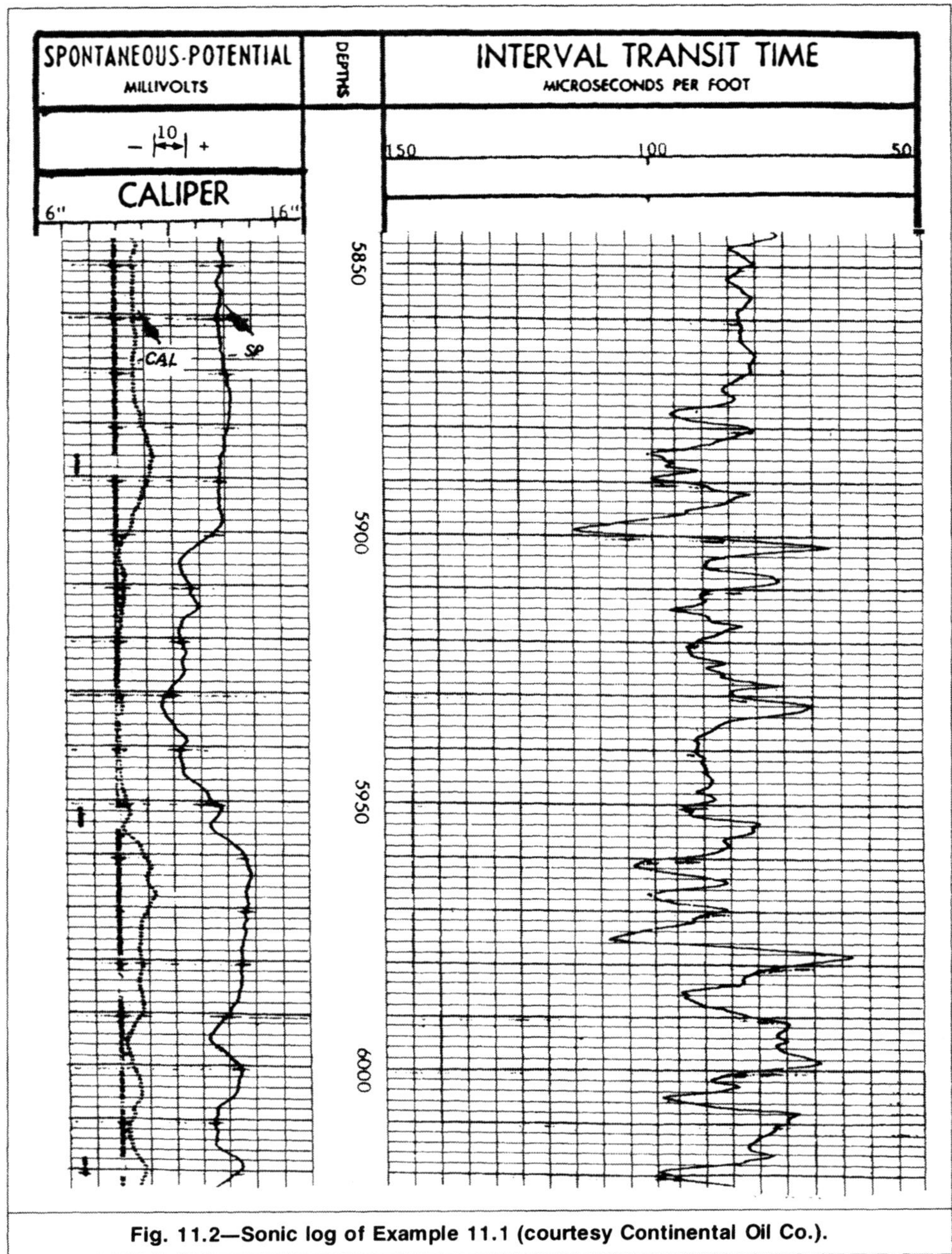

Fig. 11.2—Sonic log of Example 11.1 (courtesy Continental Oil Co.).

The correlation shows that, in the subject interval, the two logs are off-depth by 4 ft. **Fig. 11.5** shows the two logs on-depth relative to each other, with the IES log being the depth prime reference.

11.2.2 Zones Selection. After the logs are placed on-depth, the next step is to select the zones of interest. In the detection of hydrocarbons, the zones of interest are those that display permeability. The permeable beds are usually identified using the SP log. The Microlog™ is an excellent permeability indicator. The effect of mud-filtrate invasion on different resistivity tools helps indicate permeable beds. Shallow-investigation resistivity devices are most affected by invaded-zone resistivity and, in the case of freshwater-based muds, usually display an apparent resistivity that is higher than that of the deep resistivity tool. Separation between resistivity curves can be absent, however, in permeable beds in cases of extremely shallow or extremely deep invasion conditions where both shallow and deep devices investigate practically the same resistivity profile.

Thick permeable beds seldom display a constant resistivity, porosity, or other log readings. These beds must be divided into zones on the basis of log-reading variations. Zone selection is a delicate procedure because different tools have different vertical resolutions; i.e., they measure formation properties in different detail. Laterologs™ and porosity tools respond to beds as thin as 1 ft, while in resistive beds, the induction tool averages 4 ft or more at one time. Thus, the induction conductivity curve is smoother than the log of interval travel time. This is illustrated in **Fig. 11.6,** which is a tracing of the conductivity curve of Example 11.1 onto the sonic log. In addition to being smoother, the conductivity curve does not clearly show very thin beds like those indicated by arrows in Fig. 11.6.

When pairs of conductivity and interval travel-time values are picked to represent zones of interest, the zones should be picked and marked first on the smoother curve, which in this case is the conductivity curve. The Δt value reflecting the same zone investigated by the induction log is then chosen. In relatively thick beds, the choice is immediate, as in Beds 1 through 13 in Fig. 11.6. When beds include thin stringers that are not indicated on the conductivity curves, such as Beds a through d, the sonic log reading should be averaged as shown so that the responses of the two tools are equivalent. In certain cases, the problem of zone selection is more complex than discussed above. Such cases rely on the analyst's judgment, experience, and ingenuity.

After the zones are selected, the resistivity or conductivity, Δt, and other values are read and tabulated. It is good practice also to record the values on the log itself to provide a record that will not be separated from the log.

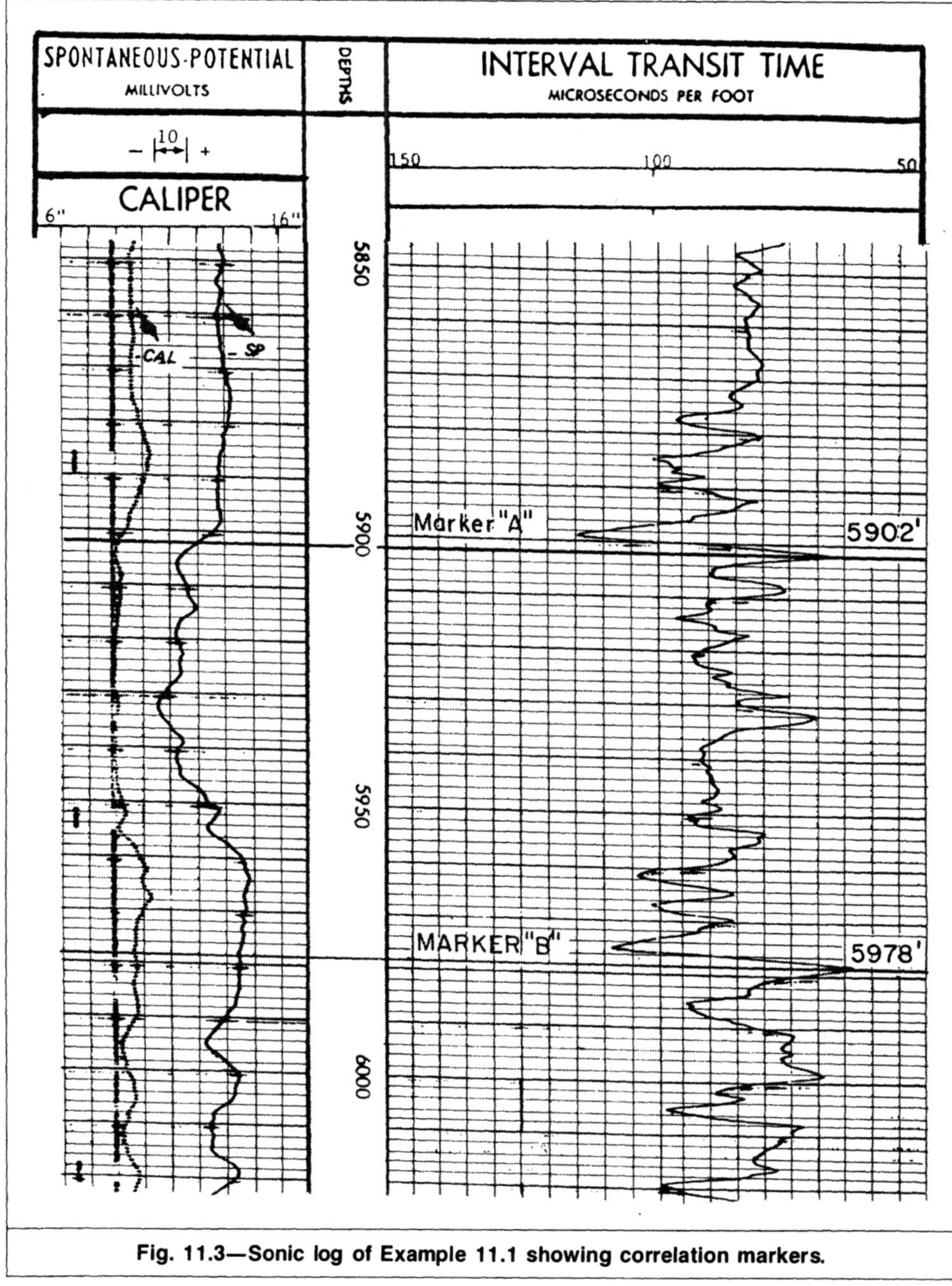

Fig. 11.3—Sonic log of Example 11.1 showing correlation markers.

11.3 Basic Concepts of Conventional Interpretation Technique

The conventional interpretation technique makes use of the following equations:

$$F=a/\phi^m, \quad (11.1)$$

$$R_o=FR_w, \quad (11.2)$$

and $S_w=(R_o/R_t)^{1/n}$. (11.3)

These equations, derived for clean formations, were discussed in detail in Chap. 1. Cases of complex lithology and shaly formations, because of their complexity and importance, are discussed in separate chapters, as are gas-bearing formations. This chapter concentrates on clean, simple-lithology oil-bearing formations.

The use of the conventional technique requires resistivity logs, a porosity log (sonic, density, or neutron), and an SP log or formation water resistivity. Resistivity logs are used to determine R_t and, if possible, R_{xo} values. The method used to determine these values is explained in Chap. 5.

The porosity-log reading is used to calculate the formation porosity with Eq. 2.50, 3.46, 3.47, or 3.62. These equations were discussed thoroughly in previous chapters. Their use assumes that the pore fluid is a liquid and requires knowledge of the matrix type; matrix density, ρ_{ma}; and matrix travel time, Δt_{ma}. Once the porosity is determined, F is calculated from Eq. 11.1 and the appropriate values of the coefficients a and m. Selection of appropriate a and m values is explained in Chap. 1.

R_w can be estimated from the SP log reading with one of the approaches described in Chap. 6. When the SP log is not available or when another value is needed to cross check, R_w can be estimated from a clean water-bearing zone believed to have the same water as the zone of interest. Because $R_t=R_o$, in water-bearing zones, Eq. 11.2, after rearranging, can be used to calculate R_w:

$$R_w=R_o/F, \quad (11.4)$$

where F is the formation factor of the water-bearing zone derived from its porosity.

For development wells where a reliable sample of formation water can be obtained, a direct measurement of R_w provides a means to verify values obtained from either the SP log or the water-bearing formation resistivity method. Water catalogs, compiled by geological societies, oil companies, etc.,[1-4] usually list chemical analyses and resistivity data for formation water collected from different fields and different producing horizons. When only chemical analysis of the formation water is available, the method discussed in

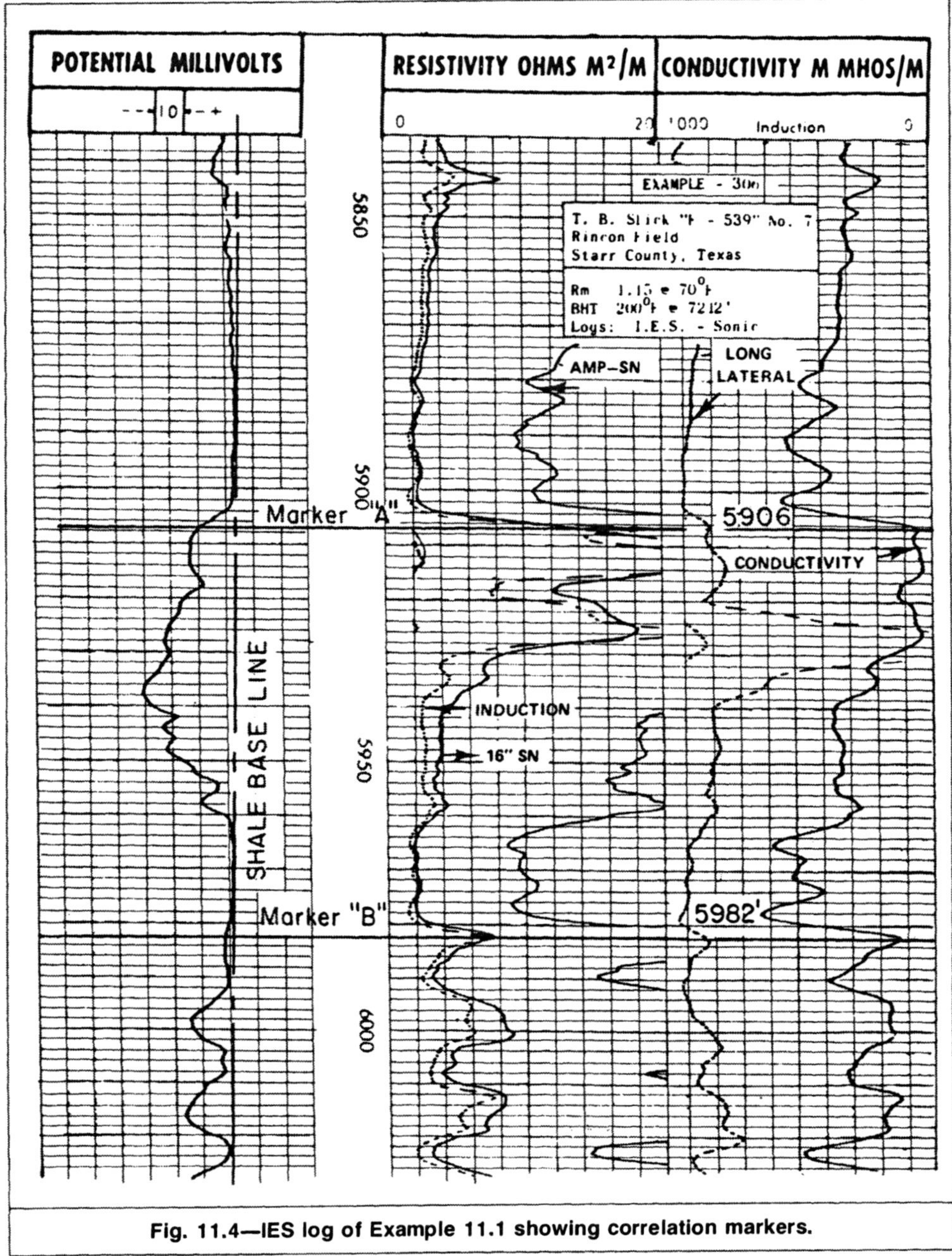

Fig. 11.4—IES log of Example 11.1 showing correlation markers.

Chap. 1 can be used to derive an equivalent NaCl concentration, which, in turn, is used to determine R_w.

Now that R_w and F values have been obtained, R_o is estimated from Eq. 11.2. R_o is the estimated resistivity of the formation of interest when fully saturated with water. Comparing R_o to the resistivity-log-derived R_t will show whether hydrocarbon is present. In hydrocarbon-bearing formations, $R_t > R_o$. The values of R_o and R_t are then used in Eq. 11.3 to estimate water saturation, S_w. The hydrocarbon saturation is, of course, equal to $(1 - S_w)$. Chap. 1 should be consulted when determining the value of the exponent n in Eq. 11.3.

Example 11.2. Figs. 11.7 and 11.8 are IES and Formation Density Compensated (FDC) logs obtained in a development well drilled to produce the oil sand located at 10,402 ft. A formation-water resistivity of 0.11 Ω · m at 60°F is listed in formation-water catalogs. Mud and temperature data listed on the well heading are as follows.

$$R_m = 1.17 \text{ at } 58°\text{F}, R_m = 0.43 \text{ at } 60°\text{F, and } R_{mc} = 1.70 \text{ at } 60°\text{F}.$$

Bottomhole temperature (BHT) = 274°F at 11,300 ft give a complete qualitative and quantitative analysis of the oil sand of interest.

Solution. The oil sand of interest is in the interval of 10,402 to 10,422 ft. The SP curve indicates a fairly clean sand. The sand displays a uniform porosity indicated by the almost-constant density log response. Hence, the sand is evaluated whole. Average log response is used for the evaluation.

Net Pay Determination. The sand is clean with no major impermeable streaks, so the net sand thickness is equal to the gross thickness, 20 ft (10,422 − 10,402 ft). No water/oil contact is present because the entire sand is resistive. The net pay is equal to the net sand thickness, 20 ft.

Porosity Determination. The density log displays an average reading of 2.25 g/cm³. The tool is automatically compensated for mudcake effect, and the caliper indicates a regular borehole. The log reading is reliable and can be used to estimate the porosity with Eq. 8.11:

$$\phi = (\rho_{ma} - \rho_b)/(\rho_{ma} - \rho_f).$$

Consistent with porosity evaluation principles of Chap. 8, values of 2.65 and 1.0 g/cm³ are assumed for ρ_{ma} and ρ_f, respectively. Then,

$$\phi = (2.65 - 2.25)/(2.65 - 1.00) = 0.24 \text{ or } 24\%.$$

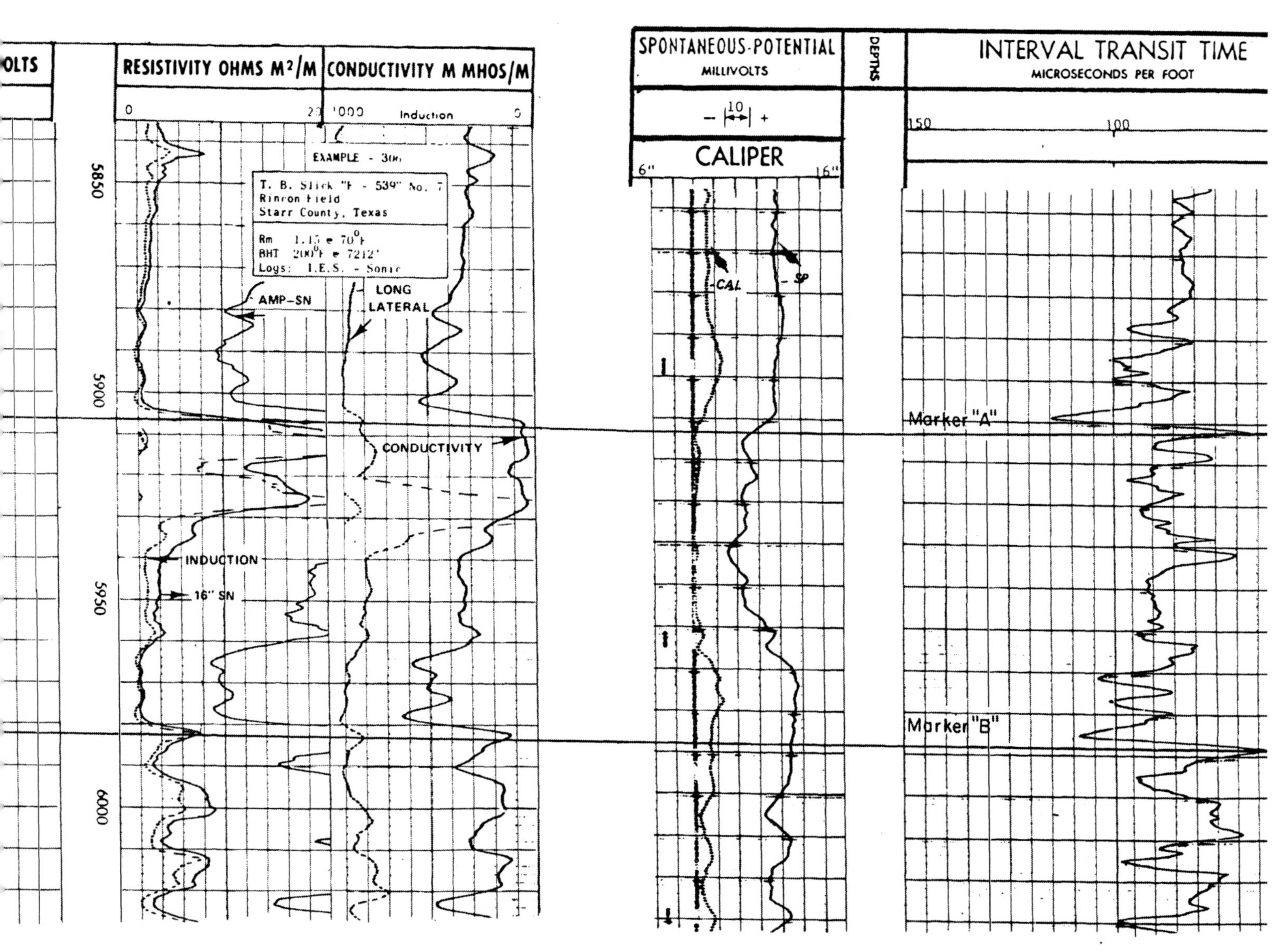

Fig. 11.5—IES and sonic logs of Example 11.1 on-depth relative to each other.

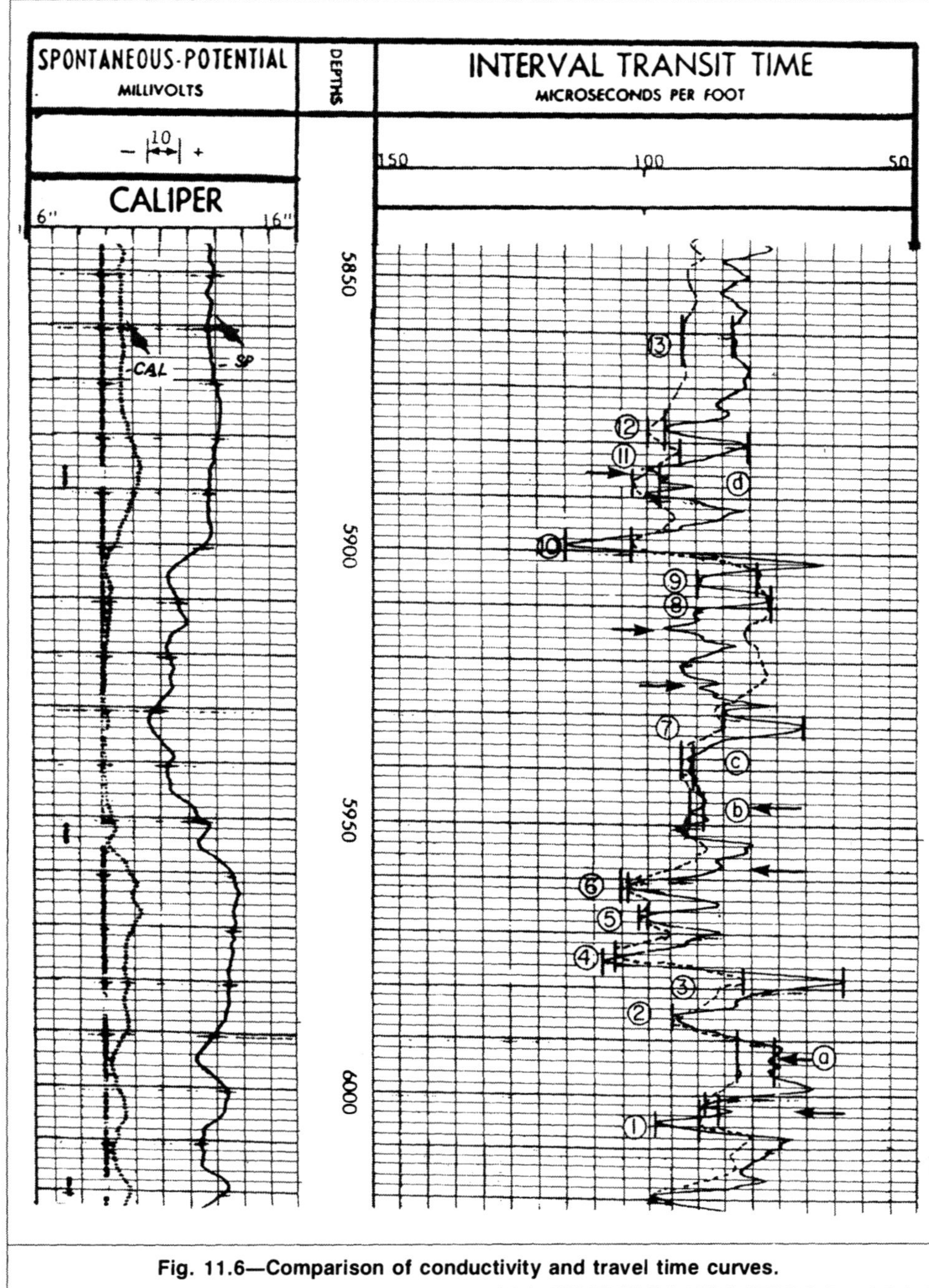

Fig. 11.6—Comparison of conductivity and travel time curves.

Formation Temperature. A linear temperature distribution and a surface temperature of 70°F are assumed. The geothermal gradient and formation temperature are calculated by Eqs. 4.15 and 4.16.

$$g_G = \frac{T_{max} - T_s}{D_{max}} \times 100 = \frac{274-70}{11{,}300} \times 100 = 1.8°\text{F}/100 \text{ ft.}$$

and $T_f = T_s + g_G \frac{D}{100} = 70 + 1.8 \frac{10{,}402}{100} = 258°\text{F}$.

Drilling-Fluid Properties. R_m and R_{mf} at formation temperature can be calculated from Eq. 1.11 or with Fig. 1.7.

$$R_m = 1.17 \frac{58+6.77}{258+6.77} = 0.29\,\Omega\cdot\text{m at } 258°\text{F}$$

and $R_{mf} = 0.43 \frac{60+6.77}{258+6.77} = 0.11\,\Omega\cdot\text{m at } 258°\text{F}$.

R_w From SP Log.
Maximum SP deflection $= -(5 \text{ div.} \times 10 \text{ mV/div.})$

$= -50$ mV.

This maximum deflection is expected to be less than the static self-potential (SSP) value because of the effect of the relatively high formation resistivity. However, the empirical chart in Fig. 6.10 indicates a negligible correction of less than 5% in this case where the diameter of invasion is expected to be small because of relatively high porosity, $h = 20$ ft, and

$$\frac{R_i}{R_m} \cong \frac{R_{SN}}{R_m} = \frac{18}{0.29} = 62.$$

Using Eqs. 6.11 and 6.21 results in

$K = 61.3 + 0.133 T_f$

$= 61.3 + 0.133(258) = 95.3$ mV

and $R_{we} = R_{mf} / 10^{SP/-K}$

$= 0.11/10^{-50/-95.3} = 0.033\,\Omega\cdot\text{m}$.

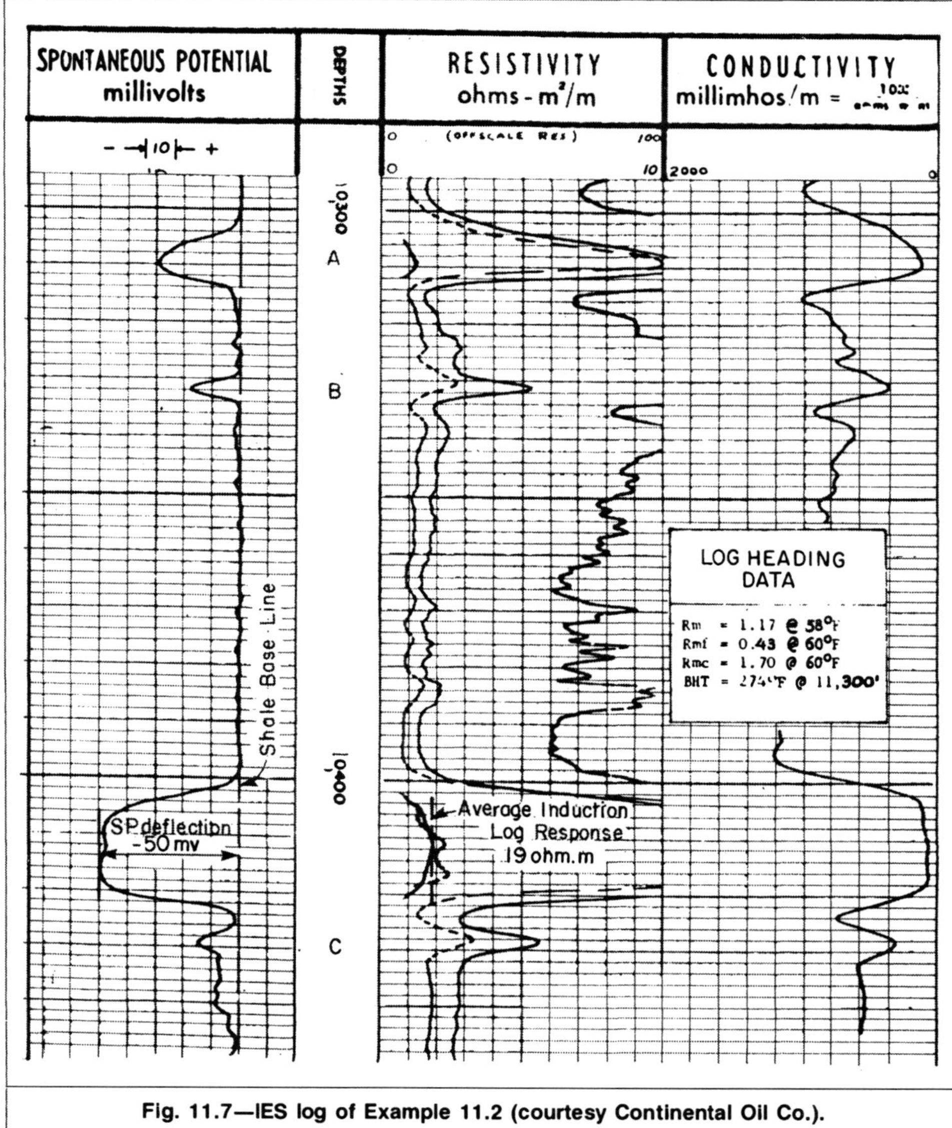

Fig. 11.7—IES log of Example 11.2 (courtesy Continental Oil Co.).

From Fig. 6.13, R_w=0.038Ω·m at 258°F.

R_w From Water Catalogs. Water catalogs indicate a resistivity of 0.11 Ω·m at 60°F, which is equivalent to 0.028 Ω·m at a formation temperature of 258°F. This value is different from that derived from the SP log response. The value of 0.028 is retained for the evaluation because it was obtained by direct measurement and is believed to be more representative. The SP response is believed to be affected by the formation resistivity at a degree much higher than indicated by the empirical chart in Fig. 6.10. The use of a low SP value in Eq. 6.21 results in a high R_w value.

R_t Determination. The deep induction tool displays an average apparent resistivity, R_a, of 19 Ω·m. Consulting Fig. 5.56 confirms the negligible borehole effect on a deep induction tool in a small-diameter borehole drilled with a low-conductivity mud. Then, borehole-effect-free apparent resistivity is

$$R_a' = R_a = 19 \Omega\cdot\text{m}.$$

On the other hand, Fig. 5.62 indicates a significant bed-thickness effect resulting in bed-thickness-effect-free apparent resistivity, R_a''=32Ω·m. Use of Fig. 5.62 was dictated by the value of the resistivity of the shale that surrounds the sand. This value is approximately 1 Ω·m.

Because of high porosity, the invasion effect is assumed to be negligible and

$$R_t = R_a'' = 32 \Omega\cdot\text{m}.$$

F and R_o Calculations. In the absence of a and m values specific to this formation, generalized Eq. 1.19 is used to estimate F:

$$F = 1.13\phi^{-1.73}$$

$$= 1.13(0.24)^{-1.73} = 13.3$$

and $R_o = FR_w$

$$= 13.3(0.028) = 0.37 \Omega\cdot\text{m}.$$

Because $R_t \gg R_o$, the zone is definitively a hydrocarbon-bearing zone.

Determination of Oil Saturation.

$$S_w = \left(\frac{R_o}{R_t}\right)^{1/2} = \left(\frac{0.37}{32}\right)^{1/2} = 0.11 \text{ or } 11\%$$

and S_o=1−0.11=0.89 or 89%.

Representativity of Results. It is often said that there are as many answers to a log interpretation problem as there are log analysts. Although this statement is somewhat exaggerated, it is quite possible that different analysts will arrive at different quantitative interpretations—i.e., different values for h, ϕ, and S_o. In this clear-cut case, all analysts should arrive at the same qualitative interpretation: the sand is hydrocarbon-bearing.

Different analysts might obtain different saturation values for several reasons. Some analysts favor a generalized relationship between F and ϕ. Some could have had bad experience with water resistivity values obtained from water catalogs, so they opt for the SP-log-derived value. Others use the induction log response as R_t without any correction, arguing that, because surrounding shales tend to decrease the apparent resistivity value and the invasion effect tends to increase its value, the two effects compensate for each other. The log analyst who uses $R_w=0.038\ \Omega\cdot m$ and $R_t \cong 19\ \Omega\cdot m$ arrives at the following S_o value:

$$S_w=(FR_w/R_t)^{1/2}$$

$$=[13.3(0.038)/19]^{1/2}=0.16 \text{ or } 16\%$$

and $S_o=1-0.16=0.84$ or 84%.

The difference between this and the previously calculated value is small (only 5%). In addition, the two values lead to the same conclusion about the high potential of the sand. In other cases, however, the difference could be larger, and the values obtained could lead to contradictory conclusions.

Earliest quantitative interpretations were based on Eq. 11.3 alone. The resistivity of an adjacent water-bearing zone is taken to be equal to R_o of the zone analyzed. This technique was used when porosity logs and core porosity were not available. Because $R_o=FR_w$, its use assumes that the water-bearing zone has the same formation factor as the hydrocarbon zone. If the two zones are not situated within the same bed, use of this technique also implies that R_w is the same. Unless data other than resistivity logs indicate that the analyzed zone is hydrocarbon-bearing, assuming that it is hydrocarbon-bearing on the basis of its relatively high resistivity only can lead to gross qualitative misinterpretation. The same is true for the adjacent zone, which is usually assumed to be fully saturated with water on the basis of only its relatively low resistivity. This practice is valid only in formations known to display uniform porosities, such as sandstones and carbonate formations with only intergranular and intercrystalline porosity. Use of this technique is discouraged in formations known to have erratic porosity distribu-

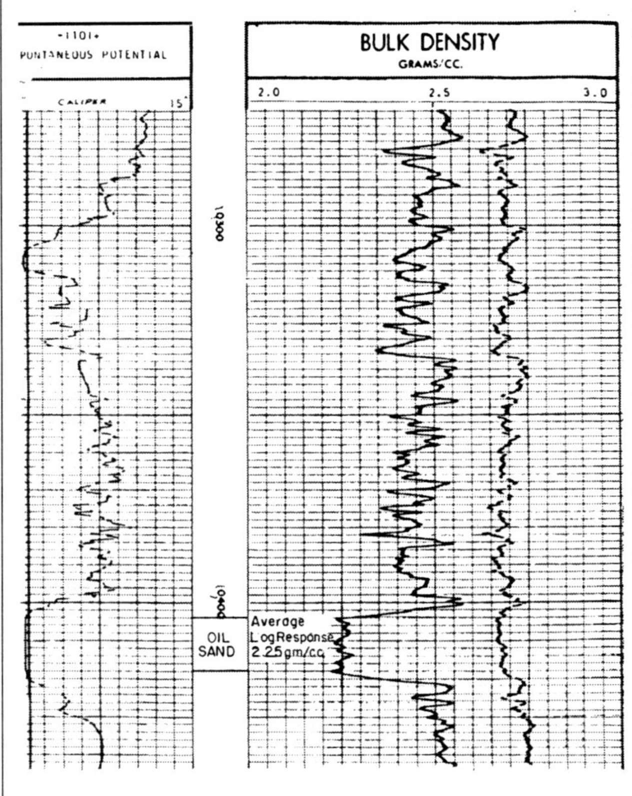

Fig. 11.8—FDC log of Example 11.2 (courtesy Continental Oil Co.).

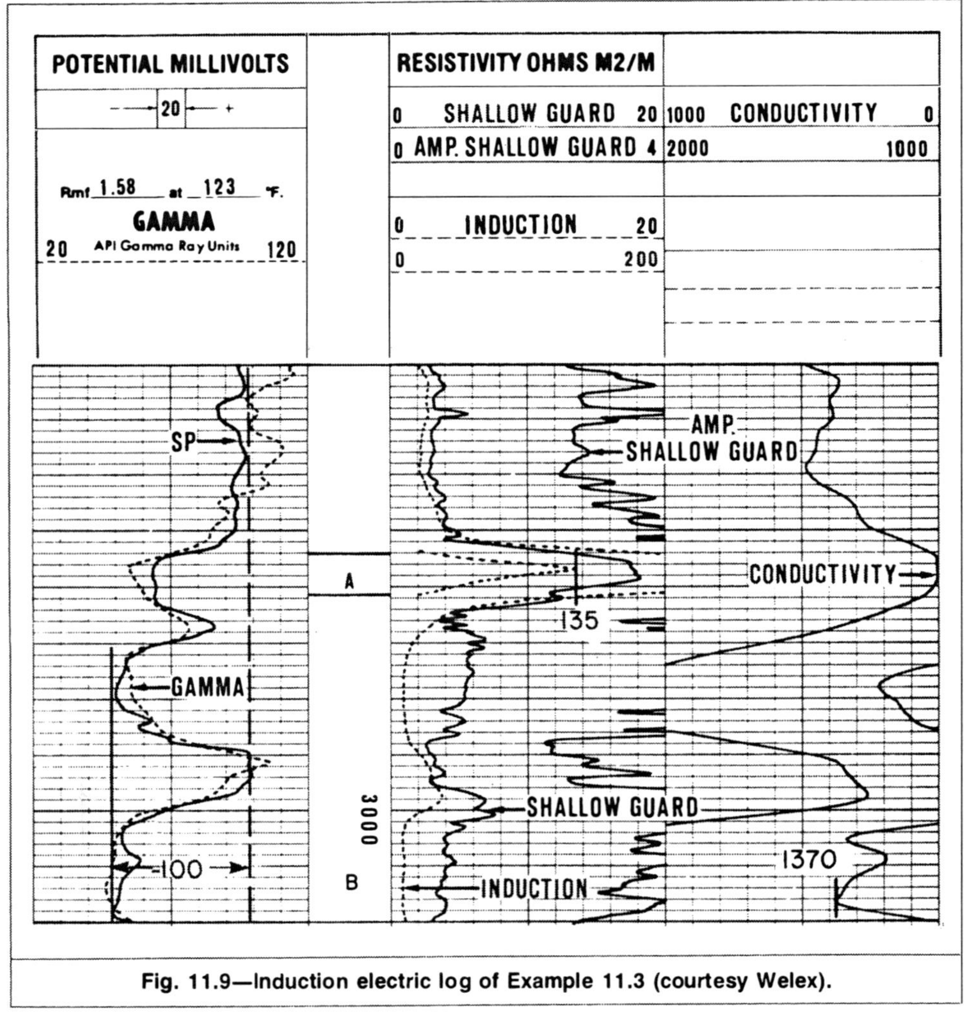

Fig. 11.9—Induction electric log of Example 11.3 (courtesy Welex).

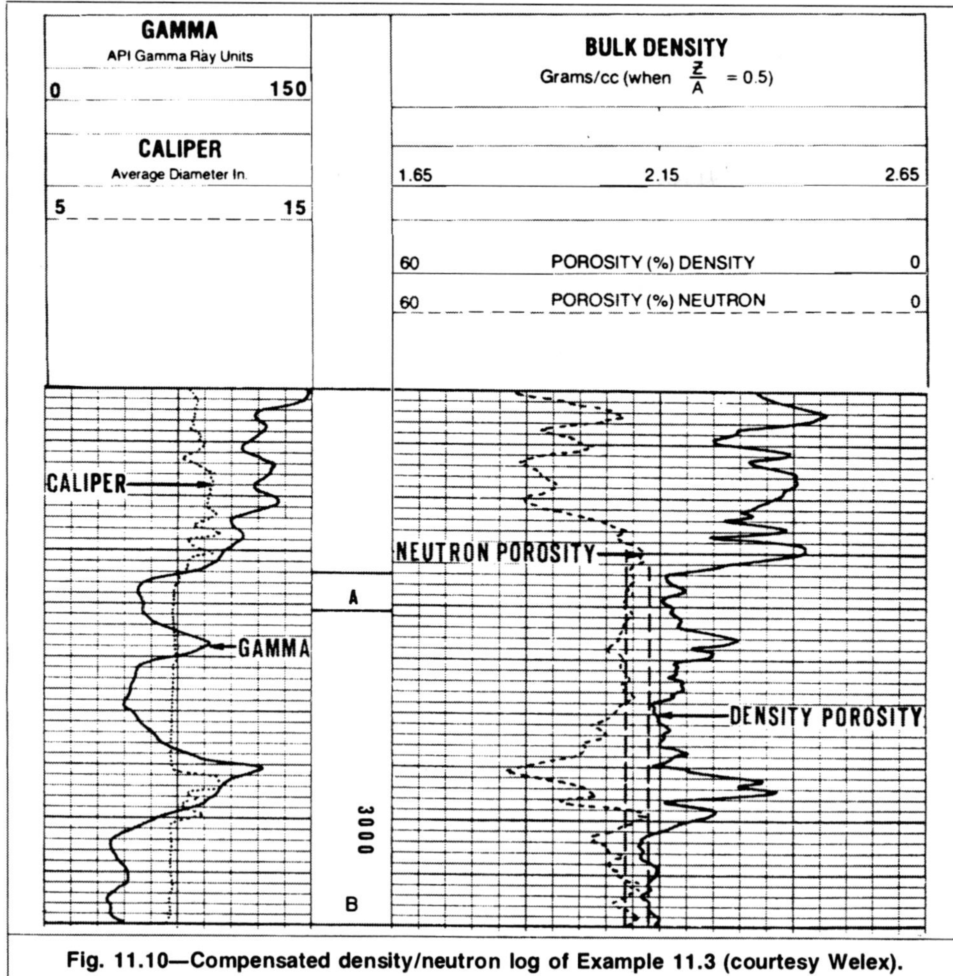

Fig. 11.10—Compensated density/neutron log of Example 11.3 (courtesy Welex).

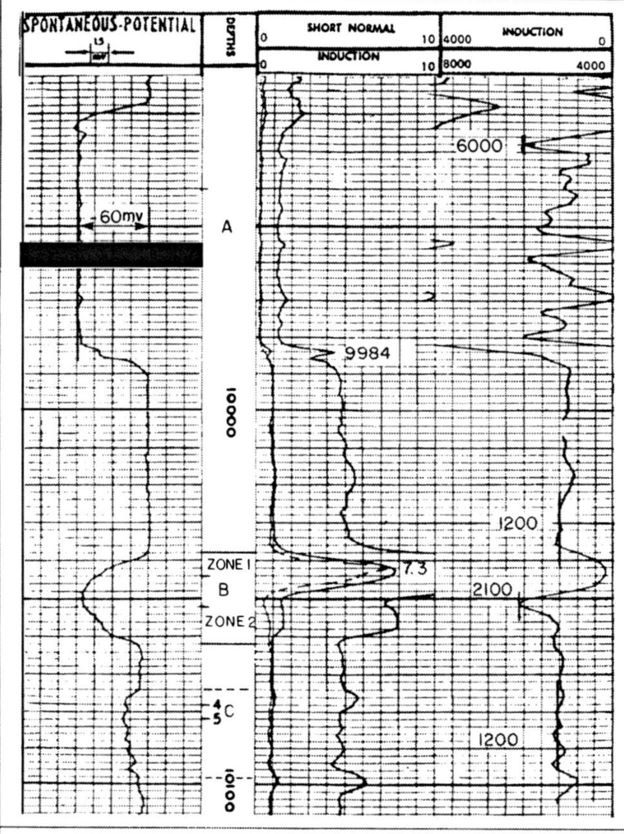

Fig. 11.11—IES log of Example 11.4 (courtesy Amoco Production Co.).

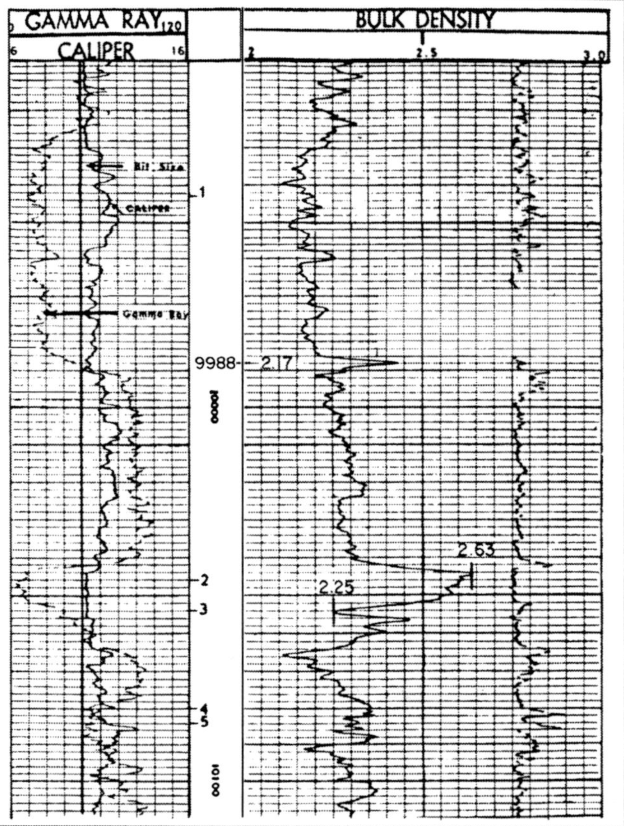

Fig. 11.12—FDC log of Example 11.4 (courtesy Amoco Production Co.).

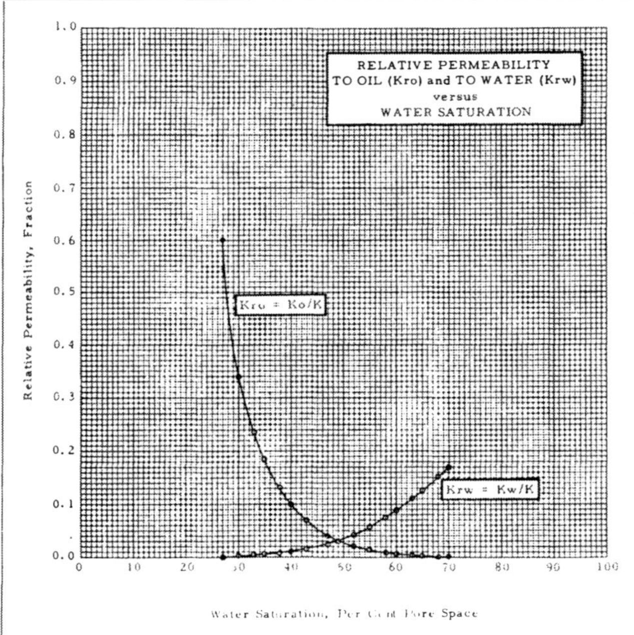

Fig. 11.13—Typical oil and water relative permeability curves (courtesy Core Laboratory Inc.).

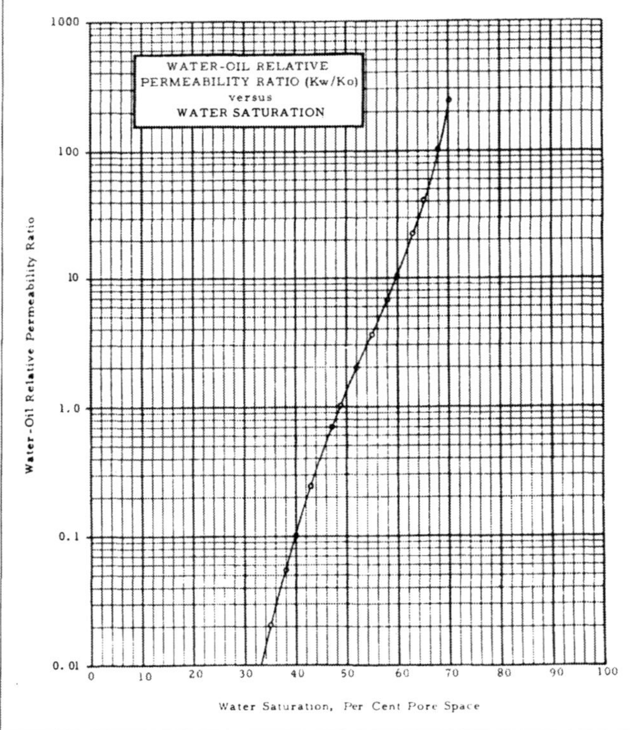

Fig. 11.14—Typical water/oil relative permeabilities ratio (courtesy Core Laboratory Inc.).

tions, such as formations with secondary porosities generated by fractures and/or vugs.

An R_o value from an adjacent bed should be used only under the following conditions.

1. No porosity tools (logs or cores) are available.
2. From data other than resistivity logs, the zone analyzed is known to be hydrocarbon-bearing.
3. The porosity is known to be uniform in the formations analyzed.
4. The formation-water resistivity is known or can be proved to be the same in both zones.
5. Both zones belong to the same rock type.

When a porosity tool displays the same response in hydrocarbon- and water-bearing intervals of the same formation, use of Eq. 11.3, which is based on resistivity values alone, is recommended. Here we know that the formation factor, formation-water resistivity, and rock type are the same. Using a log-measured R_o value avoids the uncertainties that could be introduced by Eq. 11.2. These uncertainties are generated by the models and equations used to calculate R_w from the SP log and porosity from the porosity log response. Finally, F is calculated from the porosity value with a generalized equation usually of questionable representativity.

Example 11.3. Figs. 11.9 and 11.10 illustrate the induction electric log and compensated density/neutron log obtained in a Cretaceous sand in Van Zandt County, TX. Estimate the water saturation in the hydrocarbon-bearing Zone A marked on the logs. The log heading lists the following data.

R_{mf}=2.21 at 90°F.

BHT=123°F at 3,300 ft.

Solution. The sand of interest is relatively thick, almost clean at the bottom (Zone B), and somewhat shaly at the top (Zone A). The shaliness is indicated by the increase in gamma ray reading and by the separation of the density porosity and neutron porosity curves. Shaly sand interpretation techniques described in Chap. 15 could be used. However, because of the low degree of shaliness and the uniform porosity, a reasonable estimate can be obtained from resistivity readings alone. The resistivity of Zone B, which is highly likely to be water-bearing, will replace R_o in Eq. 11.3. Then,

$R_t = 135\Omega\cdot$m,

$R_o = 1{,}000/1{,}370 = 0.73\Omega\cdot$m,

and $S_w = (0.73/135)^{1/2} = 0.07$ or 7%.

Use of log readings uncorrected for borehole, bed thickness, and invasion effects can be justified because the interpretation approach used will yield an approximate S_w value anyhow. Corrections for these effects or proof of their low values are not warranted.

Example 11.4. Figs. 11.11 and 11.12 show IES and FDC logs obtained in the 9,910- to 10,108-ft interval of a Louisiana gulf coast well where $R_{mf} = 0.062\Omega\cdot$m at BHT and BHT=200°F at 11,800 ft. Does the interval shown by the logs contain potential hydrocarbon-bearing zones?

Solution. *Analyst X.* Relying on the fact that U.S. gulf coast formations consist mainly of shale/sand series and using SP log deflection, Analyst X recognized three sand formations: Sands A through C (marked on the log). Sand A displays a conductivity as high as 6,000 m℧/m (0.17 Ω·m) in the cleanest zones. This is an extremely low resistivity, and Analyst X concluded that Sand A is a water-bearing formation. Sand C displayed a low SP deflection and a conductivity of 1,200 m℧/m (0.83 Ω·m), almost the same as adjacent shales. Analyst X concluded that Sand C is a sandy shale of no potential.

Formation B appeared to Analyst X as one sand unit with a high-resistivity zone (Zone 1) at the top and a low-resistivity zone (Zone 2) beneath it. The sand becomes shaly at the bottom of Zone 2, explaining the gradual but slight increase in resistivity. From these observations and knowing that gulf coast sands usually display uniform porosities, Analyst X concluded that Zone 1 of Formation B is hydrocarbon-bearing, displaying a resistivity of 7.3 Ω·m, and that Zone 2 of the same formation is a water-bearing zone, displaying a resistivity of 0.48 Ω·m calculated from an induction conductivity reading of 2,100 m℧/m. S_w can be estimated from Eq. 11.3, where R_o of Zone 1 is assumed to be equal to R_o of Zone 2:

$S_w = (0.48/7.3)^{1/2} = 0.26$ or 26%

and $S_h = 1 - 0.26 = 0.74$ or 74%.

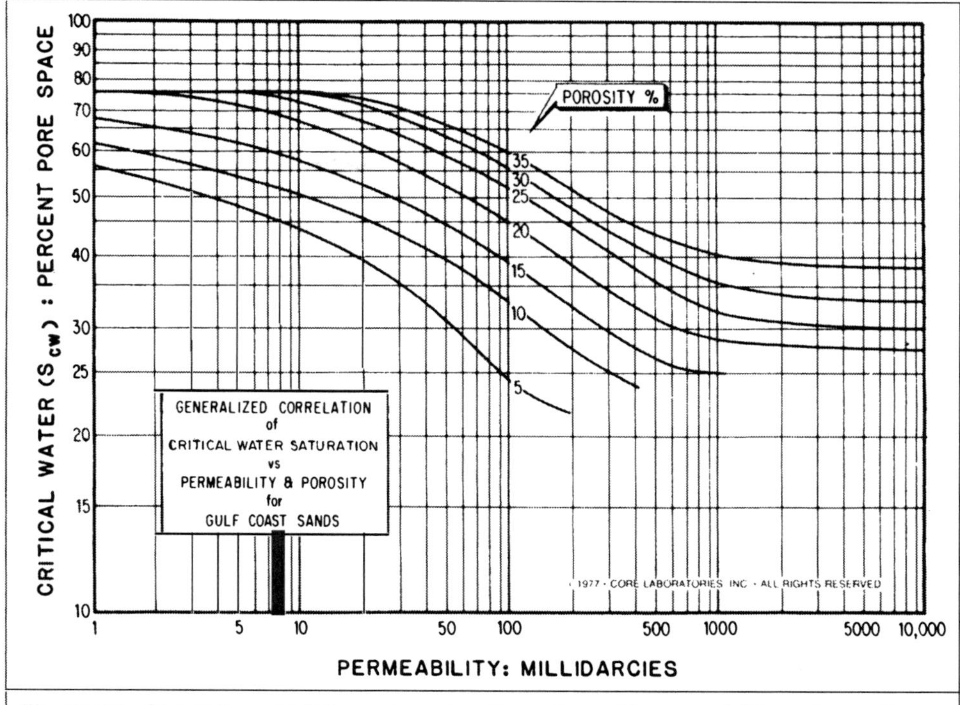

Fig. 11.15—Correlation of critical water saturations, S_{cw}, with permeability and porosity for U.S. gulf coast sands (after Ref. 9).

This is a relatively high hydrocarbon saturation, which usually indicates a zone of good potential.

Analyst X did not take into consideration the response of the density tool. This major flaw resulted, in this case, in serious misinterpretation, as demonstrated by Analyst Y.

Analyst Y. Analyst Y reached the same conclusions as Analyst X for Formations A and C. In Formation B, however, he noticed the extremely high density of Zone 1. The density reading reached a maximum of 2.63 g/cm^3. Assuming sandstone results in

$$\phi=(2.65-2.63)/(2.65-1)=0.012 \text{ or } 1.2\%.$$

This is too low to be true for U.S. gulf coast sandstone. Formation B is probably a carbonate formation. This conclusion is confirmed by the following log responses: high density response, indicating a low-porosity (tight) zone; high resistivity response caused mainly by low porosity; a caliper log reading that shows a regular borehole almost drilled to gauge; a low level of radioactivity (about 20 API units less than clean Sand A); no or negligible density log correction ($\Delta\rho=0$), which can be attributed to the absence of mudcake; and to a lesser extent, the uncommon response of the SP log.

Such carbonate formations are known to interrupt the shale/sand sequences of the gulf coast sediments. They are usually tight formations with no hydrocarbons present.

Analyst Y concluded that the interval shown is devoid of hydrocarbon zones. His interpretation is sound, but it lacks the quantitative support that might add to its credibility.

Analyst Z. In addition to the qualitative interpretation of Analyst Y, the analyst could give a quantitative interpretation in its support. Because an order of magnitude of different parameters would be sufficient in this case, Analyst Z used raw log readings without corrections. This is not serious because it can easily be shown that these corrections are negligible or will result in slight and insignificant numerical differences.

For Sand A, using Eqs. 4.15, 4.16, and 1.11,

$$G_g=[(200-70)/11{,}800]100=1.1°\text{F}/100 \text{ ft},$$

$$T_f=70+1.1(9{,}950/100)=180°\text{F},$$

and $R_{mf}=0.062[(200+6.77)/(180+6.77)]=0.069\Omega\cdot\text{m}$.

From the SP log, $E_{SSP}=-60$ mV.

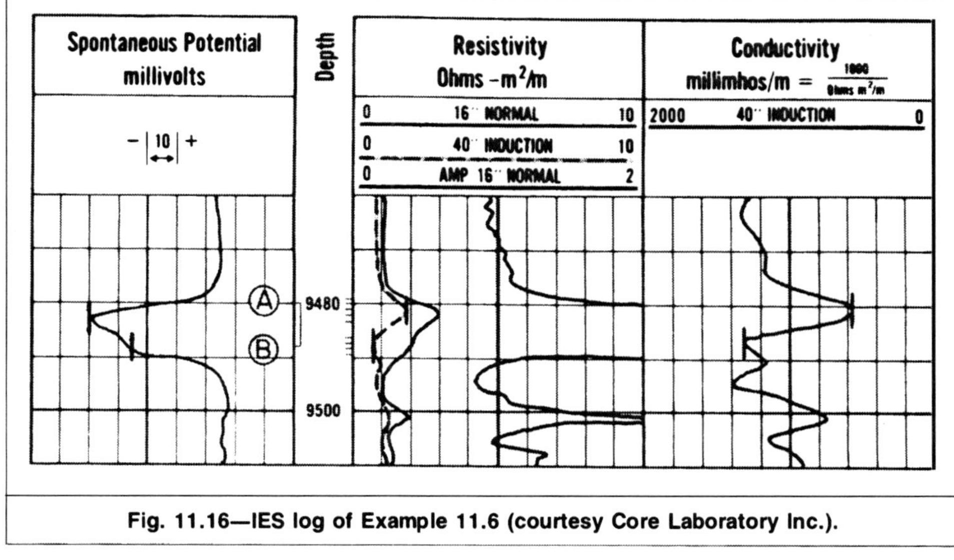

Fig. 11.16—IES log of Example 11.6 (courtesy Core Laboratory Inc.).

From Fig. 6.14,

$R_w=0.023\Omega\cdot m$ at 180°F.

Using Eqs. 2.50, 1.19, and 11.2 and an average log response of 2.17 g/cm³ gives

$\phi=(2.65-2.17)/(2.65-1)=0.29$ or 29%,

$F = 1.13/(0.29)^{1.73} = 9.6$,

and $R_o=9.6(0.023)=0.22\Omega\cdot m$.

This R_o value is of the same order of magnitude as the sand's log resistivity, confirming that it is a water-bearing formation.

To convince himself that Formation B is not potentially hydrocarbon-bearing, Analyst Z estimated the order of magnitude of R_o using the following assumptions: (1) R_w is the same as in the nearby clean sand (0.023 Ω·m); (2) the matrix density is 2.71 g/cm³, which is that of pure calcium carbonate; and (3) cementation exponent $m=2$.

Zone	R_t (Ω·m)	ρ_{log}	ϕ (%)	F	R_o	S_w (%)	Conclusion
1	7.3	2.63	4.7	457	10.5	120	Water, no potential
2	0.48	2.25	27.0	13.8	0.32	81	Water, no potential

A physically impossible S_w value of 120% was calculated because the assumptions introduced errors. This also explains why Zone 2 is qualified as water despite the calculated 19% hydrocarbon saturation.

Quantitative interpretation of Sand C, which is a shaly sand, requires use of the interpretation techniques explained in Chap. 15.

Some analysts, like Analyst Y, could argue that the calculations performed by Analyst Z are not necessary to show the absence of potential hydrocarbon zones in the section analyzed. This might be true for experienced analysts already familiar with log responses in the type of formation analyzed. A novice analyst, however, should rely on a complete and detailed qualitative and quantitative interpretation to gain experience.

11.4 Calculation of Recoverable Hydrocarbons

The amount of oil that may be recovered (reserves) from a well that encountered a formation of interest can be calculated with

$$N_R = 7,758\frac{AF_R}{B_o}\sum_{i=1}^{n} h_i\phi_i(S_o)_i, \quad \text{(11.5)}$$

where N_R=recoverable oil volume in STB, A=drainage area in acres, F_R=recovery factor, B_o=oil FVF, h_i=thickness in ft of an individual zone capable of flowing oil at rates of economic interest within the interval of interest, ϕ_i=fractional porosity, S_o=fractional oil saturation, and 7,758=number of barrels per acre-foot.

A similar equation can be written to estimate the standard cubic feet of recoverable free gas, G_R:

$$G_R = 43,560\frac{AF_R}{B_g}\sum_{i=1}^{n} h_i\phi_i(S_g)_i, \quad \text{(11.6)}$$

where S_g=fractional gas saturation, B_g=gas FVF, and 43,560=number of cubic feet per acre-foot.

The drainage area, recovery factor, and hydrocarbon FVF are determined from data other than well logs or from previous experience. B_o and B_g can be measured if fluid samples are available. They also can be estimated from available correlations.[5] To use these correlations, the formation pressure and temperature and the fluid gravity are required. Drainage area can be estimated from pressure and flow well-test analysis[6] or may be inferred from knowledge of the geologic environment. In many instances, the well spacing is substituted for drainage area, A. Identification of the producing drive mechanism allows an acceptable estimate of the recovery factor. Capillary pressure or flow test data on cores might give a hint as to the recovery factor. In certain cases, a rough estimate can be derived from log data.

The parameters h_i, ϕ_i, and $(S_h)_i$, which are inside the summation sign in Eqs. 11.5 and 11.6, are determined from log measurements by methods detailed in this and following chapters.

11.5 Concept of Critical or Cutoff Saturation

Of the three parameters ϕ, h, and $S_h(=1-S_w)$, the hydrocarbon saturation is the most important and critical value for several reasons.

1. The amount of hydrocarbon in place is directly proportional to S_h.
2. The ability of a zone to transmit hydrocarbons at an economic rate is determined by S_h.

TABLE 11.1—CORE DATA FOR EXAMPLE 11.6

Depth (ft)	Core Permeability (md)	Core Porosity (%)	Pore Saturations (%) Residual Oil	Total Water
9,480	260	34.3	9.9	60.1
9,481	239	33.9	10.1	67.4
9,482	436	34.0	10.0	68.7
9,483	256	33.5	12.2	69.8
9,486	66	30.8	7.5	70.3
9,487	91	35.1	9.4	64.4
9,488	71	30.7	4.2	76.2
9,489	77	34.9	9.2	67.8

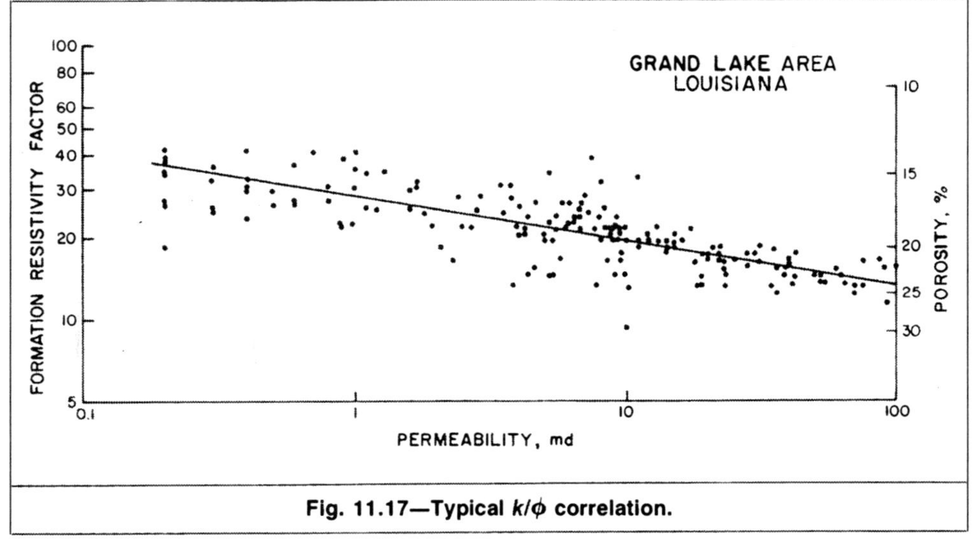

Fig. 11.17—Typical k/ϕ correlation.

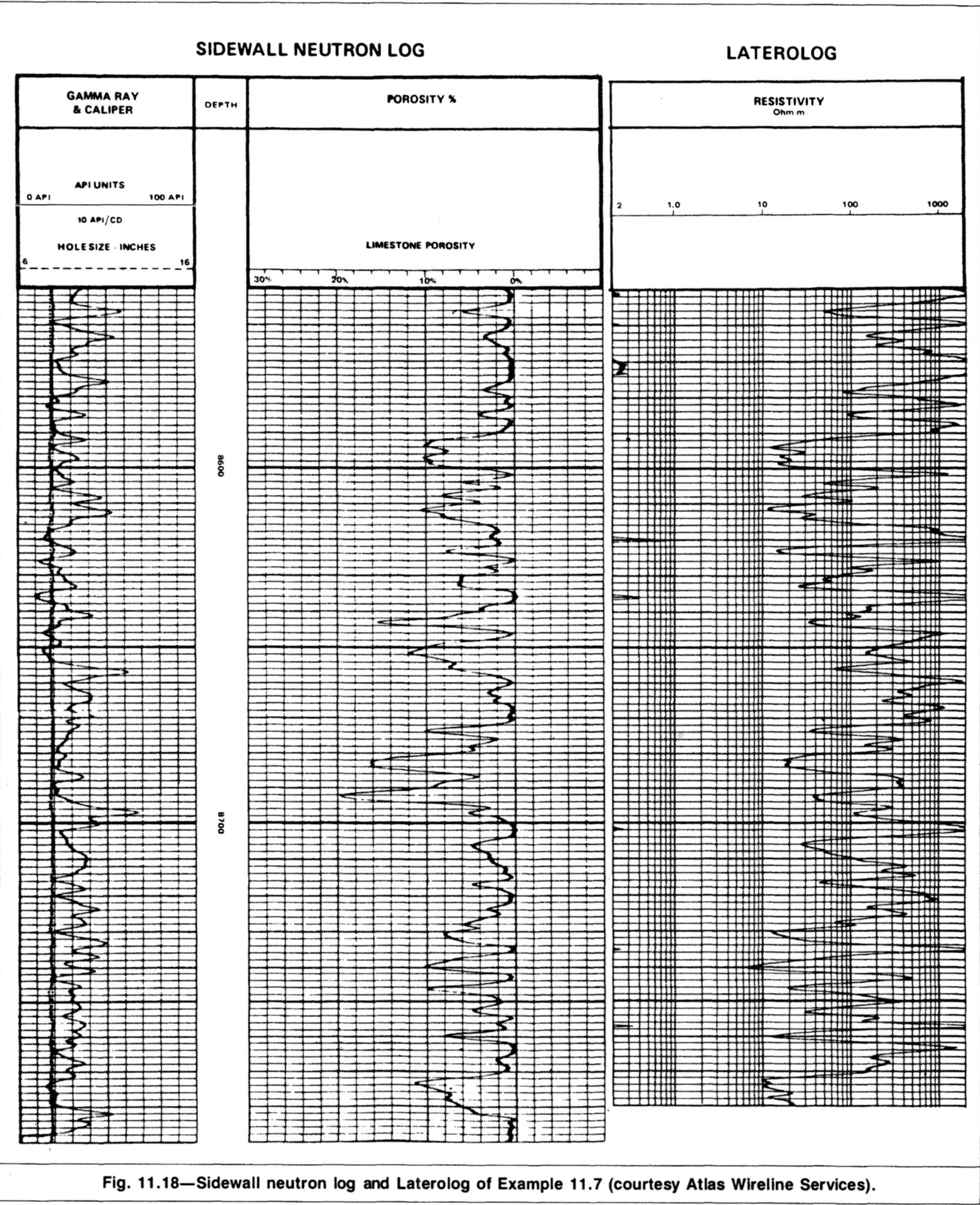

Fig. 11.18—Sidewall neutron log and Laterolog of Example 11.7 (courtesy Atlas Wireline Services).

3. Frequently, the geological and physical properties that control S_h also determine the drainage area and recovery factor.

Thus, it can be seen that, to determine the economic potential of a formation, only zones that exhibit sufficient permeability to hydrocarbons should be included in Eq. 11.5 or 11.6. All rocks contain some water, and the effective hydrocarbon permeability, k_{eh}, will be less than the rock absolute permeability, k:

$$k_{eh}=k_{rh}k, \quad (11.7)$$

where k_{rh} is the hydrocarbon relative permeability. An empirical relation exists between water saturation, hydrocarbon relative permeability, and water relative permeability, k_{rw}. **Fig. 11.13** shows k_{ro} and k_{rw} curves, and **Fig. 11.14** shows a typical k_{rw}/k_{ro} curve.

The fractional flow of water or water cut, f_w, is determined by the water saturation, which controls the hydrocarbon and water relative permeabilities, and by fluid viscosities. f_w is expressed analytically as[7]

$$f_w=1\Bigg/\left(1+\frac{k_{ro}\mu_w}{k_{rw}\mu_o}\right), \quad (11.8)$$

where μ_w and μ_o are the water and oil viscosities, respectively. Thus, as S_w approaches some "critical" or "cutoff" value, S_{cw},

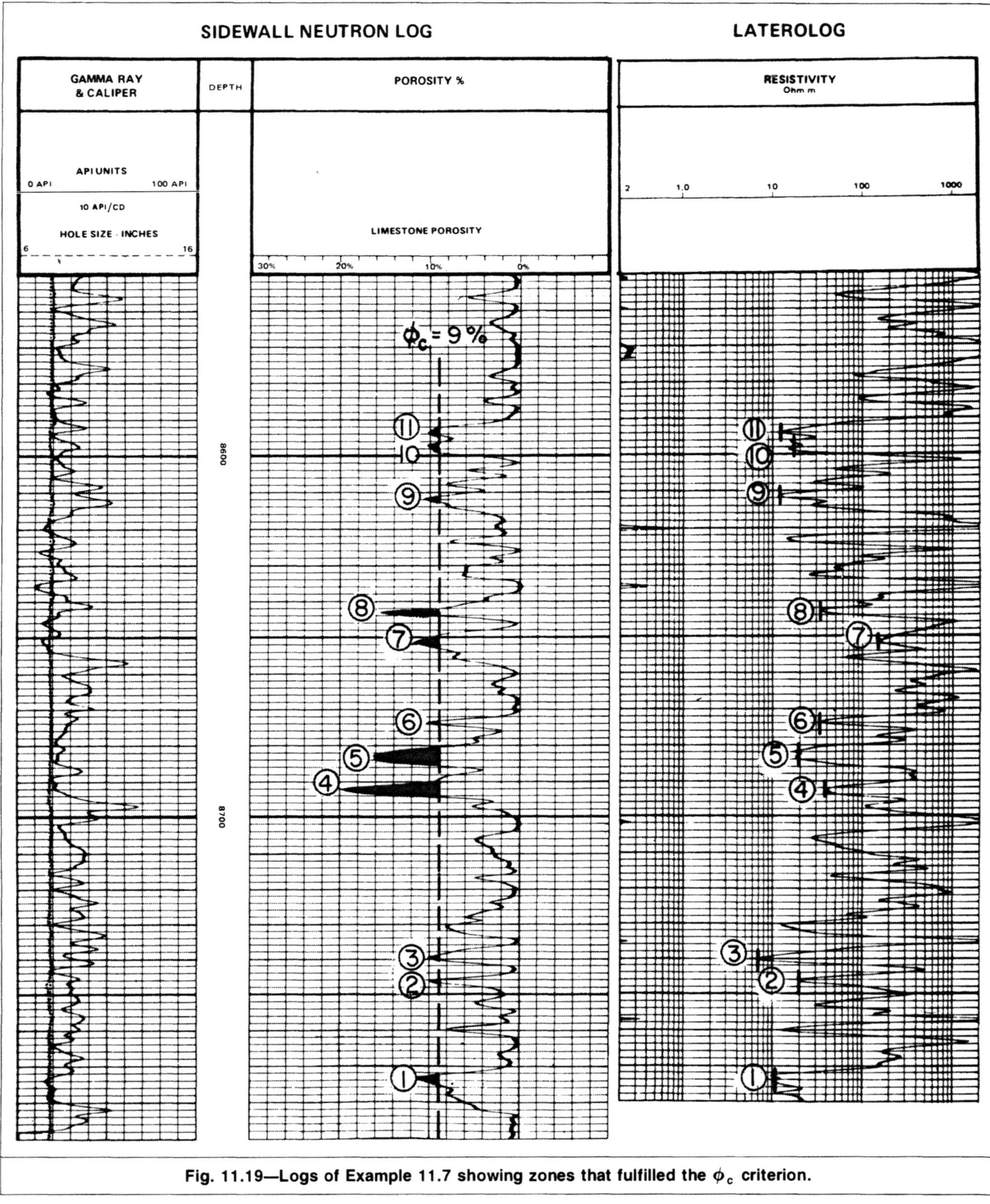

Fig. 11.19—Logs of Example 11.7 showing zones that fulfilled the ϕ_c criterion.

the ability of the rock to transmit hydrocarbons decreases rapidly, and the ability of the rock to transmit water increases rapidly. For practical purposes, this means that Eq. 11.5 should be qualified by the additional relation[8]

$$S_w < S_{cw}. \quad \text{(11.9)}$$

This means that only the zones that fulfill this inequality are included in the calculation of the recoverable oil potential. In practice, we find that S_{cw} is usually about 50% (normally within 10% saturation of 50%), but S_{cw} also has been as low as 25% and as high as 75%.

Correlations between permeability, porosity, and critical water saturation are available for different formations or can be developed from special core analysis data. **Fig. 11.15** shows a generalized correlation for gulf coast Tertiary sands.[5] By referring to such a correlation, we can estimate a maximum water saturation above which water will be produced for given permeability and porosity values. In the reservoirs used to develop the correlation in Fig. 11.15, the oil viscosity was, on average, approximately twice that of the formation water. If the reservoir oil/water viscosity ratio is considerably different from two, the S_{cw} value should be adjusted. The value should be reduced by about 5% pore space for low-gravity oil (18 to 20°API). Conversely, the value should be increased by about 5% pore space for high-gravity oil (>40°API) or 7% pore space for gas reservoirs.[9] Note that these correlations are intended to provide an immediate estimate of S_{cw} using rou-

TABLE 11.2—CALCULATION OF RECOVERABLE OIL FOR EXAMPLE 11.7

Zone	ϕ (%)	R_t (Ω·m)	$S_w=(R_w/\phi^2 R_t)^{1/2}$ (%)	$S_w<S_{cw}$	h (ft)	$h\phi(1-S_w)$ (ft)
1	11.0	11	58	no	—	—
2	12.0	19	41	no	—	—
3	12.0	7	67	no	—	—
4	19.5	40	17	yes	4	0.6474
5	16.0	19	30	yes	6	0.6720
6	10.0	35	36	yes	3	0.1920
7	12.0	150	14	yes	3	0.3096
8	15.5	35	23	yes	2	0.2387
9	10.5	12	58	no	—	—
10	10.0	17	51	no	—	—
11	10.0	13	60	no	—	—
					$\Sigma h\phi S_w=$	2.0597

tine permeability and porosity values. Greater accuracy and more reliable data may be obtained from laboratory measurements where a conventional core and sufficient time are available.

S_{cw} can be compared directly with the formation water saturation. It can also be substituted for S_w in Archie's saturation equation, Eq. 1.43, to calculate a resistivity value, R_{mp}, defined as[10]

$$R_{mp}=FR_w/(S_{cw})^n, \quad (11.10)$$

where R_{mp} is the minimum productive resistivity. If $R_t<R_{mp}$, water production is indicated.

Example 11.5. Determine the water saturation cutoff value, S_{cw}, corresponding to 50% reservoir water cut from an oil formation whose relative permeability characteristics are illustrated in Figs. 11.13 and 11.14. Assume that the oil viscosity is twice the water viscosity at formation temperature and pressure.

Also calculate the minimum productive resistivity, R_{mp}, if the formation water resistivity and the average formation resistivity factor are 0.08 Ω·m and 20, respectively.

Solution. The water cut is defined by Eq. 11.8. Solving for k_{rw}/k_{ro} results in

$$k_{rw}/k_{ro}=1/(\mu_o/\mu_w)(1/f_w-1).$$

For $f_w=0.5$ and $\mu_o/\mu_w=2$,

$$k_{rw}/k_{ro}=1/2(1/0.5-1)=0.5.$$

From Fig. 11.14, the water saturation that corresponds to this relative permeability ratio is 46%, which is the cutoff saturation.

R_{mp} is calculated by Eq. 11.10 using a saturation exponent of two:

$$R_{mp}=20(0.08)/(0.46)^2=7.6\ \Omega\cdot\text{m}.$$

This means that, for a hydrocarbon-bearing zone located within the subject formation to produce at a water cut lower than 50%, the zone should display a true resistivity in excess of 7.6 Ω·m, which translates into a water saturation of less than 46%.

Example 11.6. Fig. 11.16 shows a log section of a 10-ft Frio sand. The sand has much lower resistivity in the bottom and appears to have a water level at 9,485 ft. Core analysis summarized in **Table 11.1** indicates that the entire zone is oil-bearing. The sand was perforated from 9,480 to 9,488 ft and produced at a rate of 121 BOPD with no water production for more than 3 years. The formation water resistivity and oil gravity are 0.033 Ω·m and 37°API, respectively.

Examine the data and explain why the lower part of the sand displayed lower resistivity than the upper part, and why water was not produced from either part.

Solution. Answering these questions will probably change nothing for the subject case. These answers will, however, provide more understanding of log response in certain types of formations. Such additional information and insight help the analyst with future interpretations of log responses obtained under similar conditions.

Two zones, Zones A and B, are clearly identified on the log. Log readings and average core properties are listed below:

Zone	$\bar\phi$ (%)	$\bar k$ (md)	E_{SP} (mV)	C_t (m℧/m)	R_t (Ω·m)	S_w* (%)	S_{cw}** (%)
A	33.9	298	−45	560	1.79	37	47
B	32.9	76	−30	1,320	0.76	58	62

*$S_w=(1.13R_w/\phi^{1.73}R_t)^{1/2}$.
**From Fig. 11.15 using average core ϕ and k.

The porosities of Zones A and B are too close to cause the major difference in their resistivities. Zone B is much shalier than Zone A, as indicated by the SP log readings. This shaliness is responsible for the relatively low resistivity reading of the zone.

The water saturation in both zones is lower than the estimated critical saturation, which explains why water was not produced. Note that the use of shaly-sand interpretation techniques from Chap. 15 to calculate the water saturation of Zone B will yield a value lower than the 58% calculated with conventional equations.

11.6 Concept of Critical or Cutoff Porosity

In addition to having a large-enough relative permeability to hydrocarbons, a zone of economic interest should have sufficient absolute permeability to ensure an economical flow rate. Absolute permeability is usually inferred from porosity with empirical relations. These relations, however, have a large degree of uncertainty and vary significantly with rock type. **Fig. 11.17** is a typical k-ϕ correlation.

The k-ϕ correlation leads to the concept of critical or cutoff porosity, ϕ_c. Cutoff porosity is the minimum porosity for a rock type that will promise sufficient absolute permeability. Thus, a second qualification sometimes is added to Eq. 11.5, namely

$$\phi>\phi_c. \quad (11.11)$$

In practice, the two qualifications expressed by Eqs. 11.9 and 11.11 mean that the summation in Eq. 11.5 will be carried out for all zones within the gross interval of interest where $S_w<S_{cw}$ and $\phi>\phi_c$.

Example 11.7. Fig. 11.18 shows the Laterolog and neutron porosity logs obtained in a thick limestone formation known to have oil-bearing zones.

Estimate the recoverable oil using the following data derived from experience with the same formation type.

R_w = 0.045 Ω·m.
ϕ_c = 9%.
S_{cw} = 40%.
A = 40 acres.
F_R = 25%.
B = 1.2 RB/STB.

Solution. The recoverable oil estimated by Eq. 11.5 is qualified by the relations $\phi>\phi_c>9\%$ and $S_w<S_{cw}<40\%$. Zones that ful-

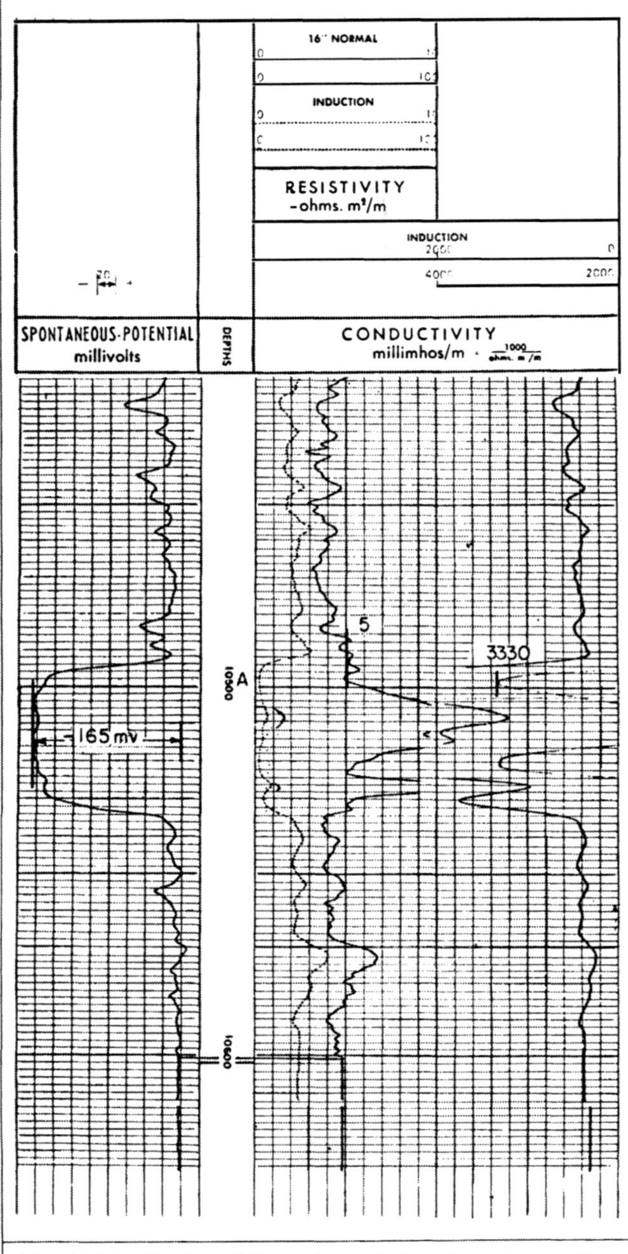

Fig. 11.20—IES log of Example 11.8 (courtesy Amoco Production Co.).

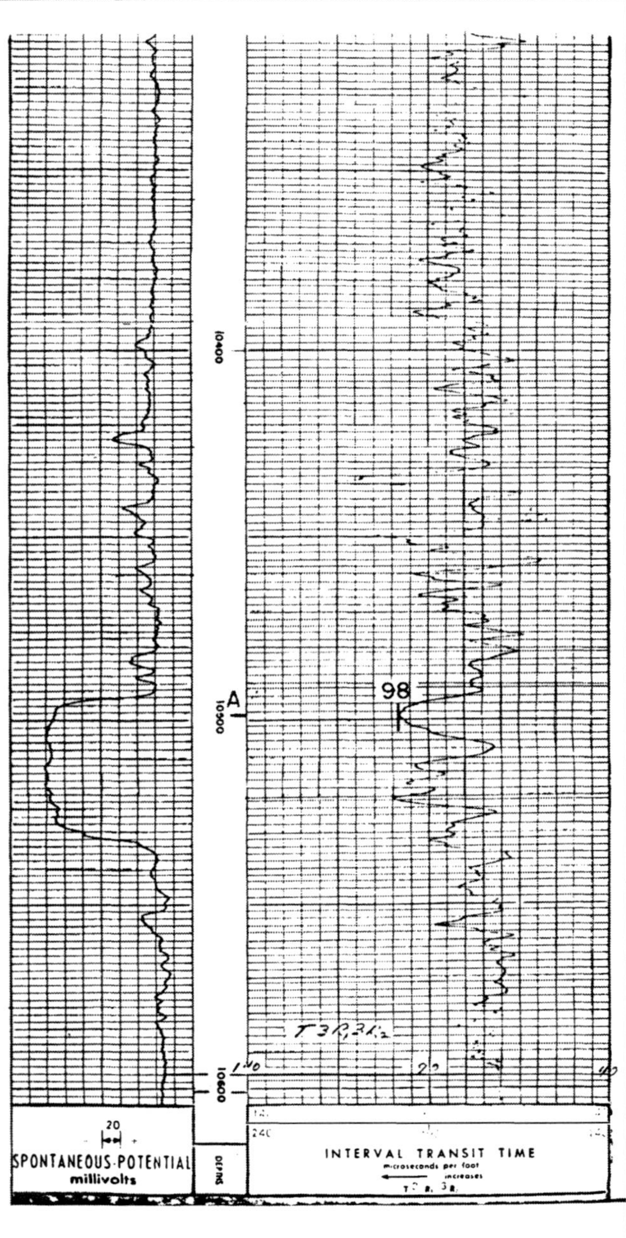

Fig. 11.21—Sonic log of Example 11.8 (courtesy Amoco Production Co.).

fill the ϕ_c criterion can be identified directly on the log, as shown in **Fig. 11.19**. **Table 11.2** lists porosity and resistivity values of these zones. Also listed are the water saturation values and the thicknesses of the zones that fulfilled the S_{cw} criterion. The product $\phi h S_o$ is computed for each zone. It adds up to 2.0597 ft. From Eq. 11.5,

$$N_R=7{,}758[40(0.25)/1.2]2.0597=133{,}000 \text{ STB}.$$

11.7 Concept of Movable Oil Saturation

The movable oil saturation, S_{mo}, is the difference between the initial oil saturation, S_{oi}, and the residual oil saturation, S_{or}, that remains after the formation is swept by water:

$$S_{mo}=S_{oi}-S_{or}. \quad (11.12)$$

S_{mo} provides a qualitative means of evaluating the producibility of a formation. A recovery factor, F_R, for an efficient waterdrive mechanism can be estimated as

$$F_R=S_{mo}/S_{oi}. \quad (11.13)$$

F_R for a depletion-drive mechanism is empirically assumed to be half that of a waterdrive's.

S_{mo} can be determined from special displacement tests. It can also be inferred from borehole measurements if it is acceptable to use mud filtrate flushing the formation that surrounds the borehole as a representation of the waterflooding mechanism. Because $S_{oi}=1-S_w$ and in the flushed zone $S_{or}=1-S_{xo}$, Eq. 11.12 can be rewritten as

$$S_{mo}=S_{xo}-S_w. \quad (11.14)$$

S_{xo} is calculated from Eq. 4.7, which assumes perfect flushing—i.e., all original formation water is removed from the flushed zone. Use of this equation requires reliable knowledge of R_{xo} and R_{mf}, which, in many cases, are difficult to obtain with sufficient accuracy.

Example 11.8. Evaluate the top zone (Zone A) of the sand shown by the IES and sonic logs in **Figs. 11.20 and 11.21** in terms of the presence of movable hydrocarbons. The sand is a Lower Tus-

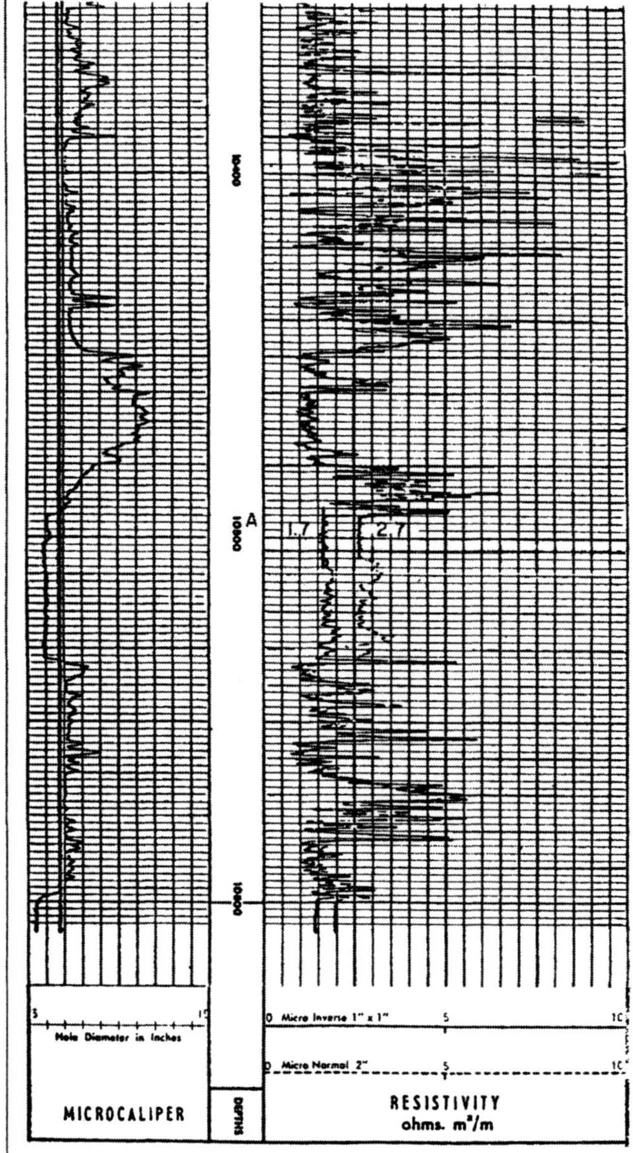

Fig. 11.22—Microlog of Example 11.9 (courtesy Amoco Production Co.).

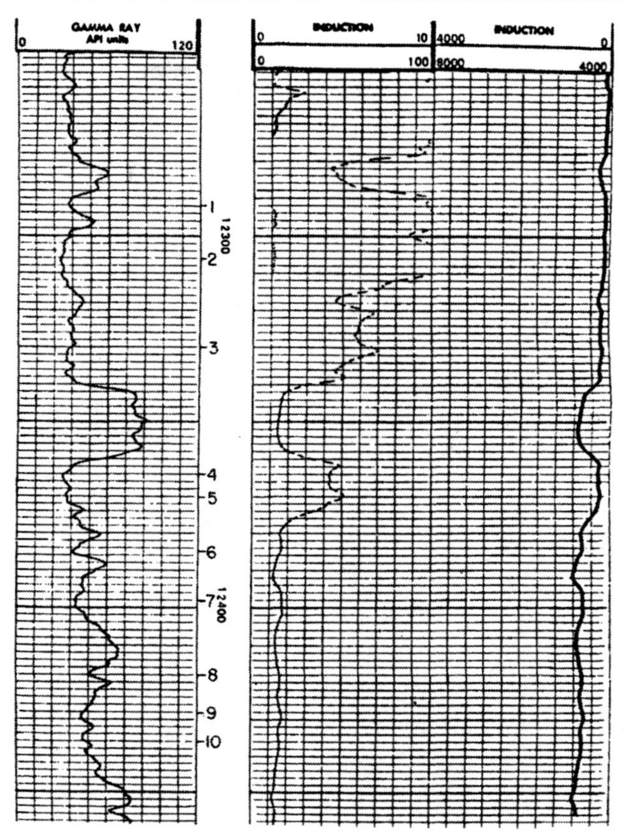

Fig. 11.23—Induction/gamma ray log of Problem 11.1 (courtesy Amoco Production Co.).

caloosa (state of Mississippi) formation that produced oil from wells drilled in other favorable structural positions.

The log heading lists the following data.

Depth reached = 10,602 ft.

Maximum recorded temperature = 216°F.

$R_m = 1.89\Omega \cdot m$ at 81°F.

$R_{mf} = 1.40\Omega \cdot m$ at 80°F.

$R_{mc} = 0.75\Omega \cdot m$ at 216°F.

Bit size = 6¾ in.

Solution. Zone A displays a relatively low resistivity:

$R_t \cong 1{,}000/3{,}330 = 0.3\ \Omega \cdot m$.

However, the SP reading of −165 mV and a Δt reading of 98 μsec/ft indicate an abnormally low water resistivity and high porosity. The possibility of oil presence, then, cannot be dismissed. Detailed quantitative interpretation has to be executed.

R_w Determination. The sand is less than 100 ft from the bottom of the hole. It is reasonable to consider that T_f = maximum recorded temperature = 216°F. The sand is thick and apparently clean, so $E_{SSP} = E_{SP} = -165$ mV.

Using the log-heading R_{mf} value in Eq. 1.11 results in

$R_{mf} = 1.4(80+6.77)/(216+6.77) = 0.55\ \Omega \cdot m$.

From Fig. 6.14,

$R_w = 0.017$ at 216°F.

According to Fig. 1.7, this value corresponds to a relatively high salinity of 180,000 ppm (NaCl equivalent).

Porosity Determination. The shale travel time displayed by the sonic log is less than 100 μsec/ft, indicating a normally compacted formation. According to the concepts discussed in Chap. 10, Eq. 10.1 can be used. According to the same concepts, the following values can be used: $\Delta t_{ma} = 55.5$ μsec/ft and $\Delta t_f = 189$ μsec/ft. Then,

$\phi = (98 - 55.5)/(189 - 55.5) = 0.32$ or 32%.

Estimation of Oil Saturation. Using Eqs. 1.19, 11.2, and 11.3 yields

$F = 1.13(0.32)^{-1.73} = 8.1$,

$R_o = 8.1(0.017) = 0.14$,

$S_w = (0.14/0.3)^{1/2} = 0.68$ or 68%,

and $S_o = 1 - 0.68 = 0.32$ or 32%.

Estimation of S_{mo}. A 68% water saturation is relatively high; however, critical water saturation values as high as 75% might exist under optimum rock properties (see Fig. 11.15). An estimation of S_{mo} might help the evaluation. The S_{mo} technique requires an R_{xo} value. A reliable R_{xo} value cannot be determined from available measurements. The short normal, however, can provide a lower limit for R_{xo} because it is expected to be affected, to some unknown degree, by the uninvaded-zone resistivity; however, because R_t is small, the effect is also small. Then,

$R_{xo} \geq R_{SN} \geq 5\Omega \cdot m$.

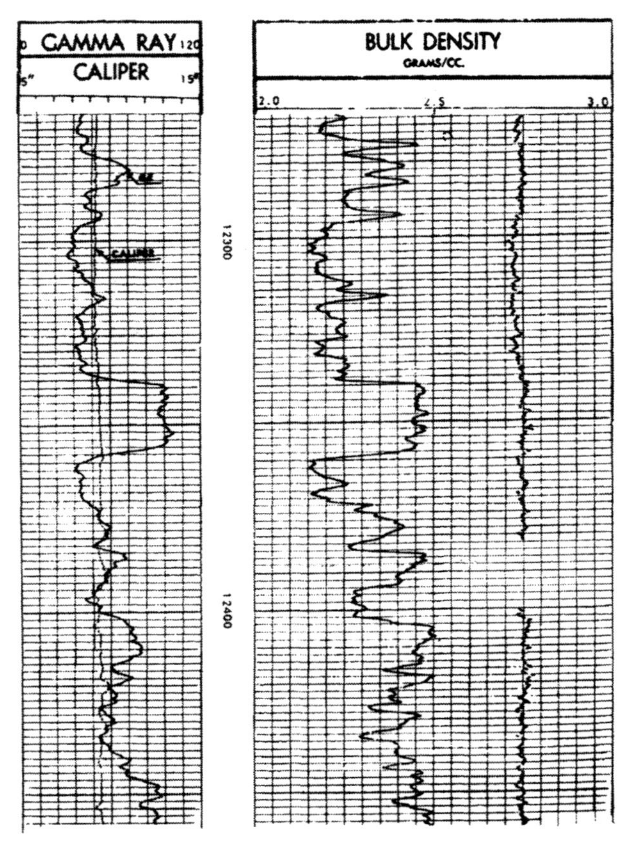

Fig. 11.24—Gamma ray/FDC log of Problem 11.1 (courtesy Amoco Production Co.).

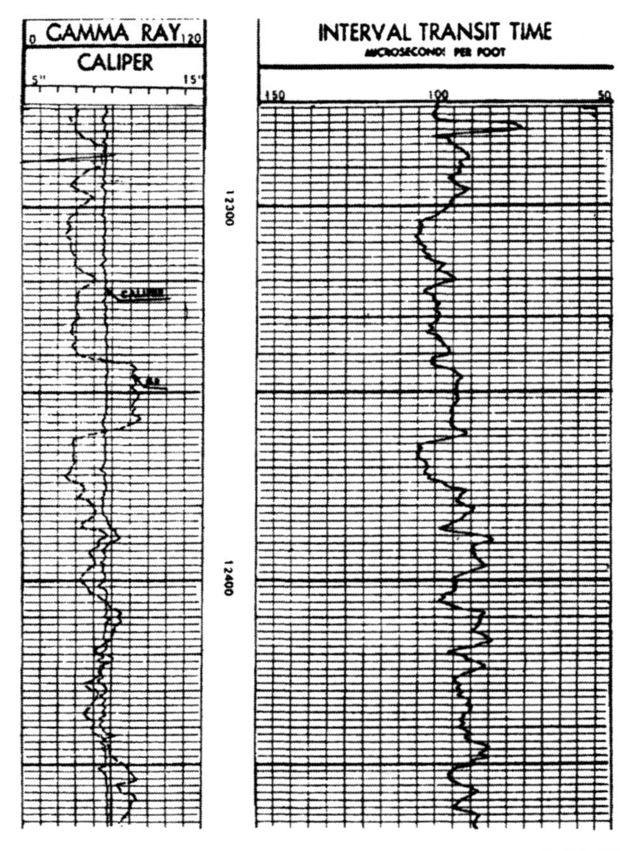

Fig. 11.25—Gamma ray/BHC sonic log of Problem 11.1 (courtesy Amoco Production Co.).

Using Eq. 4.7 gives

$$S_{xo}=(FR_{mf}/R_{xo})^{1/2} \leq (FR_{mf}/R_{SN})^{1/2}$$

$$\leq [8.1(0.55)/5]^{1/2}$$

$$\leq 0.94$$

and $S_{mo}=S_{xo}-S_w$

$$\leq 0.94-0.68$$

$$\leq 0.26 \text{ or } 26\%.$$

Although the interpretation failed to give a specific value for S_{mo}, it definitely indicated movable oil could be present. Only an actual production test could give a conclusive answer. This well was perforated from 10,494 to 512 ft, and initial production was 110 BOPD with no water.

Example 11.9. Fig. 11.22 shows the Microlog obtained in the same well and displays the sand that was the subject of Example 11.8. Review the interpretation given for Example 11.8 in light of the new data.

Solution. The Microlog should yield a fairly reliable R_{xo} value. Another attempt at S_{mo} evaluation is in order.

$$R_{1\times1}=1.7\Omega\cdot\text{m}.$$

$$R_2=2.7\Omega\cdot\text{m}.$$

$$R_{mc}=0.75\Omega\cdot\text{m}.$$

$$R_{1\times1}/R_{mc}=1.7/0.75=2.27.$$

$$R_2/R_{mc}=2.7/0.75=3.60.$$

Entering Fig. 5.70 with these values gives $R_{xo}/R_{mc}=16$ and $h_{mc}=\frac{1}{2}$ in. h_{mc} obtained from the chart is on the same order of magnitude as that indicated by the Microcaliper. The R_{xo}/R_{mc} value is slightly above the limitation of the calculation technique, which is 15 (refer to Chap. 5). With this in mind, the attempt at S_{mo} estimation is continued.

$$R_{xo}\cong 16(0.75)=12.$$

$$S_{xo}\cong[8.1(0.55)/12]^{1/2}=0.61 \text{ or } 61\%.$$

$$S_{mo}=S_{xo}-S_w$$

$$=61-68=-7\%.$$

This answer is physically impossible. Under other circumstances, zero movable oil saturation might be concluded. In this case, however, the answer is in gross error because oil has been produced from the subject sand. Several possible explanations for this error can be cited.

1. The quality of the log is poor (e.g., calibration error).
2. The conditions used to construct Fig. 5.70 are not met here.
3. The R_{mc} value is incorrect.
4. m and n values are not representative.

This is a good example of possible misinterpretation. It is rather ironic that, without the Microlog, the interpretation indicated possible movable oil, and with the Microlog, a conclusion of no movable oil could have been reached. This case is the exception because more data usually add to the flexibility and reliability of the interpretation. Log interpretation is a serious attempt to find answers for an extremely difficult and complex problem. Always remember that these answers could be inaccurate or completely wrong. The ultimate proof of the presence of recoverable hydrocarbons is the production test.

11.8 Advantages and Limitations of Conventional Interpretation Techniques

Conventional techniques have the advantage of being well-established and therefore can be discussed among practitioners relatively easily. These techniques account for several variables by direct and well-known relations and can be performed for a single zone

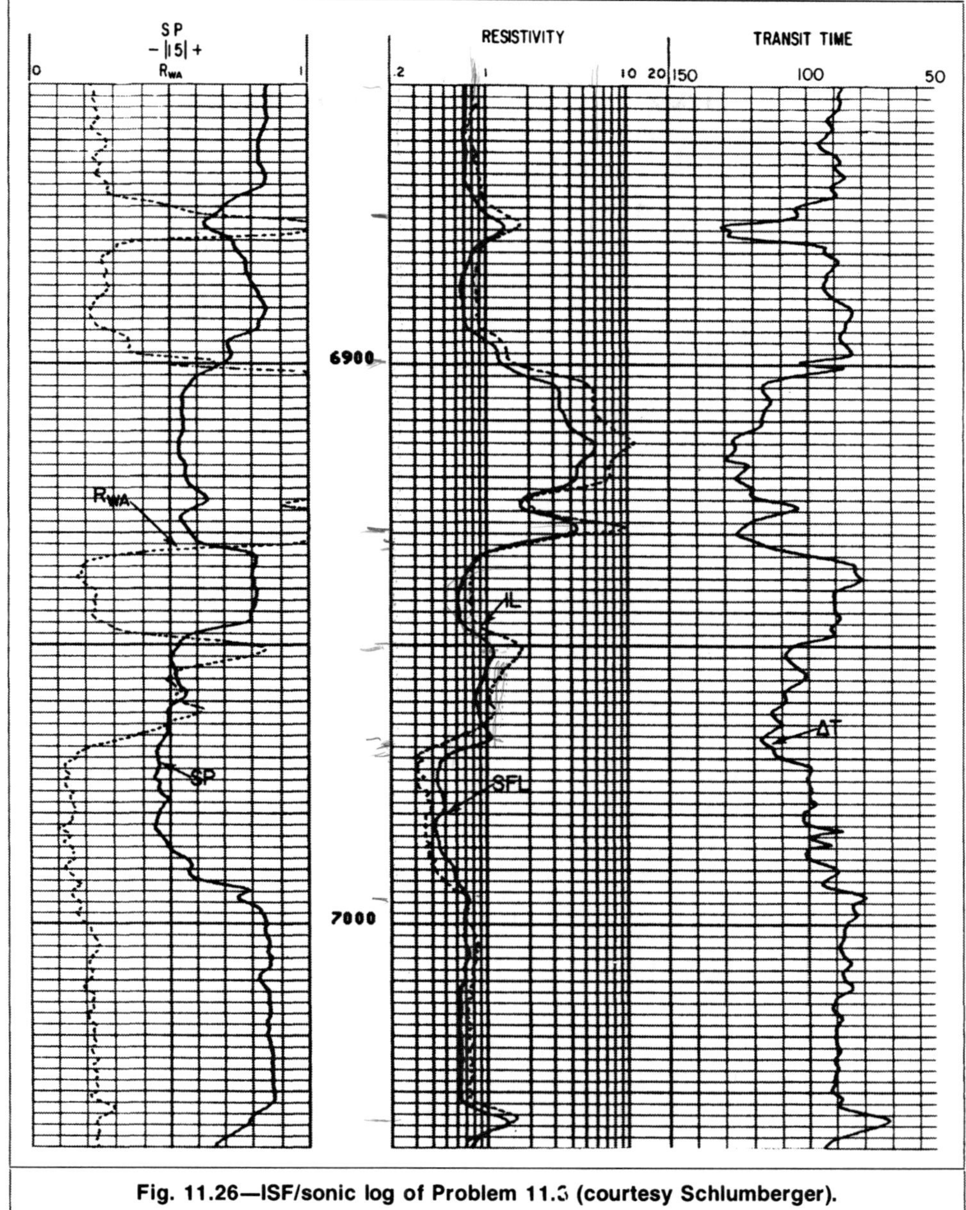

Fig. 11.26—ISF/sonic log of Problem 11.3 (courtesy Schlumberger).

alone. They have the basic disadvantage of requiring a large number of parameters: R_t, ϕ, R_w, a, m, and n. The calculated water saturation value is susceptible to errors in these parameters. Sensitivity analysis of Archie's saturation equation can be conducted easily for any unreliable parameter. One analysis method is to vary an unreliable parameter throughout a specific range and observe the corresponding change in the calculated S_w value.

Sensitivity of Archie's equation to errors in interpretation parameters increases as water saturation increases. For instance, the same percentage error in porosity will cause more change in apparent water saturation at a higher S_w value than at a lower one. The exception to this is the saturation exponent, which causes maximum error when true S_w is about 50%. Overestimation of n, R_w, and m causes overestimation of S_w. The reverse is true for ϕ and R_t. Overestimation of m causes higher errors at low porosity values. In terms of practical evaluation procedures, the conventional interpretation techniques are more sensitive to the values of m and n used when the formation porosity and saturation are near typical cutoff values. Errors in the evaluation of such marginal formations could lead the analyst to qualify an unfavorable zone as favorable or, worse yet, to qualify a favorable zone as unfavorable.

Example 11.10. Examine the sensitivity of Archie's saturation equation to the uncertainty in the saturation exponent ranging from 1.6 to 2.2. The other parameters are kept constant or varied as follows.

$R_w = 0.028 \Omega \cdot m$.

$\phi = 0.20$.

$m = 2.0$.

$R_t = 70.0$, 21, 7, 2.8, 2.1, 1.4, and 0.7.

Solution. The sensitivity analysis of the equation

$S_w = (R_w / \phi^m R_t)^{1/n}$

is shown in the table below.

R_t	S_w (%)		ΔS_w
($\Omega \cdot m$)	$n=1.6$	$n=2.2$	(%)
70.0	6	12	6
21.0	12	21	9
7.0	24	35	11
2.8	40	53	13
2.1	50	61	11
1.4	65	73	8
0.7	100	100	0

The maximum ΔS_w value occurs at the middle of the S_w scale near typical cutoff saturation values.

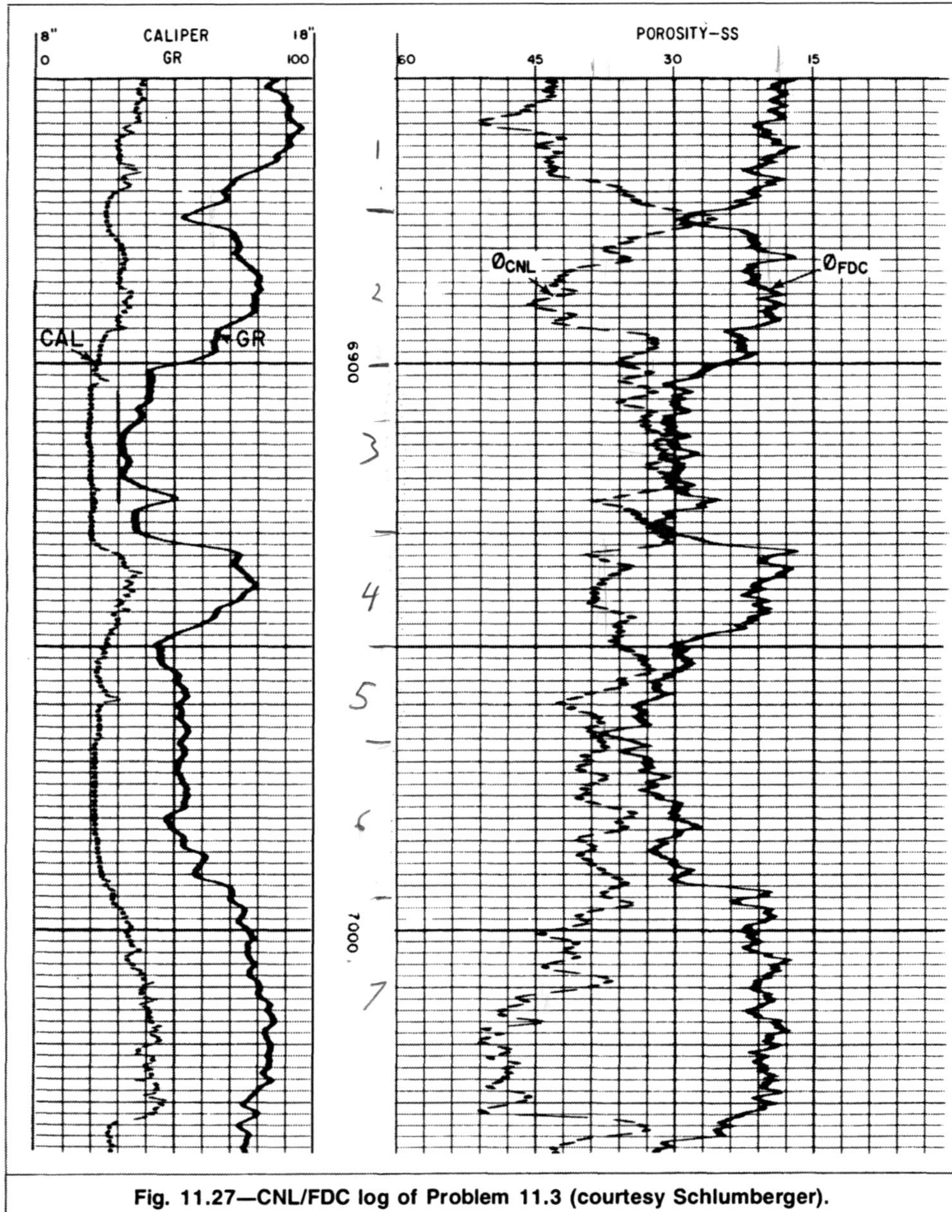

Fig. 11.27—CNL/FDC log of Problem 11.3 (courtesy Schlumberger).

The conventional interpretation technique, as any other technique, is based on both facts and assumptions. Some assumptions have been repeated so often that they are now accepted by many analysts as facts. The major assumptions are listed below in no particular order.

1. All logs are measured and recorded accurately at the well.
2. Formation waters and drilling fluids are essentially NaCl solutions.
3. The electrolyte solution in the flushed zone is all mud filtrate.
4. The mud filtrate that invades the formation at the time of drilling is the same as that sampled at the time of logging.
5. The measurement environment resembles the environment considered during construction of interpretation charts used in determining R_t.
6. Models used to derive equations for calculating porosity from density, neutron, and sonic logs represent the actual formation.
7. The saturation exponent n is 2.
8. Generalized formation-factor/porosity relationships hold true for all formations.

The purpose of listing the assumptions is not to discredit the technique but to make the user aware of the uncertainty associated with the interpretation process. It also shows that new data-gathering and/or interpretation techniques are needed to reduce the number of assumptions.

Review Questions

1. Compare exploration interpretation problems to development interpretation problems.
2. How and why would you qualify interpretation problems pertaining to recompletion of wells?
3. What are the main questions that the log analyst must answer when using logs to detect and evaluate hydrocarbon formations?
4. Is the use of well logs limited to hydrocarbon detection only? Explain.
5. Why is correlation between logs required?
6. Why do shale and low-porosity streaks qualify as correlation markers?
7. What are the precautions one should exercise while selecting zones for quantitative evaluation?
8. What are the basic concepts of conventional interpretation techniques?
9. Is it an acceptable practice to take the resistivity of an adjacent water-bearing zone to be equal to R_o of the zone analyzed?
10. Show that basing interpretation on resistivity log data alone can, in certain cases, lead to gross misinterpretation.
11. Explain why different analysts might arrive at different results for the same interpretation problem.
12. Explain why a novice log analyst always has to support his or her interpretation with quantitative analysis.

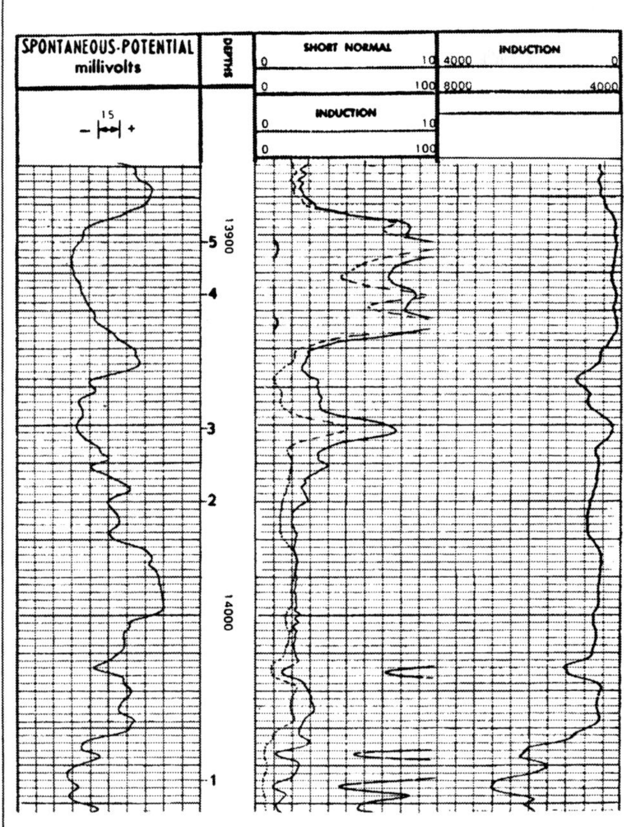

Fig. 11.28—IES of Problem 11.4 (courtesy Amoco Production Co.).

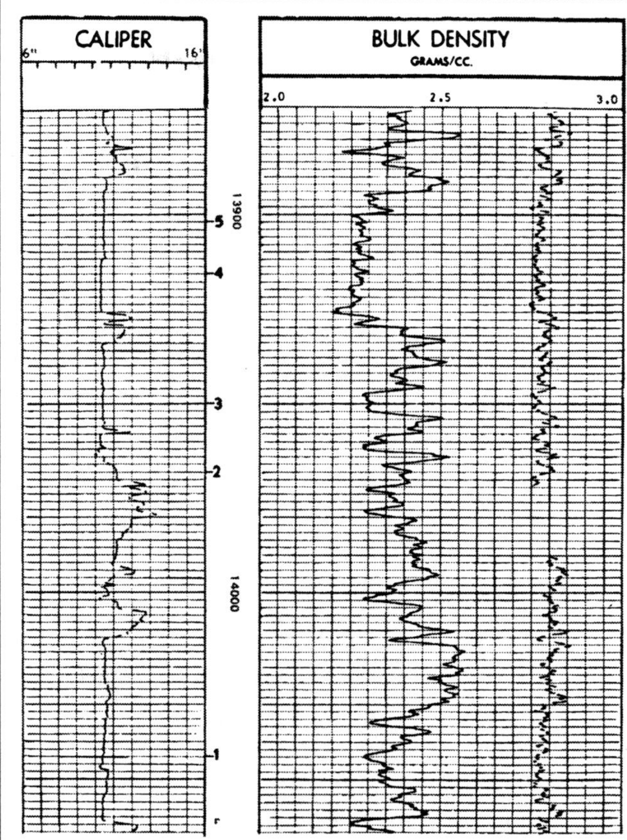

Fig. 11.29—FDC log of Problem 11.4 (courtesy Amoco Production Co.).

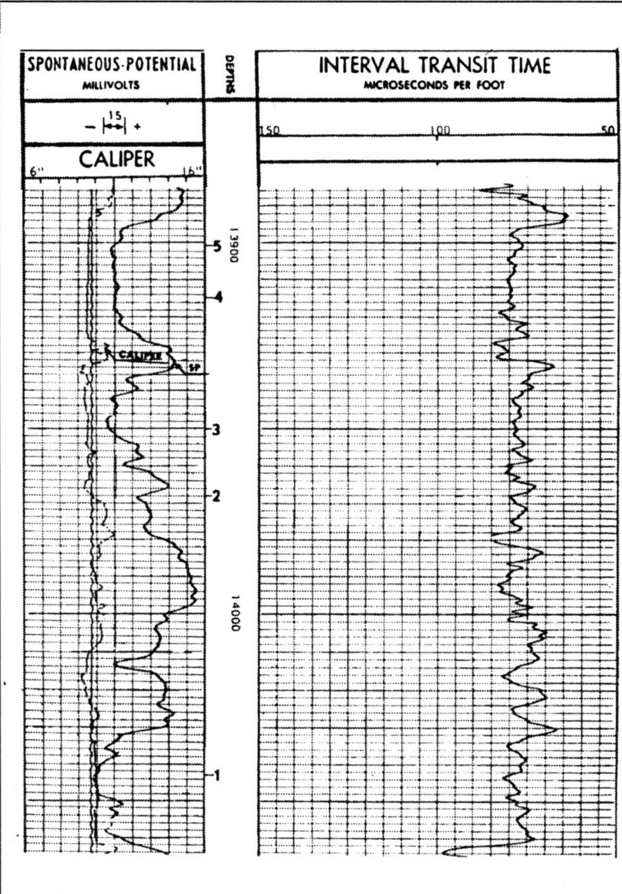

Fig. 11.30—Sonic log of Problem 11.4 (courtesy Amoco Production Co.).

13. What data are needed to estimate recoverable hydrocarbons? What portion of this information can be derived from well logs?

14. What are the practical implications of each of the following concepts: cutoff saturation, cutoff porosity, and movable oil saturation?

15. Discuss the advantages and limitations of conventional interpretation techniques.

Problems

11.1 A well is drilled with an oil-based mud. The set of logs obtained in this well consists of gamma ray/induction, gamma ray/borehole compensated (BHC) sonic, and gamma ray/density logs. **Figs. 11.23 through 11.25** show the sections of these three logs at the interval 12,274 to 12,458 ft. This interval displays thick sands known to contain oil and water. The formation-water resistivity is 0.025 Ω·m at formation temperature.

a. Are these three logs on-depth? If not, correlate between the logs using the resistivity log as the prime reference.

b. Ten relatively clean zones are marked on the resistivity log. Mark the 10 zones on the other two logs and tabulate their R_t, ρ, and Δt values.

c. Is the sonic log affected by lack of compaction?

d. Using density porosity values, estimate water saturation values of the different zones selected above.

e. Is there oil/water contact in the sands shown by the logs?

11.2 Refer to the IES and BHC sonic logs of Figs. 11.20 and 11.21.

a. The sand in the interval 10,494 to 10,535 ft appears uniform on the SP curve. However, it is composed of several zones of different porosities, resistivities, etc. Mark and label these zones on the interval transit and conductivity curves.

b. Tabulate the values of R_{ILd}, R_{SN}, and Δt of these zones.

c. Estimate the porosity and water saturation values of the zones you selected.

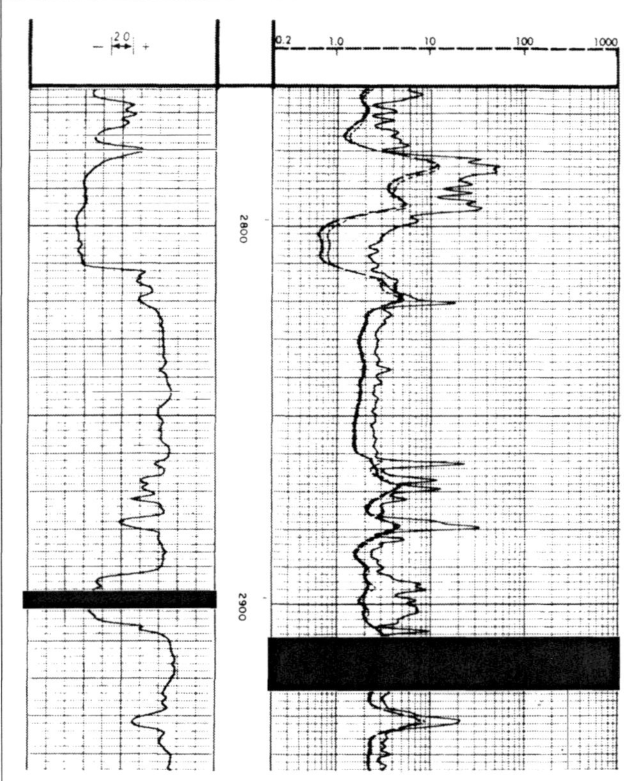

Fig. 11.31—DIL-LL8 log of Problem 11.5 (courtesy Schlumberger).

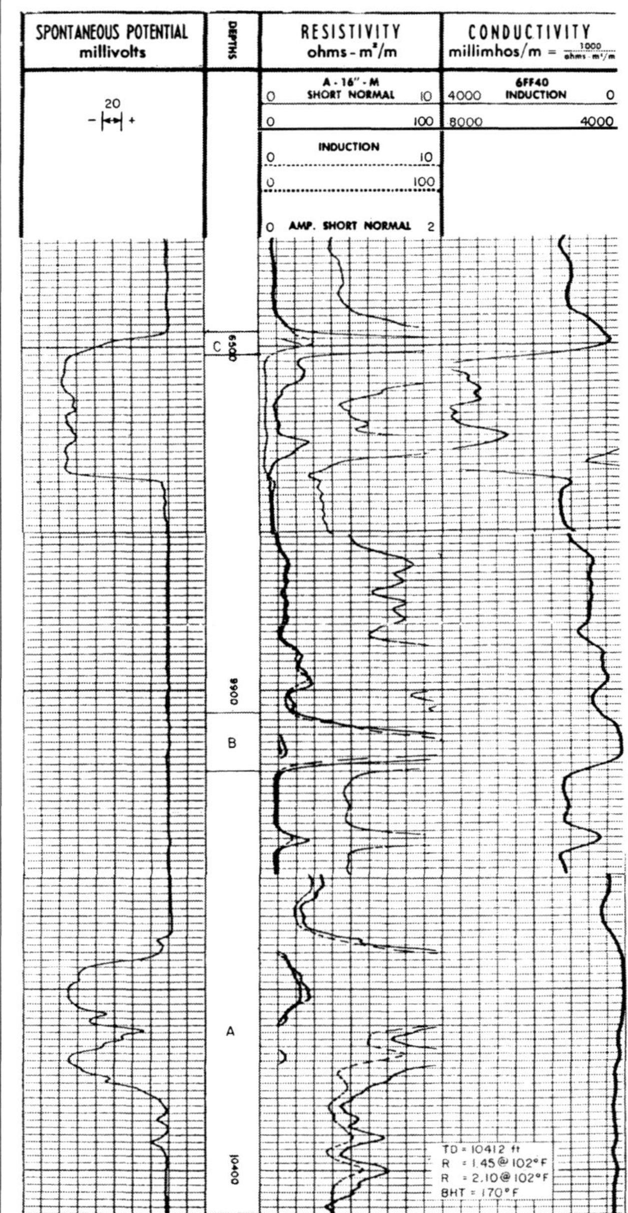

Fig. 11.32—IES log of Problem 11.6 (courtesy Continental Oil Co.).

d. Plot porosity vs. R_{SN} values. Does this plot support the assumption that the depth of mud-filtrate invasion increases as porosity decreases? Explain.

11.3 A development well is drilled to evaluate the oil sand shown by the ISF/sonic and CNL/FDC logs of **Figs. 11.26 and 11.27.** Interpretation experiences with this formation type indicate that $R_w=0.05$ Ω·m at formation temperature; that the formations are unconsolidated and hence high sonic log readings are expected; and that the sands are moderately shaly, so a good porosity value can be obtained by averaging the porosity readings of the density and neutron curves.

a. Is there oil/water contact in the sands shown by the logs?

b. Estimate the value of the summation $\Sigma\phi_i h_i (S_o)_i$ in Eq. 11.5.

11.4 In answering the following questions, refer to the IES, FDC, and sonic logs of **Figs. 11.28 through 11.30,** respectively. The logs were recorded in a 9⅞-in. hole drilled through a sand/shale series in the U.S. gulf coast. The log heading indicates that BHT=211°F at 14,000 ft, $R_m=0.35$, and $R_{mf}=0.265$ Ω·m at BHT.

a. Estimate the water saturation for Zones 1 through 5 using the resistivity log alone.

b. Estimate the water saturation for the same zones, now using all available data.

c. Compare S_w values evaluated in Part a and in Part b. Which of these two sets of values should be recommended for further analysis? Explain.

11.5 The section of the Dual Induction Laterolog 8SM (DIL-LL8) log shown in **Fig. 11.31** is obtained in a shale/sand series. The sands in the interval shown are known to display uniform porosity and water salinity.

a. Identify all hydrocarbon zones in the interval shown.

b. Estimate the water saturation value in each of these zones.

c. List and justify the assumptions used in answering Parts a and b.

11.6 **Fig. 11.32** illustrates an induction electrical log for intervals of interest of a well you are requested to evaluate. The well is operated by Company X, which produced oil from Interval A (10,342 to 10,382 ft). The well recently began producing at an economically unacceptable high WOR, and the well was shut in. Because Company X is reducing its production activity, it offered to sell the well to your company (Company Y) with the premise of completing the well in another zone located higher in the well. Company X points out two zones of possible interest: Zone B (9,908 to 9,920 ft), which displays a resistivity of 12 Ω·m, and Zone C (6,496 to 6,502 ft), which displays a resistivity of 3 Ω·m. Your Company Y will consider buying the well if Zone B or C meets the completion criterion. Completion criterion set forth by Company Y under these conditions is a hydrocarbon saturation exceeding 50%. The purchase price will be based on the net pay.

a. Would you recommend buying the well? Explain.

b. After your evaluation, based solely on the IES, you learn that Company X ran a BHC sonic log. You request and obtain that log (**Fig. 11.33**). How do the newly acquired data (sonic log) affect your previous conclusions?

c. Write a letter to your supervisor, Mr. Z, that clearly states your recommendation on the completion possibility of Zones B and C and on the purchasing of the proposed well.

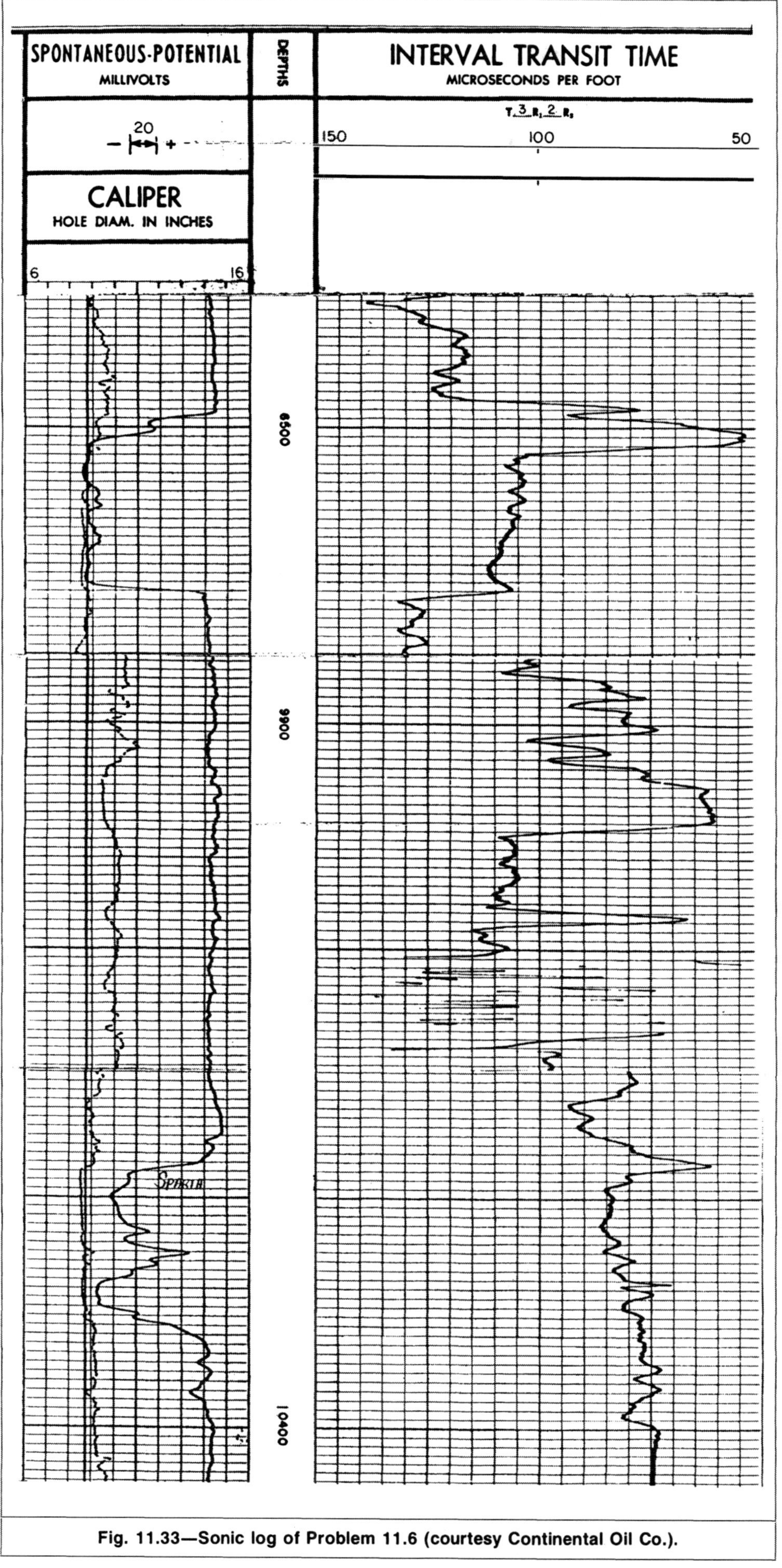

Fig. 11.33—Sonic log of Problem 11.6 (courtesy Continental Oil Co.).

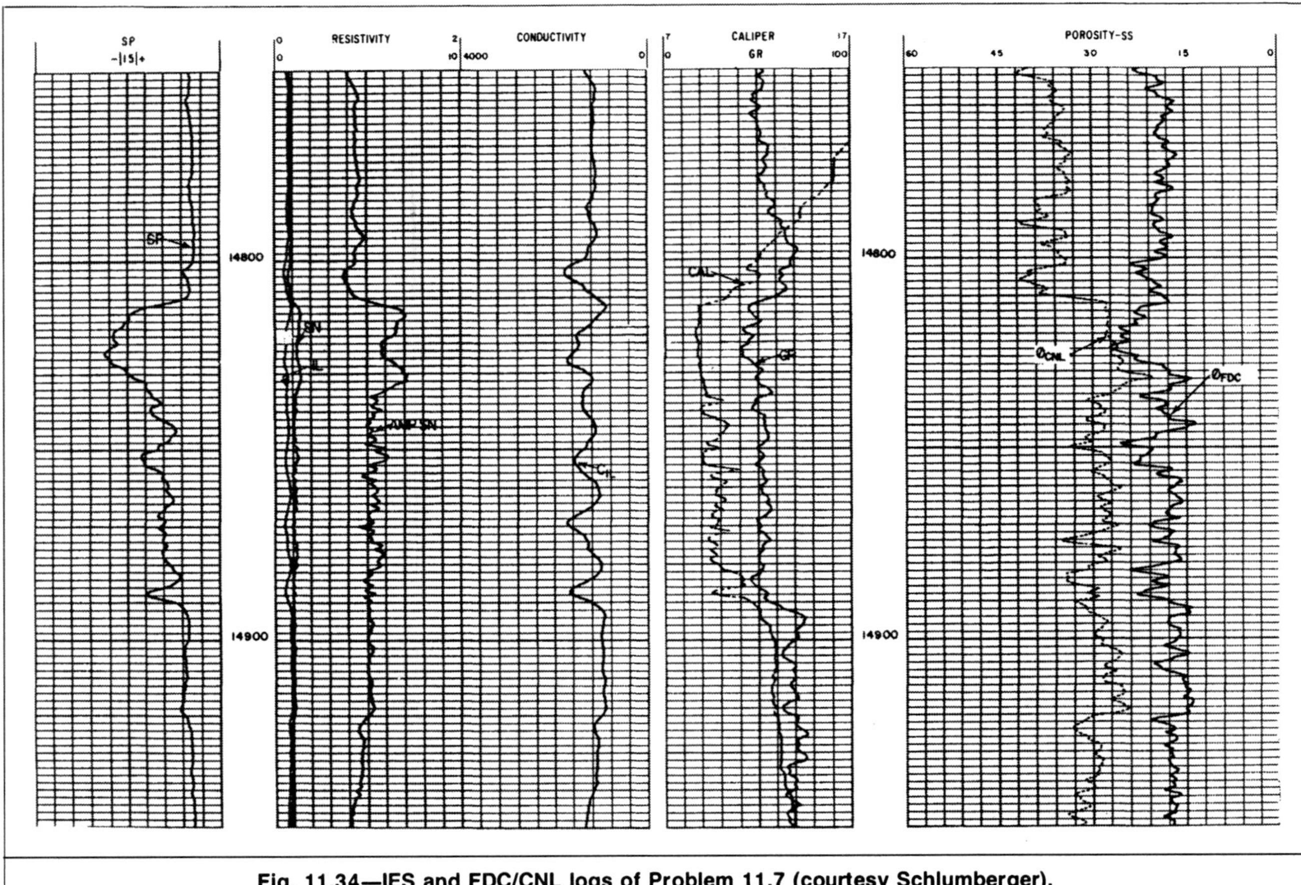

Fig. 11.34—IES and FDC/CNL logs of Problem 11.7 (courtesy Schlumberger).

11.7 The IES and FDC/CNL logs obtained in the interval 14,750 to 14,950 ft of a U.S. gulf coast well are shown in **Fig. 11.34.** The logs indicate a low-resistivity sand that appears to be somewhat clean at the top. Use the conventional interpretation approach to show that the cleanest zone of this sand contains hydrocarbon. The formation water resistivity is known to be equal to 0.02 Ω·m at formation temperature.

11.8 Using the IES, FDC, and sonic log in Figs. **11.35 through 11.37,** respectively, estimate the recoverable oil from the sand situated in the interval 11,520 to 11,597 ft. Experience with the same formation type favors the use of the following data:

$A=40$ acres.

$B_o=1.25$ RB/STB.

$F_R=35\%$.

Only zones with a shale index of less than 50% are considered for recoverable oil calculations. The IES log heading lists the following parameters.

BHT=206°F.

R_m at BHT=0.17Ω·m.

R_{mf} at BHT=0.08Ω·m.

Total depth=12,115 ft.

11.9 **Figs. 11.38 and 11.39** show sections of a gamma/guard log (focused-current resistivity log) and compensated acoustic velocity log obtained in the Krider formation encountered by a well drilled in Stevens County, KS. The Krider formation is a dolomitic limestone characterized by the following parameters.

$\Delta t_{ma}=47.6$ μsec/ft.

$m=2.1$.

$R_w=0.04$Ω·m.

Estimate the porosity and water saturation values of Zones A and B marked on the logs.

11.10 **Figs. 11.40 and 11.41** show the sections of Gamma-Guard-sidewall neutron log and the compensated acoustic velocity log obtained in the Mississippian formation encountered in a well drilled in Edwards County, KS. This formation is a limestone characterized by the following parameters.

$\Delta t_{ma}=46.5$ μsec/ft.

$m=2.1$.

$R_w=0.055$Ω·m.

a. Estimate the porosity and water saturation values of Zones A through F.

b. Calibrate the neutron log to read limestone porosity directly from the log.

11.11 **Figs. 11.42 and 11.43** show the sections of a dual induction guard log and a compensated density log obtained in the Mississippi Chat formation encountered by a well drilled in Logan County, OK. The Mississippi Chat is a carbonate formation characterized by the following parameters.

$T_f=119$°F.

$R_{mf}=0.23$Ω·m at T_f.

$R_w=0.045$Ω·m at T_f.

$\rho_{ma}=2.71$ g/cm³.

$m=1.3$.

Estimate the value of the summation $\Sigma\phi_i h_i (S_o)_i$ in Eq. 11.5.

11.12 The IES log of **Fig. 11.44** shows a Miocene sand that is oil-productive from 13,060 to 13,104 ft. The well was conventionally cored, and 3-ft sliding average permeability and porosity curves are plotted on the SP and conductivity tracks, respectively.

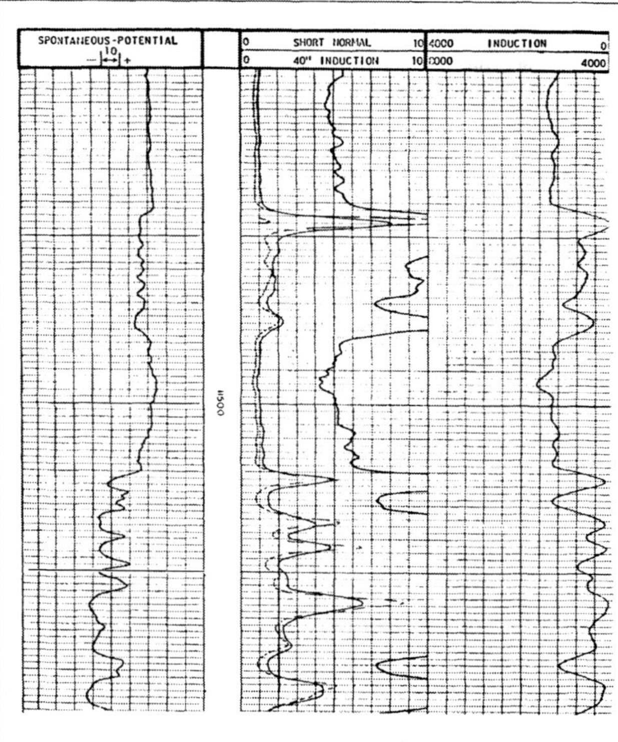

Fig. 11.35—IES for Problem 11.8 (courtesy Schlumberger).

a. Explain the probable reasons behind the loss of permeability in the middle one-third of the sand (Zone B).
b. Estimate the critical water saturation of Zones A through C.
c. Explain why Zone B is not water-productive.

11.13 A Dual Laterolog (**Fig. 11.45**), an FDC log (**Fig. 11.46**), and a proximity log/Microlog (**Fig. 11.47**) were obtained in a well drilled to evaluate a thick Devonian carbonate formation. Pertinent log-reading information is as follows.

Depth (logger)=3,129 ft.

Bit size=7⅞ in.

Mud weight=9.5 lbm/gal.

R_m=2.26 Ω·m at 80°F.

R_{mf}=2.16 Ω·m at 61°F.

R_{mc}=1.81 Ω·m at 73°F.

A special core analysis study was performed on plugs drilled from full-diameter well cores. The results of this analysis are the following: the identification and description of samples (**Table 11.3**), formation factor and resistivity index data (**Table 11.4**), waterflood test results (**Table 11.5**), and water/oil relative permeability ratio curves (**Fig. 11.48**).

A reservoir fluid study performed on an oil sample obtained from another well produced from the same formation indicates B_o=1.323 RB/STB and μ_o=0.9 cp at formation pressure of 1,239 psig and temperature of 100°F. Formation water salinity is estimated to be 180,000 ppm.

a. Determine the cementation exponent, *m*, and saturation exponent, *n*, for this formation.
b. Give your best estimate of the cutoff water saturation that corresponds to a 50% water cut.
c. Use core analyses to estimate the average irreducible water saturation.
d. Estimate the average waterflooding recovery factor from waterflood test results.

Refer to the permeable Zone A marked on the logs to answer the following questions.

e. Estimate the average density porosity. How does it compare with average core porosity? Explain the difference, if any.
f. Estimate the average water saturation. How does it compare with the irreducible value estimated in Part c? Explain the difference, if any.

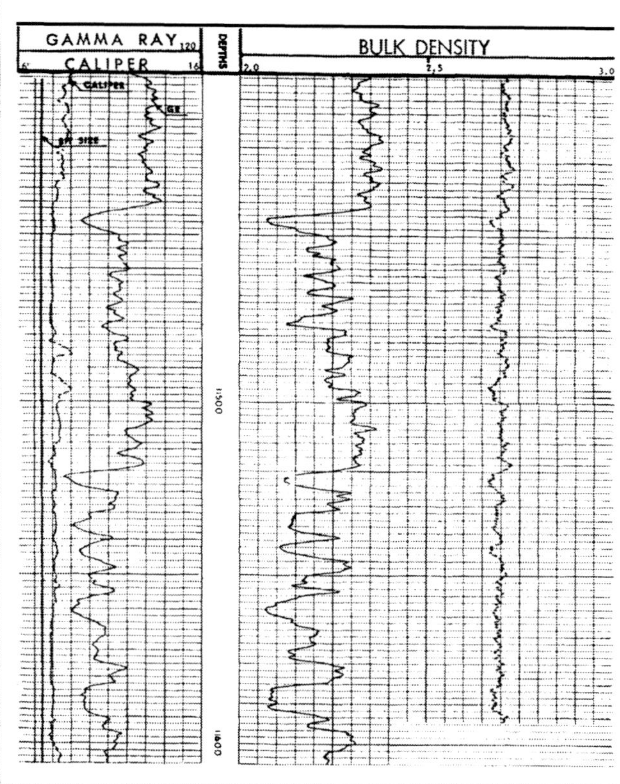

Fig. 11.36—FDC log for Problem 11.8 (courtesy Schlumberger).

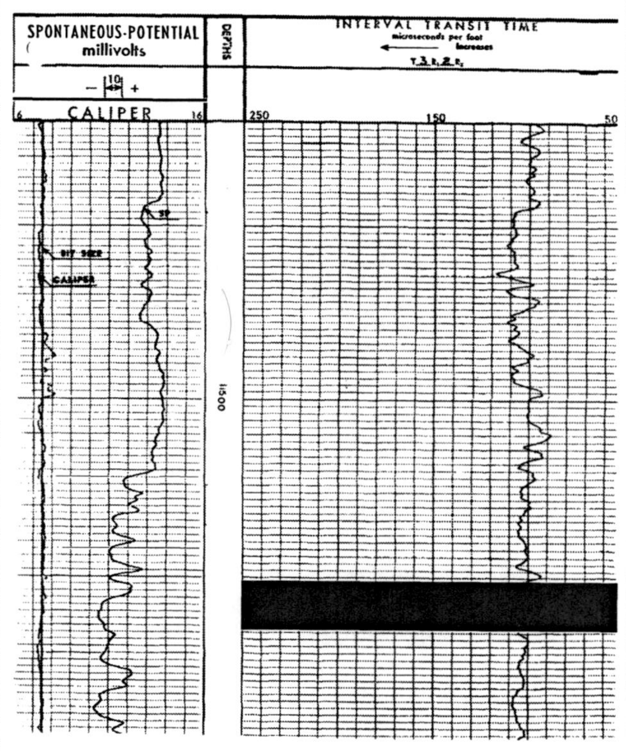

Fig. 11.37—Borehole-compensated sonic log for Problem 11.8 (courtesy Schlumberger).

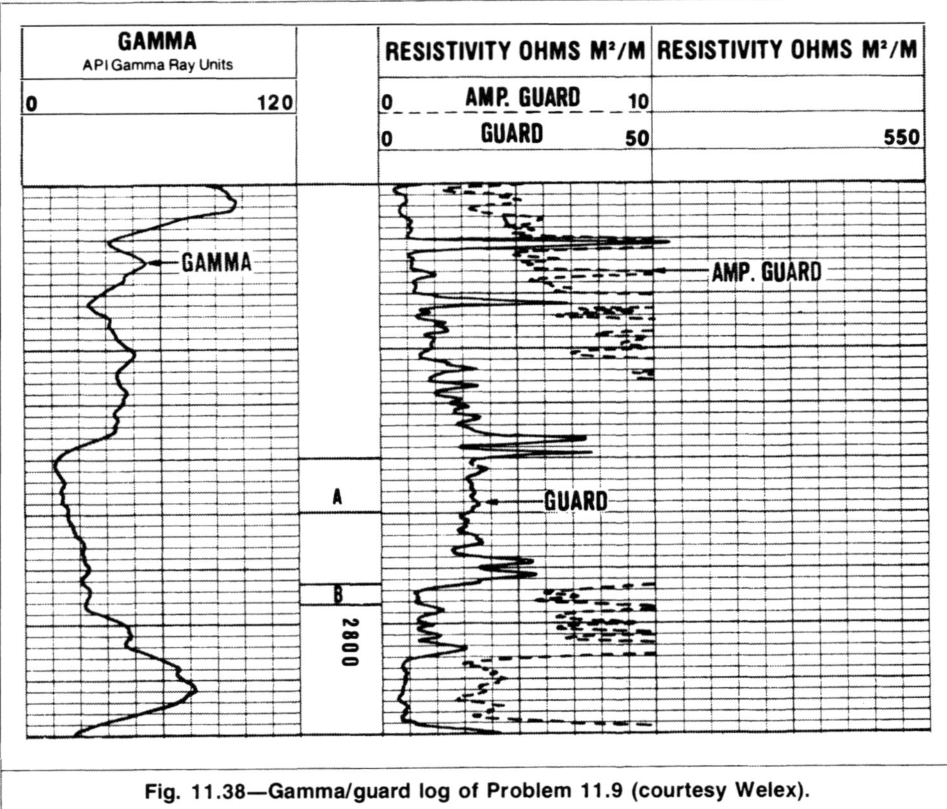

Fig. 11.38—Gamma/guard log of Problem 11.9 (courtesy Welex).

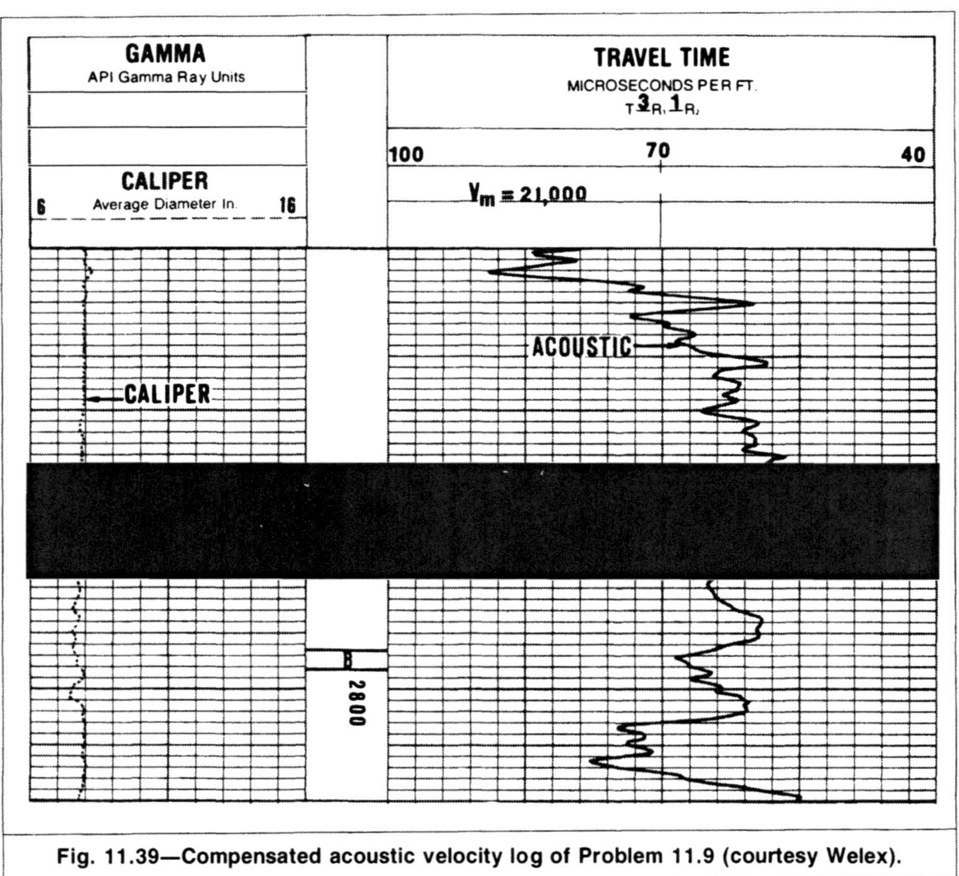

Fig. 11.39—Compensated acoustic velocity log of Problem 11.9 (courtesy Welex).

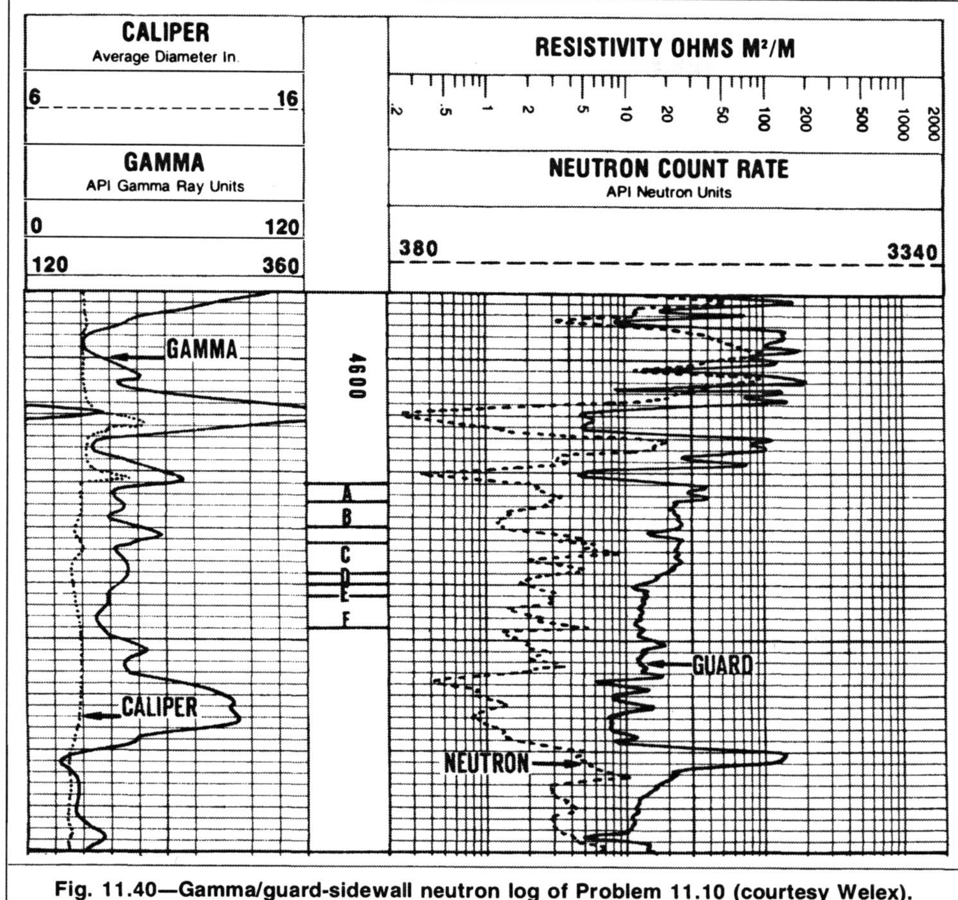

Fig. 11.40—Gamma/guard-sidewall neutron log of Problem 11.10 (courtesy Welex).

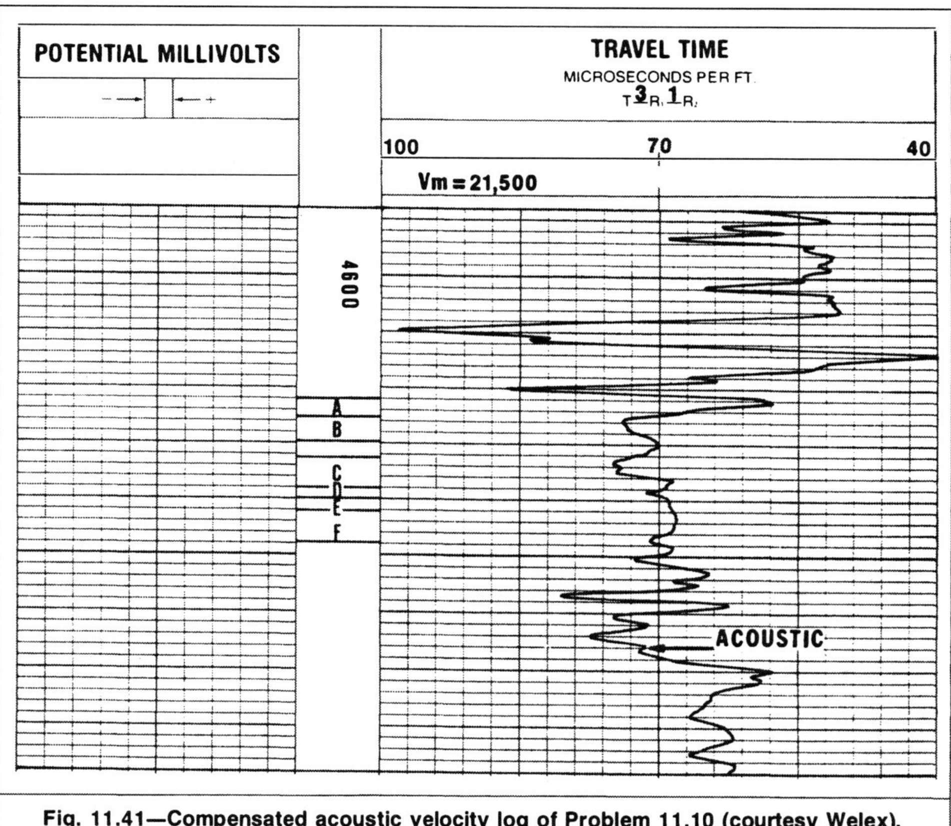

Fig. 11.41—Compensated acoustic velocity log of Problem 11.10 (courtesy Welex).

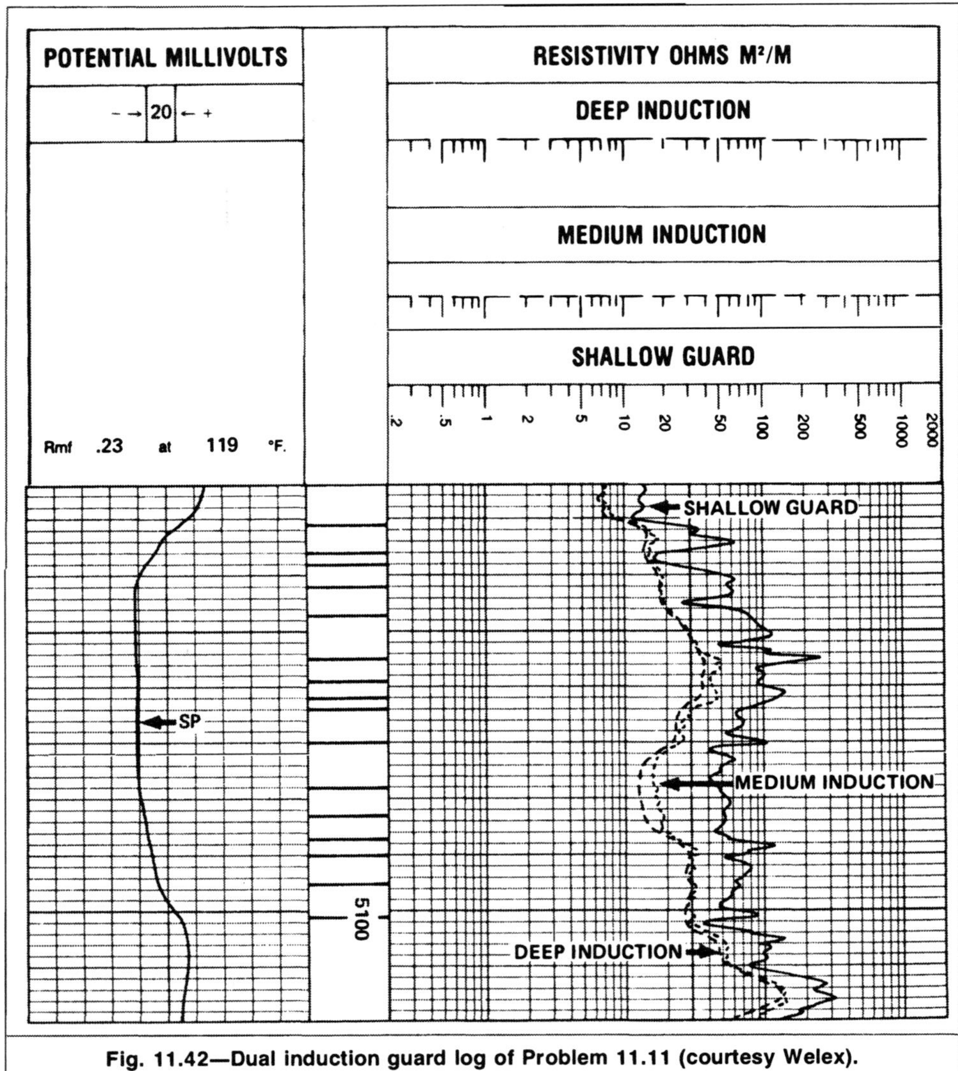

Fig. 11.42—Dual induction guard log of Problem 11.11 (courtesy Welex).

g. Does this zone meet the cutoff saturation criterion? At what water cut will it flow if perforated?

h. Estimate R_{xo} from both the Microlog and proximity log readings. Which value would you recommend for further analyses?

i. Use log data to estimate the waterflooding recovery factor. How does it compare with that obtained in Part d? Explain the difference, if any.

Repeat Parts e through i for any zone of interest you identify on the log. Considering the whole interval, complete the following.

j. On the density log, mark the zones that will meet a porosity cutoff criterion of 6%.

k. Estimate the water saturation of the zones identified in Part j. Which of these zones will have irreducible water saturation?

Nomenclature

a = coefficient in Eq. 11.1
A = drainage area, acres
B = FVF, RB/STB
C = conductivity, m℧/m
D = depth, ft
E_{SP} = measured SP, mV
E_{SSP} = theoretical SP, mV
f_w = fractional flow of water or water cut, fraction
F = formation resistivity factor
F_R = recovery factor, fraction
g_G = geothermal gradient, °F/100 ft
G_R = recoverable gas volume, scf
h = thickness, ft
k = permeability, md

TABLE 11.3—IDENTIFICATION AND DESCRIPTION OF SAMPLES, PROBLEM 11.13

Sample	Depth (ft)	Lithological Description*
1	2,915.8	Ls, lt gry, fn xln, w/sh stks
2	2,915.8	Ls, lt gry, fn xln, w/sh stks
3	2,917.4	Ls, d gry, fn xln, w/sh stks, few vugs
4	2,917.4	Ls, lt gry, fn xln, w/sh stks
5	2,919.4	Ls, lt gry, fn xln, sh stks, secondary calcite
6	2,919.4	Ls, lt gry, fn xln, sh stks, secondary calcite
10	2,927.3	Ls, d gry, fn xln, sh stks

*Key: LS = Limestone; lt = light; d = dark; gry = gray; fn = fine; xln = crystalline; w/sh stks = with shale streaks.

TABLE 11.4—FORMATION FACTOR AND RESISTIVITY INDEX DATA, PROBLEM 11.13

Sample	Porosity (%)	Formation Factor	Brine Saturation (% pore space)	Resistivity Index
1	11.6	59.0	100.0	1.00
			85.4	1.42
			29.5	8.56
			17.3	25.3
4	10.2	70.9	100.0	1.00
			93.2	1.17
			42.6	4.26
			16.7	20.2
6	6.1	153.0	100.0	1.00
			69.7	1.79
			31.8	6.41
10	2.9	603.0	100.0	1.00

k_r = relative permeability
K = temperature-dependent coefficient in Eq. 6.11, mV
m = cementation exponent
n = saturation exponent
N_R = recoverable oil volume, STB
R = resistivity, $\Omega \cdot m$
R_a' = borehole-effect-free apparent resistivity, $\Omega \cdot m$
R_a'' = bed-thickness-effect-free apparent resistivity, $\Omega \cdot m$
R_o = formation resistivity when fully saturated with water, $\Omega \cdot m$
S = saturation, fraction
S_{mo} = movable oil saturation
Δt = travel time, μsec/ft
T = temperature, °F
μ = viscosity, cp

TABLE 11.5—SUMMARY OF WATERFLOOD TEST RESULTS, PROBLEM 11.13

				Initial Conditions		Terminal Conditions		Oil Recovered	
Sample	Depth (ft)	Air Permeability (md)	Porosity (%)	Water Saturation (% pore space)	Oil Permeability (md)	Oil Saturation (% pore space)	Water Permeability (md)	Pore space (%)	Oil in place (%)
2	2,915.8	0.75	9.8	18.3	0.62	26.4	0.27	55.3	67.7
3	2,917.4	2.5	8.0	18.7	1.7	23.8	0.82	57.5	60.8
5	2,919.4	0.10	5.5	28.0	0.05	18.4	0.010	53.6	74.5

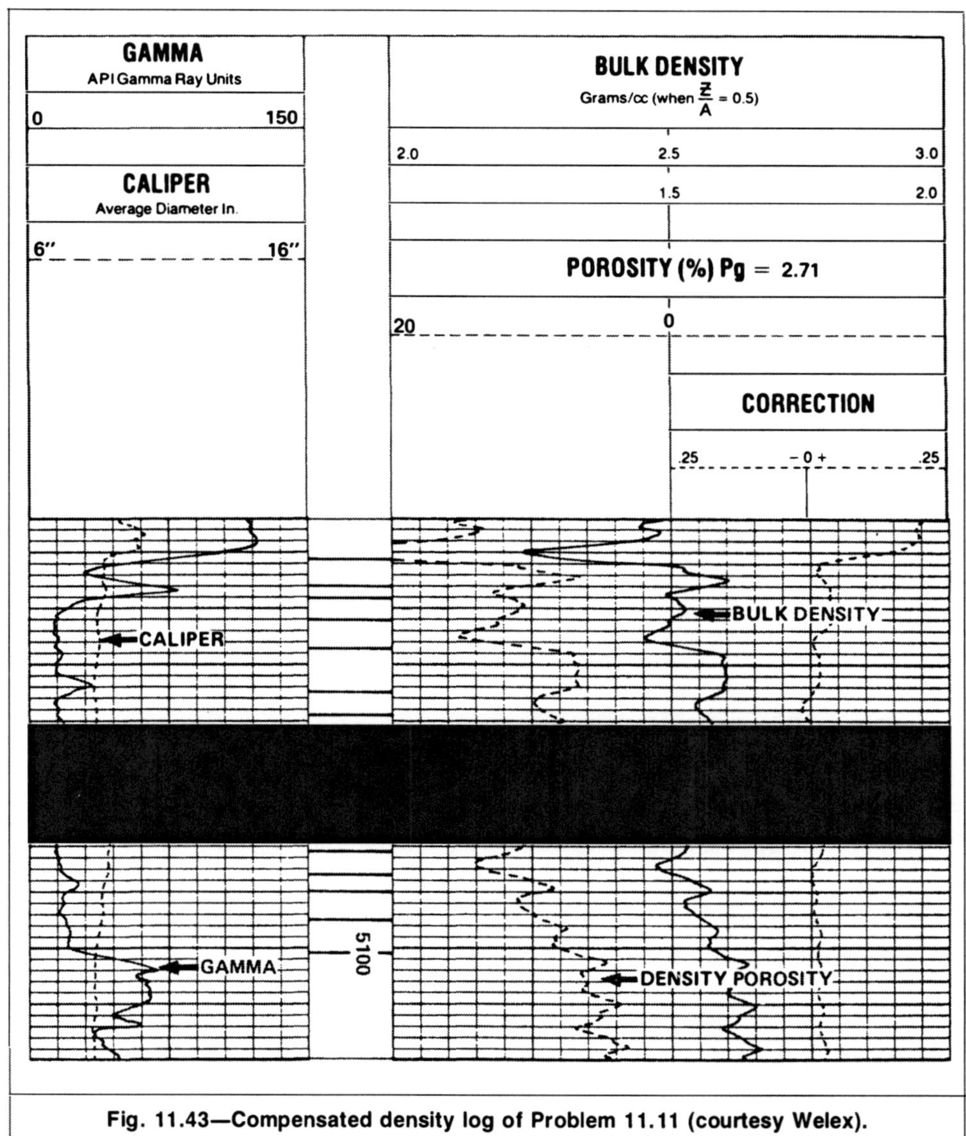

Fig. 11.43—Compensated density log of Problem 11.11 (courtesy Welex).

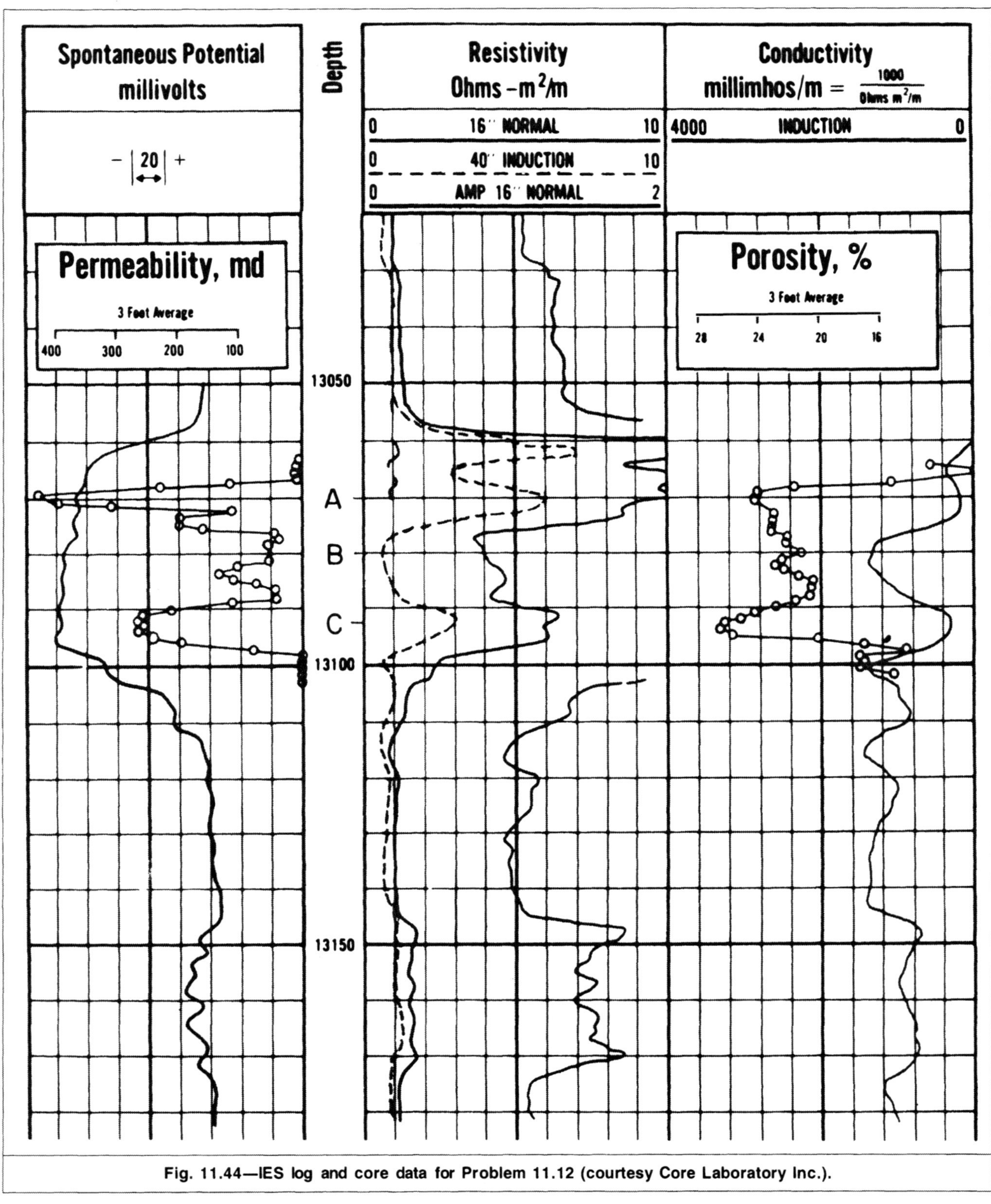

Fig. 11.44—IES log and core data for Problem 11.12 (courtesy Core Laboratory Inc.).

ρ = density, g/cm^3
ϕ = porosity, fraction

Subscripts

a = apparent
b = bulk
c = critical or cutoff value
e = effective, equivalent
f = formation
g = gas
h = borehole
h = hydrocarbon
i = invaded zone
ILd = deep induction log
m = mud
ma = matrix
max = maximum
mc = mudcake
mf = mud filtrate
mp = minimum productive
o = oil
oi = initial oil
or = residual oil
s = surface

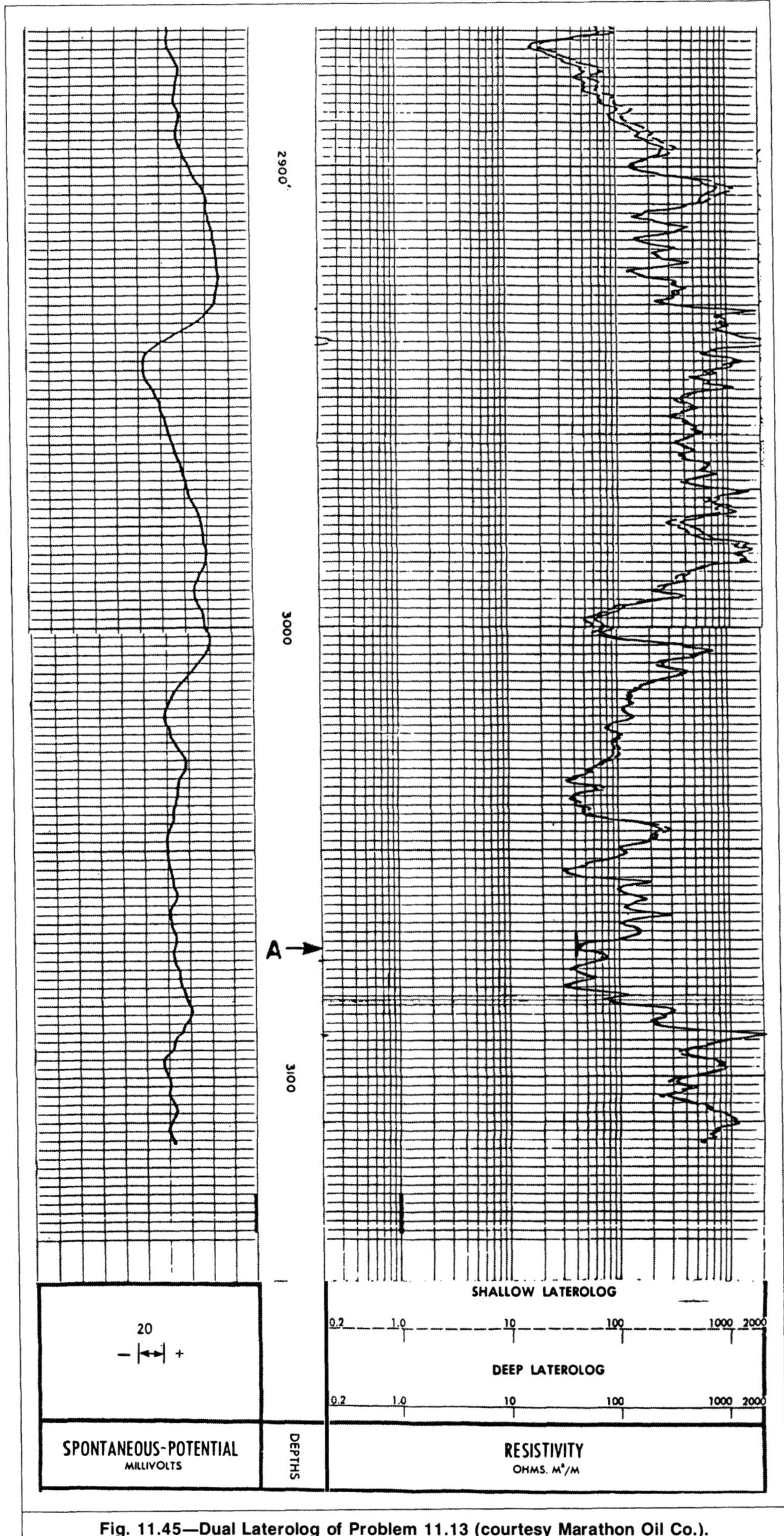

Fig. 11.45—Dual Laterolog of Problem 11.13 (courtesy Marathon Oil Co.).

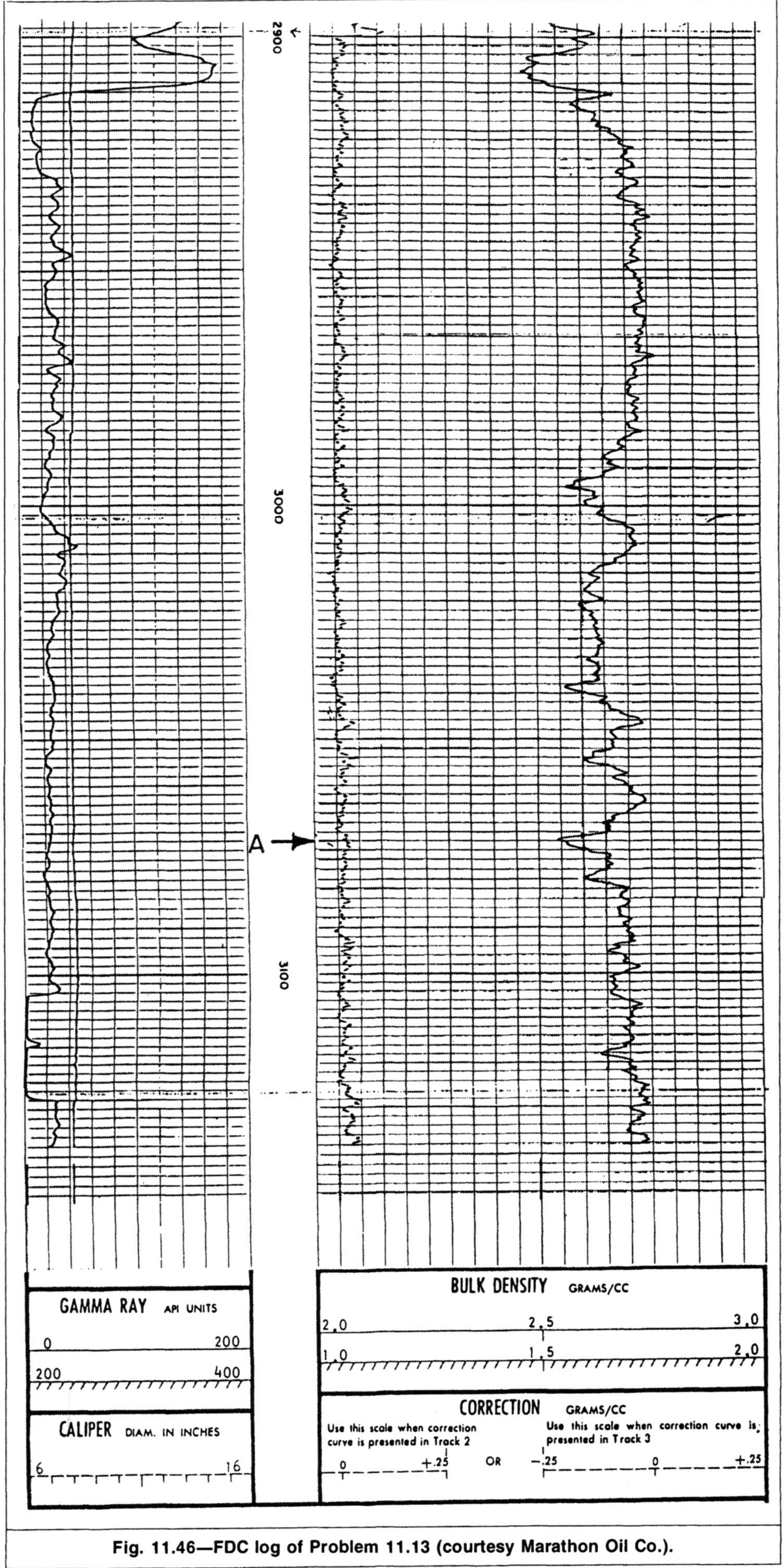

Fig. 11.46—FDC log of Problem 11.13 (courtesy Marathon Oil Co.).

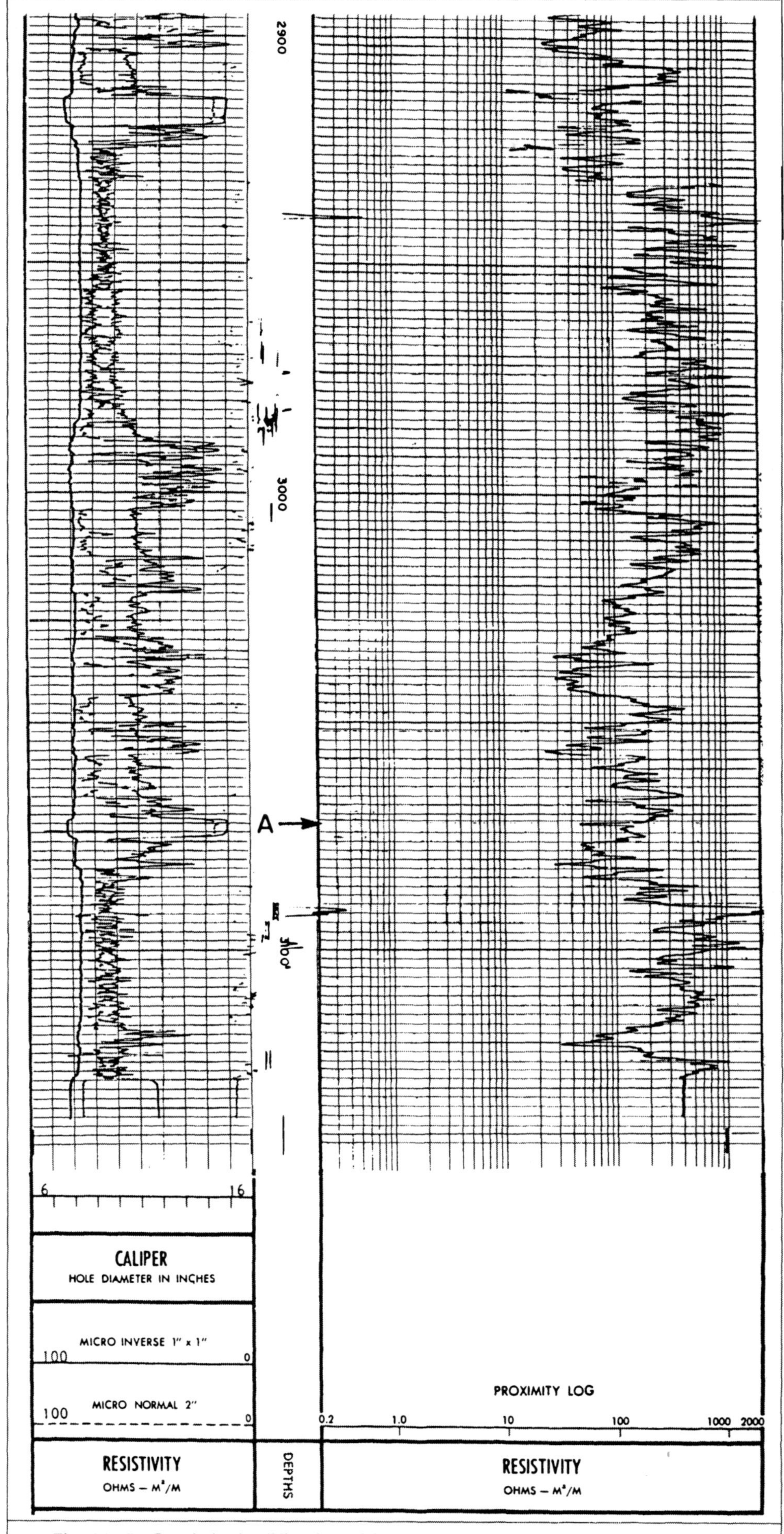

Fig. 11.47—Proximity log/Microlog of Problem 11.13 (courtesy Marathon Oil Co.).

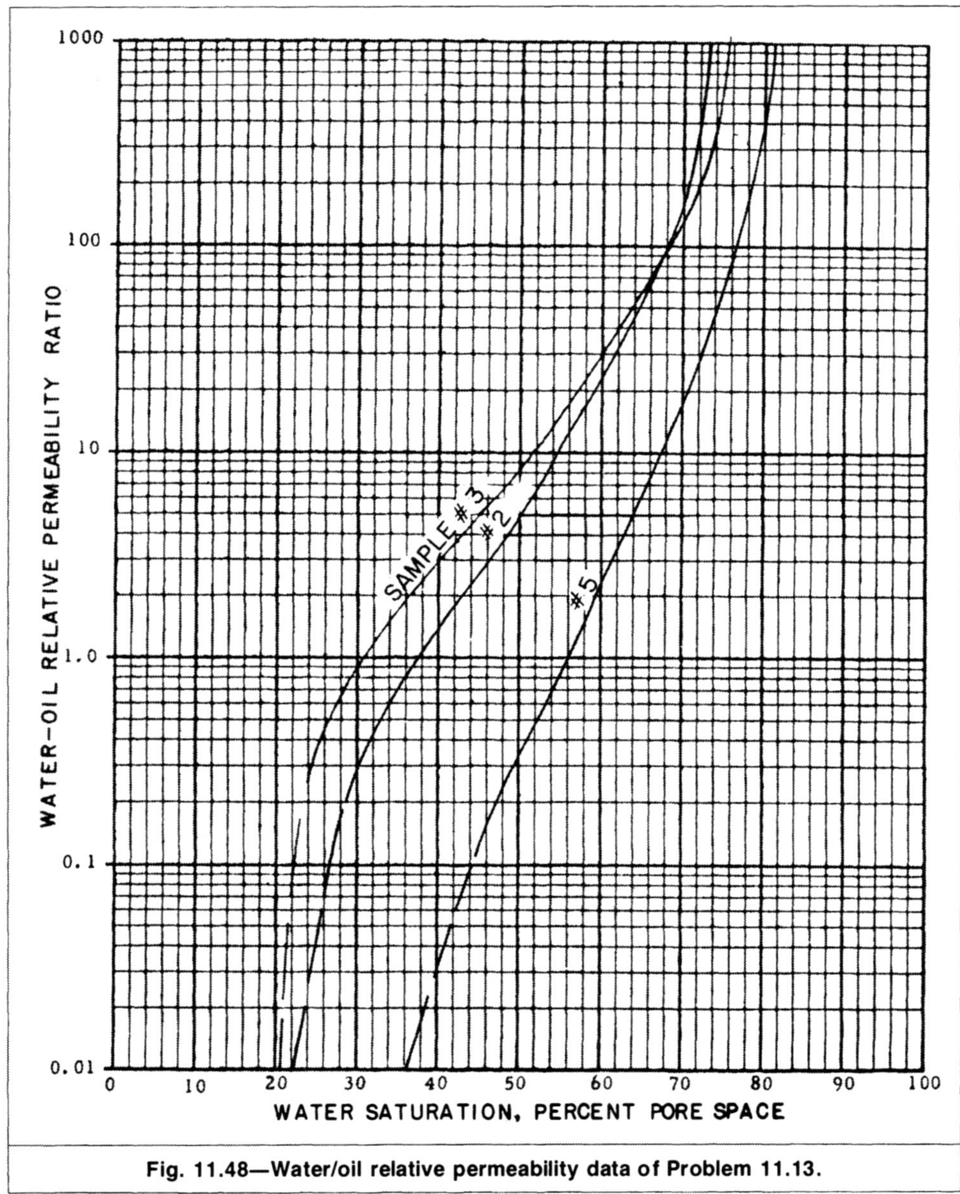

Fig. 11.48—Water/oil relative permeability data of Problem 11.13.

SN = short normal tool
t = true
w = water
xo = flushed zone

Superscript

$\bar{}$ = average

References

1. "Formation Water Resistivity Data, South Louisiana Offshore and Adjacent Areas," Lafayette Chap., SPWLA, Lafayette (Jan. 1969).
2. Taylor, R.E.: "Chemical Analyses of Ground Water for Saline-Water Resources Studies in Texas Coastal Plain Stored in National Water Data Storage and Retrieval System," USGS, Bay St. Louis, MS, U.S. Dept. of Interior Open File Report 75-79 (1975) **1,2**.
3. "Formation Water Resistivities of Canada," Canadian Well Logging Soc. (June 1978).
4. "Survey of Resistivities of Water From Subsurface Formations in Oklahoma," Oklahoma City Section, SPE (March 1975).
5. McCain, W.D. Jr.: *The Properties of Petroleum Fluids,* second edition, PennWell Publishing Co., Tulsa (1990).
6. Earlougher, R.C. Jr.: *Advances in Well Test Analysis,* second printing, Monograph Series, SPE (1977) **5**.
7. Willhite, P.G.: *Waterflooding,* Textbook Series SPE, (1986) **3**.
8. Pickett, G.R.: "Principles for Application of Borehole Measurements in Petroleum Engineering," *Log Analyst* (May-June 1969) 22-33.
9. Granberry, R.J. and Keelan, D.K.: "Critical Water Estimates for Gulf Coast Sands," *Trans.* Gulf Coast Assn. of Geologic Scientists (1977) **27**.
10. Granberry, R.J. and Wilshusen, R.C.: "Improved Interpretation of Formation Productivity by Combined Use of Core Analysis and Electric Log Data," *Trans.*, Gulf Coast Assn. of Geologic Scientists (1962) **12**.

Chapter 12
Reconnaissance Interpretation Techniques

12.1 Introduction

Reconnaissance interpretation techniques, also called "quick look," "eyeball," or "parameters" techniques, have been developed to present well-log data and calculations so that they may be scanned quickly and easily to identify zones that warrant a more detailed analysis. Reconnaissance interpretation techniques generated by service companies at the wellsite are among the most useful. They provide the elements of information needed for making decisions quickly when time is of the essence. Quick-look methods do not, however, provide a final interpretation. More and better knowledge of the formations of interest should be obtained from complete and in-depth log analysis.

Some reconnaissance interpretation methods also provide solutions to specific interpretation problems that cannot be managed by the conventional techniques of Chap. 11. Conventional interpretation techniques cannot be used in certain cases when insufficient information necessary for the calculation of water saturation is unavailable. The reconnaissance technique is used in such a case to calculate a numerical parameter that reflects water saturation but does not require the missing information for calculation.

Reconnaissance interpretation techniques are numerous. Some have general applicability while others are designed to answer specific problems encountered in localized areas or geological sections. Techniques discussed in this chapter are those that log analysts use most commonly at the wellsite: (1) the apparent water resistivity, R_{wa}, technique; (2) the R_o overlay; (3) the F_{xo}/F_s approach; (4) the R_{xo}/R_t method; and (5) the movable oil plot (MOP). Reconnaissance techniques specially designed for shaly sands and gas-bearing formations are discussed in Chaps. 15 and 16, respectively.

12.2 R_{wa} Technique

The R_{wa} technique is a quick-look technique useful for detection of hydrocarbon pay zones and estimation of water saturation and formation water resistivity. R_{wa} is defined as

$$R_{wa}=R_t/F, \quad (12.1)$$

where the true formation resistivity, R_t, is read directly off a resistivity log with deep-enough investigation to have little or no invasion effect. The formation resistivity factor, F, is derived from a porosity log, usually the sonic or the density log.

In clean, water-bearing zones, the reading of the deep resistivity tool will approximate R_o, and because $R_o=FR_w$,

$$R_{wa}=R_t/F=R_o/F=R_w. \quad (12.2)$$

It follows that, in hydrocarbon zones where $R_t>R_o$,

$$R_{wa}>R_w. \quad (12.3)$$

Simple rearrangement of Archie's saturation equation (Eq. 1.45) shows that

$$R_{wa}=R_w/S_w^2. \quad (12.4)$$

Eq. 12.4 can be used to define an apparent water saturation, S_{wa}:

$$S_{wa}=(R_w/R_{wa})^{1/2}. \quad (12.5)$$

"Apparent" is used in conjunction with this saturation value because it is derived from a reconnaissance-technique value that may be subject to uncertainties owing to the approximations used.

Simple rules of cutoff values with the parameter R_{wa}/R_w can be established for a given formation.[1] For example, productive zones are usually indicated in clean sands when the ratio $R_{wa}/R_w>3.0$ or 4.0. Pay zones are indicated in shaly sands when $R_{wa}/R_w>2.0$. **Table 12.1** lists the R_{wa}/R_w and corresponding S_w values for quick reference.

In practice, either R_{wa} values are calculated over the section of interest or an R_{wa} curve is presented on the log, as shown in **Fig. 12.1**. If R_w is available from water analysis or can be estimated from the self-potential (SP) log, R_{wa}/R_w can be calculated, and hydrocarbon zones meeting a specific cutoff saturation value can be identified.

The R_{wa} technique is still applicable when water resistivity, R_w, is unknown. However, two conditions must be fulfilled: (1) water zones must be present in the section evaluated and (2) no abrupt salinity changes can occur over the section of interest so that R_w can be considered constant. In such a case, the consistently lowest computed R_{wa} value, $(R_{wa})_{\min}$, in water-bearing zones is then selected to define R_w.

Note that abnormally high R_{wa} values can be displayed by zones other than hydrocarbon. Relatively high-resistivity shale will display a high R_{wa} value. Shale zones, however, can be distinguished easily by shale indicators, such as the SP and gamma ray logs. Coal seams display high R_{wa} because of their high resistivities and low densities. They also frequently display gas presence on the mud log. The combination of these characteristics has caused log analysts to recommend testing of such apparent "pay zones" that actually are coal seams.[1] Cycle skipping of the sonic log will also cause high R_{wa} values. This results from the anomalously high transit time, Δt, displayed by the log, which will result in an erroneously low F.

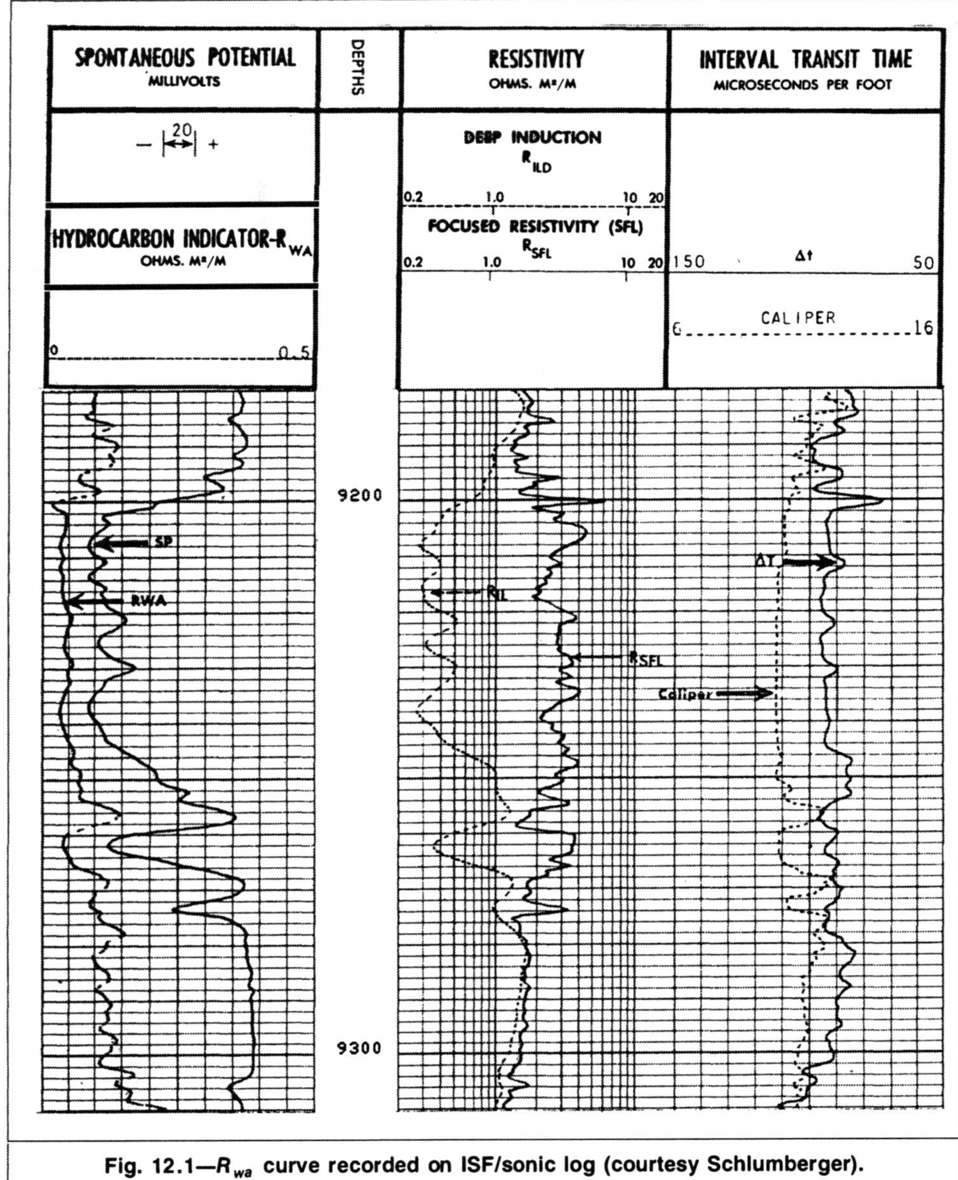

Fig. 12.1—R_{wa} curve recorded on ISF/sonic log (courtesy Schlumberger).

TABLE 12.1—R_{wa}/R_w AND CORRESPONDING S_w VALUES

R_{wa}/R_w	S_w (%)
1	100
2	71
3	58
4	50
5	45
8	35
10	32

The logs used most commonly for R_{wa} calculations are the simultaneously run induction and sonic logs. This combination works satisfactorily both in clean and shaly formations and in liquid-filled and gas-filled formations. In shaly formations, the sonic and resistivity log readings tend to cross compensate for the shale effect. The presence of shale in a formation will cause a reduction in R_t and an increase in Δt. When R_{wa} is calculated, the reduction in R_t will be canceled out by the low F obtained from the relatively high apparent porosity derived from the sonic log. If present, the gas effect could result in a high apparent sonic porosity. The derived low F, and consequently high R_{wa} value, will tend to enhance the detection of gas zones. In this case, however, S_{wa} value calculated from Eq. 12.5 will be higher than the actual saturation value.

The other two porosity logs (neutron and density) may be used. The R_{wa} values obtained, however, are not universally reliable. For example, use of the neutron log in gas-filled formations will give a pessimistic R_{wa} evaluation with the possibility that some potential pay sections may be overlooked or bypassed. The same is true for the use of the density log in liquid-filled shaly formations. Shaly but still potential zones may be overlooked because a severe reduction in R_t will not be fully compensated for by the usual slight increase in the apparent density porosity.

Corrections for invasion effect on resistivity log readings can be performed to obtain improved R_t values. For reconnaissance interpretations, such as the R_{wa} calculations, these corrections are generally not performed at the wellsite. In severe invasion, the log values could deviate too greatly from the true values of formation resistivity. When $R_w < R_{mf}$ and $R_t < R_{xo}$, deep log readings will be higher than R_t. R_{wa} evaluation could be too optimistic, resulting in the possibility of mistaking a deeply invaded water-bearing zone for a hydrocarbon zone. Abnormal invasion conditions can be checked by the use of a reconnaissance-type technique credited to Tixier *et al.*[2] This technique calls for calculation of an apparent mud filtrate resistivity, R_{mfa}, defined as

$$R_{mfa} = R_{shallow}/F, \quad \text{(12.6)}$$

where $R_{shallow}$ is read directly off a microresistivity device (e.g., the Microlaterolog, proximity log, and microspherically focused log) and F is the formation resistivity factor used in the R_{wa} calculation.

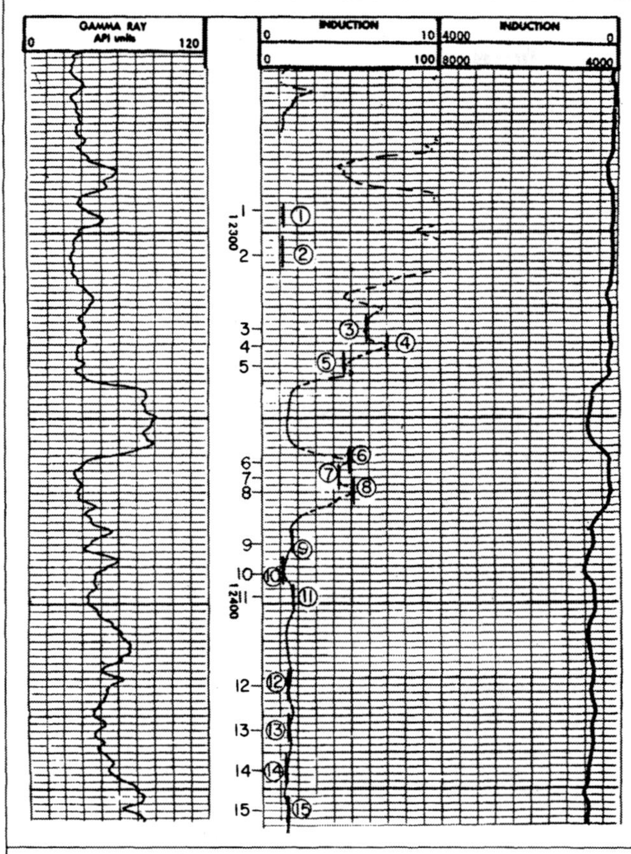

Fig. 12.2—Induction log of Example 12.1 (courtesy Amoco Oil Co.).

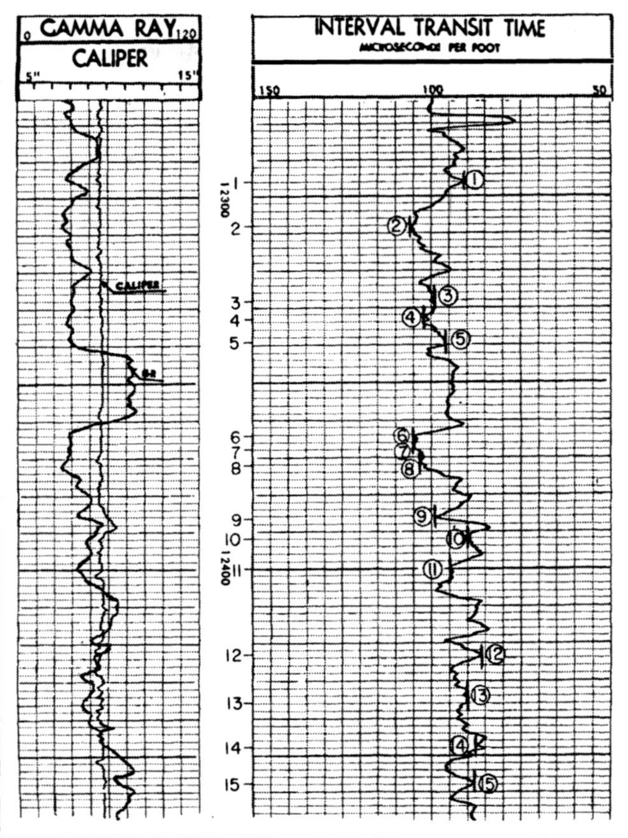

Fig. 12.3—Sonic log of Example 12.1 (courtesy Amoco Oil Co.).

R_{mfa} is equal to R_{mf} in deeply invaded water zones where $R_{\text{shallow}}=R_{xo}=FR_{mf}$. $R_{mfa}<R_{mf}$ in the case of shallow invasion where the reading of microresistivity devices will be less than R_{xo} because of the effect of the R_t zone. R_{mfa} will usually be higher than R_{mf} in oil and gas formations because of residual oil and gas. It can be seen easily from Eq. 4.7 that

$$R_{mfa}=R_{mf}/S_{xo}^2 . \quad (12.7)$$

Comparing R_{mfa} and R_{mf} provides a check for abnormal invasion conditions. For freshwater-based mud where $R_{mf}>R_w$, the following hold true.

1. If $R_{mfa}<R_{mf}$, invasion is very shallow. R_{wa} indications are probably representative.
2. If $R_{mfa}>R_{mf}$ and $R_{wa}>R_w$, R_{mfa} confirms the presence of hydrocarbon as indicated by R_{wa}.
3. If $R_{mfa}=R_{mf}$, deep invasion may be present. Favorable R_{wa} indication should be investigated further.

Of course, this checking technique requires a reliable R_{mf} value.

In addition to being used to check the validity of the R_{wa} evaluation, the R_{mfa} technique can be used to check the accuracy of the R_{mf} value recorded on the well heading. The consistently highest computed R_{mfa} values in water-bearing zones are generally very close to the real R_{mf} value. The R_{mfa} value also can be used to estimate movable-oil saturation, S_{mo}. Solving for S_{xo}, Eq. 12.7 becomes

$$S_{xo}=(R_{mf}/R_{mfa})^{1/2} . \quad (12.8)$$

Substituting the S_w and S_{xo} values from Eqs. 12.5 and 12.8 into Eq. 11.14 results in

$$S_{mo}=(R_{mf}/R_{mfa})^{1/2}-(R_w/R_{wa})^{1/2} . \quad (12.9)$$

Example 12.1. Figs. 12.2 and 12.3 are the gamma ray/induction log and the gamma ray/sonic log, respectively, obtained in a well

TABLE 12.2—LOG DATA AND CALCULATIONS FOR EXAMPLE 12.1

Zone	RIL (Ω·m)	Δt (μsec/ft)	φ (%)	F	R_{wa} (Ω·m)	R_{wa}/R_w	S_{wa} (%)	Zone's Potential
1	12	92	27	10.7	1.13	11.3	30	Good
2	12	105	37	6.3	1.91	19.1	23	Good
3	5.8	99	33	7.9	0.74	7.4	37	Good
4	7	102	35	7.0	1.00	10.0	32	Good
5	4.6	96	30	8.9	0.52	5.2	44	Good
6	5	105	37	6.3	0.79	7.9	36	Good
7	4.3	105	37	6.3	0.69	6.9	38	Good
8	5.1	103	36	6.8	0.76	7.6	36	Good
9	1.6	98	32	8.2	0.20	2.0	71	None
10	1.2	90	26	11.7	0.10	1.0	100	None
11	1.7	95	30	9.3	0.18	1.8	75	None
12	1.5	86	23	14.5	0.10	1.0	100	None
13	1.6	90	26	11.7	0.14	1.4	85	None
14	1.4	87	24	13.7	0.10	1.0	100	None
15	1.3	88	24	13.0	0.10	1.0	100	None

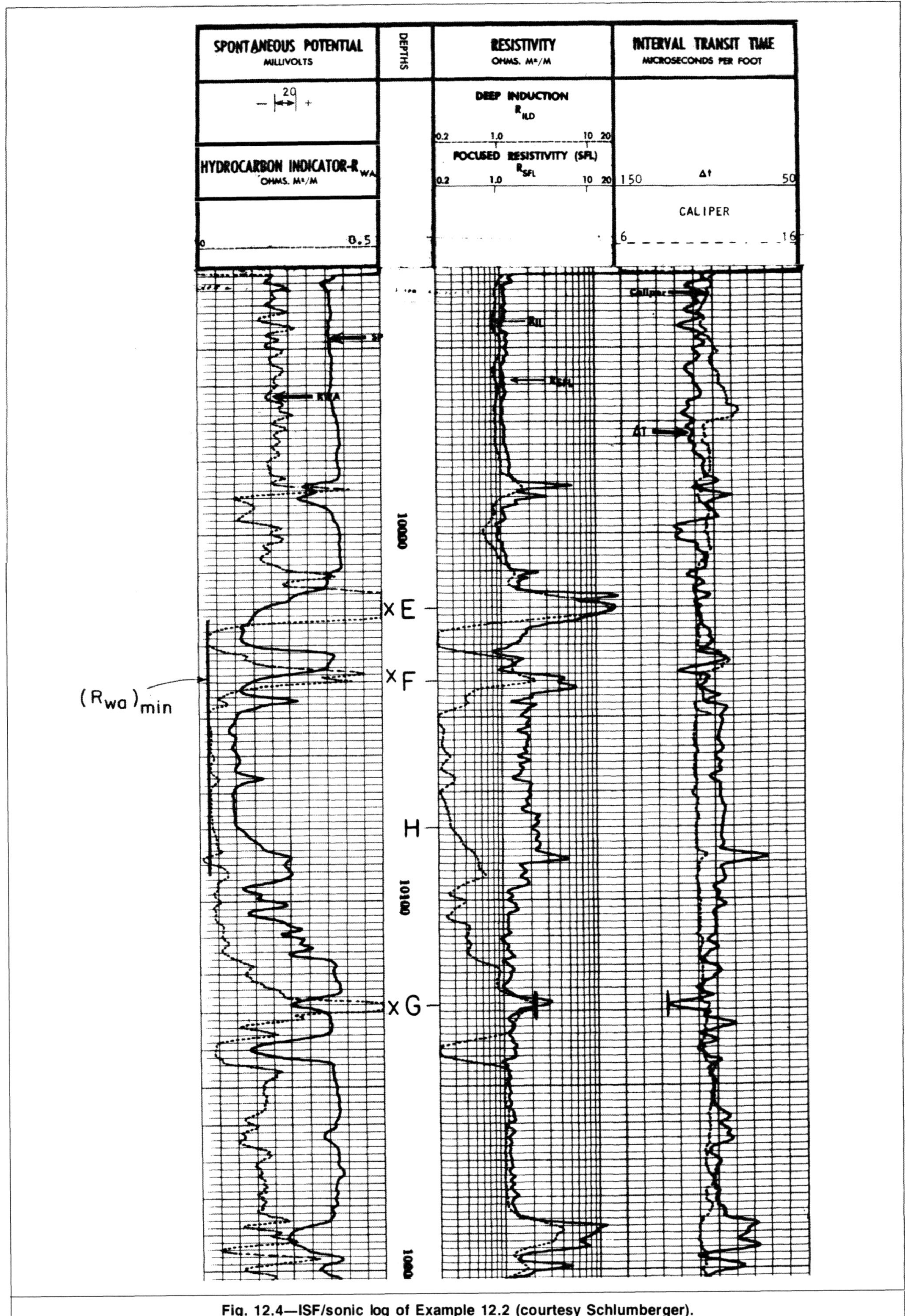

Fig. 12.4—ISF/sonic log of Example 12.2 (courtesy Schlumberger).

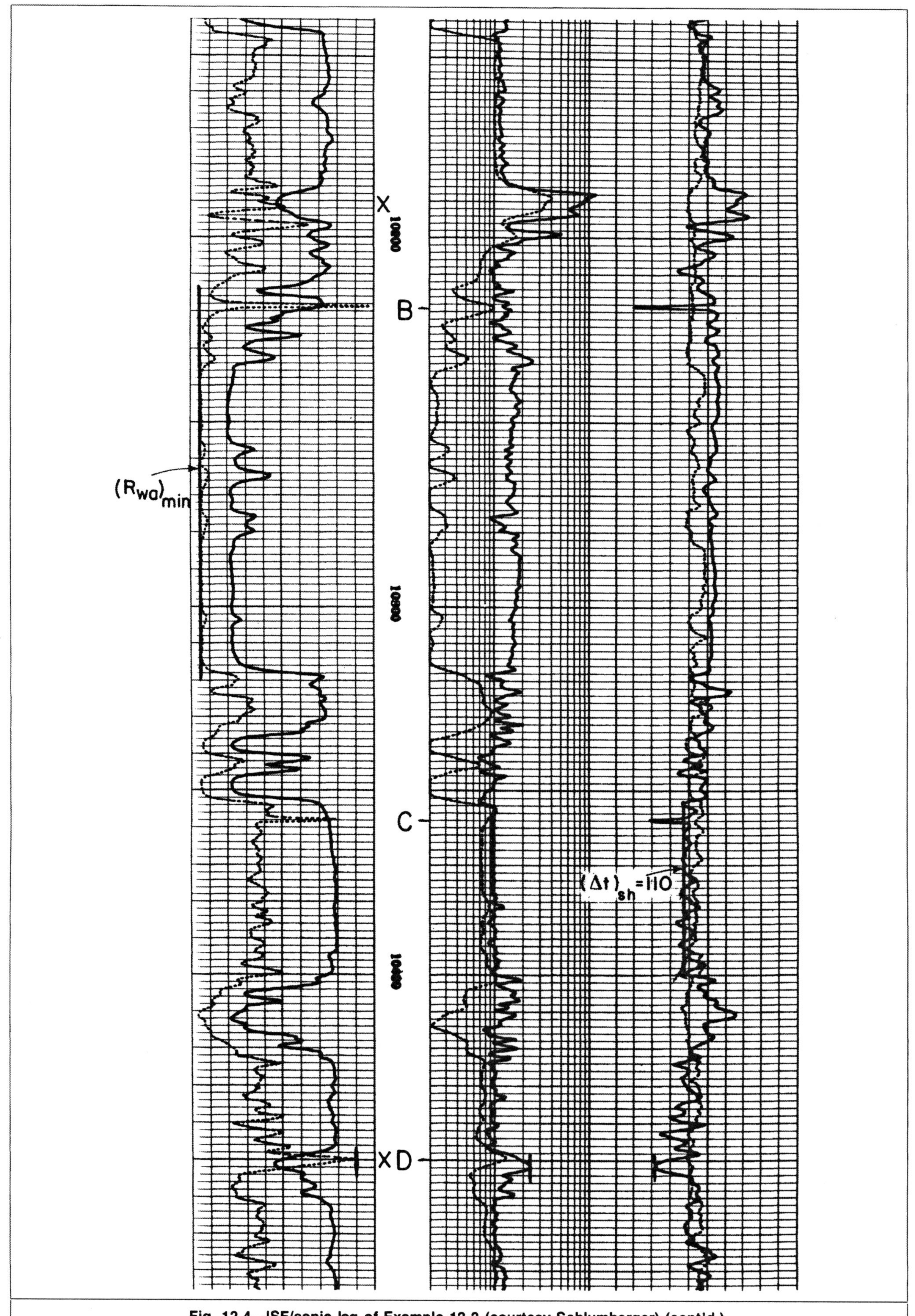

Fig. 12.4—ISF/sonic log of Example 12.2 (courtesy Schlumberger) (cont'd.).

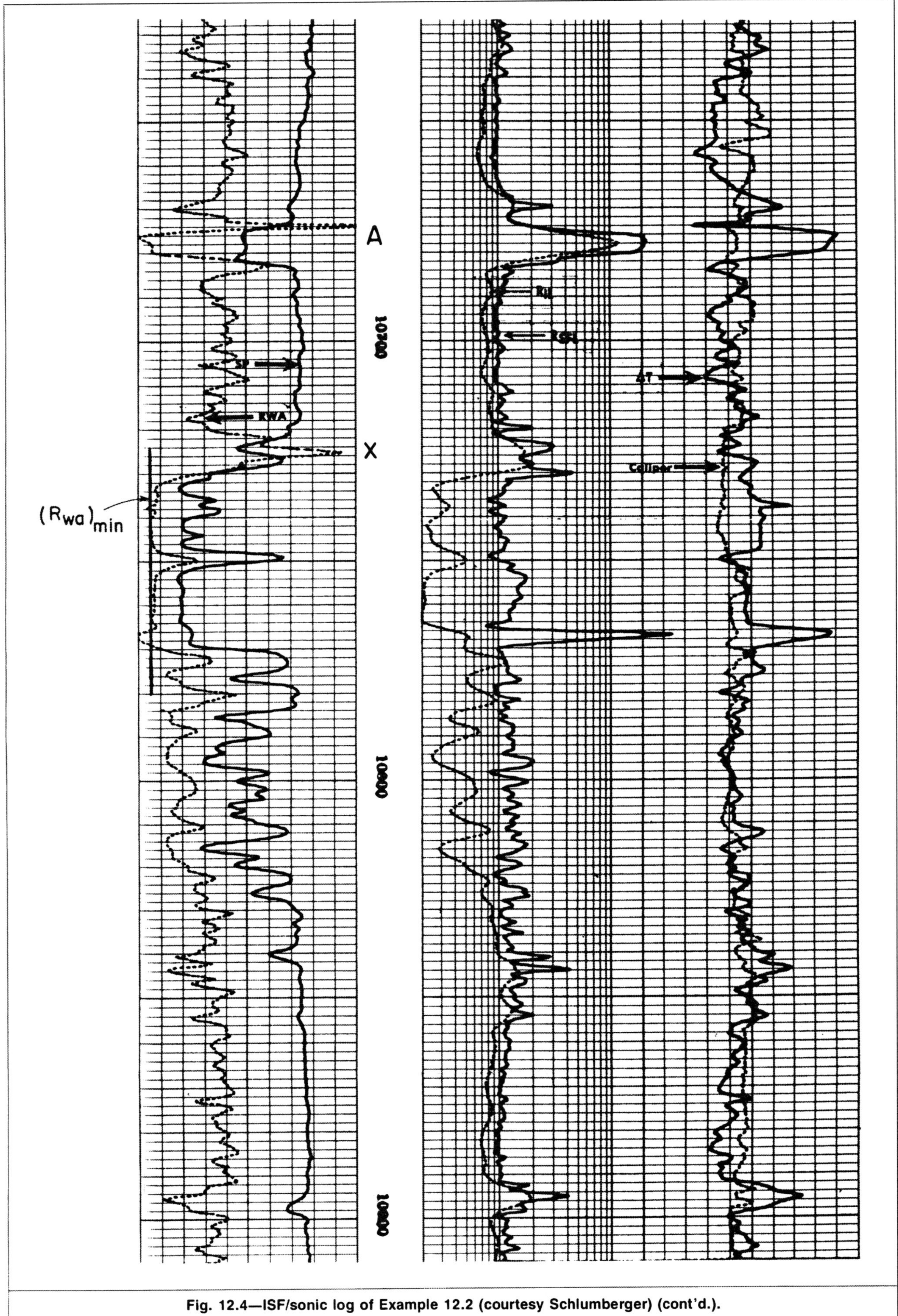

Fig. 12.4—ISF/sonic log of Example 12.2 (courtesy Schlumberger) (cont'd.).

drilled with oil-based mud. Estimate the porosity and water saturation of the 15 zones marked on the logs.

Solution. Because oil-based mud was in the borehole at the time of logging, the SP and 16-in. short-normal logs were not recorded. Also, because an SP log is not available, conventional interpretation techniques that require an R_w value cannot be used. The R_w value is needed to calculate R_o $(=FR_w)$. Because the variations in Δt indicate a nonuniform porosity, use of the R_o value from an adjacent water-bearing bed is not recommended.

Use of the R_{wa} technique to determine formation potential is in order. The porosity will be derived from the sonic log using $\Delta t_m = 55.5$ μsec/ft and $\Delta t_f = 189$ μsec/ft in Eq. 10.1. No compaction correction is needed in this case because $\Delta t_{sh} < 100$ μsec/ft. **Table 12.2** lists sonic log transit travel time, calculated porosity values, and the values of F calculated from the generalized Eq. 1.19.

The relatively high porosity range (23% to 37%) indicates shallow invasion. Thus, the reading of the deep induction tool can be considered equal to R_t. The listed values of calculated R_{wa} indicate a consistently lowest value of 0.1 Ω·m. This value is considered equal to R_w and is used to calculate R_{wa}/R_w and the S_{wa} values listed in Table 12.2. The indicated zone potential is based on the generally accepted cutoff value of $R_{wa}/R_w = 4$, which corresponds to a water saturation of 50%.

Example 12.2. The R_{wa} technique is available at the wellsite with the Induction Spherically Focused (ISF)/sonic combination as a recorded quick-look log. **Fig. 12.4** shows this log combination obtained in an Iberville Parish, LA, well. This well reached a total depth (TD) of 11,000 ft, where a maximum temperature of 162°F was recorded. Measured mud and mud filtrate resistivities are 0.9 Ω·m at 101°F and 1.30 Ω·m at 72°F, respectively. Refer to Fig. 12.4 in answering the following questions.

a. One of the conditions favorable for the use of the R_{wa} log is that R_w remains constant. Is this condition met in the interval shown in Fig. 12.4?

b. What is the consistently low R_{wa}, i.e., $[(R_{wa})_{min}]$, value displayed by the log? Compare it with the R_w value derived from the SP log.

c. Explain the relatively low R_{wa} value obtained for the resistive Zone A.

d. Discuss the relatively high R_{wa} value obtained for Zones B and C.

e. Discuss the relatively high R_{wa} values displayed by shales.

f. If the cutoff water saturation is 50%, mark the promising zones.

g. Calculate and comment on the water saturation indicated by R_{wa} evaluation of Zones D through H.

h. Use the R_{mfa} approach to check for invasion conditions in Zones E and H.

i. Calculate the residual hydrocarbon saturation in Zone E using Eq. 12.8.

Solution.

a. The SP deflection in clean sands within the interval shown in Fig. 12.4 is almost a constant −120 mV. The consistency of the SP deflection in clean sands is a good indication that R_w remains constant.

b. The R_{wa} curve displays a consistently low value of $R_{wa} = 0.03$ Ω·m. Entering Fig. 6.14 with an E_{SP} value of −120 mV, a formation temperature of about 160°F, and an R_{mf} value of 0.61 Ω·m at 160°F yields an R_w value of 0.03 Ω·m. While $(R_{wa})_{min}$ and R_w show excellent agreement in this well, this is not always the case. Differences could result because of the usual uncertainties associated with the estimation of R_w, R_{wa}, ϕ, and F.

c. The low R_{wa} value displayed by Zone A indicates that the zone does not contain hydrocarbons. The high resistivity results from the low porosity implied by a low Δt, which approaches a low value of 59 μsec/ft. Zone A is a water-bearing tight formation, probably limestone.

One advantage of using the R_{wa} evaluation as shown here is that it incorporates porosity and resistivity values. The true potential of tight zones is then clearly indicated.

d. The high R_{wa} values obtained for Zones B and C are the result of an abnormally high Δt. The character of these two deflections suggests that they are caused by cycle skipping. This is a

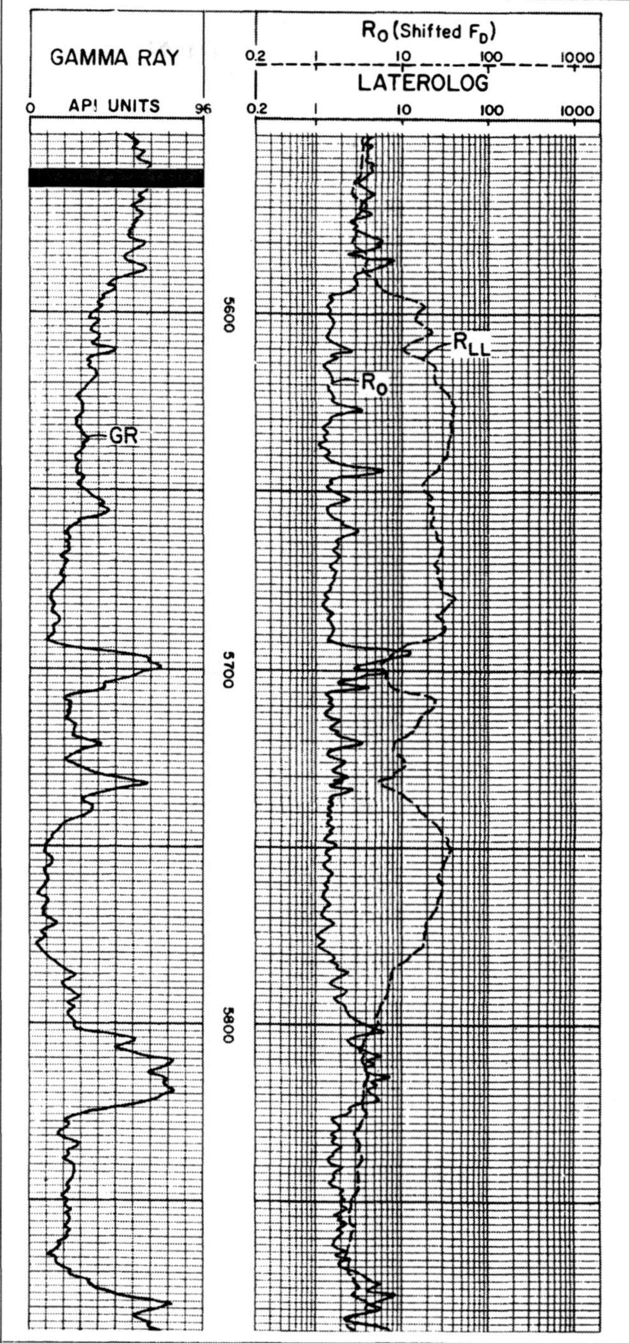

Fig. 12.5—Example of R_o log (courtesy Schlumberger).

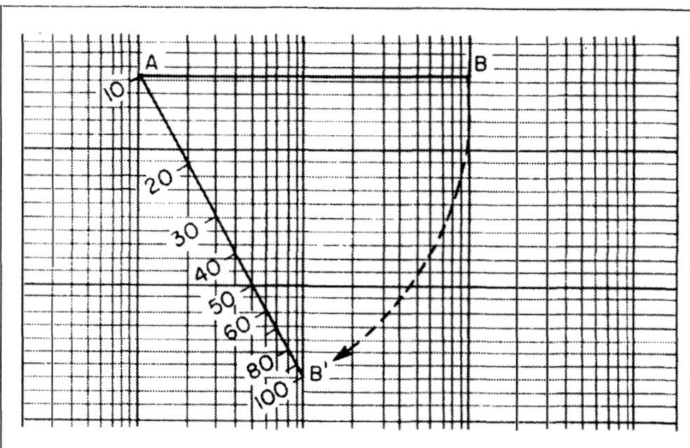

Fig. 12.6—Procedure for constructing "exponent 0.5" logarithmic scale.

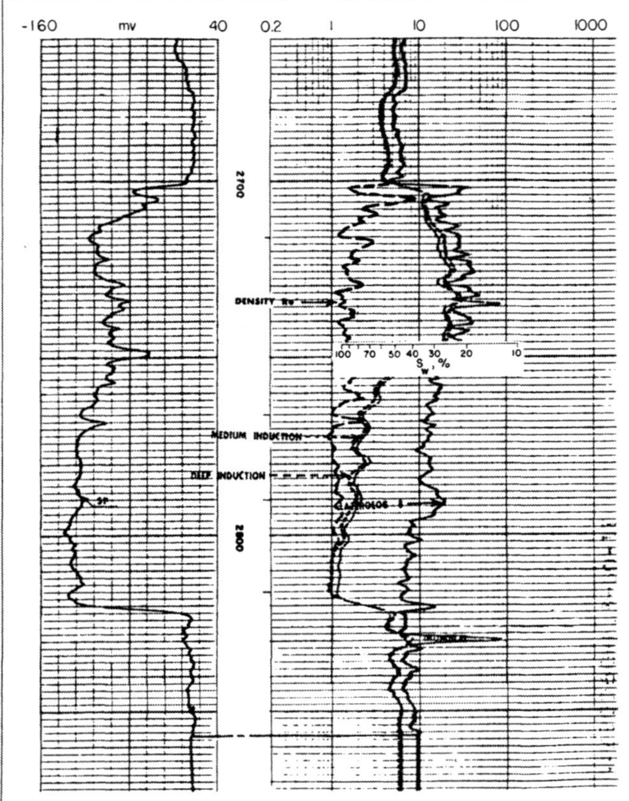

Fig. 12.7—R_o log traced into a dual induction/Laterolog 8 (courtesy Continental Oil Co.).

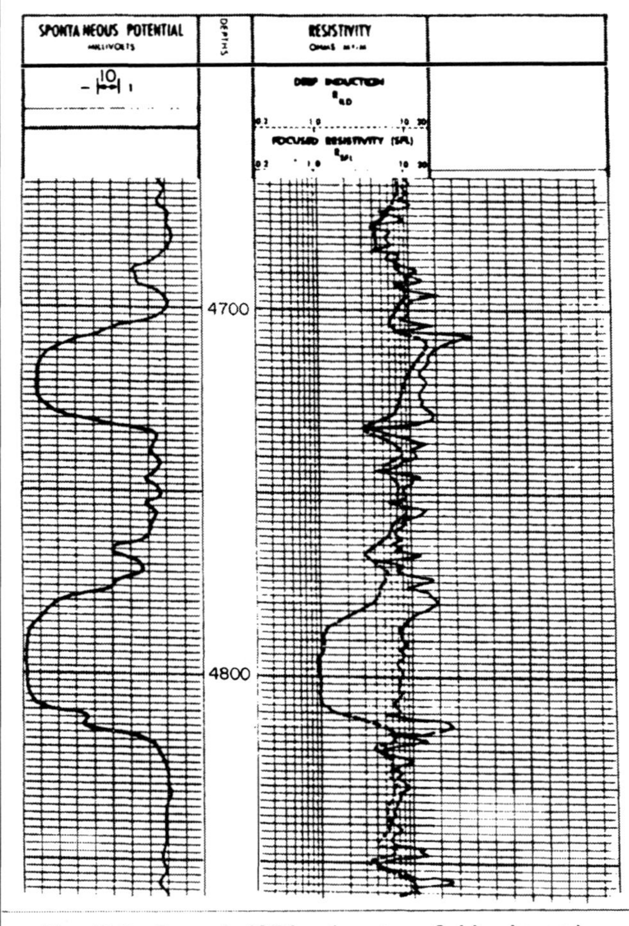

Fig. 12.8—Example ISF log (courtesy Schlumberger).

measurement anomaly and should not be mistaken for a hydrocarbon indicator.

e. The R_{wa} values for shale zones were calculated with porosity and F models developed for sands. This means that the R_{wa} value is nonrepresentative. As discussed in Chap. 1, shales are believed to contain low-salinity water called bound water. This also could explain the relatively high value of R_{wa} obtained in shales.

f. According to Table 12.1, a cutoff saturation of 50% corresponds to $R_{wa}/R_w=4$. A cutoff R_{wa} value, $(R_{wa})_c$, can then be calculated:

$$(R_{wa})_c=4(R_{wa})_{min}$$

$$=4(0.03)=0.12\ \Omega\cdot m.$$

Promising zones will then display a value of $R_{wa} \geqq 0.12$. These zones are marked by an X on the log.

g. The following table lists Δt, apparent porosity, ϕ_a, R_{wa}, and S_{wa} values for Zones D through H.

Zone	Δt (μsec/ft)	ϕ_a (%)	R_{wa} ($\Omega\cdot$m)	S_{wa} (%)
D	129	50	0.45	26
E	110	37	2.69	11
F	120	44	0.45	26
G	127	49	0.5	24
H	96	30	0.03	100

For Zones D, F, G, and H, the value of R_{wa} was read directly off the log. R_{wa} for Zone E is off-scale but can be calculated as follows:

$\Delta t_{sh}=110\ \mu$sec/ft,

$B_{cp}=1.1$,

$\Delta t=110\ \mu$sec/ft,

$R_t=17\ \Omega\cdot m$,

$\phi_a=(110-55.5)/(189-55.5)/1.1=0.37$ or 37%,

$F=1.13\ \phi_a^{-1.73}=6.3$,

and $R_{wa}=R_t/F=17/6.3=2.69$.

The high sonic porosity values obtained in Zones D through G indicate that these zones are gas-filled. The calculated R_{wa} values are exaggerated and will result in an apparent water saturation that is lower than the true saturation.

h. In the absence of a microresistivity log, the spherically focused log resistivity, R_{SFL}, will be used in place of $R_{shallow}$ in Eq. 12.6. Because $R_{xo} \geqq R_{SFL}$, the calculated R_{mfa} is expected to be on the low side.

Zone H. From Part g, $\phi=30\%$, which is a true value because the zone is 100% saturated with water. Therefore,

$F=1.13/(0.3)^2=9.1$,

$R_{SFL}=2.3\ \Omega\cdot m$,

and $R_{mfa} \geqq R_{SFL}/F$

$\geqq 2.3/9.1=0.25\ \Omega\cdot m$.

From Part b, $R_{mf}=0.61\ \Omega\cdot m$. R_{mfa} is definitely smaller than R_{mf}, which indicates very shallow invasion. This invasion condition is expected in relatively high-porosity sands.

Zone E. The porosity of Zone E calculated in Part g is an apparent porosity that bears the gas effect. Zones E and H are part of the same sand that appears to have a constant porosity. Thus, the true porosity of Zone E is equal to that of Zone H:

$\phi=30\%$,

$F=9.1$,

$R_{SFL}=21\ \Omega\cdot m$,

and $R_{mfa} \geqq 21/9.1 \geqq 2.3\ \Omega\cdot m$.

R_{mfa} is much greater than R_{mf}. Regardless of invasion conditions, R_{mfa} confirms the indication of hydrocarbons by R_{wa}.

i. $S_{xo}=(R_{mf}/R_{mfa})^{1/2}$

$=(0.61/2.3)^{1/2}=0.51$ or 51%.

$S_{gr}=49\%$.

12.3 The R_o Log

The R_o log is a graphical solution of Archie's saturation equation,

$$S_w=(R_o/R_t)^{1/2}, \quad (12.10)$$

through use of logarithmic scales. Logarithmic scaling of resistivity and porosity logs is useful for quick-look presentations because of the properties of logarithms that tend to simplify mathematical operations.

To generate the R_o log, the value of F is first derived from a porosity log. F is then multiplied by the R_w value that prevails in the interval evaluated. The resulting R_o value is plotted on a logarithmic scale together with the deepest resistivity curve of the dual induction log or Laterolog[SM], R_{deep}. **Fig. 12.5** shows an example R_o log produced with F derived from a density log and plotted with the deep Laterolog on a logarithmic scale. The logarithm of a curve reading is proportional to the distance from the unity line to the curve, then the distance separating the R_o and R_{deep} curves is proportional to (log R_o − log R_{deep}). Assuming that $R_t=R_{deep}$, Eq. 12.10 written in logarithmic form becomes

$$\log S_w=0.5(\log R_o-\log R_{deep}). \quad (12.11)$$

A separation between the R_o and R_{deep} curves means that $R_t>R_o$ and indicates the presence of hydrocarbons. The greater the separation is, the lower the S_w and the higher the hydrocarbon saturation.

A scale can be constructed to determine the value of S_w quickly from the distance between the two resistivity curves. Taking a line the length of two logarithmic cycles and scaling it to one logarithmic cycle (**Fig. 12.6**) is equivalent to taking the difference of log R_o and log R_{deep} and dividing it by two. As shown by Eq. 12.11, this is equivalent to log S_w. When the 100% mark of the S_w scale is placed on the R_o curve, the value of S_w can be read where the R_{deep} curve intersects the scale. This scale is usually prepared in the form of a transparent overlay. Such logarithmic overlay scales for conventional grids, available from service companies, are usually called exponent 0.5 scales. **Fig. 12.7** shows an R_o log derived from a density log and traced onto a dual induction/Laterolog 8[SM]. It also shows how the saturation scale can be used. When the scale is placed at a depth of 2,740 ft, a water saturation value of about 27% is indicated by the deep induction log curve.

The R_o log has the advantage of being a continuous plot of the saturation equation and can facilitate quick identification of hydrocarbon zones. However, it does not automatically compensate for changes in R_w, lithology, shaliness, and/or fluid type. To avoid possible misinterpretations, the log analyst should exhibit care in using this log.

Detailed evaluation of zones pointed out by this log is required to obtain the true potential of each zone. The water saturation value derived from the R_o log is an apparent value that could be equal to, higher than, or lower than the true value, depending on circumstances. If the sonic log is used to derive the R_o log, then the calculated saturation will be representative of the true value only in clean, consolidated, and oil-bearing zones. For example, the presence of gas would, in most cases, enhance the separation between the R_o and R_{deep} curves, but the indicated water saturation would be lower than the true value. The hydrocarbon effect also enhances the separation between the R_{deep} and R_o values derived from the density log. Use of the neutron log in gas-bearing formations gives a pessimistically high R_o value, which could result in potential gas zones being overlooked.

A reliable value of R_w must be available to generate an R_o log. If R_w is not known, F is first derived from a porosity log and plotted on a logarithmic grid. $R_o=FR_w$, so

$$\log R_o=\log F+\log R_w. \quad (12.12)$$

The F curve can be converted to an R_o curve. A shift on logarithmic grid is equivalent to multiplying or dividing. Thus, if the F curve is shifted so that it overlays the R_{deep} curve in a water zone where $R_{deep}=R_o$, the amount of the shift corresponds to R_w, and the F curve is now transformed to an R_o curve. This approach is often called "normalizing in a water zone." **Figs. 12.8 and 12.9** show an induction/spherically focused log and an F log traced on

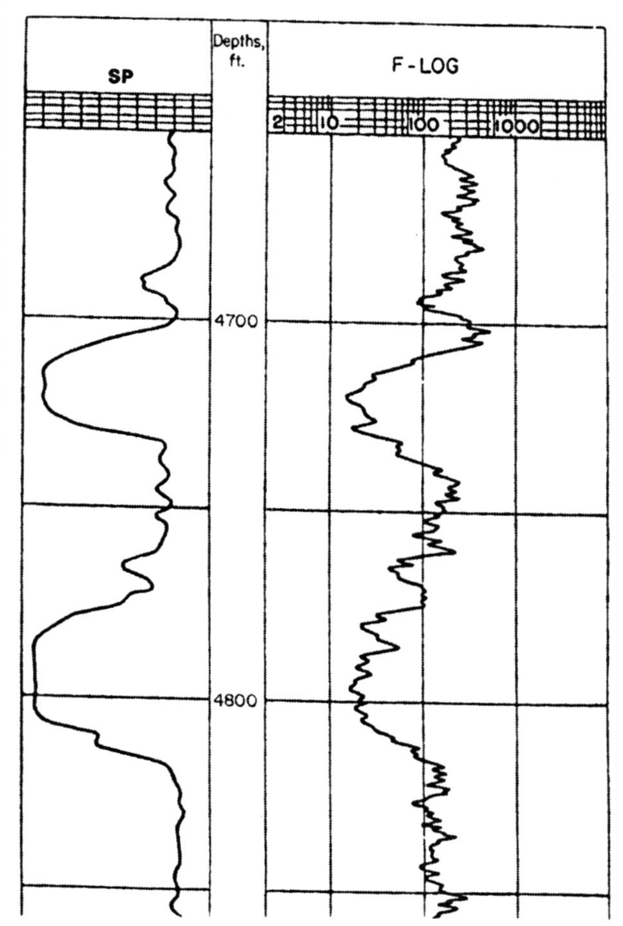

Fig. 12.9—F log for the interval shown in Fig. 12.8 (courtesy Schlumberger).

logarithmic scales. The logs show two sands in the illustrated interval. The F log was placed on the induction log and shifted until the two curves coincided in what is believed to be a water zone at 4,800 ft. **Fig. 12.10** shows the result of normalizing the F curve. The normalized F curve now represents R_o when read off the resistivity scale. The separation between the R_o and R_{deep} curves indicates the presence of hydrocarbons in the upper sand and in the top 8 ft of the lower sand.

The value of R_w also can be estimated. Fig. 12.10 shows the scale of the F log at the top and the scale of the induction log at the bottom. The F value of 10 overlays the resistivity value of 0.5, just as the F value of 100 overlays 5 on the resistivity scale. This results in an R_w value of 0.05 Ω·m:

$$R_w=R_o/F=0.5/10=0.05\ \Omega\cdot m.$$

Thus, the F curve was shifted the distance of log R_w or log (0.05) to convert it to an R_o log.

An additional advantage of the R_o curve is that it can be calibrated easily in porosity units. Using the relationship $F=\phi^{-2}$ in a logarithmic form results in

$$\log F=-2\log\phi$$

or $\log\phi=-(\log F)/2$.

This means that one cycle of the porosity scale is equivalent to two cycles of the F scale. A porosity scale can be constructed by scaling a line the length of two cycles on the F scale to one logarithmic cycle (see Fig. 12.6). Because we are using the relationship $F=\phi^{-2}$, then $F=100$ when $\phi=10\%$. By placing the 10% mark from the derived porosity scale on the R_o scale where $F=100$, we calibrate the R_o curve in porosity units (Fig. 12.10). According to this scale, the average porosity of the potential zone (4,708 to 4,732 ft) is 22%.

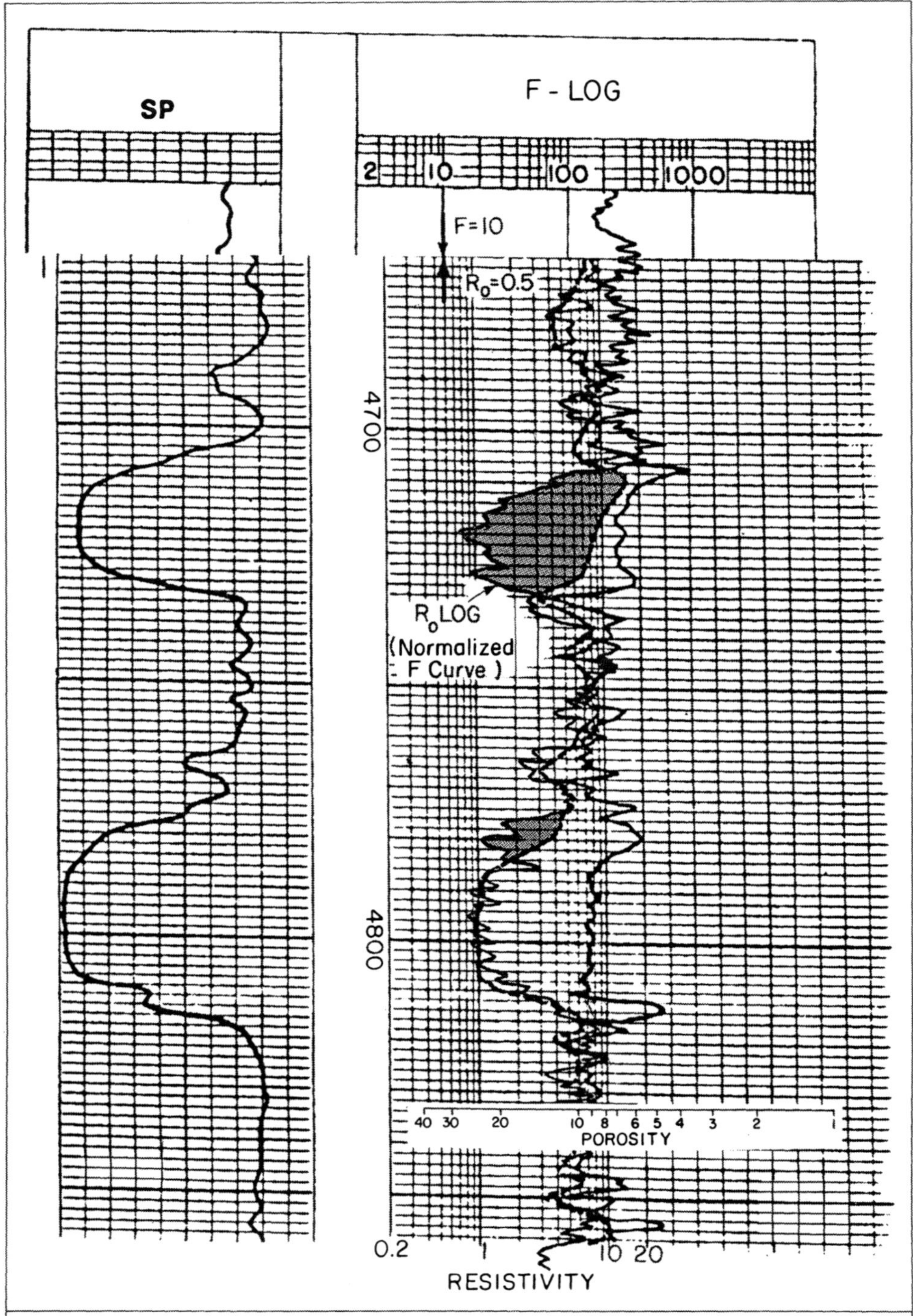

Fig. 12.10—R_o log of the interval illustrated in Figs. 12.8 and 12.9 derived by normalization of the *F* log.

12.4 The F_{xo}/F_s Approach

The F_{xo}/F_s method[2] was designed as a qualitative approach to locating oil or gas zones when R_t cannot be obtained reliably. Inability to determine R_t usually occurs in zones of deep invasion. The F_{xo}/F_s approach is based on the definition of two parameters, F_{xo} and F_s. F_{xo} is an apparent formation factor derived from a shallow resistivity log:

$$F_{xo}=R_{\text{shallow}}/R_{mf}. \quad (12.13)$$

In deeply invaded zones, R_{shallow} will approach R_{xo}:

$$R_{\text{shallow}}=R_{xo}=FR_{mf}/S_{xo}^2,$$

and F_{xo} becomes

$$F_{xo}=F/S_{xo}^2. \quad (12.14)$$

F_s is an estimate of F derived from the sonic log. The log Δt measurement is converted to sonic porosity, ϕ_s, by the appropriate response equation, and F_s is calculated from the relationship $F_s=a\phi_s^{-m}$. F_s approximates the true formation factor, F:

$$F_s=F \quad (12.15)$$

and

$$F_{xo}/F_s=1/S_{xo}^2. \quad (12.16)$$

In oil or gas formations, $S_{xo}<1$ and

$$F_{xo}/F_s>1. \quad (12.17)$$

In water-bearing formations, $S_{xo}=1$ and

$$F_{xo}/F_s=1. \quad (12.18)$$

Oil and gas formations can be identified with the above technique in deep-invasion environments by detection of residual hydrocarbons in the invaded zone. Note that the presence of gas can result in $F_s<F$. However, this will enhance the detection of gas zones.

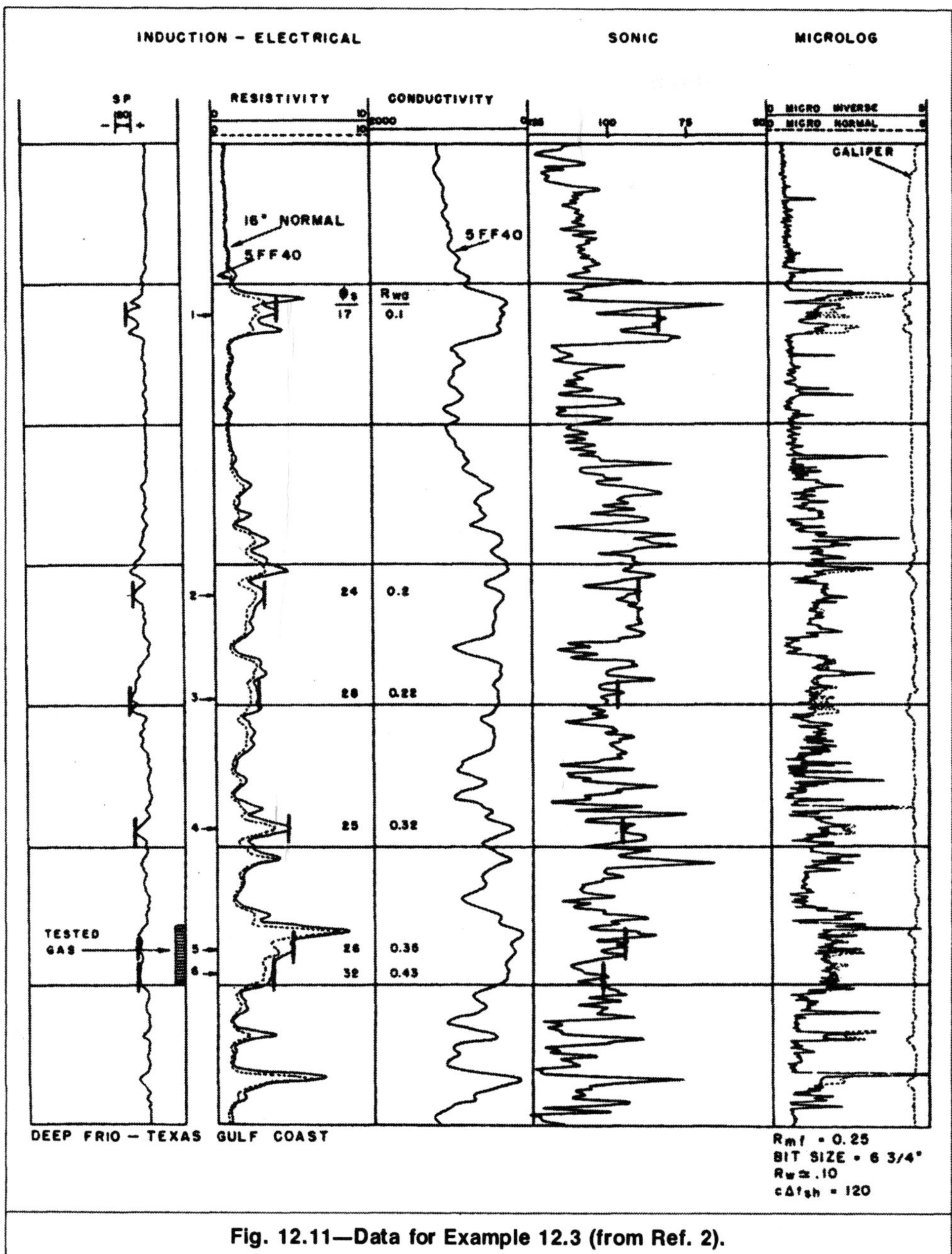

Fig. 12.11—Data for Example 12.3 (from Ref. 2).

The F_{xo}/F_s technique is valid in shaly sands because the sonic and resistivity logs are affected by the presence of shaliness in a somewhat similar manner. The technique, however, will not work properly under shallow-invasion conditions because $R_{shallow}$ does not adequately approximate R_{xo}. If the invasion conditions are such that $R_{shallow} < R_{xo}$, then the F_{xo}/F_s ratio in water-bearing formations will be less than unity. Of course, shallow invasion allows good knowledge of R_t, and other types of interpretation techniques can be used. Note also that $F_{xo}/F_s = R_{mfa}/R_{mf}$, so the technique can be used in conjunction with the evaluation of R_{wa} to guard against adverse invasion effects.

Example 12.3. Fig. 12.11 illustrates the logs made in a Deep Frio section of the Texas gulf coast. Because of shaliness and a relatively low R_{mf}/R_w ratio, little SP deflection occurs in the sands. Calculated R_{wa} values have been written on the induction electrical log for six zones of potential interest.

Level 1 shows an R_{wa} value equal to the known formation water resistivity of 0.1 Ω·m. This section is water-bearing. Only Zones 4 through 6 show an R_{wa} value at least three times greater than that of Zone 1, indicating hydrocarbons. Tests of Levels 5 and 6 show gas.

Show that the F_{xo}/F_s approach corroborates the R_{wa} value and test indications.

Solution. The shales in the interval shown display $\Delta t = 120$ μsec/ft, which indicates undercompacted formations. To calculate the sonic porosity, a compaction correction factor, B_{cp}, of 1.2 will be used in the following model:

$$\phi_s = [(\Delta t_{\log} - \Delta t_{ma})/(\Delta t_f - \Delta t_{ma})]/B_{cp},$$

where Δt_{ma} and Δt_f are 55.5 and 189 μsec/ft, respectively.

The shallowest tool available is the 16-in. short normal. The readings of the tool will be used in Eq. 12.13 to calculate F_{xo}.

F_{xo}/F_s calculations for the six zones marked on the log are tabulated below:

Zone	Δt (μsec/ft)	ϕ_s (%)	F_s (m=1.8)	R_{SN} (Ω·m)	F_{xo}	F_{xo}/F_s
1	83	17	23.9	4.0	16.0	0.67
2	94	24	13.0	3.3	13.2	1.02
3	100	28	10.0	2.8	11.2	1.12
4	96	25	11.9	4.5	18.0	1.51
5	97	26	11.4	5.5	20.0	1.75
6	107	32	7.7	3.6	14.4	1.87

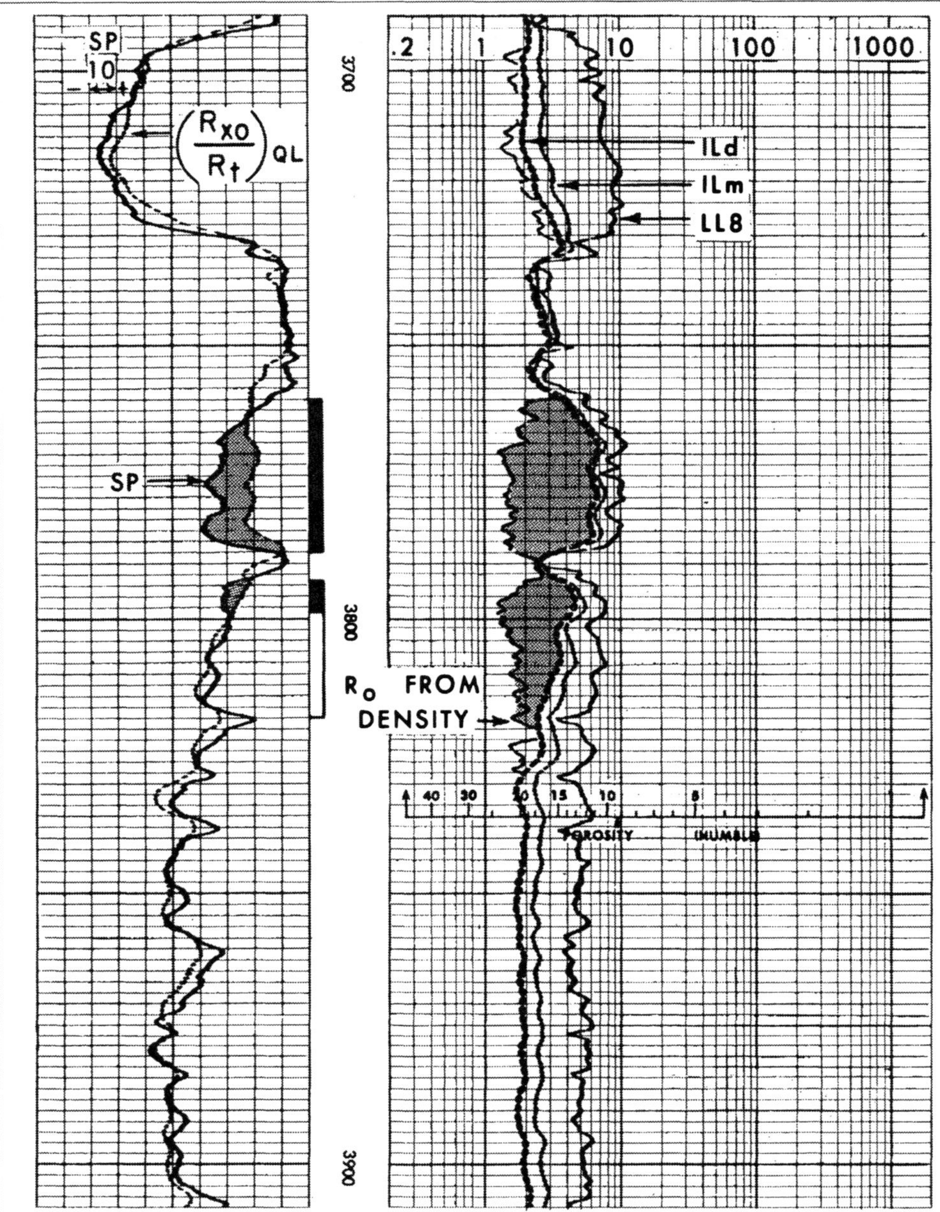

Fig. 12.12—Example $(R_{xo}/R_t)_{QL}$ curve used for comparison with SP to identify zones with movable hydrocarbons (from Ref. 3).

The ratios F_{xo}/F_s obtained for Zones 4 through 6 are substantially greater than unity. This clearly indicates that hydrocarbons are present in the invaded zones of these sections. Thus, the F_{xo}/F_s approach is in agreement with R_{wa} and test evaluations. The ratio F_{xo}/F_s is close to unity in Zones 1 through 3, indicating little or no hydrocarbon. This again is in agreement with the R_{wa} evaluations. $F_{xo}/F_s<1$ in Zone 1 because $R_{SN}<R_{xo}$.

12.5 The R_{xo}/R_t Method

Dumanoir *et al.*[3] introduced the R_{xo}/R_t method for wellsite interpretation. This technique involves computing log (R_{xo}/R_t) from either the R_{LL8}/R_{ILd} or the R_{SFL}/R_{ILd} ratio and recording it as a comparative curve with the SP. Separations between the properly scaled (R_{xo}/R_t) curve and the SP provide a quick-look location of producible hydrocarbons. The induction/spherically focused, dual induction/Laterolog 8, and dual induction/spherically focused log equipment can record the quick-look R_{xo}/R_t curve simultaneously, which is called $(E_{SP})_{QL}$ or $(R_{xo}/R_t)_{QL}$. **Fig. 12.12** shows such a recording.

The calculation of R_{xo}/R_t does not require knowledge of porosity, formation factor, or R_w. The technique is most suitable for cases where porosity is not available or cannot be determined accurately owing to complex lithology, or where known ϕ-F relationships are inappropriate. The main advantage of this technique may be that it provides a means for locating movable hydrocarbons. Because the proven presence of hydrocarbons does not necessarily mean commercial production, the $(R_{xo}/R_t)_{QL}$ curve is an appropriate companion to techniques that only indicate hydrocarbon presence, such as R_{wa}, R_o log, or F_{xo}/F_s.

The (R_{xo}/R_t) method is based on the calculation of the parameter $(E_{SP})_{QL}$ derived from the ratio R_{xo}/R_t:

$$(E_{SP})_{QL}=-K\log(R_{xo}/R_t). \quad (12.19)$$

Because $R_{xo}=FR_{mf}/S_{xo}^n$ and $R_t=FR_w/S_w^n$, then

$$R_{xo}/R_t=(S_w/S_{xo})^n(R_{mf}/R_w)$$

and Eq. 12.19 becomes

$$(E_{SP})_{QL}=-K\log[(S_w/S_{xo})^n(R_{mf}/R_w)]$$

$$=-K[\log(R_{mf}/R_w)+\log(S_w/S_{xo})^n].$$

But the measured E_{SP}, $(E_{SP})_{log}$, can be approximated by the term $-K\log(R_{mf}/R_w)$; therefore,

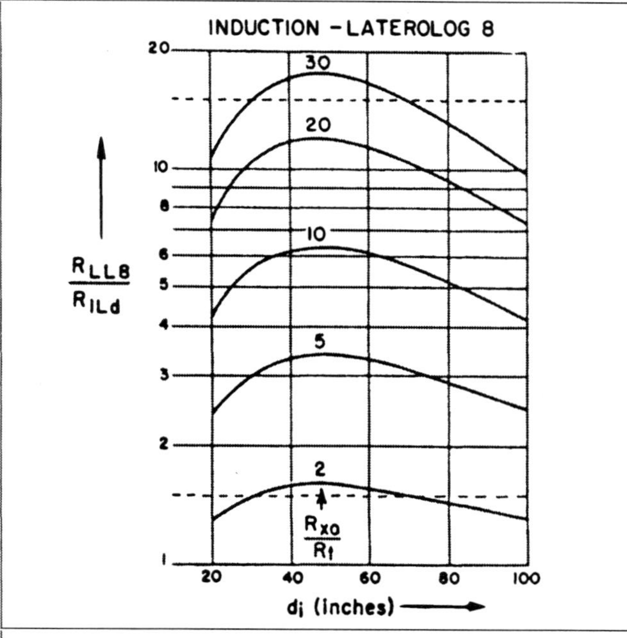

Fig. 12.13—Variations of R_{LL8}/R_{ILd} vs. d_i for various values of R_{xo}/R_t (from Ref. 3).

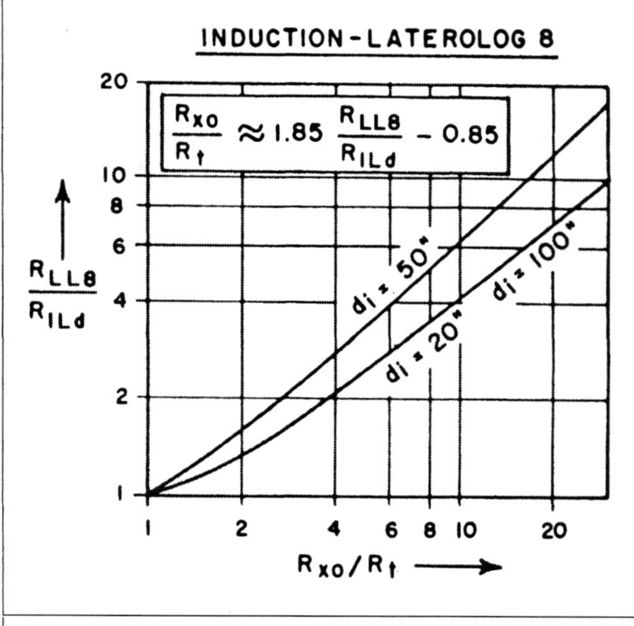

Fig. 12.14—R_{LL8}/R_{ILd} vs. R_{xo}/R_t limit curves for 20 in. $< d_i <$ 100 in.

$$(E_{SP})_{QL} = (E_{SP})_{log} - K \log(S_w/S_{xo})^n. \quad (12.21)$$

In water-bearing zones or zones with no movable hydrocarbons, $S_w = S_{xo}$ and Eq. 12.21 reduces to

$$(E_{SP})_{QL} = (E_{SP})_{log}. \quad (12.22)$$

In water-bearing zones, the E_{SP} curve and either the $(E_{SP})_{QL}$ or the $(R_{xo}/R_t)_{QL}$ curve will converge, showing no or negligible separation. An example of this is the water sand of Fig. 12.12 in the interval 3,692 to 3,731 ft. In hydrocarbon-bearing zones with movable saturation, $S_{xo} > S_w$ and $\log(S_w/S_{xo}) < 0$. Then in absolute value,

$$(E_{SP})_{QL} < (E_{SP})_{log}. \quad (12.23)$$

In hydrocarbon-bearing zones with movable saturation, the $(E_{SP})_{QL}$ or $(R_{xo}/R_t)_{QL}$ curve will separate from the measured E_{SP} curve. This is the case of the log interval from 3,760 to 3,788 ft in Fig. 12.12. The zone below is also hydrocarbon-bearing from 3,793 to 3,880 ft, as indicated by the separation between the R_o and R_{ILd} curves. Hydrocarbon production can be expected only from the top 6 ft (3,793 to 3,799 ft), as indicated by the separation between the two SP curves.

For wellsite interpretation, R_{xo}/R_t is calculated from the ratios R_{LL8}/R_{ILd} or R_{SFL}/R_{ILd} using average relationships.[3] **Fig. 12.13** shows the variations of R_{LL8}/R_{ILd} vs. d_i for various values of R_{xo}/R_t. **Fig. 12.14** is derived from Fig. 12.13 and shows the grouping of the curves for 20 in. $< d_i <$ 100 in. For this range, the curves of R_{LL8}/R_{ILd} vs. R_{xo}/R_t lie between the two limit curves shown. These curves can be approximately represented by the average relation

$$R_{xo}/R_t = 1.85(R_{LL8}/R_{ILd}) - 0.85. \quad (12.24)$$

A similar relation can be developed for the ILd/spherically focused log combination. In cases of either very shallow or abnormally deep invasions, the computed R_{xo}/R_t value will be too low, and the conclusion reached with the $(R_{xo}/R_t)_{QL}$ method will be optimistic. This can often be recognized by observing a separation between $(R_{xo}/R_t)_{QL}$ and the E_{SP} curve in water sands. This effect can be corrected for by normalizing the $(SP)_{QL}$ curve to match the SP in water sands.

The independence of the R_{xo}/R_t ratio of the porosity makes it an extremely valuable tool in complex lithology interpretation problems. In this case, accurate ϕ and subsequently accurate F and R_o values cannot be obtained. This, of course, can result in gross misinterpretation. This advantage can be illustrated best by the log in **Fig. 12.15**. The operator had no shows on the mud log while drilling the interval shown. Also, the separation between the R_o and R_{ILd} curve is small, indicating high water saturation. This is an example of a well that would have been abandoned except for the fact that a separation exists between the $(R_{xo}/R_t)_{QL}$ and E_{SP} curves, indicating movable hydrocarbons.

The well was completed in the zones marked as being perforated. Initial production was 240 BOPD with no water. The high R_o calculated results from the presence of volcanic material in the sands.[3] The matrix density of these sands is much higher than the routinely accepted value of 2.65 g/cm^3. Use of a matrix density of 2.65 g/cm^3 resulted in a low apparent porosity, which subsequently resulted in a high F and R_o.

A value of water saturation, S_w, can be derived empirically from the difference in millivolts, Δ_{SP}, between the E_{SP} and $(R_{xo}/R_t)_{QL}$ curves. According to Eq. 12.21,

$$\Delta_{SP} = -K \log(S_w/S_{xo})^n, \quad (12.25)$$

where K is expressed in terms of formation temperature, T_f, by Eq. 6.11:

$$K = 61.3 + 0.133\, T_f. \quad (12.26)$$

To determine S_w from Eq. 12.25, S_{xo} must be known. For moderate invasion and typical residual oil saturation, an empirical relation between S_{xo} and S_w is given by[4]

$$S_{xo} = S_w^{1/5}. \quad (12.27)$$

This relation is strictly empirical and may differ appreciably from the actual case. Its use here is justified by the reconnaissance nature of the evaluation. Inserting this into Eq. 12.25 and taking $n=2$ results in

$$\Delta_{SP} = -1.6\, K \log S_w. \quad (12.28)$$

Solving for S_w gives

$$S_w = 10^{\Delta_{SP}/-1.6K}. \quad (12.29)$$

Example 12.4. Refer to **Fig. 12.16** in providing the following information.

a. Using Eq. 12.19, calculate $(E_{SP})_{QL}$ at Level A. Compare the calculated value with that displayed by the log.

b. Explain the relatively high resistivity displayed at Level B.

c. Estimate S_w and S_{xo} at Level C.

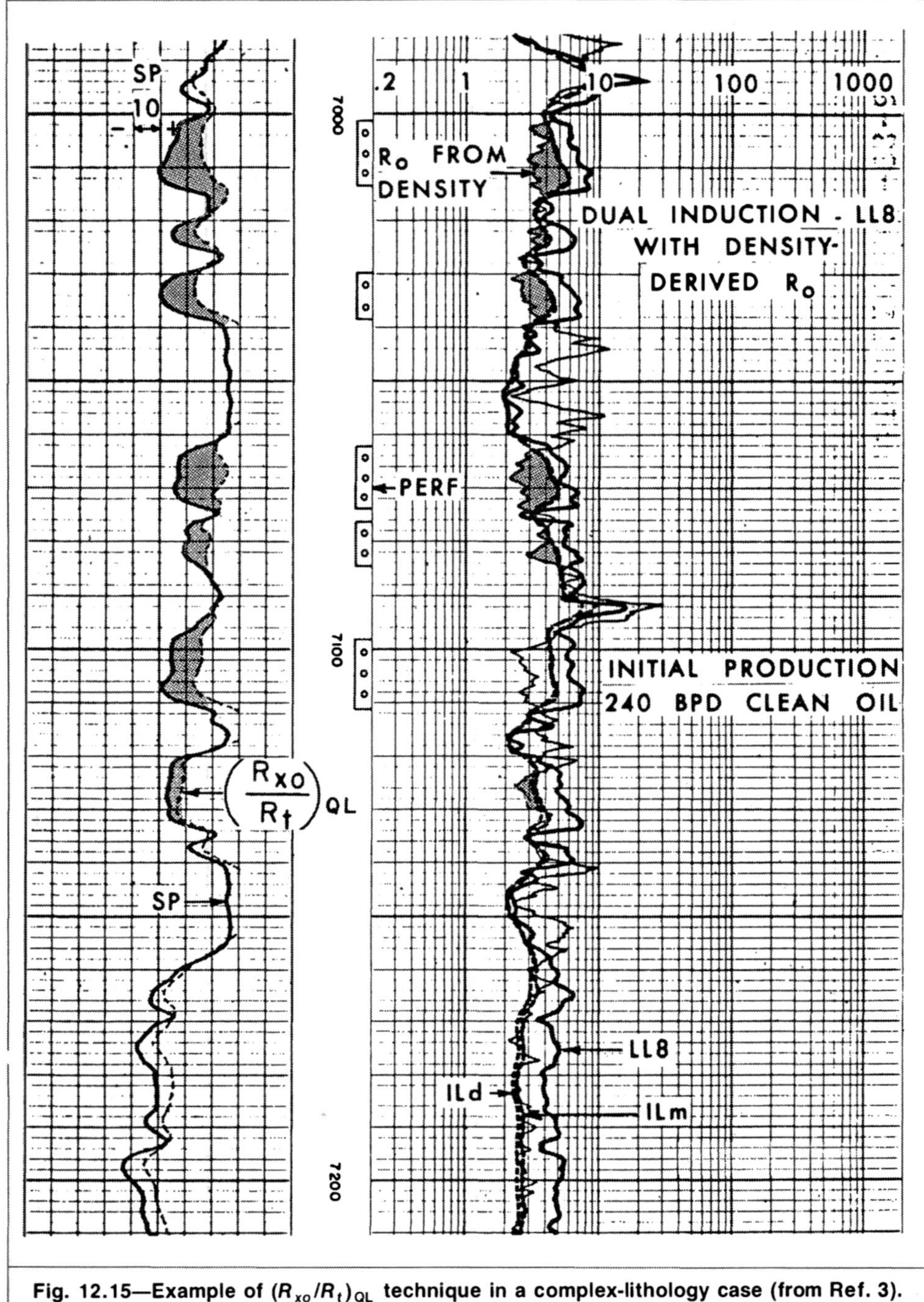

Fig. 12.15—Example of $(R_{xo}/R_t)_{QL}$ technique in a complex-lithology case (from Ref. 3).

Solution.

a. The log displays the following values at Level A: $R_{ILd}=1$ $\Omega\cdot$m, $R_{LL8}=7$ $\Omega\cdot$m. From Eq. 12.24,

$$R_{xo}/R_t=1.85(7/1)-0.85=12.1$$

Assuming a geothermal gradient of 1°F/100 ft and a surface temperature of 70°F, we estimate the formation temperature to be 117°F. K can be calculated by Eq. 12.26.

$$K=61.3+0.133(117)=76.9 \text{ mV}.$$

Now, from Eq. 12.19,

$$(E_{SP})_{QL}=-76.9 \log(12.1)=-83.3 \text{ mV}.$$

This calculated value is very close to that displayed by the log.

b. The E_{SP} and $(R_{xo}/R_t)_{QL}$ curves do not show a separation at Level B. This indicates a water-bearing zone of lower porosity than the rest of the sands. Assuming uniform porosity within the sand that encompasses Zones A and B, Zone B could be a hydrocarbon-bearing zone containing residual hydrocarbons.

c. S_w can be estimated from Eq. 12.29. From the log,

$$\Delta_{SP}=68 \text{ mV}.$$

From Part a,

$$K=76.9 \text{ mV}.$$

Thus,

$$S_w=10^{68/1.6(-76.9)}=0.28 \text{ or } 28\%.$$

The conventional approach—i.e., $(S_w)_C=(R_A/R_C)^{½}$,

$$(S_w)_C=(1/13)^{½}=0.28 \text{ or } 28\%—$$

yields the same value.

Eq. 12.29 is based on the empirical relation given by Eq. 12.27. Then,

$$S_{xo}=(0.28)^{⅕}=0.78 \text{ or } 78\%.$$

12.6 MOP's

The detection of movable hydrocarbons with the R_{xo}/R_t method requires a good-quality SP log. The technique is ineffective in high-

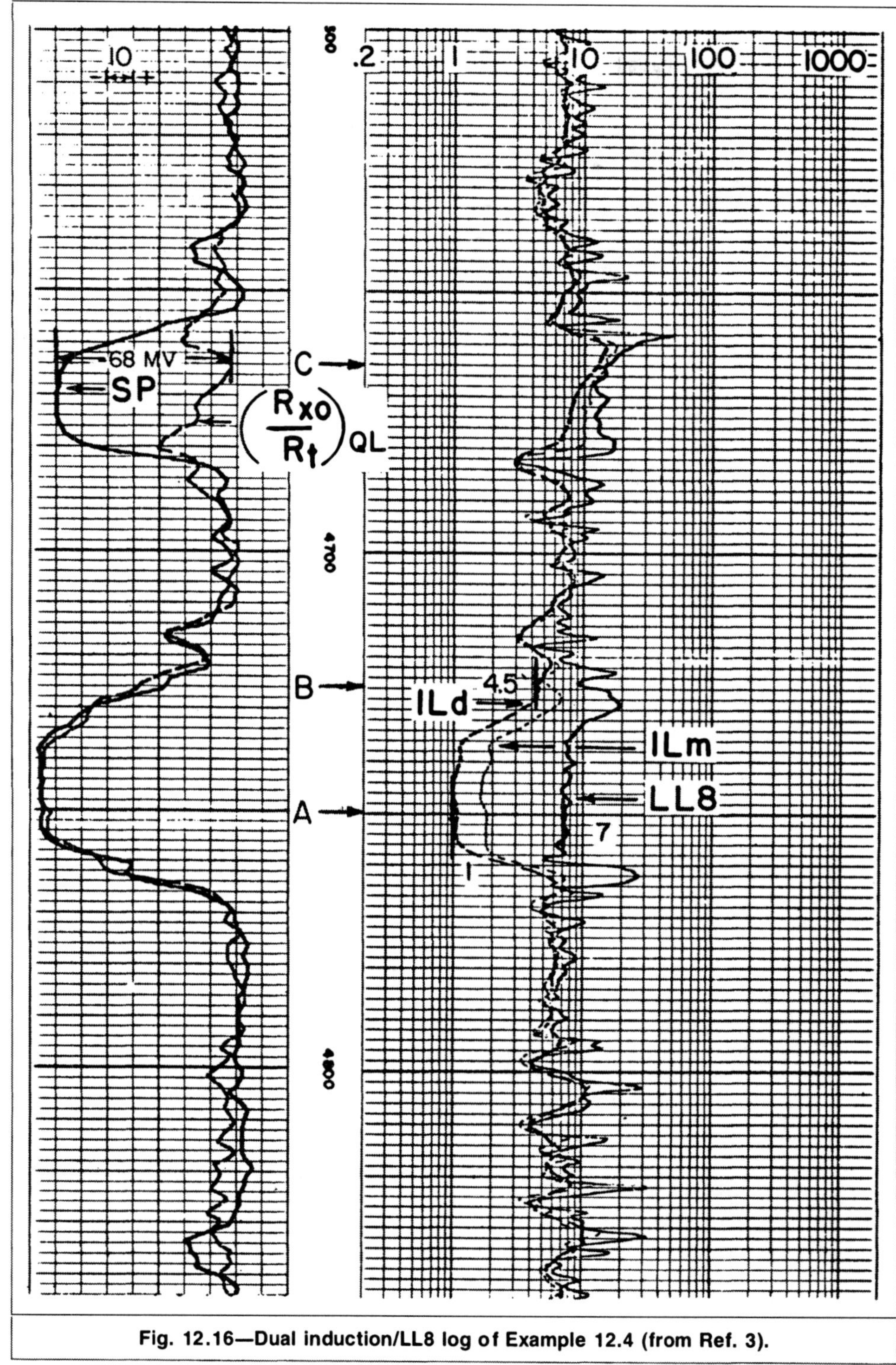

Fig. 12.16—Dual induction/LL8 log of Example 12.4 (from Ref. 3).

resistivity carbonate formations and in wells drilled with saltwater-based muds. An expanded version of the F_{xo}/F_s approach described in Sec. 12.4 yields a plot of movable oil ($F-$MOP) that is effective in locating movable hydrocarbons. The $F-$MOP compares three formation resistivity factors calculated with three different approaches. In addition to comparing F_{xo} and F_s, the plot also compares F_{xo} to a formation resistivity factor, F_R, calculated from a deep resistivity log. F_R is defined as

$$F_R=R_{deep}/R_w. \quad (12.30)$$

Assuming $R_t=R_{deep}$ gives

$$F_R=R_t/R_w=F/S_w^2. \quad (12.31)$$

F_{xo} and F_s are given by Eqs. 12.14 and 12.15. The three F curves are plotted on a logarithmic grid as shown in **Fig. 12.17**.

Combining Eqs. 12.14 and 12.30 results in

$$F_R/F_{xo}=(S_{xo}/S_w)^2. \quad (12.32)$$

It can be deduced easily from Eq. 12.32 that a separation between the F_R and F_{xo} curves gives a qualitative indication of movable hydrocarbons. A quantitative estimate of hydrocarbon movability is determined by comparing S_w and S_{xo} values that can be determined from F curves.

The ratio F_{xo}/F_s is given by Eq. 12.16, which can be expressed in logarithmic form as

$$\log S_{xo}=0.5(\log F_s-\log F_{xo}). \quad (12.33)$$

On a logarithmic grid, the separation between F_s and F_{xo} is proportional to S_{xo}. The ratio F_R/F_s can be derived from Eqs. 12.15 and 12.31:

$$F_R/F_s=1/S_w^2. \quad (12.34)$$

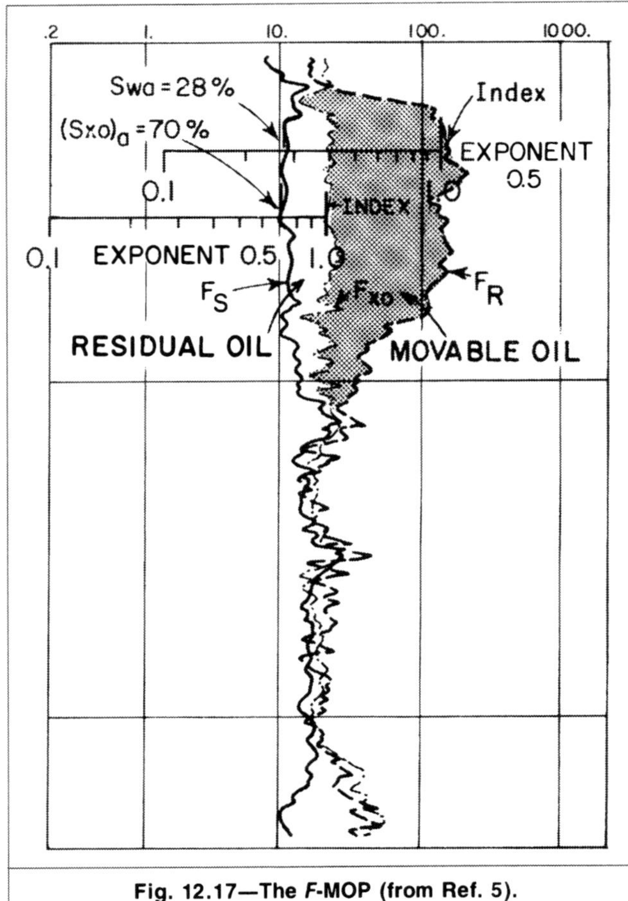

Fig. 12.17—The *F*-MOP (from Ref. 5).

In logarithmic form,

$$\log S_w = 0.5(\log F_s - \log F_R). \quad (12.35)$$

As for S_{xo}, the separation between F_s and F_R is proportional to S_w. The value of S_{mo} is calculated from S_w and S_{xo}, which are determined as shown in Fig. 12.17 by applying a logarithmic overlay scale with exponent 0.5. The F_s curve could be scaled in terms of porosity with the exponent 0.5 overlay scale. The concept and construction of this scale are discussed in Sec. 12.3.

The $F-$MOP provides a quick-look qualitative log of porosity, water saturation, residual hydrocarbons, and movable hydrocarbons. A quantitative recording of the above parameters is obtained by another version of MOP generated in terms of porosity instead of formation factor.[6] This plot compares the porosity values calculated from the sonic log, F_{xo}, and F_R. These porosities are defined as ϕ_s, ϕ_{xo}, and ϕ_R, respectively. They are recorded on a linear scale (**Fig. 12.18**).

With an F-ϕ relation of the form $F=\phi^{-2}$, Eqs. 12.13 and 12.30 yield

$$\phi_{xo} = (R_{mf}/R_{xo})^{1/2} \quad (12.36)$$

and $\phi_R = (R_w/R_t)^{1/2}$. (12.37)

Replacing R_{xo} and R_t with FR_{mf}/S_{xo}^2 and FR_w/S_w^2, respectively, and recalling that the sonic log porosity is equal to the true porosity in clean liquid-filled formations results in

$$\phi_s = \phi, \quad (12.38a)$$

$$\phi_{xo} = \phi S_{xo}, \quad (12.38b)$$

and $\phi_R = \phi S_w$. (12.38c)

These expressions show that ϕ_{xo} and ϕ_R are the water bulk volumes in the invaded and noninvaded zones of the formation, respectively.

The significance of the relative values of the three porosities can be summarized as follows. In water-bearing zones,

$$\phi_s = \phi_{xo} = \phi_R. \quad (12.39)$$

In hydrocarbon-bearing zones with no movable saturation,

$$\phi_s > \phi_{xo} = \phi_R. \quad (12.40)$$

In hydrocarbon-bearing zones with movable saturation,

$$\phi_s > \phi_{xo} > \phi_R. \quad (12.41)$$

Eqs. 12.38 indicate that

$$\phi_s - \phi_R = \phi(1 - S_w) \quad (12.42)$$

and $\phi_{xo} - \phi_R = \phi(S_{xo} - S_w) = \phi S_{mo}$. (12.43)

In addition to indicating the presence of movable hydrocarbons qualitatively, the separation between the ϕ_s and ϕ_R curves and between the ϕ_{xo} and ϕ_R curves is a quantitative measure of the bulk volume of the oil in place (OIP) and of the movable hydrocarbons, respectively. In addition to being used as a quick-look technique,

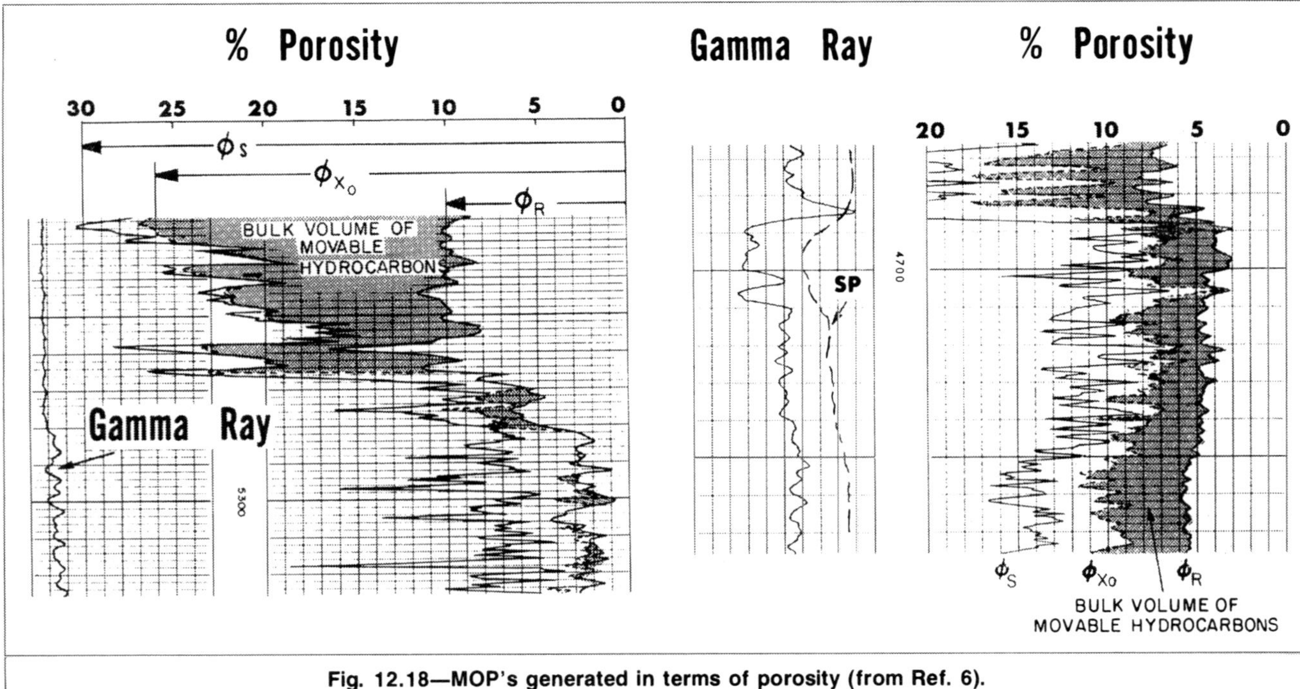

Fig. 12.18—MOP's generated in terms of porosity (from Ref. 6).

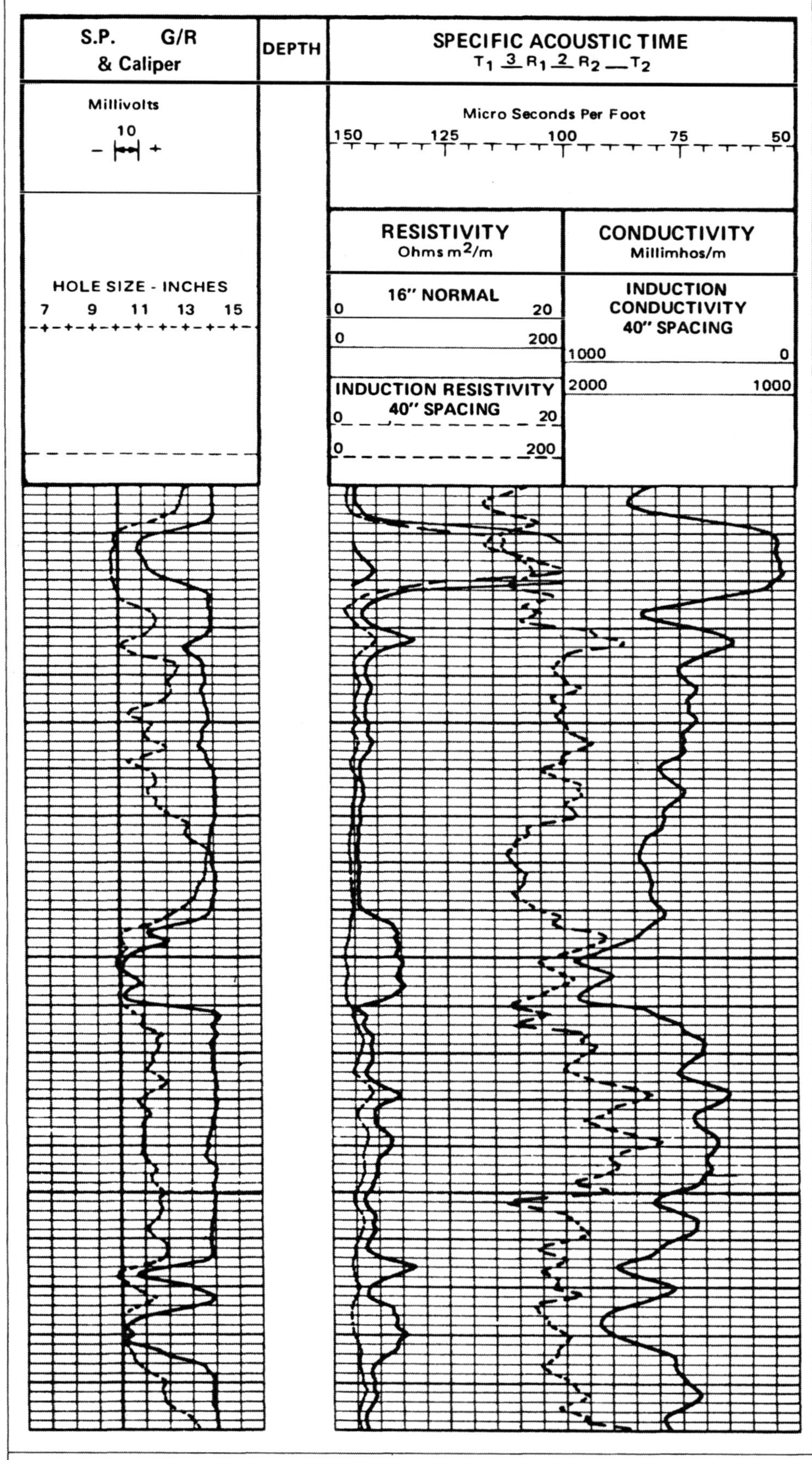

Fig. 12.19—Induction Electrolog/Acoustilog for Example 12.5 (courtesy Atlas Wireline Services).

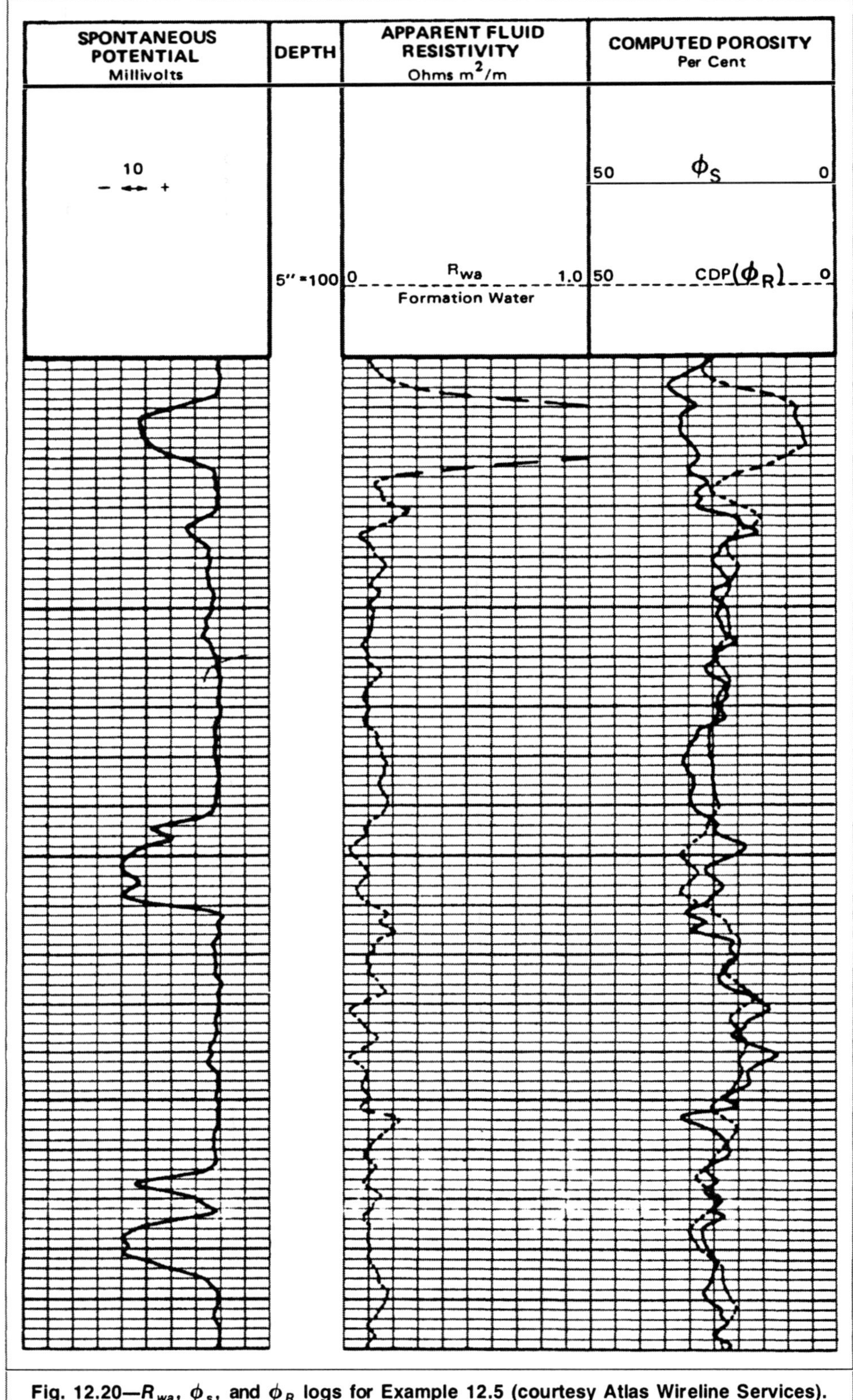

Fig. 12.20—R_{wa}, ϕ_s, and ϕ_R logs for Example 12.5 (courtesy Atlas Wireline Services).

the porosity MOP is used in the presentation of an in-depth computer-processed interpretation.[5]

Example 12.5. Fig. 12.19 is a combination induction Electrolog™/Acoustilog™ from a south Texas oil well. This log was processed by a computer to obtain the R_{wa}, ϕ_s, and ϕ_R logs in **Fig. 12.20.**

a. Do the logs show any oil zones?
b. Estimate the OIP (B_o=1.2).
c. Does the zone contain movable oil?

Solution.

a. The logs show a 12-ft-thick hydrocarbon-bearing zone at the top of the interval illustrated. The presence of hydrocarbon is indicated both by the high R_{wa} value, relative to that of lower sands, and by the separation between ϕ_s and ϕ_R.

b. OIP is given by

N=7,758$h\phi(1-S_w)/B_o$, STB/acre.

From the computed porosity logs $\overline{(\phi_s)}$=30% and $\overline{(\phi_R)}$=7%. Because $\phi_x-\phi_R=\phi(1-S_w)$, as indicated by Eq. 12.42,

N=7,758(12)(0.3−0.07)/1.2=17,800 STB/acre.

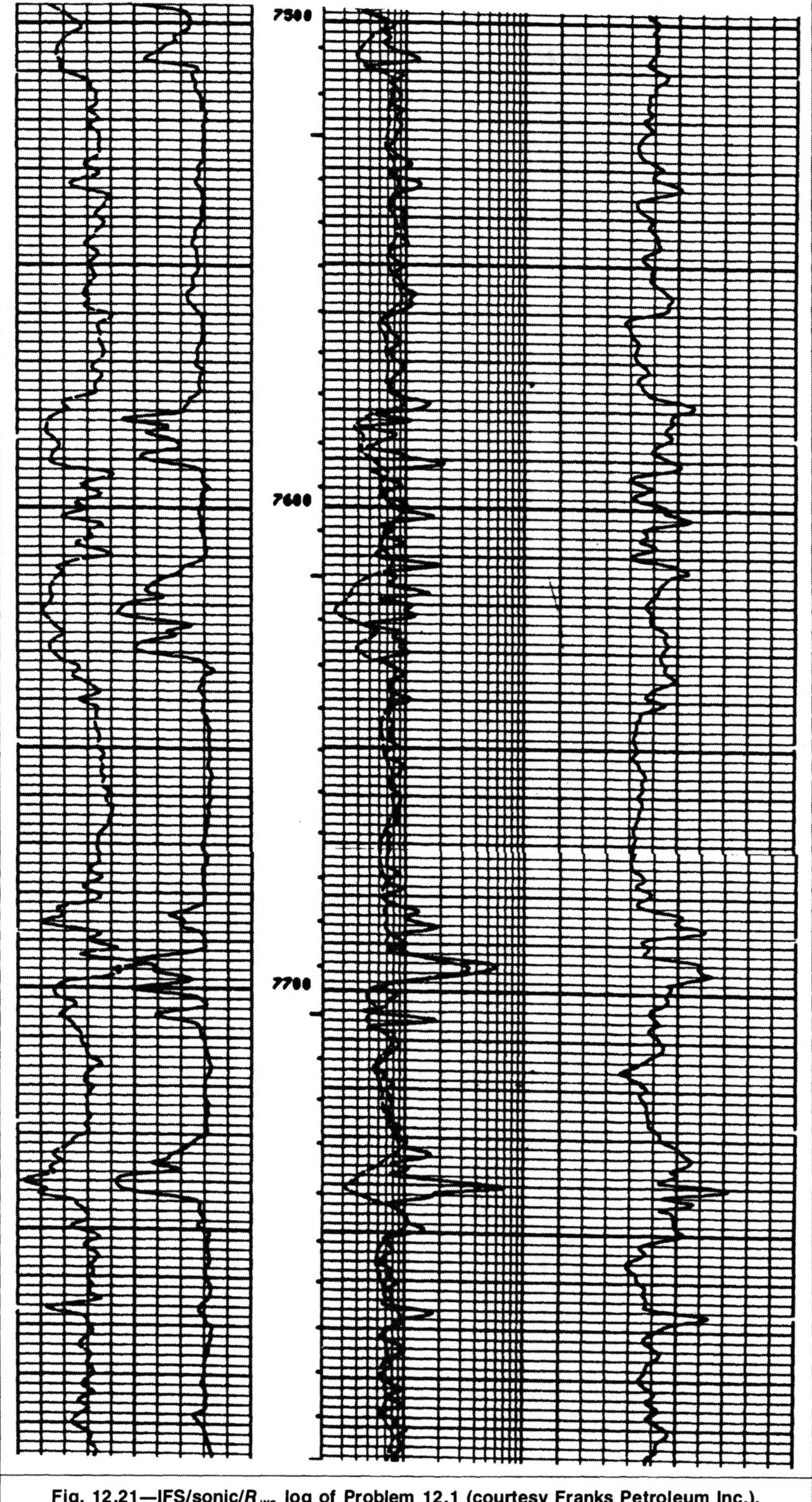

Fig. 12.21—IFS/sonic/R_{wa} log of Problem 12.1 (courtesy Franks Petroleum Inc.).

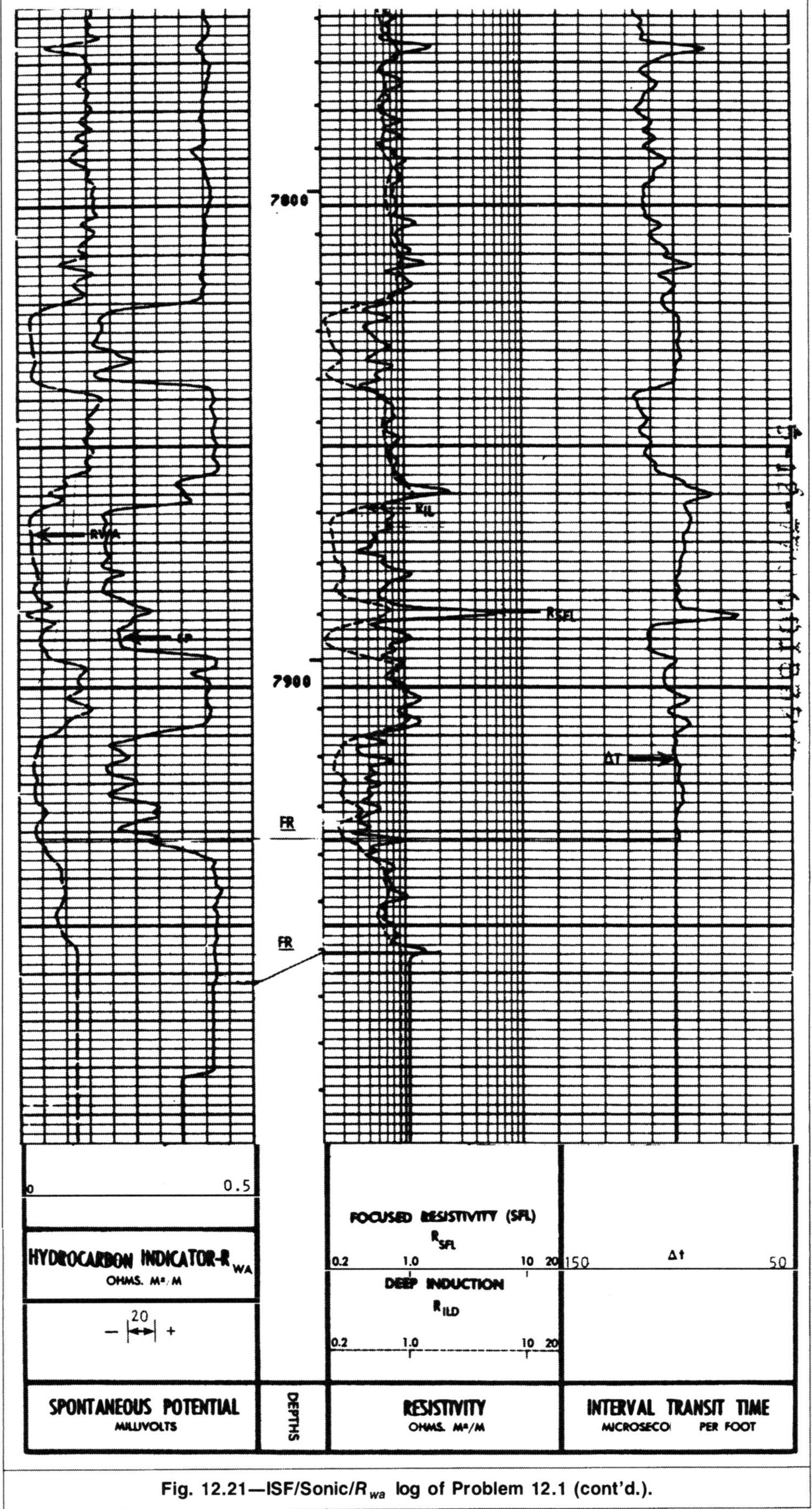

Fig. 12.21—ISF/Sonic/R_{wa} log of Problem 12.1 (cont'd.).

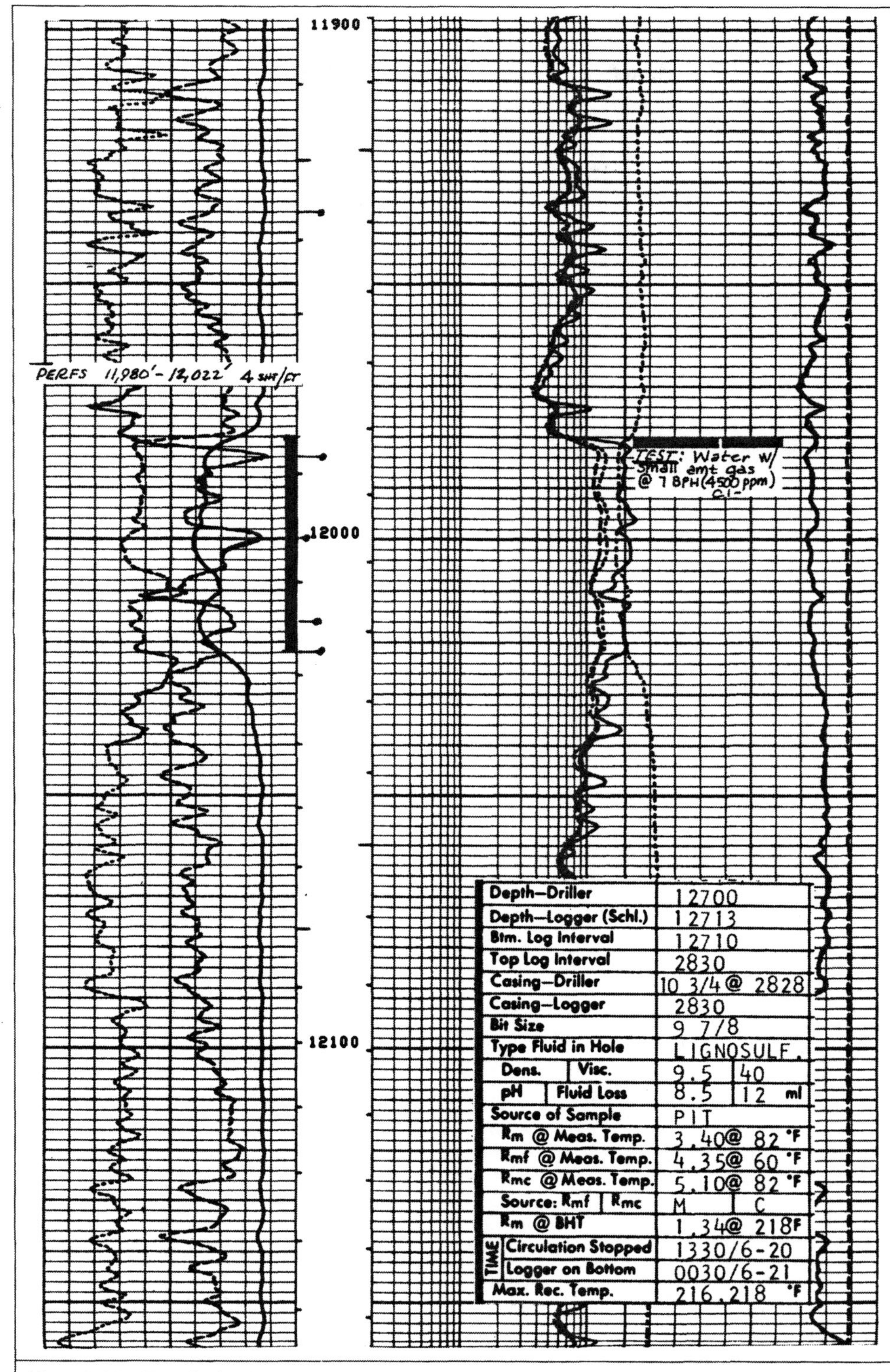

Fig. 12.22—Dual induction/borehole-compensated sonic/R_{wa} log for Problem 12.2 (courtesy Amoco Oil Co.).

c. Because the logs do not show ϕ_{xo}, a direct indication of movable oil is not available. However, from the ϕ_s and ϕ_R readings, the oil saturation can be estimated to be 77%:

$\phi_R=\phi_s S_w$,

$S_w=\phi_R/\phi_s=0.07/0.3=0.23$ or 23%,

so $S_o=0.77$ or 77%.

With this relatively high oil saturation, oil movability is expected.

Review Questions

1. What is the main purpose of reconnaissance interpretation techniques?

2. What data are required to perform R_{wa} evaluation? What parameter is needed for conventional interpretation but not necessary for estimating S_w with the R_{wa} technique? Can this parameter be obtained from R_{wa} calculation?

3. Give examples of zones other than hydrocarbon-bearing zones that will display relatively high R_{wa} values. Explain why and how these zones are distinguished from hydrocarbon zones.

4. Why are the induction and sonic logs most commonly used for R_{wa} calculations? Can the density or neutron log be used instead of the sonic log? Explain.

5. Explain the effect of deep mud-filtrate invasion on the R_{wa} evaluation. How can the effect of deep-invasion conditions be recognized?

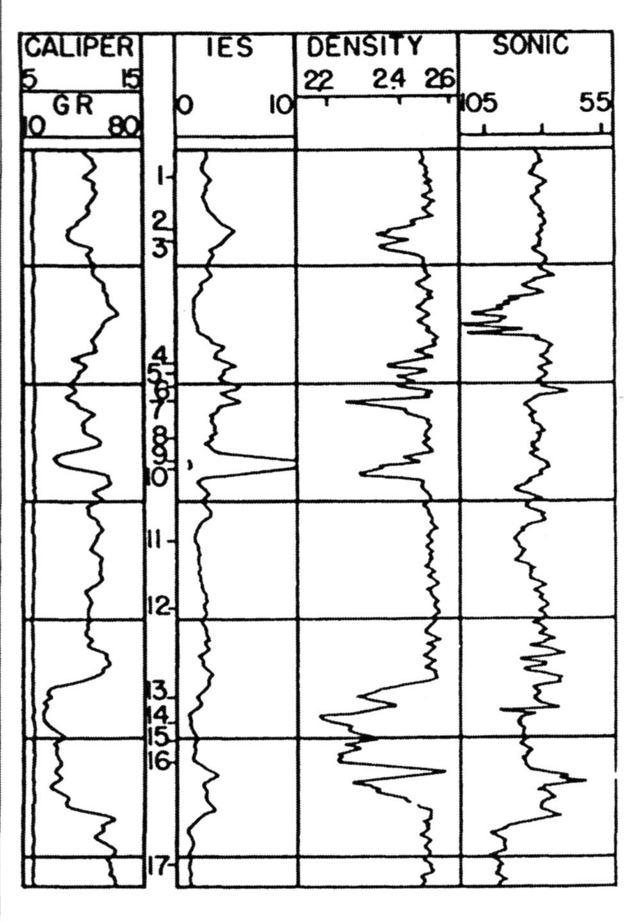

Fig. 12.23—Log data for Problem 12.3 (courtesy Schlumberger).

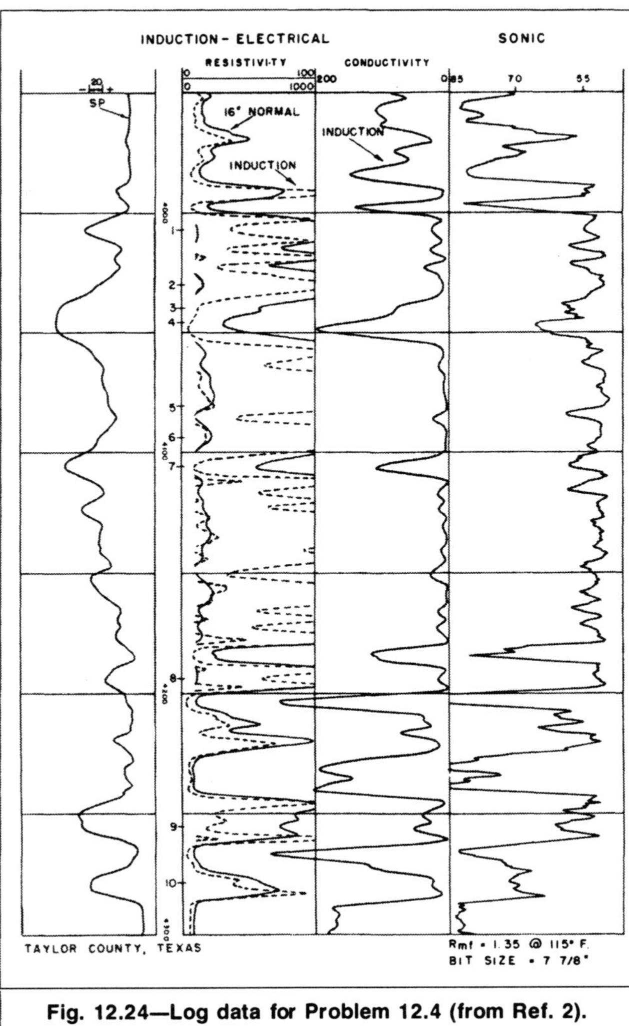

Fig. 12.24—Log data for Problem 12.4 (from Ref. 2).

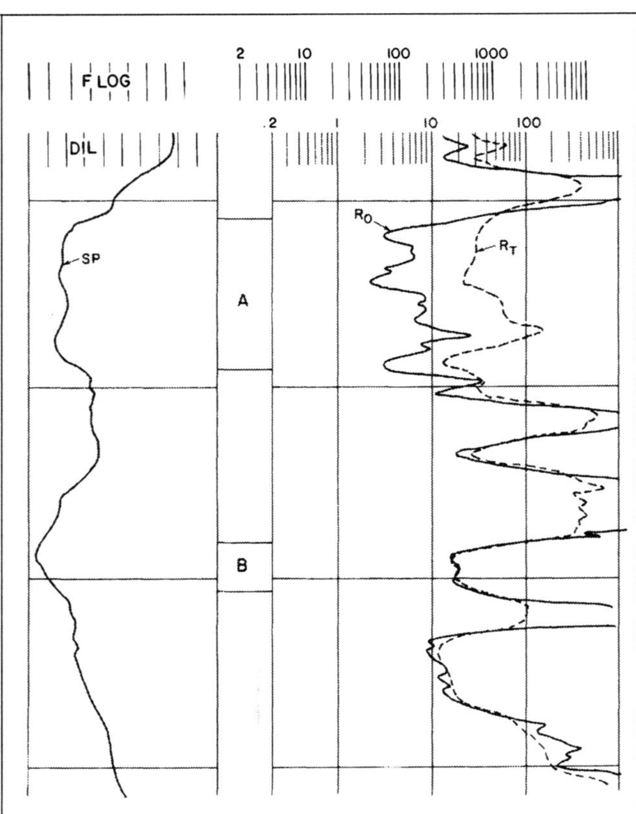

Fig. 12.25—Log data for Problem 12.6 (courtesy Schlumberger).

6. If petrophysical models for clean liquid-filled formations are used, discuss the response of the R_{wa} log in the following cases: clean, consolidated, gas-bearing sands; clean, unconsolidated, gas-bearing sands; shaly sands; and clean, liquid-filled carbonate.
7. The R_{wa} approach is effective when R_w is constant in the interval studied. How can this condition be checked?
8. Does the consistently lowest value of R_{wa}, $(R_{wa})_{min}$, necessarily equal the R_w value derived from the SP log? Explain.
9. List conditions for which the water saturation value calculated with Eq. 12.5 may differ considerably from the actual saturation.
10. What data are needed to generate the R_o log? What is the advantage of this log?
11. Explain how an F log is converted to an R_o log.
12. Explain how a logarithmic exponent 0.5 scale is constructed.
13. What are the advantages of each of the following reconnaissance interpretation techniques: the F_{xo}/F_s approach, the R_{xo}/R_t method, and the MOP.
14. Show that $F_{xo}/F_s = R_{mfa}/R_{mf}$.
15. What are the definitions of F_R, F_{xo}, and F_s? How are they related to ϕ_R, ϕ_{xo}, and ϕ_s?
16. Compare the concept, advantages, and limitations of both the R_{wa} and the R_{xo}/R_t techniques.
17. What is the fundamental difference between the $F-$MOP and the $\phi-$MOP?

Problems

12.1 Scan **Fig. 12.21,** which shows 400 ft of an ISF/sonic/R_{wa} log obtained in a south Louisiana well.

a. At what depth does the interval shown contain hydrocarbon zone(s)?
b. Give the gross and net thicknesses of the potential sand(s).
c. If the cutoff saturation is 50%, give the depth and thickness of the pay zone(s).

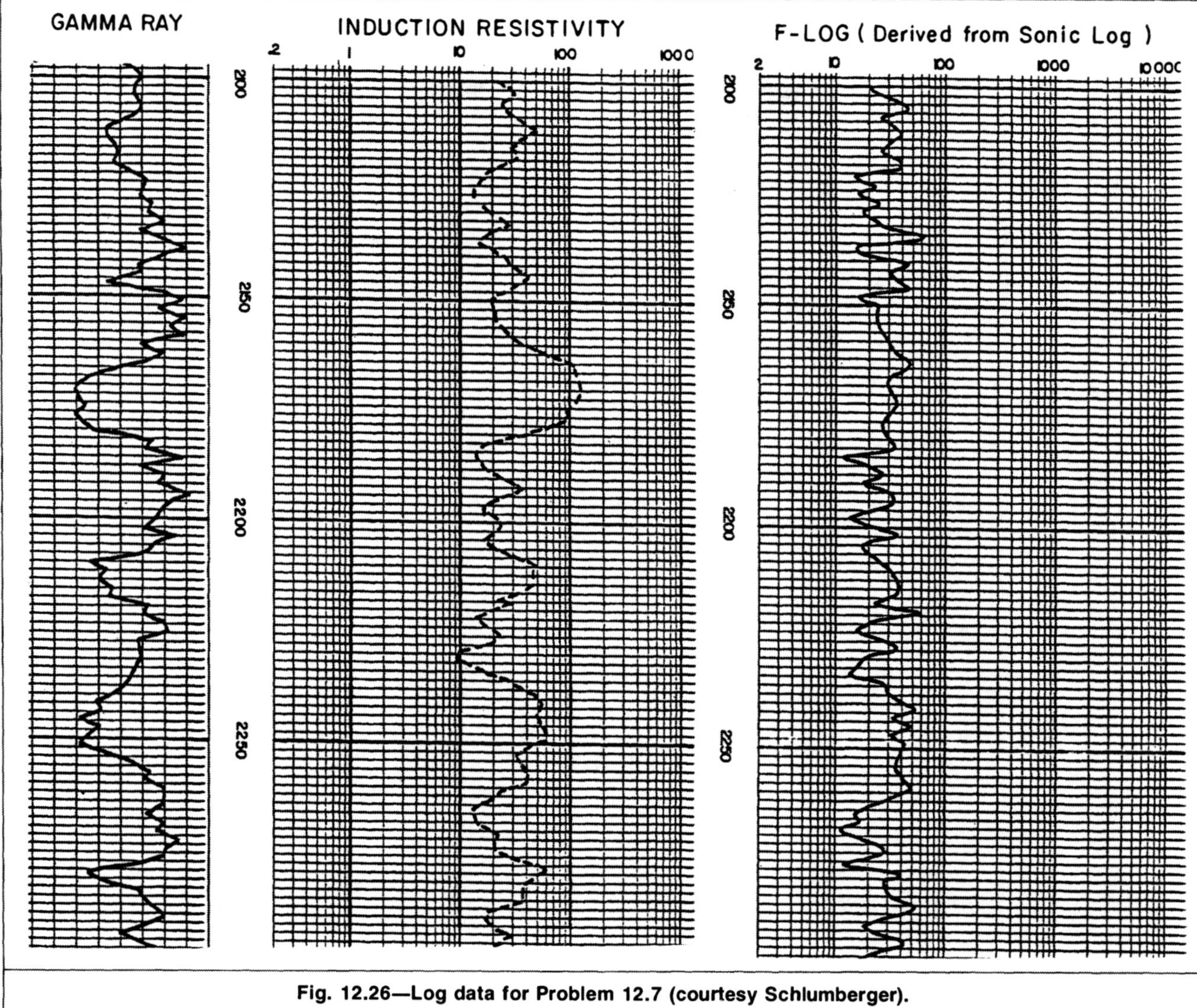

Fig. 12.26—Log data for Problem 12.7 (courtesy Schlumberger).

d. Estimate the average porosity and water saturation of the pay zone(s).
e. Is the above evaluation final? If it is not, which interpretation technique is warranted for finalization.

Note: initial production for the pay zone was 10 BOPD.

12.2 **Fig. 12.22** shows the dual induction/borehole-compensated sonic/R_{wa} log for the interval of 11,900 to 12,160 ft. The interval shown contains a 42-ft-thick sand that displays a relatively high resistivity and high R_{wa}. The sand was perforated and tested. It produced 4,500-ppm-Cℓ water with a small amount of gas.

a. Explain why the R_{wa} value is high.
b. Could the test result be predicted with conventional interpretation techniques? Explain.

12.3 Induction/gamma ray, density, and sonic logs were recorded in oil-based mud of a U.S. gulf coast well that penetrates a sand/shale series. Tabulated below are log values for each of the levels numbered on the log of **Fig. 12.23.**

a. Calculate R_{wa} for each level using sonic log porosity.
b. Repeat the above calculations using density log porosity.
c. R_{wa} density differs slightly from R_{wa} sonic in certain zones (e.g., Zones 5, 6, 13, and 14). Explain the reason for this slight difference.
d. R_{wa} density differs drastically from R_{wa} sonic in Zones 1, 8, 11, 12, and 17. Explain the reason for this drastic difference.
e. What is the most probable value of formation water resistivity?
f. If the cutoff saturation is 50%, what are the probable pay zones?

Level	R_{IL} ($\Omega \cdot m$)	ρ_b (g/cm^3)	Δt (μsec/ft)
1	2.6	2.51	77.5
2	4.8	2.32	81
3	4.0	2.33	79.5
4	5.0	2.345	79.5
5	4.0	2.39	79.5
6	5.0	2.41	79.5
7	5.8	2.24	87
8	3.0	2.49	82
9	11.0	2.32	79.5
10	10.0	2.27	82.5
11	1.65	2.50	90
12	2.3	2.53	79.5
13	1.8	2.26	83.5
14	1.25	2.195	90
15	1.75	2.32	87.5
16	1.50	2.22	87.5
17	1.25	2.49	94

12.4 **Fig. 12.24** illustrates the logs of a well surveyed in fresh mud in the Strawn section, Taylor County, TX. Ten zones of interest are marked, and log data pertaining to these zones are listed below.

Zone	R_{IL} ($\Omega \cdot m$)	R_{SN} ($\Omega \cdot m$)	Δt (μsec/ft)
1	37.5	125	54.5
2	100	200	50.5

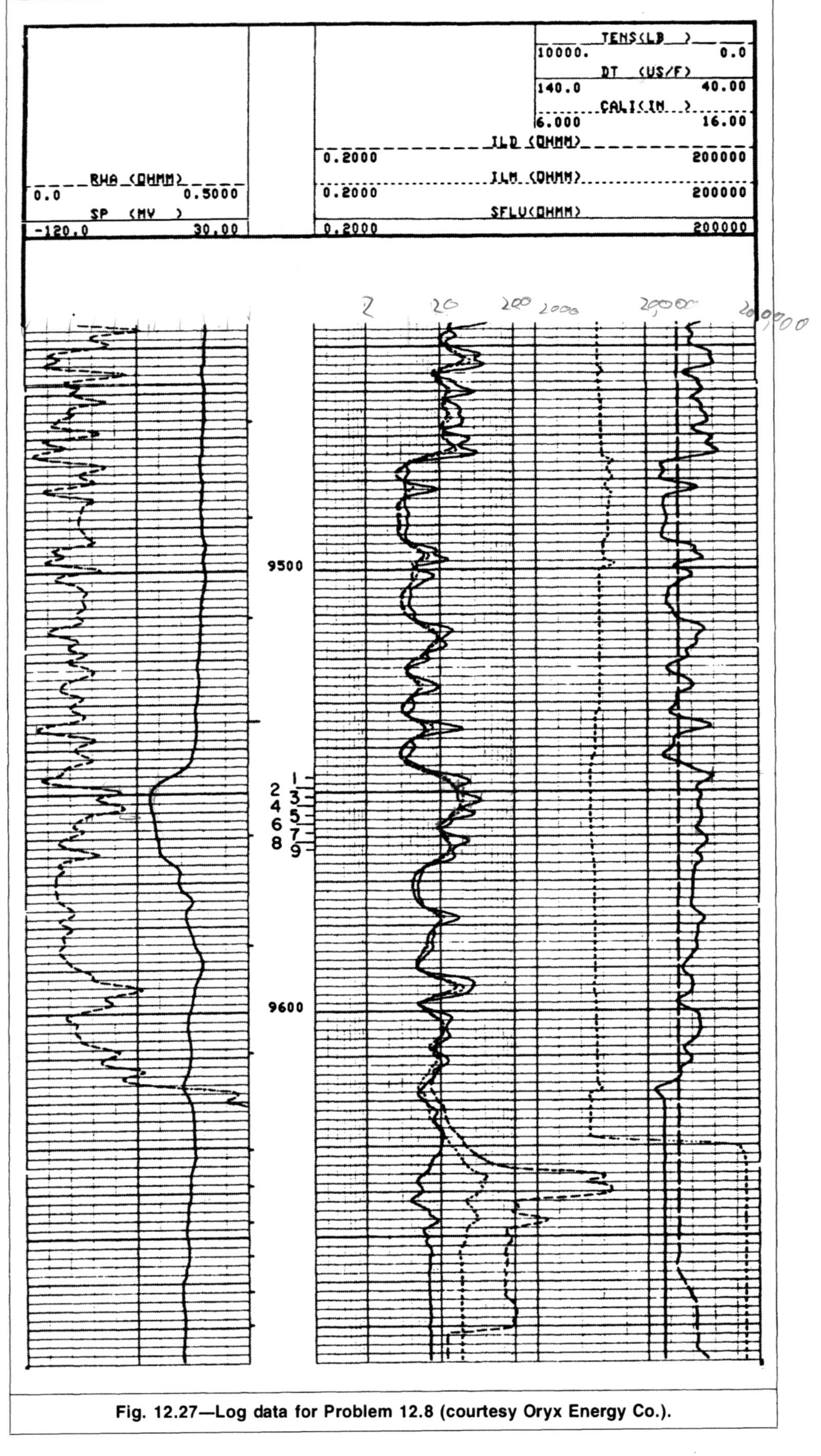

Fig. 12.27—Log data for Problem 12.8 (courtesy Oryx Energy Co.).

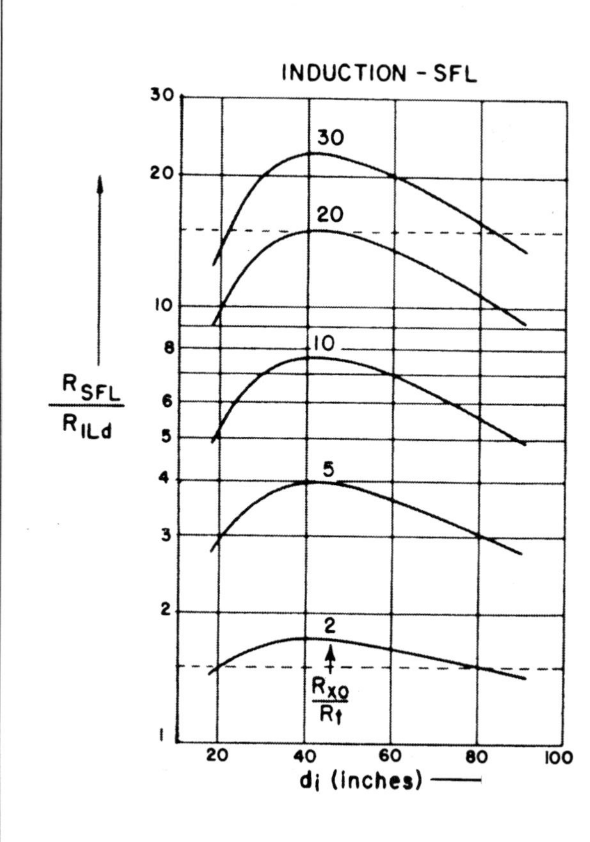

Fig. 12.28—Variations of R_{SFL}/R_{ILd} vs. d_i for various values of R_{xo}/R_t (from Ref. 3).

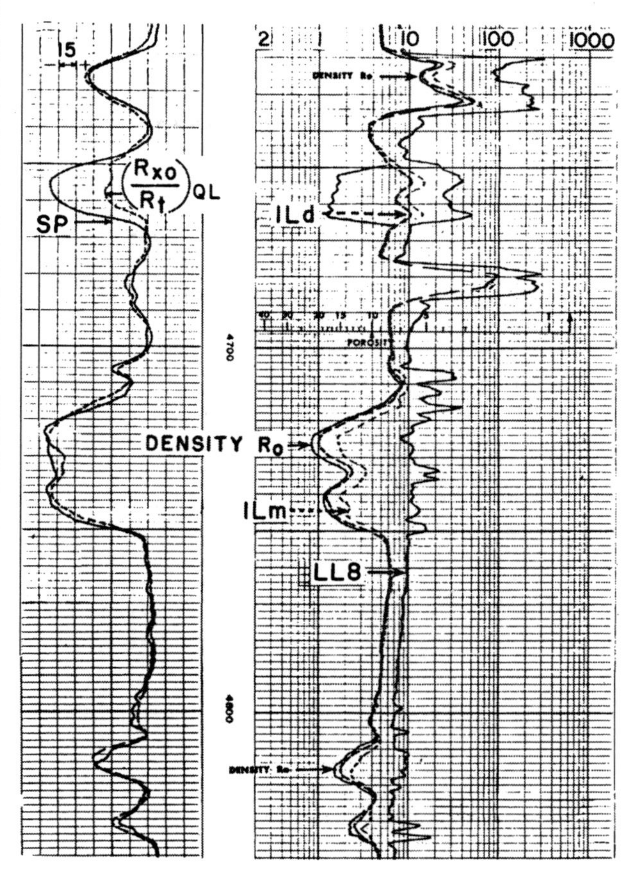

Fig. 12.29—Log data for Problem 12.10 (from Ref. 3).

3	12.5	50	58.5
4	5	22.5	65.0
5	230	380	49.5
6	175	310	51
7	9.5	40	58
8	60	150	52
9	17.5	75	61
10	40	60	70.5

Additional data available are as follows

$T_f = 115°F$.

$R_{mf} = 1.35\ \Omega \cdot m$ at T_f.

Bit size $= 7\frac{7}{8}$ in.

$\Delta t_{ma} = 47.6\ \mu sec/ft$.

Use the R_{wa} technique to point out potential hydrocarbon zones. Because the sonic log response indicates low porosity and possible deep invasion, use the R_{mfa} approach to support the R_{wa} evaluation.

12.5 Construct a logarithmic exponent 0.5 scale that matches the grid of Fig. 12.5. Use the scale to determine the minimum water saturation indicated by the log.

12.6 **Fig. 12.25** illustrates an R_o log. The log is derived from an F log that is normalized in Water Zone B.
a. Does Zone A contain hydrocarbons? Explain.
b. Give the average water saturation in the top 5 ft of Zone A.
c. Calibrate the R_o curve in porosity units; then determine the average porosity of the top 5 ft of Zone A.
d. What is the value of R_w in this interval?

12.7 The gamma ray/induction log and F log of a 400-ft interval are shown in **Fig. 12.26.**
a. Trace the F log on transparent tracing paper. Overlay the paper on the induction log and normalize it to derive an R_o log. Trace the induction curve onto the tracing paper.

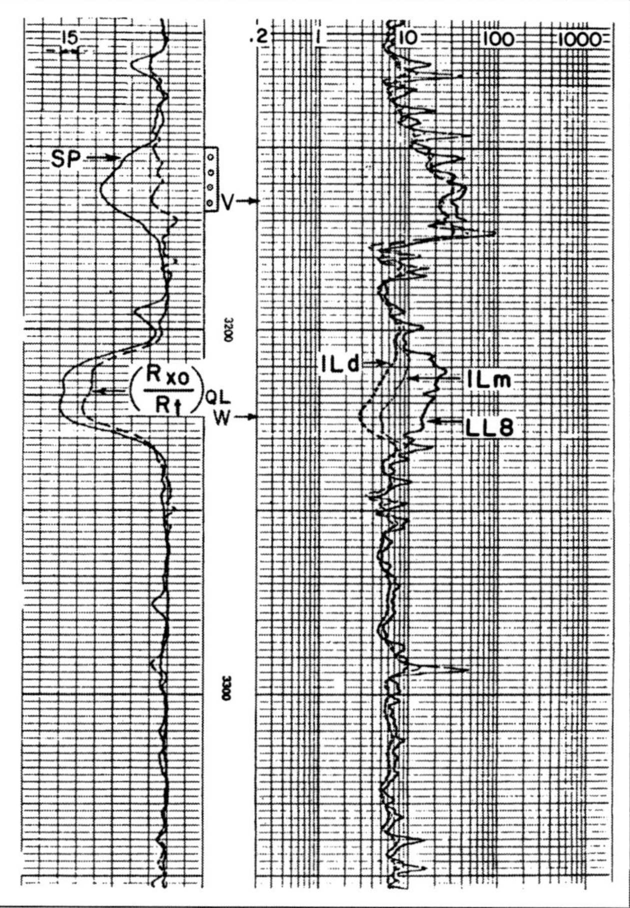

Fig. 12.30—Log data for Problem 12.11 (from Ref. 3).

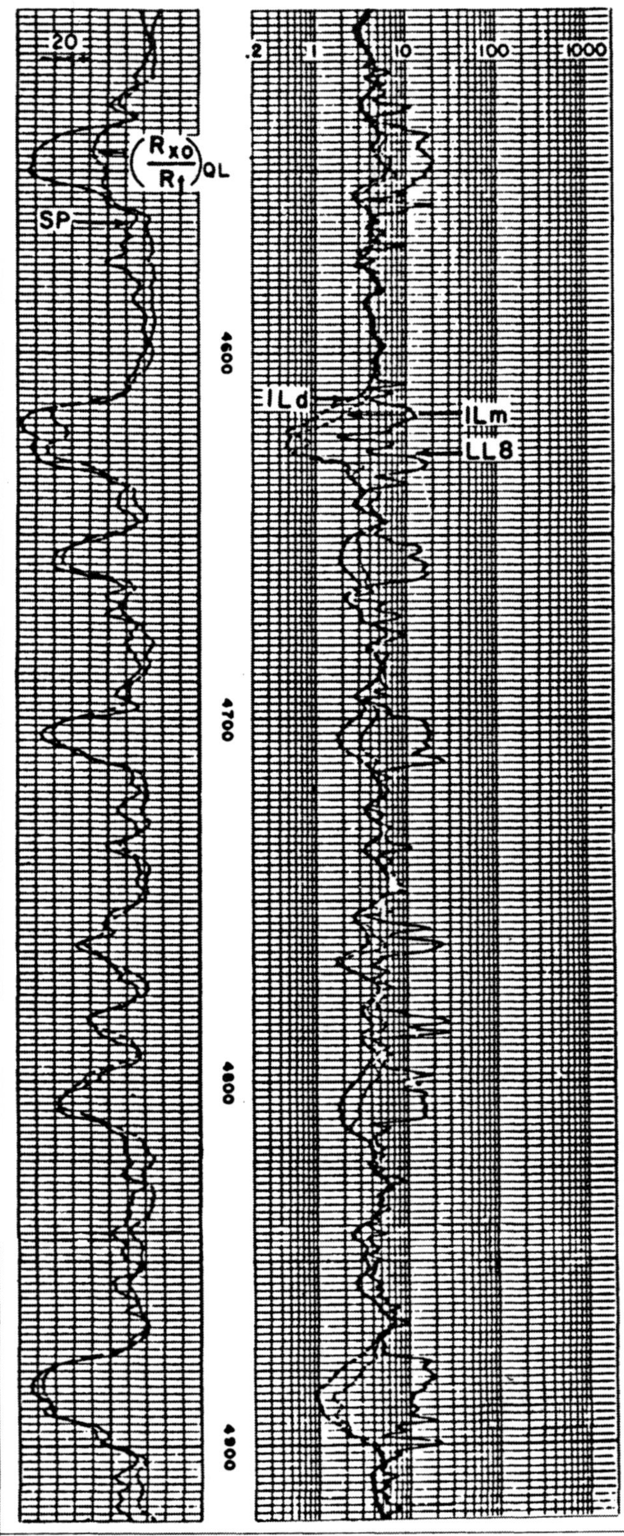

Fig. 12.31—Log data for Problem 12.12 (from Ref. 3).

b. Mark hydrocarbon zone(s) on the tracing paper. Determine the net pay, average porosity, and average water saturation of the hydrocarbon zone(s).

c. Determine R_w.

12.8 The ISF/sonic log section of **Fig. 12.27** shows a low-porosity sand at 9,550 ft. Pertinent wellhead information includes the following.

Total depth=9,663 ft.

Maximum recorded temperature=230°F.

R_m at 73°F=1.0 Ω·m.

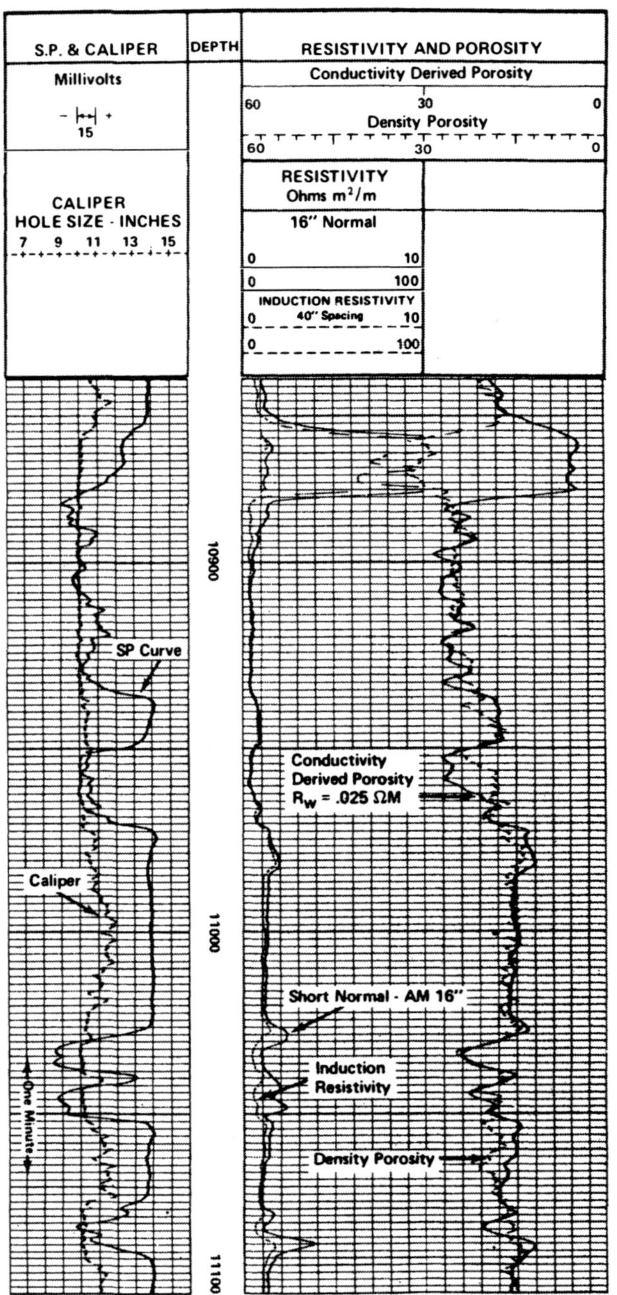

Fig. 12.32—Log for Problem 12.14 (courtesy Atlas Wireline Services).

R_{mf} at 73°F=0.76 Ω·m.

a. Estimate R_w from the SP log.

b. Using the R_{wa} curve, identify the potential zones. Cutoff saturation is 50%.

c. The low porosity and possibility of deep invasion could result in a rather optimistic R_{wa} evaluation. Use the F_{xo}/F_s approach to confirm the findings in Part b. Perform calculations for the nine intervals marked on the log.

12.9 **Fig. 12.28** shows the variation of R_{SFL}/R_{ILd} vs. d_i for various values of R_{xo}/R_t. Use the data displayed by this chart to generate a graph similar to Fig. 12.14 that gives the relation between R_{xo}/R_t and R_{SFL}/R_{ILd}. Find the average relation, in the form of Eq. 12.24, between these two ratios.

12.10 The logs of **Fig. 12.29** show a potential zone at 4,650 ft. The potential of the zone is indicated by the separation between R_o and R_t and between the SP and $(R_{xo}/R_t)_{QL}$ logs. The zone is a gas-bearing limy sand.

a. Using the R_o log, determine the average value of S_w in the subject zone.
b. Using the $(R_{xo}/R_t)_{QL}$ information, determine the average value of S_w in the same zone.
c. Explain the good agreement between the two S_w values obtained above.
d. Estimate S_{xo}.

12.11 The logs obtained over an interval of interest in a north Louisiana well are shown in **Fig. 12.30**. The logs show two sands. The top sand (3,150 to 3,170 ft) has been perforated and tested for gas; the lower sand (3,206 to 3,228 ft) was found to be wet.
a. Show how the $(E_{SP})_{QL}$ value at Levels V and W were calculated.
b. Explain why the calculated $(E_{SP})_{QL}$ falls short of the measured, E_{SP}, value in a wet zone.
c. What do you suggest to avoid misinterpretation that might result in a similar situation?

12.12 Scan the dual induction/Laterolog 8/$(R_{xo}/R_t)_{QL}$ log of **Fig. 12.31**.
a. Indicate the zones containing movable hydrocarbon saturation.
b. Identify the zones that should be evaluated further before a final conclusion can be drawn. Explain why the $(R_{xo}/R_x)_{QL}$ is not conclusive in these zones.

12.13 The following values were obtained in a well drilled through carbonate formations with saltwater-based mud.

Zone	R_{LL3} (Ω·m)	R_{MLL} (Ω·m)	ϕ_s (%)
1	40	3.7	10
2	12	12	15
3	8	1.8	15
4	12	4.7	20
5	10	5.3	15
6	32	10.8	12

Additional available information is as follows.

$m=2$.

$R_{mf}=R_w=0.03$ Ω·m.

Mudcake is thin.

a. Using the quick-look MOP algorithm, determine the movable oil and water saturations for each zone.
b. Is the indication given by Part a conclusive in Zone 2? If not, what plan of action should be recommended?

12.14 **Fig. 12.32** shows a combination induction Electrolog/Densilog with conductivity-derived porosity (ϕ_R). The logs show a potential oil sand at the top. Estimate the OIP to be assigned to this zone. Use a drainage area of 40 acres and oil shrinkage factor of 0.8.

Nomenclature

B = FVF, RB/STB
B_{cp} = compaction correction factor
d_i = invasion diameter, in.
E_{SP} = spontaneous potential, mV
F = formation resistivity factor
K = temperature-dependent coefficient in Eqs. 6.11, mV
N = volume of oil in place, STB
R = resistivity, Ω·m
S = saturation, fraction
T_f = formation temperature, °F
Δt = transit time, μsec/ft
ϕ = porosity, fraction
ρ = bulk density, g/cm³

Subscripts

a = apparent
c = cutoff value
f = fluid
gr = residual gas
ILd = Deep Induction log
LL3 = Laterolog-3 tool
LL8 = Laterolog-8 tool
log = log response
ma = matrix
mf = mud filtrate
min = minimum
MLL = microlaterolog tool
mo = movable oil
o = oil
QL = quick look
R = resistivity
s = sonic
SFL = spherically focused log
sh = shale
SN = short normal log
t = true
w = water
xo = flushed zone

Superscript

$\bar{}$ = average

References

1. *Well Logging and Interpretation Techniques, The Course for Home Study*, Dresser Atlas Inc., Houston (1982).
2. Tixier, M.P., Alger, R.P., and Tanguy, D.R.: "New Developments in Induction and Sonic Logging," *Trans.*, AIME (1960) **219**, 362–74.
3. Dumanoir, J.L., Hall, J.D., and Jones, J.M.: "R_{xo}/R_t Methods for Wellsite Interpretation," *Proc.*, SPWLA 13th Annual Logging Symposium, Tulsa (May 7–17, 1972) paper R.
4. *Log Interpretation, Vol. I—Principles*, Schlumberger Ltd., Houston (1972) Chap. 15.
5. *Log Interpretation, Vol. II—Applications*, Schlumberger Ltd., Houston (1974) Chap. 3.
6. Millican, M.L., Raymer, L.L., and Alger, R.P.: "Well Site Recording of Movable Oil Plot," *Proc.*, SPWLA Annual Logging Symposium, Midland (1964) Paper F.

Chapter 13
Pattern-Recognition Interpretation Techniques (Crossplotting)

13.1 Introduction

Basic borehole measurements provide an array of recorded or calculated parameters. The parameters usually available are true formation resistivity, R_t; flushed zone resistivity, R_{xo}; formation bulk density, ρ_b, or formation density porosity, ϕ_D; formation acoustic interval transit time, Δt; formation neutron porosity, ϕ_N; self-potential deflection next to the formation, E_{SP}; and formation natural gamma-ray radioactivity, γ.

Pattern-recognition interpretation techniques are based on plotting one of these parameter vs. another. The quantities plotted could also be the ratio of or the difference between two parameters. The plot is known as a "crossplot." A pattern or trend representing zones having the same value of a specific formation property is first recognized on the crossplot. "Normal" trends usually represent water-bearing formations, liquid-filled formations, clean formations, etc. The zones that fall off a normal trend are abnormal in that they are, respectively, hydrocarbon-bearing, gas-bearing, shaly, etc. In addition to this qualitative information, deviation of the abnormal zone from the normal trend will provide quantitative information, such as hydrocarbon saturation and shale content. The unknown formation property of interest to the interpreter will determine the parameters to be plotted and the type of crossplot.

Crossplotting is a very useful interpretation technique. An ambiguous relationship between two parameters can be clarified by crossplotting the parameters. Also, a not-so-obvious solution can be made more obvious by crossplotting. The technique's basic limitation is that it requires a statistically significant number of zones. Crossplotting can also be tedious and time-consuming. However, the advent of taped and digitized logs and the increase of computer use in interpretation greatly enhanced the use of crossplotting.[1] This chapter discusses crossplots devised for the detection of hydrocarbon-bearing zones and the estimation of hydrocarbon saturation in the detected zones. Crossplots used to determine lithology and porosity and to evaluate shaly and gas-bearing formations are discussed in later chapters.

13.2 Nonlinear-Resistivity/Linear-Porosity (Hingle) Crossplot

13.2.1 Concepts of the Hingle Plot. A.T. Hingle[2] introduced the concept of plotting resistivity vs. porosity. The method is based on Archie's basic petrophysical model rearranged and plotted on special grid-type paper. The rearranged Archie model given by Eq. 1.44 is

$$(R_t)^{-1/m}=(S_w^n/aR_w)^{1/m}\phi. \quad \text{(13.1)}$$

Considering zones of constant R_w and the same lithology (i.e., constant a, m, and n), a plot of R_t vs. ϕ yields a family of nonlinear trends. However, these trends can be made into straight lines by plotting $(R_t)^{-1/m}$ instead of R_t vs. porosity. Eq. 13.1 can, in effect, be expressed in the form

$$y=c\phi, \quad \text{(13.2)}$$

where $y=(R_t)^{-1/m}$ .... (13.3)

and $c=(S_w^n/aR_w)^{1/m}$. .... (13.4)

As **Fig. 13.1** shows, Eq. 13.2 describes a set of straight lines fanning out from a common point of origin, ($\phi=0$, $y=0$), which corresponds to ($\phi=0$, $R_t=\infty$). This point of origin physically represents the matrix. Each of the lines corresponds to a specific value of parameter c. Because a,m,n, and R_w are constants, the change in c reflects a change in S_w. The top line, designated by c_o, corresponds to $S_w=1$ and represents water-bearing zones. The equation of this line is a special form of Eq. 13.1 in which $S_w=1$ and $R_t=R_o$. It can be expressed as

$$(R_o)^{-1/m}=(aR_w)^{-1/m}\phi. \quad \text{(13.5)}$$

The top line is called the water or R_o trend. The other lines represent $S_w<1$, with S_w decreasing clockwise.

The technique is not limited to measurements that furnish porosity explicitly, such as the neutron log. It is also applicable to measurements that provide quantities proportional to porosity, such as the sonic and density logs. Sonic-log porosity defined by Eq. 10.1 can be expressed as

$$\phi=\frac{\Delta t-\Delta t_{ma}}{\Delta t_f-\Delta t_{ma}}$$

$$=\frac{\Delta t}{\Delta t_f-\Delta t_{ma}}-\frac{t_{ma}}{\Delta t_f-\Delta t_{ma}}$$

$$=\alpha\Delta t-\beta, \quad \text{(13.6)}$$

where α and β are coefficients reflecting matrix and fluid properties. Substituting Eq. 13.6 into Eq. 13.2 results in

$$y=c(\alpha\Delta t-\beta). \quad \text{(13.7)}$$

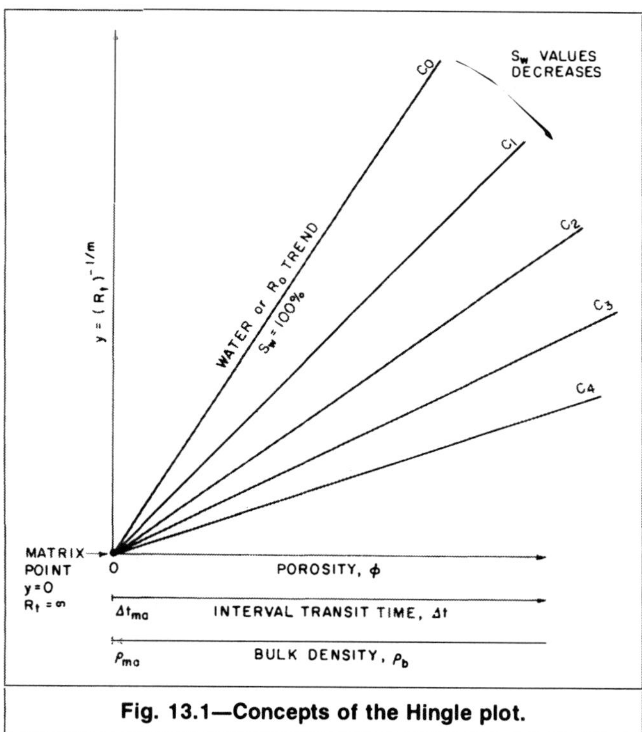

Fig. 13.1—Concepts of the Hingle plot.

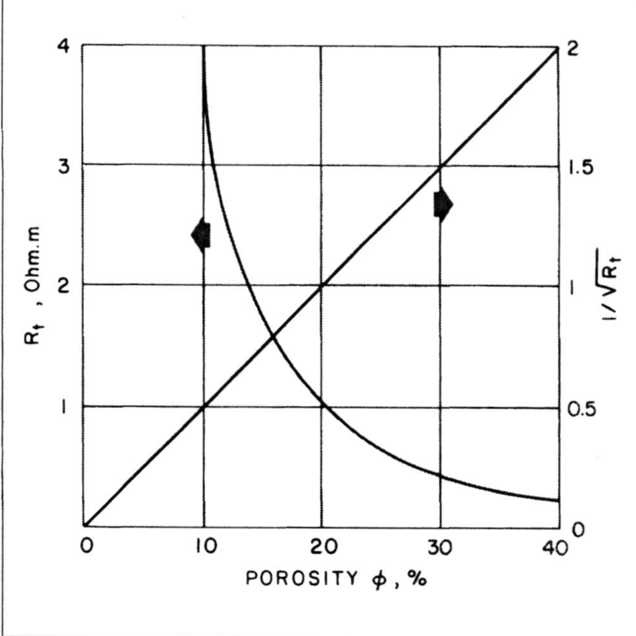

Fig. 13.2—Graphical representation of the R_t-ϕ relationship of Example 13.1.

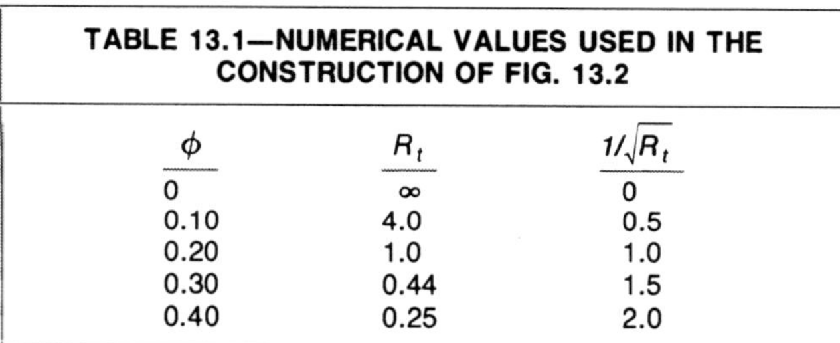

TABLE 13.1—NUMERICAL VALUES USED IN THE CONSTRUCTION OF FIG. 13.2

ϕ	R_t	$1/\sqrt{R_t}$
0	∞	0
0.10	4.0	0.5
0.20	1.0	1.0
0.30	0.44	1.5
0.40	0.25	2.0

Density-log porosity defined by Eq. 8.11 can also be expressed as

$$\phi=\frac{\rho_{ma}-\rho_b}{\rho_{ma}-\rho_f}$$

$$=\frac{\rho_{ma}}{\rho_{ma}-\rho_f}-\frac{\rho_b}{\rho_{ma}-\rho_f}$$

$$=\kappa-\delta\rho_b, \quad \text{(13.8)}$$

where κ and δ are also coefficients that reflect matrix and fluid properties.

Substituting Eq. 13.8 into Eq. 13.2 yields

$$y=c(\kappa-\delta\rho_b). \quad \text{(13.9)}$$

Eqs. 13.8 and 13.9 each describes a set of straight lines fanning out of a point that represents the matrix. The coordinates of the matrix point are ($\Delta t=\Delta t_{ma}$, $R_t=\infty$) for the sonic log and ($\rho_b=\rho_{ma}$, $R_t=\infty$) for the density log. As Fig. 13.1 shows, the other concepts of the plot are unchanged.

Example 13.1. Using linear scales, plot the trend of R_t vs. ϕ that represents water-bearing formations ($S_w=1$) characterized by the following parameters: $R_w=0.04\Omega\cdot$m, $m=n=2$, and $a=1$. How can this trend be made into a straight line?

Solution. Substituting the given values of S_w, m, n, and a into Eqs. 13.2 through 13.4 results in

$$c=1/\sqrt{R_w},$$

$$y=1/\sqrt{R_t},$$

and $1/\sqrt{R_t}=\phi/\sqrt{R_w}$.

For the given value of R_w,

$$1/\sqrt{R_t}=5\phi,$$

or $R_t=1/25\phi^2$.

Fig. 13.2, a plot of R_t vs. ϕ, represents this relation. The trend is nonlinear, but it can be made into a straight line if we plot $1/\sqrt{R_t}$ instead of R_t vs. ϕ. The linear trend is also shown by Fig. 13.2. **Table 13.1** gives the values used to construct Fig. 13.2.

13.2.2 Hingle Crossplot Special Grid. To simplify the construction of the Hingle crossplot, the R_t scale can be adjusted in advance. This will eliminate the need to calculate $(R_t)^{-1/m}$. The adjustment of the R_t scale depends, of course, on the value of m or, in other words, on the F-ϕ relationship. **Figs. 13.3 and 13.4** show two Hingle crossplot grids with adjusted resistivity scales using the relationships $F=0.62/\phi^{2.15}$ and $F=1/\phi^2$, respectively. These two grids are usually available in logging-company chart books.[3] Other charts are available[4] or can be constructed, but these two are the most commonly used. The grid of Fig. 13.3 usually is used for sandstone lithology, the other grid for carbonate lithology. Note that the resistivity scale on this special grid type can be multiplied by any constant value to obtain a scale that covers a different range of values.

Example 13.2. Construct the vertical resistivity scale of a Hingle crossplot grid that represents $m=2$. Use a 5-in. distance to represent a resistivity range of 1 to $\infty\Omega\cdot$m. Show that the scale can be converted to represent a range of 5 to $\infty\Omega\cdot$m simply by multiplying by 5.

Solution. The resistivity scale is adjusted by calculating the parameter y in terms of R_t and m. In this case, $y=1/\sqrt{R_t}$. **Table 13.2** lists the values of y for selected values of R_t. It can be seen that a y range of 0 to 1 corresponds to a resistivity range of ∞ to 1. Because y varies linearly with ϕ (Eq. 13.2), the 5-in. distance can be subdivided linearly in terms of y. The distance, L, to a specific y value is given by $L=5y$ in. The L values are also listed in Table 13.2. To facilitate the plotting process, the y value on the scale is replaced by its corresponding R_t value, as **Fig. 13.5** shows.

If the resistivity range of interest is 5 to ∞, instead of 1 to ∞, and if full use of the 5-in. distance is required, the resistivity values

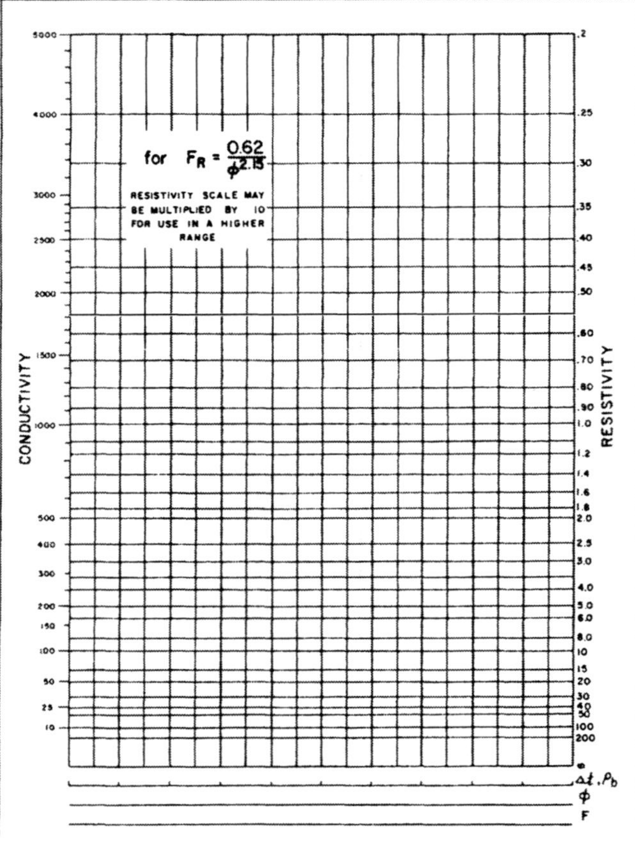

Fig. 13.3—Resistivity/porosity Hingle crossplot for sandstones (courtesy Schlumberger).

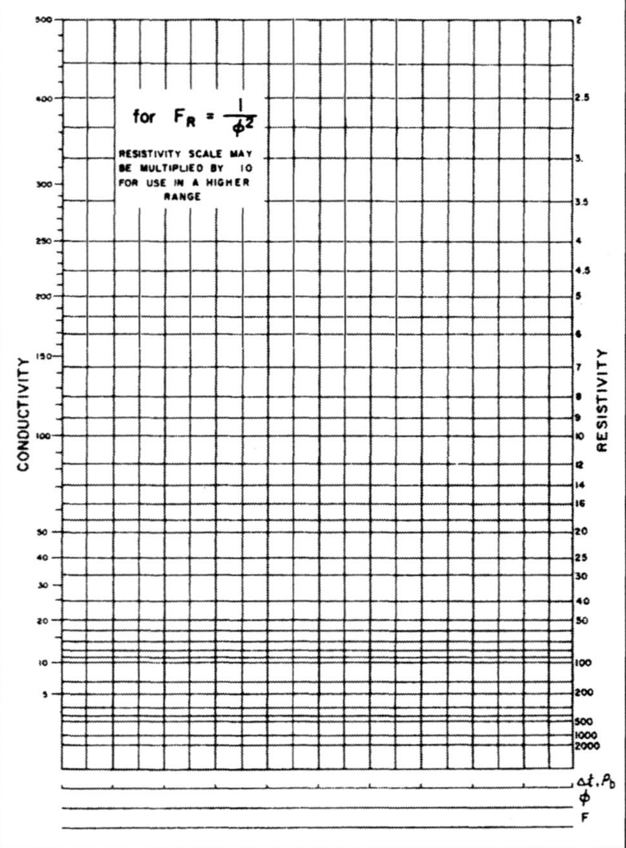

Fig. 13.4—Resistivity/porosity Hingle crossplot for carbonates (courtesy Schlumberger).

can be multiplied by five to restrict the scale's range without modifying its property. This is possible because the y' range of 0 to 0.4772 that corresponds to the R_t range of 5 to ∞ when scaled over a distance of 5 in. gives the same results as previously. The values of R_t', y', and L' for this case are listed in Table 13.2.

TABLE 13.2—SCALING OF A HINGLE PLOT GRID, EXAMPLE 13.2

R_t	$y=1/\sqrt{R_t}$	L (in.)	$R_t'=R_t 5$	$y'=1/\sqrt{R_t'}$	L' (in.)
1	1	5	5	0.4472	5
2	0.7071	3.536	10	0.3162	3.536
10	0.4472	2.236	25	0.2	2.236
20	0.2236	1.118	100	0.1000	1.118
50	0.1414	0.707	250	0.0632	0.707
100	0.1000	0.500	500	0.0447	0.500
1,000	0.0316	0.158	5,000	0.01414	0.158
∞	0	0	∞	0	0

13.2.3 Plotting Procedure. The Hingle Crossplot is prepared as follows.

1. Raw data from logs are acquired and tabulated. This requires correlation between logs followed by zone selection. Concepts of correlation and zone selection are discussed in Sec. 11.2.

2. The plotting procedure starts with the selection or the construction of the proper grid. The grid to be used is determined by the value of m believed to be representative of the prevailing lithology.

3. If the preprinted values of the R_t scale do not cover the desired range of resistivity, or if the desired range of resistivity occupies a small interval on the scale, the scale can be adjusted by multiplying the values by any constant value. The multiplier could be smaller or greater than unity.

4. The linear x axis is scaled in terms of the available porosity log parameter: Δt, ρ_b, or ϕ_N. Because it is customary to have the porosity increase to the right, Δt also must increase to the right; however, ρ_b should decrease in this direction. The scale should be selected properly so that the matrix point and the highest expected porosity value fall within the graph paper.

5. R_t is plotted vs. log data (ϕ_N, Δt, or ρ_b). It is recommended to designate each data point by a number or letter symbol.

Example 13.3. Prepare a Hingle crossplot using the data listed in **Table 13.3** for a well drilled through limestone in south Texas.[5]

Solution. Because hard rock conditions prevail and no specific value of m is recommended, the data are plotted on the grid of Fig. 13.4, (m=2). Because the minimum resistivity listed in Table 13.3 in 2 Ω·m, there is no need to change the resistivity scale.

Fig. 13.6 is a Hingle crossplot displaying the resistivity and density data of Table 13.3. The x axis is scaled in bulk density from 2.52 to 2.72 g/cm^3, with values increasing to the left to locate the matrix point left of the scale, as is customary. The minimum measured density is 2.53 and the matrix density of limestone is 2.71, so all points, including the matrix point, fall on the graph paper.

13.2.4 Interpretation Procedure. The data displayed on a Hingle crossplot, such as **Fig. 13.7A**, are interpreted as follows.

1. The trend that represents water-bearing formations is recognized. Because water-bearing formations display lower resistivities for specific porosity values, the water trend is drawn through the most northwesterly points on the crossplot—i.e., Points 1 through 4 in **Fig. 13.7B.** If the matrix property ρ_{ma} or Δt_{ma} is known, the water trend can be force-fitted through the matrix point.

2. If the matrix is not known, the water trend is extrapolated to intersect with the x axis at the matrix point. This is how ρ_{ma} or Δt_{ma} is determined. The construction of the water trend as ex-

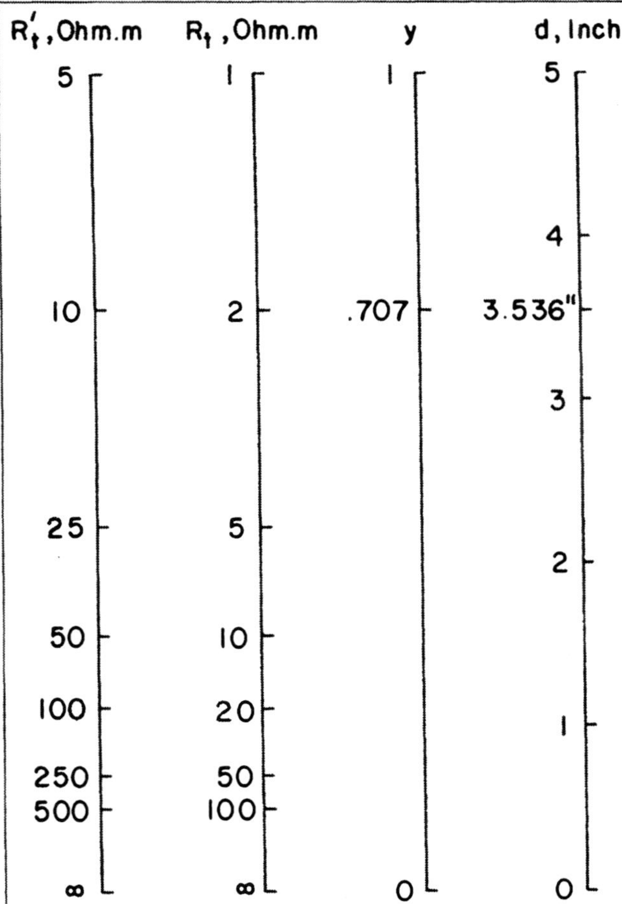

Fig. 13.5—Vertical resistivity scale of a Hingle crossplot grid ($m=2$).

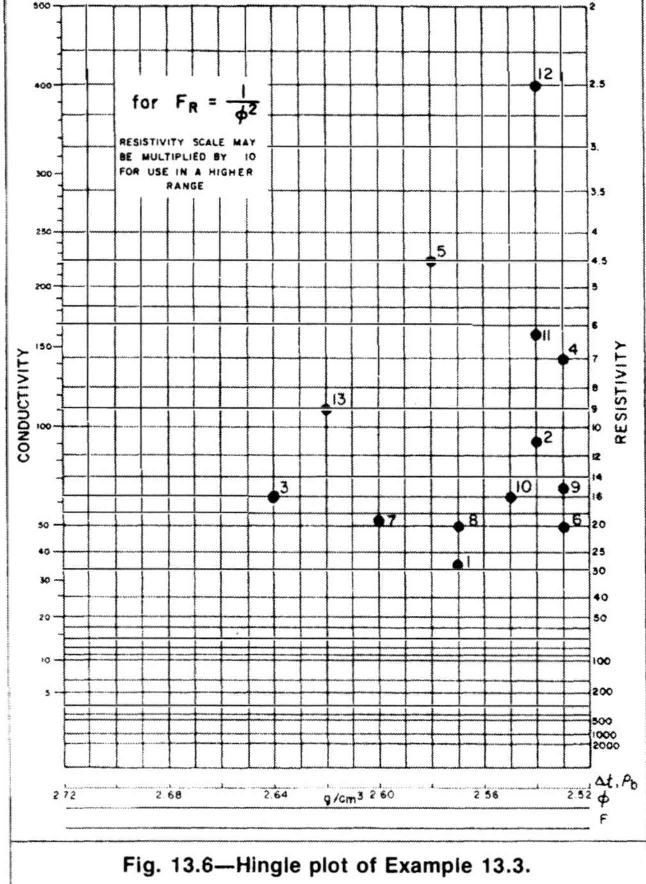

Fig. 13.6—Hingle plot of Example 13.3.

TABLE 13.3—DENSITY AND RESISTIVITY DATA REPRESENTING CARBONATE ROCK IN SOUTH TEXAS (DATA FOR EXAMPLES 13.3 AND 13.6)

Zone	Depth (ft)	ρ_b (g/cm³)	R_t (Ω·m)
1	7,160	2.57	28.0
2	7,168	2.54	11.0
3	7,174	2.64	16.0
4	7,180	2.53	7.0
5	7,185	2.58	4.5
6	7,199	2.53	20.0
7	7,212	2.60	19.0
8	7,230	2.57	20.0
9	7,241	2.53	15.0
10	7,247	2.55	16.0
11	7,264	2.54	6.3
12	7,269	2.54	2.5
13	7,275	2.62	9.0

plained above assumes that a relatively large number of data points were taken from the logs within obvious water zones that ideally have a nice spread of porosities. Knowing R_w and ρ_{ma} or Δt_{ma}, we can construct the water trend without a large number of data points representing water zones.

3. The known or derived ρ_{ma} or Δt_{ma} is used with Eq. 8.11 or 10.1 for proper scaling of the x axis in terms of porosity. The x axis also can be scaled in terms of the formation resistivity factor with the relationship $F=a\phi^{-m}$. The value of m should be the same as that used to construct the Hingle-plot grid.

4. Formation water resistivity can be estimated from the water trend. Any point (e.g., Point A in Fig. 13.7B) lying on the water trend is selected. The y coordinate of such a point is R_o. Its x-axis coordinate is F or a parameter that can be converted easily to F. The value of R_w is simply equal to R_o/F.

5. Points 5 through 10 in Fig. 13.7B lie off the established water trend and represent hydrocarbon zones. This is true, provided that R_w and lithology are the same for all data points considered. The water saturation of a hydrocarbon zone can be estimated from the ratio of the R_t value for that zone to the value of resistivity R_o at the point of intersection with the water trend of a vertical line drawn from the point of interest (e.g., Zone 5 in Fig. 13.7B). The water trend represents zones that have the same R_w and lithology as the hydrocarbon zones; thus, the R_o value read on the vertical line of constant porosity passing through the hydrocarbon zone of interest is the value of the zone of interest's resistivity when fully saturated with water.

6. The estimation of water saturation values for numerous hydrocarbon zones can be simplified by plotting lines that represent S_w values other than 100%. This is done by designating the R_t values that correspond to other S_w values on a vertical line that represents a specific R_o value. For quick reference, **Table 13.4** shows how the ratio R_t/R_o corresponds to S_w. This is based on Archie's equation $S_w^2=R_o/R_t$. For example, the line representing $S_w=50\%$ corresponds to $R_t=4R_o$. This line can be constructed by joining the $R_t=4R_o$ point and the matrix point (**Fig. 13.7C**).

7. The crossplot provides a quick look at potential zones by plotting lines that represent the cutoff saturation and cutoff porosity (**Fig. 13.7D**). It can easily be seen that, with the illustrated cutoff values, only Zones 5 and 7 through 9 are potential zones. Zone 10 does not meet the minimum porosity criterion, and Zone 6 does not meet the minimum hydrocarbon saturation criterion.

If the density or sonic log is used, the above qualitative interpretation is valid regardless of the type of hydrocarbon. However, the saturation values derived from the plot are true only for oil zones. For gas zones, unless the porosity parameter used is free from the gas effect, the water saturation value derived from the plot will be lower than the true value. If the neutron log is used, the opposite effect occurs.

The above qualitative and quantitative interpretation assumes that the zones of interest and those on the reference water trend have

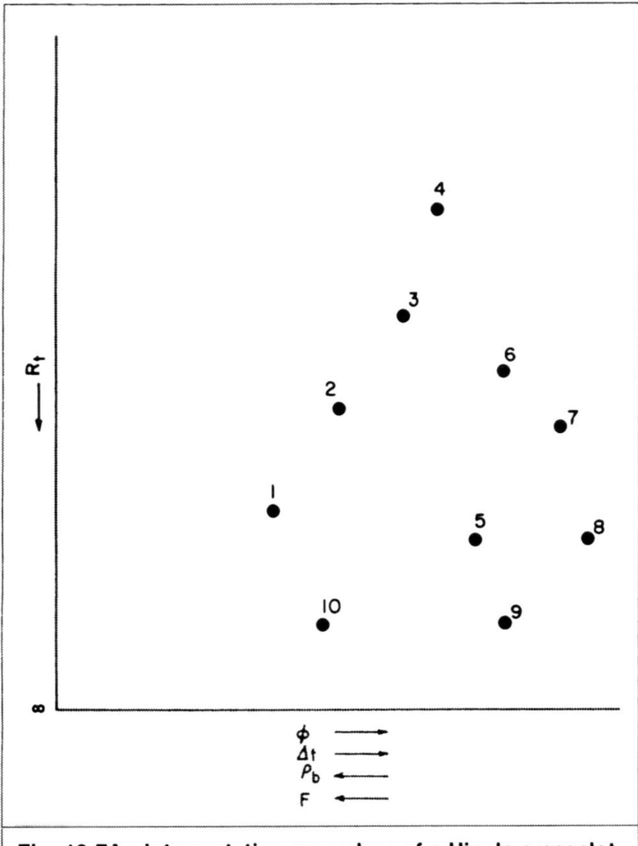

Fig. 13.7A—Interpretation procedure of a Hingle crossplot.

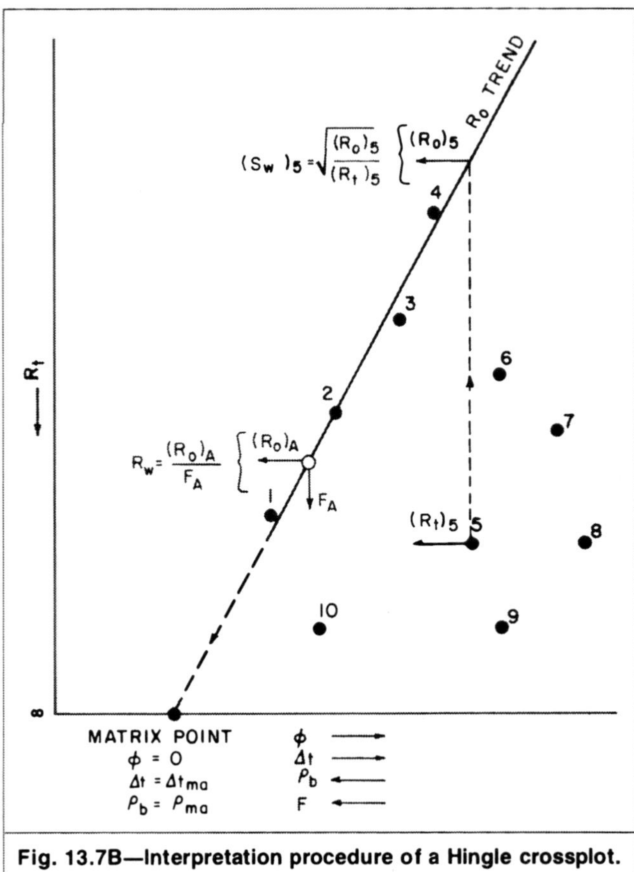

Fig. 13.7B—Interpretation procedure of a Hingle crossplot.

the same R_w, lithology, and matrix properties. It also assumes that the points used to establish the reference line are fully (i.e., 100%) saturated with water. However, as explained in Sec. 13.2.5, under favorable conditions, effects of changes or presence of more than one value for any number of these parameters can be predicted, recognized, and accounted for during interpretation.

Example 13.4. Make a complete qualitative and quantitative interpretation of the Hingle plot of Fig. 13.6.

a. Define the hydrocarbon-bearing zones.
b. Scale the x axis in terms of porosity and formation factor.
c. Estimate the formation water resistivity.

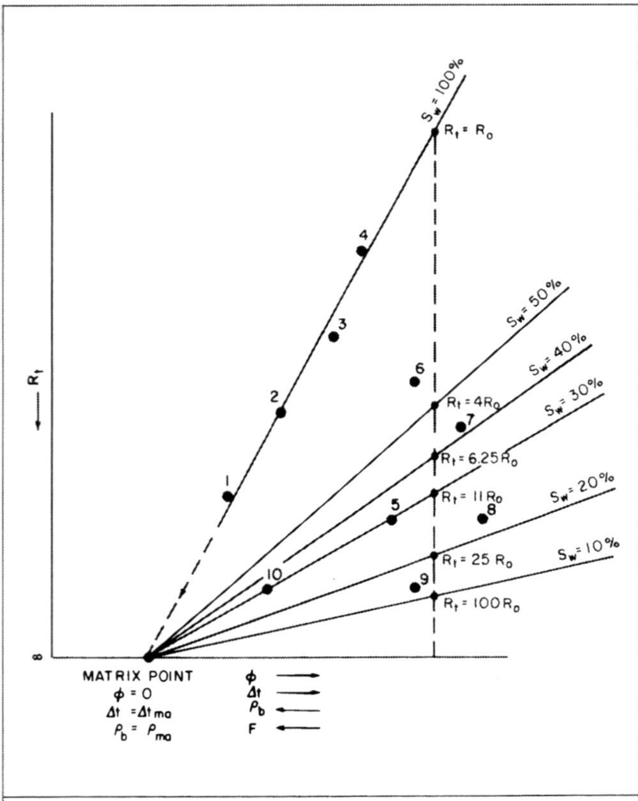

Fig. 13.7C—Interpretation procedure of a Hingle crossplot.

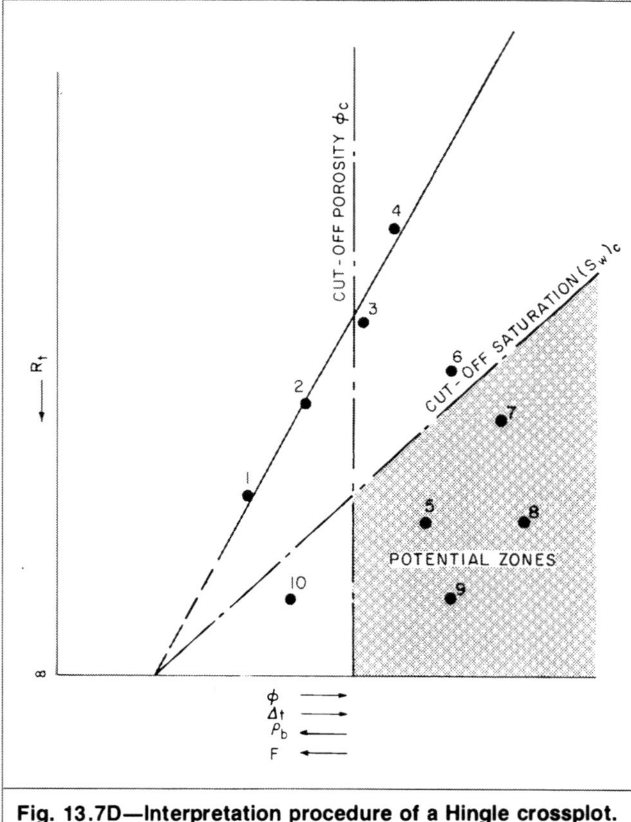

Fig. 13.7D—Interpretation procedure of a Hingle crossplot.

TABLE 13.4—PAIRING OF S_w AND R_t/R_o VALUES

S_w (%)	R_t/R_o
100	1.00
90	1.23
80	1.56
70	2.04
60	2.78
50	4.00
40	6.25
30	11.11
20	25.00
10	100.00

d. List the zones that meet the cutoff criteria $\phi>9\%$ and $S_w<50\%$. Also, list the porosities and water saturations of these zones.

Solution.

a. To detect hydrocarbon zones, the water trend has to be defined. The water trend for this group of points passes through Points 3, 5, 12, and 13 (see **Fig. 13.8**). This is not a large number of points; however, when extrapolated, the water trend intersects the x axis at a point that corresponds to a believable matrix density of 2.71 g/cm^3. The remaining points (Points 1, 2, 4, and 6 through 11) fall off the now-defined water trend and contain hydrocarbons.

b. The x axis is scaled in terms of porosity by calculating ρ_b values that correspond to round porosity values with the following form of Eq. 8.11:

$$\rho_b=\phi\rho_f+(1-\phi)\rho_{ma}.$$

Assuming that $\rho_f=1$ g/cm^3 and $\rho_{ma}=2.71$ g/cm^3 yields

$$\rho_b=2.71-1.71\phi.$$

The following values also result.

ϕ (%)	ρ_b (g/cm^3)
0	2.710
5	2.625
10	2.540

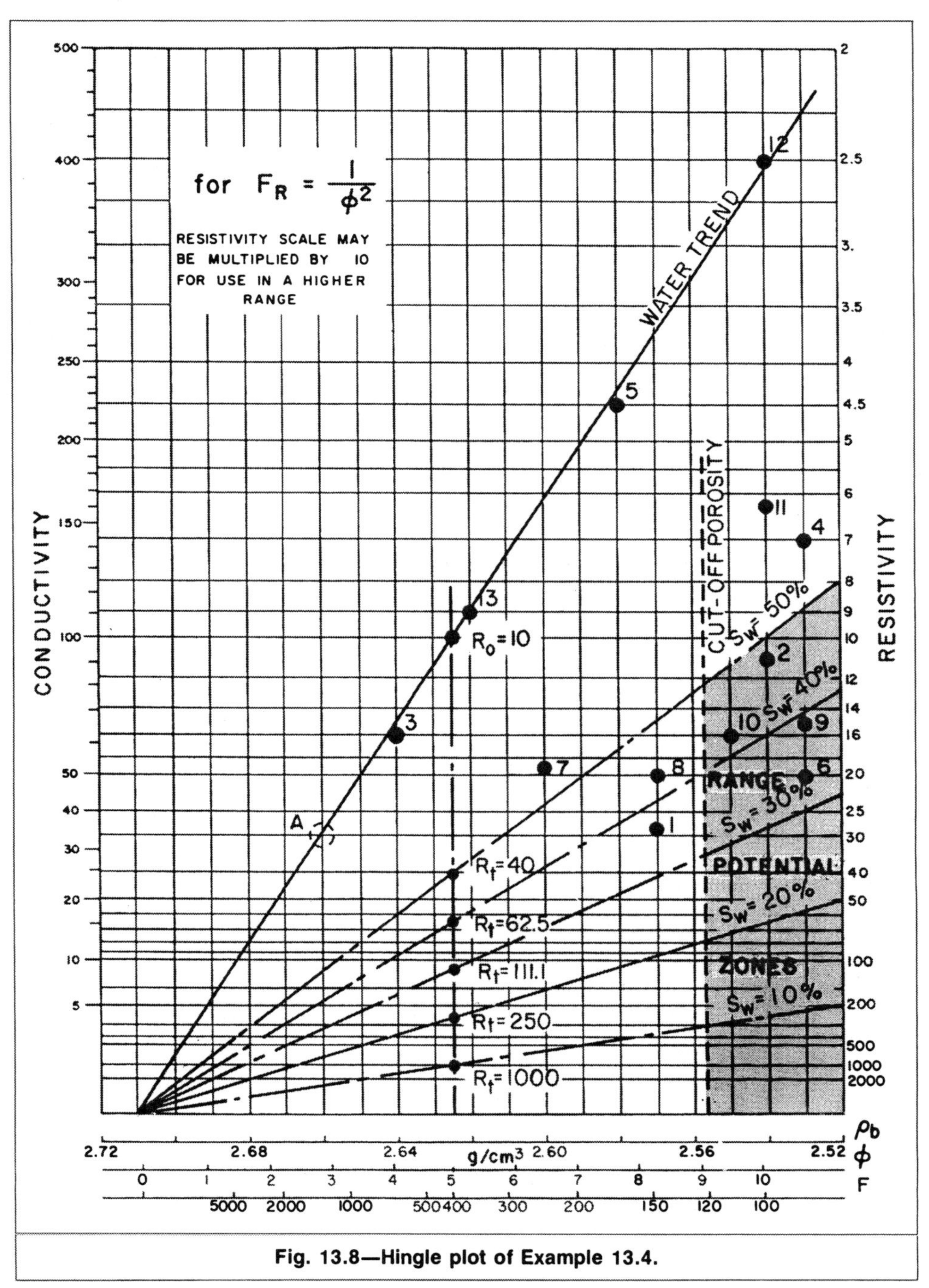

Fig. 13.8—Hingle plot of Example 13.4.

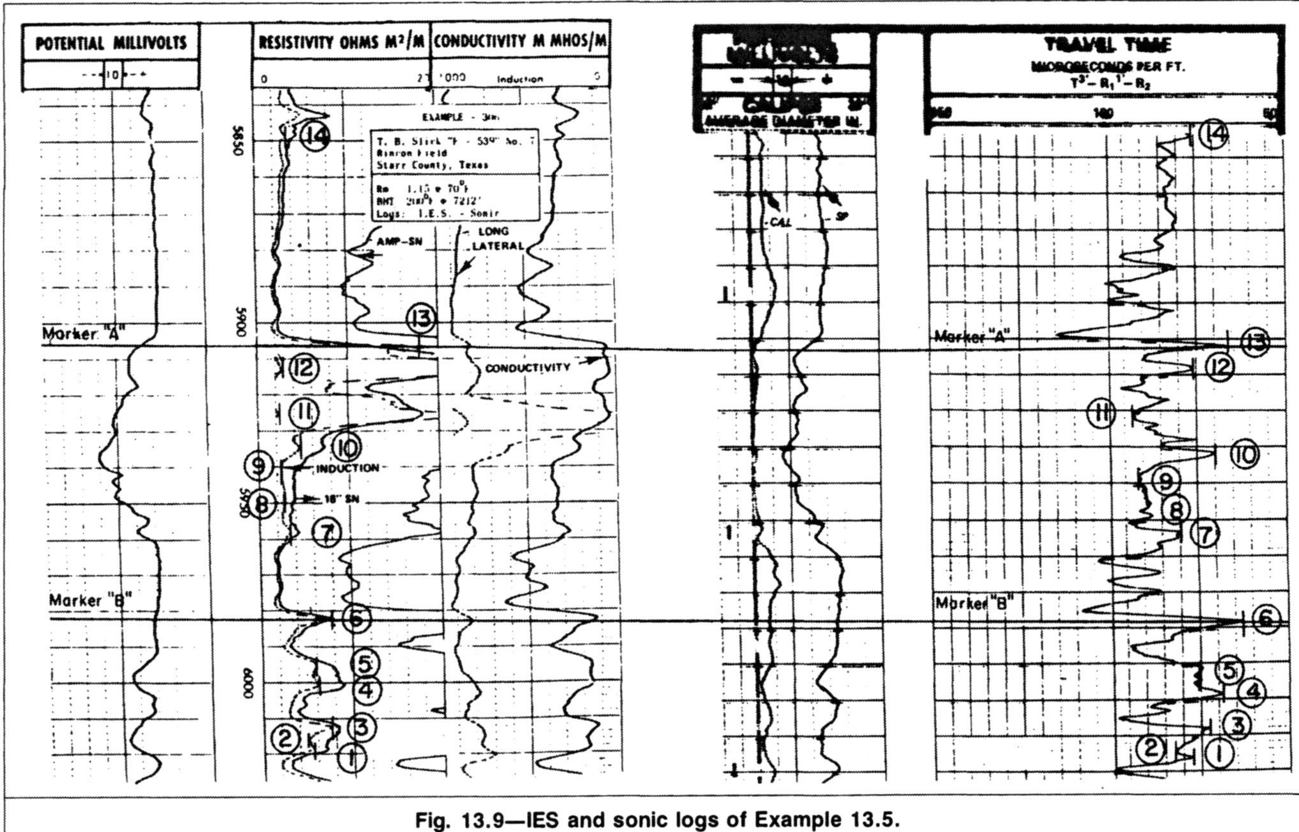

Fig. 13.9—IES and sonic logs of Example 13.5.

The x axis also can be scaled in terms of F by calculating porosity values that correspond to round F values from the relationship $F=\phi^{-2}$.

F	ϕ (%)
100	10.0
200	7.1
300	5.8
400	5.0
500	4.5
1,000	3.2
2,000	2.2
5,000	1.4

c. Select a point located on the water trend. For Point A,

$R_o=30$.

$\rho_b=2.66$ g/cm^3.

$\phi=(2.71-2.66)/(2.71-1)=0.029$.

$F=\phi^{-2}=(0.029)^{-2}=1{,}170$.

$R_w=R_o/F$

$=30/1{,}170=0.026\Omega\cdot$m.

d. Lines representing water saturations of 10%, 20%, 30%, 40%, and 50% are plotted with values from Table 13.4.

e. A cutoff porosity of 9% corresponds to a cutoff bulk density of

$\rho_b=2.71-1.71(0.09)=2.556$ g/cm^3.

The hatched area in Fig. 13.8 is bounded by a constant-porosity line of 9% and a constant-S_w line of 50% that corresponds to the cutoff porosity and saturation values, respectively. A zone should fall within the shaded area to be considered potential. Only Zones 2, 6, 9, and 10 are potential zones. The porosities and water saturations of these four zones can be read directly off the crossplot.

Zone	ϕ (%)	S_w (%)
2	10.0	48
6	10.5	34
9	10.5	39
10	9.5	42

13.2.5 Advantages and Limitations of the Hingle Crossplot. Like any other interpretation technique, the Hingle crossplot has several advantages and limitations. The two advantages follow.

1. The plot provides a quick means to evaluate qualitatively a long interval for the presence of significant hydrocarbon saturations. Once the plot is made, such parameters as R_w, ρ_{ma}, and Δt_{ma} can be varied easily without tedious recalculations.

2. A quantitative measure of water saturation can be derived without knowledge of R_w and without having a calibration for the porosity tool (i.e., ρ_{ma} and Δt_{ma}). Under favorable conditions, the recognized water trend can be used to obtain values of R_w, ρ_{ma}, and Δt_{ma}.

Limitations include the following.

1. To establish a representative water trend, water zones displaying a relatively wide range of porosity must be present in the section analyzed.

2. The cementation exponent, m, must be known or assumed.

3. The parameter $(R_t)^{-1/m}$ has to be calculated first or a special graph paper is needed.

4. The petrophysical models on which the plot's concept is based are for clean, consolidated, and liquid-filled formations. Handling of shaly sands, unconsolidated formations, and gas zones requires certain precautions.

5. Formation-water resistivity and lithology must remain fairly constant over the interval analyzed.

Changes in R_w and/or lithology either can show potential zones where none exist or can mask the presence of potential zones. This is not a weakness peculiar to the Hingle crossplot but is a manifestation of the basic ambiguity inherent in all interpretation techniques. The following example shows how such ambiguity is handled in practice. Note that there is no predetermined procedure to detect those anomalies. Recognizing changes in R_w and/or lithology will largely depend on the situation and on the ingenuity and experience of the analyst.

Example 13.5. Refer to the IES and sonic log of Fig. 11.5 to provide the following information.

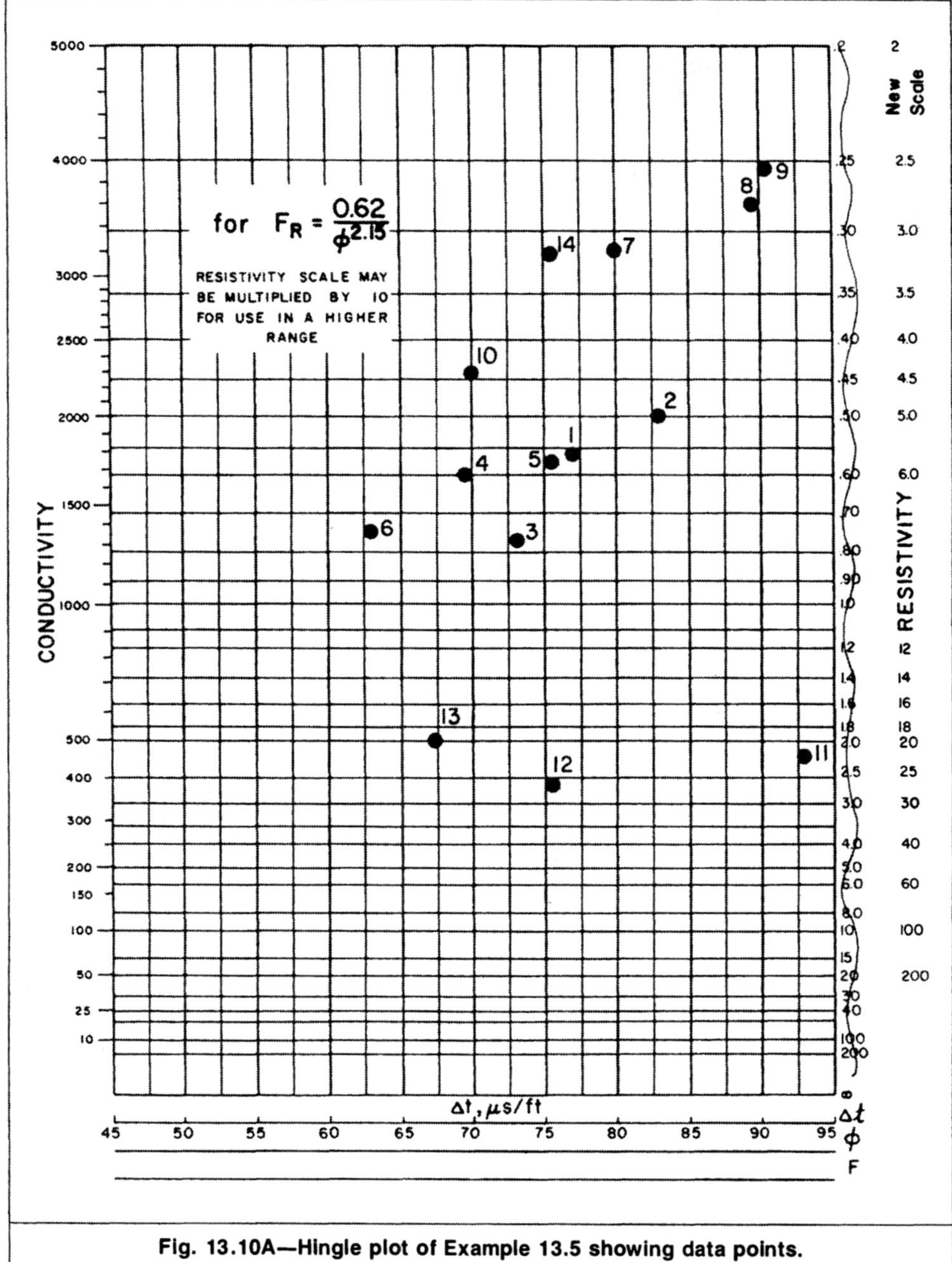

Fig. 13.10A—Hingle plot of Example 13.5 showing data points.

a. Mark and label the zones of interest on the interval transit and induction curves. Tabulate the values of R_t and Δt of these zones.

b. Prepare a Hingle plot using the data prepared in Step a.

c. Trace the most representative water trend onto the plot.

Solution.

a. Thirteen zones of interest are marked and labeled on the sonic and induction curves of **Fig. 13.9.** R_t and Δt of these zones are tabulated below.

Zone	R_t ($\Omega \cdot$m)	Δt (μsec/ft)
1	5.6	77
2	5.0	83
3	7.6	73
4	6.0	69
5	5.8	76
6	7.4	63
7	3.2	80
8	2.8	89
9	2.6	91
10	4.4	70
11	22.0	93
12	26.0	76
13	19.4	67
14	3.2	76

b. Because the lithology of interest in the interval shown is sandstone, Fig. 13.3 used to construct the Hingle plot. The horizontal scale is selected to cover the range of 45 to 95 μsec/ft. With this scale, the sandstone matrix point and the 13 data points fall within the boundaries of the graph paper. The values on the resistivity scale were multiplied by ten so that a better plot could be obtained. **Fig. 13.10A shows the Hingle plot.**

c. According to the interpretation procedure discussed in Sec. 13.2.4, a linear trend passing through the most northwesterly points on the crossplot will represent the water trend. Trend A in **Fig. 13.10B** is drawn through the most northwesterly points (Points 6, 10, and 14) and yields a matrix travel time of 37.5 μsec/ft. This value, which is not representative of sandstone, should alert the analyst to the presence of a lithology other than sandstone. This conclusion is supported by the fact that Zone 6 displays a low porosity and an undefined SP deflection, which are not characteristic of sandstone.

A more representative water trend for sandstone formations present in the interval analyzed, Trend B, is constructed as shown by assuming Δt_{ma}=55 μsec/ft.

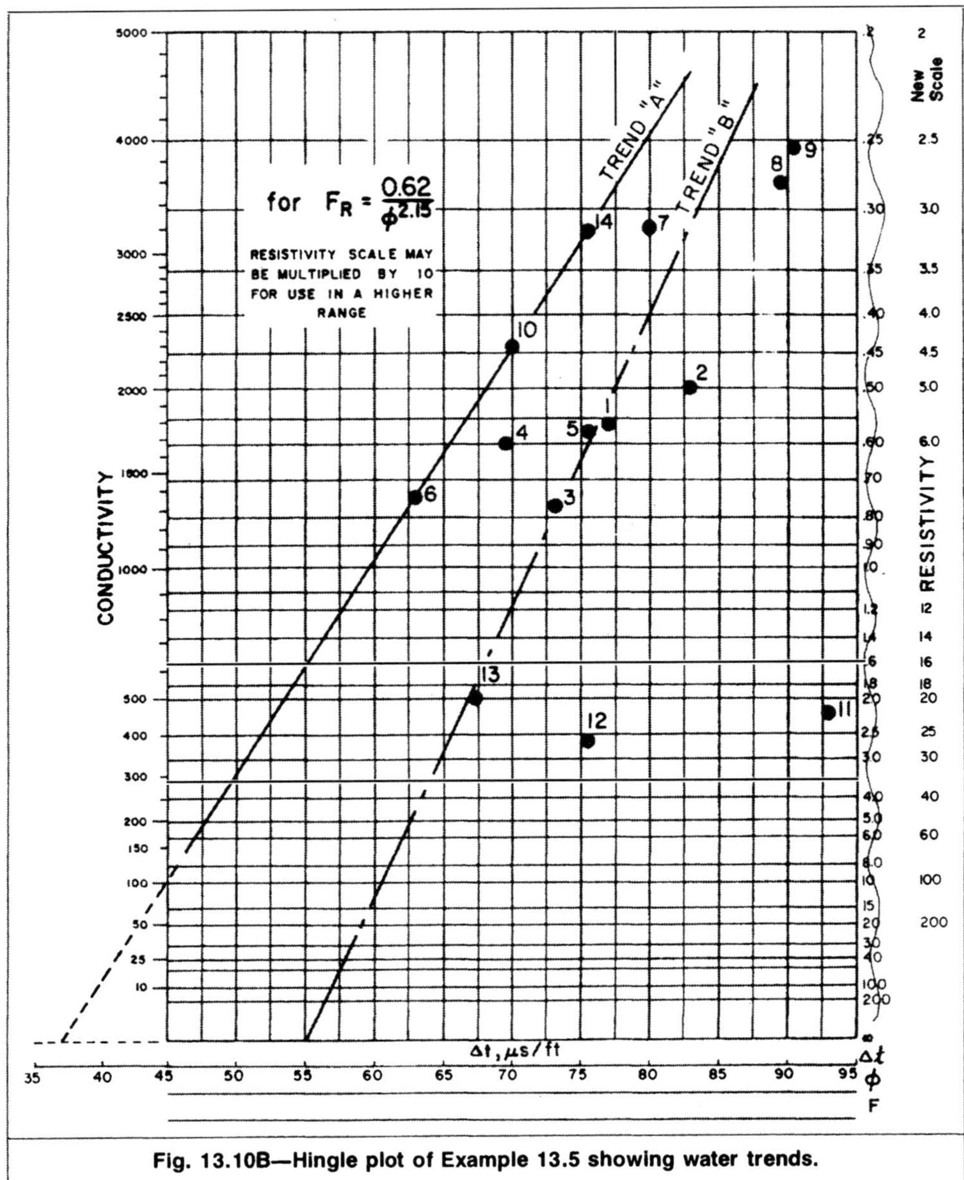

Fig. 13.10B—Hingle plot of Example 13.5 showing water trends.

13.3 Log R_t vs. Log ϕ Pickett Crossplot

13.3.1 Concept of Pickett Crossplot. Pickett[6,7] introduced the concept of a log-log crossplot of porosity vs. resistivity. The technique is based on manipulation of three basic equations ($F=\phi^{-m}$, $R_o=R_wF$, and $R_t=R_oS_w^{-n}$) to obtain

$$\log R_t=-m\log\phi+\log R_w-n\log S_w. \quad (13.10)$$

Considering zones of constant R_w and the same lithology (i.e., same m and n), a plot of ϕ vs. R_t on log-log paper yields a family of parallel linear trends (**Fig. 13.11**). Each line corresponds to a specific S_w value. The lowest line corresponds to the highest value of S_w, or 100%. This line, called the water or R_o trend, is expressed by a special form of Eq. 13.10:

$$\log R_o=-m\log\phi+\log R_w \quad (13.11)$$

$$\text{or } \log\phi=-1/m(\log R_o-\log R_w). \quad (13.12)$$

The slope of this line is $-1/m$. It intercepts the $\phi=1$ (or 100%) line at a resistivity value equal to R_w.

The technique is applicable to measurements that allow either the explicit calculation of porosity or the derivation of a quantity from a log that is proportional to porosity. Appropriate response equations for sonic and density logs can be expressed as

$$\rho=\rho_{ma}-A\phi \quad (13.13)$$

$$\text{and } \Delta t=\Delta t_{ma}+B\phi, \quad (13.14)$$

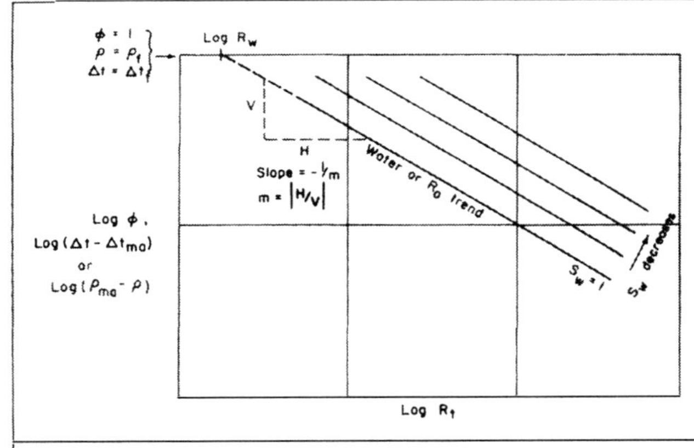

Fig. 13.11—Concepts of the Pickett plot.

where A and B are coefficients reflecting matrix and fluid properties. According to Eqs. 8.11 and 10.1,

$$A=\rho_{ma}-\rho_f \quad (13.15)$$

$$\text{and } B=\Delta t_f-\Delta t_{ma}. \quad (13.16)$$

Solving Eqs. 13.13 and 13.14 for ϕ and substituting them into Eq. 13.11 leads to

$$\log R_o=-m\log(\rho_{ma}-\rho)+m\log A+\log R_w \quad (13.17)$$

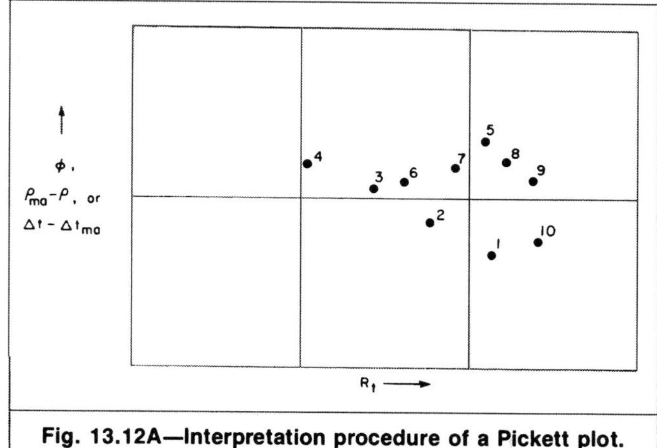

Fig. 13.12A—Interpretation procedure of a Pickett plot.

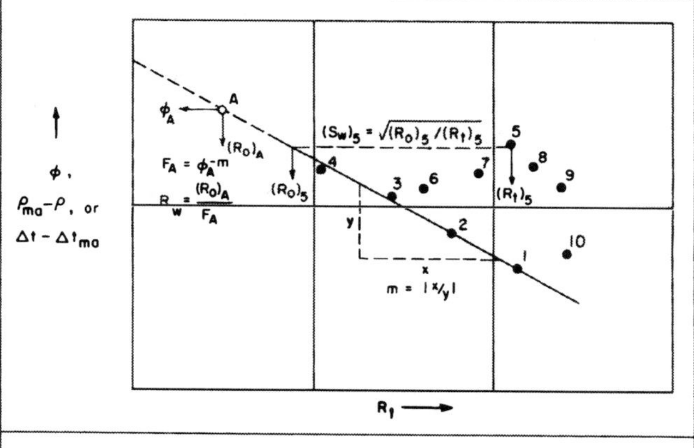

Fig. 13.12B—Interpretation procedure of a Pickett plot.

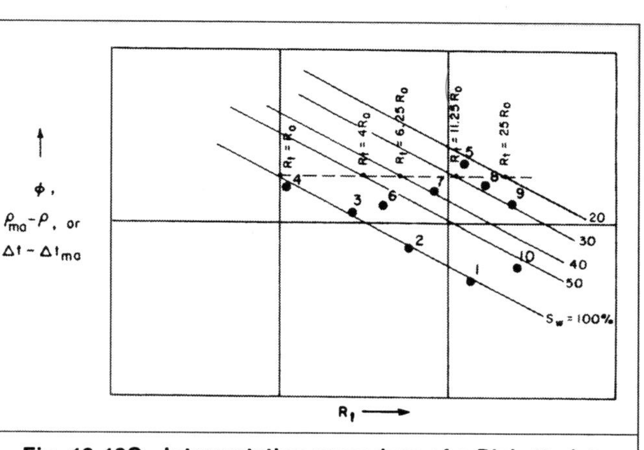

Fig. 13.12C—Interpretation procedure of a Pickett plot.

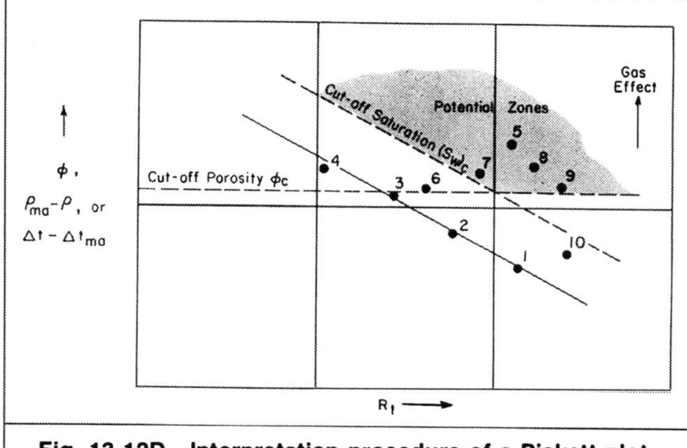

Fig. 13.12D—Interpretation procedure of a Pickett plot.

and $\log R_o = -m \log(\Delta t - \Delta t_{ma}) + m \log B + \log R_w$. . . . .(13.18)

Eqs. 13.17 and 13.18 show that a log-log plot of R_t vs. $(\Delta t - \Delta t_{ma})$ or $(\rho_{ma} - \rho)$ yields linear trends whose slopes are proportional to m. The R_o trend intercepts the line that corresponds to $\rho = \rho_f$ or $\Delta t = \Delta t_f$ at $R_t = R_w$.

13.3.2 Plotting and Interpretation Procedure. A Pickett crossplot is prepared as follows.

1. Proper log-log paper is selected. A 2- ×3-cycles paper is usually required.
2. Resistivity values, R_t, are plotted vs. ϕ, $(\rho_{ma} - \rho)$, or $(\Delta t - \Delta t_{ma})$ data. Conventionally, R_t is plotted on the abscissa and ϕ, $(\rho_{ma} - \rho)$, or $(\Delta t - \Delta t_{ma})$ on the ordinate (**Fig. 13.12A**). The points are numbered to avoid confusion. The value of ρ_{ma} or Δt_{ma} is estimated from knowledge of the lithology in the interval analyzed.
3. If the interval analyzed contains water zones, then the straight line drawn through the most southwesterly points defines the water trend, or the $S_w = 100\%$ line. Points 1 through 4 in Fig. 13.12A define the water trend.
4. The cementation exponent of the $F = \phi^{-m}$ relationship can be determined from the slope of the water trend because the slope is equal to $-1/m$.
5. The R_t value at which this line intersects the $\phi = 100\%$ line will represent R_w. If the intercept is off the page or if the ordinate is scaled in $(\rho_{ma} - \rho)$ or $(\Delta t - \Delta t_{ma})$ values, the coordinates of any point on the water trend can be used to calculate R_w. The x coordinate of such a point is R_o, and the y coordinate is a parameter (ϕ, $\rho_{ma} - \rho$, or $\Delta t - \Delta t_{ma}$) that can be easily converted to F. The value of R_w is simply equal to R_o / F.
6. If none of the intervals analyzed are 100% water-bearing and if R_w is known, then that value of R_w and an engineering estimate of m are used to construct the 100% S_w line.
7. Points 5 through 10 in **Fig. 13.12B** lie off the established water trend and represent hydrocarbon zones. This is true provided that R_w and the lithology are the same for all data points considered. The water saturation of a hydrocarbon zone can be estimated from the ratio of the zone's R_o value to the resistivity R_t value. The resistivity R_o value for this ratio is taken where R_o and the water trend of a horizontal line drawn from the point of interest (e.g., Zone 5 in Fig. 13.12B) intersect. Because the water trend represents zones with the same R_w and lithology as the hydrocarbon zones, the R_o value read on the horizontal line of constant porosity passing through the hydrocarbon zone of interest is that of the zone of interest when fully saturated with water.
8. Water saturation values for numerous zones can be estimated simply by constructing lines that represent S_w values other than 100% (**Fig. 13.12C**). To construct additional water saturation lines on this graph, a horizontal line is drawn to the right, beginning at the point where the 100% S_w line crosses a "1" ($R_t = R_o = 0.1$, 1, 10, etc.). The R_t values that correspond to different water saturations are marked on this line. The values are listed in Table 13.4.
9. The crossplot provides a quick look at potential zones by plotting lines that represent the cutoff saturation and porosity (**Fig. 13.12D**). It can easily be seen that, with the illustrated cutoff values, only Zones 5 and 7 through 9 are potential zones. Zone 10 does not meet the minimum porosity criterion, and Zone 6 does not meet the minimum hydrocarbon saturation criterion.

As for the Hingle plot, the above qualitative and quantitative interpretation assumes that the zones of interest and those on the reference water trend have the same R_w, lithology, and matrix properties. It is also assumed that the points used to establish the reference line are actually fully saturated with water. However, under favorable conditions, effects of changes or the presence of more than one value for any number of these parameters can be predicted, recognized, and accounted for during interpretation.

For a gas-bearing formation, the use of a sonic or density log is recommended. The plot will help detect gas-bearing zones. However, unless the porosity log used is free from the gas effect, the

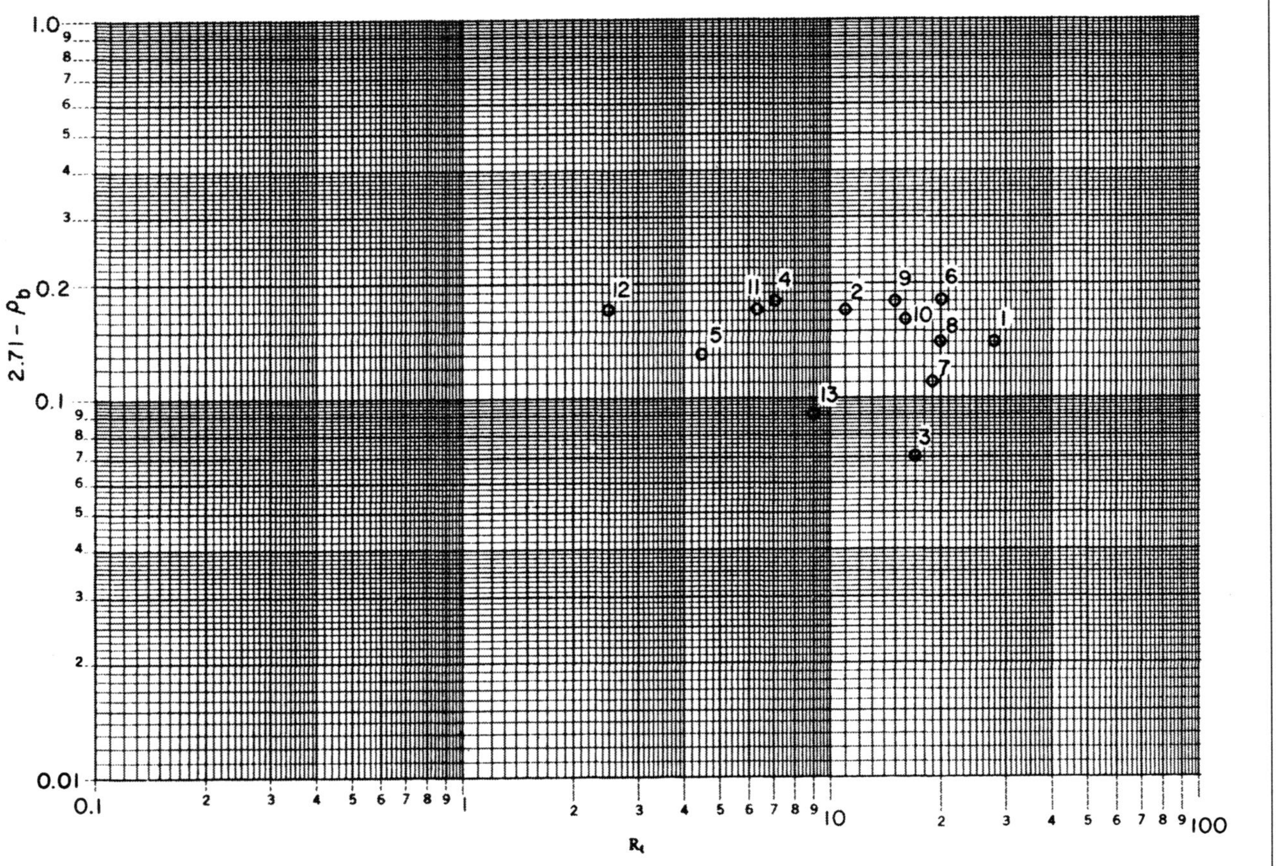

Fig. 13.13A—Pickett plot of Example 13.6.

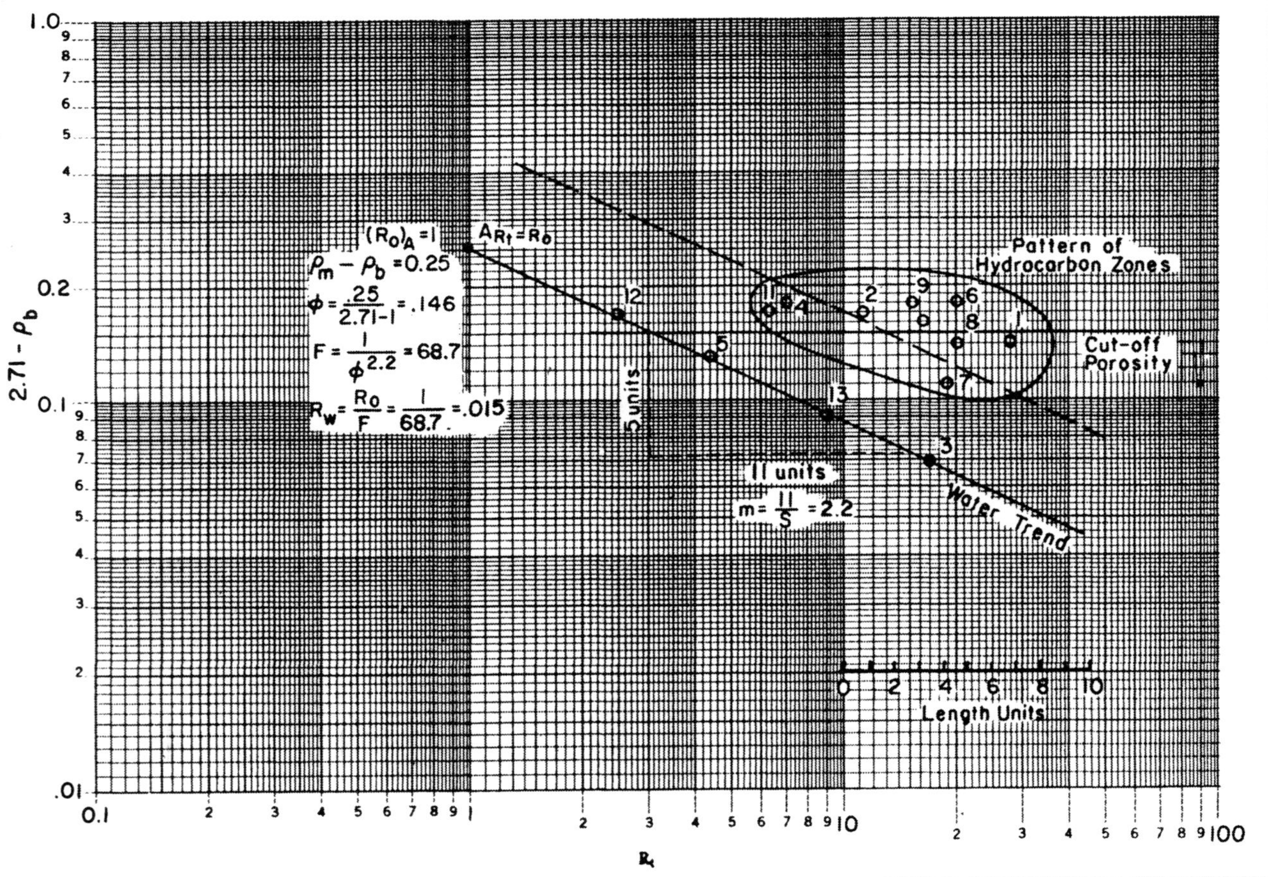

Fig. 13.13B—Pickett plot of Example 13.6.

TABLE 13.5—DENSITY-RESISTIVITY DATA REPRESENTING CARBONATE ROCK IN SOUTH TEXAS (DATA FOR EXAMPLE 13.6)

Zone	Depth (ft)	ρ_b (g/cm³)	R_t (Ω·m)	$2.71-\rho_b$ (g/cm³)
1	7,160	2.57	28.0	0.14
2	7,168	2.54	11.0	0.17
3	7,174	2.64	16.0	0.07
4	7,180	2.53	7.0	0.18
5	7,185	2.58	4.5	0.13
6	7,199	2.53	20.0	0.18
7	7,212	2.60	19.0	0.11
8	7,230	2.57	20.0	0.14
9	7,241	2.53	15.0	0.18
10	7,247	2.55	16.0	0.16
11	7,264	2.54	6.3	0.17
12	7,269	2.54	2.5	0.17
13	7,275	2.62	9.0	0.09

water saturation value derived from the plot will be lower than the true value.

Example 13.6.

a. Prepare a Pickett crossplot using the data listed in Table 13.3.
b. Define the hydrocarbon-bearing zone.
c. Estimate the cementation exponent.
d. Estimate the formation water resistivity.
e. List the porosity and water saturation for the zones that meet cutoff porosity and saturation values of 9% and 50%, respectively.
f. Compare the qualitative and quantitative information obtained from the Pickett crossplot to that obtained previously from the Hingle crossplot (Example 13.4).

Solution

a. The lithology is known to be limestone, so a matrix density of 2.71 g/cm³ is used to calculate the value of $(\rho_m-\rho_b)$ for each of the 13 zones. These values are listed in **Table 13.5.** A 2- ×3-cycles log-log graph paper is used to prepare the plot. The 2-cycles axis is scaled in terms of $(\rho_{ma}-\rho_b)$ values and covers the range of 0.01 to 1 g/cm³. The 3-cycles axis is scaled in terms of R_t and covers the resistivity range of 0.1 to 100 Ω·m. With these ranges, all data points fall within the boundaries of the graph paper, as shown in **Fig. 13.13A.**

b. Fig. 13.13A shows a well-defined linear pattern formed by Zones 3, 5, 12, and 13. This pattern is taken to represent the water trend. Fig. 13.13A also shows a second pattern formed by points falling off the water trend. If formation water resistivity and lithology are the same for all zones considered, then these zones are hydrocarbon-bearing zones. This pattern and the water trend are well-identified in **Fig. 13.13B.** The slope of the water trend and any point lying on the trend are used to determine m and R_w, respectively.

c. Slope $=-1/m=5$ units/-11 units.

$m=11/5=2.2.$

d. Using the coordinates of Point A gives

$R_o=1$ Ω·m,

$\rho_{ma}-\rho_b=0.25$ g/cm³,

$\phi=(\rho_{ma}-\rho_b)/(\rho_{ma}-\rho_f)=0.25/1.71=0.146$ or 15%,

$F=1/\phi^{2.2}=65,$

and $R_w=R_o/F=1/65=0.0154$ Ω·m.

e. The value of $\rho_{ma}-\rho_b$ that corresponds to the cutoff porosity is first calculated:

$0.09=(\rho_{ma}-\rho_b)/(2.71-1),$

$\rho_{ma}-\rho_b=0.15.$

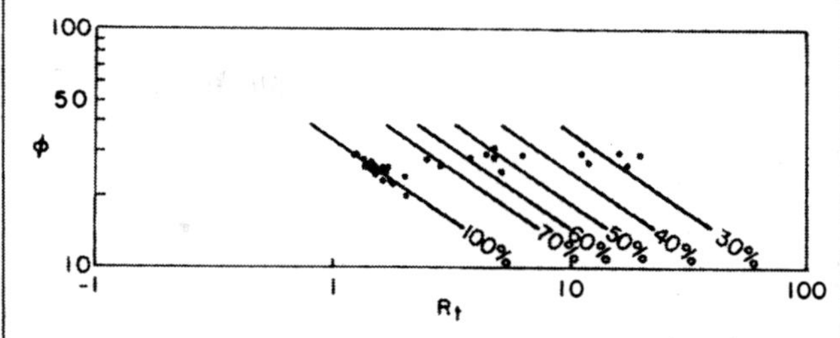

Fig. 13.14A—Pickett plot showing presence of a transition zone.

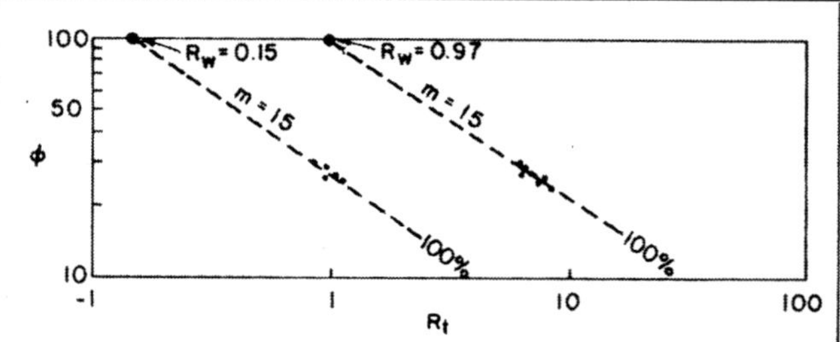

Fig. 13.14B—Pickett plot showing parallel grouping indicating change in R_w.

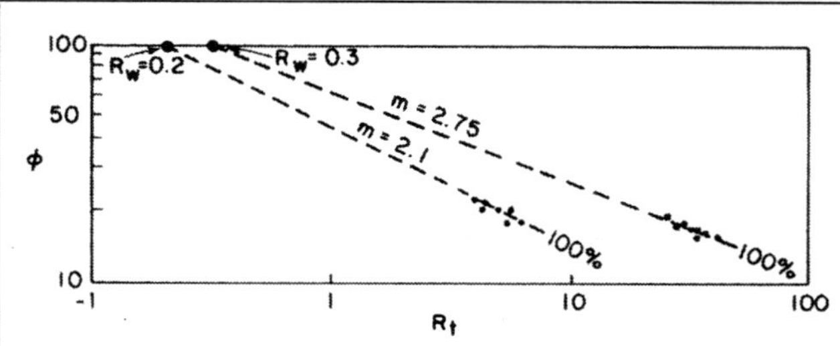

Fig. 13.14C—Pickett plot indicating presence of two different lithologies.

The two lines that represent the cutoff porosity and saturation are added to the plot (Fig. 13.13B).

Only Zones 2, 6, 9, and 10 meet the cutoff porosity and saturation criteria. The porosity and saturation values of these zones are listed below. R_o values are obtained from Fig. 13.13B.

Zone	ρ_b (g/cm³)	ϕ (%)	R_t Ω·m	R_o Ω·m	S_w (%)
2	2.54	10.0	11.0	2.5	48
6	2.53	10.5	20.0	2.1	32
9	2.53	10.5	15.0	2.1	37
10	2.55	9.0	16.0	2.8	42

f. Except for the value of R_w, the qualitative and quantitative data obtained from the Hingle plot of Fig. 13.8 and the Pickett plot of Fig. 13.13 are very similar. Both plots identify the same four zones as potential zones.

Two considerably different R_w values were derived from the two plots. Values of 0.026 and 0.015 Ω·m were obtained from the Hingle and Pickett plots, respectively. The difference is attributed to the value of m used to calculate F from porosity values. In preparing the Hingle plot of Fig. 13.8, we assumed that $m=2$. The Pickett plot of Fig. 13.13 shows that, for this rock type, $m=2.2$ is more representative.

A better agreement between the R_w values obtained from Hingle and Pickett plots can be realized with the iterative technique discussed in Sec. 13.5.

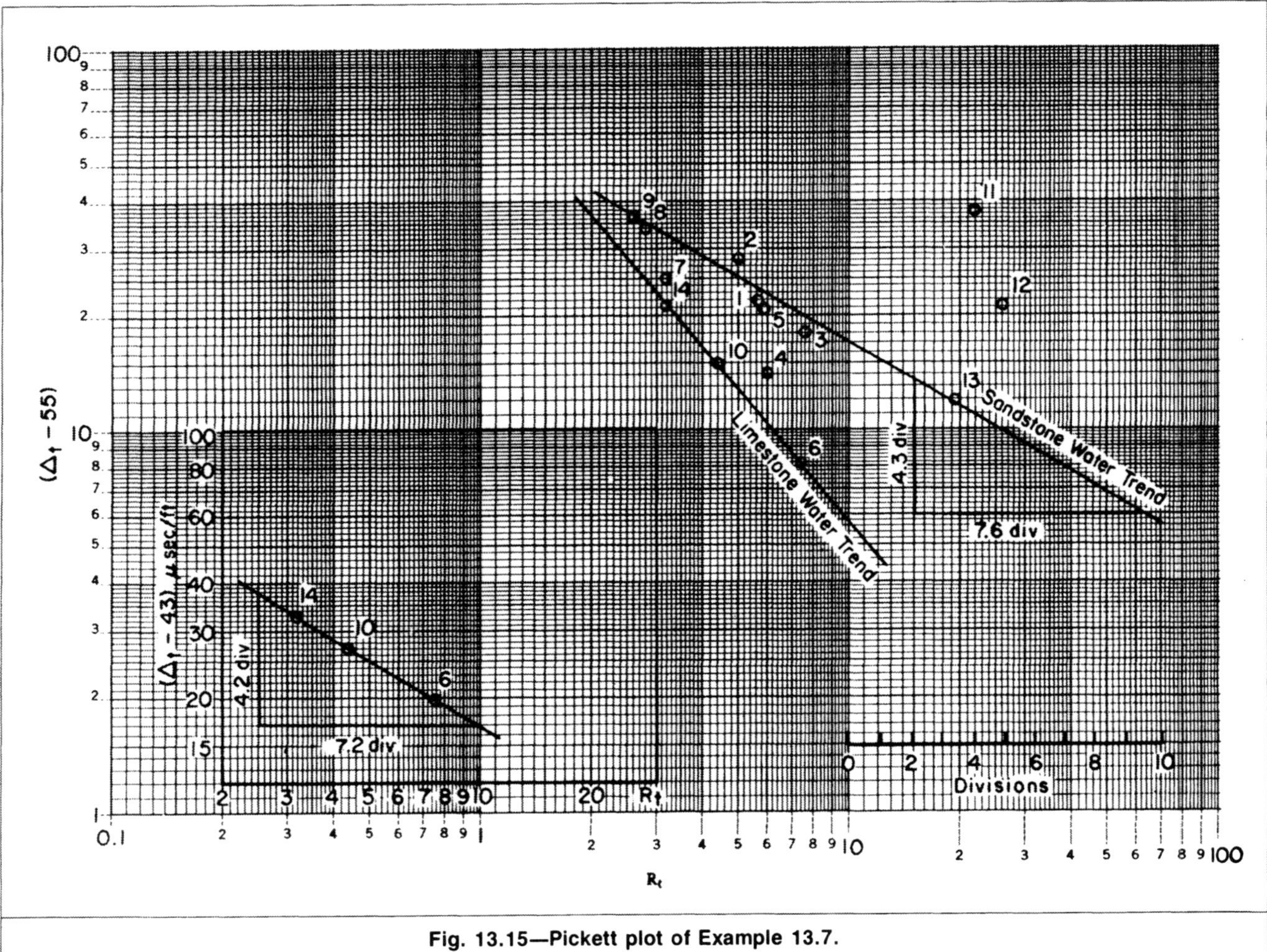

Fig. 13.15—Pickett plot of Example 13.7.

13.3.3 Advantages and Limitations of the Pickett Plot. The Pickett crossplot has the same advantages and limitations of the Hingle plot except for the following.

1. The F/ϕ relationship does not have to be known or assumed; in fact, m can be computed from the graph.
2. The graph paper used to construct the plot is readily available.
3. The matrix properties (Δt_{ma} *or* ρ_{ma}) have to be known or assumed. However, in certain cases where a large number of data points are available, these properties can be derived by trial-and-error. This approach is explained in Sec. 13.3.4.

The Pickett plot helps the log analyst make a number of observations.[8]

1. A lateral spread of data points (**Fig. 13.14A**) usually indicates the presence of a transition zone.
2. Parallel groupings of data points (**Fig. 13.14B**) indicate changes in R_w.
3. Nonparallel straight-line groupings (**Fig. 13.14C**) indicate different lithologies, resulting in different values of m.

Example 13.7. Prepare a Pickett plot using the data of Example 13.5. The predominant lithology present in the interval analyzed is sandstone; $\Delta t_m = 55$ μsec/ft.

a. Like the Hingle plot, does the Pickett plot indicate the presence of more than one lithology?

b. Determine the value of m for sandstone present in the interval analyzed.

c. If the other lithology present in the interval analyzed is dolomite ($\Delta t_{ma} = 43$ μsec/ft), determine the value of m.

Solution.

a. The values of $(\Delta t - 55)$ are calculated and plotted vs. R_t on a log-log paper, as shown in **Fig. 13.15.**

Zone	R_t (Ω·m)	Δt (μsec/ft)	$(\Delta t - 55)$ (μsec/ft)	$(\Delta t - 43)$ (μsec/ft)
1	5.6	77	22	—
2	5.0	83	28	—
3	7.6	73	18	—
4	6.0	69	14	—
5	5.8	76	21	—
6	7.4	63	8	20
7	3.2	80	25	—
8	2.8	89	34	—
9	2.6	91	36	—
10	4.4	70	15	27
11	22.0	93	38	—
12	26.0	76	21	—
13	19.4	67	12	—
14	3.2	76	21	33

As with the Hingle plot of Example 13.5, two trends representing the water-bearing zones of the two lithologies present in the interval analyzed are indicated by the Pickett plot (Fig. 13.15).

b. m is calculated from the slope of the sandstone water trend:

$m = 7.6/4.3 = 1.7$.

c. Although a linear trend is indicated for limestone lithology, m cannot be determined accurately from this trend because it is developed with a value of $\Delta t_{ma} = 55$ μsec/ft, which is not that of dolomite. To determine the value of m, another plot is prepared with $\Delta t_{ma} = 43$ μsec/ft.

The slope of the new trend shown in the insert of Fig. 13.10 can be used to derive the value of m:

$m = 7.2/4.2 = 1.71$.

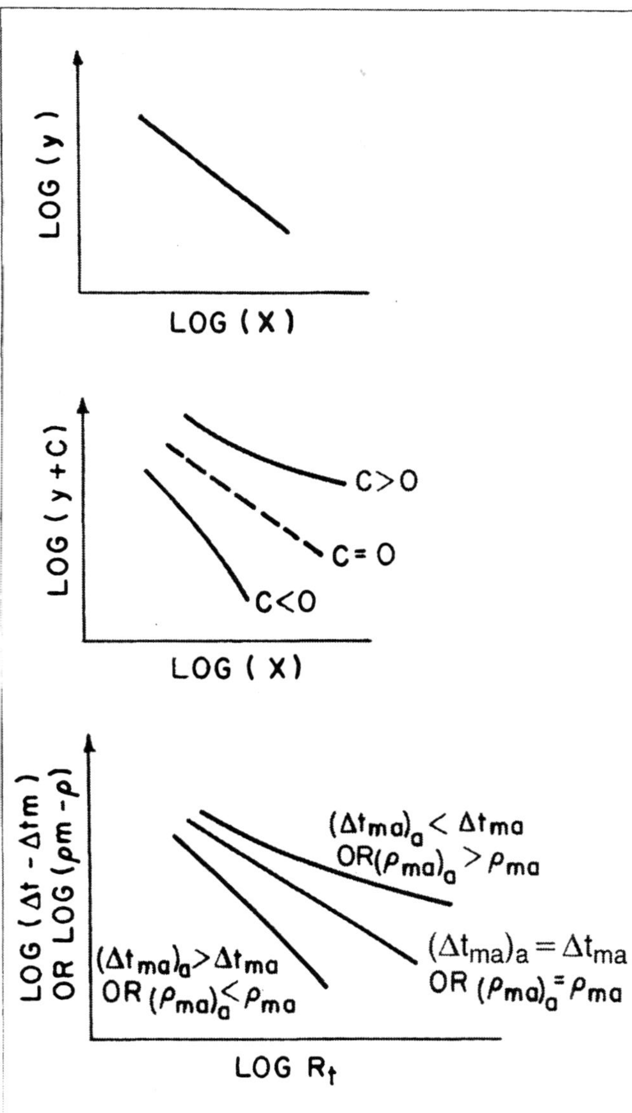

Fig. 13.16—Effect of the selected matrix properties on the linearity of the Pickett plot.

13.3.4 Determination of Matrix Properties With the Pickett Plot. If two quantities, y and x, are exponentially related,

$$y=ax^b, \quad (13.19)$$

then $$\log y=\log a+b \log x \quad (13.20)$$

and a log-log plot of y vs. x is a straight line. If the paired arrays of y or x are changed by a constant, c, the log-log plot of $(y+c)$ vs. x or y vs. $(x+c)$ will be curvilinear. The curvature will be concave or convex, depending on whether c is positive or negative (**Fig. 13.16**).

Eqs. 13.17 and 13.18 can be expressed in the form of Eq. 13.20:

$$\log R_o=\log a+b \log(\Delta t-\Delta t_{ma}) \quad (13.21)$$

and $$\log R_o=\log a+b \log(\rho_{ma}-\rho). \quad (13.22)$$

Then, if a plot of R_t vs. $(\rho_{ma}-\rho)$ is prepared from the log data of ρ and R_t for a series of water-saturated zones, a straight line is obtained only for the correct difference $(\rho_{ma}-\rho)$. For $(\rho_{ma}-\rho)$ values to be correct, the assumed value of ρ_{ma} has to be the true matrix density value. If the plot is curvilinear, then the assumed ρ_{ma} is incorrect. Other values of ρ_{ma} are then used in a trial-and-error approach until the plot of R_t vs. $(\rho_{ma}-\rho)$ results in a straight line for water-bearing zones. The same trial-and-error approach applies to the Δt measurement when the matrix value Δt_{ma} is established.

For this trial-and-error approach to work, a relatively large number of points representing water-bearing zones have to be available. These points also must be spread over a wide range of resistivity and porosity values.

Example 13.8. The table below lists pairs of R_o and Δt values observed in water-bearing zones of a limy sandstone formation. Use the Pickett plot to determine the most representative value of Δt_{ma}.

R_o (Ω·m)	Δt (μsec/ft)
0.5	89
1.0	76
2.0	67
4.0	61
9.0	56.5
18.5	54

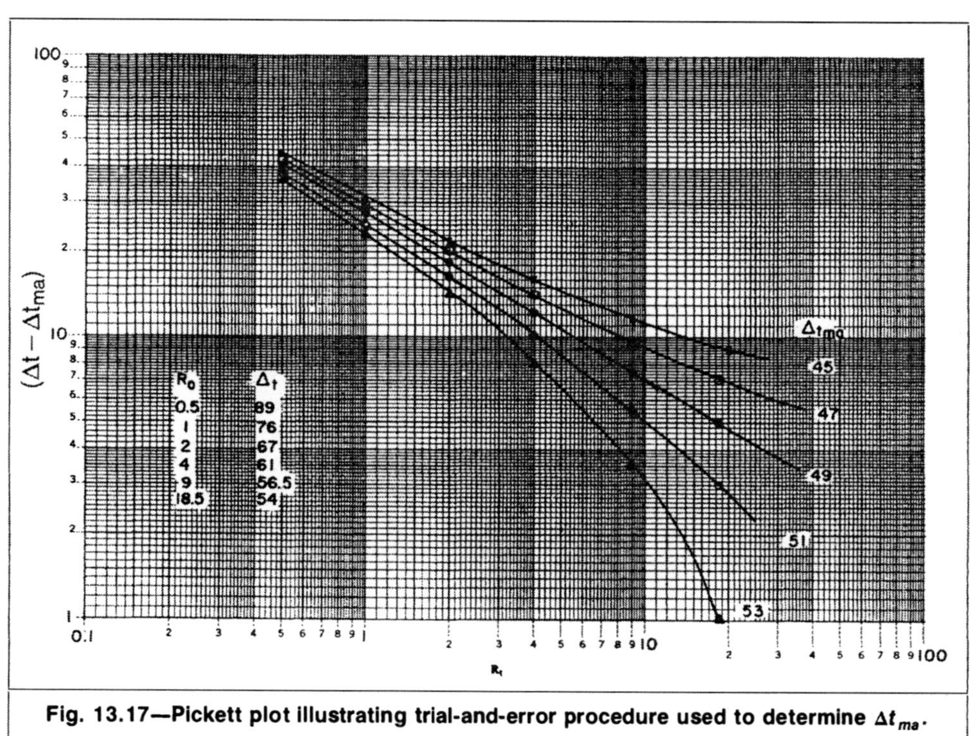

Fig. 13.17—Pickett plot illustrating trial-and-error procedure used to determine Δt_{ma}.

Solution. Several reasonable values (listed below) of Δt_{ma} are used to prepare the Pickett plot of **Fig. 13.17.**

R_o ($\Omega\cdot$m)	Δt (μsec/ft)	$\Delta t-\Delta t_{ma}$ (μsec/ft) $\Delta t_{ma}=45$ (μsec/ft)	$\Delta t_{ma}=47$ (μsec/ft)	$\Delta t_{ma}=49$ (μsec/ft)	$\Delta t_{ma}=51$ (μsec/ft)	$\Delta t_{ma}=53$ (μsec/ft)
0.5	89	44	42	40	38	36
1	76	31	29	27	25	23
2	67	22	20	18	16	14
4	61	16	14	12	10	8
9	56.5	11.5	9.5	7.5	5.5	3.5
18.5	54	9	7	5	3	1

The matrix value of 49 μsec/ft results in the most rectilinear trend. It should be then considered the most representative value.

13.3.5 Effects of Logging Tools' Miscalibration on Pickett Crossplot. Three types of errors can result from tool miscalibration[6]: "zero" errors, where the apparent log response is equal to the true log response plus or minus a constant; "sensitivity" errors, where the apparent log response is equal to a constant times the true log response; and "combination" errors, where the apparent log response is equal to a constant plus the product of another constant and the true log response. When applied to a sonic log, these three types of errors result in

$$\Delta t_a=C_1+\Delta t \quad (13.23)$$

(zero error),

$$\Delta t_a=C_2\Delta t \quad (13.24)$$

(sensitivity error), and

$$\Delta t_a=C_1+C_2\Delta t \quad (13.25)$$

(combination error), where Δt_a=apparent log response, Δt=true value that should have been recorded, and C_1, C_2=constants.

These conditions are applied assuming that the plot of $(\Delta t-\Delta t_{ma})$ vs. R_t is the interpretation approach used and that several values of apparent matrix interval travel time, $(\Delta t_{ma})_a$, are attempted until the water-bearing trend is straightened, as explained in the previous section. When a zero error exists, the value plotted is $[\Delta t_a-(\Delta t_{ma})_a]$, which, according to Eq. 13.23, can be written as

$$\Delta t_a-(\Delta t_{ma})_a=C_1+\Delta t-(\Delta t_{ma})_a. \quad (13.26)$$

However, the water-bearing trend can be a straight line only if $\Delta t_a-(\Delta t_{ma})_a$ is equal to the difference between the true log response, Δt, and the true matrix travel time, Δt_{ma}:

$$\Delta t_a-(\Delta t_{ma})_a=C_1+\Delta t-(\Delta t_{ma})_a=\Delta t-\Delta t_{ma} \quad (13.27)$$

$$\text{or } (\Delta t_{ma})_a=C_1+\Delta t_{ma}. \quad (13.28)$$

This means that the water-bearing trend was straightened by use of an assumed $(\Delta t_{ma})_a$, which, although not equal to the true value, compensated for the miscalibration error.

When a sensitivity error exists, the value plotted $[\Delta t_a-(\Delta t_{ma})_a]$ can be expressed according to Eq. 13.24 as

$$\Delta t_a-(\Delta t_{ma})_a=C_2\Delta t-(\Delta t_{ma})_a. \quad (13.29)$$

Only a value of $(\Delta t_{ma})_a=C_2\Delta t_{ma}$ will result in a rectilinear water trend because

$$C_2\Delta t-(\Delta t_{ma})_a=C_2(\Delta t-\Delta t_{ma}) \quad (13.30)$$

$$\text{and } \log C_2(\Delta t-\Delta t_{ma})=\log C_2+\log(\Delta t-\Delta t_{ma}). \quad (13.31)$$

The straightened water trend, however, is shifted on the plot by the constant factor log C_2. This process does not eliminate the error. However, because data points that represent hydrocarbon-bearing zones also are shifted by log C_2, they still will be off the water-bearing trend, and the correct values of water saturation are calculated from the plot. Although m also is correct, an apparent formation water resistivity, which is equal to the product of C_2 and the true formation water resistivity, R_w, is derived from the plot.

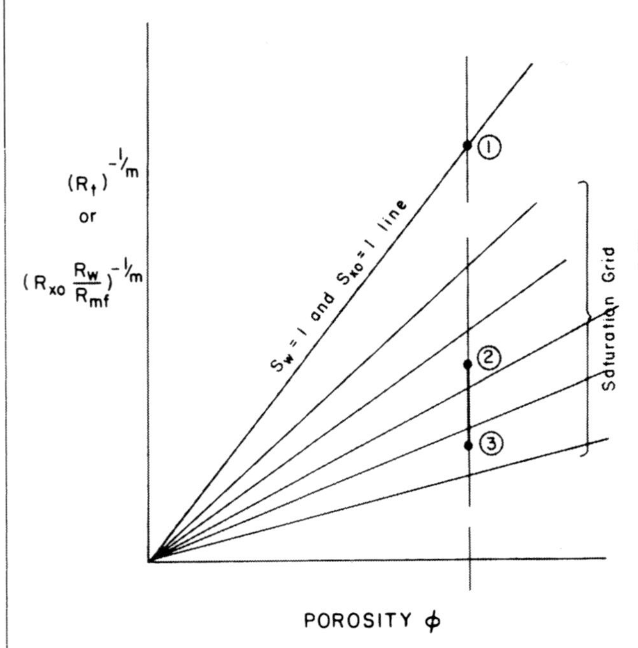

Fig. 13.18—Concepts of S_{mo} determination with a Hingle plot.

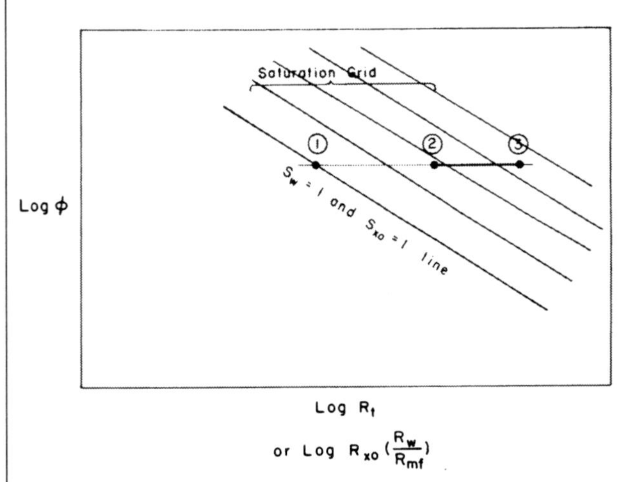

Fig. 13.19—Concepts of S_{mo} determination with a Pickett plot.

By applying the same analysis, we can show that, where a combination error exists, the water trend is straightened by use of an apparent matrix travel time, $(\Delta t_{ma})_a$, equal to $C_1+C_2\Delta t_{ma}$. Then,

$$\Delta t_a-(\Delta t_{ma})_a=C_1+C_2\Delta t-[C_1+C_2(\Delta t_{ma})_a]$$

$$=C_2(\Delta t-\Delta t_{ma}). \quad (13.32)$$

This means that the zero error is compensated for, and the sensitivity error results in a shift of all data points by a constant factor, log C_2. Correct values of S_w and m are derived from the crossplot.

This analysis also can be used to show that, for a Pickett plot, using the values of $(\rho_{ma}-\rho)$ to straighten the water trend results in compensation for zero errors. The sensitivity errors will shift all the data points by the same amount.

If a zero error exists in the resistivity log, such as

$$R_a=C_1+R_t, \quad (13.33)$$

and if the apparent log response, R_a, is used as R_t in the Pickett crossplot, the water trend will be a curve. The curvature is most

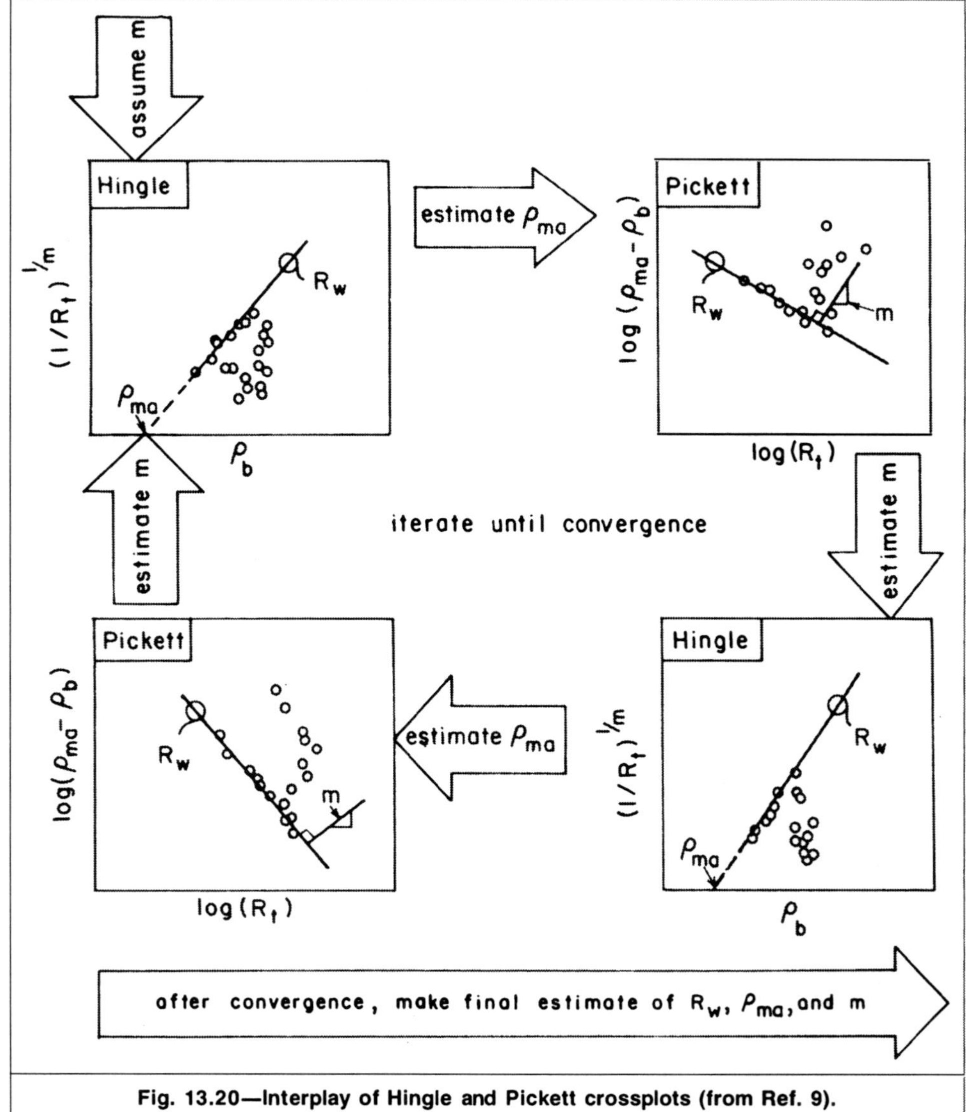

Fig. 13.20—Interplay of Hingle and Pickett crossplots (from Ref. 9).

marked in the lower resistivity range. The same type of error in porosity tools results in a curvature in the higher resistivity range, making it possible to distinguish a porosity tool miscalibration from a resistivity tool miscalibration. The curvature can be eliminated by guessing the value of C_1 that results in R_t—i.e., $R_a - C_1$.

A sensitivity error in the resistivity log results in

$$R_a = C_2 R_t. \quad (13.34)$$

However, because

$$\log R_a = \log C_2 + \log R_2, \quad (13.35)$$

a sensitivity error simply results in a shift of the data points on the resistivity scale by a constant factor, log C_2. Correct values of S_w still can be derived from the plot.

A combination error in the resistivity tool can be reduced to a sensitivity error by finding the C_1 that straightens the water-bearing trend.

If a statistically significant number of data points is available, the Pickett plot interpretation approach circumvents all types of errors that might exist in readings from resistivity and porosity tools. In general, the conventional interpretation techniques do not circumvent these miscalibration errors.

13.4 Flushed-Zone Resistivity/Porosity Crossplots

When R_{xo} values are available, the two types of resistivity/porosity crossplots described previously can be adapted to determine S_{xo} and movable oil saturation, S_{mo}, values. Eqs. similar to Eqs. 13.1 and 13.10 can be written for R_{xo} and S_{xo}:

$$(R_{xo})^{-1/m} = (S_{xo}^n / aR_{mf})^{1/m} \phi \quad (13.36)$$

and

$$\log R_{xo} = -m \log \phi + \log R_{mf} - n \log S_{xo}. \quad (13.37)$$

This means that a plot of R_{xo} vs. porosity that follows the procedures of the Hingle or Pickett plot will make it possible to determine the S_{xo} values of the plotted zones. Comparison of S_{xo} and S_w values will indicate zones where oil is movable and the amount that can be moved.

S_{mo} can be determined graphically from the R_t/ϕ Hingle or Pickett crossplots. Because

$$R_{xo} = FR_{mf}/S_{xo}^n \text{ and } R_t = FR_w/S_w^n,$$

then

$$R_t = R_{xo}(R_w/R_{mf})(S_{xo}/S_w)^n. \quad (13.38)$$

Consequently, for water-bearing zones where $S_{xo} = S_w$,

$$R_t = R_{xo}(R_w/R_{mf}), \quad (13.39)$$

and for oil-bearing zones where $S_{xo} > S_w$,

$$R_t < R_{xo}(R_w/R_{mf}). \quad (13.40)$$

To determine S_{mo} graphically, the R_{xo} value first is normalized by the resistivity contrast between formation water and mud filtrate. This normalization results in the value for $(R_{xo}R_w/R_{mf})$. The normalized values then are entered into the R_t vs. ϕ crossplot.

According to Eq. 13.39, the $S_w = 1$ and $S_{xo} = 1$ lines assume the same position on the Hingle and Pickett plots (**Figs. 13.18 and 13.19**, respectively). The true resistivity and normalized R_{xo} values for a specific water-bearing zone assumes the same position on the plot (e.g., Zone 1). The true resistivity for an oil zone of the same porosity as Zone 1 assumes Position 2. According to Eq. 13.40, its normalized R_{xo} value will assume Position 3.

The separation between Positions 2 and 3 qualitatively indicates the presence of movable oil in the specific zone. The values of saturation read off the saturation grid at Positions 2 and 3 indicate S_w and S_{xo}, respectively. S_{mo} is, of course, the difference between these two values—i.e., $(S_{xo}-S_w)$.

13.5 Interplay of Hingle and Pickett Crossplots

In Secs. 13.2 and 13.3, two resistivity vs. porosity crossplots were examined. In a Hingle plot one must assume the cementation exponent, m, and plot $R_t^{-1/m}$ vs. a porosity parameter for the zones of interest. One may establish the R_o trend and extrapolate it to determine the matrix point. The Pickett plot requires the choice of a matrix point, calculation of $(\rho_{ma}-\rho_b)$ or $(\Delta t-\Delta t_{ma})$, and plotting of this value vs. (R_t) on log-log paper. From this plot, one can estimate m from the negative inverse of the slope of the R_o trend.

An interpretation approach that combines the Hingle and Pickett plots was proposed in Refs. 9 and 10; **Fig. 13.20** illustrates this approach. An initial m is assumed and applied to a Hingle plot from which an initial matrix point can be estimated. This matrix point is used in the Pickett plot, from which a new m can be derived. Again, this is substituted into the Hingle plot, and the cycle is repeated until successive matrix points and cementation exponents are equal. Once convergence is reached, the R_o trend of either plot can easily be used to estimate the formation water resistivity.

The interplay of the Hingle and Pickett crossplots can be conducted more efficiently with an interactive computer system with graphics capabilities.

Example 13.10. The data of Table 13.3 were used to construct a Hingle plot (Example 13.4) and a Pickett plot (Example 13.6). These two plots display the following information.

	Hingle Plot	Pickett Plot
m	2 (assumed)	2.2
ρ_{ma}, g/cm³	2.71	2.71 (assumed)
R_w, Ω·m	0.026	0.015

Use the iterative technique discussed in this section to obtain the most representative values for m, ρ_{ma}, and R_w.

Solution. The value of $m=2.2$ yielded by the Pickett plot is used to prepare a second Hingle plot (**Fig. 13.21**). The water trend shown by the plot is extrapolated to the matrix value of 2.71 g/cm³ as in the other two plots.

The value of R_w is estimated from the coordinate of Point 12 as follows:

$R_o=2.5$ Ω·m,

$\rho_b=2.54$ g/cm³,

$\phi=(2.71-2.54)/(2.71-1)=0.099$ or 10%,

$F=1/\phi^{2.2}=161$,

and $R_w=1.5/161\cong 0.015$ Ω·m.

The values of m, ρ_{ma}, and R_w of the second Hingle/Plot are the same as those displayed by the Pickett plot. Convergence is then attained, and the most representative values of m, ρ_{ma}, and R_w are $m=2.2$, $\rho_{ma}=2.71$ g/cm³, and $R_w=0.015$ Ω·m.

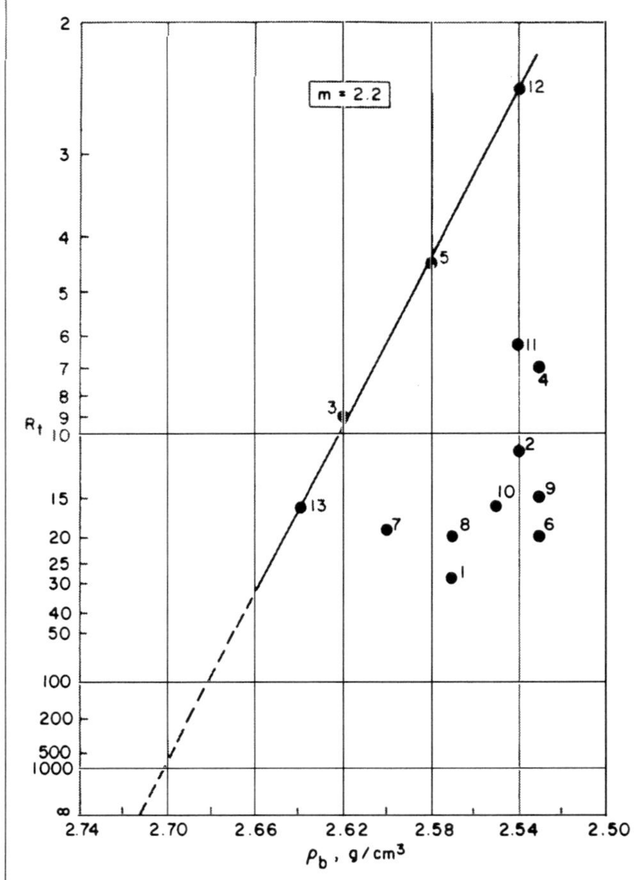

Fig. 13.21—Hingle plot of Example 13.10 prepared with $m=2.2$.

13.6 Summary

Crosspotting techniques encompass the conventional and reconnaissance interpretation techniques while offering a significant advantage over these methods. The crossplotting approach has five principal advantages.

1. A long log section can be evaluated quickly for the presence of significant hydrocarbon saturations without the time-consuming process of calculating water saturation for different zones by the basic equations.
2. Crossplotting gives the log analyst a quick idea of value ranges, a better feel for the data with which he or she is working, and insight into current shortcomings of the data and interpretation techniques, which may lead to the acquisition of more data and the development of a new interpretation approach.
3. A measure of water saturation can be derived without knowledge of m or R_w, without the calibration for the porosity tool (ρ_{ma}, Δt_{ma}, etc), and in the presence of logging calibration errors.
4. Under favorable conditions, the recognized patterns can be used to obtain values of R_w, m, Δt_{ma}, and ρ_{ma}. Also, rock types can be distinguished.
5. Once a crossplot has been made, parameters such as R_w and m can be varied easily without tedious recalculations.

The crossplotting pattern-recognition technique is one of the most powerful interpretation approaches. Whenever log analysis is expanded beyond the use of conventional interpretation techniques, crossplotting should be considered as an evaluation aid.

Review Questions

1. What is the basic concept of the pattern-recognition interpretation technique?
2. Is the pattern-recognition interpretation technique restricted only to porosity/resistivity crossplots?
3. What factors determine the parameters to be crossplotted and the type of crossplot?
4. What are the advantages of the pattern-recognition interpretation technique over conventional techniques?
5. What are the basic limitations of the crossplotting interpretation technique?

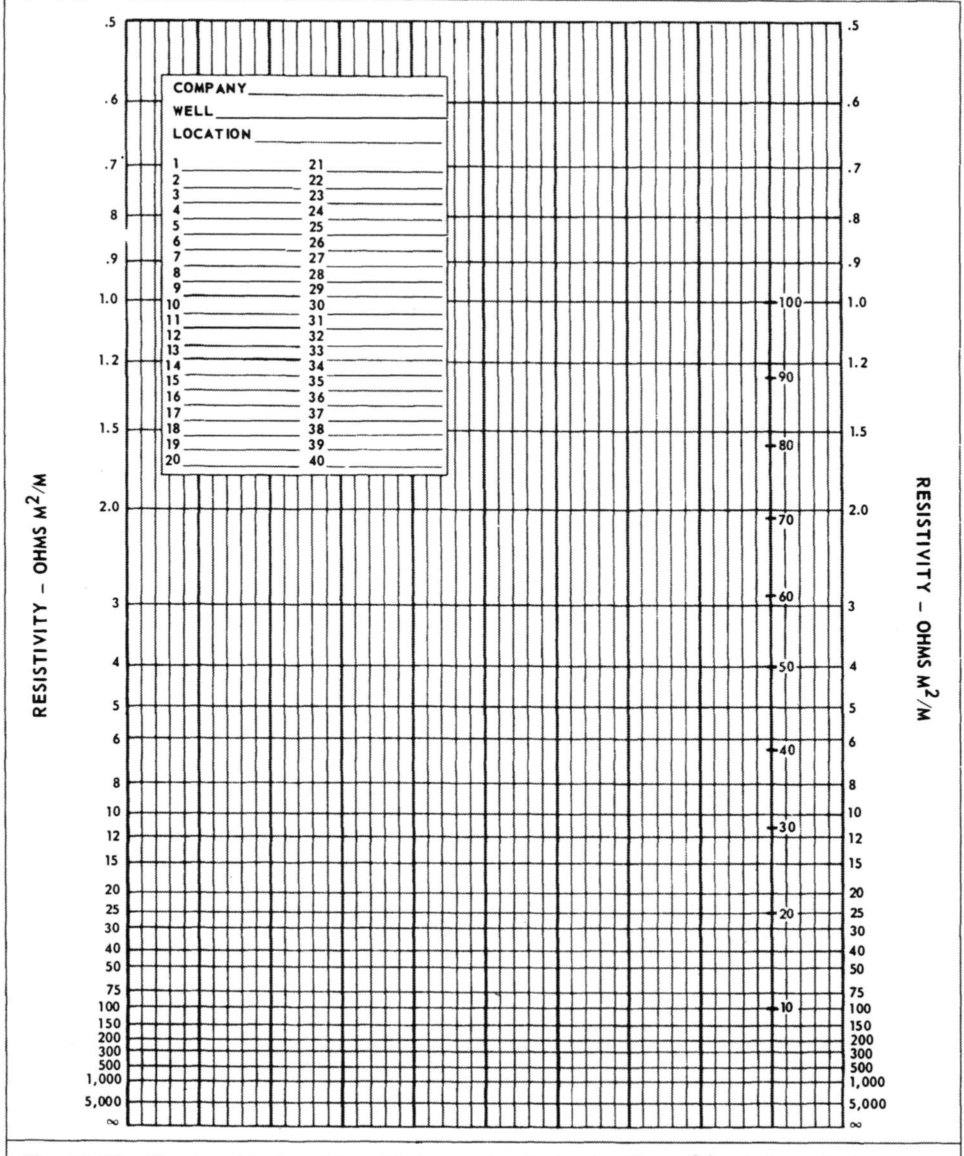

Fig. 13.22—Hingle grid of problem 13.1, constructed using $F=\phi^{-2.4}$ relationship (courtesy Gearhart Industries Inc.).

6. What are the specific advantages and limitations of the Hingle crossplot?

7. What are the specific advantages and limitations of the Pickett crossplot?

8. How are the Hingle and Pickett crossplots similar? How do they differ?

9. How can the Hingle and Pickett crossplots be interplayed for additional advantages of the crossplotting technique?

10. Why is the neutron/resistivity crossplot not recommended in cases of gas-bearing formations?

11. What are the two basic assumptions inherent in the concept of the Hingle and Pickett crossplots?

12. How can changes in formation water resistivity and lithology within the interval analyzed be detected?

13. Does a data point situated off the water trend on the porosity/resistivity crossplot always indicate a hydrocarbon zone?

14. How are matrix properties determined with a Pickett crossplot?

15. How does the Pickett crossplotting technique account for zero, sensitivity, and combination errors that result from tool miscalibration?

Problems

13.1 a. On the Hingle grid of **Fig. 13.22,** plot the water trend that represents formations where $\rho_{ma}=2.8$ g/cm^3 and $R_w=0.02$ Ω·m.

b. On the same grid, plot lines that represent cutoff saturation and porosity of 50% and 10%, respectively. The saturation exponent, n, is 2.2.

13.2 On a 2-×3-cycle log-log paper, prepare a porosity/resistivity Pickett crossplot that features the following.

a. A water trend representing formations characterized by $m=2$ and $R_w=0.02$ Ω·m.

b. A water trend representing formations characterized by $m=2$ and $R_w=0.08$ Ω·m.

c. A water trend representing formations characterized by $m=1.5$ and $R_w=0.02$ Ω·m.

d. An oil trend characterized by $R_w=0.02$ Ω·m, $m=2$, and $S_w=50\%$.

e. An oil trend characterized by $R_w=0.02$ Ω·m, $m=1.5$, $\phi=10\%$, and $S_w=57\%$.

13.3 **Figs. 13.23 and 13.24** illustrate the Laterolog (LL3) and sidewall neutron logs obtained over an Abu Reef limestone section.

a. Can you, by qualitative visual inspection of the logs alone, detect the presence or absence of hydrocarbon zones?

b. Identify at least 40 zones for evaluation. Mark and label the zones on the gamma ray, resistivity, and porosity curves.

c. Read and tabulate the resistivity and porosity values of each zone.

d. Prepare a Hingle crossplot using the data prepared in Part c. What value of m did you consider?

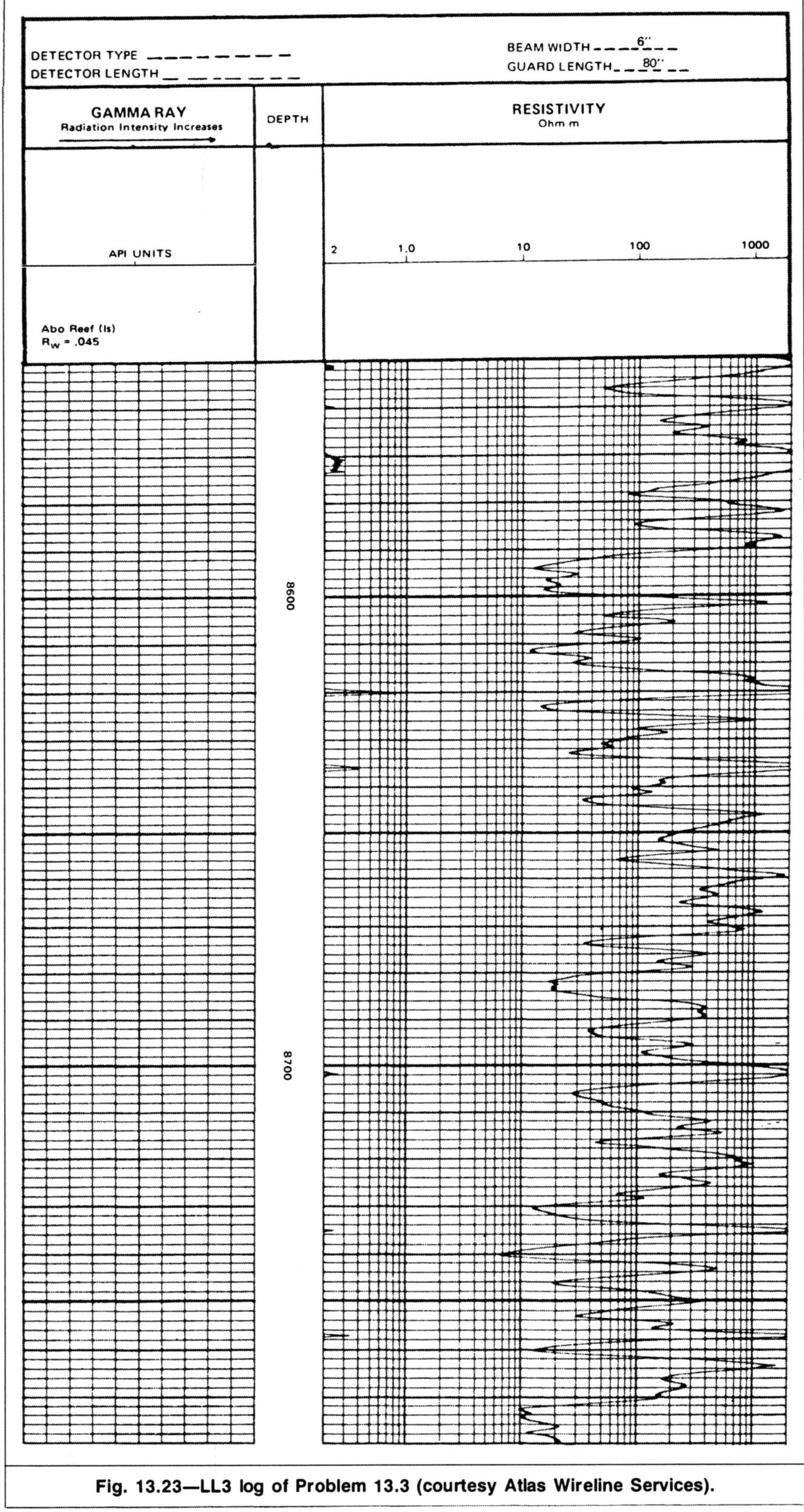

Fig. 13.23—LL3 log of Problem 13.3 (courtesy Atlas Wireline Services).

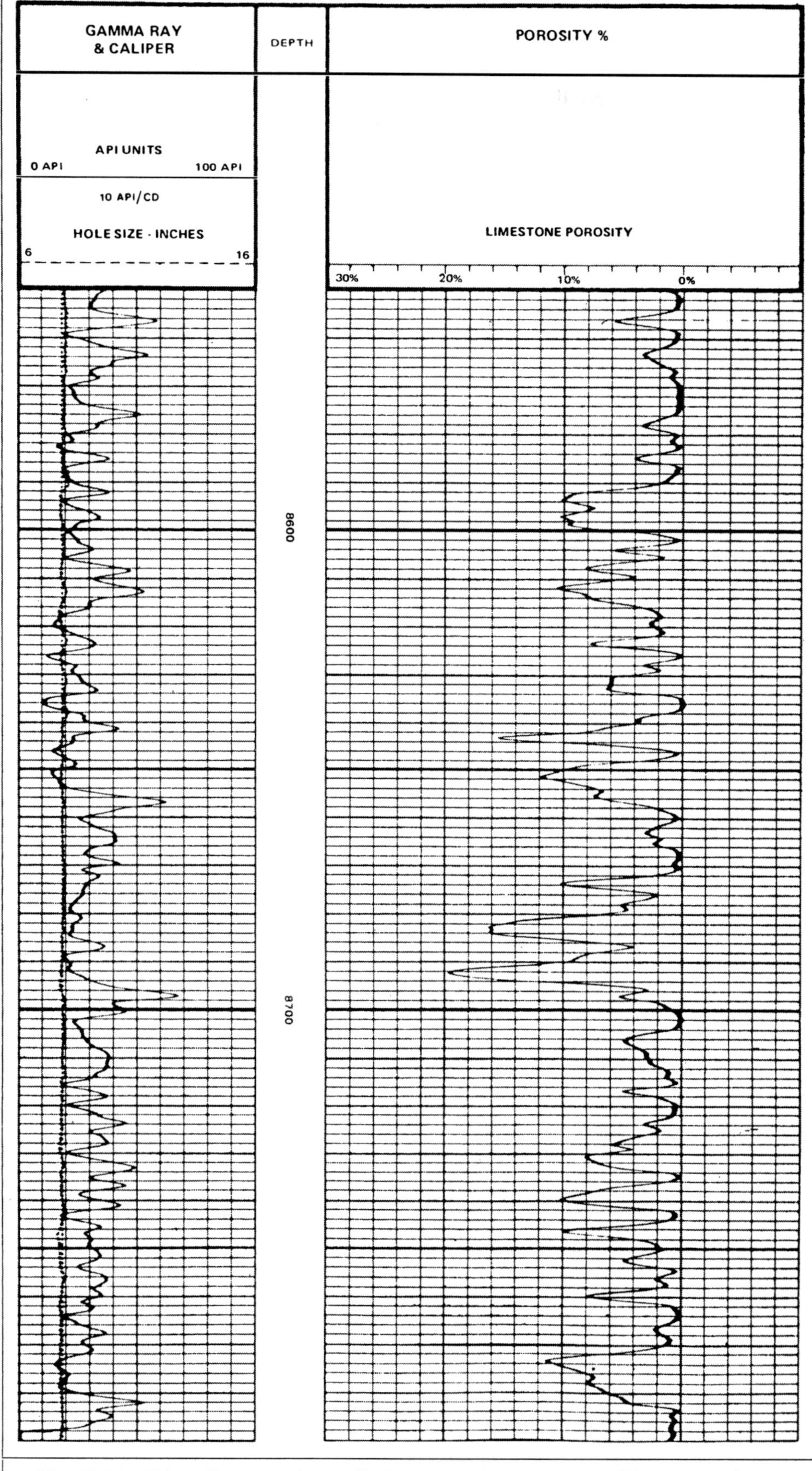

Fig. 13.24—Sidewall neutron log of Problem 13.3 (courtesy Atlas Wireline Services).

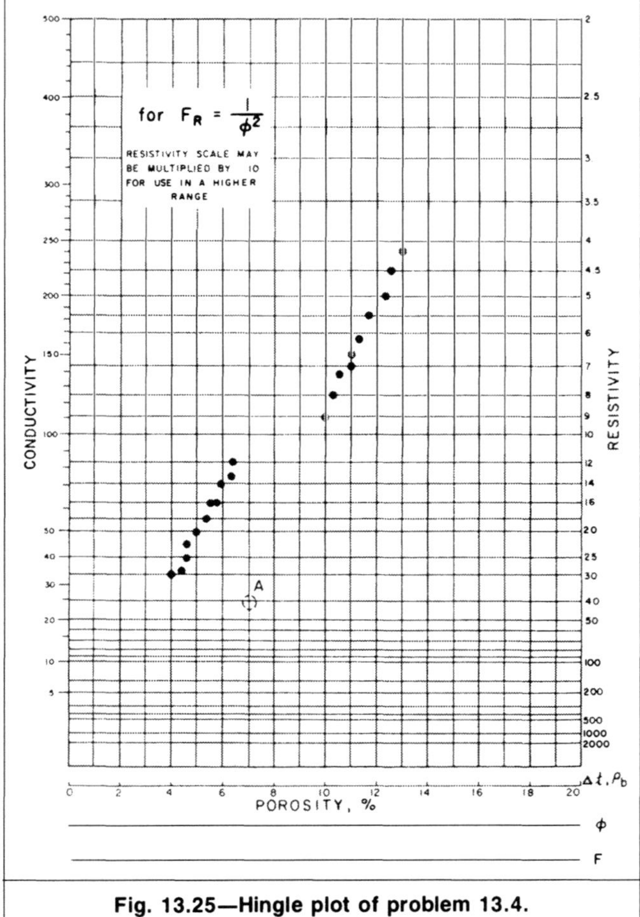

Fig. 13.25—Hingle plot of problem 13.4.

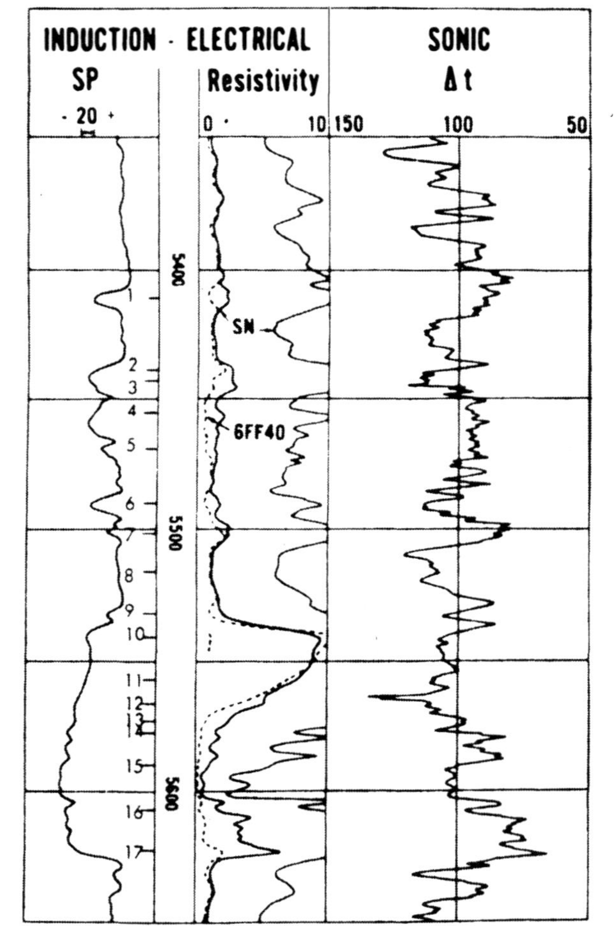

Fig. 13.26—Logs of Problem 13.6 (courtesy Schlumberger).

e. Assuming that the illustrated section contains some zones that are 100% water-bearing, plot the optimum "R_o trend" and estimate the corresponding R_w value. Also, plot the 50%, 40%, 30%, 20%, and 10% water saturation lines. On the Hingle plot, circle the potential zones if ϕ_c and S_{wc} are 12% and 50%, respectively.

f. Repeat Part e, assuming that the section shown does not contain any 100% water-bearing zones. Water catalogs list a value of 0.045 Ω·m for R_w of the subject formation.

g. How do the results obtained in Part f differ from those obtained in Part e?

h. Repeat Parts d through f using a Pickett plot.

i. Compare the results obtained using the Pickett plot to those obtained using the Hingle plot.

13.4 **Fig. 13.25** shows the Hingle plot prepared for permeable zones present in a 200-ft interval. The R_t values are derived from the dual induction/SFL log, and the porosity values are derived from the bulk density log using a matrix density value of 2.71 g/cm^3. If the subject interval is known to contain two types of lithology, explain the two lithology types and determine the value of R_w in each lithology type. If hydrocarbons are present in the subject interval, the hydrocarbons are expected to be oil. Does Zone A shown on the plot contain oil? If yes, what is the oil saturation?

13.5 The following values were read from the induction and sonic logs recorded through a sand/shale section.

Zone	Conductivity (m℧/m)	Δt (μsec/ft)
1	400	95
2	150	96
3	370	83
4	520	85
5	50	58
6	140	76
7	130	67

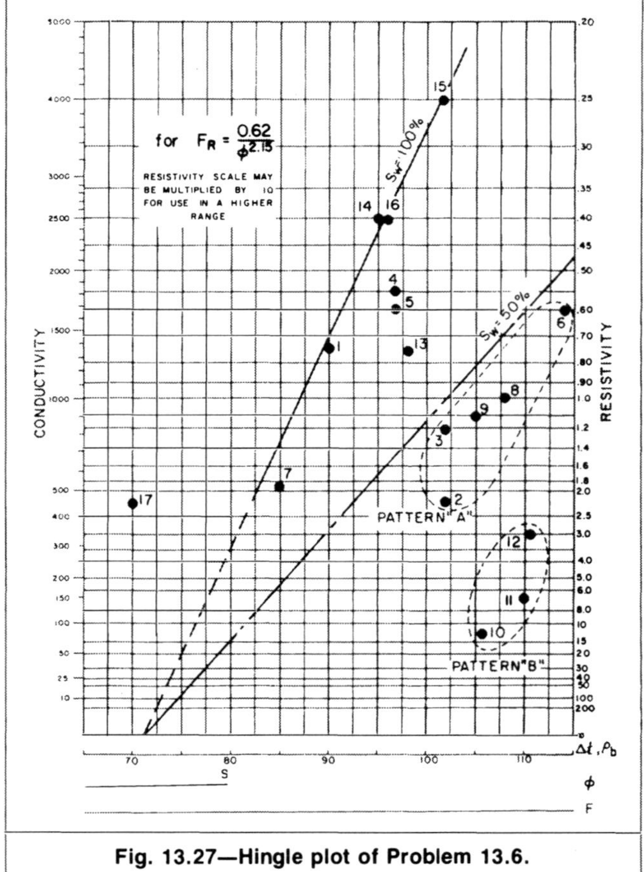

Fig. 13.27—Hingle plot of Problem 13.6.

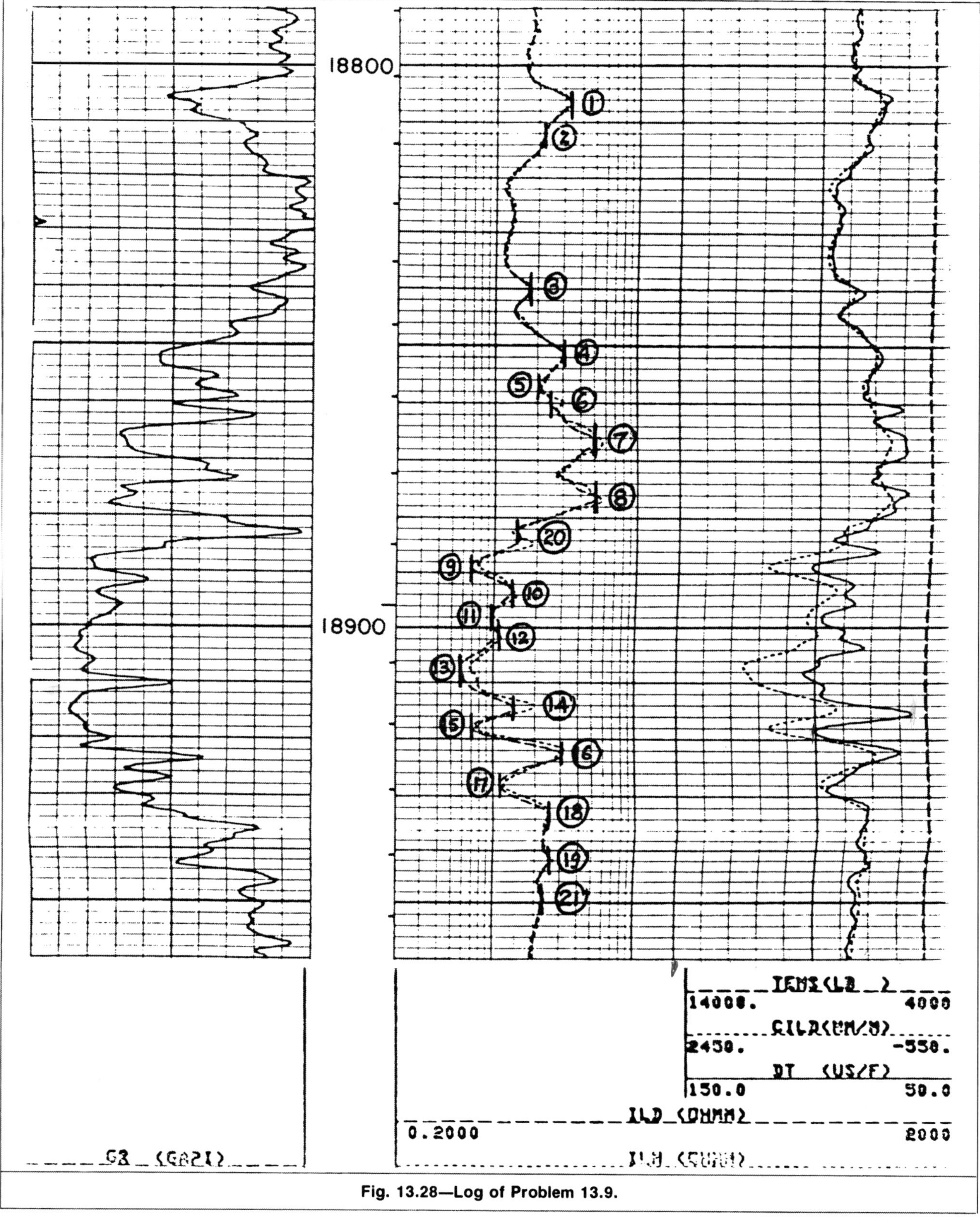

Fig. 13.28—Log of Problem 13.9.

8	100	80
9	100	55
10	200	69
11	150	90
12	1,050	86
13	1,500	91

a. Prepare a Hingle plot using $F=0.62/\phi^{2.15}$.

b. Prepare a Pickett plot using $\Delta t_{ma}=55.5$ μsec/ft.

c. Does the section analyzed contain lithologies other than sandstone?

d. Determine R_w and Δt_{ma} from the Hingle plot.

e. Determine R_w and m from the Pickett plot.

f. According to the Hingle and Pickett plots, which zones exhibit hydrocarbon saturations higher than 50%?

13.6 **Fig. 13.26** shows logs recorded through a sand/shale section. **Fig. 13.27** is a Hingle crossplot of induction resistivity vs. sonic travel time prepared using the data of the numbered zones.

a. Why was Point 17 not considered when determining the water trend?

b. How can you explain the high matrix travel time of 71 μsec/ft indicated by the plot?

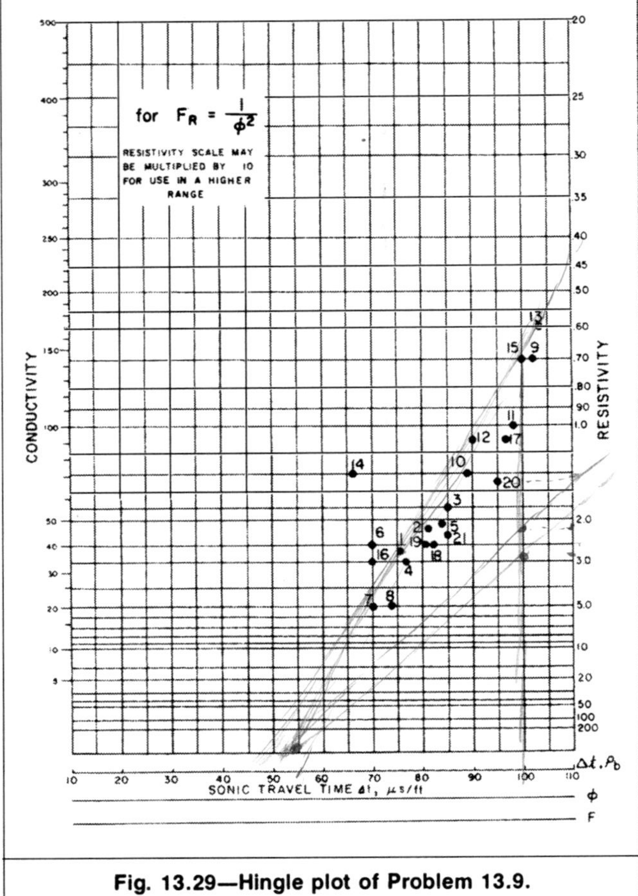

Fig. 13.29—Hingle plot of Problem 13.9.

c. Are the water saturation values indicated by the plot for zones within Patterns A and B representative? If not, are they on the high or low side?

13.7 The following data were obtained in a well where a conventional core, induction log, and density log were run for an interval of interest.

Zone	Depth (ft)	ϕ_{core}	ρ_b (g/cm^3)	R_t ($\Omega \cdot$m)
1	4,180	14.0	2.43	13.5
2	4,186	17.0	2.39	8.8
3	4,193	18.0	2.37	7.8
4	4,197	20.0	2.34	6.0
5	4,203	17.5	2.38	8.2
6	4,209	19.0	2.36	11.0
7	4,212	15.0	2.47	90.0
8	4,220	17.0	2.44	100.0
9	4,225	17.5	2.43	14.0
10	4,230	13.5	2.49	23.0
11	4,234	12.5	2.51	28.0

a. Prepare a plot of ρ_b vs. ϕ_{core}.

b. Prepare a Pickett crossplot (ϕ_{core} vs. R_t).

c. How many rock types are present in this interval? Determine the following for each rock type: most probable lithology, m and R_w.

d. Indicate the hydrocarbon zones, and determine their water saturation values.

13.8 Tabulated below are values read from logs obtained through a limestone formation.

a. Prepare a Pickett crossplot.

b. Determine m and R_w from the crossplot.

c. Determine the water saturation and porosity of potential zones.

Zone	R_{ILd} ($\Omega \cdot$m)	Δt (μsec/ft)
1	90	55
2	105	52
3	80	52
4	55	54
5	120	49
6	8	52
7	5	54
8	1.9	58
9	4	55
10	2.4	59
11	1.6	64
12	2.3	58
13	0.8	63.5
14	1.3	60

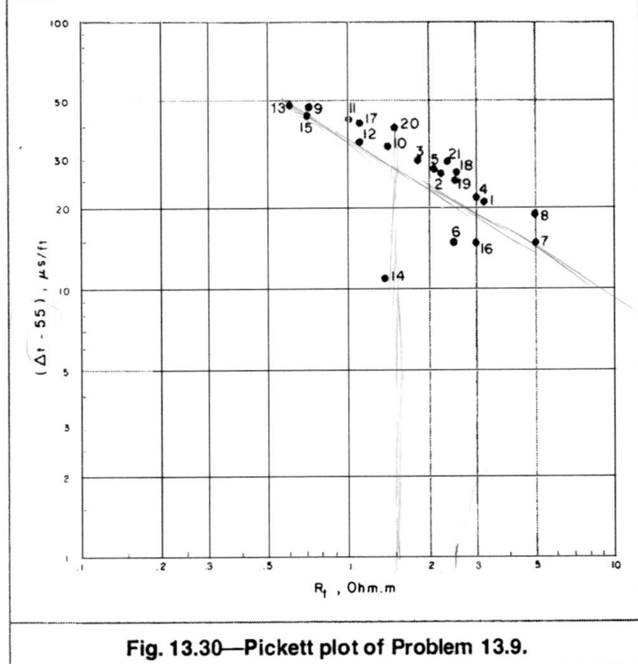

Fig. 13.30—Pickett plot of Problem 13.9.

13.9 An exploration well was drilled to test sands in the interval 18,790 to 18,960 ft shown in **Fig. 13.28**. Only a dual induction/sonic log was obtained. The openhole condition precludes the acquisition of additional data (logs, cores, formation tester, drillstem test, etc.). Further testing will require running a casing, which, given the formation depth, represents a substantial investment.

Twenty-one zones, marked on the log, were selected for the analyses. Hingle and Pickett plots were prepared and are shown in **Figs. 13.29 and 13.30.**

a. Are lithologies other than sandstone present in the interval analyzed?

b. Is the analyst justified in using $m=2$ and $\Delta t_{ma}=55$ μsec/ft in constructing the two crossplots?

c. Does the interval analyzed contain potential hydrocarbon zones?

d. If experience with this formation indicates that $R_w=0.06$ $\Omega \cdot$m at formation temperature, will this affect the conclusion you reached in Part c, and how do you explain the difference between this R_w value and the value obtained from the crossplots?

e. In the form of a memorandum, give your recommendation as to what to do next with this well.

Nomenclature

a = coefficient in Archie's F-ϕ relationship
E_{SP} = self-potential, mV
F = formation resistivity factor
L = distance, in.
m = cementation exponent in Archie's F-ϕ relationship
n = saturation exponent
R = resistivity, $\Omega \cdot$m
S = saturation, fraction
S_{mo} = Movable Oil Saturation

Δt = formation acoustic interval transit time, μsec/ft
α,β = coefficients in Eq. 13.6
γ = formation natural gamma-ray radioactivity
κ,δ = coefficients in Eq. 13.9
ρ = density, g/cm^3
ϕ = porosity, fraction

Subscripts

a = apparent
b = bulk
c = cutoff
D = formation density
f = fluid
ma = matrix
mf = mud filtrate time
N = neutron
o = oil
t = true
w = water
xo = flushed zone

References

1. McFadzen, T.B.: "Cross-Plotting. A Neglected Technique In Log Analysis," *Proc.*, SPWLA 14th Annual Logging Symposium, Lafayette (1973) paper Y.
2. Hingle, A.T.: "The Use of Logs in Exploration Problems," paper presented at the Soc. of Exploration Geophysicists 29th Intl. Annual Meeting, Los Angeles, Nov. 1959.
3. *Log Interpretation Charts,* Schlumberger Technical Services, Houston (1991) 134–35.
4. *Formation Evaluation Data Handbook,* Gearhart-Owen Wireline Services, Ft. Worth, TX (1978) 81–84.
5. Fertl, W.H.: "Hingle Crossplot Speeds Long-Interval Evaluation," *Oil & Gas J.* (Jan. 15, 1979) 114–16.
6. Pickett, G.R.: "Pattern Recognition as a Means of Formation Evaluation," *Proc.*, SPWLA 14th Annual Logging Symposium, Lafayette (1973) paper A.
7. Pickett, G.R.: "A Review of Current Techniques for Determination of Water Saturation from Logs," *JPT* (Nov. 1966) 1425–33.
8. Lang, W.H. Jr.: "Porosity-Resistivity Crossplots Can Help To Evaluate Formations," *Oil & Gas J.* (Nov. 29, 1976) 94–98.
9. Gael, T.B.: "Estimation of Petrophysical Parameters by Crossplot Analysis of Well Log Data," MS thesis, Louisiana State U., Baton Rouge (May 1981).
10. Hilchie, D.W.: *Applied Openhole Log Interpretation for Geologist and Engineers,* D.W. Hilchie Inc., Golden, CO (1978).

Chapter 14
Interpretation in Complex Lithologies

14.1 Introduction

Accurate porosity determination is necessary for effective log interpretation. When the lithology of a clean, liquid-filled formation is known or can be assumed with reasonable accuracy, representative porosity values can be derived from the reading of a porosity log. However, porosity determination becomes more involved when the lithology is not known or when it consists of two or more minerals of unknown proportions. Most reservoir rocks are composed of one of two main minerals and may contain various amounts of clay. The most common binary mixtures associated with carbonate rocks are limestone/dolomite, limestone/silica, dolomite/silica, and dolomite/anhydrite. For sands, silica/dolomite, silica/limestone, and silica/heavy mineral (such as pyrite or siderite) are likely combinations. Ternary mixtures of these minerals and more complex compositions also occur.[1]

Log-derived porosity values ϕ_D, ϕ_N, and ϕ_S can be expressed in general terms as[2]

ϕ_D or $\phi_N = f$(matrix, total porosity, shale type and amount, type and amount of fluids in pore space) . . . . (14.1)

and $\phi_S = f$(matrix, primary porosity only, degree of formation compaction, shale type and amount, type and amount of fluids in pore space). . . . . (14.2)

Although all three porosity device responses are affected by porosity, matrix, and fluid parameters, the effects on the devices are different. These differences can be resolved with a graphical technique usually called "litho/porosity crossplots," and both porosity and matrix characteristics can be defined. Two types of crossplots are used. The "dual-mineral crossplots" presume that the matrix consists of two minerals. The other type treats the case of three-mineral, or ternary mixtures.

In addition to porosity computation in complex lithologies, the litho-porosity technique has many applications in formation evaluation, including: secondary porosity detection; gas detection; lithology determination for stratigraphic and environmental studies; and detection and evaluation of mineral deposits, such as sulfur, potash, coal, oil shale, and certain metallic minerals.[3] For these applications, certain combinations of the three porosity tools are more appropriate than others. Various combinations of porosity tools are discussed in this chapter. Except as indicated, nonshaly, liquid-saturated formations are considered in the discussion. Applications of litho/porosity crossplots in shaly formation evaluation and gas detection are addressed in detail in Chaps. 15 and 16, respectively.

14.2 Lithology/Porosity Interpretation With Two Porosity Tools

14.2.1 Neutron/Density Combination. In clean, liquid-filled formations, Eq. 14.1 reduces to

ϕ_D or $\phi_N = f$(matrix, porosity) . . . . (14.3)

For the density tool, Eq. 14.3 can be expressed in terms of bulk density:

$$\rho_b = \rho_{ma} - \phi(\rho_{ma} - \rho_f). \quad (14.4)$$

When a porosity index is derived from the neutron log and when the matrix is assumed to have the same properties as a water-filled limestone, Eq. 14.3 reduces to

$$\phi_N = f(\phi), \quad (14.5a)$$

or $$\phi = f(\phi_N), \quad (14.5b)$$

where ϕ_N and ϕ are the apparent and true porosities, respectively. Eq. 14.5 is expressed graphically for the Compensated Neutron Log (CNL^{SM}) by the curves of **Fig. 14.1.** Combining Eqs. 14.4 and 14.5 results in

$$\rho_b = \rho_{ma} - (\rho_{ma} - \rho_f) f(\phi_N). \quad (14.6)$$

For water-filled limestone where $\rho_f = 1$ g/cm^3, $\rho_{ma} = 2.71$ g/cm^3, and $\phi_N = \phi$, Eq. 14.6 becomes

$$\rho_b = 2.71 - 1.71\phi_N. \quad (14.7)$$

Eq. 14.7 represents a straight line if ρ_b and ϕ_N are crossplotted on linear scales, as shown by **Fig. 14.2a.** This straight line that defines the water-filled limestone trend can be scaled in porosity units. The vertical scale that represents ρ_b can also be scaled in apparent limestone density porosity. Scaling is done with Eq. 8.11 where the values of 2.71 and 1 g/cm^3 are substituted for ρ_{ma} and ρ_f, respectively. As can easily be seen, the limestone line is one of equal neutron and density porosities. **Table 14.1** lists the ϕ_D, ϕ_N, and ρ_b values that correspond to different values of ϕ. For water-saturated sandstones where $\rho_{ma} = 2.65$ g/cm^3 and $\rho_f = 1$ g/cm^3, **Table 14.2** gives values of ϕ_N, ϕ_D and ρ_b that correspond to different values of ϕ. The values of ϕ_N are obtained from Fig. 14.1; the values of ρ_b are calculated with Eq. 14.4 (i.e., $\rho_b = 2.65 - 1.65\phi$); and the ϕ_D values are obtained with Eq. 8.11 [i.e., $\phi_D = (2.71 - \rho_b)/1.71$]. Crossplotting ρ_b vs. ϕ_N yields the water-saturated sandstone trend, which is easily graduated in true porosity values (**Fig. 14.2b**).

The water-saturated dolomite trend can be constructed similarly. **Table 14.3** lists the values necessary for plotting the dolomite trend.

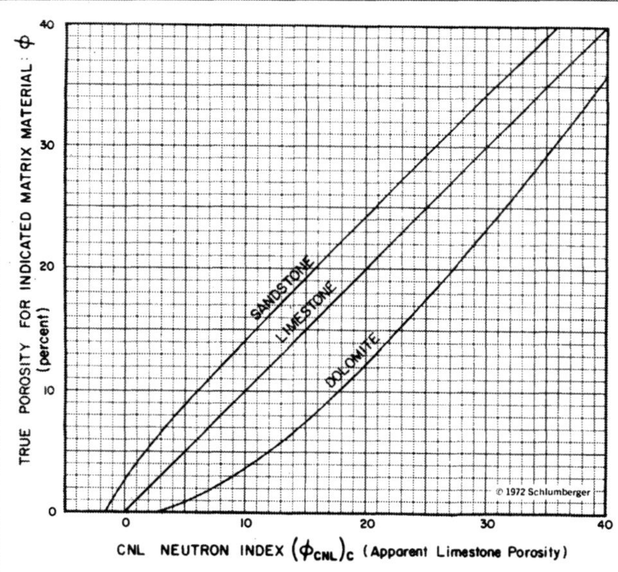

Fig. 14.1—CNL neutron porosity index derived with limestone matrix vs. true porosity for common reservoir minerals (courtesy Schlumberger).

Note that the three trends shown in Fig. 14.2B converge to a common point ($\rho_b = 1$ g/cm^3, $\phi_N = 100\%$). This convergence represents the process of fluid (fresh water in this case) progressively filling the pore space.

Dual-mineral litho/porosity crossplots are usually included in well-logging companies' chart books. **Fig. 14.3 and 14.4,** similar to the crossplot of Fig. 14.2B, are for Schlumberger's Formation Density Log/Compensated Neutron Log (FDC/CNL) and Formation Density Log/Sidewall Neutron Porosity Log (FDC/SNP) combinations, respectively. These charts are constructed for clean, fully liquid-saturated formations and for boreholes filled with fresh water or freshwater-based mud (i.e., $\rho_f = 1.0$ g/cm^3). Similar charts are available for cases when the borehole is filled with salt water or saltwater-based mud. In addition to sandstone, limestone, and dolomite trends, Figs. 14.3 and 14.4 show the trend for anhydrite and single points for zero-porosity minerals (salt, sulfur, etc.).

Because the density/neutron crossplot is based on two equations, it can, in principle, be used to solve for two unknowns—e.g., the porosity and the proportion of two known minerals. It sometimes is possible to deduce a third unknown, such as the presence of an additional mineral, with some expertise of the formations analyzed.

With the assumption that the formation is clean and liquid-filled, if the data that pertain to a zone of interest are plotted on one of the three major lithology lines or on one of the single zero-porosity points, then the zone's matrix can be considered a single-mineral unit. If the zone is plotted between two lithology lines, then it is a combination of at least two minerals and the porosity is determined from the value of an isoporosity line that connects the two lithology lines. The relatively large separation between the sandstone, limestone, and dolomite lines indicates good resolution for these three common lithologies. Evaporites like anhydrite and salt can also be identified. For example, consider Zone P, for which the FDC log indicates 2.46 g/cm^3 and the CNL shows an apparent limestone porosity of 21%. According to Fig. 14.3, if the zone is known or assumed to be a limestone/dolomite mixture, Zone P would consist of about 60% limestone and 40% dolomite and would have a porosity of roughly 18%. The two unknowns solved for are the porosity and the proportion of the two known or assumed minerals.

Zone P also could be a cherty (sandy) dolomite because it is located between the sandstone and dolomite lines. If so, it would consist of about 60% dolomite and 40% sand and would have a porosity of about 18.3%.

This example illustrates that an error in choosing the two minerals that make up the matrix will not result in a large error in determining the porosity value. Because of this unique feature, a point (ϕ_N, ρ_b) interpreted in terms of any binary combination of the four common minerals (silica, limestone, dolomite, and anhydrite) yields essentially the same porosity.[1]

Example 14.1. Listed below are the FDC log bulk densities and the SNP log apparent limestone porosities for different zones. Assuming that, if porous, the formation is liquid-filled and that its matrix consists of no more than two minerals, determine the lithology, the proportions of the minerals in the rock, and the porosity for each zone.

Zone	ρ_b (g/cm^3)	ϕ_{SNP} (%)
A	2.24	22
B	2.68	13
C	2.94	2
D	2.02	4
E	2.30	20
F	2.50	20

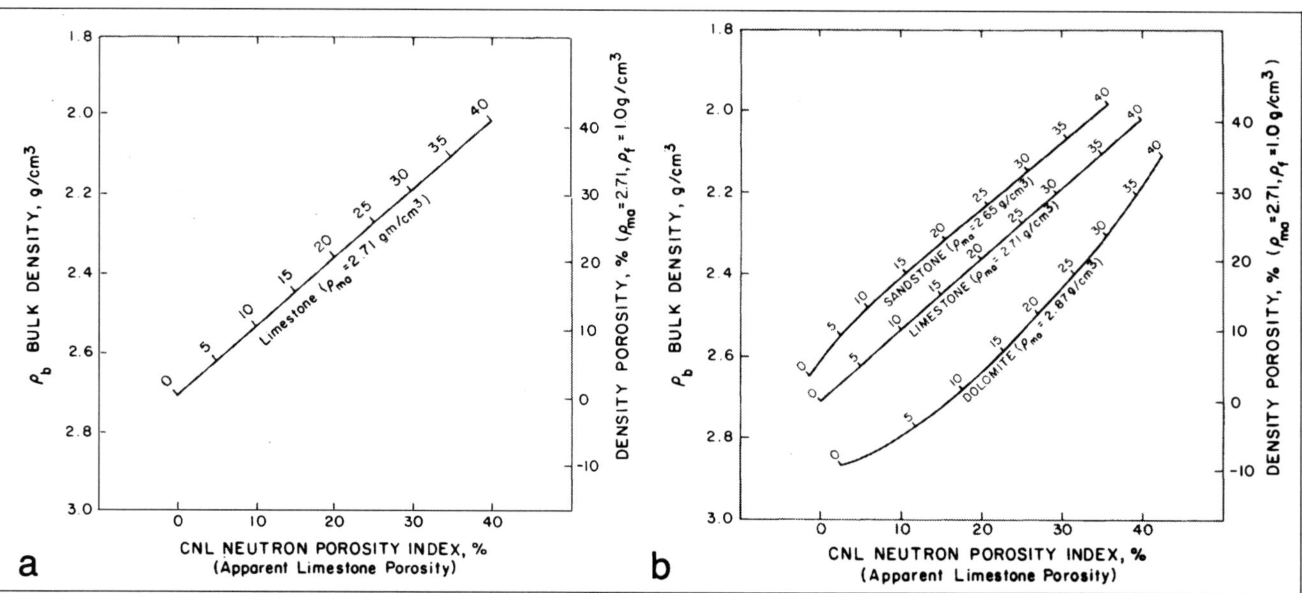

Fig. 14.2—(a) Density/neutron dual-mineral crossplot showing water-saturated limestone line; (b) density/neutron dual-mineral crossplot showing water-saturated sandstone, limestone, and dolomite lines.

Solution. *Zone A.* The point that represents Zone A lies on the sandstone trend when plotted on the FDC/SNP litho/porosity chart of Fig. 14.4. If the formation is liquid-filled, Zone A is 100% sand and has a porosity of about 25%.

Zone B. According to the stated assumptions, this zone is 100% dolomite because it lies on the dolomite trend. The porosity of Zone B is roughly 10%.

Zone C. According to Fig. 14.4, Zone C is most probably anhydrite. The porosity can be estimated from Eq. 8.11:

$$\phi=(2.98-2.94)/(2.98-1)=0.02 \text{ or } 2\%.$$

Zone D. This zone is salt. The point that represents the zone does not fall exactly on the salt point of the chart, probably because of the effect of borehole conditions on the density log response.

Zone E. The point that represents this zone lies between the sandstone and limestone lines in Fig. 14.4. According to its location, the zone is mostly sand. The second lithology is most probably limestone. Zone E is then a limy sandstone with about 70% sand and 30% limestone. The porosity is roughly 22%.

Zone F. This zone is limy or cherty dolomite. In either case, the porosity is about 18%. If assumed to be limy dolomite, Zone F then consists of 75% dolomite and 25% limestone.

Because Figs. 14.3 and 14.4 are constructed with the assumption that the formations are clean and liquid-filled, the presence of gas and/or shale in the formation will render the intrepretation more complex but not impossible. Use of litho/porosity charts to evaluate a shaly and/or gas-bearing formation is discussed in detail in Chaps. 15 and 16.

Quick visual lithological interpretation can be obtained by a graphical comparison of porosity logs by use of porosity overlays, where the two porosity logs are recorded on the same track with the same porosity scale. **Fig. 14.5** illustrates this technique. Both neutron and density porosity curves are recorded in limestone porosity units on the right track. If the fluids that fill the formations are known or can be assumed to be liquids, then the separation, or the lack thereof, between the two curves can be interpreted in terms of lithology.

A lack of separation, as in Interval A in Fig. 14.5 where $\phi_N \cong \phi_D$, indicates that both logs are presenting the true porosity and that the lithology is limestone. A "negative" separation, where $\phi_N > \phi_D$, usually corresponds to shale, as seen in Interval B in Fig. 14.5. A sandstone such as Interval C is characterized by a "positive" separation, where $\phi_N < \phi_D$. In such a case, both ϕ_N and ϕ_D are apparent porosities. The true porosity can be determined from ϕ_N and ϕ_D by use of the appropriate density/neutron litho/porosity crossplot. The true porosity can be determined from ϕ_N alone with Fig. 14.1. It can also be determined from ϕ_D alone by first back-calculating the bulk density reading and then substituting the calculated value and the matrix density of sandstone into Eq. 8.11.

The presence of gas and/or shale in the formation will affect the relative position of the porosity curves on the overlay. These effects are discussed in Chaps. 15 and 16. Note that other logs, such as the gamma ray log, should be used when available to confirm the lithology suggested by the overlay.

Example 14.2. Fig. 14.6 shows an FDC/CNL porosity overlay. The porosities are given in limestone porosity units. Examine the overlay, and determine the lithology and porosity of zones G, H, and J. If porous, the formations are known to be liquid-filled.

Solution. *Zone G.* Because $\phi_N=\phi_D$, the lithology is limestone. This equality also indicates that both logs are reading the true porosity, which is about 24%. This zone plots on the limestone line of the dual-mineral porosity chart of Fig. 14.3. The gamma ray and caliper log responses concur with the interpretation.

Zone H. The separation of the two porosity curves ($\phi_N=0$, $\phi_D=-16\%$) and the negative value of the density porosity strongly indicate that the porosity readings are apparent values. Crossplotting the two values in Fig. 14.3 clearly indicates that Zone H is a nonporous anhydrite. Once again, the gamma ray and caliper logs confirm this interpretation.

TABLE 14.1—BULK DENSITY VS. POROSITY FOR LIMESTONES

$\phi=\phi_N=\phi_D$ (%)	ρ_b (g/cm^3)
0	2.710
5	2.625
10	2.539
15	2.454
20	2.368
25	2.283
30	2.197
35	2.112
40	2.026

TABLE 14.2—APPARENT NEUTRON AND DENSITY POROSITY FOR SANDSTONES

ϕ (%)	ϕ_N (From Fig. 14.1) (%)	ρ_b $(=2.65-1.65\phi)$ (g/cm^3)	ϕ_D $[=(2.71-\rho_b)/1.71]$ (%)
0	−1.5	2.650	3.6
5	2.7	2.568	8.6
10	6.0	2.485	13.6
15	10.9	2.403	18.6
20	15.8	2.320	23.6
25	20.8	2.238	28.6
30	25.8	2.155	33.6
35	30.8	2.073	38.6
40	35.8	1.990	43.6

TABLE 14.3—APPARENT NEUTRON AND DENSITY POROSITY FOR DOLOMITES

ϕ (%)	ϕ_N (From Fig. 14.1) (%)	ρ_b $(=2.87-1.87\phi)$ (g/cm^3)	ϕ_D $[=(2.71-\rho_b)/1.71]$ (%)
0	2.5	2.870	−9.4
5	11.9	2.777	−3.9
10	17.8	2.683	1.6
15	22.6	2.590	7.0
20	27.1	2.500	12.3
25	31.5	2.403	18.0
30	35.5	2.309	23.5
35	39.4	2.216	28.9
40	42.6	2.122	34.4

Zone J. The separation shown by the overlay in this zone where $\phi_N > \phi_D$ could result from the presence of shale and/or a mineral heavier than limestone. The low gamma ray response and the absence of hole enlargement on the caliper log eliminate shale effect as the cause of the separation. Plotting Zone J in Fig. 14.3 ($\phi_N=20\%$, $\phi_D=9\%$) suggests that the zone is limy dolomite with about 73% dolomite and 27% limestone. The porosity is about 14.5%.

The position of Point J on the crossplot can also suggest the presence of silica. However, given the environment suggested by the logs, this is not likely. What is likely, however, is the presence of anhydrite and/or other evaporites in addition to limestone and dolomite. This cannot be ascertained with available data.

14.2.2 Sonic/Neutron Combination. Like the neutron/density combination, a litho/porosity crossplot can be constructed for the sonic/neutron combination. **Fig. 14.7** shows a sonic/CNL crossplot. The crossplot shows that, like the density/neutron combination, the resolution between the sandstone, limestone, and dolomite lithology is good, and errors in choice of lithology pairs among these minerals will have a negligible effect on the estimated porosity value. Note that the crossplot was constructed with the assumption that the formations were compacted by use of a weighted-average transform, Eq. 10.1, to relate sonic travel transit time to porosity.

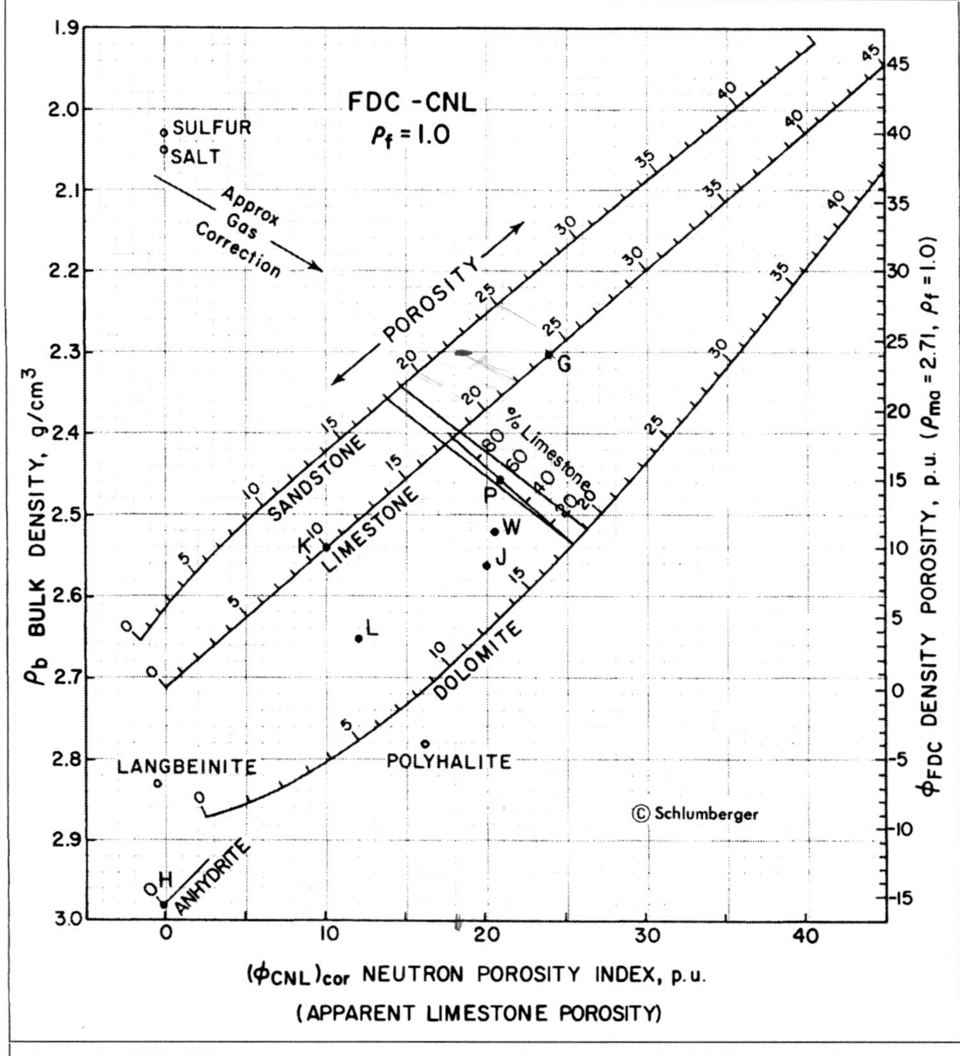

Fig. 14.3—FDC/CNL dual-mineral crossplot, case of fresh water, liquid-filled boreholes (courtesy Schlumberger).

A sonic/neutron dual-mineral crossplot based on an empirically observed transform is available.[4] Because of this additional assumption, the sonic/neutron combination is less effective than the density/neutron combination in lithology determination.

The crossplot also assumes that only primary porosity exists. This fact, however, can be used to detect the presence of secondary porosity. The concept relies on the fact that a sonic log generally is oblivious to secondary porosity (i.e., vuggy porosity and fractures). Thus, on a sonic/neutron crossplot, zones with secondary porosities will be plotted off the correct lithology trend, indicating an apparent porosity that is less than the total porosity. Detection of secondary porosity requires knowledge of the lithology, which can be obtained from core information or determined from the density/neutron litho/porosity crossplot. The magnitude of the secondary porosity can be reflected by computing a secondary porosity index (SPI). SPI is defined as the difference between the total porosity value, ϕ, determined from the density/neutron crossplot and the apparent porosity determined from sonic log, ϕ_S:

$$I_{sp}=\phi-\phi_S. \quad (14.8)$$

Example 14.3. Listed below are the FDC, CNL, and sonic log responses in two carbonate formations. Show that both zones contain secondary porosities, and estimate the SPI.

Zone	ρ_b (g/cm^3)	ϕ_N (Limestone Matrix) (%)	Δt (μsec/ft)
K	2.54	10	54
L	2.65	12	51

Solution. Zones K and L are plotted on the FDC/CNL crossplot of Fig. 14.3 and the sonic/CNL crossplot of Fig. 14.7. These two charts suggest the following information.

	FDC/CNL Crossplot		Sonic/CNL Crossplot	
Zone	Lithology	Porosity (%)	Lithology	Porosity (%)
K	100% limestone	10	50% limestone 50% dolomite	6
L	50% limestone 50% dolomite	8	100% dolomite	5

The discrepancy between the information obtained from these two charts probably results from the presence of secondary porosity, which is ignored by the sonic tool. If the information obtained from the FDC/CNL crossplot is assumed to be correct, the lithology, total porosity, and SPI of the two zones are as follows.

Zone	Lithology	Total Porosity (%)	Sonic Porosity (%)	SPI (%)
K	100% Limestone	10	5	5
L	50% Limestone 50% Dolomite	8	4	4

$\Delta t_{ma}=45.5$ μsec/ft and $\Delta t_f=189$ μsec/ft were used to determine the sonic porosity of Zone L.

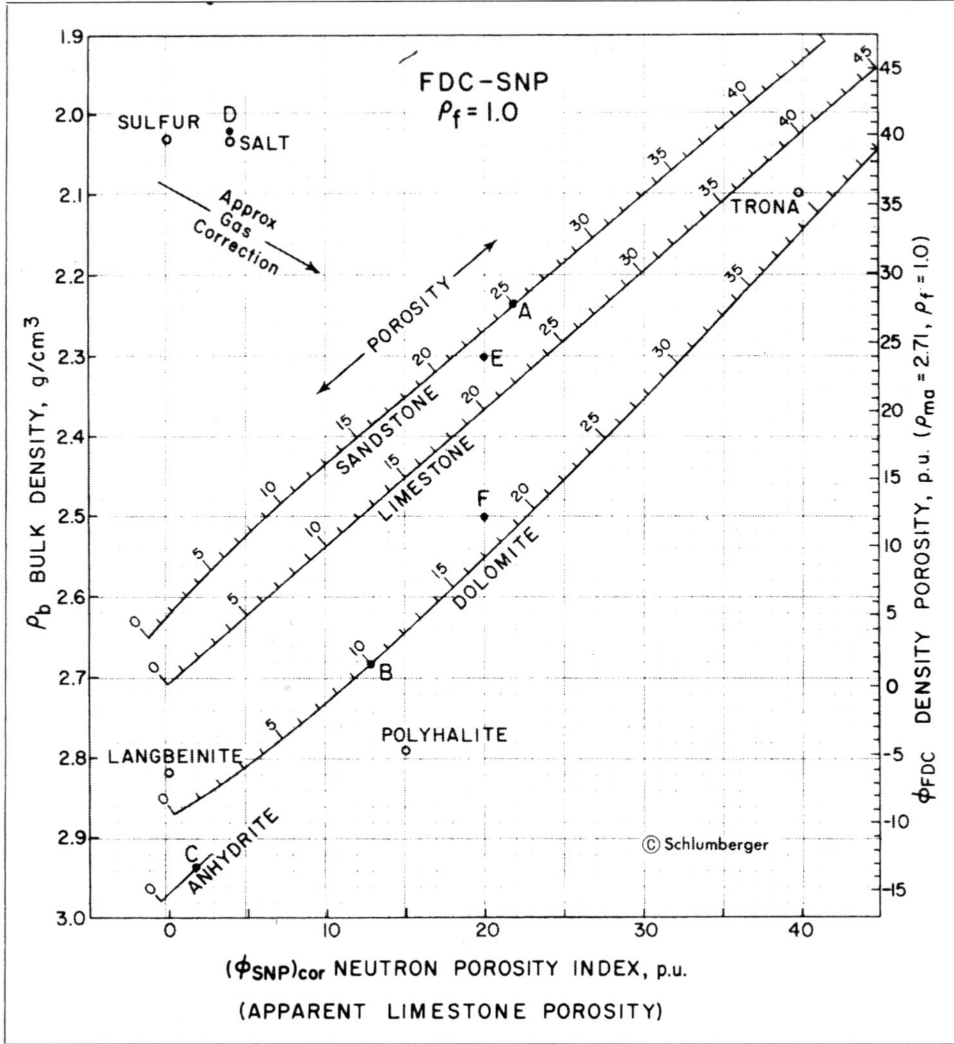

Fig. 14.4—FDC/SNP dual-mineral crossplot, case of fresh water, liquid-filled boreholes (courtesy Schlumberger).

14.2.3 Sonic/Density Combination. Fig. 14.8 shows the sonic/FDC Crossplot. This crossplot clearly has poor resolution for common reservoir rock lithology. The sonic/density crossplot, however, has good resolution for determining the existence of evaporites (such as salt, gypsum, and anhydrite) and nonmetallic minerals (like sulfur). This good resolution results from the wide separation between the points that represent these minerals in Fig. 14.8.

14.3 Lithology/Porosity Interpretation With Three Porosity Tools

When the three porosity tools are available, a set of four equations can be written for cases when the formation matrix is composed of three minerals.[5,6] Three of these equations represent the three different responses of each tool; the fourth is a material-balance equation.

$$\Delta t = \phi \Delta t_f + (1-\phi)(V_1 \Delta t_{ma1} + V_2 \Delta t_{ma2} + V_3 \Delta t_{ma3}), \quad \ldots . (14.9)$$

$$\phi_N = \phi(\phi_N)_f + (1-\phi)[V_1(\phi_N)_{ma1} + V_2(\phi_N)_{ma2} + V_3(\phi_N)_{ma3}], \quad \ldots (14.10)$$

$$\rho_b = \phi \rho_f + (1-\phi)(V_1 \rho_{ma1} + V_2 \rho_{ma2} + V_3 \rho_{ma3}), \quad \ldots \ldots (14.11)$$

and $1 = V_1 + V_2 + V_3$, ... (14.12)

where ϕ = fractional porosity; V_1, V_2, V_3 = volume fractions of the three minerals; ρ_f, $(\phi_N)_f$, Δt_f = fluid values; and $(\Delta t_{ma})_i$, $[(\phi_N)_{ma}]_i$, $(\rho_{ma})_i$ = matrix values.

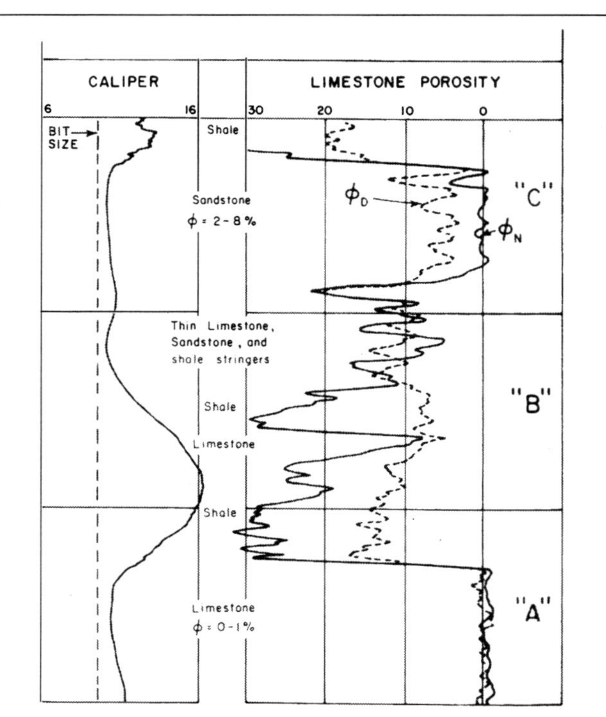

Fig. 14.5—FDC/CNL porosity overlay (courtesy Schlumberger).

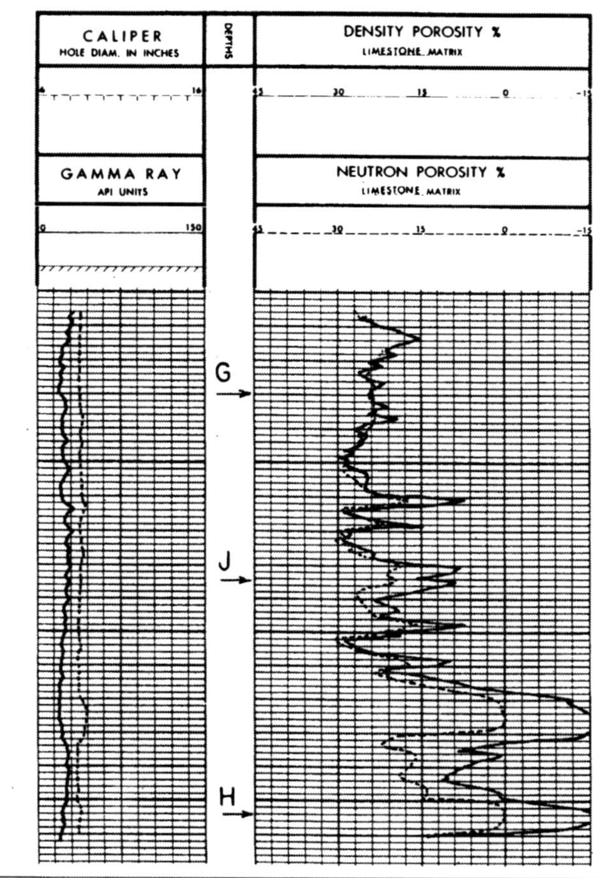

Fig. 14.6—FDC/CNL porosity overlay of Example 14.2 (courtesy Schlumberger).

These equations can be solved simultaneously for four unknowns, usually ϕ, V_1, V_2, and V_3. The solution, called the tri-porosity solution, assumes that the effective parameters of a mineral mixture are determined by linear combination of the parameters of the minerals' matrices. It also assumes that the formations are compacted, clean, and liquid-filled without secondary porosity. When there is a sandstone/limestone/dolomite mixture and a FDC/CNL/sonic combination, Eqs. 14.9 through 14.12 become

$$\rho_b = 1.0\phi + (1-\phi)(2.65V_{ss} + 2.71V_{ls} + 2.87V_{dol}), \quad (14.13)$$

$$\phi_{CNL} = 1.0\phi + (1-\phi)(0.015V_{ss} + 0.0V_{ls} + 0.025V_{dol}), \quad (14.14)$$

$$\Delta t = 189\phi + (1-\phi)(55.5V_{ss} + 47.5V_{ls} + 43.5V_{dol}), \quad (14.15)$$

and $$1 = V_{ss} + V_{ls} + V_{dol}. \quad (14.16)$$

Lithology interpretation can be facilitated by use of graphical techniques like the *M/N* plot and the matrix identification (MID) plot. The concept of these plots is based on defining parameters that are independent of porosity. Thus, each mineral is represented on the plot by a unique point regardless of its porosity.

The true advantage of the graphical solution over the analytical one is that the graphical solution makes detecting the presence of gas, shale, or secondary porosity easier. Secondary porosity, shaliness, and gas will shift the position of the points with respect to their true lithologies.

If the response of tools other than the neutron, density, and sonic logs can be expressed in terms of volume fraction of the formation constituents, more than four equations can be written. The set of equations can then be solved for more unknowns. This approach can be used in cases when more than three minerals are present.

14.3.1 The M/N Plot. On a neutron/density litho/porosity crossplot, the 0% to 100% porosity range for a specific mineral can be represented as illustrated in **Fig. 14.9A.** That graph shows a matrix point $[\rho_{ma}, (\phi_N)_{ma}]$ that corresponds to 0% porosity and a fluid point $[\rho_f, (\phi_N)_f]$ that corresponds to 100% porosity. The slope of the line joining these two points is used to identify the mineral. The parameter N, which is related to the slope, is defined as[3]

$$N = \frac{(\phi_N)_f - (\phi_N)_{ma}}{\rho_{ma} - \rho_f}. \quad (14.17)$$

Similarly, the slope of the line joining the matrix and fluid points on the sonic/density crossplot of **Fig. 14.9B** can be used to identify a specific mineral. A parameter M, which is related to the slope, is defined as[3]

$$M = 0.01(\Delta t_f - \Delta t_{ma})/(\rho_{ma} - \rho_f). \quad (14.18)$$

The factor 0.01 is introduced arbitrarily to make the values of M compatible in magnitude with the values of N.

Eqs. 14.17 and 14.18 indicate that M and N are independent of porosity. In fact, they depend only on the fluid and matrix characteristics. Log data (Δt, ρ_b, and ϕ_N) displayed in a formation composed of the mineral considered in this discussion will be plotted along the lines that join matrix and fluid points (see Figs. 14.9A and 14.9B). N and M also can be defined as

$$N = [(\phi_N)_f - \phi_N]/(\rho_b - \rho_f) \quad (14.19)$$

and $$M = 0.01(\Delta t_f - \Delta t)/(\rho_b - \rho_f). \quad (14.20)$$

These definitions are possible only because the porosity response is assumed to be linear between the matrix and fluid points. This is a good assumption for most minerals. However, for some minerals, dolomite in particular, the neutron log exhibits nonlinearity with porosity (Figs. 14.1 through 14.4). This nonlinearity is more pronounced in the low porosity range. In these cases, several porosity ranges are considered. N is computed in each range using an approximate $(\phi_N)_{ma}$ value. **Table 14.4** lists the porosity ranges, matrix coefficients, and M and N values of dolomite and other common minerals. The fluid coefficients Δt_f, ρ_f, and $(\phi_N)_f$ used in computing M and N are 189 μsec/ft, 1.00 g/cm^3, and 1.00 for a freshwater-based mud and 185 μsec/ft, 1.1 g/cm^3, and 1.00 for a saltwater-based mud.

The tabulated values of M are plotted vs. N to generate the *M/N* plot used for mineral identification. As **Fig. 14.10** shows, the pure mineral points fall at definite locations regardless of porosity. Each mineral is represented by a pair of points, one for fresh mud and one for salt mud. Dolomite is represented by three pairs of points, each for a specific porosity range. The sandstone (silica) is represented by two pairs of points, each representing a specific degree of rock compaction.

The *M/N* plot also displays arrows to indicate the direction along which points will move away from their true lithology locations owing to the effect of gas, secondary porosity, or shale. No unique shale point exists on the *M/N* plot because shales tend to vary in their characteristics. Most shales, however, will be situated below the line that joins the silica and anhydrite points.

To identify the lithology of a specific zone, the ρ_b, Δt, and ϕ_N values displayed by the three porosity tools are read. The values of N and M are computed with Eqs. 14.19 and 14.20. Δt_f, ρ_b, and $(\phi_N)_f$ are selected according to the mud type. Next, the value of M is plotted vs. N in the *M/N* plot of Fig. 14.10. If the zone is composed of one mineral, the zone will be plotted on the point that represents that mineral. If the zone consists of a mixture of two minerals, it will be plotted on the line joining these two minerals. In the case of a matrix composed of three minerals, the zone will be plotted within the triangle described by these three mineral points.

The location of the point on a line or within a triangle that represents three different mineral points will determine the lithology fraction. These fractions are used to compute average or apparent matrix density, which, in turn, is used with the bulk and fluid densities to calculate porosity. The apparent matrix density is expressed as

$$(\rho_{ma})_a = \sum_{i=1}^{n} V_i(\rho_{ma})_i \quad (14.21)$$

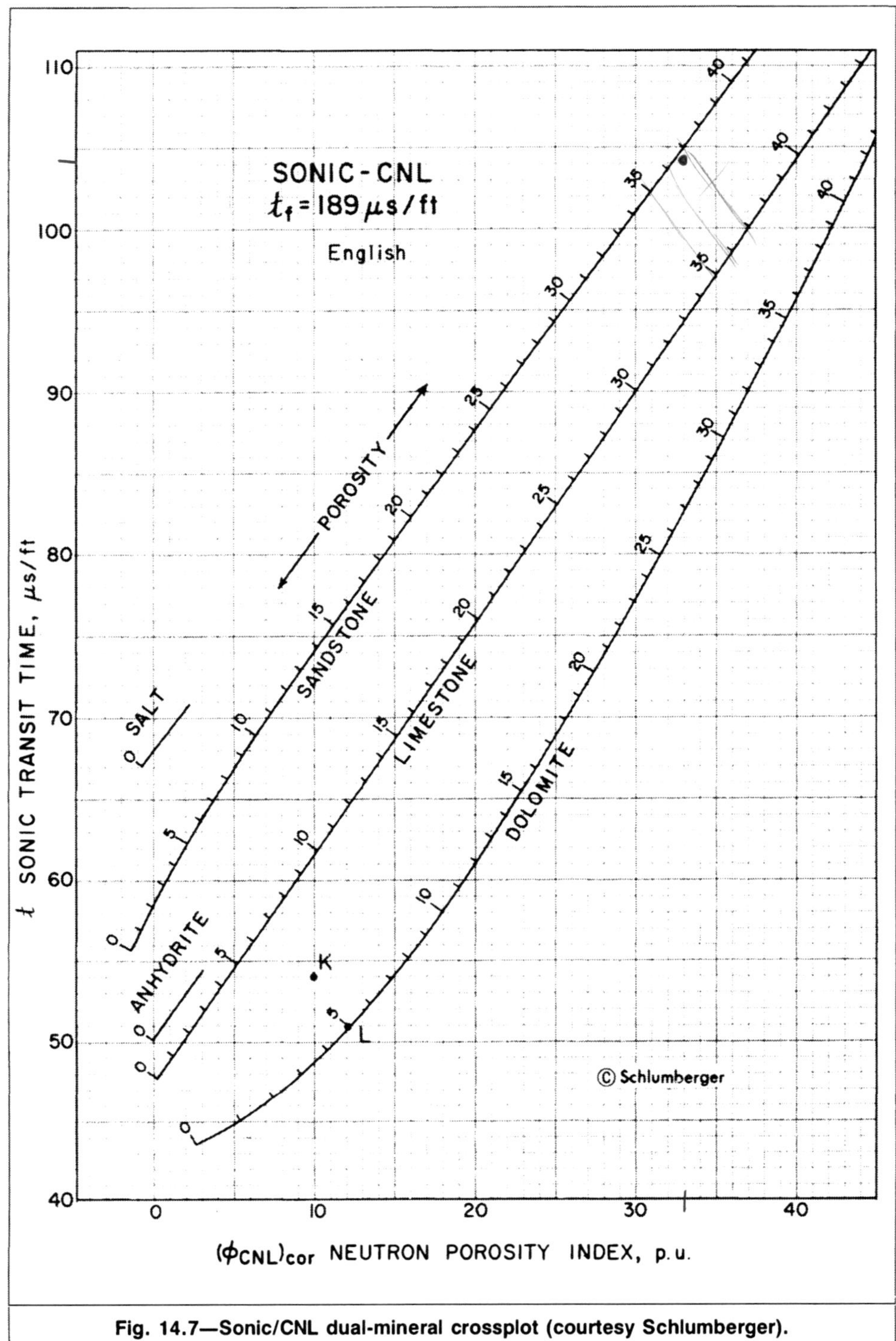

Fig. 14.7—Sonic/CNL dual-mineral crossplot (courtesy Schlumberger).

where V_i = the lithology fractions ($\sum_{1}^{n} V_i = 1$) and $(\rho_{ma})_i$ = matrix densities of the different minerals present in the zone.

Similar equations can be written for the sonic and neutron information.

Geologic age, depositional environment, diagenesis, and other detailed geologic information help determine the probable minerals present in the interval studied. In the absence of such detailed information, a standard set of lithology triangles (**Fig. 14.11**) is used to interpret the *M*/*N* plot. The triangles represent the mineral combinations most likely to occur in shale-free carbonate and evaporite rocks.[3]

Example 14.4. Deposition environment in a section of interest indicates that either one or a combination of silica, calcium carbonate, and dolomite is the most probable lithology. Using the *M*/*N* plot, determine the lithology, lithology fraction, and porosity of the following four zones.

Zone	ρ_b (FDC Log) (g/cm³)	Δt (Sonic Log) (μsec/ft)	ϕ_N (SNP log, Limestone Matrix) (%)
A	2.34	80.5	16
B	2.81	48.5	5
C	2.71	54	7
D	2.60	59	6

The section was drilled with freshwater-based mud.

Solution. The *M* and *N* values for the different zones were computed with Eqs. 14.19 and 14.20 and are listed below. $\rho_f = 1$ g/cm³, $(\phi_N)_f = 1$, and $\Delta t_f = 189$ μsec/ft are used because the mud base is fresh water. Also listed below is the most likely lithology

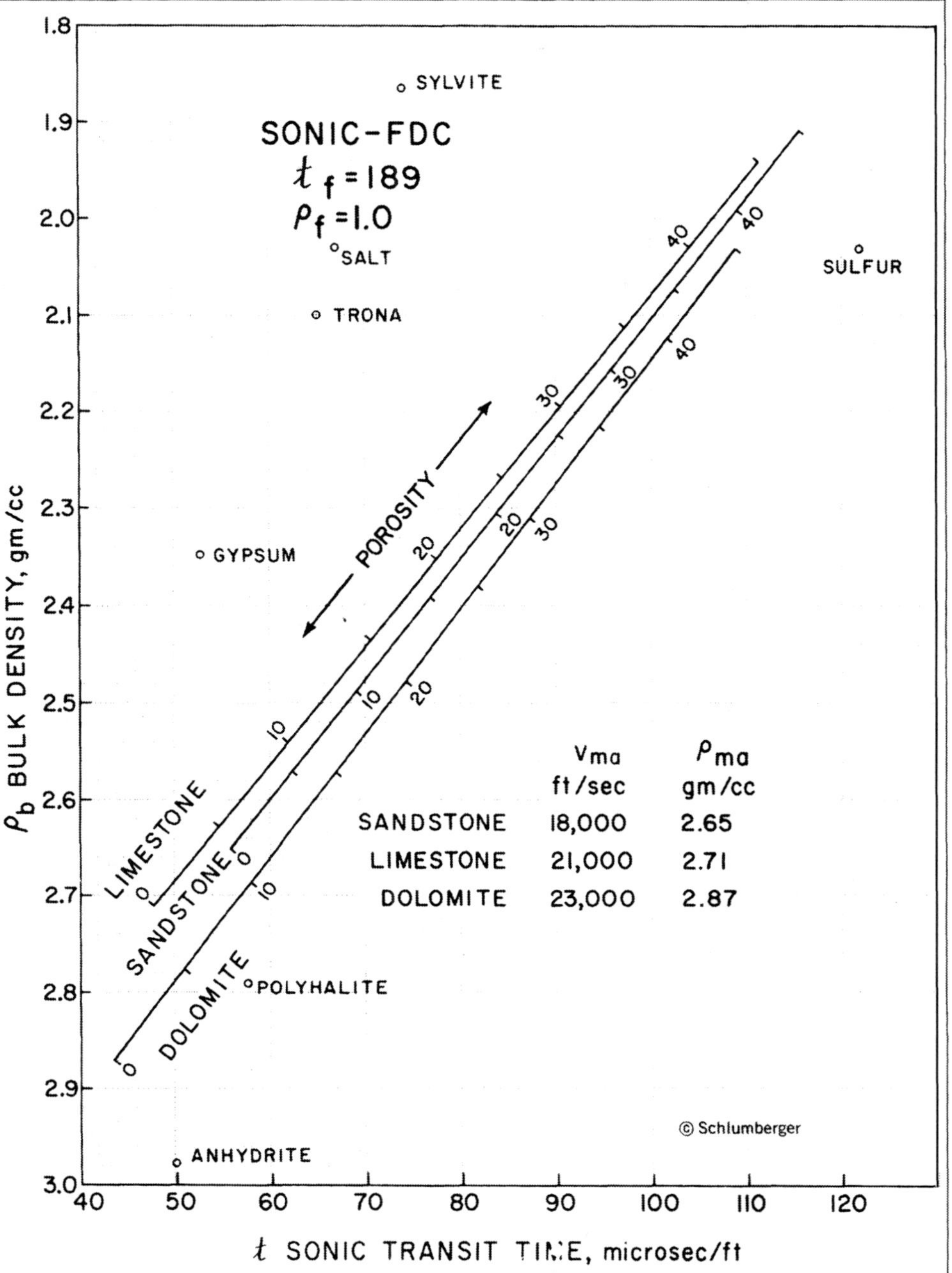

SONIC-FDC

$t_f = 189$

$\rho_f = 1.0$

	V_{ma} ft/sec	ρ_{ma} gm/cc
SANDSTONE	18,000	2.65
LIMESTONE	21,000	2.71
DOLOMITE	23,000	2.87

© Schlumberger

Fig. 14.8—Sonic/density log (FDC) dual-mineral chart (courtesy Schlumberger).

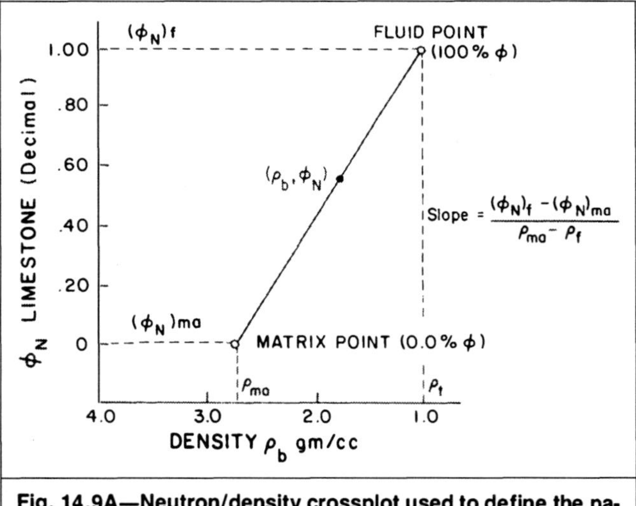

Fig. 14.9A—Neutron/density crossplot used to define the parameter *N*.

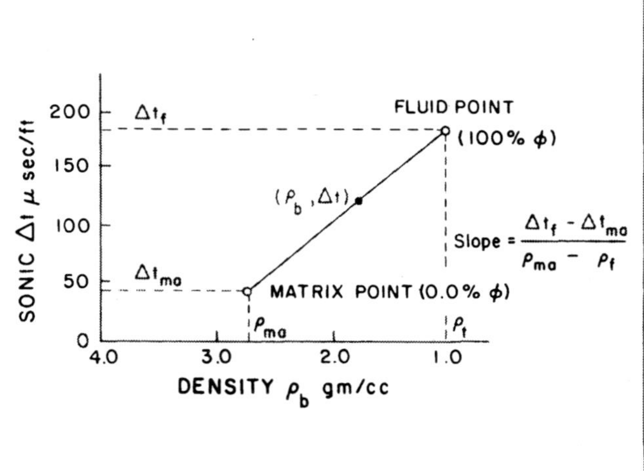

Fig. 14.9B—Sonic/density crossplot used to define the parameter *M*.

TABLE 14.4—MATRIX COEFFICIENTS AND M AND N VALUES FOR SOME COMMON MINERALS (FROM REF. 3)

	Matrix Coefficients			Salt Mud $\Delta t_f = 185$ μsec/ft, $\rho_f = 1.10$ g/cm³		Fresh Mud $\Delta t_f = 189$ μsec/ft, $\rho_f = 1.00$ g/cm³	
Mineral	Δt_{ma} (μsec/ft)	ρ_{ma} (g/cm³)	$(\phi_{SNP})_{ma}$	M	N	M	N
Silica (1) (V_{ma} = 18,000 ft/sec)	55.5	2.65	−0.035	0.835	0.669	0.810	0.628
Silica (2) (V_{ma} = 19,500 ft/sec)	51.2	2.65	−0.035	0.862	0.669	0.835	0.628
$CaCO_3$	47.6	2.71	0.00	0.854	0.621	0.827	0.585
Dolomite (1) (ϕ = 5.5% to 30%)	43.5	2.87	0.035	0.800	0.544	0.778	0.513
Dolomite (2) (ϕ = 1.5% to 5.5% and >30%)	43.5	2.87	0.02	0.800	0.554	0.778	0.524
Dolomite (3) ϕ = 0.0% to 1.5%	43.5	2.87	0.005	0.800	0.561	0.778	0.532
Anhydrite	50.0	2.98	0.00	0.718	0.532	0.702	0.505
Gypsum	52.0	2.35	0.049	1.060	0.408	1.015	0.378
Salt	67.0	2.05	0.04	1.240	1.010	1.16	0.914

of each zone. The lithology is deduced by considering the location that the zone assumes on the M/N plot of Fig. 14.11.

Zone	M	N	Most Likely Lithology
A	0.806	0.629	Silica (V_{ma} = 18,000 ft/sec)
B	0.776	0.525	Dolomite (dense)
C	0.789	0.544	Dolomite/limestone/silica
D	0.813	0.588	Limestone/silica/dolomite

TABLE 14.5—M AND N VALUES OF EXAMPLE 14.4

Mineral	M	N
Silica (1)	0.810	0.628
Calcite	0.827	0.585
Dolomite (3)	0.778	0.532

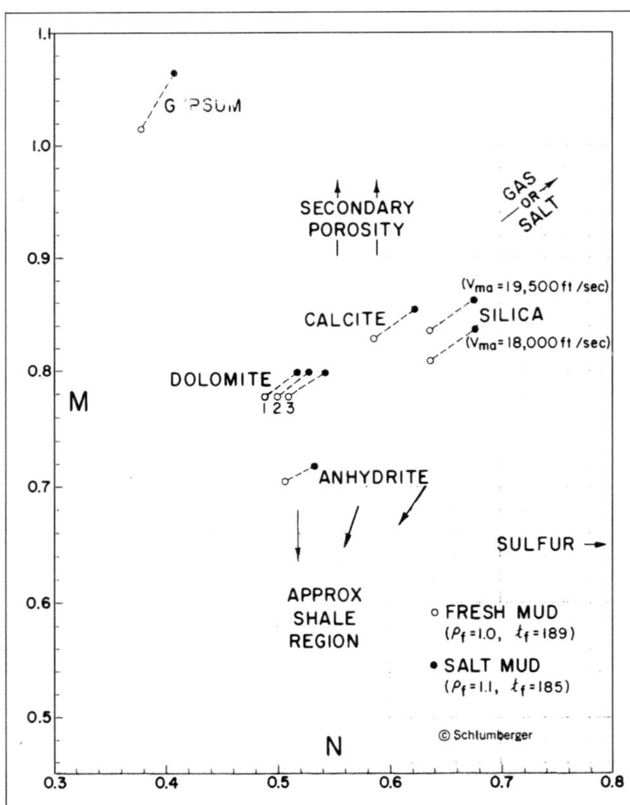

Fig. 14.10—M/N plot used for mineral identification (CNL is used) (courtesy Schlumberger).

The lithology fractions and porosity of each zone can be determined as follows.

Zone A. The matrix defined by Fig. 14.11 is 100% silica. The porosity can be determined from Eq. 14.21 or 8.11 using a matrix density of 2.65 g/cm³.

$\phi = (2.65 - 2.34)/(2.65 - 1) = 0.188$ or 19%.

The porosity can also be calculated both from the SNP reading with neutron porosity equivalence curves similar to those of Fig. 14.1 and from the sonic log information:

$\phi = (80.5 - 55.5)/(189 - 55.5) = 0.187$ or 19%.

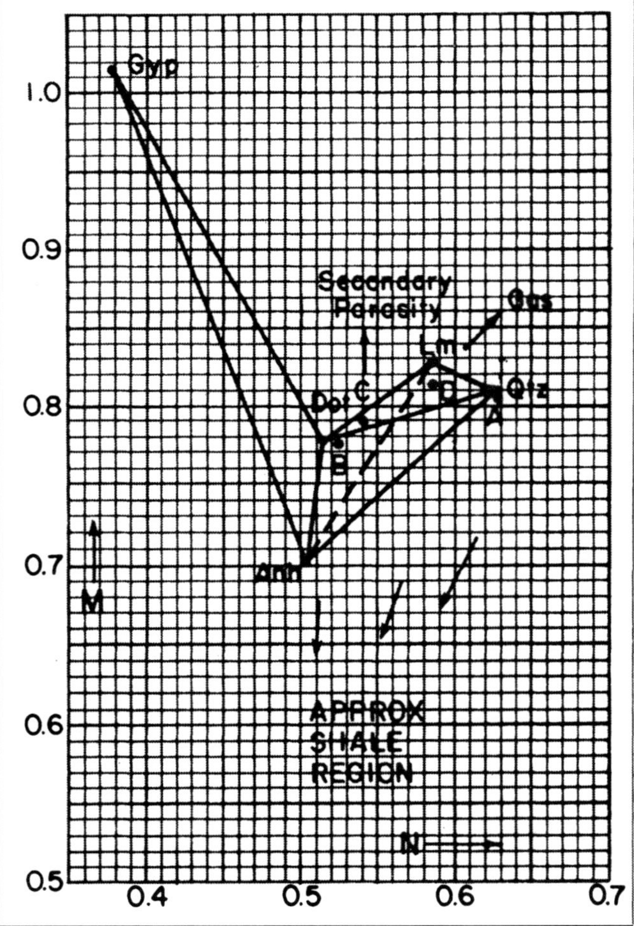

Fig. 14.11—M/N plot showing stardard lithology triangles (SNP is used) (courtesy Schlumberger).

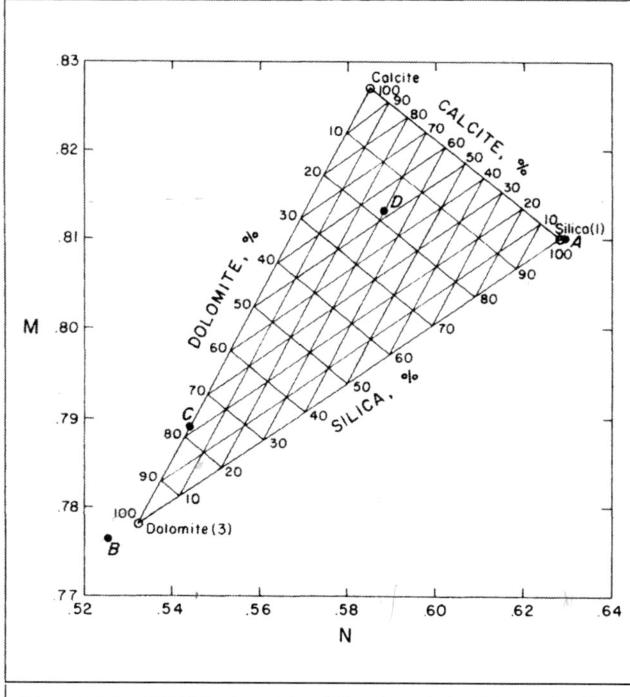

Fig. 14.12—*M/N* lithology identification plot of Example 14.4.

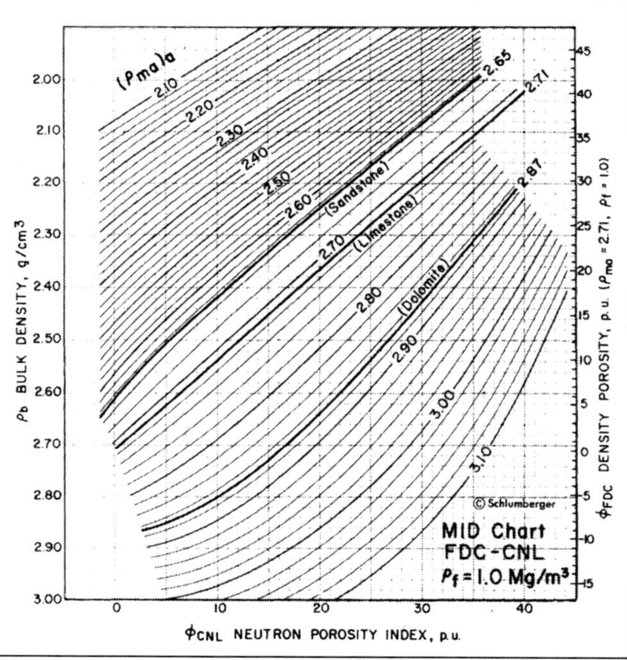

Fig. 14.13—FDC/CNL MID chart for fresh mud scaled in $(\rho_{ma})_a$ values (courtesy Schlumberger).

Zone B. Zone B is identified as dense dolomite because it lies to the right of the point on the chart representing low-porosity dolomite. The porosity can be estimated as in the previous case from either density log information,

$\phi=(2.87-2.81)/(2.87-1)=0.032$ or 3%,

or sonic log information,

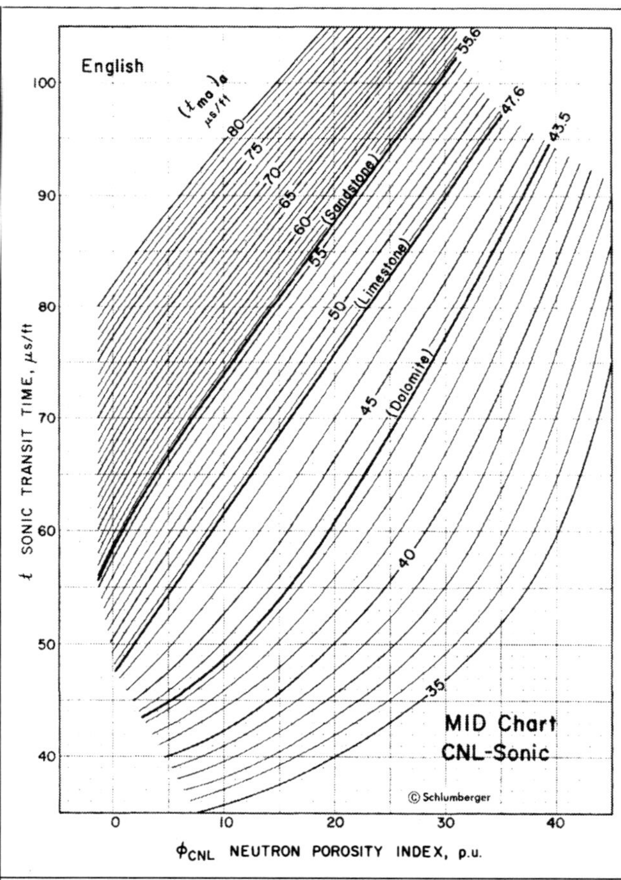

Fig. 14.14—CNL/Sonic MID chart for fresh mud scaled in $(\Delta t_{ma})_a$ (courtesy Schlumberger).

$\phi=(48.5-43.5)/(189-43.5)=0.034$ or 3%.

The lithology fractions of zones composed of three minerals are determined graphically. Isolithology lines are first constructed within the three-minerals triangle. The coordinates of the zone relative to the isolithology-lines network give the lithology fractions. For better accuracy, the silica/limestone/dolomite triangle is constructed on a larger scale (**Fig. 14.12**). Silica type (1) and dolomite type (3) points are used. Silica (1) and dolimite (3) are the lithology of Zones A and B, respectively. Thus, we can conclude that they are the silica and dolomite types most likely to exist in the interval analyzed. As can be seen, the lithology type of other zones within the interval of interest is used to an advantage. If this information is missing, additional assumptions will be required to select the silica and dolomite points. The *M* and *N* values of the three minerals are obtained from **Table 14.5.**

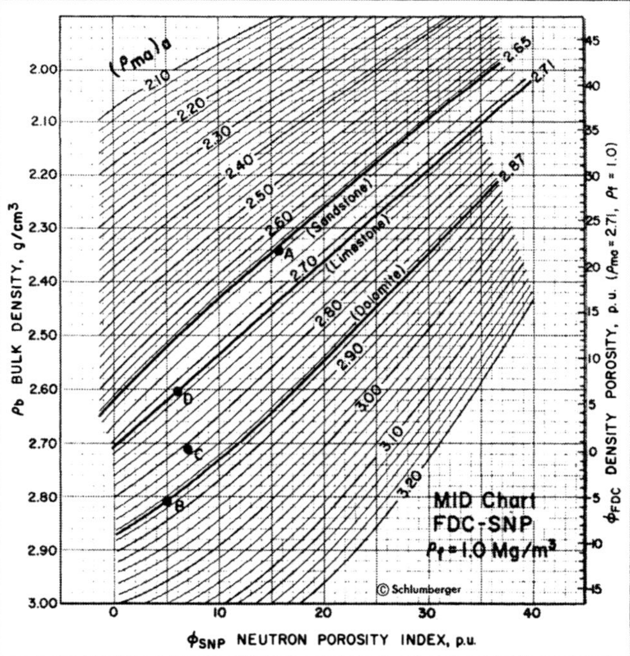

Fig. 14.15—FDC/SNP MID chart for fresh mud scaled in $(\rho_{ma})_a$ (courtesy Schlumberger).

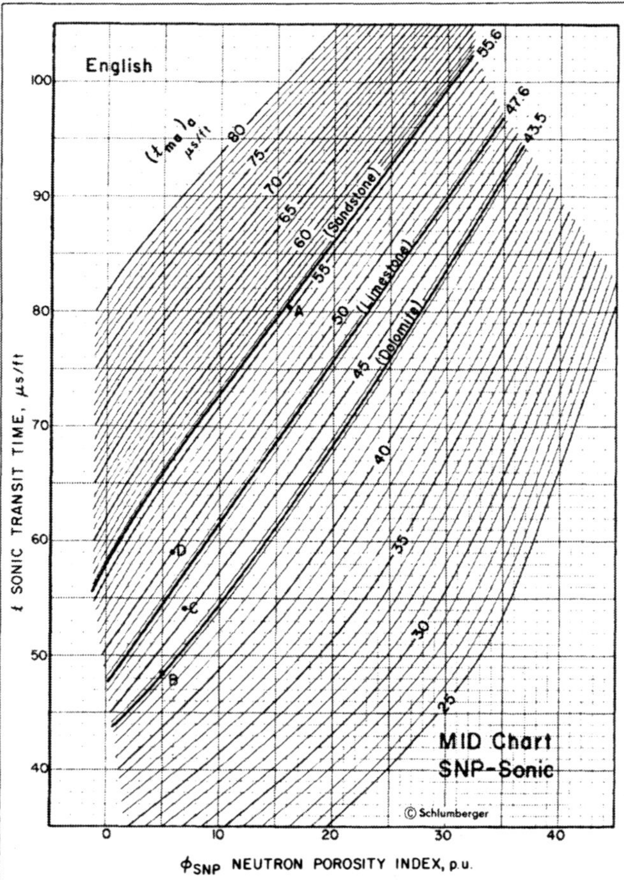

Fig. 14.16—SNP/Sonic MID chart for fresh mud scaled in $(\Delta t_{ma})_a$ (courtesy Schlumberger).

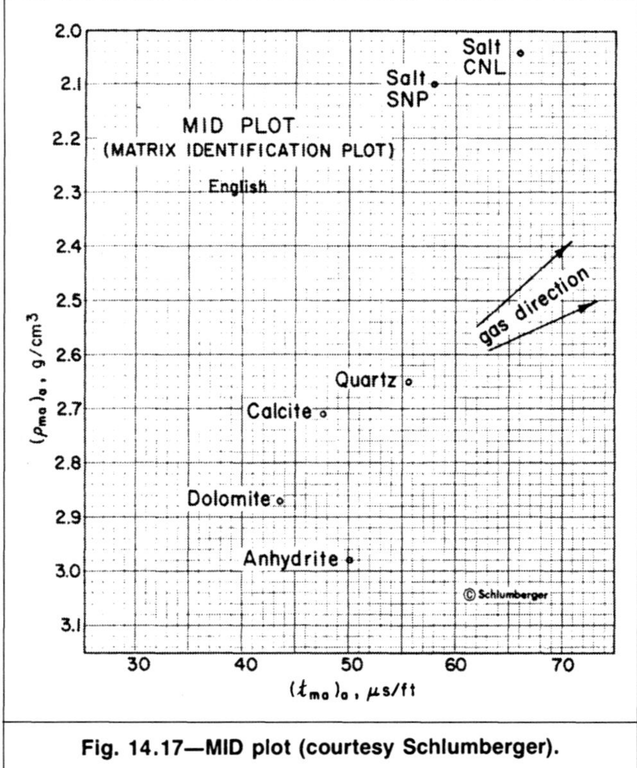

Fig. 14.17—MID plot (courtesy Schlumberger).

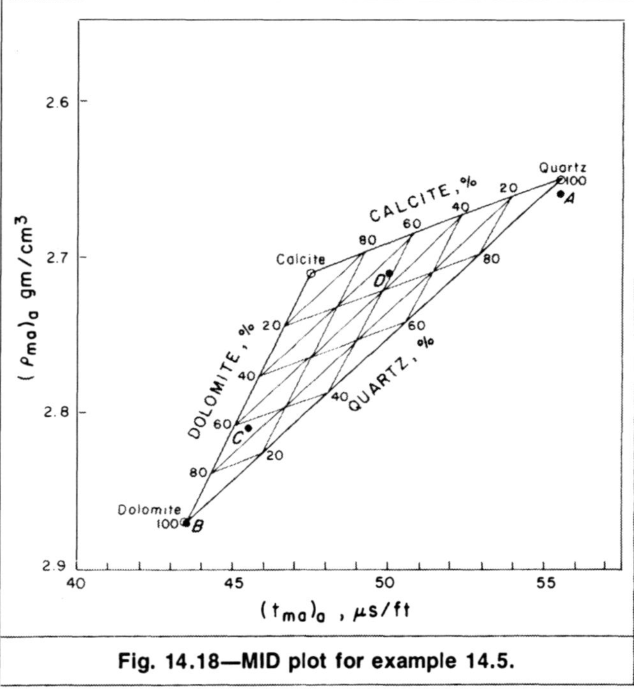

Fig. 14.18—MID plot for example 14.5.

Note that, in Fig. 14.12, Point A does not fall exactly on top of the silica (1) point, probably because of errors introduced when the logs were read. Also, Point B does not fall exactly on the dolomite (3) point. In addition to log-reading errors, this misalignment also could result because the dolomite points occupy approximate positions on the *M/N* plot.

Zone C. According to Fig. 14.12, which is based on dense dolomite, Zone C appears to contain only dolomite and limestone. The lithology fractions are 78% dolomite and 22% limestone. Zone C consists of only two minerals, so the lithology fractions and porosity can be determined best from the lithology/porosity chart of Fig. 14.4. The chart indicates lithology fractions of 67% and 33% for dolomite and limestone, respectively. The porosity is roughly 5.5%.

Zone D. According to Fig. 14.12, the zone consists of 52% calcite, 30% silica, and 18% dolomite. Using Eqs. 14.21 and 8.11 gives

$$(\rho_{ma})_a = 0.52(2.71) + 0.30(2.65) + 0.18(2.87)$$
$$= 2.72 \text{ g/cm}^3$$

and $\phi = (2.72-2.60)/(2.72-1.0) = 0.07$ or 7%.

The lithology fractions and porosity based on Fig. 14.12 are only estimates. Their accuracy is largely affected by assumptions used to construct the *M/N* plot.

Lithology also can be identified with an *A/K* plot. This plot is similar in concept and use to the *M/N* plot because *A* is defined as the reciprocal of *N*, and *K* is defined as the ratio *M/N*.

14.3.2 MID Plot. The *M/N* plot has certain disadvantages. Although the computation of *M* and *N* is not complex, it becomes tedious when done by hand over a long log interval. Also, *M* and *N* lack direct physical meaning; they cannot be related to known rock properties. In addition, the multiplicity of matrix points on the *M/N* plot makes the plot subject to certain ambiguities.

The MID plot was devised to overcome these disadvantages.[7] The principle of the MID plot is similar to that of the *M/N* plot: two porosity-independent parameters—apparent matrix density, $(\rho_{ma})_a$, and apparent matrix sonic travel time, $(\Delta t_{ma})_a$—are cross-plotted. These two parameters are obtained with litho/porosity cross-plots similar to those of Figs. 14.2, 14.3, and 14.7. Fig. 14.2B shows three lithology lines that represent sandstone, limestone, and dolomite. The matrix densities of these lithologies are 2.65, 2.71, and 2.87 g/cm³, respectively. The data used to construct these three trends came from actual tool responses. Trends representing hypothetical lithologies exhibiting apparent matrix densities other than 2.65, 2.71, or 2.87 g/cm³ can be constructed by interpolation above the sandstone line and below the dolomite line. The construction technique was chosen to give good porosity reso-

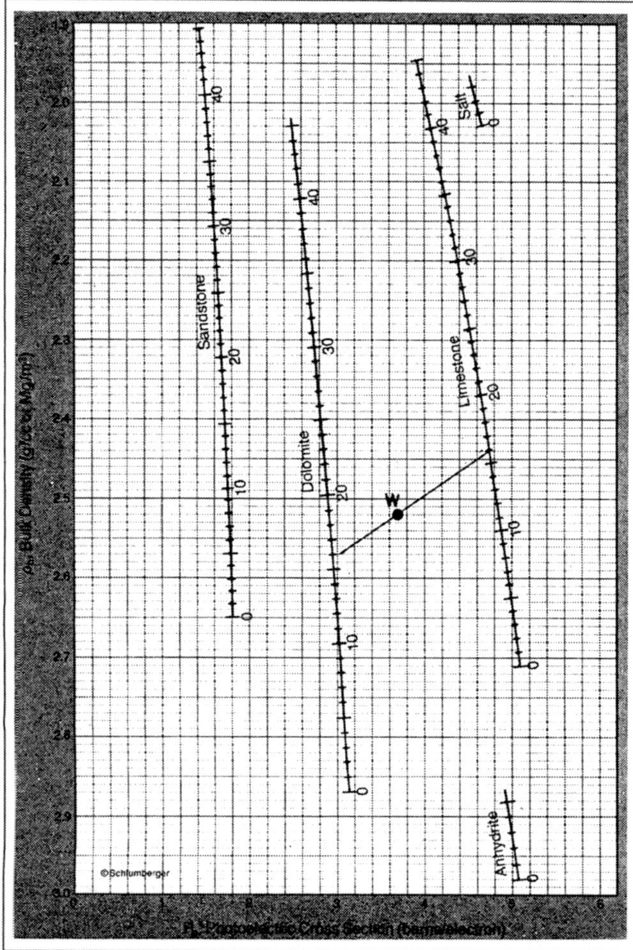

Fig. 14.19—ρ_b/P_e dual-mineral crossplot, case of fresh water, liquid-filled holes (courtesy Schlumberger).

lution because determining porosity usually is the main objective.[7] **Fig. 14.13** shows a density/neutron crossplot constructed in this manner and scaled in $(\rho_{ma})_a$ values. **Fig. 14.14** shows a sonic/neutron crossplot similarly scaled in values of $(\Delta t_{ma})_a$. Figs. 14.13 and 14.14 are for CNL tools and fresh mud. **Figs. 14.15 and 14.16** show charts for SNP tools and fresh mud. Charts for salt mud are included in service companies' chart books.[4]

The MID plot itself is shown in **Fig. 14.17.** This plot is similar to the *M/N* plot and is used in the same manner. First, the values of $(\rho_{ma})_a$ and $(\Delta t_{ma})_a$ are determined from the appropriate charts with the readings of the sonic, density, and neutron logs. Next, the MID plot is entered. The proximity of the zone to mineral points (its position either on a line joining two mineral points or within a triangle formed by three mineral points) will identify the lithology. The lithology fractions can be estimated as with the *M/N* plot.

Also, as for the *M/N* plot, the presence of secondary porosity produces displacements that are parallel to the sonic-sensitive axis. The displacement is along the $(\Delta t_{ma})_a$ axis toward smaller values. The presence of gas will displace points along the direction shown on the plot. Shales and shaly formations tend to be plotted along a southwest trend. Shaliness is identified best by plotting some shale points to establish the shale trend.

Example 14.5. Determine the lithologies, lithology fractions, and porosities of the four zones of Example 14.4 using the MID plot. Compare the results obtained from the MID plot and the *M/N* plot.

Solution. The MID coordinates obtained from Figs. 14.15 and 14.16 with the porosity data of the four zones are listed below. The four zones are also marked in Figs. 14.15 and 14.16.

	Data			MID Coordinates	
Zone	ρ_b (g/cm^3)	Δt (μsec/ft)	Φ_{SNP} (%)	$(\rho_{ma})_a$ (g/cm^3)	$(\Delta t_{ma})_a$ (μsec/ft)
A	2.34	80.5	16	2.66	55.5
B	2.81	48.5	5	2.87	43.5
C	2.71	54.0	7	2.81	45.5
D	2.60	59.0	6	2.71	50.0

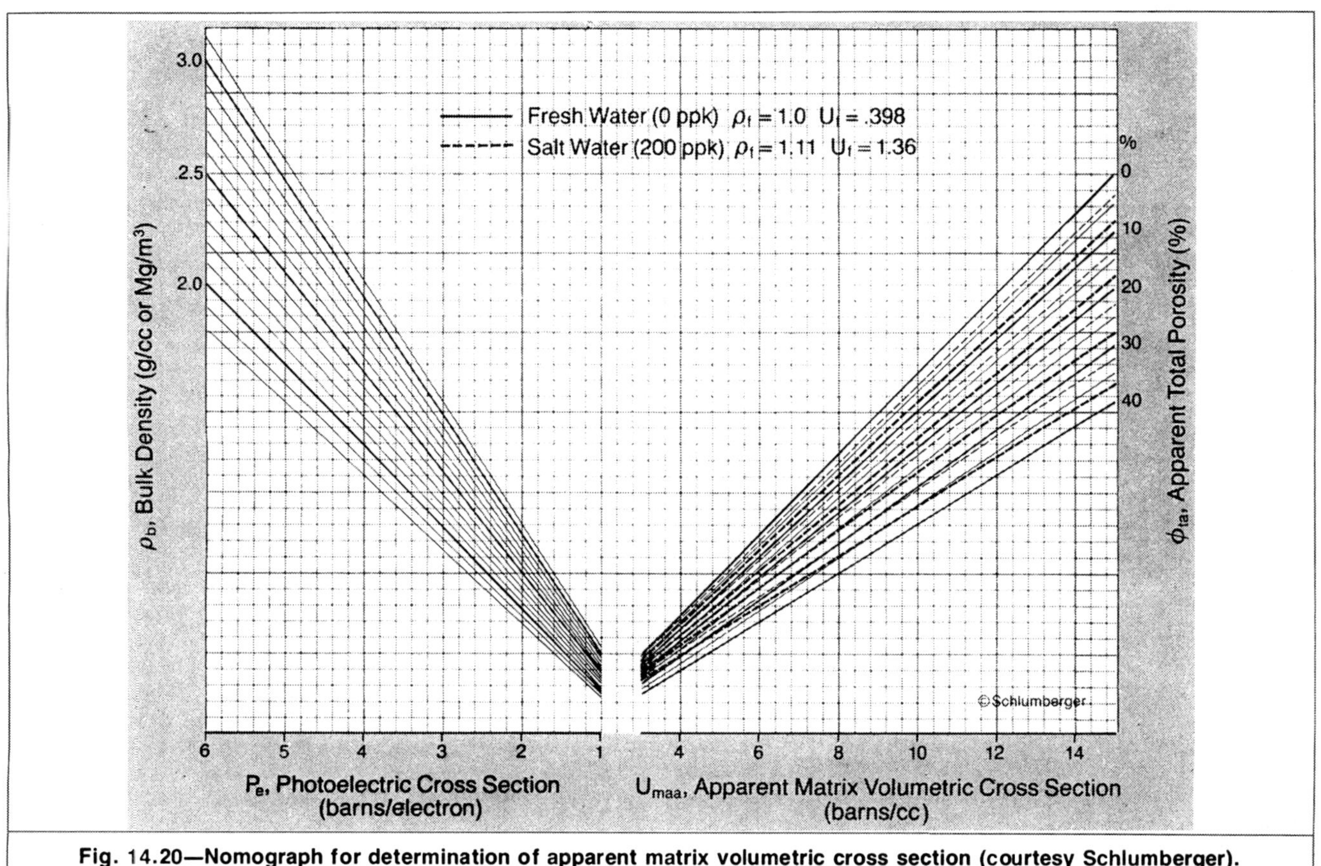

Fig. 14.20—Nomograph for determination of apparent matrix volumetric cross section (courtesy Schlumberger).

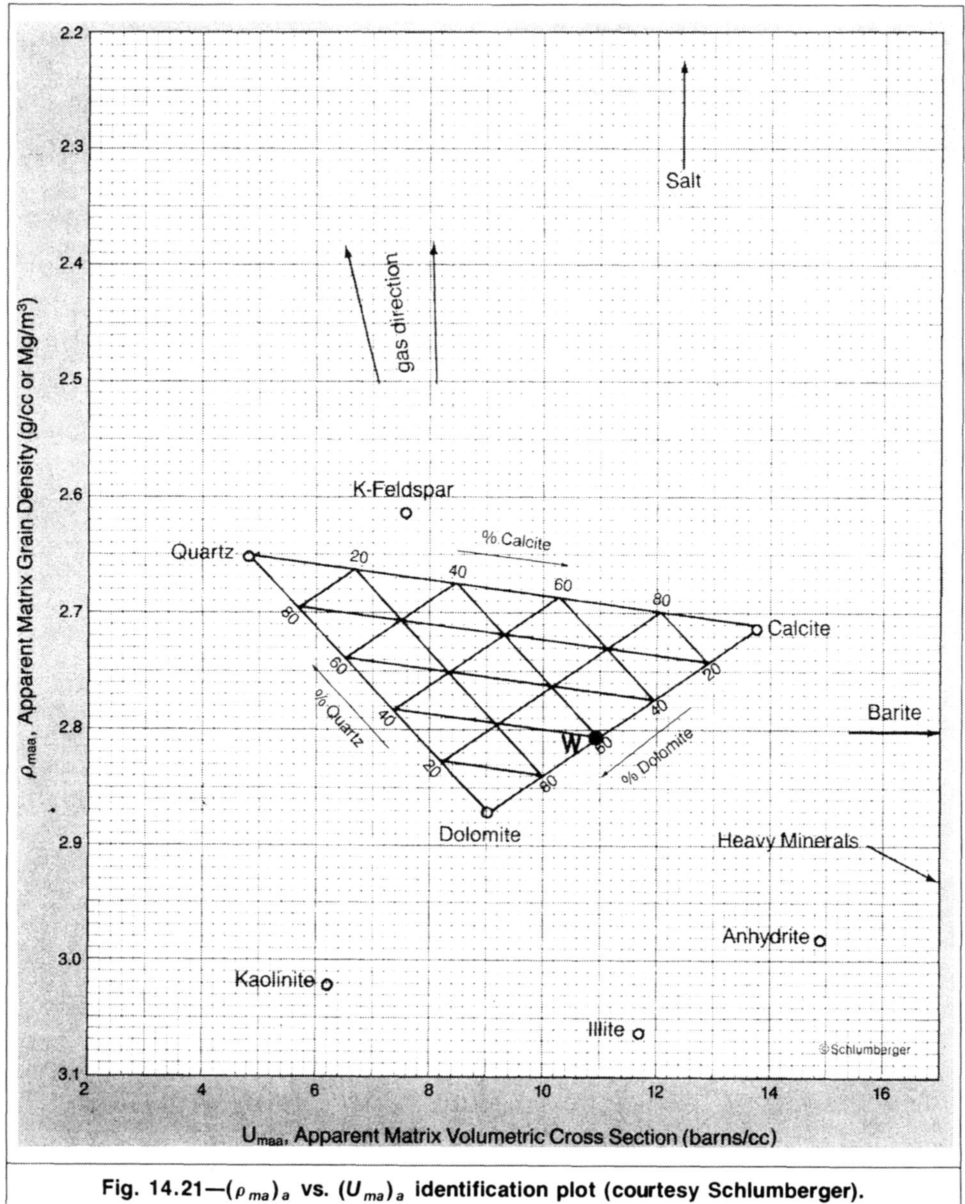

Fig. 14.21—$(\rho_{ma})_a$ vs. $(U_{ma})_a$ identification plot (courtesy Schlumberger).

$(\rho_{ma})_a$ and $(\Delta t_{ma})_a$ are entered on the MID plot in **Fig. 14.18.** The MID plot indicated the following lithologies and porosities for the four zones:

Zones	Lithology	Porosity (Eq. 8.11) (%)
A	100% sandstone	19
B	100% dolomite	3
C	64% dolomite 30% limestone 6% sandstone	5.5
D	50% limestone 38% sandstone 12% dolomite	6.5

The *M/N* and MID plots supply the same information for Zones A and B. This is expected because the zones are made up of a single lithology. Note that Zone B fell on top of the dolomite point of the MID plot. This results because the MID plot counteracts the ambiguities in the *M/N* plot caused by nonlinear neutron response.

The MID and *M/N* plots indicate slightly different lithologies for Zones C and D, mainly because of the approximation inherent in each plot. Because the MID plot calls for less approximation, it permits for more representative determinations. Porosity values estimated for each zone from the two plots are very close. Note that the determination of $(\rho_{ma})_a$, needed for porosity calculations, is an integral part of the interpretation procedure with the MID plot. Calculation of $(\rho_{ma})_a$ is an additional step for the M/N plot.

14.4 Lithology/Porosity Interpretation With the Litho-Density Tool

The Litho-Density[SM] log adds new possibilities to lithology identification. As detailed in Chap. 8, the Litho-Density tool provides bulk density, ρ_b, and P_e measurements. P_e is the index of the effective photoelectric absorption cross-section of the formation. Two parameters, ρ_e and U, are derived from ρ_b and P_e with Eqs. 2.44, 2.47, and 8.15. ρ_e and U are the electron density index and the effective photoelectric absorption cross-section index per unit volume, respectively.

The P_e measurement is strongly related to the matrix lithology. In a simple lithology, it can be used alone as a matrix indicator with the lithology curve of Fig. 8.19. The P_e measurement can also be used in combination with the bulk density, ρ_b, to analyze two-mineral matrices and to determine porosity. This is done with the dual-mineral crossplot of **Fig. 14.19.** This chart solves for porosity and lithology fractions with the following equations[8]:

$$\rho_b = \phi\rho_f + V_1\rho_1 + V_2\rho_2, \quad (14.22)$$

$$P_e\rho_e = \phi U_f + V_1 U_1 + V_2 U_2, \quad (14.23)$$

and $1 = \phi + V_1 + V_2$, $\quad (14.24)$

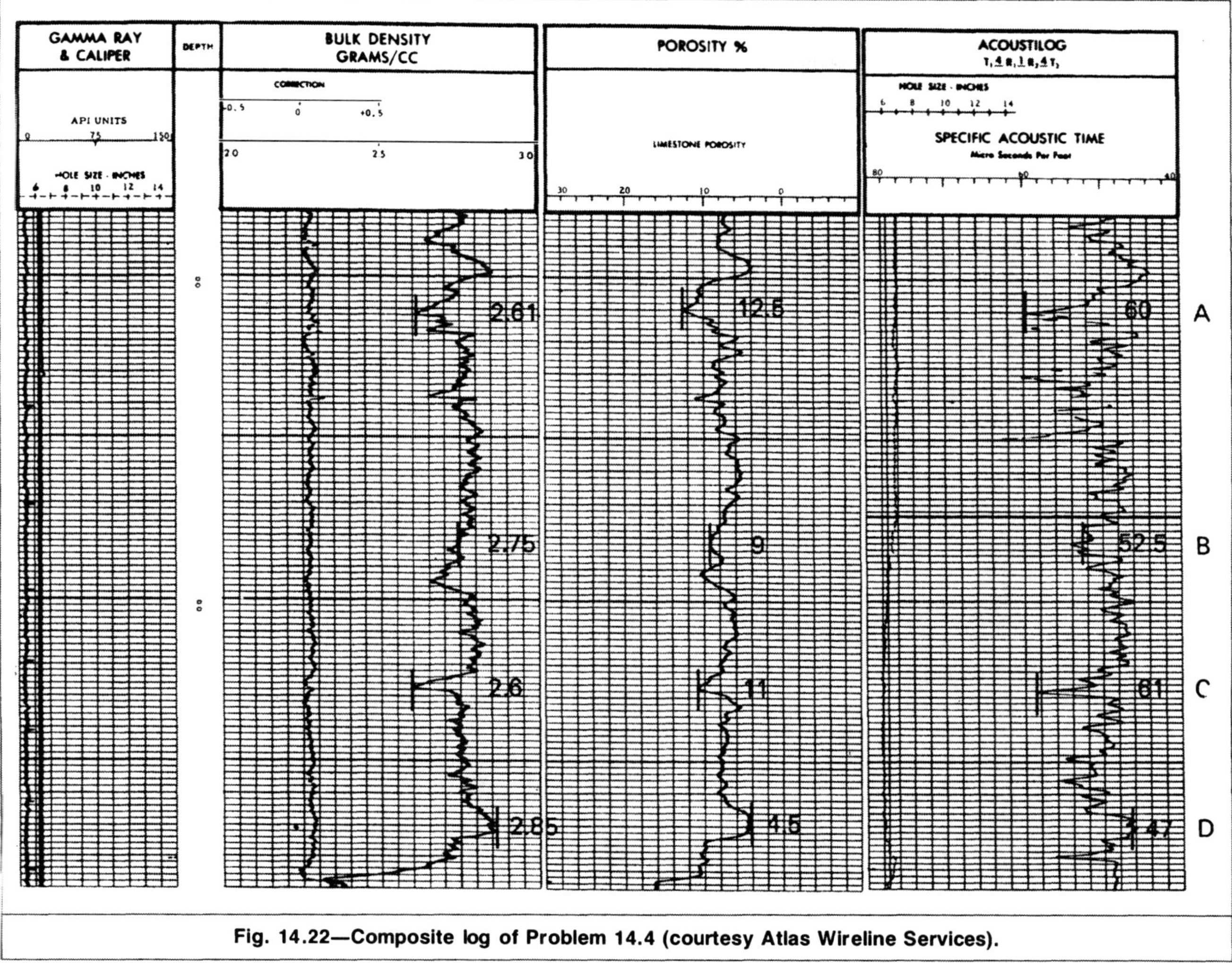

Fig. 14.22—Composite log of Problem 14.4 (courtesy Atlas Wireline Services).

where ρ_f = fluid density (assumed = 1 g/cm^3), U_f = fluid volumetric cross section (assumed U_f = 0.398), $\rho_{1,2}$ = matrix density of minerals 1,2, $U_{1,2}$ = volumetric cross section of Minerals 1,2, and $V_{1,2}$ = bulk volume fraction of Minerals 1,2.

For other dual-mineral crossplots, the two minerals known or assumed to be present in the matrix must be selected. The chart is entered using ρ_b and P_e of the zone of interest. The distance from the point to the pure-mineral trends determines the relative proportions of the minerals in the matrix. The porosity is read by interpolation between isoporosity lines. If the neutron porosity, ϕ_N, is available for the zone of interest, a porosity value can be derived from the ρ_b/ϕ_N dual-mineral crossplot (e.g., Fig. 14.3). Comparing the two porosity values determined from the two crossplots can result in more accurate interpretations.[8] If the porosity values are equal, the choice of minerals is correct, and the porosity is liquid-filled as assumed. If the two porosity values are different, the choice of another pair of minerals may be warranted to reconcile the two porosity determinations. The presence of gas also could cause disagreement between the two porosity values.

The P_e measurement also is used in combination with bulk density and neutron porosity information to analyze more complex lithologies—e.g, determination of the relative proportions of three minerals present in a matrix. This is accomplished through a comparison of apparent matrix grain density, $(\rho_{ma})_a$, and apparent volumetric cross section, $(U_{ma})_a$. The apparent matrix density is determined from neutron and density data and the appropriate MID chart of Figs. 14.13 through 14.16. The apparent volumetric cross section is calculated with

$$(U_{ma})_a = (P_e \rho_e - \phi U_f)/(1-\phi). \quad (14.25)$$

Porosity is estimated from Eq. 14.21. $(U_{ma})_a$ also can be determined graphically with the nomograph of **Fig. 14.20.** To use the nomograph, enter with the value of P_e, proceed vertically to ρ_b and then horizontally to the porosity value, and finally move vertically downward to define $(U_{ma})_a$.

The values of $(\rho_{ma})_a$ and $(U_{ma})_a$ are then entered in the ordinate and abscissa of the MID plot of **Fig. 14.21.** As with *M/N* and MID plots, rock mineralogy is identified by the position of the plotted data point relative to the points labeled on the plot. The ternary diagram in Fig. 14.21 represents a mixture of quartz, calcite, and dolomite. Matrix proportions for any three minerals can be determined by[8]

$$(\rho_{ma})_a = V_1\rho_1 + V_2\rho_2 + V_3\rho_3, \quad (14.26)$$

$$(U_{ma})_a = V_1U_1 + V_2U_2 + V_3U_3, \quad (14.27)$$

$$\text{and } 1 = V_1 + V_2 + V_3, \quad (14.28)$$

where $\rho_{1,2,3}$ = matrix densities of Minerals 1 through 3, $U_{1,2,3}$ = volumetric cross sections for Minerals 1 through 3, and $V_{1,2,3}$ = proportions of Minerals 1 through 3 in the formation matrices.

Fig. 14.21 is based on a three-mineral analysis in a clean, liquid-filled formation. Arrows indicate directions along which displacement will take place owing to the presence of gas, heavy minerals, salt, or barite in the formation. The plot also illustrates the location of different clay minerals. The natural gamma ray spectrometry and Litho-Density measurements can provide insight into clay mineralogy.

Example 14.6. Log measurements in carbonate Formation *W* are ρ_b = 2.52 g/cm^3, ϕ_N = 20.5% (CNL, limestone matrix assumed), and P_e = 3.65 barns/electron. The drilling fluid is a freshwater-based mud.

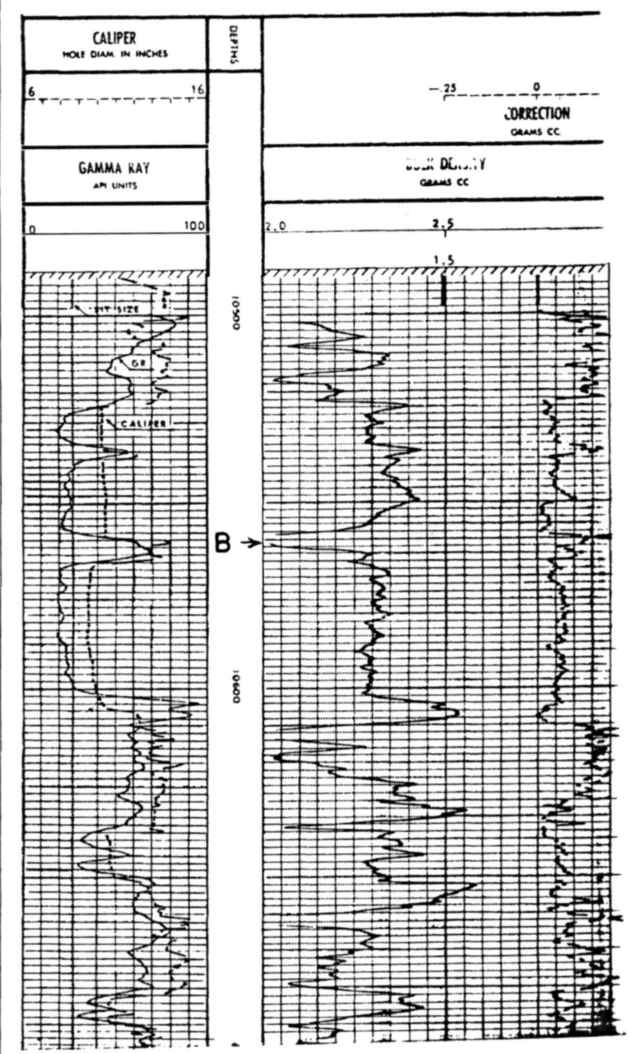

Fig. 14.23—Bulk density log of Problem 14.6 (courtesy Marathon Oil Co.).

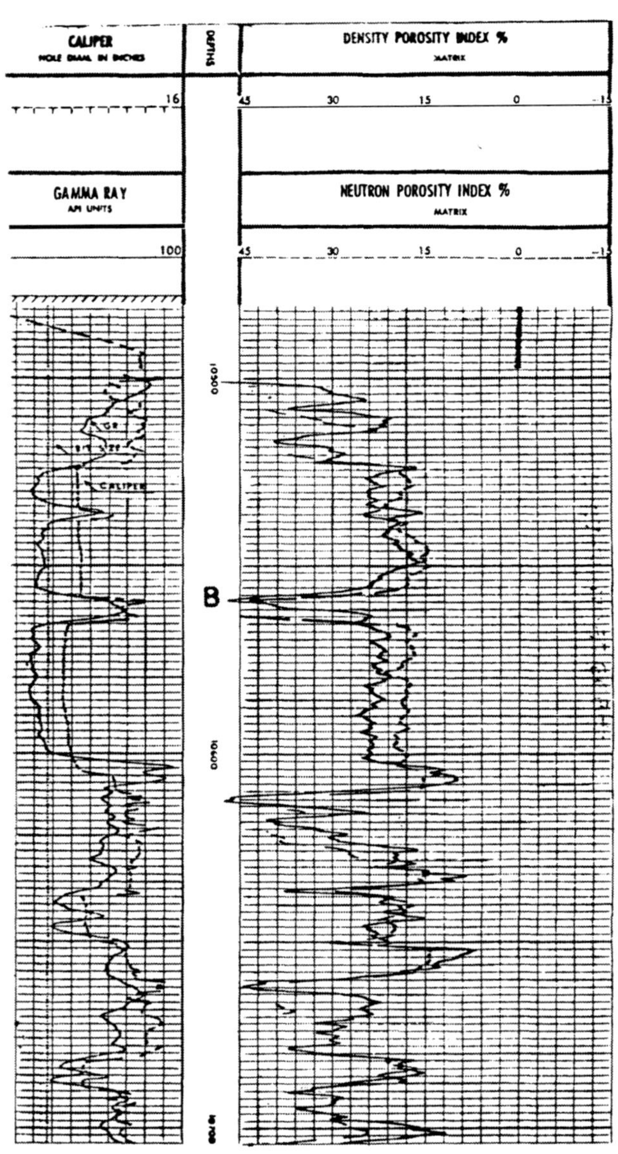

Fig. 14.24—FDC/CNL porosity log of Problem 14.6 (courtesy Marathon Oil Co.).

a. Using the dual-mineral crossplot of Fig. 14.19 determine the mineralogy and porosity of the formation.

b. Corroborate the results of Part a with two different approaches.

Solution.

a. The location of Formation W on the ρ_b/P_e dual-mineral crossplot is shown in Fig. 14.19. This location indicates that the zone is limy dolomite with 16% porosity. The mineral proportions are about 57% dolomite and 43% limestone.

b. The results of Part a can be corroborated with the ρ_b/ϕ_N dual-mineral crossplot of Fig. 14.3 and/or the $(\rho_{ma})_a/(U_{ma})_a$ MID plot.

The position of Data Point W in Fig. 14.3 confirms the interpretation of Part a. It indicates the same porosity value of 16% and similar proportions of limy dolomite lithology.

To use the MID plot of Fig. 14.21, $(\rho_{ma})_a$ and $(U_{ma})_a$ are first evaluated. $(\rho_{ma})_a$ is obtained from the MID chart of Fig. 14.13 using ρ_b and ϕ_N values. It also can be calculated from Eq. 8.11, which is solved for $(\rho_{ma})_a$:

$$(\rho_{ma})_a=(\rho_b-\phi)/(1-\phi)$$

$$=(2.52-0.16)/(1-0.16)=2.81 \text{ g/cm}^3.$$

$(U_{ma})_a$ is calculated with Eqs 8.9 and 14.25 as follows.

$$\rho_e=(2.52+0.1883)/1.0704=2.53 \text{ electrons/unit volume.}$$

$$(U_{ma})_a=[3.65(2.53)-0.16(0.398)]/(1-0.16)$$

$$=10.92 \text{ barns/unit volume.}$$

Formation W displays the above $(\rho_{ma})_a$ and $(U_{ma})_a$ values, so it falls on the calcite/dolomite line as shown in Fig. 14.21. The MID plot corroborates the above interpretation; it indicates a limy dolomite of 60% dolomite and 40% calcite.

Review Questions

1. Under what conditions can formation porosity be estimated reasonably from a single-porosity log?

2. Why and how do combinations of two or three porosity logs provide better information about formation lithology than a single log?

3. How many logs are required to give a good estimate of porosity and mineralogy of a limy sandstone?

4. What type of separation [i.e., none ($\phi_D=\phi_N$), positive ($\phi_D>\phi_N$), or negative ($\phi_D<\phi_N$)] will be observed on the FDC/CNL porosity overlay next to each of the following zones? The two porosity curves are derived assuming a liquid-filled limestone matrix.

a. Liquid-filled sandstone.
b. Liquid-filled dolomite.
c. Liquid-filled limestone.
d. Liquid-filled limestone with secondary porosity.
e. Shale.
f. Anhydrite.
g. Salt.

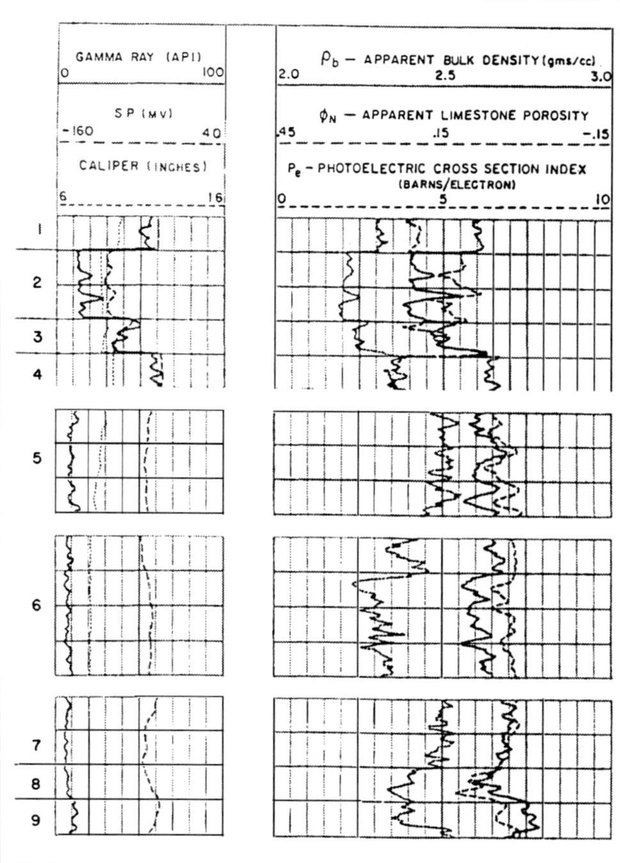

Fig. 14.25—Litho-Density composite log of Problem 14.13 (courtesy Schlumberger).

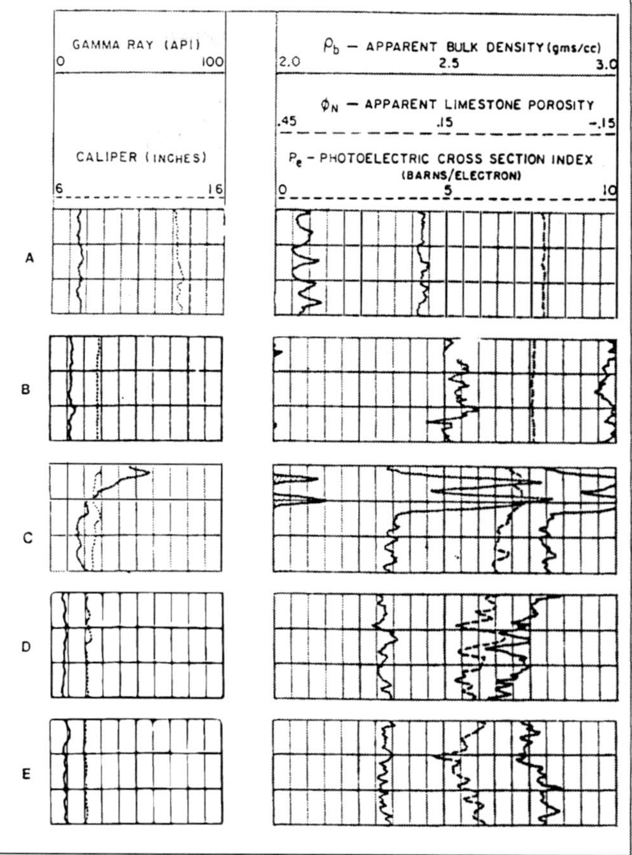

Fig. 14.26—Litho-Density composite log of Problem 14.14 (courtesy Schlumberger).

h. Sulfur.
i. Polyhalite.
j. Langbeinite.

5. How does the presence of secondary porosity, shale, or gas in the formation show on the different dual-mineral plots, the *M/N* plot, and the MID plot?

6. How can the presence and the magnitude of secondary porosity be determined from the different litho/porosity charts?

7. Why does the sonic/density dual-mineral crossplot have a poor resolution for common reservoir rocks? For what application is this crossplot most suited?

8. What is the set of equations used in the tri-porosity method of lithology determination for the case of a FDC/SNP/sonic combination and anhydrite/gypsum/salt formation drilled with saltwater-based mud?

9. Why are the parameters *M* and *N* independent of porosity?

10. Why does the *M/N* plot display four silica points and six dolomite points?

11. What factors determine the probable minerals present in a formation of interest? What is meant by standard lithology triangles?

12. What are the advantages of the MID plot over the *M/N* plot?

13. Why and how does the Litho-Density tool add a new dimension to lithology identification?

14. What is the set of equations used in the Litho-Density MID plot method of lithology determination for a sandstone/limestone matrix?

Problems

14.1 a. Generate tables similar to Tables 14.1 through 14.3 for the sonic/neutron (CNL) combination.
b. Use the tabulated values to construct a sonic/neutron dual-mineral litho/porosity crossplot.
c. Compare your crossplot to that of Fig. 14.7.

14.2 Assuming that the formation is liquid-filled ($\rho_f = 1.0$ g/cm^3) and its matrix consists of a single mineral, indicate the lithology and porosity for each set of data.

TABLE 14.6—DATA FOR PROBLEM 14.11

Point	ρ_b (g/cm^3)	ϕ_{SNP} (%)	Δt (μsec/ft)
1	2.60	11	75
2	2.50	8	62
3	2.63	10	71.5
4	2.63	7.5	59
5	2.60	5.5	58
6	2.63	12.5	62
7	2.57	7	62
8	2.65	8	60
9	2.60	8.5	62
10	2.61	8	62
11	2.60	7	61
12	2.60	8	60

TABLE 14.7—DATA FOR PROBLEM 14.12

Point	ρ_b (g/cm^3)	ϕ_{SNP} (%)	Δt (μsec/ft)
1	2.55	11.5	65
2	2.87	4	46
3	2.79	5.5	55
4	2.77	11	62
5	2.81	6	46
6	2.74	12	58
7	2.76	6	50
8	2.55	23	69
9	2.81	4	52
10	2.82	7.5	55
11	2.71	10	55
12	2.47	24	67

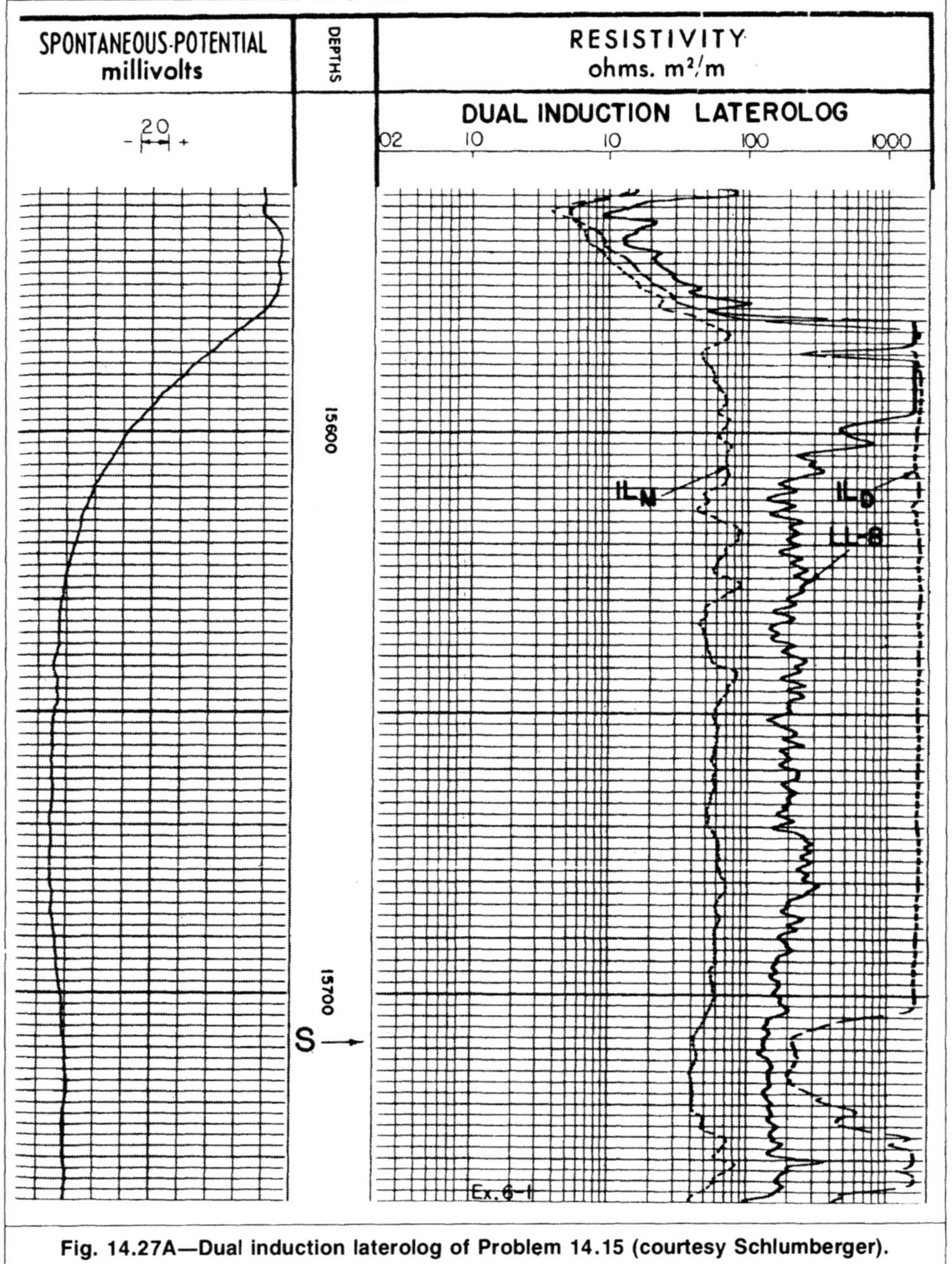

Fig. 14.27A—Dual induction laterolog of Problem 14.15 (courtesy Schlumberger).

Zone	ρ_b (g/cm^3)	ϕ_{CNL} (Limestone Matrix) (%)
A	2.20	30
B	2.78	12
C	2.94	2
D	2.17	25
E	2.04	0

14.3 Assuming that the formation is liquid-filled ($\rho_f = 1.0$ g/cm^3) and its matrix consists of two common minerals, indicate the most probable lithology and porosity for each set.

Zone	ρ_b (g/cm^3)	ϕ_{SNP} (Limestone Matrix) (%)
V	2.20	26
W	2.70	10
X	2.54	10
Y	2.92	0
Z	2.40	20

14.4 **Fig. 14.22** shows a composite log for ≈200 ft of a carbonate formation. The log recorded by Atlas Wireline Services consists of the following curves: gamma ray, caliper, bulk density, SNP, and Acoustilog^{SM}.

a. Using the density/neutron combination, determine the mineralogies and porosities of Zones A through D.

b. Repeat Part a using the Acousticlog/neutron combination.

c. Compare the results obtained in Parts a and b. Explain the differences, if any.

d. Using information from all three logs and the *M/N* approach, determine the lithologies and porosities of Zones A through D.

e. Which of the above interpretations would you recommend retaining for further evaluation of the zones and why?

Note: For accurate evaluation, dual-mineral charts and the *M/N* plot developed for Atlas Wireline Services' tools should be used.

14.5 Consider a zone for which the FDC log shows a bulk density of 2.58 g/cm^3 and the SNP log indicates an apparent limestone porosity of 11%. Determine the porosity, lithology, and matrix density assuming the following situations.

a. The zone matrix consists of a combination of limestone and dolomite.

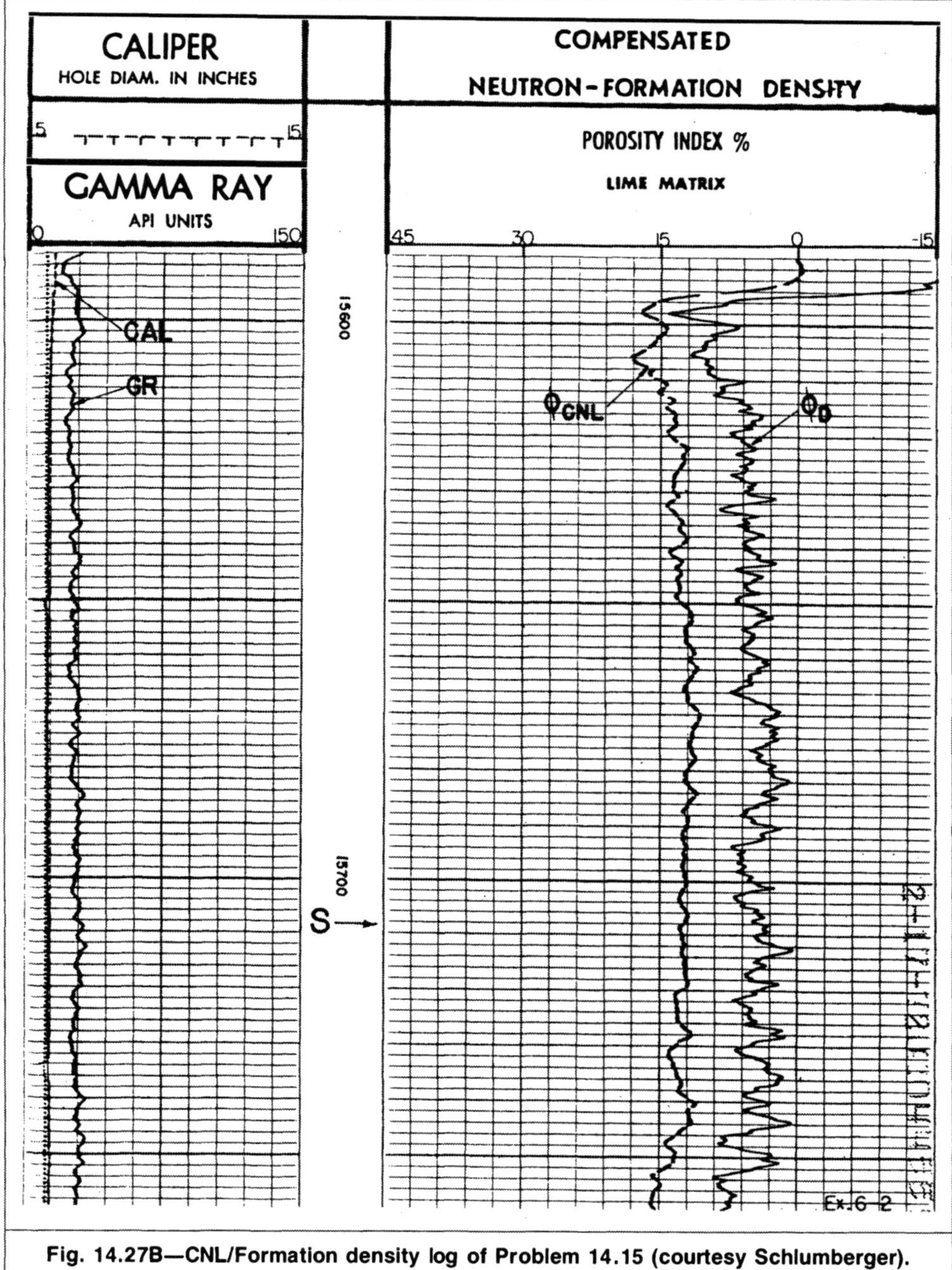

Fig. 14.27B—CNL/Formation density log of Problem 14.15 (courtesy Schlumberger).

b. The matrix is a combination of dolomite and silica.
c. The matrix is a combination of silica and anhydrite.
d. The four common minerals (silica, limestone, dolomite, and anhydrite) are present in this zone, but their exact compositions are not specified. How accurately can the matrix density and porosity be determined?

14.6 **Figs. 14.23 and 14.24** show the bulk density, density porosity index, and neutron porosity index logs obtained in the 10,500- to 10,700-ft interval of a well drilled near Lafayette, LA.

a. What matrix was assumed in determining the density and neutron porosity indices?
b. Determine the lithology of the formation displayed by the logs in the interval 10,566 to 10,601 ft.
c. Explain the relatively high density porosity and neutron porosity readings of Zone B.

14.7 The following values were obtained from sonic and CNL logs recorded through a liquid-filled carbonate formation: $\Delta t = 57$ μsec/ft, and ϕ_{CNL} (apparent limestone porosity) = 10%.

a. Determine the apparent lithology and porosity of the zone.
b. If sidewall cores show the zone to be clean limestone, what is the probable reason that the plotted point does not fall on the limestone line? What is the probable porosity of the zone? Assume that the log readings are good and are not affected by borehole size, mudcake, etc.

14.8 The following data were obtained for three zones of interest.

Zone	Δt (μsec/ft)	ϕ_{CNL} (Apparent Limestone Porosity) (%)
1	98	34
2	93	30
3	95.5	32

a. Using the appropriate chart, determine the lithologies and porosities of the three zones. Geological environment does not favor the presence of dolomites.
b. Visual inspection of the cores obtained from these zones indicates that the lithology is clean sandstone. How can you explain the interpretation obtained in Part a?

14.9 a. Prepare a bulk density/SNP crossplot that shows the sulfur, gypsum, and anhydrite points. Matrix properties of these minerals are as given below.

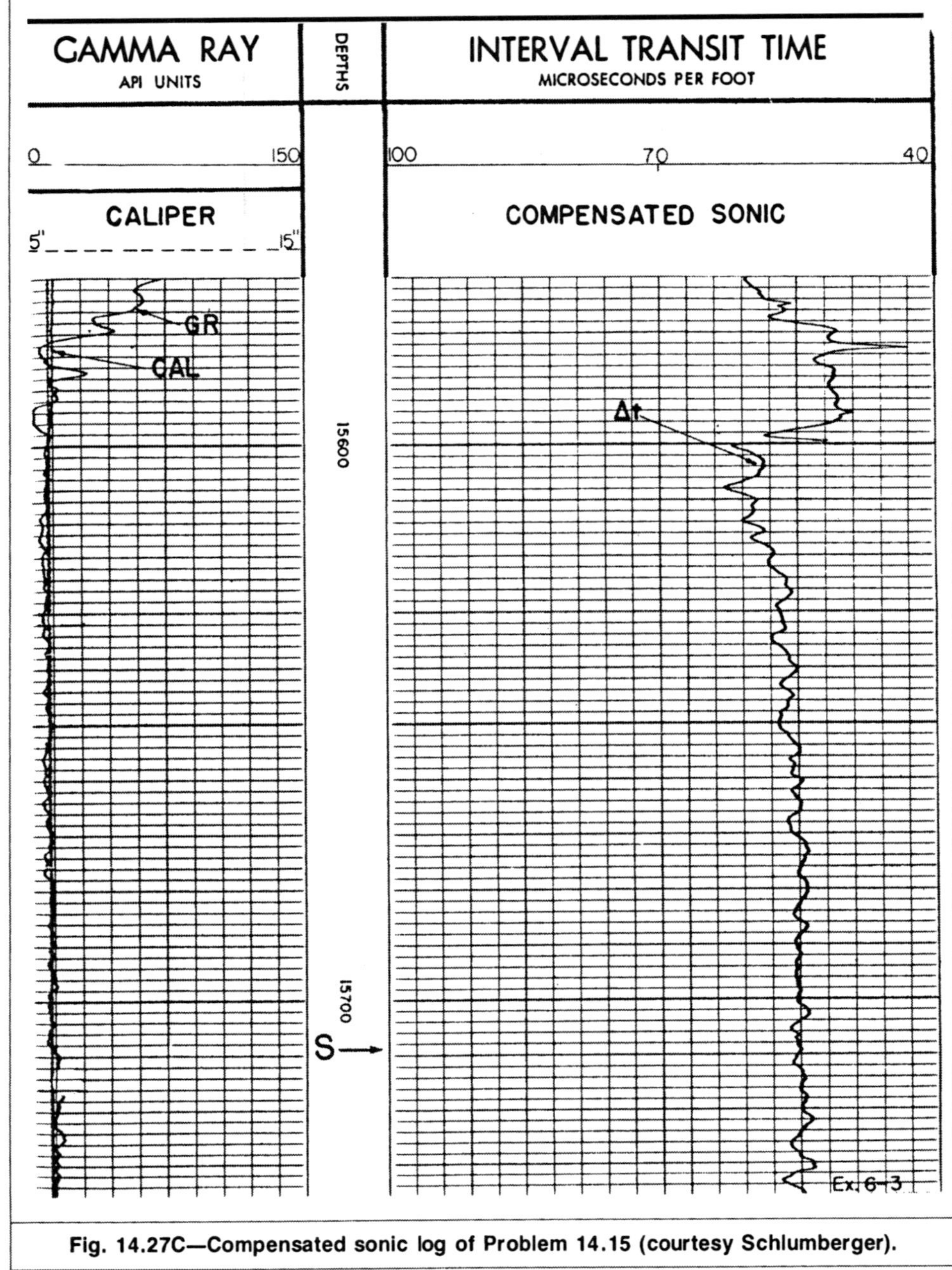

Fig. 14.27C—Compensated sonic log of Problem 14.15 (courtesy Schlumberger).

Mineral	Bulk Density (g/cm^3)	SNP Porosity (Limestone Matrix) (%)
Sulfur	2.030	0
Gypsum	2.351	4.9
Anhydrite	2.977	0

b. Determine the SNP limestone porosity index and bulk density of Zone A, which is composed of 42% gypsum, 40% anhydrite, and 18% sulfur.

c. What is the true porosity of zone A?

d. What is the true porosity and lithology of a single-lithology formation that displays the same SNP (limestone porosity) and bulk density as Zone A?

14.10 a. Prepare an MID plot that shows the calcite, dolomite, and anhydrite points. Also, construct a ternary diagram that shows the 20%, 40%, 60%, and 80% composition lines of each of the three minerals.

b. In a clean, liquid-saturated formation drilled with salty mud, the following log values were obtained: $\rho=2.75$ g/cm^3, $\phi_{SNP}=7.5\%$ (limestone matrix), and $\Delta t=58$ $\mu sec/ft$. If the matrix is known to be composed of limestone, dolomite, and anhydrite, determine the proportions of these minerals. Also, estimate the matrix porosity.

14.11 **Table 14.6** lists three porosity log values taken over a section in a Utah well.

a. Prepare an MID plot, an *M/N* plot, or both.

b. What is the most probable lithology present in this section?

c. Are any of the zones shaly?

d. Determine the mineralogies and porosities of Zones 5 through 7 and 12.

14.12 **Table 14.7** lists three porosity log values over a 350-ft section in a west Texas well.

a. Prepare an MID plot, an *M/N* plot, or both.

b. What is the most probable lithology present in this section?

c. If production from this section is associated with the presence of secondary porosity, list the possible potential zones.

d. If sidewall cores show that Zone 12 is a pure dolomite, determine its secondary porosity.

14.13 **Fig. 14.25** is a composite of several Litho-Density logs from an area with formations containing limestone, dolomite, sand,

and shale. Determine the lithologies and porosities of the nine zones marked on the logs.

14.14 **Fig. 14.26** is a composite of several Litho-Density logs from an area with carbonate and evaporite formations. Determine the lithologies and porosities of the five zones marked on the logs.

14.15 **Figs. 14.27A, 14.27B, and 14.27C** show, respectively, the dual induction/Laterolog^{SM}, FDC/CNL, and borehole-compensated logs obtained from a section of a well drilled in the southeastern U.S.

a. Give a general description of the lithology present in the interval shown.

b. For Zone S, give a detailed interpretation in terms of lithology, lithology fractions, porosity, porosity type, hydrocarbon type, and saturation. Well heading information is as follows.

R_m=0.534 Ω·m at 100°F.
R_{mf}=0.348 Ω·m at 79°F.
Bottomhole=276°F at 15,760 ft.
Bit size=5⅞-in.
Mud weight=10.7 lbm/gal.

Nomenclature

I_{sp} = secondary porosity index
P_e = photoelectric absorption cross section, barns/electron
R = resistivity, Ω·m
Δt = transit time, μsec/ft
U_f = photoelectric absorption cross section, barns/cm^3
v = velocity, ft/sec
V = volume, fraction
ρ = density, g/cm^3
ϕ = porosity, fraction

Subscripts

a = apparent
b = bulk
CNL = Compensated Neutron Log
dol = dolomite
D = density log
f = fluid
ls = limestone
ma = matrix
N = neutron log
ss = sandstone
S = sonic log
SNP = Sidewall Neutron Porosity tool

References

1. Poupon, A., Hayle, W.R., and Schmidt, A.W.: "Log Analysis in Formations With Complex Lithologies," *JPT* (Aug. 1971) 995–1005; *Trans.*, AIME, **251.**
2. Ferlt, W.H.: "Lithology, Other Effects on Porosity Logs," *Oil & Gas J.* (March 12, 1979) 68–70.
3. Burke, J.A., Campbell, R.L. Jr., and Schmidt, A.W.: "The Litho-Porosity Crossplot," *Trans.*, SPWLA 10th Annual Logging Symposium, Houston (May 1969) paper Y.
4. *Chart Book,* Schlumberger, Houston (1978).
5. *Log Interpretation Vol. 1: Principles,* Schlumberger, Houston, Chap. 12 (1972).
6. Savre, W.C.: "Determination of More Accurate Porosity and Mineral Composition in Complex Lithologies with the Use of Sonic, Neutron and Density Surveys," *JPT* (Sept. 1963) 945–59.
7. Clavier, C. and Rust, D.H.: "MID Plot: A New Lithology Technique," *Log Analyst* (Nov.–Dec. 1976) **XVII,** No. 6, 16–24.
8. Gardner, J.S. and Dumanoir, J.L.: "Litho-Density Log Interpretation," *Trans.*, SPWLA 21st Annual Logging Symposium, Lafayette (July 1980) paper N.

Chapter 15
Log Interpretation of Shaly Formations

15.1 Introduction

The evaluation of shaly formations (i.e., formations containing clay minerals) has long been a difficult task. Clay minerals affect all well-logging measurements to some degree. The shale effects have to be considered during evaluation of such reservoir parameters as porosity and water saturation. Although considerable research has been devoted to studying shaly formations, most of these effects are not fully understood, and perfect and universal models describing them remain illusive. For example, models that express shaly sand resistivity are deficient or require knowledge of parameters that cannot be determined practically. These models and the electrical properties of shaly sands in general were presented in Chap. 1. This chapter focuses on the practical aspects of shaly-sand interpretation.

Fig. 15.1 shows idealized responses of the self-potential (SP), the deep resistivity, and the three porosity logs in a hypothetical clean sand. If shale is added to this sand, the tool response is displaced toward the normal shale response. The degree of displacement increases as shale content, V_{sh}, increases. In the U.S. gulf coast or similar formations characterized by high porosity and formation water salinity, the shale resistivity, R_{sh}, usually exceeds that of a water-bearing zone, R_o. The presence of shale in the sand tends to reduce the true resistivity, R_t, of hydrocarbon-bearing zones and to increase the value of R_o. This can affect both quantitative and qualitative interpretations. Quantitatively, a nonrepresentative high S_w value is calculated if clean formation models are used. The potential of the zone is then underestimated or completely masked. A high V_{sh} might even encumber visual detection of hydrocarbon zones.

As Fig. 15.1 also shows, the presence of shale affects the response of the three porosity tools. Using clean formation models in the quantitative interpretation results in overestimation of porosity values.

The interpretation problem in shaly formations is in calculating porosity and saturation values free from the shale effect. Because the shale effect depends on shale content, the estimation of V_{sh} is of prime importance. Qualitatively, V_{sh} indicates whether the formation can be considered clean or shaly. This determines the type of model or approach to use in the interpretation. Quantitatively, V_{sh} is used to estimate the shale effect on log responses and, if needed, to correct them to the clean formation responses.

Because shale affects every logging tool to some degree, numerous methods have been developed to indicate the presence and to estimate the content of shale.[1] The most often used methods involve the SP, gamma ray, and porosity logs.

15.2 Shale Content From the SP Log

V_{sh} from the SP log can be estimated with the following linear relationship:

$$V_{sh}=1-E_{PSP}/E_{SSP}, \quad\quad (15.1)$$

where E_{PSP} is the shale response in the shaly zone of interest and E_{SSP} is the shale response in an adjacent clean, thick zone that contains the same water salinity as the zone of interest.

Use of this method should be restricted to cases where the SP is of good quality and other shale indicators are absent. Several factors, such as salinity changes, R_w/R_{mf} contrast, and hydrocarbon content, affect the estimation of V_{sh} from the SP log. The presence of hydrocarbon in a formation reduces the reading of the SP log, so Eq. 15.1 tends to give an upper limit for shale content.

15.3 Shale Content From the Gamma Ray Log

The shale volume is related to a shale index, I_{sh}:

$$I_{sh}=(\gamma_{\log}-\gamma_c)/(\gamma_{sh}-\gamma_c), \quad\quad (15.2)$$

where $\gamma_{\log}$=gamma ray response in the zone of interest, γ_c= average gamma ray response in the cleanest formations, and γ_{sh}=average gamma ray responses in shales.

It is customary to assume that $V_{sh}=I_{sh}$. This assumption, however, tends to exaggerate the shale volume. Empirical relationships were found to be more reliable. Several empirical relationships were developed for different geologic ages and areas. The most notable correlations were developed by Larionov,[2] Stieber,[3] and Clavier *et al.*[4] **Fig. 15.2** illustrates these correlations, which also can be expressed analytically.

For tertiary rocks, the Larionov equation is

$$V_{sh}=0.083(2^{3.7I_{sh}}-1), \quad\quad (15.3)$$

the Stieber equation is

$$V_{sh}=I_{sh}/(3-2I_{sh}), \quad\quad (15.4)$$

and the Clavier *et al.* equation is

$$V_{sh}=1.7-[3.38-(I_{sh}+0.7)^2]^{½}. \quad\quad (15.5)$$

For older rocks, the Larionov equation is

$$V_{sh}=0.33(2^{2I_{sh}}-1). \quad\quad (15.6)$$

An empirical equation can be developed specifically for a formation or geologic unit of interest, as discussed in Example 15.5.

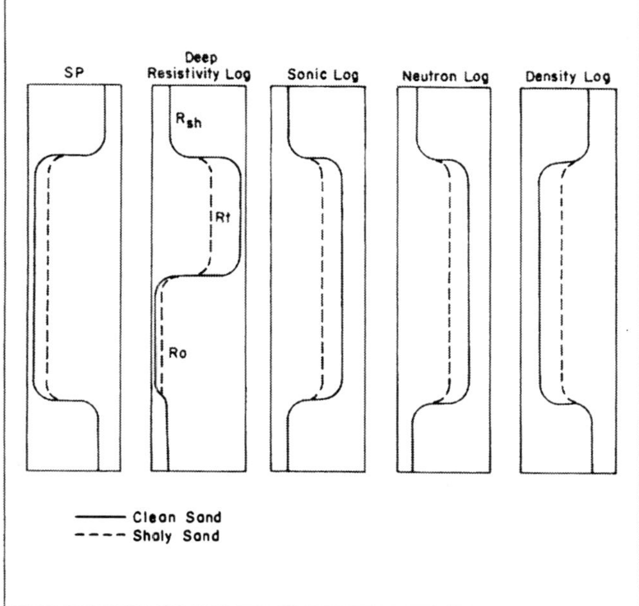

Fig. 15.1—Idealized response of different tools in hypothetical clean and shaly sands.

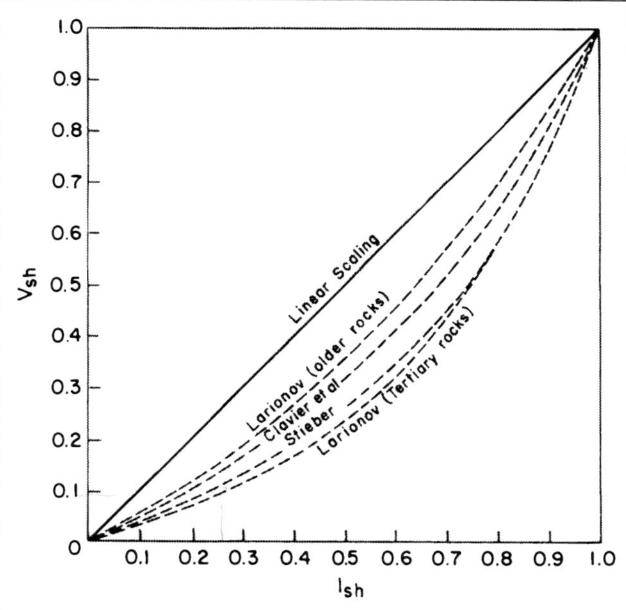

Fig. 15.2—Empirical correlations relating shale content, V_{sh}, to gamma ray shale index, I_{sh}.

Example 15.1. Figs. 15.3 and 15.4 show a section of IES and FDC logs obtained in a 9,100-ft-deep well drilled offshore Louisiana. R_m at a bottomhole temperature (BHT) of 156°F was 0.34 Ω·m.

a. Using the SP curve, determine the shale content of Zone A in the logs.

b. List the assumptions implicit in the procedure used in Part a.

c. Using the gamma ray curve, determine the shale content of Zone A.

d. List the assumptions implied in the procedure used in Part c.

e. Compare the values calculated in Parts a and c, and recommend a V_{sh} value.

Solution

a. Zone A displays an E_{PSP} value of −47 mV. The cleanest zone in the interval shown is Zone B, which displays an E_{SSP} value of −80 mV. Using Eq. 15.1 gives

$$V_{sh}=1-[(-47)/(-80)]=0.41 \text{ or } 41\%.$$

b. Assumptions implied in Part a are that Zone B is a clean formation, Zones A and B have the same formation water salinity, and the effects of bed-thickness, formation resistivity, and invasion R_w/R_{mf} contrast on the SP deflection in Zone A are negligible.

c. The γ_{log}, γ_c, and γ_{sh} values obtained in Zones A through C are 44, 28, and 92 API units, respectively. Using Eq. 15.2 yields

$$I_{sh}=(44-28)/(92-28)=0.44 \text{ or } 44\%.$$

Because the sediments offshore Louisiana are Tertiary, V_{sh} can be estimated from Eq. 15.3, 15.4, or 15.5. Using the Larionov equation gives

$$V_{sh}=0.083[2^{3.7(-44)}-1]=0.17 \text{ or } 17\%.$$

The Stieber equation results in

$$V_{sh}=-44/[3-2(0.44)]=0.21 \text{ or } 21\%.$$

The Clavier equation yields

$$V_{sh}=1.7-[3.38-(0.44+0.7)^2]^{1/2}$$

$$=0.257 \text{ or } 26\%.$$

As can be seen, the three empirical equations recommended for Tertiary rocks give different values (see Fig. 15.2). Larionov's equation gives a value on the low side, Clavier *et al.*'s equation gives a value on the high side, and Stieber's equation gives a value in between.

d. Assumptions implied in the procedure used in Part c are that Zone B is a clean formation, Zone C is a representative shale zone, and V_{sh} can be related to I_{sh} by one of the given empirical equations.

e. As previously mentioned, the SP log method yields a high V_{sh} value. Given the nature of the assumptions in Parts b and d, a V_{sh} value derived from the gamma ray log should be preferred over one derived from the SP log.

The derived gamma ray value ranges from 17% to 26%, depending on the empirical equation used. Use of a low value of 17%, a high value of 26%, or an average value of 21% depends on the spirit of the evaluations. Later sections will show that, the higher the value of V_{sh} used, the higher the correction for shale effect.

Experience has shown that the gamma ray index approach provides a reasonable estimate for V_{sh}, provided that the formation does not contain radioactive minerals that are not associated with shales (e.g., uranium). When uranium is present, use of the gamma ray spectral log is recommended. This log displays potassium (K^{40}) and thorium (Th) contents. The shale content can be approximated by potassium or thorium indices:

$$(I_{sh})_K=[(C_K)_{log}-(C_K)_c]/[(C_K)_{sh}-(C_K)_c] \quad \text{.......... (15.7)}$$

$$\text{and } (I_{sh})_{Th}=[(C_{Th})_{log}-(C_{Th})_c]/[(C_{Th})_{sh}-(C_{Th})_c], \quad \text{... (15.8)}$$

where $(C_K)_{log}$ and $(C_{Th})_{log}$ are the K^{40} and Th curve readings in the zone of interest, $(C_K)_c$ and $(C_{Th})_c$ are the K^{40} and Th curve readings in the cleanest formation, and $(C_K)_{sh}$ and $(C_{Th})_{sh}$ are the K^{40} and Th curve readings in shales.

Potassium and thorium contents vary for different types of clays. The use of $(I_{sh})_K$ or $(I_{sh})_{Th}$ alone to determine V_{sh} could lead to incorrect values. Illite, glauconite, biotite, and muscovite have high-potassium and low-thorium concentrations. On the other hand, kaolinite, montmorillonite, bentonite, and bauxite are high in thorium but low in potassium. These clays tend to form a hyperbolic pattern on a crossplot of potassium content vs. thorium content. The product index of potassium and thorium is virtually independent of clay type.[5] Hence, it can be used to define a more representative shale index:

$$(I_{sh})_{KTh}=[(C_K C_{Th})_{log}-(C_K C_{Th})_c]/[(C_K C_{Th})_{sh}-(C_K C_{Th})_c]. \quad \text{.......... (15.9)}$$

Example 15.2. Fig. 15.5 illustrates a section of a natural gamma ray spectrometry log obtained in a Gulf of Mexico well. Examine this log section and estimate the shale index of Zones V and X using the total gamma ray curve and spectrometry information.

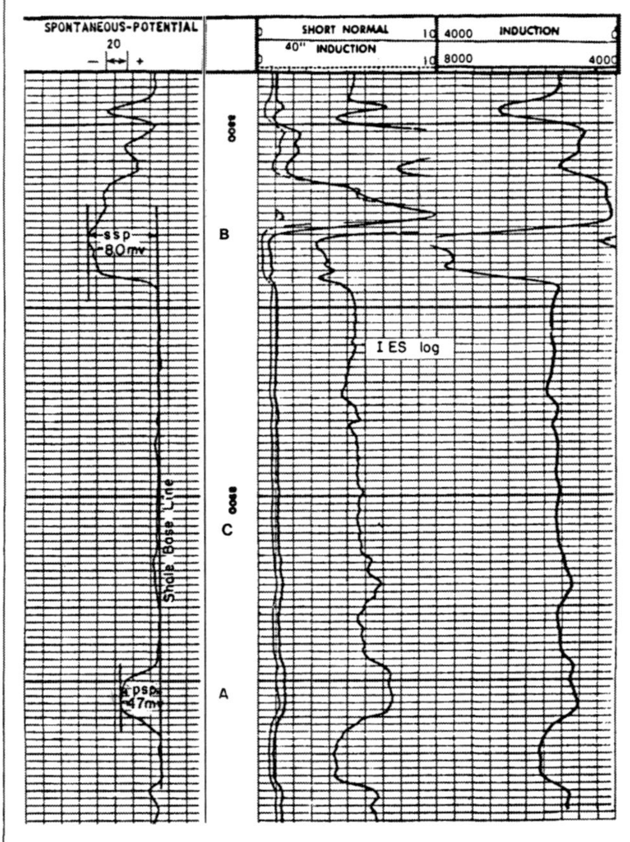

Fig. 15.3—IES log of Example 15.1 (courtesy Amoco Production Co.).

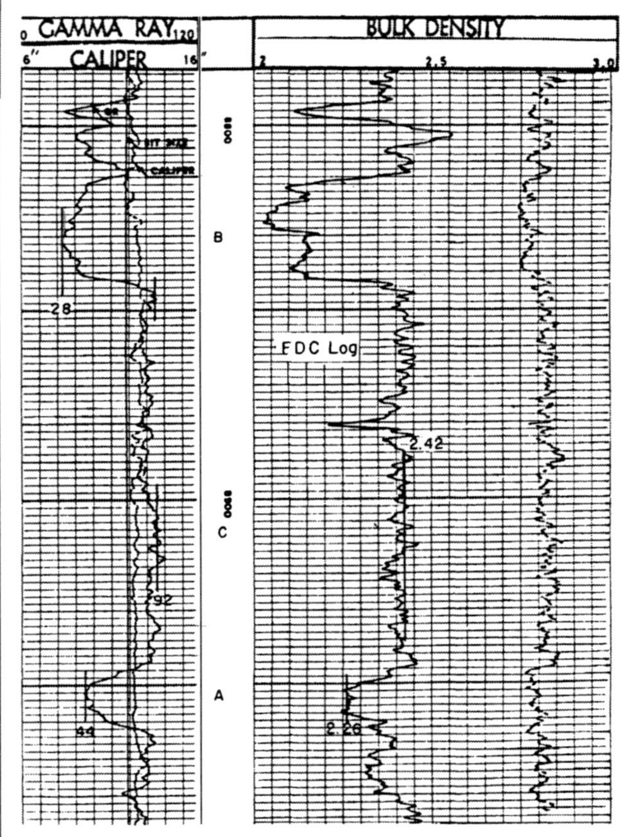

Fig. 15.4—FDC log of Example 15.1 (courtesy Amoco Production Co.).

Solution. *Shale Parameters.* The interval 8,805 to 8,830 ft appears to contain the purest shale. For this interval, $\gamma_{sh}=100$ API units, $(C_K)_{sh}=2.9$ log divisions, and $(C_{Th})_{sh}=3.7$ log divisions.

Clean Sand Parameters. The total gamma ray and potassium readings for clean sand are selected from the interval 8,436 to 8,450 ft, which displays the minimum total radioactivity in the section. The minimum thorium content is selected from the interval 8,405 to 8,416 ft, which does not display the lowest total gamma ray response because of the uranium content. In this interval, $\gamma_c=33$ API units, $(C_K)_c=1.2$ log divisions, and $(C_{Th})_c=0.8$ log divisions.

Shale Index for Zones V and X. The log readings in Zones V and X and the calculated shale index values are tabulated below:

	Zone V	Zone X
γ_{log}, API units	64	62
$(C_K)_{log}$, log divisions	2.2	1.6
$(C_{Th})_{log}$, log divisions	2.4	1.8
I_{sh} (gamma ray, Eq. 15.2), %	46	43
$(I_{sh})_K$ (Eq. 15.7), %	59	24
$(I_{sh})_{Th}$ (Eq. 15.8), %	55	35
$(I_{sh})_{KTh}$ (Eq. 15.9), %	44	20

Each of the four approaches used yielded a different I_{sh}. This points out clearly the inherent difficulty in estimating this important interpretation parameter. Fertl[5] suggested that the product index of potassium and thorium, $(I_{sh})_{KTh}$, is the most reliable index derived from the spectrometry log. The 44% value calculated for $(I_{sh})_{KTh}$ of Zone V compares well with the 46% value of I_{sh} calculated from the total gamma ray response. This is not surprising because no uranium is present in the zone. On the other hand, the 1.6 ppm uranium present in Zone X explains the much higher I_{sh} calculated from the total gamma ray curve. From this analysis, the best representative I_{sh} values to be retained for further evaluation are 44% and 20% for Zones V and X, respectively.

15.4 Porosity Logs in Shaly Formations

Porosity log response can be expressed in general by Eq. 14.1 or 14.2. In a shaly liquid-filled formation assumed to have normal compaction and no secondary porosity, these two equations reduce to

$$\text{tool response}=f(\text{matrix, porosity, shale content}). \quad (15.10)$$

The presence of shale complicates the interpretation of the tool response because of the diverse characteristics of shales and the different responses of each porosity tool to the shale content. On the density porosity log, shales display low to moderate porosity values. On the sonic and neutron logs, shales display moderate to relatively high porosity values.

The tool response expressed by Eq. 15.10 depends on the porosity, shale content, and the fluid, shale, and matrix properties. **Fig. 15.6** shows schematically the fractions and densities of the different constituents of a shaly formation. The response of the density tool, ρ_b, in such a formation can be expressed as

$$\rho_b=\phi\rho_f+V_{sh}\rho_{sh}+(1-\phi-V_{sh})\rho_{ma}, \quad (15.11)$$

where ρ_b, ρ_f, ρ_{sh}, and ρ_{ma} are the bulk, fluid, shale, and matrix densities, respectively.

Rearranging Eq. 15.11 results in

$$\rho_{ma}-\rho_b=\phi(\rho_{ma}-\rho_f)+V_{sh}(\rho_{ma}-\rho_{sh}). \quad (15.12)$$

Dividing by $(\rho_{ma}-\rho_f)$ yields

$$(\rho_{ma}-\rho_b)/(\rho_{ma}-\rho_f)$$
$$=\phi+V_{sh}[(\rho_{ma}-\rho_{sh})/(\rho_{ma}-\rho_f)]. \quad (15.13)$$

Eq. 15.13 can be expressed as

$$\phi_D=\phi+V_{sh}(\phi_D)_{sh}, \quad (15.14)$$

where $\phi_D=(\rho_{ma}-\rho_b)/(\rho_{ma}-\rho_f)$ ... (15.15)

and $(\phi_D)_{sh}=(\rho_{ma}-\rho_{sh})/(\rho_{ma}-\rho_f)$. ... (15.16)

ϕ_D is the density porosity displayed by the log in the formation of interest. This value is calculated by assuming a certain ρ_{ma} that corresponds to a clean formation (e.g., 2.65 g/cm^3 in sandstones).

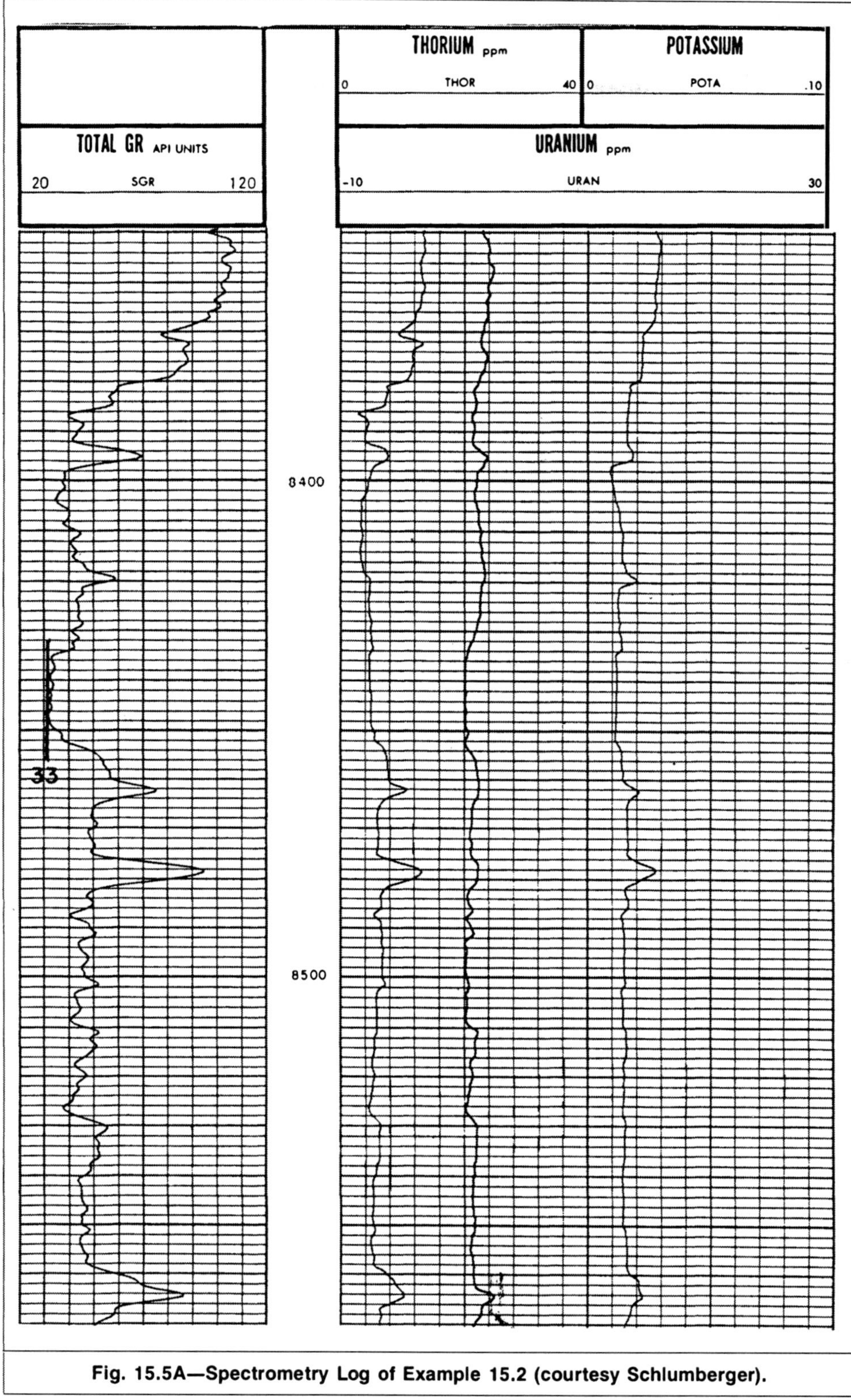

Fig. 15.5A—Spectrometry Log of Example 15.2 (courtesy Schlumberger).

The fluid type is assumed to be liquid, and ρ_f is chosen accordingly. For a freshwater-based mud, $\rho_f = 1$ g/cm^3. Because of these assumptions, ϕ_D is an apparent porosity. It is equal to the true porosity, ϕ, only in clean, liquid-filled formations.

The term $(\phi_D)_{sh}$ is the apparent density porosity displayed by the tool if placed in a 100% shale formation having the same characteristics as the shale present in the formation of interest. For practical reasons, shales adjacent to the formation of interest are assumed to be similar to the shale present within it.

Equations like Eq. 15.14 can be derived in a similar manner for the neutron and sonic logs:

$$\phi_N = \phi + V_{sh}(\phi_N)_{sh} \quad \text{(15.17)}$$

and $\phi_S = \phi + V_{sh}(\phi_S)_{sh}$, ...(15.18)

where ϕ_N and ϕ_S are the apparent porosities displayed by the neutron and sonic logs, respectively, in a shaly formation, and $(\phi_N)_{sh}$ and $(\phi_S)_{sh}$ are the apparent porosities displayed by the neutron and sonic logs, respectively, in adjacent shales.

The true porosity, ϕ, representing the effective PV of a shaly formation can be estimated from any of the porosity tools by

$$\phi = \phi_a - V_{sh}(\phi_a)_{sh}, \quad \text{(15.19)}$$

where ϕ_a is the apparent porosity displayed by the tool or calculated from raw data (i.e., ρ_b, Δt), assuming that the formation is clean and liquid-filled, and $(\phi_a)_{sh}$ is the apparent porosity dis-

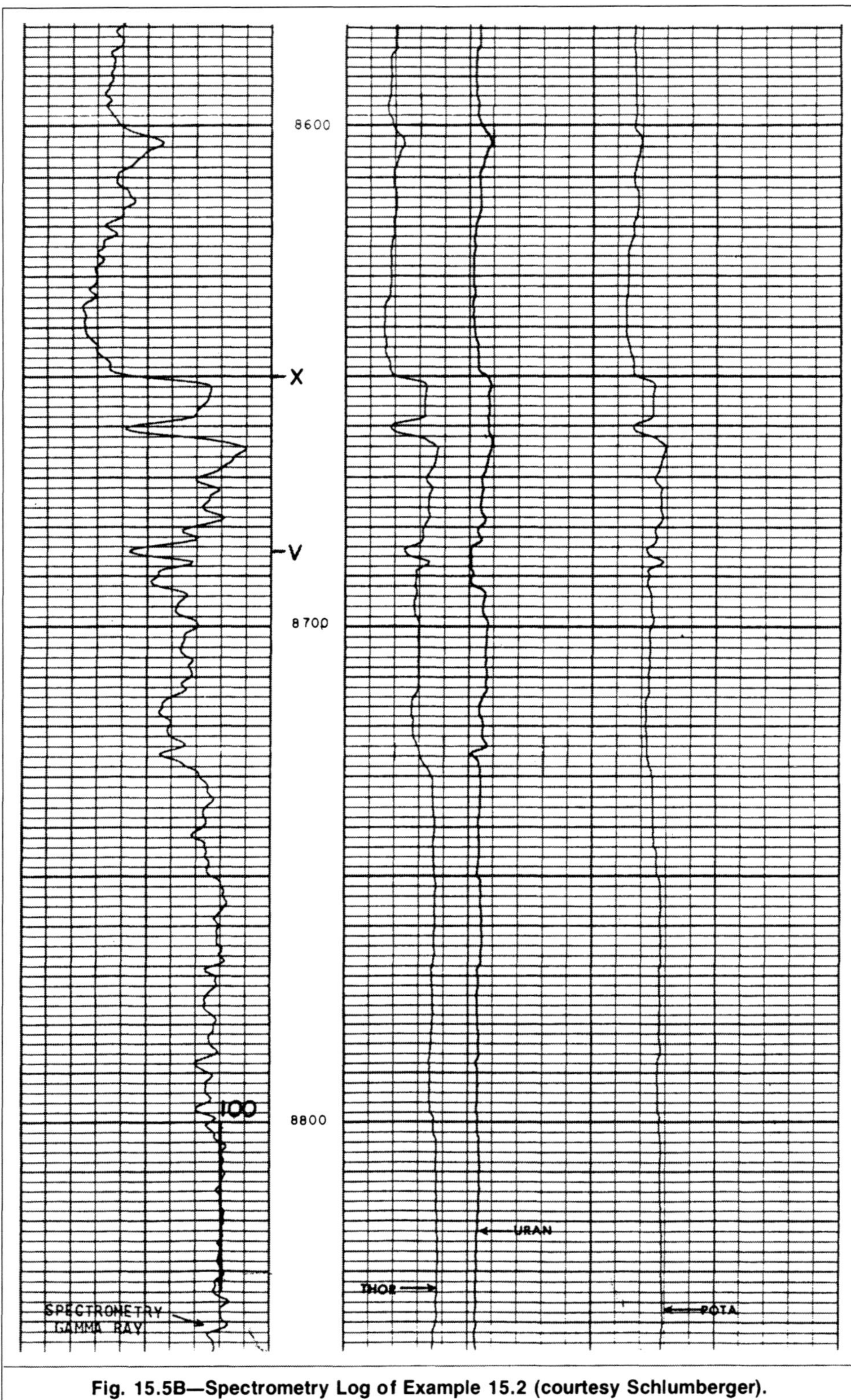

Fig. 15.5B—Spectrometry Log of Example 15.2 (courtesy Schlumberger).

played by the tool or calculated in adjacent shales from raw data, assuming the formation is clean and liquid-filled. The term $[V_{sh}(\phi_a)_s]$ is regarded as a correction term.

Example 15.3. Refer to Figs. 15.3 and 15.4 and answer the following.

a. Calculate the apparent density porosity, ϕ_D, of Zone A.

b. Calculate the apparent density porosity of shales adjacent to Zone A.

c. Estimate the true effective porosity of Zone A.

d. List the assumptions implied in the procedure used in Part c.

e. Comment on the validity of the porosity estimate.

Solution

a. The bulk density curve displays a value of 2.26 g/cm^3 in Zone A. Using Eq. 15.15 and setting ρ_{ma}=2.65 g/cm^3 and ρ_f=1 g/cm^3 yields

ϕ_D=(2.65−2.26)/(2.65−1)=0.236 or 24%.

b. The bulk density curve displays an average value of 2.42 g/cm^3 in shales adjacent to Zone A. Using Eq. 15.16 yields $(\phi_D)_{sh}$=(2.65−2.42)/(2.65−1)=0.139 or 14%.

c. An average V_{sh} value of 21% is estimated for Zone A in Example 15.1. Using Eq. 15.19 results in

ϕ=0.236−0.21 (0.139)

=0.207 or 21%.

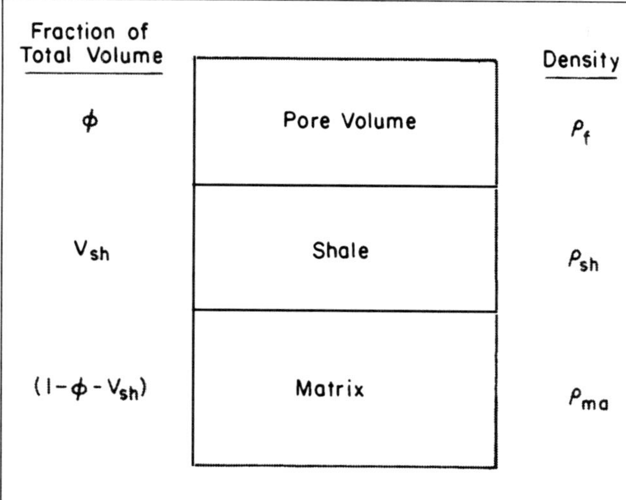

Fig. 15.6—Schematic of the fractions and properties of shaly formation constitutents.

d. Assumptions implied in the procedure used to estimate the effective porosity of Zone A are that $\rho_{ma}=2.65$ g/cm^3 and $\rho_f=1$ g/cm^3. All the assumptions implied in the estimation of V_{sh} are also implied here (see the solution of Example 15.1). Finally, it is assumed that adjacent shale and shale within Zone A are of similar properties.

e. The calculation of ϕ_D in Part a is based on reasonable assumptions. The correction term $[V_{sh}(\phi_D)_{sh}]$ contains the more questionable assumptions. The magnitude of the correction is relatively small, about 0.03 porosity units (3%). A large error in one or more of the parameters needed to calculate this term might translate into only a small error in the value of effective porosity. For example, as seen in Example 15.1, $V_{sh}=21\pm5\%$. This uncertainty in V_{sh} yields a maximum deviation of only 1% in the porosity value. Hence, an effective porosity of 21% for Zone A can be considered a good estimate.

15.5 Effective Porosity and Shale Content From Porosity Logs

Effective porosity of shaly formations can be calculated, as shown in the previous section, from the response of any porosity tool. This

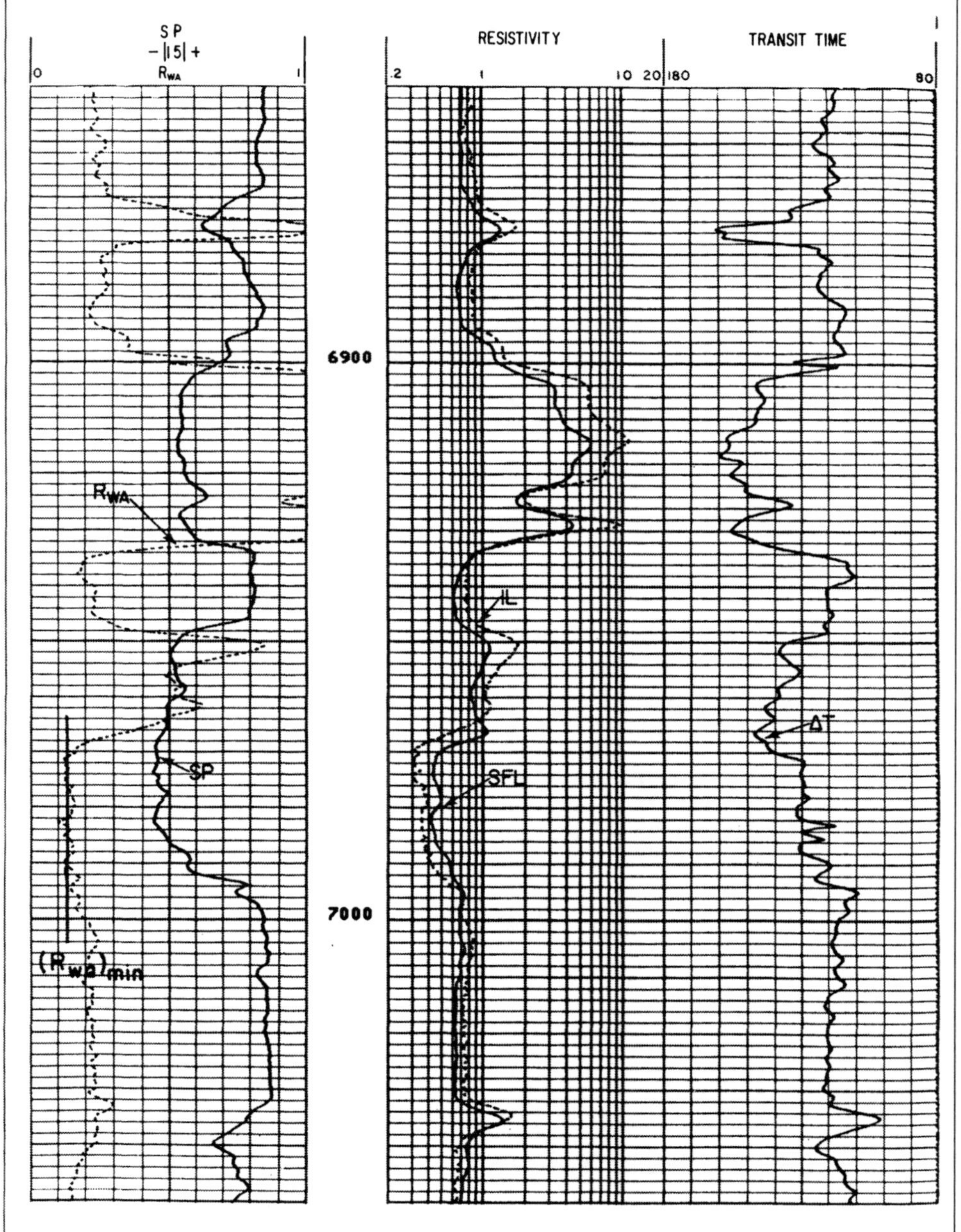

Fig. 15.7—ISF/sonic log of Example 15.4 (courtesy Schlumberger).

calculation, however, requires knowledge of V_{sh}. V_{sh} is estimated with one of the methods described in Sec. 15.3; the gamma ray log approach is used the most. Experience has shown that, when one porosity tool is used to calculate ϕ, the density log provides the best estimate. Because $(\phi_D)_{sh}$ is much smaller than $(\phi_N)_{sh}$ and $(\phi_S)_{sh}$, the correction term $[V_{sh}(\phi_a)_{sh}]$ of Eq. 15.19 is less sensitive to an incorrect estimation of V_{sh}.

To avoid misinterpretation caused by a nonrepresentative V_{sh} value estimated from the gamma ray log, an independent evaluation of V_{sh} is necessary. This can be done with two porosity tool responses. In the absence of known V_{sh}, Eq. 15.19 contains two unknowns, ϕ and V_{sh}. However, Eq. 15.19 can be written for each of the porosity tools. Having two porosity curves allows two equations with the form of Eq. 15.19 to be set up. Because the sonic tool depends on compaction and secondary porosity, in addition to porosity, shale, and matrix, density and neutron responses are used to set up the system of two equations to be solved simultaneously for ϕ and V_{sh}:

$$\phi_D=\phi+V_{sh}(\phi_D)_{sh} \quad \text{(15.20a)}$$

and $$\phi_N=\phi+V_{sh}(\phi_N)_{sh}. \quad \text{(15.20b)}$$

Example 15.4. Figs. 15.7 and 15.8 display the sections of ISF/sonic and FDC/CNL logs obtained over an interval containing Tertiary shales and sands.

a. Explain which sand shown in this interval is the cleanest.

b. Can the SP curve be used to estimate the shale content V_{sh}? Explain.

c. Using the gamma ray curve, estimate the V_{sh} content of Zones 1 through 3.

d. Estimate $(\phi_D)_{sh}$, $(\phi_N)_{sh}$, and $(\phi_S)_{sh}$. Compare their orders of magnitude. What effect will this have on the shale correction?

e. Estimate V_{sh} for Zones 1 through 3 independent of gamma ray response.

f. Compare and comment on V_{sh} values calculated in Parts c and e.

Solution.

a. The cleanest sand shown in the interval of interest is Sand X, especially the 10-ft zone located at 6,911 to 6,921 ft. It is clean because it displays the lowest gamma ray deflection. Also, both the density and neutron curves display the same porosity value ($\approx$30%). According to Eq. 15.19, ϕ_N and ϕ_D can be equal if $V_{sh}=0$.

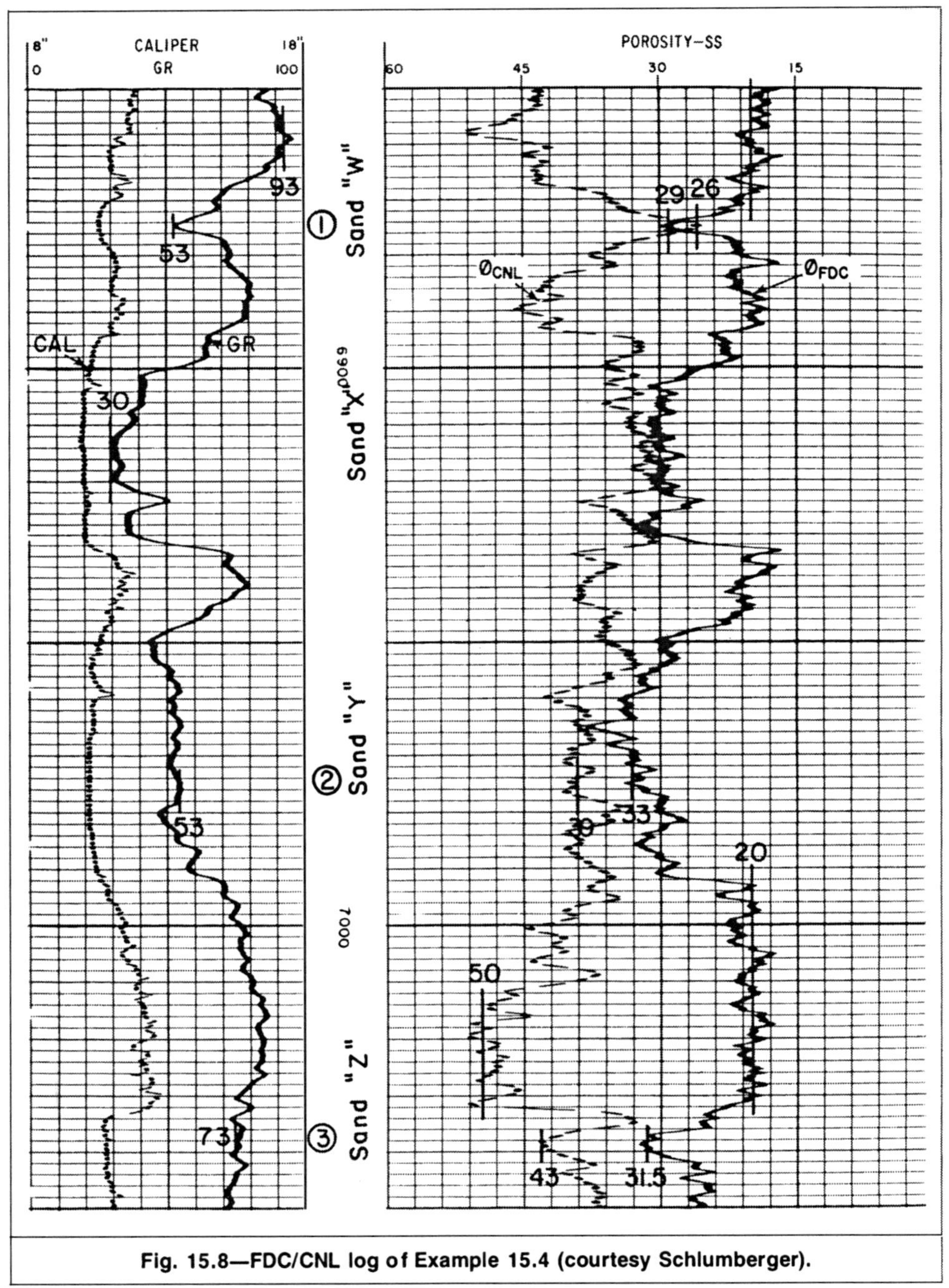

Fig. 15.8—FDC/CNL log of Example 15.4 (courtesy Schlumberger).

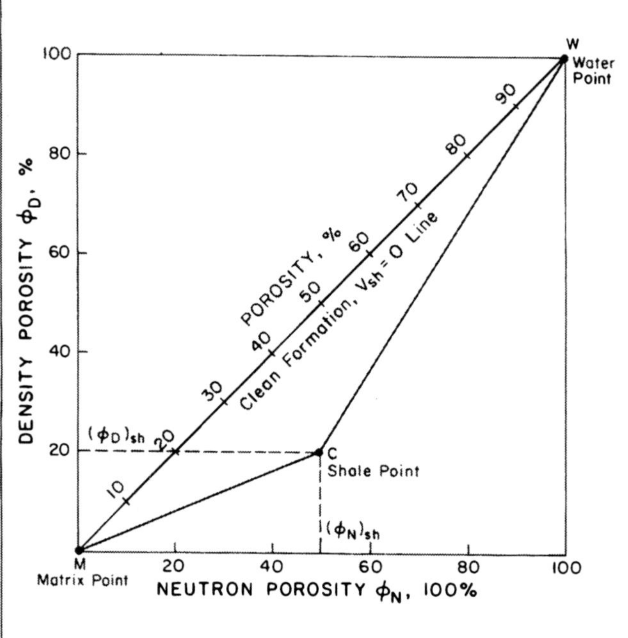

Fig. 15.9—Density/neutron crossplot showing water, matrix, and shale points.

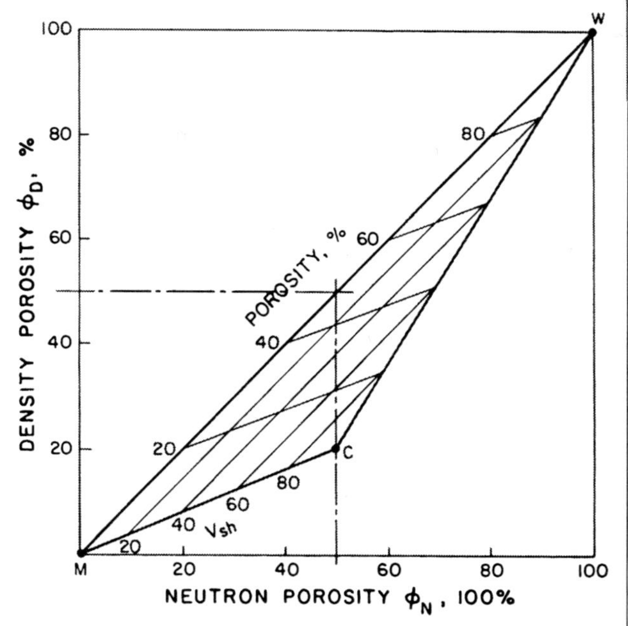

Fig. 15.10—Density/neutron crossplot showing isoporosity and isoshale-content lines.

b. The FDC/CNL log indicates that Sand Y is shaly. Sand Y displays a gamma ray response that is greater than that of the clean Sand X. The separation between the ϕ_D and ϕ_N curves occurring in Sand Y also tends to support this conclusion. Sand Y, however, displays the highest SP value because the SP deflection in Sand X is considerably reduced owing to high resistivity. Hence, the E_{SSP} in clean sand cannot be determined with satisfactory accuracy. The SP log is not suitable for V_{sh} estimation.

c. γ_c=30 API units and γ_{sh}=93 API units.

Zone	γ_{log} (API units)	I_{sh} (Eq. 15.2) (%)	V_{sh} (Eq. 15.4) (%)
1	53	37	16
2	53	37	16
3	73	68	41

d. The following average values can be read in shales present in the interval of interest: $(\phi_D)_{sh}$=20%, $(\phi_N)_{sh}$=50%, and $(\Delta t)_{sh}$=120 μsec/ft. From Eq. 10.1,

$$(\phi_S)_{sh}=(120-55.5)/(189-55.5)=0.48 \text{ or } 48\%.$$

The shale density porosity is much smaller than the shale neutron and sonic porosities. This results in a high shale effect on both the neutron and sonic porosity curves. In analytical terms, the $[V_{sh}(\phi_a)_{sh}]$ term of Eq. 15.19 is large for the sonic and neutron logs. Thus, the density log is preferred if used alone in estimating the effective porosity.

e. The simultaneous solution of Eqs. 15.20 yields the following.

Zone	ϕ_D (%)	ϕ_N (%)	ϕ (%)	V_{sh} (%)
1	29	26	31	−10
2	33	39	29	20
3	31.5	43	24	38

f. The V_{sh} values calculated in Parts c and e are listed below.

Zone	V_{sh} From gamma ray log (%)	V_{sh} From ϕ_D/ϕ_N (%)
1	16	−10
2	16	20
3	41	38

It is evident that the negative V_{sh} value calculated in Zone 1 is in error. The model used should then be questioned. The model used, Eqs. 15.20, assumes that the formations are liquid-filled. As indicated by the R_{wa} curve, Zone 1 contains hydrocarbon. If this hydrocarbon is gas, then the model is not valid. As discussed in Chap. 16, the responses of the density and neutron logs are consistent with the presence of gas.

Given the diverse asumptions involved in the V_{sh} estimation (see Examples 15.1 and 15.3), we can conclude reasonably that the V_{sh} estimates from the gamma ray log and the ϕ_D/ϕ_N reading are in agreement for Zones 2 and 3.

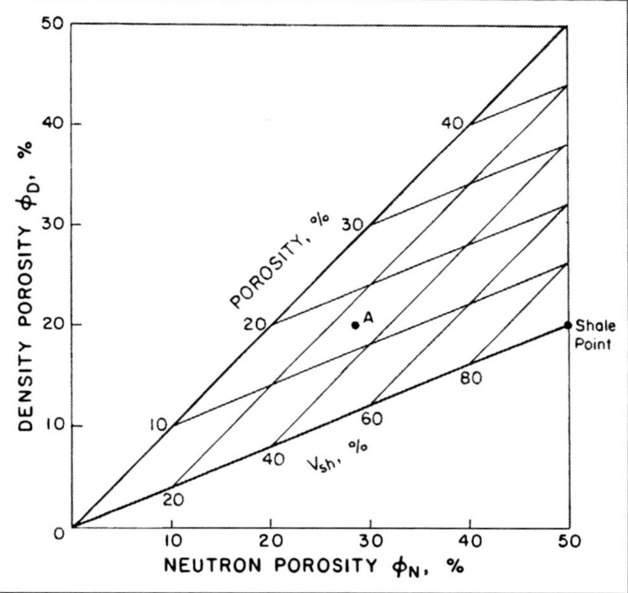

Fig. 15.11—Density/neutron crossplot over the useful range of porosity.

15.6 Density/Neutron Crossplot for Shaly Formation

Effective porosity and shale content by simultaneous solution of Eqs. 15.20 can be determined graphically by a crossplot. The general advantages of graphical solution are discussed in Chap. 13. The specific advantage of solving Eqs. 15.20 graphically is the possi-

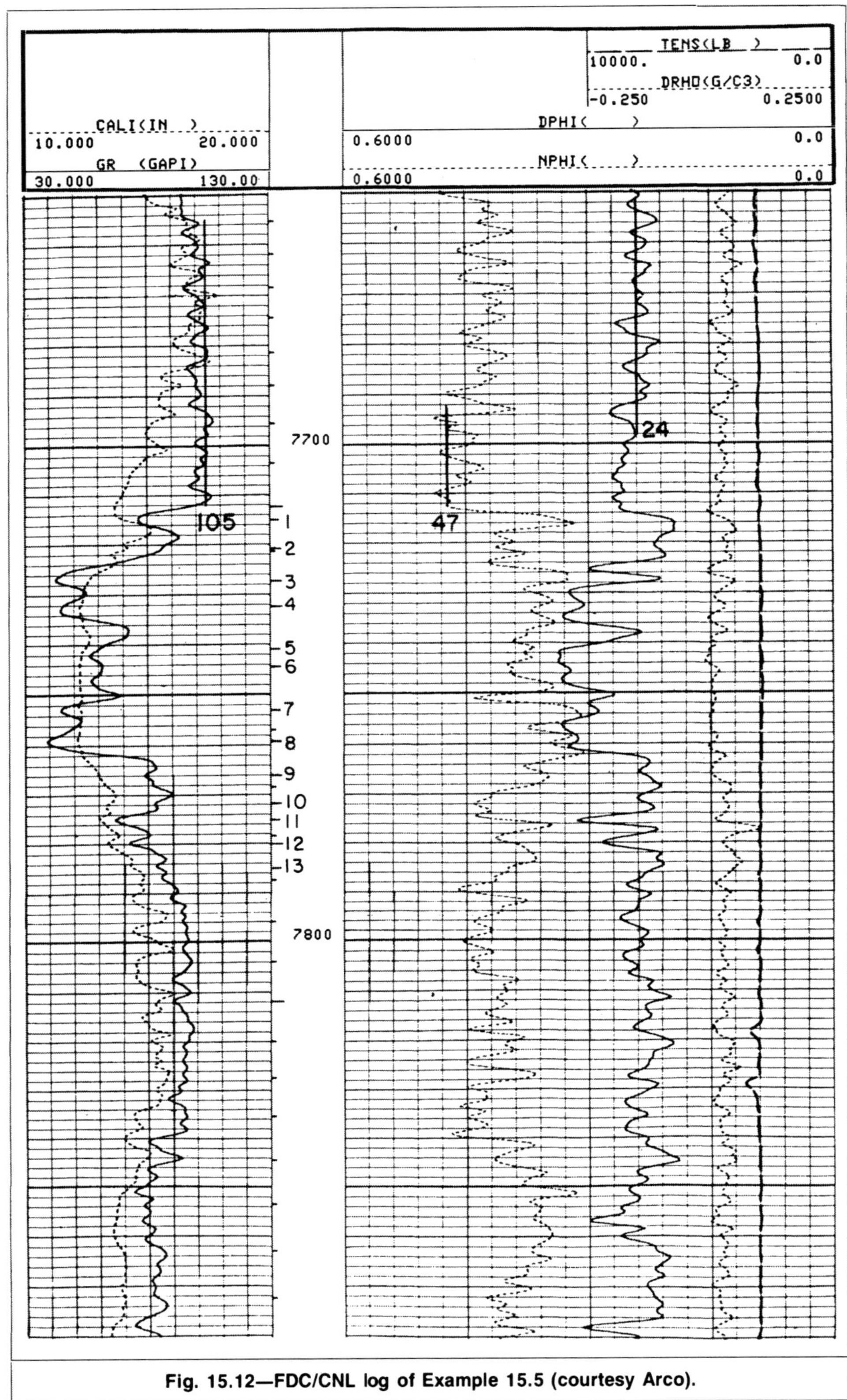

Fig. 15.12—FDC/CNL log of Example 15.5 (courtesy Arco).

bility of detecting abnormal conditions, such as a change in lithology or the presence of gas.

Fig. 15.9 shows a density-porosity/neutron-porosity crossplot for a specific case of shale properties. Three distinct points (Points W, M, and C) are shown. Point W represents the water or fluid point where $\phi_D=\phi_N=100\%$. Point M represents the matrix point. If the density and neutron tools are calibrated properly in terms of the existing matrix type, then $\phi_D=\phi_N=0\%$. Point C represents the shale point. The coordinates of Point C, $[(\phi_N)_{sh}, (\phi_D)_{sh}]$, correspond to the average shale responses in the interval analyzed.

Line $\overline{MW}$ is a 45° line that joins Points M and W. Hence, it represents a clean formation line or $V_{sh}=0$ line. Line $\overline{MW}$ can be scaled in true porosity, as shown in Fig. 15.9. Points representing ϕ_D and ϕ_N values obtained in clean formations will fall on the $\overline{MW}$ line, and their positions on the line indicate the porosity value. A point that represents shaly formations falls within the triangle. Its position, relative to the sides of the triangle, gives the values of ϕ and V_{sh}. A network of isoporosity and isoshale-content lines (**Fig. 15.10**) makes this determination possible. Isoporosity lines are parallel to the line that joins matrix and shale points. Isoshale-content lines are 45° lines. Because porosity values seldom exceed 50%, only the southwest quarter of Fig. 15.10 can be drawn (**Fig. 15.11**) to make full use of the graph paper. Point A represents a shaly formation that has values of $\phi=15\%$ and $V_{sh}=30\%$. Note

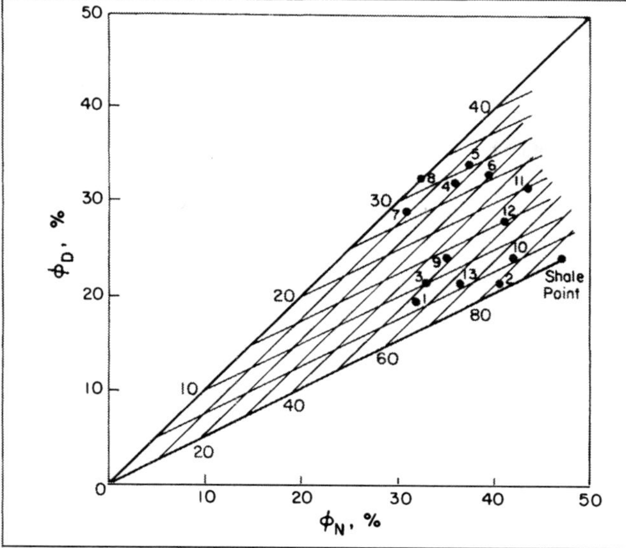

Fig. 15.13—Density/neutron crossplot of Example 15.5.

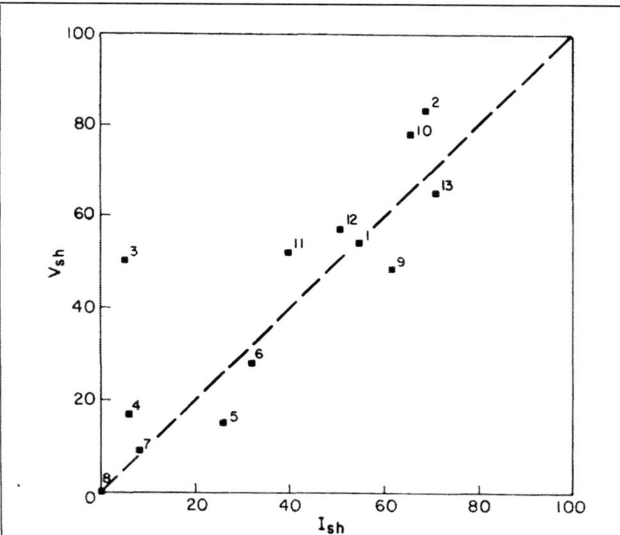

Fig. 15.14—V_{sh} estimated from the density/neutron crossplot vs. I_{sh} calculated from gamma ray data, Example 15.5.

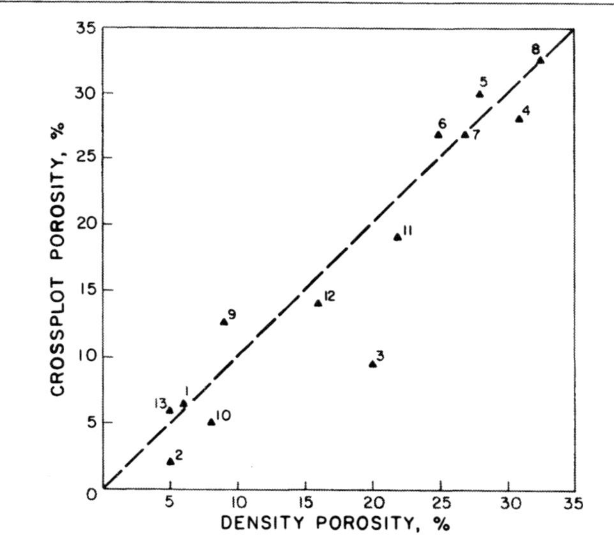

Fig. 15.15—Neutron/density crossplot porosity vs. density porosity, Example 15.5.

that the shale point plotted in Figs. 15.9 through 15.11 correspond to $(\phi_D)_{sh}$=20% and $(\phi_N)_{sh}$=50%. These coordinates vary from well to well and have to be determined for each case.

Example 15.5. Fig. 15.12 shows the FDC/CNL log section obtained in an offshore Texas well. The log shows several shaly sand zones.

a. Choose gamma ray, neutron porosity, and density porosity values for the different sand zones marked on the log.

b. Select the shale parameters, γ_{sh}, $(\phi_N)_{sh}$, and $(\phi_D)_{sh}$.

c. Select the clean formation parameter, γ_c.

d. Calculate I_{sh} for each zone using gamma ray information.

e. Estimate V_{sh} and ϕ for each zone using the density/neutron crossplot.

f. Plot V_{sh} values obtained in Part e vs. I_{sh} values calculated in Part d. Comment on the correlation, or lack thereof, between these two parameters. If a correlation exists, discuss its practical application.

g. Estimate the effective porosity of the different zones using Eq. 15.19 and the density log data.

h. Plot the porosity value estimated in Part e vs. those estimated in Part f. Comment on the correlation, or lack thereof, between these two parameters.

Solution.

a. Gamma ray, ϕ_N, and ϕ_D values for the 13 zones marked in Fig. 15.12 are listed below.

Zone	Gamma Ray (API units)	ϕ_N (%)	ϕ_D (%)
1	76	32	19.5
2	85	40.5	21.5
3	43	33	21.5
4	44	36	32
5	57	37.5	34
6	61	39.5	33
7	45	31	29
8	40	32.5	32.5
9	80	35	24
10	83	42	24
11	66	43.5	31.5
12	73	41	28
13	86	36.5	21.5

b. The representative shale parameters selected are γ_{sh}=105 API units, ϕ_N=47%, and ϕ_D=24%.

c. Zone 8 is indicated to be clean because $\phi_N \approx \phi_D$ and because it displays the lowest gamma ray reading. Therefore, γ_c=40.

d. The I_{sh} values calculated with Eq. 15.2 are tabulated below.

Zone	I_{sh} Eq. 15.2 (%)	V_{sh} Crossplot (%)	ϕ Crossplot (%)
1	55	54	6.5
2	69	83	1.7
3	5	50	9.5
4	6	17	28
5	26	15	30
6	32	28	27
7	8	9	27
8	0	0	32.5
9	62	48	12.5
10	66	78	5
11	40	52	19
12	51	57	14
13	71	65	6

e. **Fig. 15.13**, the density/neutron crossplot, shows the 13 selected zones. The values of V_{sh} and ϕ estimated from this graph appear above.

f. **Fig. 15.14** is a plot of V_{sh} estimated from the density/neutron crossplot vs. I_{sh}. I_{sh} is calculated from gamma ray data. The plot suggests that $V_{sh} \approx I_{sh}$. If more data points are used and a better correlation is obtained, this method can be very useful. It can be used to determine a regional or a field empirical correlation that

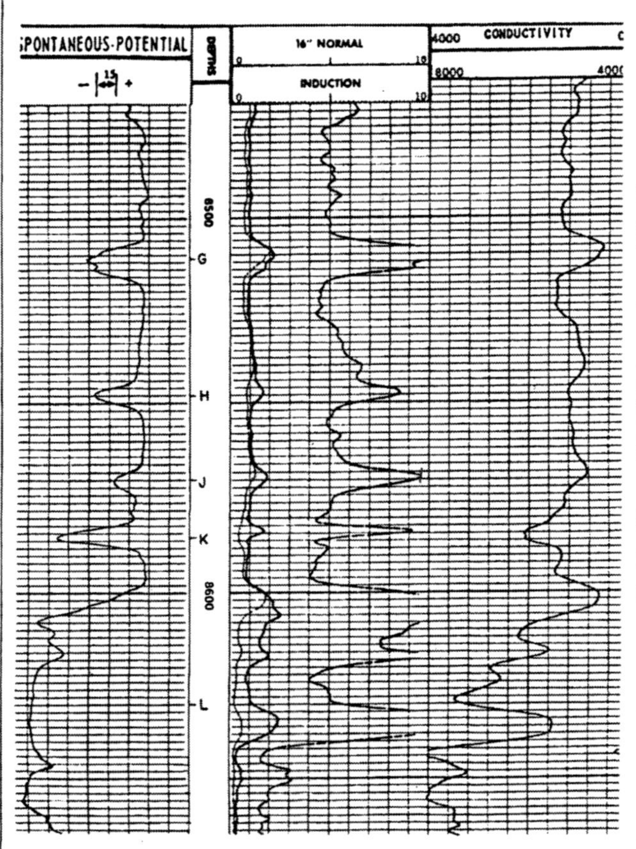

Fig. 15.16—IES log of Example 15.6 (courtesy Amoco Production Co.).

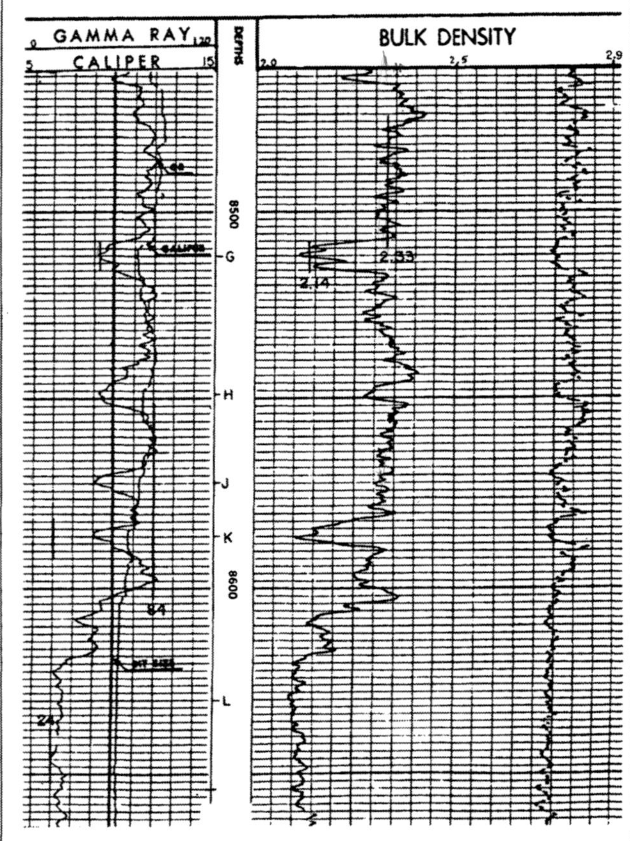

Fig. 15.17—FDC log of Example 15.6 (courtesy Amoco Production Co.).

relates I_{sh} to V_{sh}. Reliance on universal equations, Eqs. 15.3 through 15.6, is then eliminated.

g. Using ϕ_D, $(\phi_D)_{sh}$, and $V_{sh}=I_{sh}$ in Eqs. 15.20 results in the porosity values listed below.

Zone	ϕ (Eqs. 15.20) (%)
1	6
2	5
3	20
4	31
5	28
6	25
7	27
8	32.5
9	9
10	8
11	22
12	16
13	5

h. **Fig. 15.15** plots neutron/density crossplot porosity vs. that determined in Part g. A reasonable correlation exists between the two parameters, suggesting that, in the absence of the neutron log, the density log data alone can provide reasonable porosity values.

15.7 Water Saturation Determination in Shaly Formations

The interpretation problem of water saturation determination in shaly formations still lacks a satisfactory solution.[6] A wide variety of S_w models currently are used routinely to evaluate shaly sands.[1] Each model can provide a significantly different S_w value. None is universally accepted by log analysts.[6] A true solution to the shaly sand problem requires a sound scientific theory yielding an S_w model capable of a universally consistent predictive performance. It also requires that the shale term in that model be expressed in terms of log-derivable parameters. These requirements are not met by any of the S_w equations routinely used today.[6]

The evolution of shaly sand resistivity models is detailed in Chap. 1. In this chapter devoted to practical applications, two models are retained. They are the Fertl and Hammack[7] equation and the CYBERLOOK model.[8]

15.7.1 Fertl and Hammack Equation. For practical application, particularly if reservoir characteristics are similar to those in the U.S. gulf coast area, the Fertl and Hammack equation can be expressed as

$$S_w = (0.81R_w/\phi^2R_t)^{1/2} - (V_{sh}R_w/0.4\phi R_{sh}). \quad \text{(15.21)}$$

This equation is retained because it treats the shale effect as a correction term subtracted from the clean sand term. The correction term, ΔS_w, is expressed as

$$\Delta S_w = V_{sh}R_w/0.4\phi R_{sh}. \quad \text{(15.22)}$$

Eq. 15.21 has the advantage of pointing out the practical aspects of the shale effect. First, treating a shaly sand as clean will underestimate the potential of a hydrocarbon formation because high S_w values will be calculated. Second, using an inflated V_{sh} will produce exactly the opposite effect—i.e., overestimation of the potential of a hydrocarbon formation.

Example 15.6. The IES and FDC logs in **Figs. 15.16 and 15.17** were obtained in an interval of a U.S. gulf coast well drilled with freshwater-based mud. BHT is 165°F at 8,700 ft. Measured R_{mf} is 0.69 Ω·m at 85°F. Estimate the average porosity and water saturation of Sand G.

Solution. The SP and gamma ray curves clearly indicate that Sand G is shaly. A shaly formation interpretation approach is then used.

Determination of V_{sh}. Eq. 15.2 is used to determine I_{sh}: $\gamma_{log}=54$ API units, $\gamma_c=24$ API units, $\gamma_{sh}=84$ API units, and $I_{sh}=(54-24)/(84-24)=0.50$ or 50%.

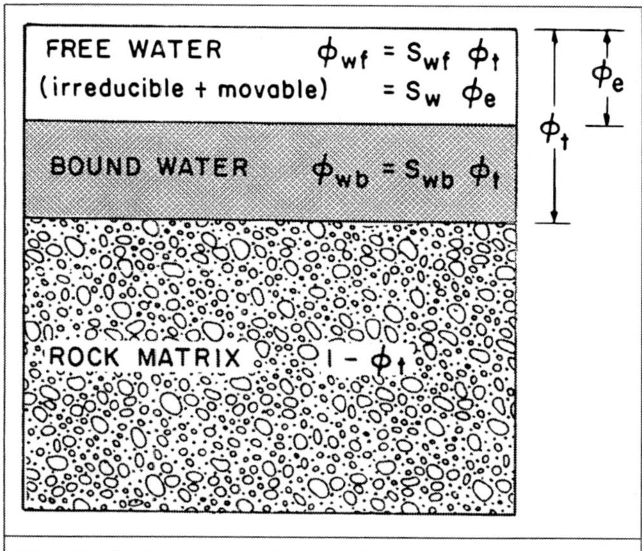

Fig. 15.18—The dual-water model of a water-bearing shaly formation.

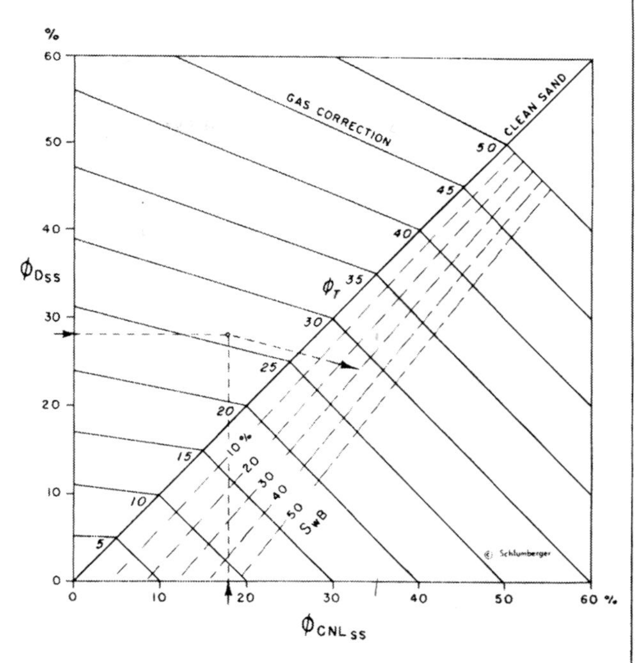

Fig. 15.19—Total porosity determination from neutron and density logs for U.S. gulf coast sandstones (from Ref. 9).

Because gulf coast sediments are of the Tertiary Age, then

$V_{sh}=0.083[2^{3.7(0.50)}-1]=0.216$ or 22%.

Eq. 15.3 is used to estimate V_{sh}. This equation results in a minimum V_{sh}, which in turn gives a minimum ΔS_w. The estimation of formation potential is then conservative.

Determination of ϕ.

$\phi_D=(2.65-2.14)/(2.65-1)=0.309$ or 31%.

$(\phi_D)_{sh}=(2.65-2.33)/(2.65-1)=0.194$ or 19%.

$\phi=0.309-(0.22)(0.194)$

$=0.266$ or 27%.

Determination of R_w. There is an evident correlation between the gamma ray and SP deflections in the different sands. This correlation indicates that the change in SP deflection is caused mainly by shaliness and bed-thickness effect and not by a change in water salinity. This means that R_w remains the same over the interval shown. A best estimate of R_w is then obtained using data that pertain to the cleanest sand, Sand L:

$E_{SSP}=-92$ mV,

$T_f=165°$F,

and $R_{mf}=0.69(85+6.77)/(165+6.77)$

$=0.37\ \Omega\cdot$m at T_f.

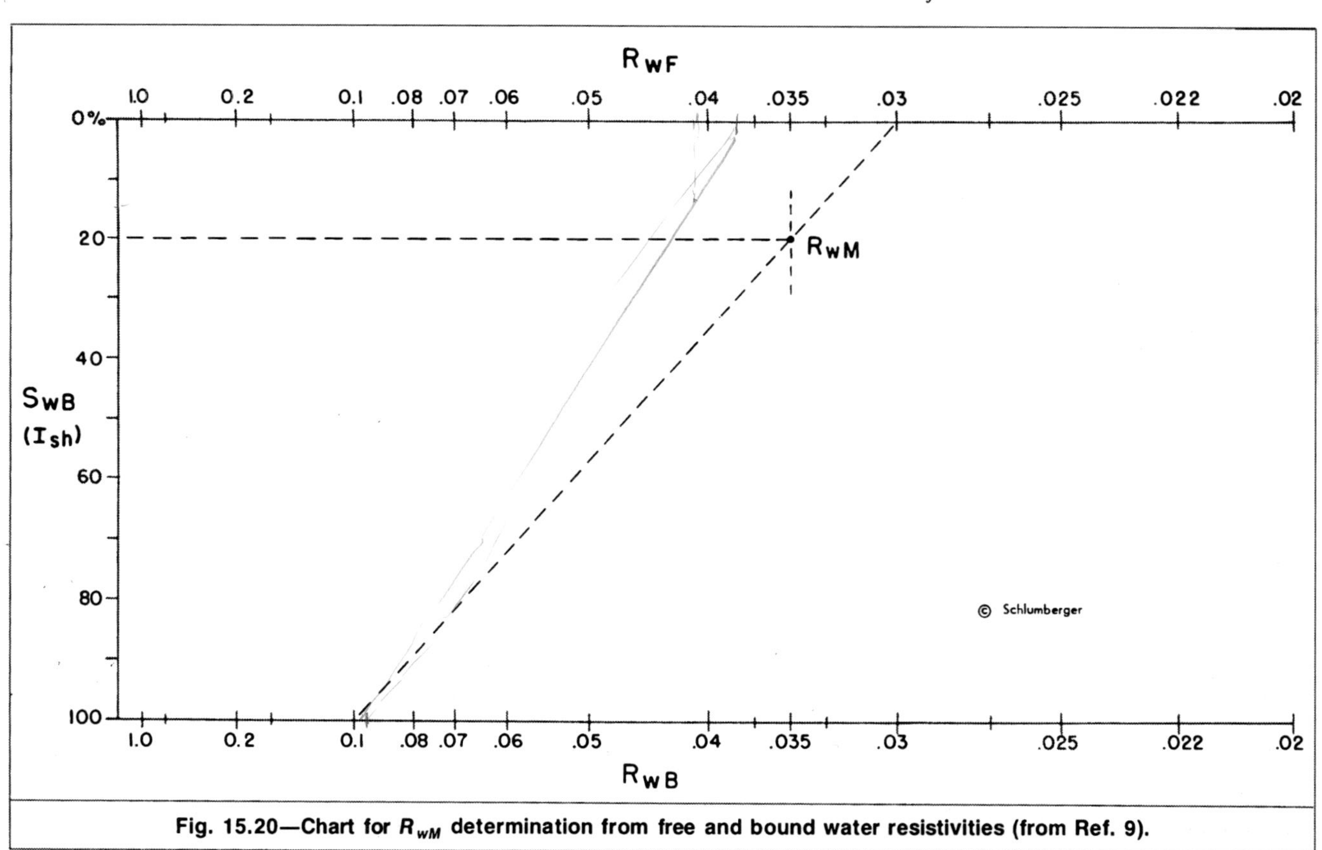

Fig. 15.20—Chart for R_{wM} determination from free and bound water resistivities (from Ref. 9).

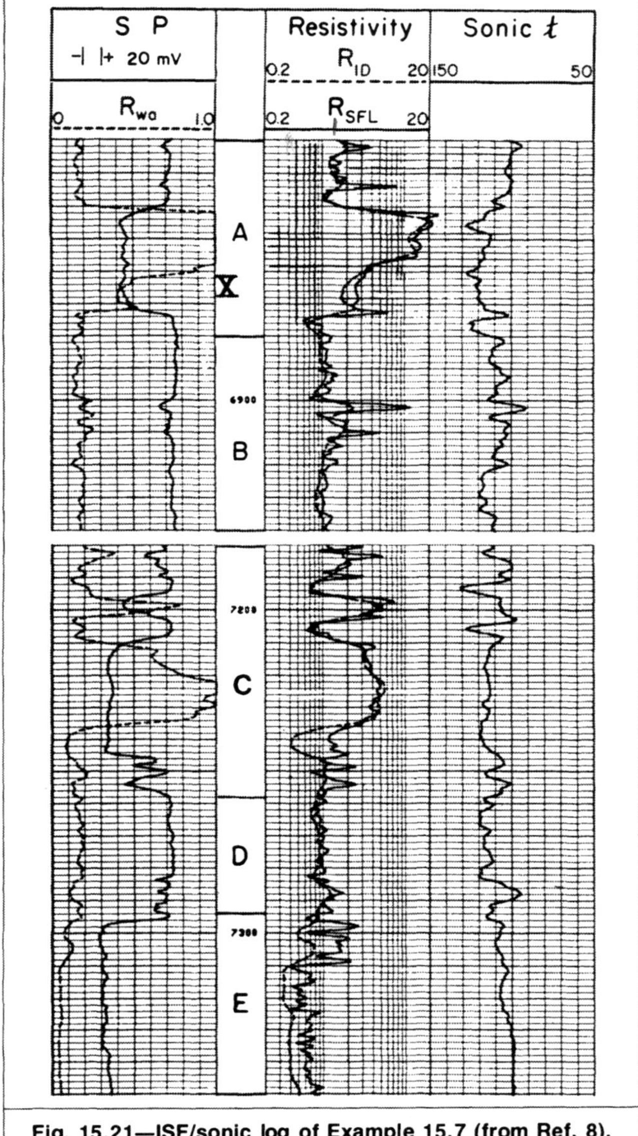

Fig. 15.21—ISF/sonic log of Example 15.7 (from Ref. 8).

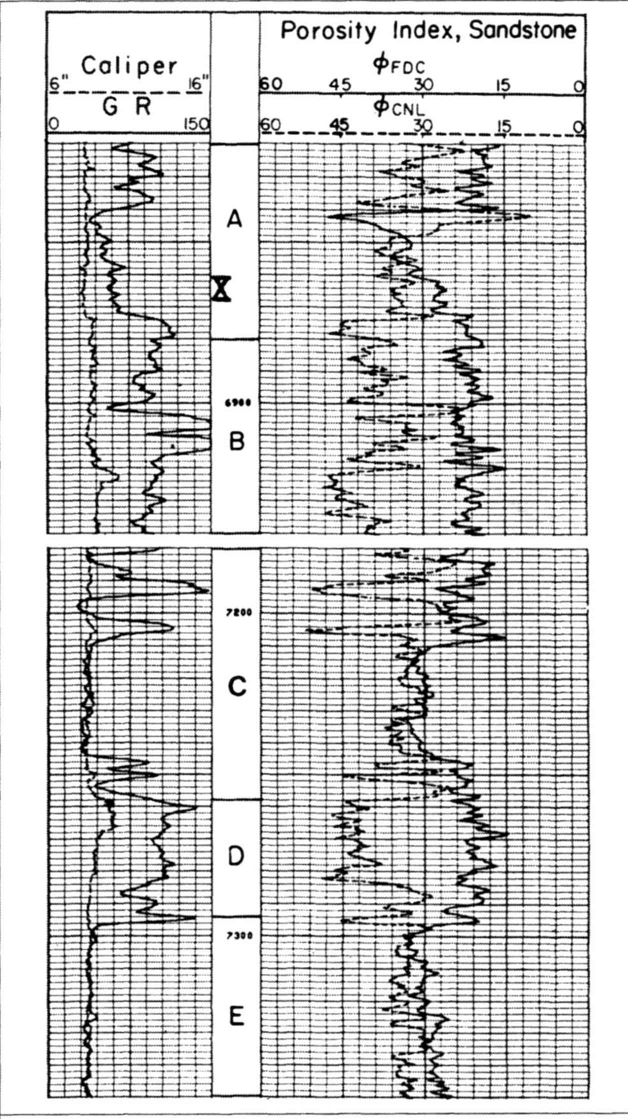

Fig. 15.22—FDC/CNL log of Example 15.7 (from Ref. 8).

From Fig. 6.14, $R_w=0.035\ \Omega\cdot m$ at T_f. A close value also is obtained from the empirical chart of Fig. 6.19.

Determination of S_w.

$R_t=2\ \Omega\cdot m.$

$R_{sh}=1{,}000/1{,}300=0.77\ \Omega\cdot m.$

$$S_w=\left[\frac{0.81(0.035)}{(0.27)^2(2)}\right]^{1/2}-\frac{0.22(0.035)}{0.4(0.27)(0.77)}$$

$$=0.44-0.09$$

$$=0.35 \text{ or } 35\%.$$

15.7.2 CYBERLOOK Water Saturation Model. Inspired by the concept of the dual-water model, Best *et al.*[8] introduced a water saturation model that is the basis of the CYBERLOOK wellsite computer process. The model is based on a water-bearing shaly "wet" resistivity, R_o, which is compared with log-measured resistivity, R_t, to detect the presence of hydrocarbons and to estimate their contents. **Fig. 15.18** shows the model used to define R_o. It is considered to be a shaly formation that behaves as a clean formation containing two types of water: bound water and free water.

Bound water is associated with shales. It occupies a bulk volume fraction, ϕ_{wB}, and has a conductivity C_{wB}. Free water is not bound to shale. It is called free water to distinguish it from the bound water; this terminology does not imply producibility. The free water includes the formation irreducible water. It occupies a fraction of the bulk volume that is equal to ϕ_{wF}, and has a conductivity C_{wF}. The total porosity, ϕ_t, which represents the bulk volume fraction of the formation occupied by free and bound water, is

$$\phi_t=\phi_{wB}+\phi_{wF}. \qquad (15.23)$$

The free and bound water saturation can then be defined as

$$S_{wF}=\phi_{wF}/\phi_t, \qquad (15.24)$$

$$S_{wB}=\phi_{wB}/\phi_t, \qquad (15.25)$$

and $S_{wF}+S_{wB}=1. \qquad (15.26)$

In the CYBERLOOK model, the formation factor is defined as

$$F=1/\phi_t^2. \qquad (15.27)$$

The conductivity C_o of the water-bearing shaly formation is

$$C_o=\phi_t^2 C_{wM}, \qquad (15.28)$$

where C_{wM} is the conductivity of the free and bound water mixture. C_{wM} is a weighted-average conductivity expressed as

$$\phi_t C_{wM}=\phi_{wB}C_{wB}+\phi_{wF}C_{wF}, \qquad (15.29)$$

or considering Eqs. 15.24 through 15.26,

$$C_{wM}=S_{wB}C_{wB}+(1-S_{wB})C_{wF}. \qquad (15.30)$$

Then

$$C_o=\phi_t^2[S_{wB}C_{wB}+(1-S_{wB})C_{wF}], \qquad (15.31)$$

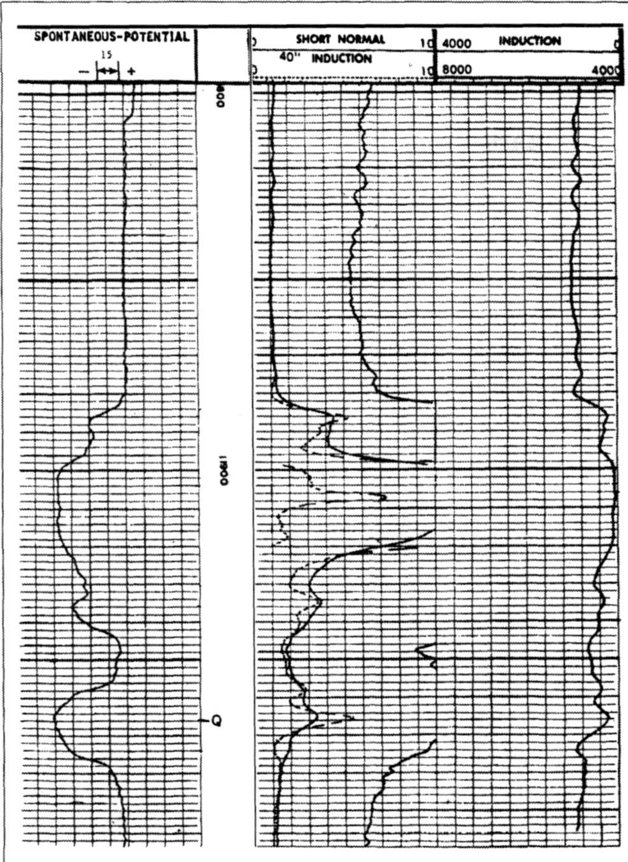

Fig. 15.23—IES log of Problem 15.1 (courtesy Amoco Production Co.).

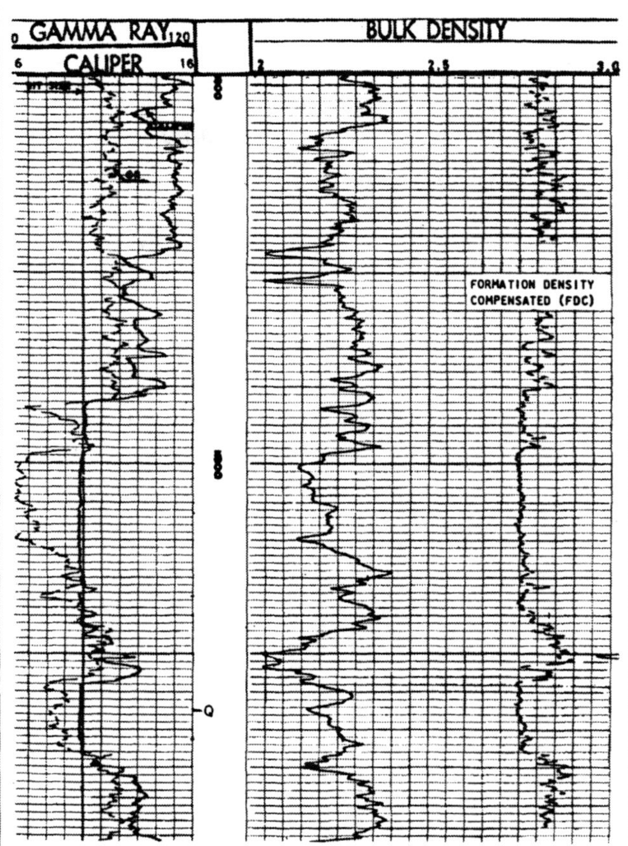

Fig. 15.24—Density log of Problem 15.1 (courtesy Amoco Production Co.).

and in resistivity terms,

$$R_o = R_{wM}/\phi_t^2, \quad (15.32)$$

where R_{wM} is the water mixture resistivity given by

$$R_{wM} = R_{wF}R_{wB}/[S_{wB}R_{wF} + (1-S_{wB})R_{wB}]. \quad (15.33)$$

R_{wF} and R_{wB} are the free and bound water resistivities, respectively.

The CYBERLOOK water saturation is calculated with Archie's model for a clean formation:

$$S_w = (R_o/R_t)^{1/2}. \quad (15.34)$$

The effective porosity, ϕ_e, which represents pore space containing only free water and possible hydrocarbons, is obtained by removing the bulk volume fraction of bound water, ϕ_{wB}, from total porosity, ϕ_t. Therefore,

$$\phi_e = \phi_t - \phi_{wB}, \quad (15.35)$$

and from Eq. 15.25,

$$\phi_e = \phi_t(1-S_{wB}). \quad (15.36)$$

The estimation of ϕ_e and S_w with Eqs. 15.33, 15.32, 15.36, and 15.34 requires knowledge of S_{wB}, ϕ_t, R_{wF}, and R_{wB}.

Determination of S_{wB}. S_{wB} is computed from I_{sh} with an experimental transform called the "S" curve. The shape of the "S" curve is controlled by porosity.[8] These transforms are not published. As a first approximation,

$$S_{wB} = V_{sh}. \quad (15.37)$$

Determination of ϕ_t. It has been found that porosity from the neutron/density crossplots of Chap. 14 provides a good value of ϕ_t. The total porosity also can be set as a first approximation, equal to the average of density and neutron porosities:

$$\phi_t = (\phi_N + \phi_D)/2. \quad (15.38)$$

If light hydrocarbons present in the formation are expected to affect the porosity logs, **Fig. 15.19** is used to estimate ϕ_t. This empirical chart was constructed for U.S. gulf coast sands and incorporates a correction factor determined by the expected grain density adjusted for shaliness.[9] To determine ϕ_t, the chart is entered with porosity from the neutron and density logs in sandstone units. Water or oil sands will intercept on or below the clean sand line, depending on their shaliness. Total porosity is determined by interpolating between the isoporosity lines that originate from the porosity values on the clean sand line. Clean gas-bearing sands will intercept above the clean sand line, and total porosity is determined as with water or oil sands. Shaly gas-bearing formations may plot above, on, or below the clean sand line. Total porosity is determined by moving parallel from the intercept to the gas correction lines and then to the intersection of S_{wB}. Total porosity is then read from the isoporosity lines previously discussed.

Determination of R_{wF}, R_{wB}, and R_{wM}. R_{wF} and R_{wB} are estimated with the apparent water resistivity approach, R_{wa}, in a clean water-bearing formation and in a 100% shale, respectively. Therefore,

$$R_{wF} = R_o\phi^2 \quad (15.39)$$

and $$R_{wB} = R_{sh}(\phi_t)_{sh}^2. \quad (15.40)$$

The total shale porosity $(\phi_t)_{sh}$ is calculated from Eq. 15.38:

$$(\phi_t)_{sh} = [(\phi_N)_{sh} + (\phi_D)_{sh}]/2. \quad (15.41)$$

R_{wF} and R_{wB} can be taken from the R_{wa} curve if available. R_{wM} is then calculated from Eq. 15.33 or **Fig. 15.20.** To use the chart, the R_{wF} and R_{wB} values are marked on the top and bottom scales. A line is then drawn to join the two water resistivity values. S_{wB} is entered onto the left side of the chart, and a horizontal line is drawn. R_{wM} is read at the intersection of the two constructed lines, as shown in Fig. 15.20.

Example 15.7. Figs. 15.21 and 15.22 show sections of an ISF/sonic log and a FDC/CNL log from a well in south Texas. This interval contains a series of Frio sands and shales. Determine the porosity and water saturation of the zone situated from 6,864 to 6,870 ft.

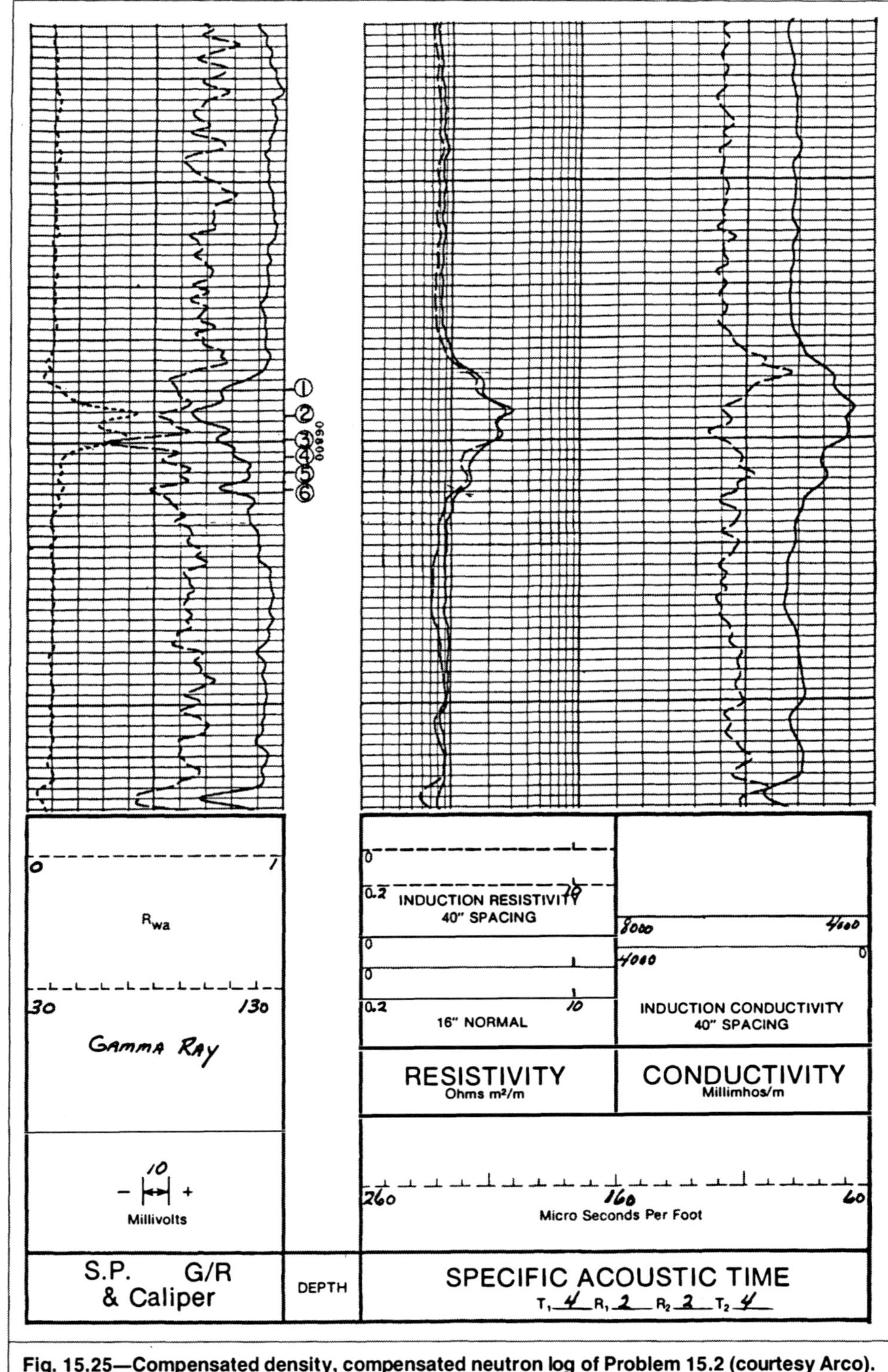

Fig. 15.25—Compensated density, compensated neutron log of Problem 15.2 (courtesy Arco).

Solution. A "quick-look" examination of the logs indicates that the section can be divided into five intervals.

1. Interval A is a hydrocarbon-bearing sand that has a gas cap, as indicated by the separation of the neutron and density porosity curves. The hydrocarbon presence is indicated by the R_{wa} curve. The 6,864- to 6,870-ft zone of interest is moderately shaly. As discussed in Chap. 16, the shale content present in this zone is not enough to offset and exceed the gas effect if present. The zone, then, is oil-bearing.
2. Intervals B and D are mostly shale with some thin sands. Shale parameters can be obtained easily from these two intervals.
3. Interval C contains clean oil-bearing sands.
4. Interval E contains a reasonably clean water-bearing sand.

Determination of S_{wB}. S_{wB} is taken to be equal to V_{sh}. $\gamma_{log}=60$ API units; $\gamma_c=30$ API units (from Interval C); and $\gamma_{sh}=105$ API units (from Interval D). From Eq. 15.2,

$$I_{sh}=(60-30)/(105-30)=0.40 \text{ or } 40\%,$$

and from Eq. 15.3,

$$V_{sh}=0.083\ [2^{3.7(0.4)}-1]$$

$$=0.149 \text{ or } 15\%.$$

Let $S_{wB}=V_{sh}=15\%$.

Determination of ϕ_t. In the zone of interest, $\phi_N=35\%$ and $\phi_D=27\%$. According to Eq. 15.38, ϕ_t can be approximated as

$$\phi_t=(35+27)/2=31\%.$$

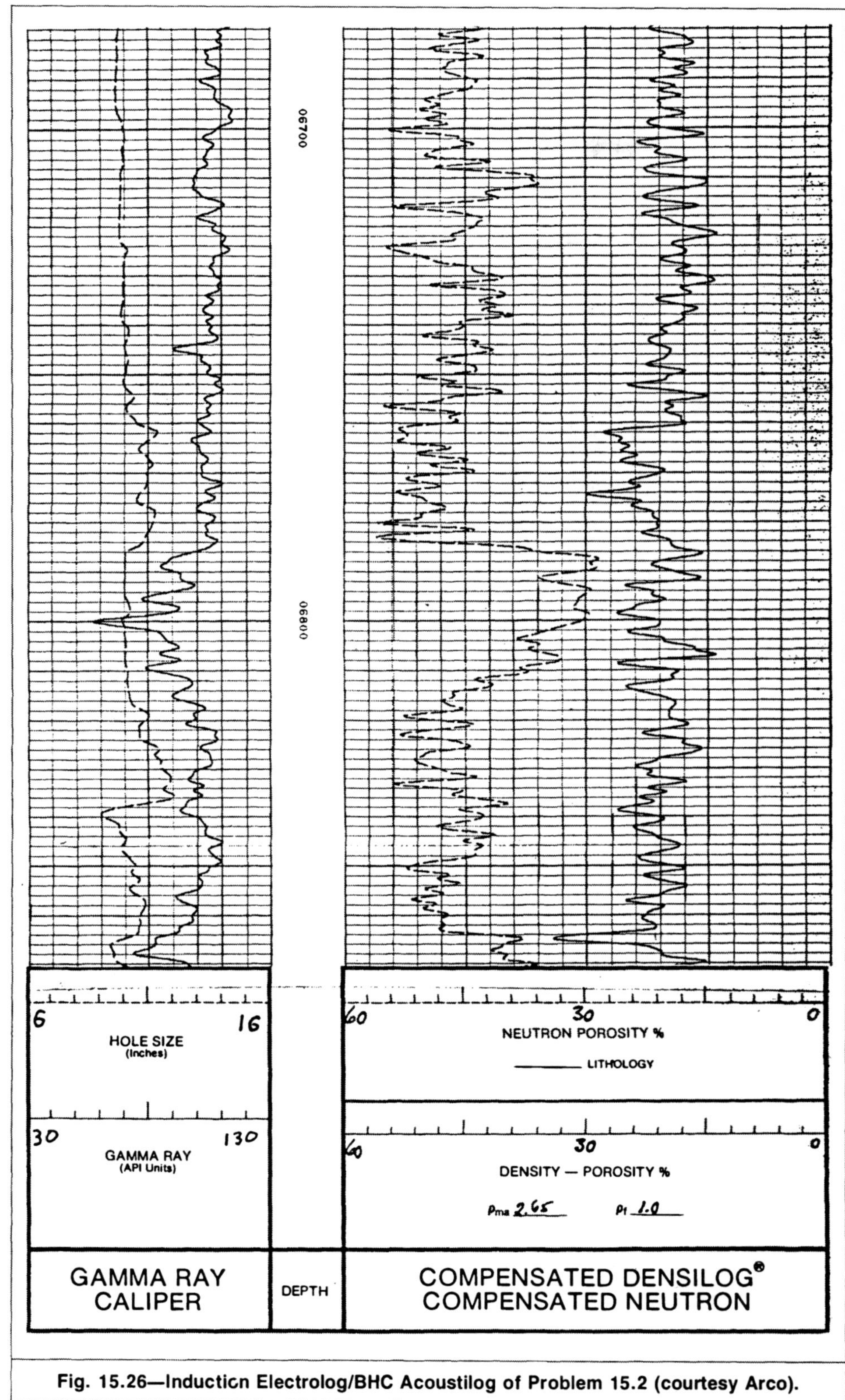

Fig. 15.26—Induction Electrolog/BHC Acoustilog of Problem 15.2 (courtesy Arco).

Fig. 15.19 yields the same value.

Determination of R_{wM}. R_{wF} and R_{wB} can be calculated with Eqs. 15.39 and 15.40. From the average readings in clean sands of Interval E, $\phi=0.31$, $R_o=0.35\ \Omega\cdot m$, and $R_{wF}=0.35(0.31)^2=0.04\ \Omega\cdot m$. From the average readings in the shale Interval D, $R_{sh}=0.9\ \Omega\cdot m$, $(\phi_t)_{sh}=(42+22)/2=32\%$, and $R_{wB}=0.9(0.32)^2=0.092\ \Omega\cdot m$. Then, from Eq. 15.33,

$$R_{wM}=0.034(0.092)/[0.15(0.034)+(1-0.15)(0.092)]$$

$$=0.038\ \Omega\cdot m.$$

Determination of ϕ_e. From Eq. 13.36,

$$\phi_e=0.31\ (1-0.15)$$

$$=0.26 \text{ or } 26\%.$$

Determination of S_w. Using Eqs. 15.32 and 15.34 gives

$$R_o=0.038/(0.31)^2=0.4\ \Omega\cdot m,$$

$$R_t=1.8,$$

and $S_w=(0.4/1.8)^{1/2}$

$$=0.47 \text{ or } 47\%.$$

Review Questions

1. How does the presence of shale affect resistivity logs in high-porosity formations that contain high-salinity formation water?
2. How does the presence of shale and hydrocarbons affect the SP and gamma ray log responses?
3. Why is the SP log not recommended as a shale indicator in all cases?
4. Can a crossplot of SP response vs. gamma ray values be used to detect the presence of hydrocarbons in shaly sands? Explain.
5. In what environment is the use of the spectral gamma ray log preferred over the total gamma ray log?
6. If only one porosity log is used to estimate the porosity of shaly formations, the density log should be the one. Do you agree or disagree with this statement? Explain.
7. Why is the sonic log usually not used with another porosity tool to set up a system of two equations to be solved simultaneously for porosity and shale content of shaly formations?
8. What are the advantages of the density/neutron crossplot over the analytical solution of Eqs. 15.20?
9. Why is the interpretation problem of S_w determination in shaly formations still without a satisfactory solution?
10. Why is it the Fertl and Hammack V_{sh} model retained for S_w estimation?
11. What is the basis of the CYBERLOOK water saturation model? Discuss.

Problems

15.1 Examine the IES and density logs of **Figs. 15.23 and 15.24,** and then answer the following questions. These logs were obtained in a U.S. gulf coast well.
   a. Using the gamma ray curve, estimate the shale content of Zone Q.
   b. Explain why in this case the SP curve cannot be used to estimate shale content.
   c. Using density log data, estimate the effective porosity of Zone Q.
   d. If R_w=0.07 Ω·m at formation temperature, use the Fertl and Hammack equation to estimate S_w of Zone Q.

15.2 **Figs. 15.25 and 15.26** show a section of a compensated density and an Induction Electrolog BHC Acoustilog recorded in an offshore Louisiana well. The section includes a shaly sand divided into six zones. A clean, nearby sand displays the properties γ=40 API units and R_{wa}=0.03 Ω·m.
   a. Using the gamma ray curve, estimate the V_{sh} value for the six zones in Fig. 15.26.
   b. Estimate V_{sh} and ϕ for each of the six zones using the density/neutron crossplot.
   c. Plot V_{sh} for Part a vs. V_{sh} estimated in Part b. Comment on the correlation, or lack thereof, between these two parameters.
   d. Using the Fertl and Hammack equation, estimate S_w for the six zones.
   e. Using the CYBERLOOK model, estimate S_w for the six zones. Compare the saturation values to those obtained in Part d.

Nomenclature

C = conductivity, m℧/m
C_K = potassium concentration, wt%
C_{Th} = thorium concentration, ppm
E_{PSP} = SP log response in the shaly zone of interest, mV
E_{SP} = measured self-potential, mV
E_{SSP} = theoretical self-potential, mV
F = formation resistivity factor
I_{sh} = shale index, fraction
R = resistivity, Ω·m
R_o = resistivity of a water-bearing formation, Ω·m
S = saturation
ΔS_w = water saturation correction term, fraction
T_f = formation temperature, °F
Δt = interval transit time, μsec/ft
V_{sh} = shale content, fraction
γ = gamma ray radioactivity, API units
γ_c = average gamma ray response in the cleanest formations, API units
γ_{log} = gamma ray log response, API units
γ_{sh} = average gamma ray response in shales, API units
ϕ = porosity, fraction
ρ = density, g/cm³

Subscripts

a = apparent
b = bulk
c = clean
D = density tool
e = effective
f = fluid
K = potassium
KTh = potassium and thorium
log = log response
m = mud
ma = matrix
mf = mud filtrate
N = neutron tool
sh = shale
S = sonic tool
t = total
Th = thorium
w = water
wa = apparent water
wB = bound water
wF = free water
wM = free and bound water mixture

References

1. *Well Logging and Interpretation Techniques. The Course for Home Study,* Dresser Atlas, Dresser Industries Inc., Houston (1982).
2. Larionov, V.V.: "Borehole Radiometry," *Nedra,* Moskwa (1969).
3. Stieber, S.J.: "Pulsed Neutron Capture Log Evaluation in the Louisiana Gulf Coast," paper SPE 2961 presented at the 1970 SPE Annual Meeting, Houston, Oct. 4-7.
4. Clavier, C., Hoyle, W.R., and Meunier, D.: "Quantitative Interpretation of Thermal Neutron Decay Time Logs: Part I—Fundamentals and Techniques," *JPT* (June 1971) 743-55; *Trans.,* AIME, **251**.
5. Fertl, W.H.: "Clay Analysis in Shaly Sands, Part VI: Gamma Ray Spectral Logging a New Evaluation Frontier," *World Oil* (Oct. 1983).
6. Worthington, P.F.: "The Evolution of Shaly-Sand Concepts in Reservoir Evaluation," *Log Analyst* (Jan.-Feb. 1985).
7. Fertl, W.H. and Hammack, G.W.: "A Comparative Look at Water Saturation Computations in Shaly Pay Sands," *Proc.,* 12th SPWLA Annual Logging Symposium (May 1971) paper R.
8. Best, D.L., Gardner, J.S., and Dumanoir, J.L.: "A Computer-Processed Wellsite Log Computation," *Proc.,* 19th SPWLA Annual Logging Symposium (June 1978) paper Z.
9. "A Guide to Wellsite Interpretation for the Gulf Coast," Schlumberger, Houston.

Chapter 16
Evaluation of Gas-Bearing Formations

16.1 Introduction

Previous chapters emphasized identification and evaluation of oil-bearing formations. This chapter treats gas-bearing formations. Generally speaking, gas-bearing formations can be identified easily unless they contain a substantial amount of shale. Quantitative log interpretation of gas formations, however, requires special consideration of the effects of gas on porosity logs' responses.

Porosity logs (density, neutron, and sonic) have relatively shallow radii of investigation. Gas saturation near the wellbore in all types of formations causes an increase in density log porosity and a decrease in neutron log porosity. The presence of gas results in an appreciable increase in sonic log porosity only in poorly compacted sands, which is the case of most shallow sands and some abnormally pressured formations. Using density or sonic log porosity values uncorrected for gas effect usually results in a low apparent value of the formation factor, F. This, in turn, results in a low apparent R_o value. Consequently, clean gas formations will be identified easily as hydrocarbon zones by use of an interpretation technique, such as the apparent water resistivity, R_{wa}, and porosity resistivity crossplots. The hydrocarbon saturation estimated under these conditions usually will be exaggerated. Use of uncorrected neutron log porosity produces the opposite effect. A low apparent porosity results in a high F, which, in turn, results in a high estimated R_o value. Subsequently, a gas zone can pass undetected.

Three factors determine the response of porosity logs in gas-bearing formations: the actual porosity, the gas saturation, and the degree of shaliness. Because of both gas and shale effects, more than one porosity log is required to evaluate gas formations. Two porosity logs are generally sufficient provided that the shale content, V_{sh}, can be reasonably estimated from the gamma ray log. The density/neutron combination usually is recommended for the evaluation. The sonic log usually is not recommended because, in addition to gas and shale effects, compaction and secondary porosity effects can be present. These additional unknowns could encumber gas identification.

16.2 Effect of Gas on the Neutron Log Response

As discussed in Chap. 9, the neutron tool responds to hydrogen content in the formation probed. When gas is present within the zone investigated by the tool, the counting rate at the detector increases. Because neutron tools are calibrated to read the true porosity in a clean lithology saturated with fresh water, the log reading in a gas formation is an apparent porosity that is much lower than the true one. This can be expressed as

$$(\phi_N)_g < \phi, \quad \text{(16.1)}$$

where $(\phi_N)_g$ is the neutron porosity reading in a clean gas-bearing formation and ϕ is the formation's true porosity.

This allows the neutron log to be used in the detection of gas zones provided that true porosities of these zones are known or can be estimated.

In a formation of uniform lithology and porosity, the neutron log alone can indicate the gas/liquid (gas/oil or gas/water) contacts, as shown in **Fig. 16.1**. This identification is possible because of the apparent porosity contrast between gas- and liquid-filled zones.

As discussed in Sec. 15.3 and indicated by Eq. 15.17, the neutron log displays a relatively high porosity in shaly formations. The shale effect is opposite that of the gas effect. The presence of shale in a gas-bearing formation makes gas detection more difficult. When shale and gas effects completely offset each other, a shaly gas-bearing formation may, on the neutron log, look just like a clean liquid-filled formation. For example, the gas effect of Zone C in Fig. 16.1 is almost entirely masked by the shale effect.

16.3 Effect of Gas on the Density Log Response

As detailed in Chap. 8, the response of the density tool is proportional to the number of electrons per unit volume, which, in turn, is related to the density of the formation constituents—i.e., lithology and fluids. The presence of gas in the zone investigated by the density tool will decrease the recorded bulk density. If the density tool is calibrated to read true porosity in a clean lithology saturated with fresh water, the log reading in a gas formation is an apparent porosity that is much higher than the true formation porosity. This can be expressed as

$$(\phi_D)_g > \phi, \quad \text{(16.2)}$$

where $(\phi_D)_g$ is the density porosity reading in a clean gas-bearing formation, and ϕ is the formation's true porosity.

This allows the density log to be used to detect gas zones provided that the true porosities of the zones are known or can be estimated. In a formation of uniform lithology and porosity, the density log alone can indicate the gas/liquid contact (**Fig. 16.2**). This identification is possible because of the bulk density contrast between the gas- and liquid-filled zones.

Caution should be exercised when the density log is used in irregular boreholes where sections of the enlarged borehole diameter are shorter than the tool itself. In this case, pockets of drilling fluids will separate the tool from the formation as the tool rests on the

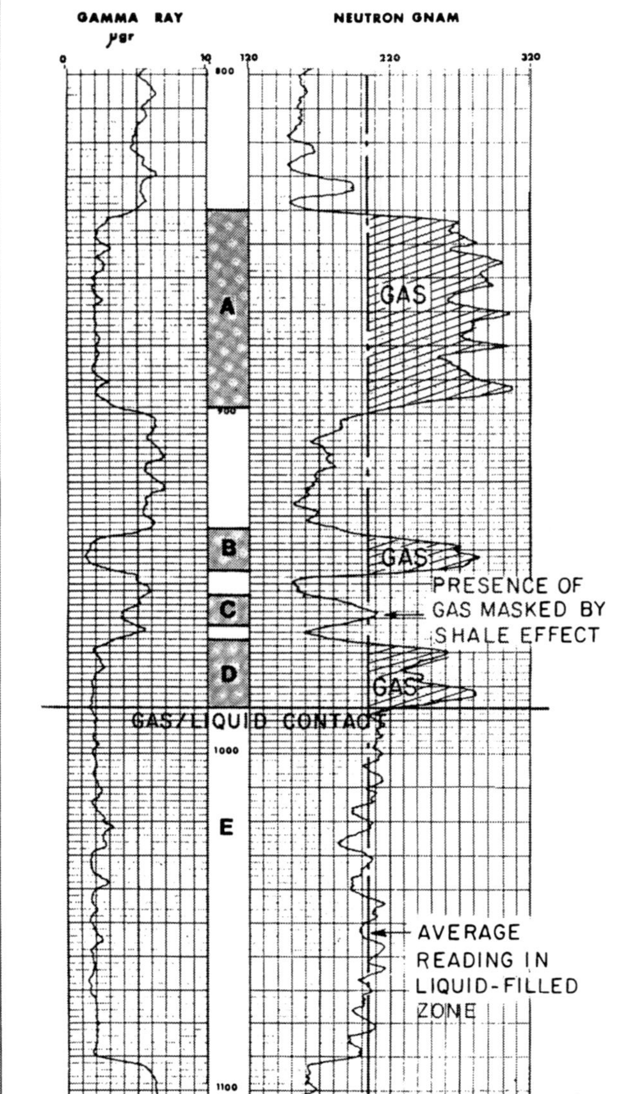

Fig. 16.1—Gas-zone identification on a neutron log in a formation of uniform lithology and porosity.

shoulders of the thin enlarged zone. The tool reading will be highly affected by the relatively low-density drilling fluid. Such a borehole effect is not entirely removed by the compensation technique incorporated into the log reading. A low bulk density resulting from borehole irregularity could be misinterpreted as gas effect. **Fig. 16.3** shows a density log recorded in an irregular borehole. The irregularity can be seen easily on the caliper log. The resemblance of the density and caliper log profiles (especially in Zones A through C) indicates that the low bulk density reading is caused by borehole irregularity. This low bulk density reading should not be confused for gas effect. Note, however, that the true bulk density of the formation is still undetermined. The porosity and fluid type cannot be deduced from the density log.

As shown in Chap. 15, shale presence in a formation affects the reading of the density log. The shale effect depends on the magnitude of the shale bulk density relative to the formation's clean and liquid-filled bulk density. As in Zones A and E in Fig. 16.2, the presence of shale tends to decrease the gas effect, making gas detection in a shaly formation more challenging.

16.4 Effect of Gas on Sonic Log Response

Limited data are available on the effect of gas on acoustic velocity and attenuation. Laboratory tests show that the presence of a compressible fluid in the pore space of unconsolidated sands greatly increases acoustic travel time and reduces amplitude.[1,2] Where the sand has some degree of consolidation, the increase in travel time caused by the gas is lessened or completely erased. With high matrix compressibility, it is probable that the gas effect is magnified because the fluid would then greatly affect the overall bulk compressibility.[3]

The presence of a transit travel time anomaly would indicate that there are gas-bearing formations. The absence of such an anomaly, however, is inconclusive. The gas effect on transit travel time in uncompacted formations is not quantitatively predictable. **Fig. 16.4** shows an example of the gas effect. The presence of a hydrocarbon zone on top of a water zone is indicated by the R_{wa} curve. The sonic log displays a uniform interval transit time response, indicating uniform porosity. An anomaly owing to the presence of gas is clearly seen within the hydrocarbon zone. For the anomaly to exist, the sand should be uncompacted. The uncompaction is empirically confirmed by a reading of 120-μsec/ft travel time in an adjacent shale formation. Using Eq. 10.3, we can estimate the compaction factor, B_{cp}, of this sand to be 1.2.

The dependence of the gas anomaly on the degree of formation compaction makes the sonic log unsuitable for gas detection in most cases. As with neutron and density logs, the presence of shale in the formation further complicates the problem.

16.5 Visual Gas Detection With Porosity Overlays

A porosity overlay is a presentation of two porosity logs recorded on the same porosity scale. For example, both logs are calibrated in terms of limestone matrix and freshwater formation fluid. In liquid-filled formations, separations between the two curves are interpreted in terms of lithology (see Fig. 14.5). In uniform lithology, when the logs are calibrated in terms of the same lithology, the separations between the two curves are interpreted in terms of fluid types (liquid or gas).

Fig. 16.5 shows an overlay of neutron and density porosity curves (FDC/CNL combination). The two logs are recorded in terms of sandstone porosity units. The separation between the two curves at the top of the sand shown by the log, where $\phi_D \gg \phi_N$, is due to the presence of gas. As mentioned, apparent porosities from neutron and density logs are affected oppositely by gas. Because neutron and density curves will also separate as a result of lithology, and because shaliness may also affect the degree of separation, other logs should be used to confirm the presence of gas.

The sand shown in Fig. 16.5 averages a porosity of 29%. In the gas zone, the neutron log displays an apparent porosity lower than the true porosity and the density log displays an apparent porosity higher than the true porosity. This is in agreement with Eqs. 16.1 and 16.2.

The degree of separation between the two curves depends on the saturation and density of the gas. The shape of the separation depends on the saturation profile, which, in turn, depends on the drilling-fluid invasion profile. The reason for this dependency is that the neutron log, especially the dual-spacing neutron log (CNL), investigates more deeply in the formation than the density log does. For both deep invasion and extremely shallow invasion where the two tools are investigating essentially the same saturation profile, one curve will mirror-image the other, as in Fig. 16.5. In shallow to moderate invasion, the two logs will not mirror-image each other because they are investigating different saturation profiles (see **Fig. 16.6**).

Visual gas detection also is possible with the sonic/neutron overlay. When the sonic log is used, however, complications resulting from compaction and/or secondary porosity effects must be considered.

16.6 Porosity Determination in Gas-Bearing Formations

When a porosity log response is converted to porosity, a lithology type is assumed. Also assumed is the fluid type, which is usually taken to be fresh water. Hence, porosity logs indicate an apparent porosity value in gas zones. Estimation of the actual formation porosity in gas zones requires expressions of tool response in such zones.

Assuming that the density tool investigates the invaded zone—i.e., the invasion extends beyond the zone investigated by the tool—the log response can be expressed as

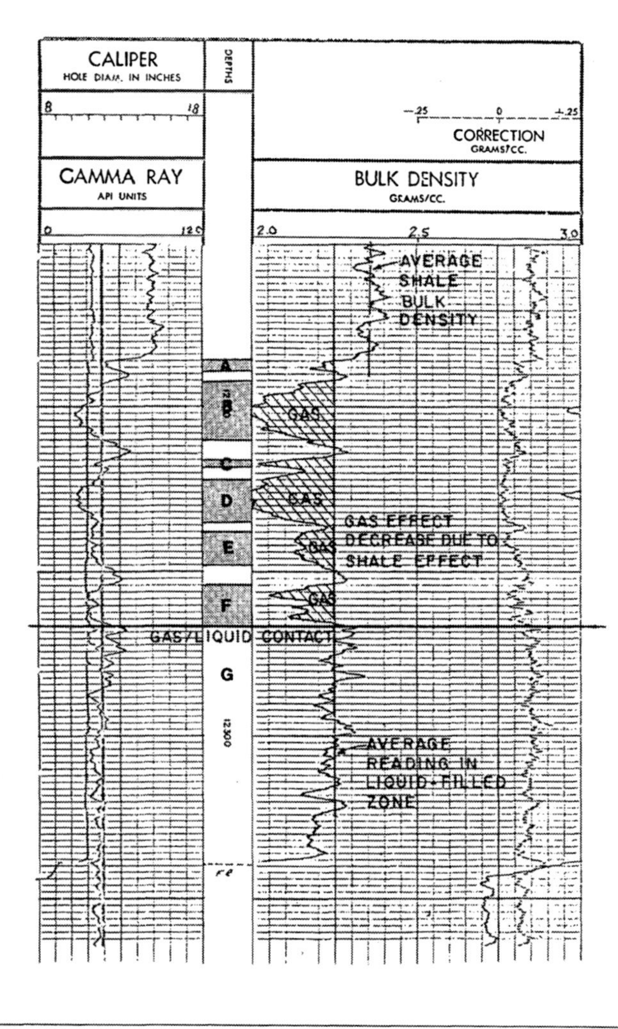

Fig. 16.2—Gas-zone identification on a density log in a formation of uniform lithology and porosity.

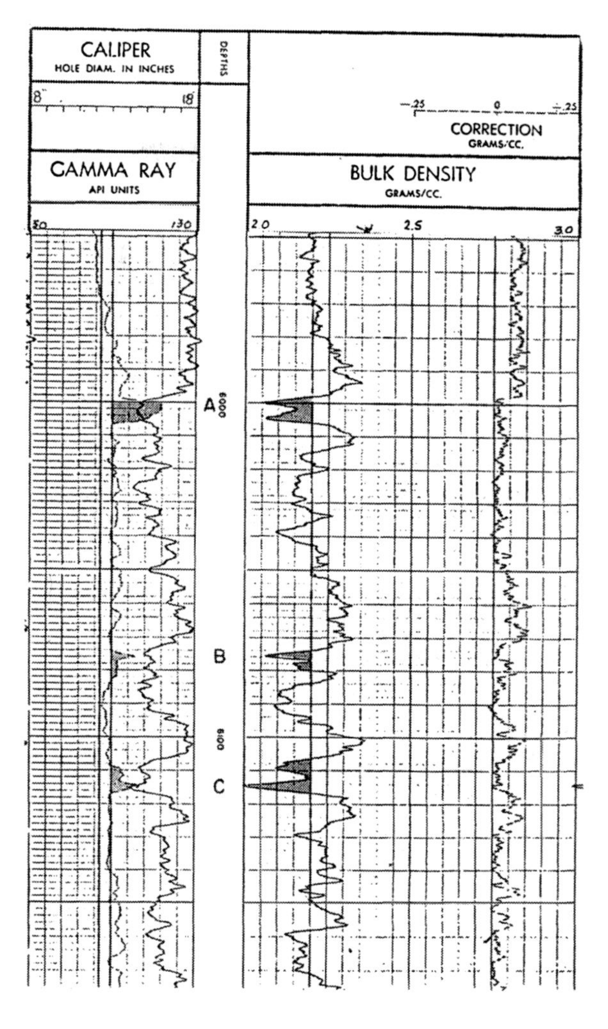

Fig. 16.3—Density log recorded in an irregular section of a borehole.

$$\rho_b=(1-\phi)\rho_{ma}+\phi[S_{xo}\rho_{mf}+(1-S_{xo})\rho_g], \quad (16.3)$$

where ρ_b=log reading, g/cm³; ϕ=true porosity, fraction; ρ_{ma}= matrix density, g/cm³; S_{xo}=mud filtrate saturation in the invaded zone, fraction; ρ_{mf}=mud filtrate density, g/cm³; and ρ_g= apparent gas density as seen by the density log, g/cm³.

Eq. 16.3 also can be written as

$$\phi_D=\phi[S_{xo}+(\phi_D)_g(1-S_{xo})], \quad (16.4)$$

where ϕ_D is the density porosity defined as

$$\phi_D=(\rho_{ma}-\rho_b)/(\rho_{ma}-\rho_f) \quad (16.5)$$

and $(\phi_D)_g$ is the apparent gas porosity as seen by the density log. $(\phi_D)_g$ is defined as

$$(\phi_D)_g=(\rho_{ma}-\rho_g)/(\rho_{ma}-\rho_{mf}). \quad (16.6)$$

Gas density depends on pressure, temperature, and gas composition. **Fig. 16.7** shows the gas gravity as a function of pressure and temperature for a gas mixture that is slightly heavier than methane with an equivalent composition of $C_{1.1}H_{4.2}$.

ρ_{mf} can be approximated by[2]

$$\rho_{mf} = 1 + 0.73n, \quad (16.7)$$

where n is the fractional salinity and ρ_{mf} is in g/cm³.

If the invasion is assumed to extend beyond the zone investigated by the tool, the neutron log response can be expressed as

$$\phi_N=(1-\phi)H_{ma}+\phi[S_{xo}H_{mf}+(1-S_{xo})H_g], \quad (16.8)$$

where ϕ_N is the reading of the log (fraction) and H_{ma}, H_{mf}, and H_g are the hydrogen indices of the matrix, mud filtrate, and gas, respectively.

H_{mf} is expressed as[4]

$$H_{mf}=(1-n)\rho_{mf}, \quad (16.9)$$

and H_g is approximated by[4]

$$H_g=9.0[(4.0-2.5\rho_g)/(16.0-2.5\rho_g)]\rho_g. \quad (16.10)$$

Fig. 16.7 shows the hydrogen index of a hypothetical gas $C_{1.1}H_{4.2}$. Additional empirical equations approximating the hydrogen indices are given in Ref. 5.

Pure sandstone, limestone, and dolomite do not contain hydrogen. Subsequently,

$$H_{ma}=0. \quad (16.11)$$

If the formation is drilled with freshwater-based mud, Eqs. 16.7 and 16.9 become

$$\rho_{mf}=1 \quad (16.12)$$

and $H_{mf}=1$, $\quad (16.13)$

and in a low-pressure environment, gas density can be considered null:

$$\rho_g=0. \quad (16.14)$$

Eq. 16.10 becomes

$$H_g=0. \quad (16.15)$$

With the simple case represented by Eqs. 16.11 through 16.15, Eqs. 16.3, 16.4, and 16.8 reduce to

$$\rho_b=(1-\phi)\rho_{ma}+\phi S_{xo}, \quad (16.16)$$

$$\phi_D=\phi[S_{xo}+\rho_{ma}(1-S_{xo})/(\rho_{ma}-1)], \quad (16.17)$$

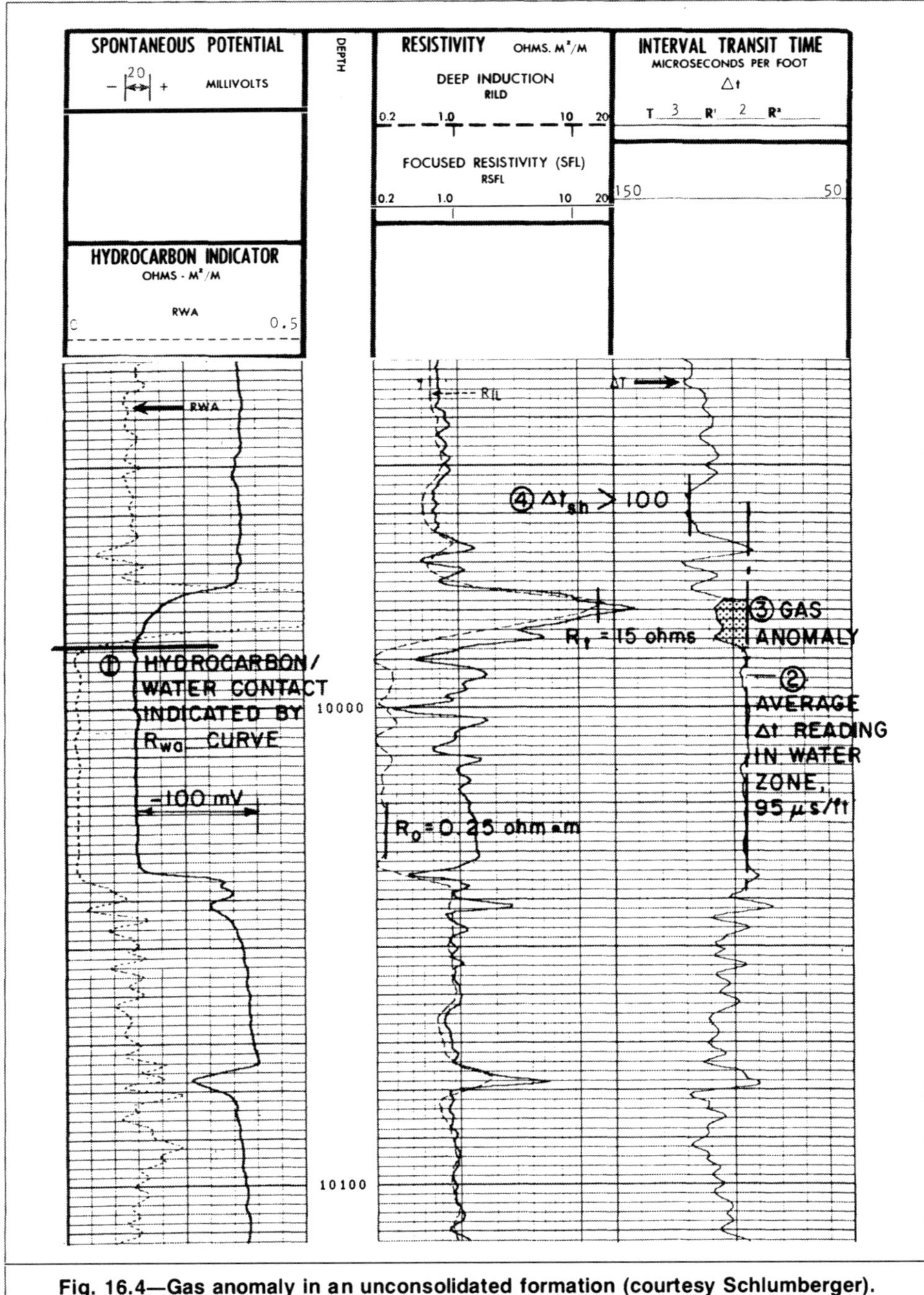

Fig. 16.4—Gas anomaly in an unconsolidated formation (courtesy Schlumberger).

and $\phi_N = \phi S_{xo}$. . . . . (16.18)

Solving for ϕ yields

$$\phi = [(\rho_{ma} - \rho_b) + \phi_N]/\rho_{ma} \quad (16.19)$$

or $\phi = [(\rho_{ma} - 1)/\rho_{ma}]\phi_D + (\phi_N/\rho_{ma})$. . . . . (16.20)

Use of Eq. 16.19 or 16.20 to estimate ϕ depends on whether ρ or ϕ_D is the available density log response. Once ϕ is calculated, S_{xo} can be estimated with Eq. 16.18 as expressed in the following form:

$$S_{xo} = \phi_N/\phi. \quad (16.21)$$

Fig. 16.8 is a graphical presentation of Eqs. 16.16 through 16.18 for a sandstone formation where $\rho_{ma} = 2.65$ g/cm³.

Investigating the effect of hydrocarbons on the neutron and density logs, Gaymard and Poupon[5] have shown that, to a good approximation, regardless of the value of ρ_g,

$$\phi^2 = \frac{\phi_N^2 + \phi_D^2}{2[1 + 0.12(1 - S_{xo})]^2[1 + 0.5n(1 - S_{xo})]^2}. \quad (16.22)$$

If the filtrate is assumed to be fresh mud (i.e., $n=0$), Eq. 16.22 simplifies to

$$\phi^2 = \frac{\phi_N^2 + \phi_D^2}{2[1 + 0.12(1 - S_{xo})]^2}. \quad (16.23)$$

To evaluate porosity with Eq. 16.23, a value of S_{xo} has to be assumed. Fortunately, this assumption is not critical. For the range of reasonable and practical values of S_{xo}, Eq. 16.23 can be simplified further to

$$\phi = [(\phi_N^2 + \phi_D^2)/2]^{1/2}. \quad (16.24)$$

16.6.1 Excavation Effect. The above simplified approaches neglect the excavation effect.[6] When properly calibrated for formation lithology, the neutron log is assumed to respond only to the hydrogen present in formation water and hydrocarbons. Consider the two formations in **Fig. 16.9**. Formation a has a 15% porosity, which is 100% water-saturated; Formation b has a 30% porosity, which is 50% water-saturated and 50% saturated with low-pressure gas. The difference between the two formations is as simple as if matrix

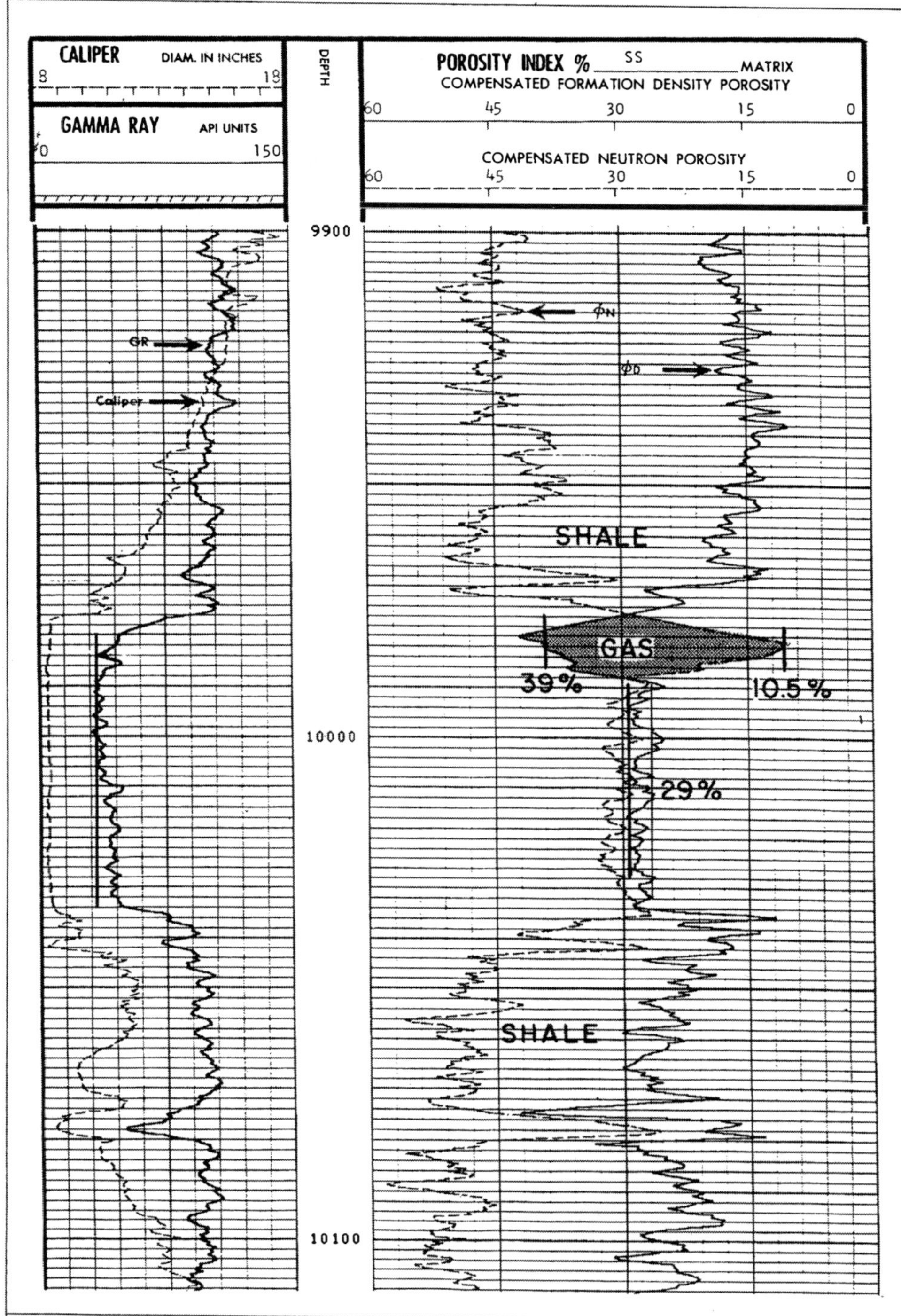

Fig. 16.5—Neutron/density porosity overlay showing separation in gas zones (courtesy Schlumberger).

material amounting to 15% of formation volume had been excavated and low-density gas substituted in its place. Both formations contain the same bulk volume fraction of water. Because a low-pressure gas has a hydrogen index that is very close to zero, the neutron log responds to the hydrogen of the water, which is the same for both cases. Equating the neutron log response for the two formations amounts to ignoring the effect of the "excavated" matrix rock on neutron slowing and diffusion, neutron capture, etc. The response is expected to be different in the two cases. This difference is called the excavation effect.

Mathematical investigations were made to estimate the magnitude of the excavation effect.[6] They were based on computations of neutron slowing lengths under different conditions of porosity, saturation, and matrix lithology. The excavation effect was found to depend on hydrocarbon saturation, fluid hydrogen index porosity, and matrix lithology. **Fig. 16.10** shows the excavation effect for sandstone, limestone, and dolomite. A simplified equation representing the curves of Fig. 16.10 is

$$\Delta\phi_{Nex}=K(2\phi^2 S_{wH}+0.04\phi)(1-S_{wH}), \quad (16.25)$$

where $K=(\rho_{ma}/2.65)^2$ ..... (16.26)

and $\Delta\phi_{Nex}$ is the excavation effect, which should be added to the neutron log porosity reading that does not already contain the correction.

S_{wH} is the equivalent saturation based on hydrogen content or pore fluids:

$$S_{wH}=S_{xo}H_{mf}+(1-S_{xo})H_g. \quad (16.27)$$

When the mud filtrate is fresh ($H_{mf}=1$) and the gas pressure is low ($H_g=0$), $S_{wH}=S_{xo}$. When the matrix is sand, $K=1$.

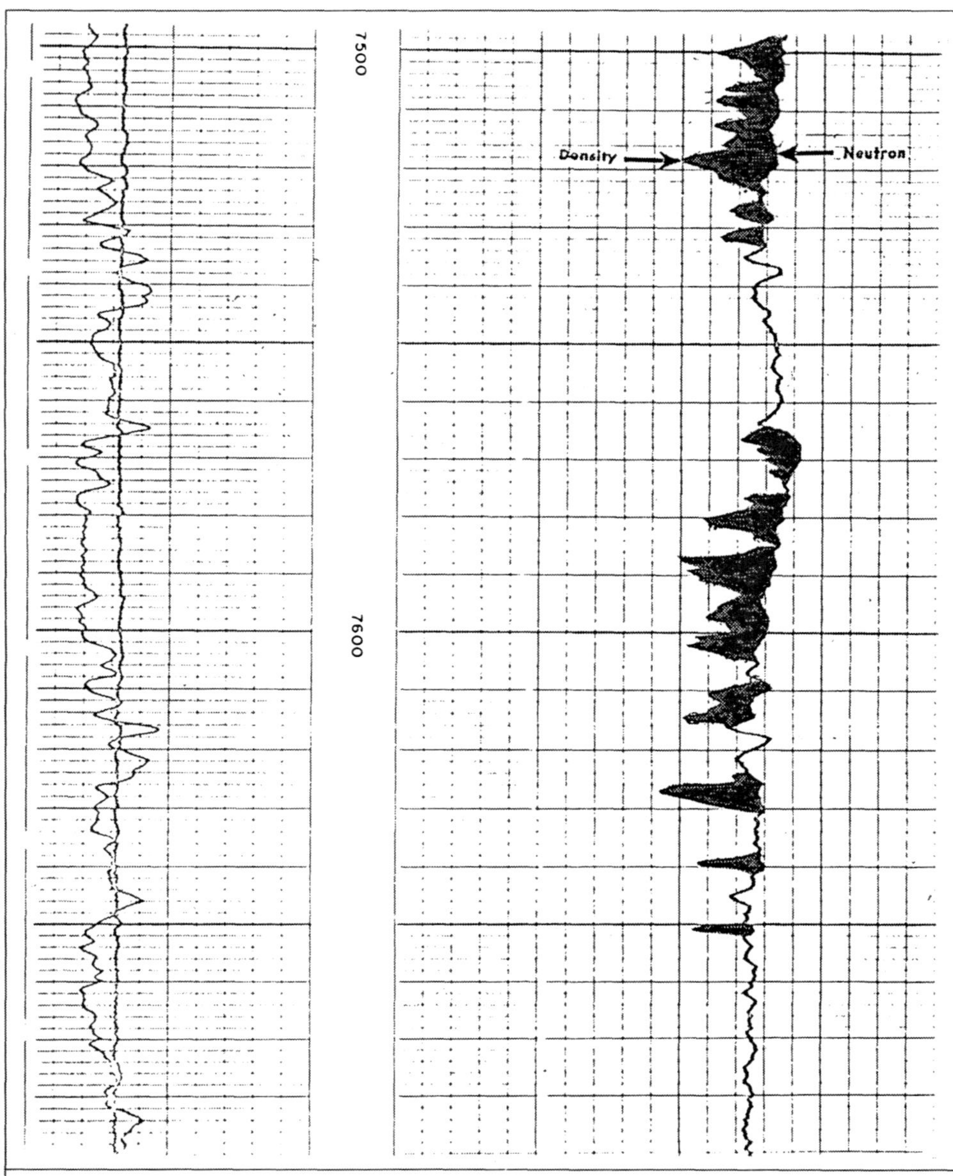

Fig. 16.6—Neutron/density porosity overlay showing separation in moderately invaded gas zones.

The excavation effect correction is added to the neutron log porosity reading:

$$\phi_{Nc}=\phi_N+\Delta\phi_{Nex}, \quad \text{(16.28)}$$

where ϕ_{Nc} is the neutron log porosity corrected for the excavation effect.

Eqs. 16.3 through 16.10 and Eqs. 16.25 through 16.28 can be used to solve for ϕ and S_{xo} if ϕ_N and ρ readings in a gas zone are known. This, of course, is a trial-and-error approach and therefore lends itself to solution by computer. Charts similar to Fig. 16.8 also can be constructed for any set of mud filtrate, gas, and matrix densities. Figs. **16.11 through 16.13** are charts constructed for sandstone and fresh mud filtrate for gas densities of 0, 0.12, and 0.25, respectively.[7]

Example 16.1. Neutron and density log readings in a clean, gas-bearing sandstone formation are 5% and 2.0 g/cm^3, respectively. Assuming the gas is low density and the filtrate is fresh mud, determine ϕ and S_{xo} with and without inclusion of the excavation effect.

Solution. Apparent density porosity, ϕ_D, can be calculated as

$$\phi_D=(\rho_{ma}-\rho_b)/(\rho_{ma}-\rho_f)$$

$$=(2.65-2.0)/(2.65-1.0)=0.39 \text{ or } 39\%.$$

Hence, $\phi_D \gg \phi_N$, which confirms the presence of gas in the formation.

ϕ and S_{xo} Determination Excluding the Excavation Effect. Using Eqs. 16.19 and 16.20 results in

$$\phi=[(2.65-2)+0.05]/2.65$$

$$=0.264 \text{ or } 26\%$$

and $S_{xo}=0.05/0.264$

$$=0.189 \text{ or } 19\%.$$

The same values of ϕ and S_{xo} can be obtained from the chart of Fig. 16.8.

ϕ and S_{xo} Determination Including the Excavation Effect. Entering the chart of Fig. 16.11 with the values ϕ_N=5% and ϕ_D=39% yields the values ϕ=26% and S_{xo}=23%.

In this case, neglecting the excavation effect appears to affect only the estimation of S_{xo}. A porosity value of 26% is obtained either with or without accounting for the excavation effect.

Example 16.2. In a clean, gas-bearing sand, neutron and density logs display porosity readings of 10% and 20%, respectively. Investigate the effect of neglecting gas density on porosity determination.

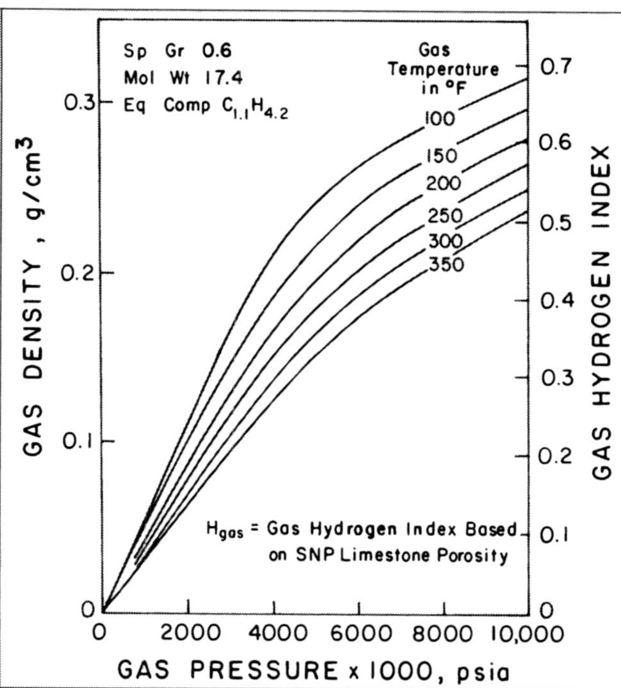

Fig. 16.7—Gas density and hydrogen index as functions of pressure and temperature for a gas mixture slightly heavier than methane ($C_{1.1}H_{4.2}$) (courtesy Schlumberger).

Solution. The following values were obtained from the charts of Figs. 16.11 through 16.13 with ϕ_N=10% and ϕ_D=20%.

ρ_g (g/cm³)	ϕ (%)	S_{xo} (%)
0	16.4	69
0.12	15.9	60
0.25	15.3	30

These values suggest that for ϕ_N and ϕ_D values of 10% and 20%, respectively, the gas density has little effect on the porosity values. The effect on S_{xo}, however, is considerable. This observation is also true for most pairs of ϕ_N and ϕ_D values.

Example 16.3. Investigate the effect of lithology on the evaluation of ϕ and S_{xo} in a gas-bearing formation using neutron and density log readings.

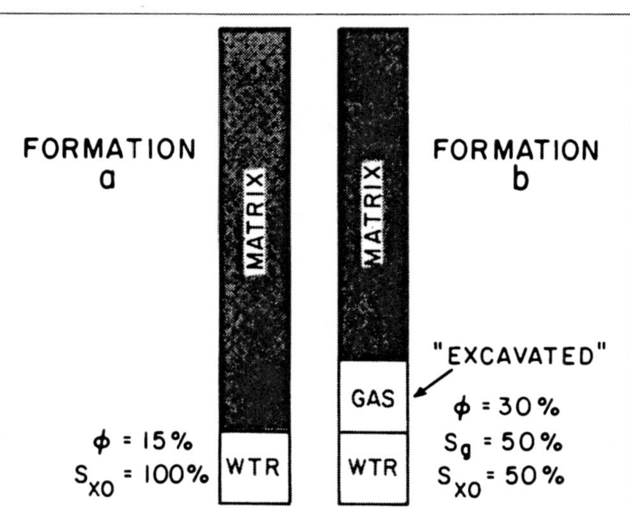

Fig. 16.9—Two formations that contain the same hydrogen content but yield different neutron log response owing to the excavation effect (after Ref. 6).

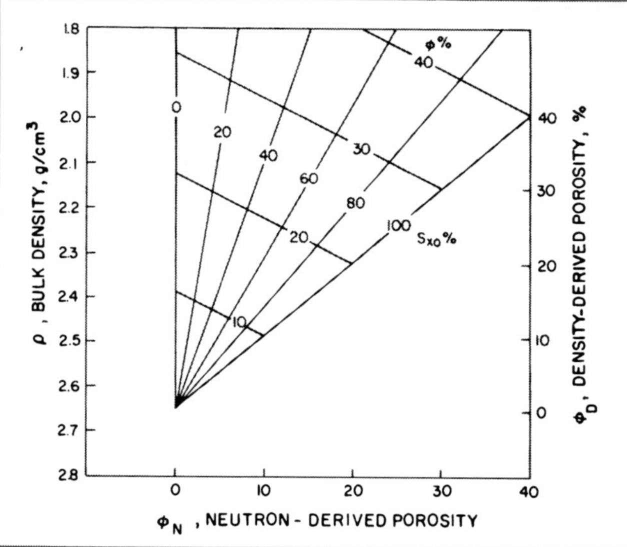

Fig. 16.8—Graphical solution for ϕ and S_{xo} as functions of ρ_b and ϕ_N for simplified case of a gas-bearing formation.

Solution. Let ϕ_N=10% and ϕ_D=20% (both porosity values are based on sandstone matrix), and let the lithology be a limy sandstone of unknown proportions. Values of ϕ and S_{xo} will be calculated from Eqs. 16.19 through 16.21, first assuming the formation to be 100% sandstone and then assuming the formation to be 100% limestone.

Sandstone Lithology. Using Eq. 16.20 gives

ϕ=[(2.65−1)/2.65]0.2+(0.1/2.65)

=0.162 or 16%.

From Eq. 16.21,

S_{xo}=0.1/0.162

=0.616 or 62%.

Limestone Lithology. From Fig. 14.1, ϕ_N=6% (apparent limestone porosity). Using Eqs. 16.19 and 16.21 yields

ϕ=[(2.71−2.32)+0.06]/2.71

=0.166 or 17%

and S_{xo}=0.06/0.166

=0.361 or 36%.

These values suggest that the type of lithology has little effect on the estimated porosity value. The effect on the S_{xo} value, however, is considerable.

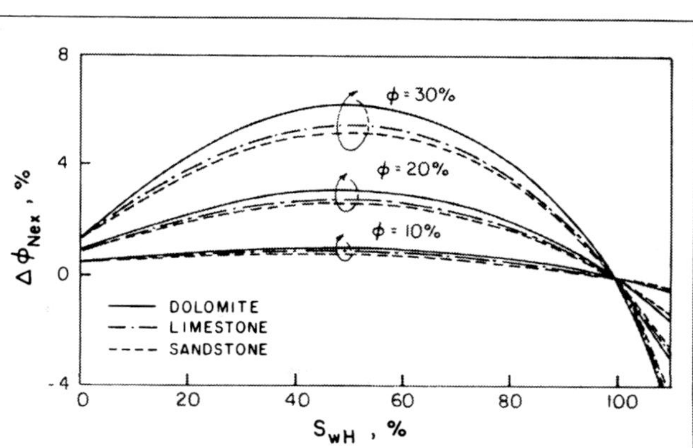

Fig. 16.10—Correction for excavation effect as a function of S_{wH} for different lithologies (after Ref. 6).

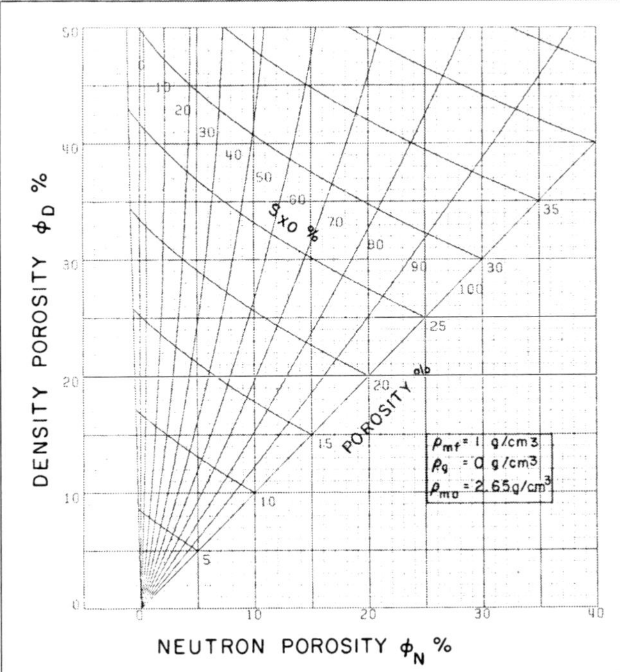

Fig. 16.11—Neutron/density crossplot for ϕ and S_{xo} determination in a gas-bearing formation, $\rho_g=0$ (from Ref. 7).

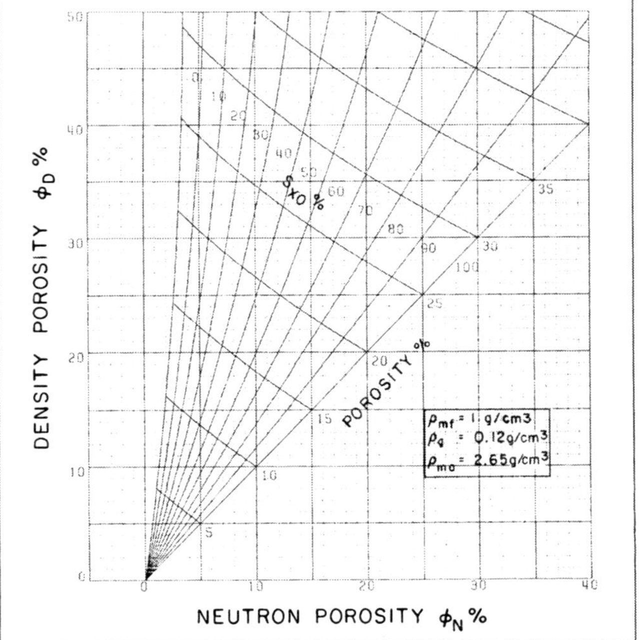

Fig. 16.12—Neutron/density crossplot for ϕ and S_{xo} determination in a gas-bearing formation, $\rho_g=0.12$ g/cm^3 (from Ref. 7).

The above treatment of porosity determination in a gas-bearing formation is valid only when both neutron and density tools are investigating the same fluid saturation profile. This is the case in deep invasion and extremely shallow invasion. Under such invasion conditions, the two porosity curves will mirror-image each other, as seen in Fig. 16.5. In the case of extremely shallow invasion, the S_{xo} value derived from these calculations will approach the S_w value in the uninvaded zone. Shallow to moderate invasion profiles will complicate the evaluation because the neutron and density tools will be investigating different saturation profiles. This case usually is obvious on the log because the neutron and density curves do not mirror-image each other, as illustrated by Fig. 16.6.

In such a case, the density log can be used alone to estimate the actual porosity. Rearranging Eqs. 16.16 and 16.17 yields

$$\phi=(\rho_{ma}-\rho_b)/(\rho_{ma}-S_{xo}\rho_{mf})$$

$$\approx(\rho_{ma}-\rho_b)/(\rho_{ma}-S_{xo}) \quad \text{.......}(16.29)$$

and $\phi=[\phi_D(\rho_{ma}-\rho_{mf})]/(\rho_{ma}-S_{xo}\rho_{mf})$

$$\approx[\phi_D(\rho_{ma}-1)]/(\rho_{ma}-S_{xo}). \quad \text{.......}(16.30)$$

Use of Eq. 16.29 or 16.30 requires a reasonable value of S_{xo} to be assumed. This approach is better than incorporating the neutron log reading in the calculation, which, in this case, will result in a porosity value that is too low.

Example 16.4. The ISF/sonic and FDC/CNL logs of Figs. 16.4 and 16.5 were obtained in a well in Cameron Parish, LA. The log reading lists the following data.

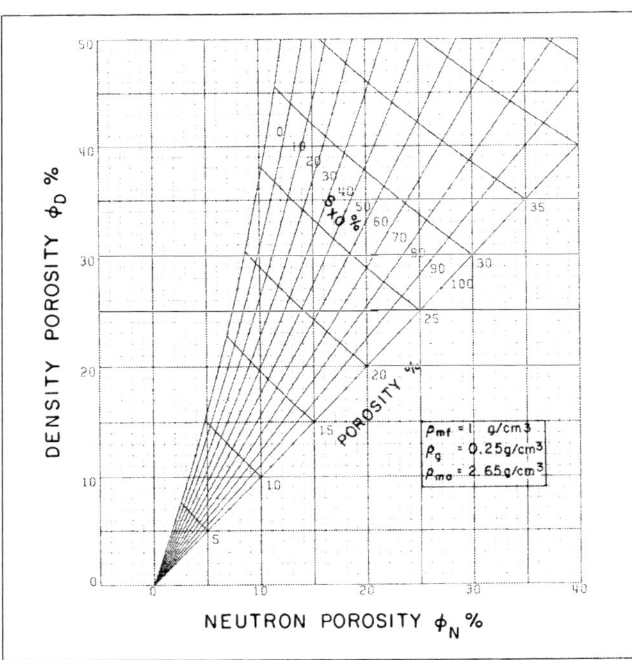

Fig. 16.13—Density crossplot for ϕ and S_{xo} determination in a gas-bearing formation, $\rho_g=0.25$ (from Ref. 7).

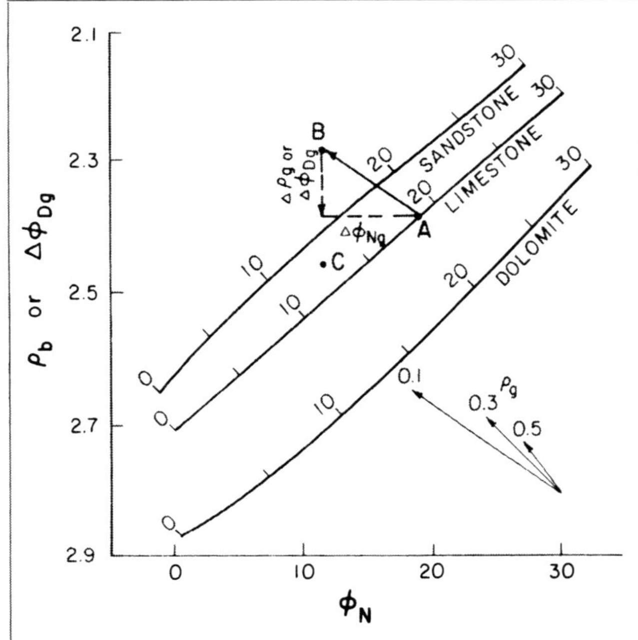

Fig. 16.14—Effect of gas on lithology determination from lithology/porosity crossplot (after Ref. 4).

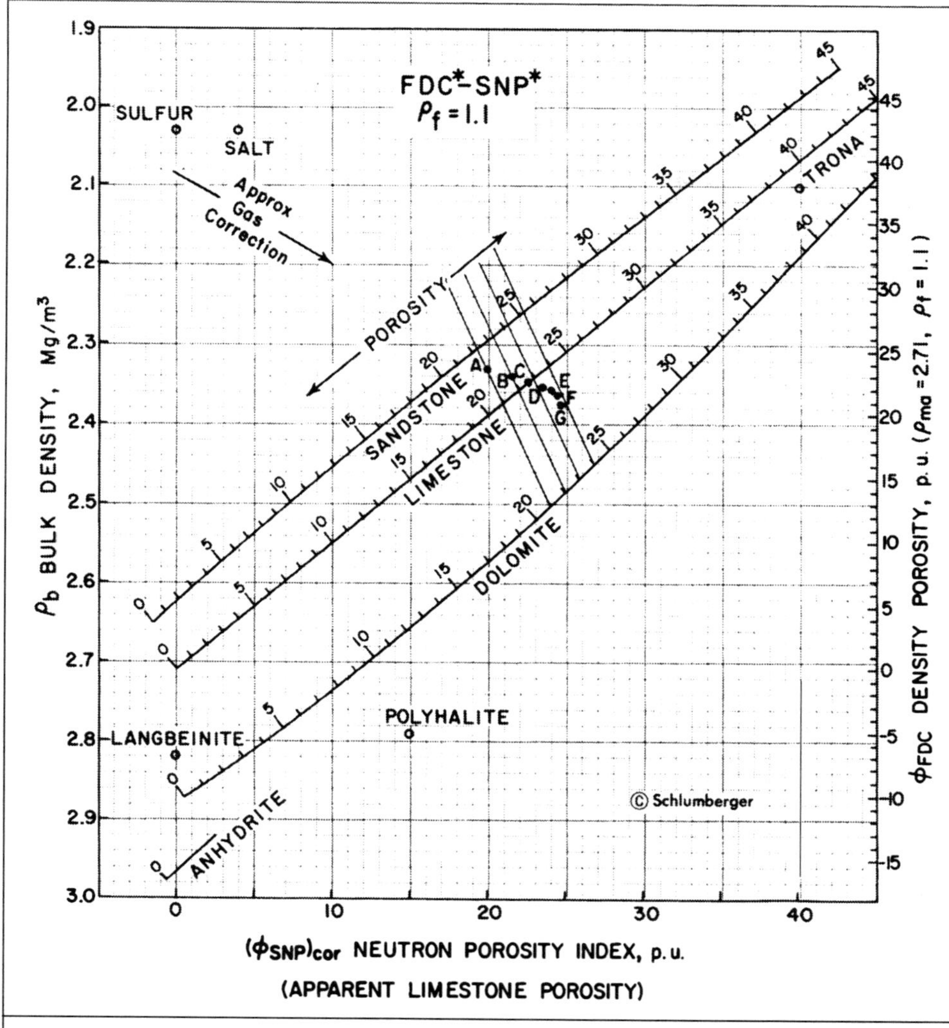

Fig. 16.15—FDC/SNP dual-mineral crossplot, case of salt water, liquid-filled borehole (courtesy Schlumberger).

Total depth	11,420 ft
Bit size	8½ in.
Mud density	10.2 lbm/gal
R_m	0.54 Ω·m at 114°F
R_{mf}	0.41 Ω·m at 74°F
Maximum recorded temperature	195°F

Estimate the porosity of the gas-bearing sand shown in the logs.

a. Take into consideration the excavation effect and gas density.

b. Use an approximate approach.

c. Use other possible approaches.

What conclusion concerning the determination of porosity in gas-bearing formations can be drawn from this example?

Solution.

a. Log readings in the gas zone situated in the interval of 9,975 to 9,987 ft are ϕ_D=39% and ϕ_N=10.5%. Formation temperature is estimated from well-heading information in the following manner:

$$g_G=[(195-70)/11{,}420]100=1.1°\text{F}/100 \text{ ft},$$

$$T_f=70+1.1(9{,}980/100)$$

$$=179°\text{F}.$$

With the assumption that the mud hydrostatic head is indicative of formation pressure,

$$\text{mud pressure gradient}=0.052\ (10.2)$$

$$=0.53 \text{ psi/ft},$$

$$\text{and formation pressure}=0.53\ (9{,}980)$$

$$=5{,}281 \text{ psig}.$$

From Fig. 16.7, ρ_g=0.21 g/cm³. Using Fig. 16.13, which includes excavation effect and is constructed for relatively high-density gas, gives

$$\phi=26\%.$$

b. Using Eq. 16.20 yields

$$\phi=[(2.65-1)/2.65](0.39)+(0.105/2.65)$$

$$=0.282 \text{ or } 28\%.$$

From Eq. 16.24,

$$\phi=\{[(0.39)^2+(0.105)^2]/2\}^{½}$$

$$=0.286 \text{ or } 29\%.$$

c. The sand that contains the gas zone appears to be of uniform porosity. From a sonic log reading in a water zone,

$$\phi=(95-55.5)/(189-55.5)$$

$$=0.296 \text{ or } 30\%.$$

The slight separation between the density and neutron curves indicates that the water zone is slightly shaly. Taking an average value of the two curves gives ϕ=29%.

This work shows that the approximate approach is sufficiently accurate for determining a gas-bearing formation's porosity. Given the agreement between the value obtained in the water zone and the approximate technique, the best estimate of the porosity of the gas zone is 29%.

16.7 Saturation Determination in Gas-Bearing Formations

Because gas has the same effect on resistivity logs as oil, water saturation values can be estimated with Archie's equation:

$$S_w^n = aR_w/\phi^m R_t. \quad (16.31)$$

The porosity in Eq. 16.31, however, is the actual formation porosity derived from log readings, as explained in Sec. 16.6.

Saturation values of gas-bearing zones derived from reconnaissance techniques, such as R_{wa} and R_o overlay methods, are inexact because uncorrected log porosity readings are used. In the R_{wa} technique where the sonic log usually is used to estimate the formation resistivity factor, F, the gas effect could result in an inflated porosity, which, in turn, would give an apparent low value of F. The calculated value of R_{wa} is then exaggerated. The result is a clear R_{wa} anomaly in the gas zone. Using Eq. 12.5 to calculate water saturation, however, results in a value that is too low. The same result will occur when the density log is used to generate the R_o log. The opposite effect, a water saturation value that is too high, results if the neutron log is used to generate the F value needed for calculating R_{wa} and R_o.

In the crossplotting techniques described in Chap. 13, the presence of gas within the zone investigated by the density or sonic log will cause the point that represents the gas zone to shift away from the water trend. The zone is detected more easily, but the hydrocarbon saturation value derived from the plot is higher than the actual value. The opposite effect results when the neutron log is used.

Once reconnaissance or crossplotting techniques determine that a zone is hydrocarbon, the hydrocarbon type should be investigated. If the hydrocarbon is gas, porosity should be determined first with the approach discussed in Sec. 16.6 before a representative saturation value is calculated.

Example 16.5. Determine gas saturation in the hydrocarbon zone displayed by the ISF/sonic and FDC/CNL logs of Figs. 16.4 and 16.5. Well-heading data are given in Example 16.4.

Solution. From the analysis given in Example 16.4, $\phi=29\%$ and $T_f=179°F$. R_w can be calculated from self-potential (SP) log data,

$E_{SSP}=-100$ mV.

From Eq. 1.11,

R_{mf} at $T_f=0.178\ \Omega\cdot m$.

From Fig. 1.38,

$R_w=0.021\ \Omega\cdot m$.

Using Archie's relationships gives

$F=0.81/(0.29)^2=9.6$

and $R_o=9.6\ (0.021)=0.2\ \Omega\cdot m$.

From the ISF log,

$R_t=15\ \Omega\cdot m$,

$S_w=(0.2/15)^{1/2}$

$=0.116$ or 12%,

and $S_g=88\%$.

The sand appears to be of uniform lithology and porosity, so R_o can be obtained directly from the water zone underlying the gas zone:

$R_o=0.25\ \Omega\cdot m$,

$S_w=(0.25/15)^{1/2}$

$=0.129$ or 13%,

and $S_g=87\%$.

The latter approach is the better one because it uses only raw log data. No possible deviations, then, are introduced in evaluations of ϕ, F, and R_w.

16.8 Gas Effect on Lithology/Porosity Crossplots

On the lithology/porosity crossplots discussed in Chap. 14, gas-bearing zones assume nonrepresentative positions. On a density/neutron crossplot, for example, a liquid-filled limestone porosity will assume Position A in **Fig. 16.14**. The presence of gas in a zone of the same lithology and porosity results in a shift upward and to the left. This shift from Position A to B is almost parallel to the isoporosity lines. The porosity of a gas zone can then be approximated by a direct reading from the chart. However, the lithology indication from the crossplot can be in error. For example, the porosity of a gas zone that assumes Position C in the crossplot is about 13%. Its lithology could be limestone or a limestone/dolomite mixture, depending on the shift caused by the gas effect.

A gas correction is needed to deduce the correct lithology.[8] The gas correction consists of shifting a point that represents a gas zone to a position that represents a liquid-filled point of the same porosity[9], e.g., shifting Point B in Fig. 16.14 to Position A. The vertical and horizontal components of Vector AB are the gas effects on the density log, $\Delta\rho_g$ or $\Delta\phi_{Dg}$, and on the neutron log, $\Delta\phi_{Ng}$, respectively. The gas effect $\Delta\rho_g$ or $\Delta\phi_{Dg}$ on the density log is defined by gas-dependent terms in Eqs. 16.3 and 16.4. They are

$$\Delta\rho_g=\phi(1-S_{xo})(\rho_{mf}-\rho_g) \quad (16.32)$$

and $$\Delta\phi_{Dg}=\phi(1-S_{xo})(1-\phi_{Dg}). \quad (16.33)$$

ϕ_{Dg} is defined in terms of ρ_g by Eq. 16.6.

Similarly, the gas effect on the neutron log, $\Delta\phi_{Ng}$, is defined by the gas-dependent term in Eq. 16.8. Considering the excavation effect gives

$$\Delta\phi_{Ng}=\phi(1-S_{xo})(H_{mf}-H_g)+\Delta\phi_{Nex}, \quad (16.34)$$

where H_{mf}, H_g, and $\Delta\phi_{Nex}$ are defined by Eqs. 16.9, 16.10, and 16.25.

To proceed with the gas effect correction, a first-trial value of ϕ can be determined from the crossplot or from Eq. 16.19 or 16.20. A first-trial value of S_{xo} can be estimated from Eq. 16.21. The first-trial values of ϕ and S_{xo} are used to compute the density and neutron gas effects from Eqs. 16.32 and 16.33. These hydrocarbon effects are used to determine the coordinates of the point representing the formation being investigated. The corrected location of the formation on the crossplot indicates a better value of ϕ.

The last value of ϕ is used again to compute S_{xo} and then to iterate the calculations of the correction. Further iterations can be made until no appreciable shift of the zone's position on the chart can be produced.

The arrows in the lower right side of Fig. 16.14 show the approximate direction and magnitude of the gas shift for three gas densities. These shifts are calculated from Eqs. 16.32 through 16.34, assuming that the filtrate is fresh mud (i.e., $\rho_{mf}=1$ and $H_{mf}=1$), that the excavation effect is negligible, and that $\phi(1-S_{xo})=0.15$.

Example 16.6. The readings of the FDC and SNP logs in a clean, gas-bearing formation with a lithology composed of limestone and dolomite are 2.33 g/cm^3 and 20% apparent limestone porosity, respectively. Determine the porosity and lithology fractions of the formation if filtrate salinity is 150,000 ppm, formation temperature is 150°F, and formation pressure is 2,000 psia.

Solution. From Eq. 16.7, the mud filtrate density is

$\rho_{mf}=1+0.73(0.15)$

$=1.1$ g/cm^3.

Crossplotting the values of ρ_b and ϕ_N on **Fig. 16.15** shows that Point A, which represents the formation, lies above the limestone line. This position unequivocally confirms that Point A assumes a nonrepresentative position because of the presence of gas. A gas correction is then required to determine ϕ and lithology proportions of limestone and dolomite.

First-trial values are $\phi=0.21$ from the crossplot, and $S_{xo}=0.20/0.21=0.95$ from Eq. 16.21.

Values of ρ_g, H_{mf}, H_g, and $\Delta\phi_{Nex}$ are as follows. From Fig. 16.7,

$$\rho_g=0.1.$$

From Eq. 16.9,

$$H_{mf}=1.1(1-0.15)$$
$$=0.935.$$

From Eq. 16.10,

$$H_g=9.0\{[4-2.5(0.1)]/[16-2.5(0.1)]\}0.1$$
$$=0.214.$$

From Eq. 16.27,

$$S_{wH}=0.95(0-0.935)+0.05(0.214)$$
$$=0.09.$$

From Eq. 16.25,

$$\Delta\phi_{Nex}=(2.71/2.65)^2[2(0.21)^2(0.9)+0.04(0.21)]0.1$$
$$=0.009.$$

Values of gas effects $\Delta\rho_g$ and $\Delta\phi_{Ng}$ are from Eqs. 16.32 and 16.34, respectively.

$$\Delta\rho_g=0.21(0.05)(1.1-0.1)=0.01 \text{ g/cm}^3.$$

$$\Delta\phi_{Ng}=0.21(0.05)(0.935-0.214)+0.009$$
$$=0.016.$$

The two correction values locate Point B in Fig. 16.15. Because the point is still located above the limestone line, additional iterations are required. Values of additional iterations are tabulated below.

	Trial						
	1	2	3	4	5	6	7
ϕ	0.21	0.22	0.225	0.232	0.235	0.238	0.238
S_{xo}	0.952	0.909	0.889	0.862	0.851	0.840	0.840
S_{wH}	0.901	0.869	0.855	0.836	0.828	0.820	0.820
$\Delta\phi_{Nex}$	0.009	0.0127	0.0145	0.0171	0.0182	0.0193	0.0193
$\Delta\rho_g$	0.01	0.020	0.0250	0.0320	0.035	0.0380	0.0467
$\Delta\phi_{Ng}$	0.0163	0.0271	0.0325	0.0401	0.0434	0.0467	0.0467
Location	B	C	D	E	F	G	G

Further iterations would not change the results. The location of Point G is that of a liquid-bearing zone that has the same lithology and porosity as Zone A. Point G indicates that the lithology is 60% limestone and 40% dolomite. The porosity is about 24%.

16.9 Shaly Gas-Bearing Formations

The difference of porosity values recorded by the neutron and density logs where $\phi_D>\phi_N$ is used as a direct method of gas detection. When the formation investigated is shaly, the separation between the two curves decreases because shale effects on both logs are opposite that of gas. This can be seen in **Fig. 16.16**, which shows a thick, gas-bearing sand in the interval 9,405 to 9,582 ft. A large separation appears in Intervals A and D where the formation is practically clean as shown by the gamma ray curve. The separation narrows in Zone B where the formation is more shaly. When the shale content is large enough, the separation will disappear completely or even reverse (i.e., $\phi_D<\phi_N$). Note that in Zone A the separation is still modulated by the shale content. In fact, a good correlation exists between the gamma ray log response and the magnitude of the separation.

Gas may be detected in shaly formations by use of one of several techniques presented in Secs. 16.9.1 through 16.9.4.

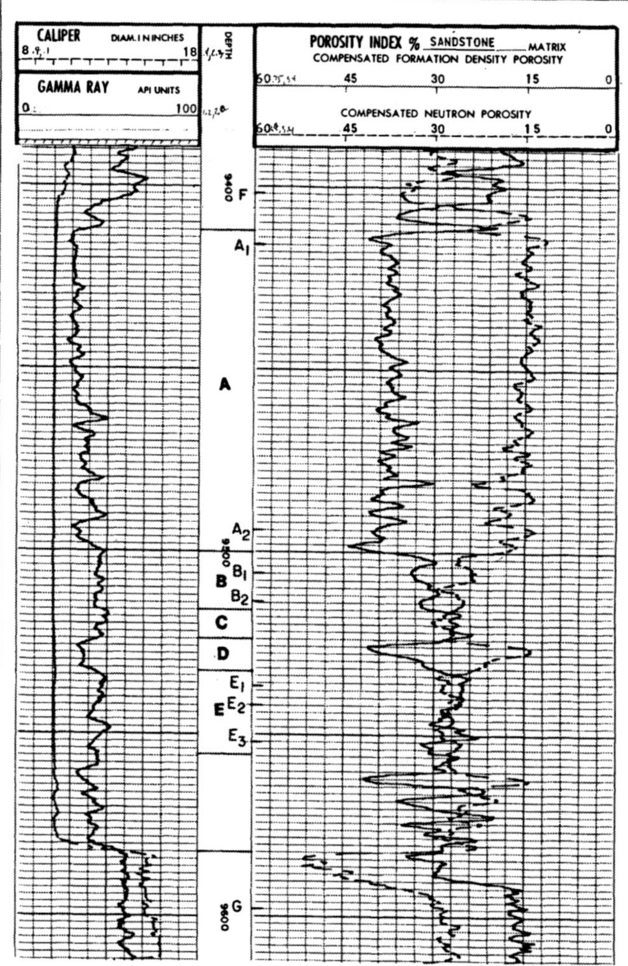

Fig. 16.16—FDC/CNL porosity overlay showing various degrees of curve separation in a gas formation (courtesy Schlumberger).

16.9.1 Gas Detection in Shaly Formations With a Neutron/Density Porosity Overlay. This quick-look technique can be used to detect gas in zones of high shale content, where the separation between the two curves disappears or even reverses because the shale effect exceeds the gas effect. With this technique, the neutron log is shifted relative to the density log to make the two porosity curves track in known shaly water sands. Gas-bearing zones then become more obvious from crossovers of the shifted neutron and density curves where the density porosity reading is higher than the now-shifted neutron porosity. **Fig. 16.17** shows an example application of the technique. The dual induction laterolog (DIL) indicates a potential zone in the lower interval where the LL8 and ILd separate. The neutron and density curves show no appreciable separation, suggesting that the formation is liquid-bearing. This interpretation is incorrect, however, because the formation appears to be quite shaly on the SP and gamma ray logs. Normalizing the two curves in a shaly water-bearing zone results in a separation characteristic of gas presence. With this quick-look technique, an attempt is made to remove the shale effect empirically, thus accentuating the gas effect. One application of this method requires the presence, in the section analyzed, of a water zone with approximately the same shaliness as the hydrocarbon zones.

16.9.2 Gas Detection in Shaly Formations With a ($\phi_N-\phi_D$) vs. Gamma Ray Crossplot. As detailed in Sec. 15.3, the neutron and density log responses in liquid-filled shaly formations are expressed as

$$\phi_N=\phi+V_{sh}\phi_{Nsh} \quad \text{......} (16.35)$$

and $$\phi_D=\phi+V_{sh}\phi_{Dsh} \quad \text{......} (16.36)$$

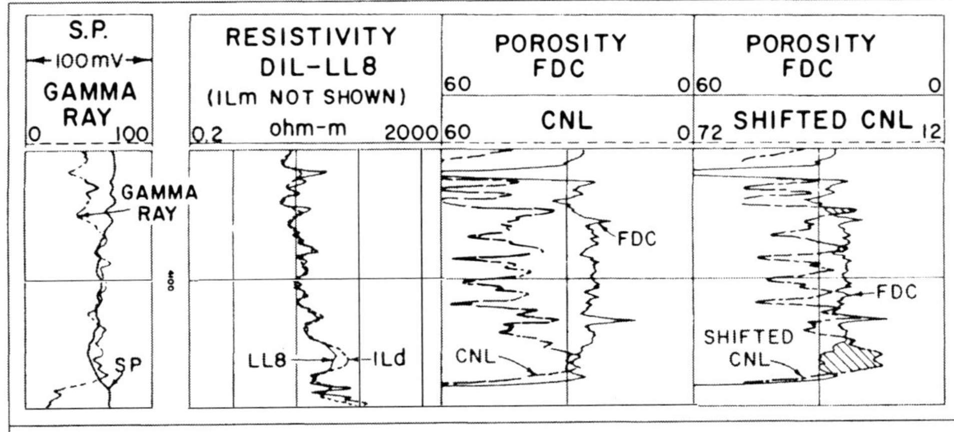

Fig. 16.17—Log example illustrating the quick-look shifted neutron and density log overlay technique (from Ref. 9).

Subtracting Eq. 16.36 from Eq. 16.35 results in

$$\phi_N - \phi_D = V_{sh}(\phi_{Nsh} - \phi_{Dsh}). \quad \ldots\ldots (16.37)$$

If the gamma ray responds to the amount of shale present in the formation, a plot of gamma ray reading vs. the difference between the neutron and density log values $\phi_N - \phi_D$ results in a straight line on coordinate paper,[10] as shown in **Fig. 16.18**. The slope of the line is determined by shale properties in the interval analyzed. The x intercept is determined by the gamma ray log response in a shale-free formation present in the interval analyzed.

Because of lithology variation, borehole effects, etc., data from clean, water-bearing and oil-bearing formations will plot in an area of low gamma ray value and little or no difference in neutron and density porosities (Fig. 16.18). Likewise, shale will plot in an area of high gamma ray readings and large $\phi_N - \phi_D$ values. Because of local effects on the gamma ray tool and the particular neutron tool type involved, instead of a unique straight line, a linear trend can be seen between clean, liquid-bearing formations and shale.[11]

The presence of gas in a formation will shift its point on the crossplot downward from the liquid-filled formation trend, with the cleaner zones showing negative $\phi_N - \phi_D$ values and the more shaly zones showing positive $\phi_N - \phi_D$ values.

Example 16.7. Prepare a crossplot of $\phi_N - \phi_D$ vs. gamma ray that shows selected zones within Intervals A through E in Fig. 16.16. Adjacent shales display the following average values.

$\phi_{Dsh} = 16\%$.
$\phi_{Nsh} = 38\%$.
$\gamma_{sh} = 85$ API units.
$\gamma_{clean} = 30$ API units.

Using the crossplot, determine the fluid type of Zones F and G situated at 9,402 and 9,599 ft, respectively.

Solution.

Zone	GR (API units)	ϕ_N (%)	ϕ_D (%)	$(\phi_N - \phi_D)$	Identification
A_1	30	12	41	−29	Clean/gas
A_2	30	13.5	40	−26.5	Clean/gas
B_1	45	24	34	−10	Shaly/gas
B_2	44	25	32	− 7	Shaly/gas
C	50	30	27	+ 3	Shaly/gas
D	35	14	41	−27	Clean/gas
E_1	47	28.5	25	+ 3.5	Shaly/gas
E_2	45	26	26	0	Shaly/gas
E_3	43	25.5	31	− 5.5	Shaly/gas
F	65	36	21.5	+15.5	Shaly/gas
G	60	28.5	17	+11.5	Shaly/gas

The crossplot of $\phi_N - \phi_D$ vs. gamma ray in **Fig. 16.19** shows gas-bearing Zones A through E in the expected location below the line that joins clean, liquid-bearing sand and shale points. The cleaner the zone, the farther off the line it is.

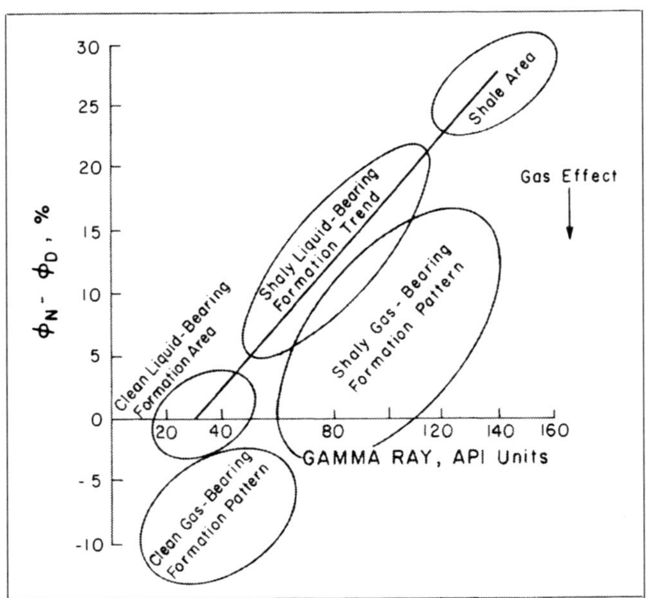

Fig. 16.18—Schematic of crossplot of ($\phi_N - \phi_D$) vs. gamma ray.

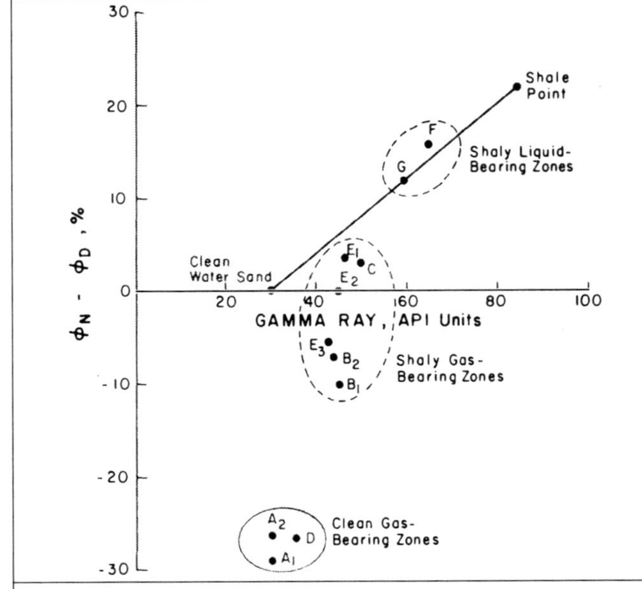

Fig. 16.19—($\phi_N - \phi_D$) vs. gamma ray crossplot of Example 16.7.

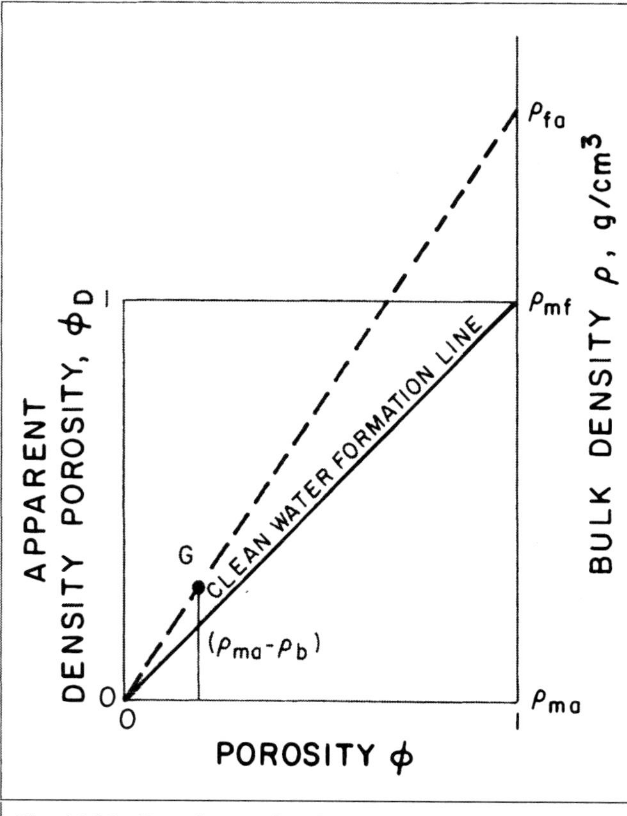

Fig. 16.20—Porosity vs. density tool response showing apparent fluid density, ρ_{fa}, of Gas Zone G (after Ref. 12).

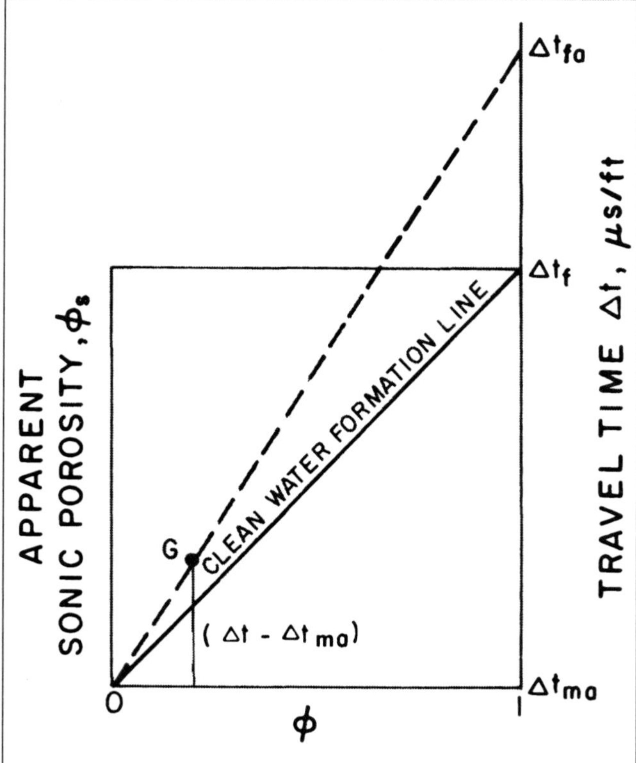

Fig. 16.21—Porosity vs. sonic tool response showing apparent fluid interval transit travel time, Δt_{fa}, of Gas Zone G (after Ref. 12).

Zones F and G are located on the line; hence, they are both liquid-filled.

16.9.3 Gas Detection in Shaly Formations With the Fluid Identification Plot. The fluid identification (FID) plot was devised mainly as an aid in identifying gas presence in shaly sands.[12] The concept of the FID plot is similar to that of the matrix identification (MID) plot discussed in Chap. 14. The MID plot method assumes a liquid content and solves for the lithology. The FID plot method, on the other hand, assumes that the lithology is known and solves for the fluid type.

The solid line of **Fig. 16.20** shows a plot of ϕ vs. the reading of the density tool in clean, water-bearing formations. Gas-bearing Formation G will plot above the line for the clean, water-bearing formation because $\phi_D > \phi_N$ in gas zones. An apparent fluid density, ρ_{fa}, is defined at the intersection of the 100% porosity line and the extension of a line that joins the origin and Point G. The slope of the dotted line can be expressed as

$$A=(\rho_{ma}-\rho_{fa})/1=(\rho_{ma}-\rho_b)/\phi. \quad (16.38)$$

The apparent fluid density can then be defined in terms of bulk density by

$$\rho_{fa}=\rho_{ma}-[(\rho_{ma}-\rho_b)/\phi] \quad (16.39)$$

and in terms of apparent density porosity by

$$\rho_{fa}=\rho_{ma}-(\phi_D/\phi)(\rho_{ma}-\rho_{mf}). \quad (16.40)$$

Similarly, an apparent fluid interval transit travel time, Δt_{fa}, of a gas zone can be derived as shown in **Fig. 16.21**. The fluid apparent travel time is defined algebraically in terms of travel time by

$$\Delta t_{fa}=\Delta t_{ma}+(\Delta t-\Delta t_{ma})/\phi \quad (16.41)$$

and in terms of apparent sonic porosity computed on the basis of compacted formation by

$$\Delta t_{fa}=\Delta t_{ma}+(\phi_S/\phi)(\Delta t_{mf}-\Delta t_{ma}). \quad (16.42)$$

The FID plot, which is a crossplot of ρ_{fa} vs. Δt_{fa}, will have the characteristic illustrated in **Fig. 16.22**. Clean, liquid-bearing sands, shaly, gas-bearing sands, and shales should occupy the positions shown. A light oil-bearing zone will plot between clean, liquid-bearing sands and clean gas sands. Its position depends on oil density and residual saturation. Note that the scaling of the Δt_{fa} axis of Fig. 16.22 is arbitrary because sonic log response will vary drastically from one geologic unit to another.

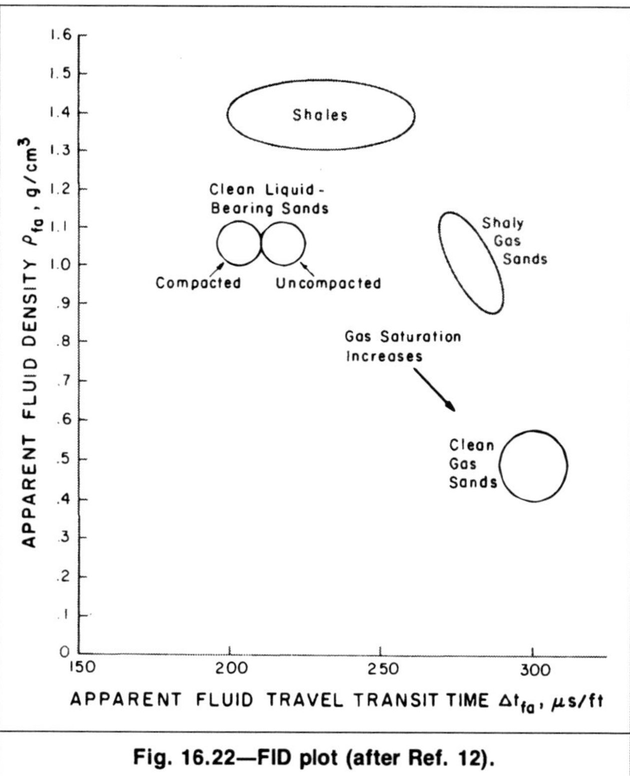

Fig. 16.22—FID plot (after Ref. 12).

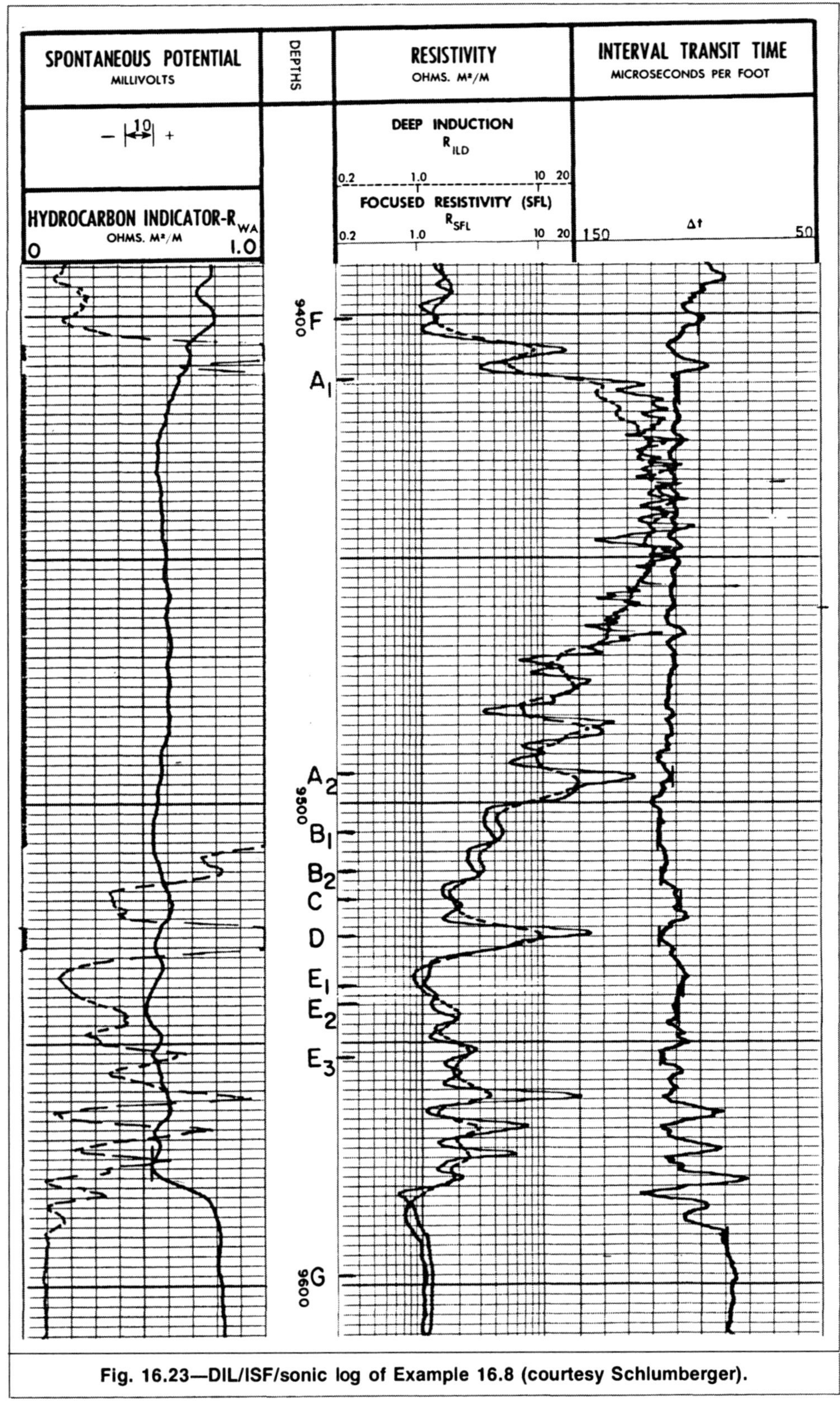

Fig. 16.23—DIL/ISF/sonic log of Example 16.8 (courtesy Schlumberger).

To help with fluid-type identification, Eqs. 16.39 and 16.41 are used to calculate ρ_{fa} and Δt_{fa} for several zones within the interval of interest. The ϕ needed for the calculation can be approximated with Eq. 16.24. A crossplot of ρ_{fa} vs. Δt_{fa} is constructed. Analysis of the crossplot should result in some or all of the patterns in Fig. 16.22.

Example 16.8. Fig. 16.23 illustrates a section of a DIL/ISF/sonic log for the same interval in the FDC/CNL log of Fig. 16.16. Using the FID plot, determine the fluid type in Zones A through G. Does the fluid identification agree with that obtained in Example 16.7 and illustrated by Fig. 16.19?

Solution.

Zone	ϕ_N (%)	ϕ_D (%)	Δt (μsec/ft)	ϕ (%)	ρ_{fa} (g/cm^3)	Δt_{fa} (μsec/ft)	Zone Identification
A_1	12	41	110	30	0.41	236	Clean/gas
A_2	13.5	40	113	30	0.44	248	Clean/gas
B_1	24	34	117	29	0.74	264	Shaly/gas
B_2	25	32	115	29	0.81	263	Shaly/gas
C	30	27	110	28.5	1.09	246	Shaly/gas
D	14	41	117	30.5	0.44	256	Clean/gas

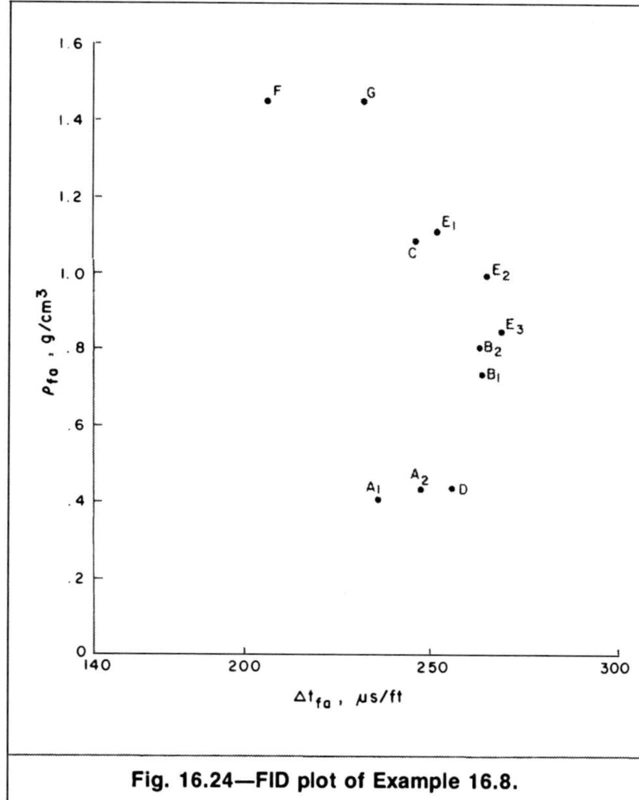

Fig. 16.24—FID plot of Example 16.8.

E_1	28.5	25	108	27	1.11	251	Shaly/gas
E_2	26	26	110	26	1.0	265	Shaly/gas
E_3	25.5	31	116	28	0.85	269	Shaly/gas
F	36	21.5	100	30	1.45	206	Shale
G	28.5	17	97	23	1.45	232	Shale

Fig. 16.24 is a crossplot of ρ_{fa} vs. Δt_{fa} (or the FID plot) for values tabulated above. **Fig. 16.25** shows the interpretation of the plot. A point that represents a clean, water-bearing formation is added to the plot at the coordinates 189 μsec/ft and 1 g/cm³ to help with pattern recognition. Tabulated above are identifications of zones according to the interpretation of the FID plot. This identification agrees with that obtained from Fig. 16.19, except for Zones F and G. The interpretation of Fig. 16.19 shows these two zones as being very shaly liquid-bearing sands; however, the FID plot identifies these zones as shales. The difference in the two interpretations is irrelevant because the zones have no potential. The difference results from the selection of the coordinates of the shale point in Fig. 16.19.

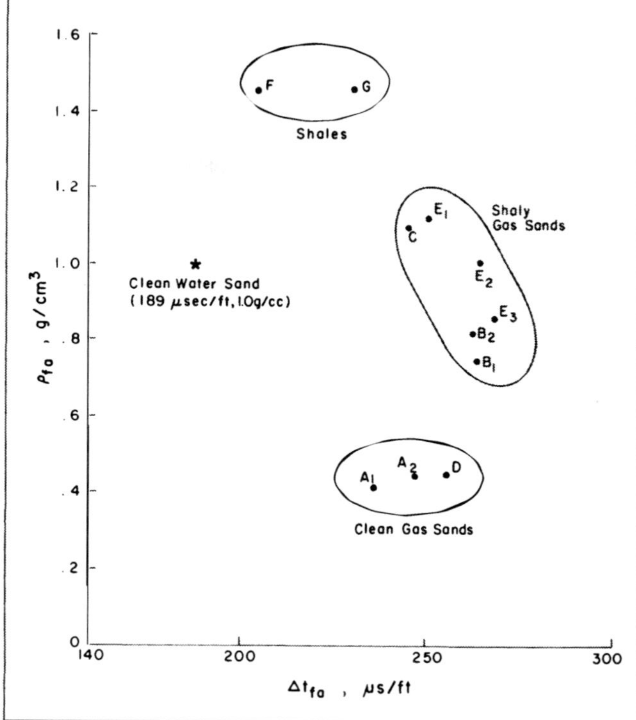

Fig. 16.25—FID plot of Example 16.8 with different patterns recognized.

16.9.4 Gas Detection in Shaly Formations With the Density/Neutron Crossplot. Fig. 16.26 is a schematic of the density/neutron porosity crossplot discussed in Sec. 15.5. A clean, gas-bearing formation represented by Point X is easily identifiable on this crossplot because it clearly falls above the clean sand line. With a shaly zone, however, such as Zone Y, two interpretations are possible. Zone Y could be a shaly liquid-filled formation characterized by

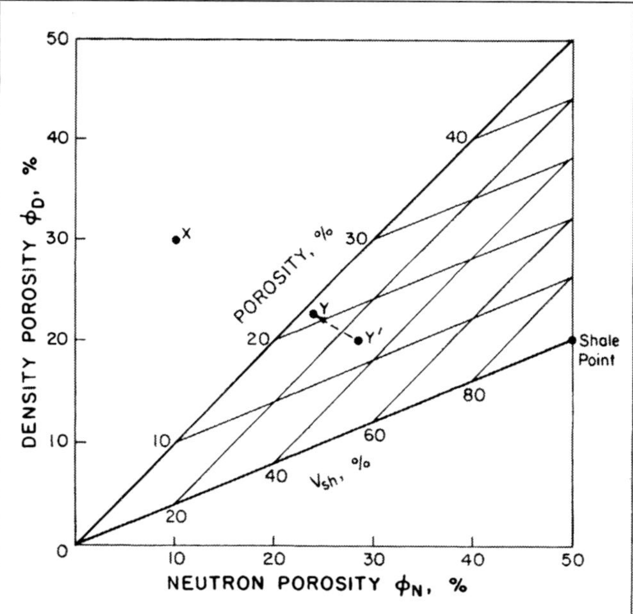

Fig. 16.26—Density/neutron crossplot showing clean gas and shaly gas zones.

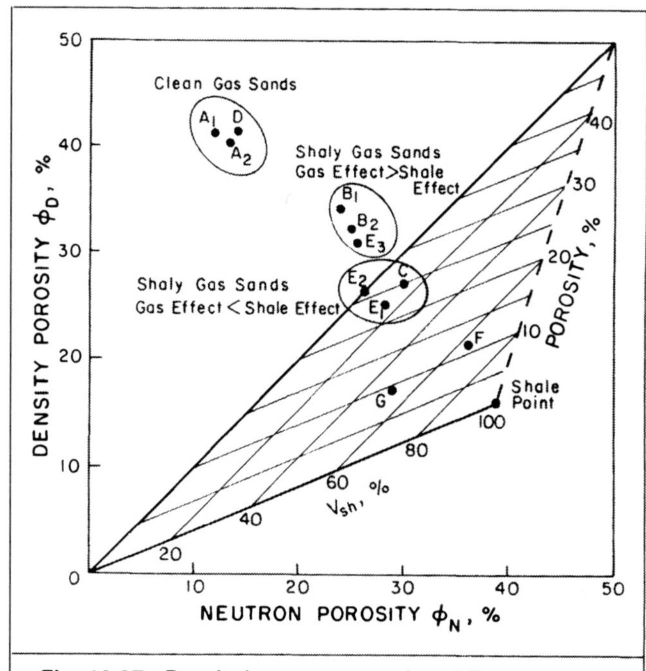

Fig. 16.27—Density/neutron crossplot of Example 16.9.

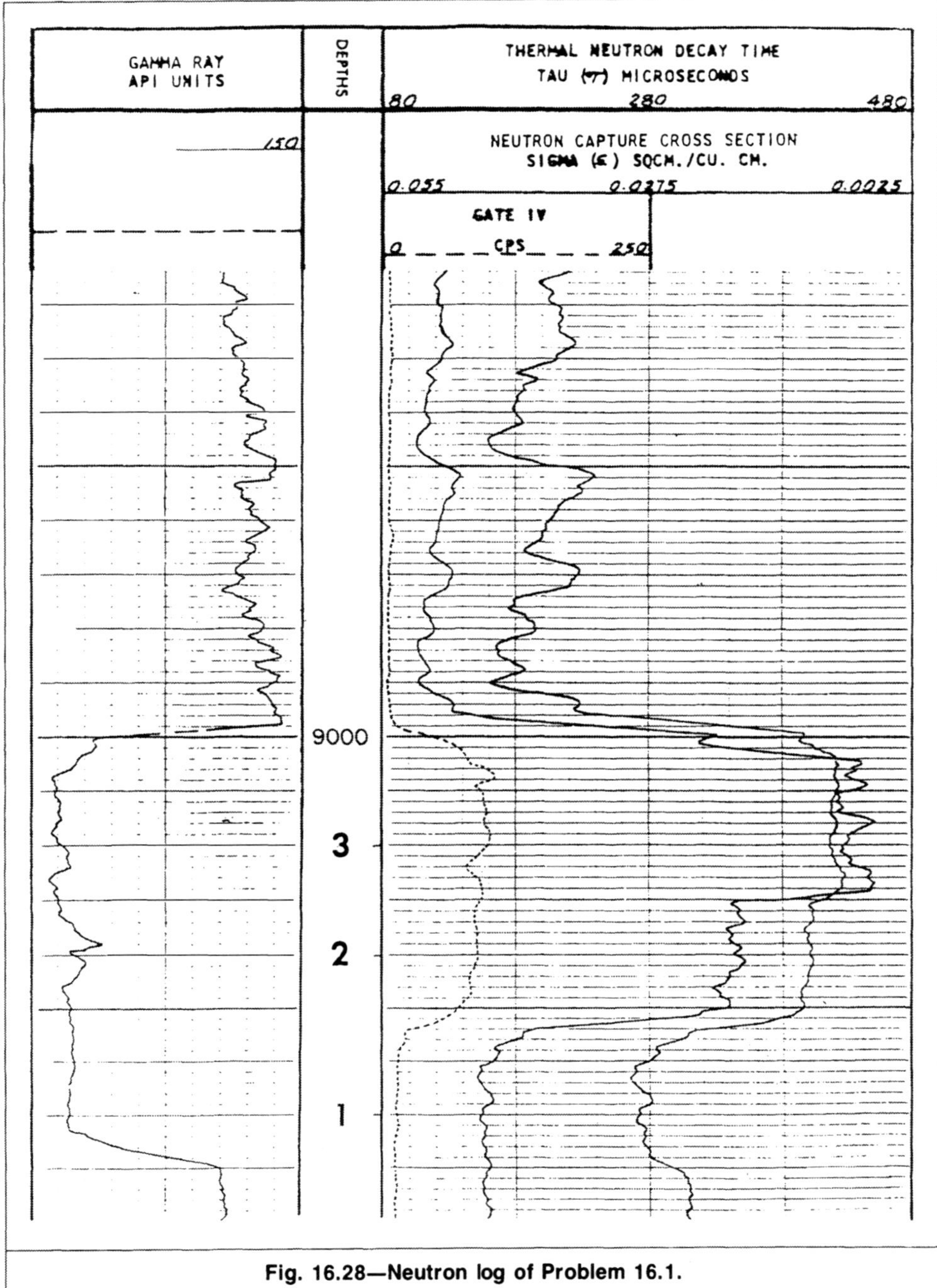

Fig. 16.28—Neutron log of Problem 16.1.

V_{sh}=5% and ϕ=21%. Zone Y could also be a much more shaly, less porous gas-bearing formation. Because of the gas effect on both the density and neutron readings, its position is shifted from Point Y', which represents the liquid-filled point of the same porosity.

If an independent and reliable measure of V_{sh} is available, the presence of gas can be confirmed because

$$(V_{sh})_{crossplot} < V_{sh}. \quad \text{(16.43)}$$

Correct porosity values can also be calculated. Shale-effect-free values of neutron and density porosities, ϕ_N' and ϕ_D', respectively, can be calculated by

$$\phi_N' = \phi_N - V_{sh}\phi_{Nsh} \quad \text{(16.44)}$$

and $\phi_D' = \phi_D - V_{sh}\phi_{Dsh}$. ......(16.45)

If V_{sh}, ϕ_{Dsh}, and ϕ_{Nsh} values are representative, then

$$\phi_D' > \phi_N'. \quad \text{(16.46)}$$

This shows the effect of gas.

The values of ϕ_N' and ϕ_D' corrected for shale effect are then used to calculate the porosity value with the method explained in Sec. 16.6.

If preparation of the density/neutron porosity crossplot is not convenient, $(V_{sh})_{crossplot}$ values can be obtained by solving Eqs. 15.14 and 15.17 simultaneously.

Example 16.9. Using the zones marked on the FDC/CNL and ISF/sonic logs of Figs. 16.16 and 16.23, show that the neutron/density porosity crossplot technique can be used to identify gas zones. For the gas zones identified, calculate the porosity and gas saturation. The log heading lists the following information.

Total depth	10,004 ft
Bit size	9⅞ in.
Mud density	12 lbm/gal
R_m	0.33 Ω·m at 74°F
R_{mf}	0.19 Ω·m at 74°F
Maximum recorded temperature	175°F

Solution. *Gas Zone Identification.* **Fig. 16.27** is a density/neutron crossplot prepared for the interval of interest. Values used to prepare the plot and analysis are tabulated below.

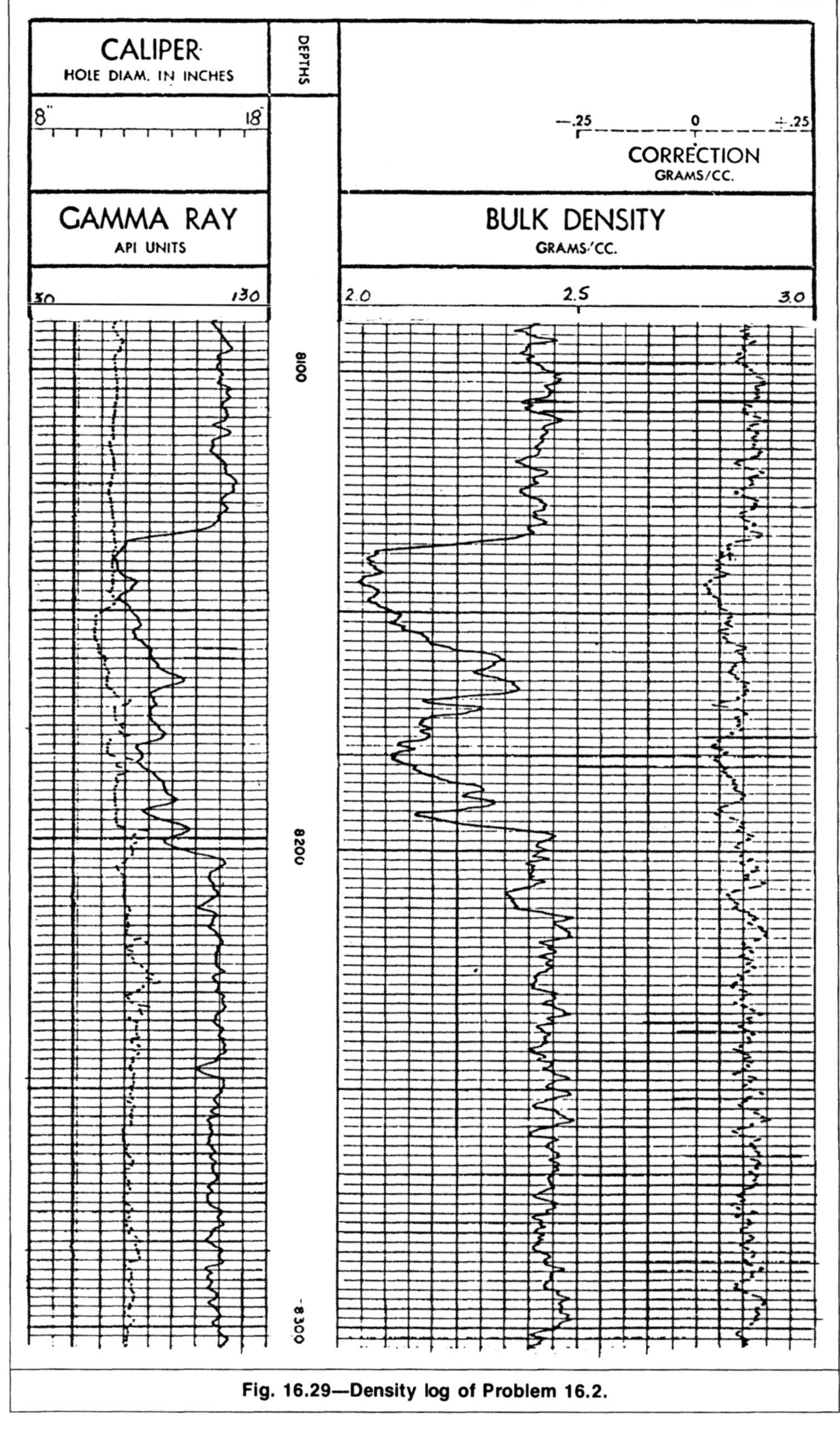

Fig. 16.29—Density log of Problem 16.2.

Zone	ϕ_N (%)	ϕ_D (%)	Gamma Ray (API units)	Crossplot ϕ (%)	Crossplot V_{sh} (%)	Gamma Ray V_{sh} (%)	Fluid Type
A_1	12	41	30	62	—	0	Gas
A_2	13.5	40	30	59	—	0	Gas
B_1	24	34	45	41	—	27	Gas
B_2	25	32	44	37	—	25	Gas
C	30	27	50	25	14	36	Gas*
D	14	41	35	61	—	9	Gas
E_1	28.5	25	47	22	16	31	Gas*
E_2	26	26	45	26	0	27	Gas*
E_3	25.5	31	43	35	—	24	Gas
F	36	21.5	65	11	66	64	Liquid**
G	28.5	17	60	9	52	55	Liquid**

$*(V_{sh})_{crossplot} < (V_{sh})_{gamma\ ray}$.
$**(V_{sh})_{crossplot} = (V_{sh})_{gamma\ ray}$.

For clean and moderately shaly sands, the points that represent the zone are clearly situated off the clean, water-bearing sand line. They can easily be identified as gas zones. Fluid identification in zones that fall within the plot (Zones C, E_1, and E_2) is not possible until the $(V_{sh})_{crossplot}$ is compared with $(V_{sh})_{gamma\ ray}$. Such

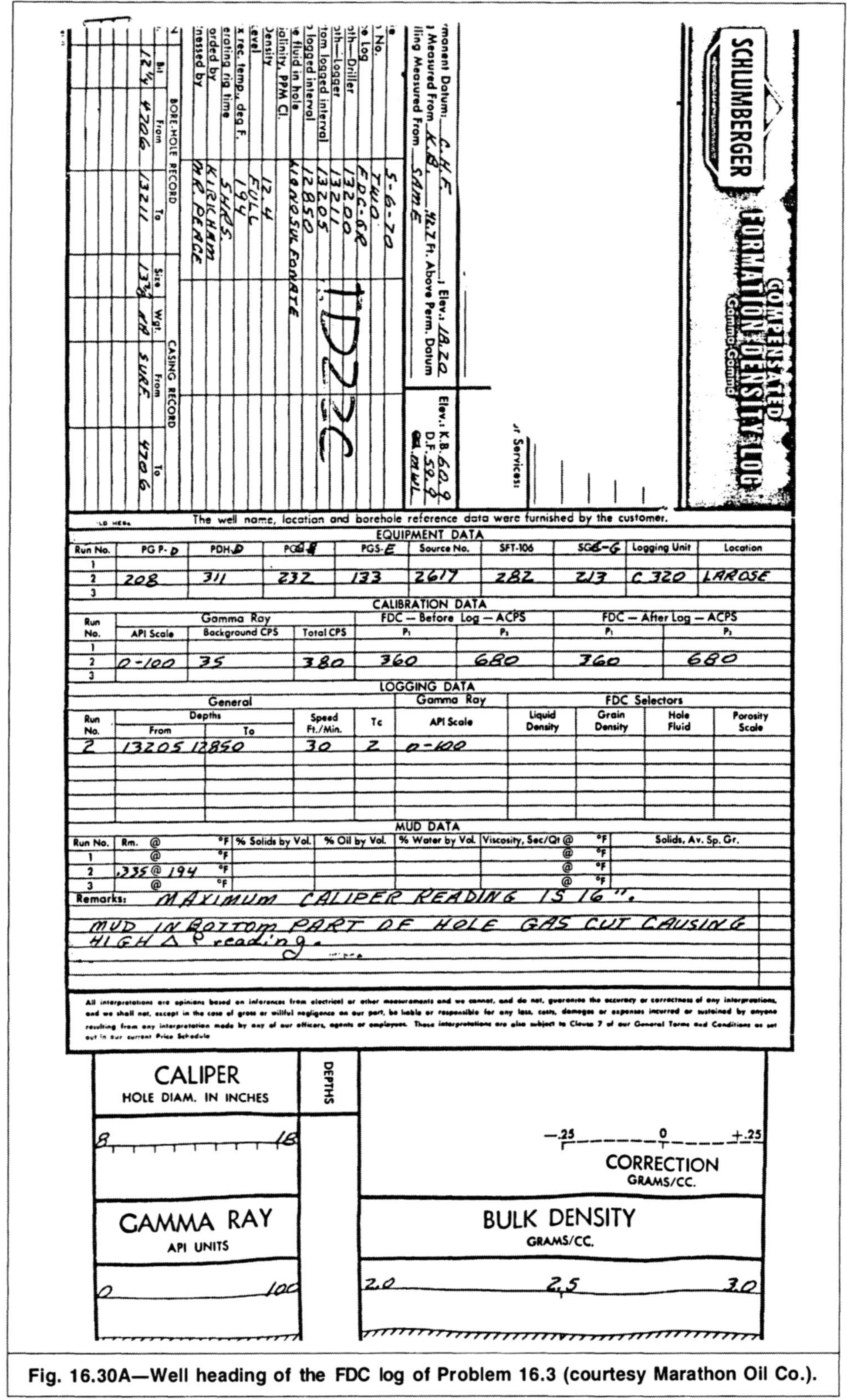

SCHLUMBERGER

COMPENSATED FORMATION DENSITY LOG
Gamma-Gamma

The well name, location and borehole reference data were furnished by the customer.

EQUIPMENT DATA

Run No.	PG P-D	PDH-D	PG[illegible]	PGS-E	Source No.	SFT-106	SGB-G	Logging Unit	Location
1									
2	208	311	232	133	2617	282	213	C 320	LAROSE
3									

CALIBRATION DATA

Run No.	Gamma Ray			FDC — Before Log — ACPS		FDC — After Log — ACPS	
	API Scale	Background CPS	Total CPS	P_1	P_2	P_1	P_2
1							
2	0-100	35	380	360	680	360	680
3							

LOGGING DATA

Run No.	General: Depths From	General: Depths To	Speed Ft./Min.	Tc	Gamma Ray: API Scale	FDC Selectors: Liquid Density	Grain Density	Hole Fluid	Porosity Scale
2	13205	12850	30	2	0-100				

MUD DATA

Run No.	Rm. @ °F	% Solids by Vol.	% Oil by Vol.	% Water by Vol.	Viscosity, Sec/Qt @ °F	Solids, Av. Sp. Gr.
1	@ °F				@ °F	
2	.335 @ 194 °F				@ °F	
3	@ °F				@ °F	

Remarks: MAXIMUM CALIPER READING IS 16".
MUD IN BOTTOM PART OF HOLE GAS CUT CAUSING HIGH Δρ reading.

All interpretations are opinions based on inferences from electrical or other measurements and we cannot, and do not, guarantee the accuracy or correctness of any interpretations, and we shall not, except in the case of gross or willful negligence on our part, be liable or responsible for any loss, costs, damages or expenses incurred or sustained by anyone resulting from any interpretation made by any of our officers, agents or employees. These interpretations are also subject to Clause 7 of our General Terms and Conditions as set out in our current Price Schedule

Fig. 16.30A—Well heading of the FDC log of Problem 16.3 (courtesy Marathon Oil Co.).

comparison indicates that the $(V_{sh})_{crossplot}$ is much lower than the actual shale content; hence, gas is present in these zones. Note that liquid is indicated as the fluid type in Zones F and G because, for all practical purposes, $(V_{sh})_{crossplot}=(V_{sh})_{gamma\ ray}$.

Porosity Determination. Porosity log readings are corrected with Eqs. 16.44 and 16.45. The porosity is then calculated either from Eq. 16.20 or 16.24 or from Fig. 16.13.

Zone	ϕ_N' (%)	ϕ_D' (%)	ϕ From Eq. 16.20 (%)	ϕ From Eq. 16.24 (%)	ϕ From Fig. 16.13 (%)
A_1	12	41	30	30	28
A_2	13.5	40	30	29	28
B_1	13.6	29.6	23.5	23	22.5
B_2	15.3	27.9	23	22.5	22.5
C	16.2	21.2	19	19	19
D	10.5	39.5	29	29	27
E_1	16.8	20.1	19	18.5	19
E_2	15.6	21.6	19	19	18.5
E_3	16.5	27.2	23	22.5	22.5

Shale-effect-free density porosity is now much larger than shale-effect-free neutron porosity because only the gas effect remains.

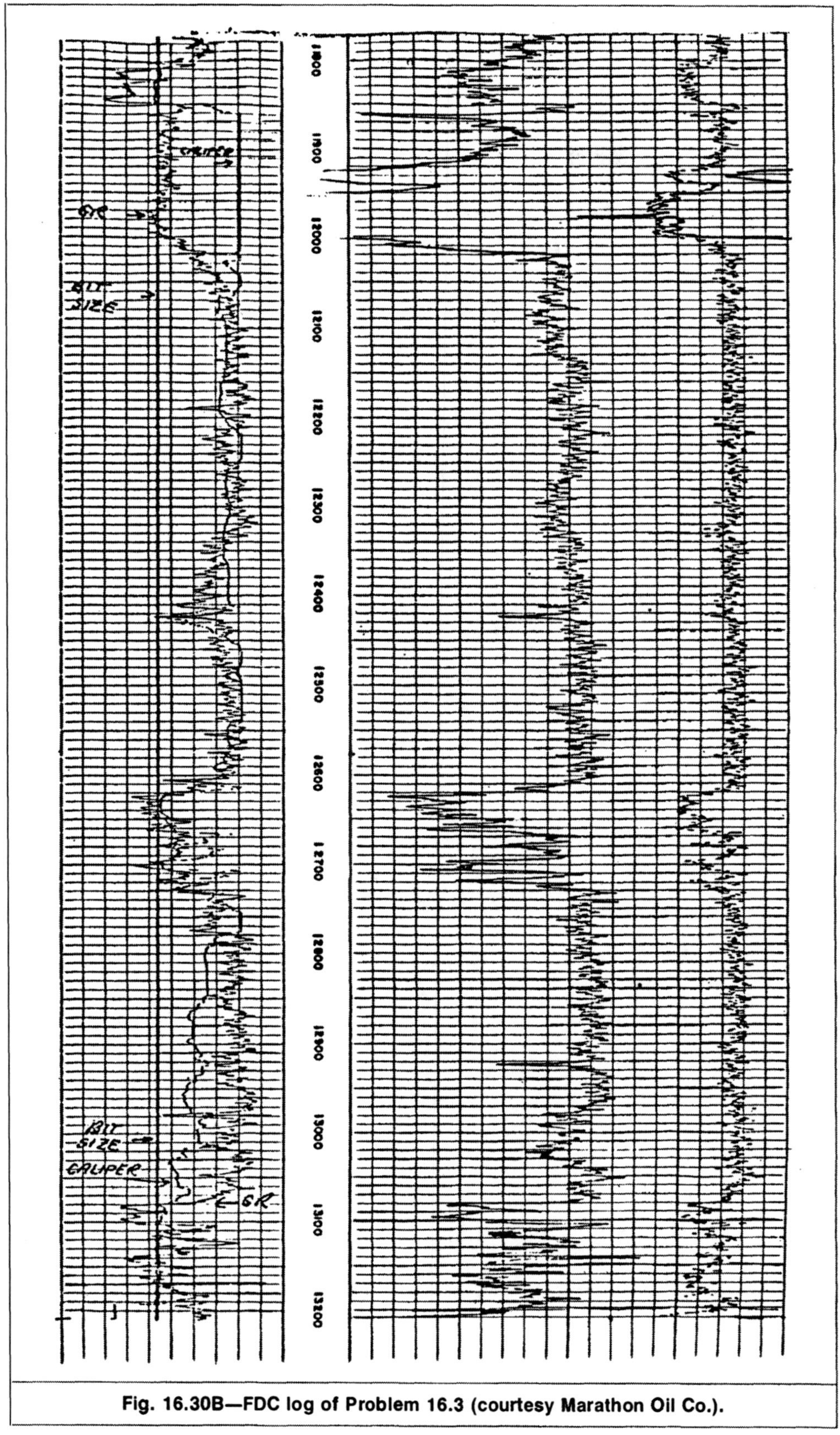

Fig. 16.30B—FDC log of Problem 16.3 (courtesy Marathon Oil Co.).

Porosity estimates from Eqs. 16.16 and 16.24 and from Fig. 16.13 are in close agreement.

Gas Saturation Determination. From log reading information,

$g_G=[(174-70)/10{,}004]100=1.04°F/100$ ft,

$T_f=70+1.04(9{,}500/100)=168°F$,

and $R_{mf}=0.19[(74+6.77)/(168+6.77)]=0.088\ \Omega\cdot m$ at 168°F.

From the resistivity log, $E_{SSP}=-30$ mV, $R_{sh}=1.2\ \Omega\cdot m$, and $R_{sh}/R_{mf}=13.6$.

From Fig. 6.19, $R_{mf}/R_w=2.1$ and $R_w=0.042\ \Omega\cdot m$ at 168°F. The same R_w value is obtained from Fig. 6.14. The water saturation can be estimated with Eq. 15.21. The saturation values are tabulated below.

Zone	R_t ($\Omega\cdot m$)	ϕ (%)	V_{sh} (%)	S'_w* (%)	ΔS_w** (%)	S_w† (%)	S_g (%)
A_1	30	28	0	11	0	11	89
A_2	20	28	0	14	0	14	86
B_1	4.7	22.5	27	36	10	26	74

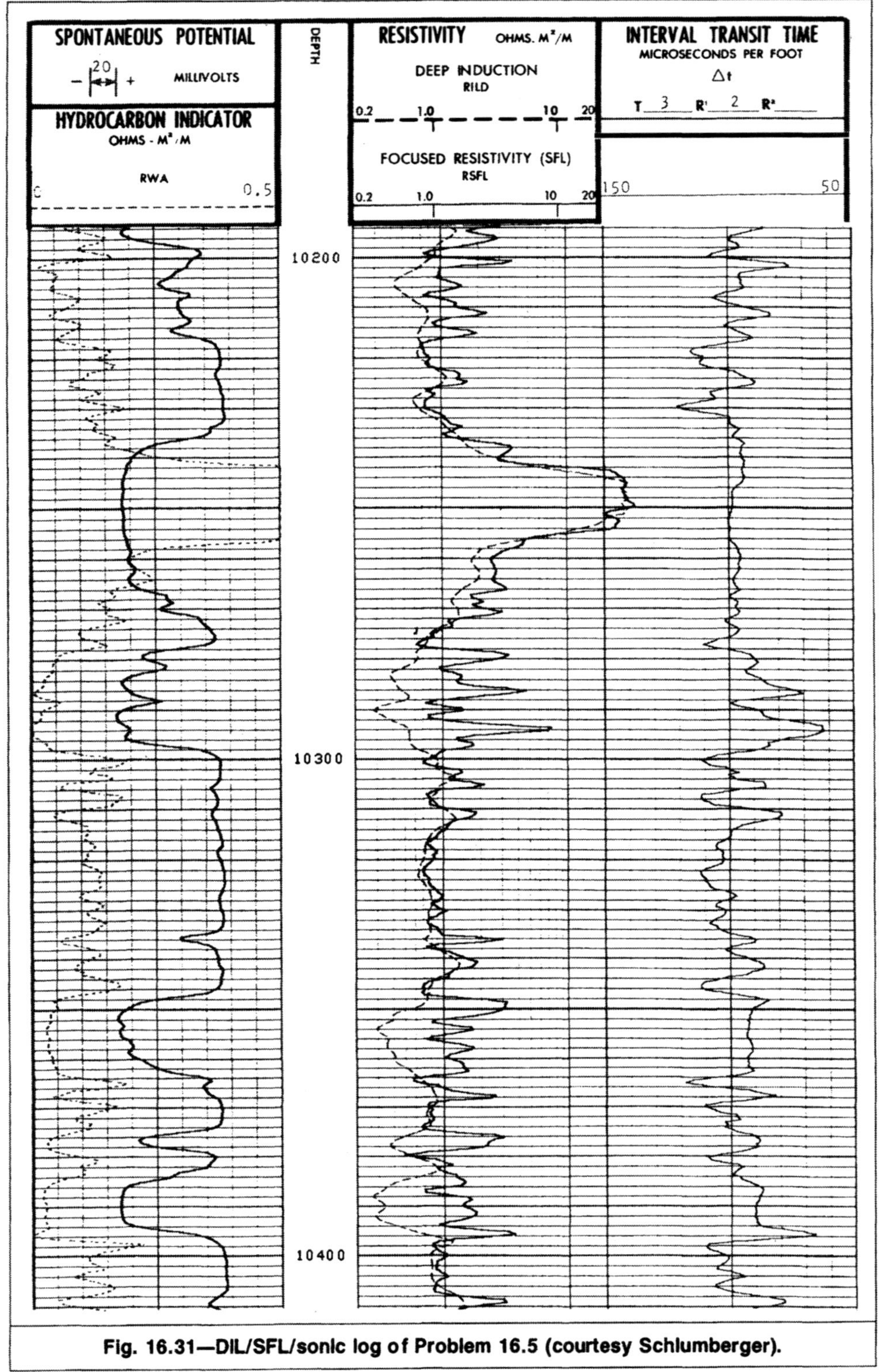

Fig. 16.31—DIL/SFL/sonic log of Problem 16.5 (courtesy Schlumberger).

B_2	3.2	22.5	25	45	10	35	65	
C	1.7	19	36	73	16	57	43	
D	9.5	27	9	21	3	18	82	
E_1	0.85	19	31	106	14	92	8	
E_2	1.5	18.5	27	77	12	66	34	
E_3	2.5	22.5	24	50	9	41	59	

*$S'_w = (0.81 R_w / \phi^2 R_t)^{1/2}$.
**$\Delta S_w = (V_{sh} R_w)/0.4\, \phi R_{sh}$.
†$S_w = S'_w - \Delta S_w$.

Review Questions

1. How does the presence of gas in a formation affect the response of the neutron, density, and sonic logs?
2. What parameters determine the response of porosity tools in a gas-bearing formation?
3. How many porosity logs are optimally required to evaluate a gas-bearing formation? Explain the reason for this number.
4. Why is the absence of gas anomaly on the sonic log inconclusive?
5. Under what conditions can the neutron or density log alone indicate the gas/liquid contact in a formation?
6. Why does the presence of clay in the formation render gas detection more difficult?
7. Explain the concept of gas zone identification with the neutron/density porosity overlay.
8. What parameters affect the magnitude and shape of the separation between the neutron and density porosity curves in a clean, gas-bearing formation?

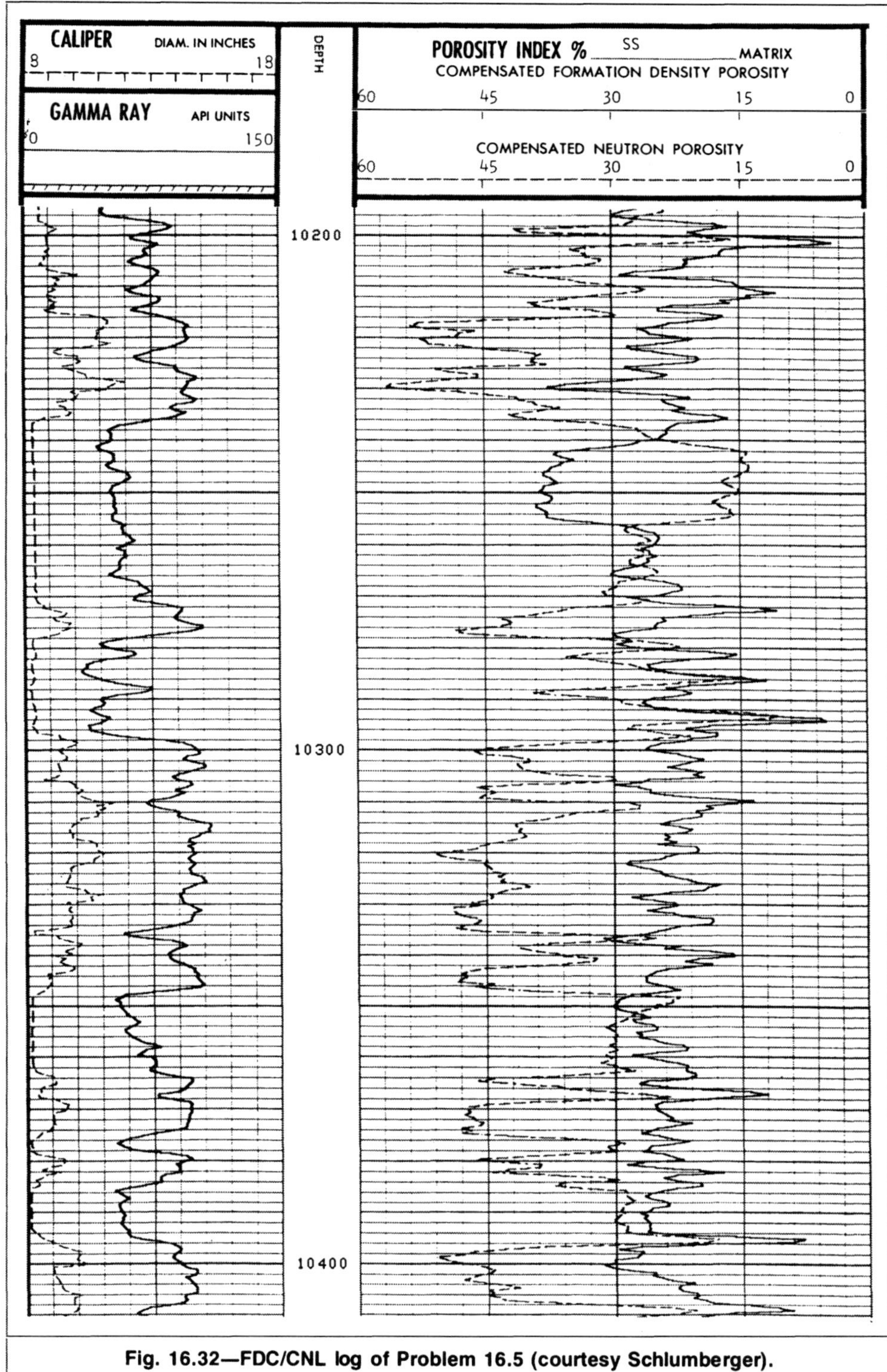

Fig. 16.32—FDC/CNL log of Problem 16.5 (courtesy Schlumberger).

9. What are the assumptions implied in Eqs. 16.19 and 16.24 that are used to approximate the porosity value in gas-bearing formations? Which of these two equations has fewer assumptions?

10. What is meant by excavation effect?

11. What is the difference between the charts of Figs. 16.8 and Fig. 16.11?

12. Can the values of porosity and deep resistivity displayed by logs in a gas-bearing formation be substituted directly into Archie's saturation equation? Explain.

13. How does the presence of gas affect the following interpretation methods: R_{wa}, R_o overlay, porosity/resistivity crossplots, and lithology/porosity crossplots.

14. Show how the terms $\Delta\rho_g$, $\Delta\phi_{Dg}$, and $\Delta\phi_{Ng}$, given by Eqs. 16.32 through 16.34, can be derived from Eqs. 16.3, 16.4, and 16.8, respectively.

15. Do the neutron and density porosity curves always show a separation in gas-bearing formations? Explain.

Problems

16.1 **Fig. 16.28** shows a neutron log recorded in a cased hole. The upper sand shown by the log is an oil reservoir with a gas cap. Can the gas/oil contact be determined from this log alone by mere visual inspection?
What quantitative interpretation approach can be suggested to improve on qualitative determination of the gas/oil contact?

16.2 The density log in **Fig. 16.29** shows a 60-ft-thick sand. The resistivity log (not shown) displays a relatively high reading of the deep resistivity, indicating the presence of hydrocarbon in the sand. Is it possible to tell from the density log alone the type of hydrocarbon—i.e., gas, oil, or both?

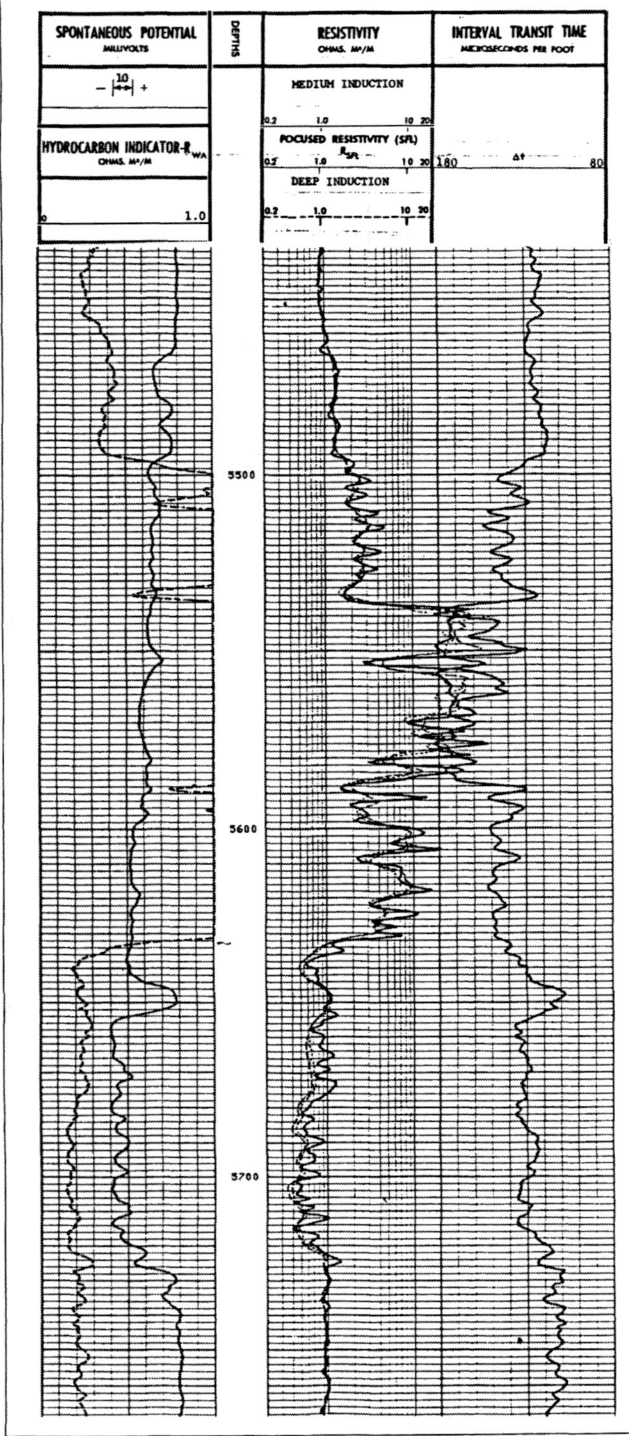

Fig. 16.33—DIL/SFL/sonic log of Problem 16.9 (courtesy Schlumberger).

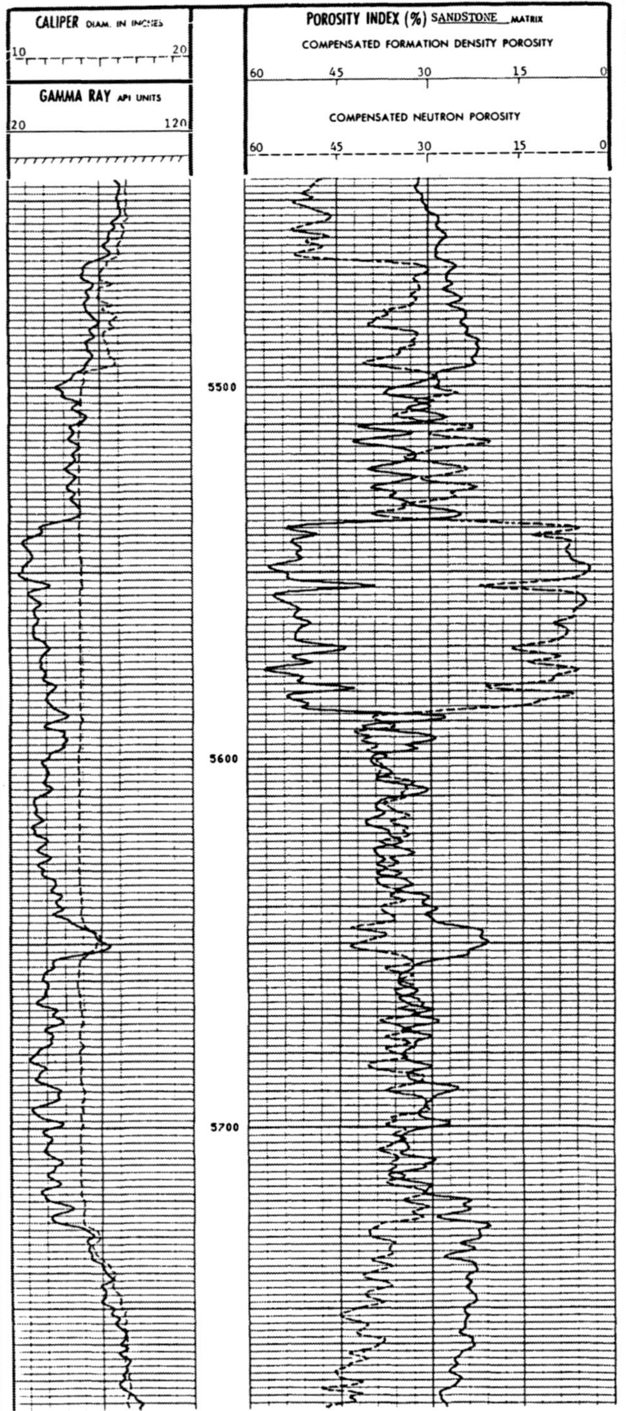

Fig. 16.34—Simultaneous FDC/CNL log of Problem 16.9 (courtesy Schlumberger).

16.3 **Figs. 16.30A and 16.30B** show a density log. If sands encountered by this borehole exhibit a porosity of about 20% and if gas-bearing sands are present in this well, mark these gas sands on the log. Explain your interpretation approach.

16.4 Construct a neutron/density porosity crossplot for limestone similar to that of Fig. 16.8.

16.5 **Figs. 16.31 and 16.32** show an attractive zone in the interval of 10,240 to 10,256 ft. Give a complete evaluation of the zone in terms of (a) hydrocarbon presence, (b) hydrocarbon type, (c) lithology, (d) porosity, and (e) fluid saturations. Log heading lists the following data.

Total depth	11,420 ft
Bit size	8½ in.
Mud density	10.2 lbm/gal
R_m at 114°F	0.54 Ω·m
R_{mf} at 74°F	0.41 Ω·m
Maximum recorded temperature	195°F

16.6 The readings of the FDC and CNL logs in a clean hydrocarbon-bearing formation are 2.3 and 10% apparent limestone porosity, respectively. Determine the hydrocarbon type, porosity, and lithology if filtrate salinity is 20,000 ppm, formation temperature is 250°F, and formation pressure is 4,000 psia.

16.7 The arrows on the lower right side of Fig. 16.14 represent approximate hydrocarbon shifts for various values of gas density. Explain how these shifts were calculated.

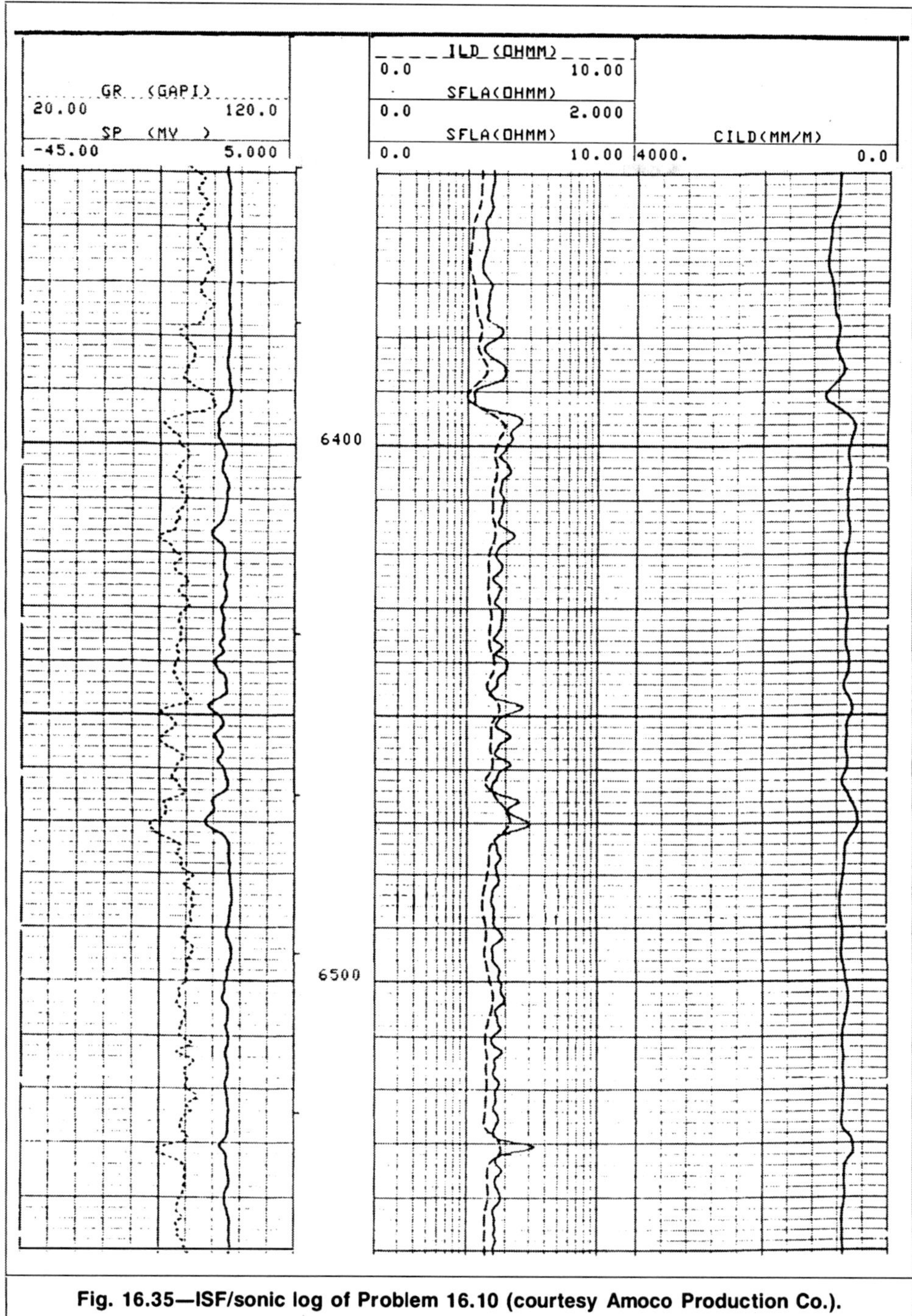

Fig. 16.35—ISF/sonic log of Problem 16.10 (courtesy Amoco Production Co.).

16.8 The FDC/CNL log of Fig. 14.24 shows a separation in the interval of 10,566 to 10,602 ft, where $\phi_D > \phi_N$. Does this interval contain gas? Explain.

16.9 The DIL/SFL/sonic and FDC/CNL logs of **Figs. 16.33 and 16.34** show a thick sand between 5,468 and 5,726 ft. Considering the response of the different tools, the sand can be divided into five major intervals.

Interval A—5,468 to 5,494 ft.
Interval B—5,494 to 5,531 ft.
Interval C—5,535 to 5,588 ft.
Interval D—5,590 to 5,632 ft.
Interval E—5,632 to 5,726 ft.

Identify the fluid type in each interval. Select and determine the porosity and fluid saturation in several zones within each interval. Log heading lists the following information.

Location	Gulf of Mexico
Total depth	7,180 ft
Bit size	12¼ in.
Mud type	Ligno
Mud density	13.0 lbm/gal
R_m	0.79 Ω·m at 78°F
R_{mf}	0.32 Ω·m at 78°F
Maximum recorded temperature	140°F

16.10 The ISF/sonic and FDC/CNL logs of **Figs. 16.35 and 16.36** show a poorly developed sand in the interval 6,394 to 6,472 ft. The sand is very shaly and displays a very low resistivity value. A production test yielded a substantial gas flow rate. Do the log data correlate with the test results? Explain. Log heading lists the following data.

Location	Offshore Louisiana
Total depth	6,816 ft
Bit size	8¾ in.
Mud density	11 lbm/gal
R_m	0.6 at 74°F
R_{mf}	0.42 at 74°F
Maximum recorded temperature	136°F

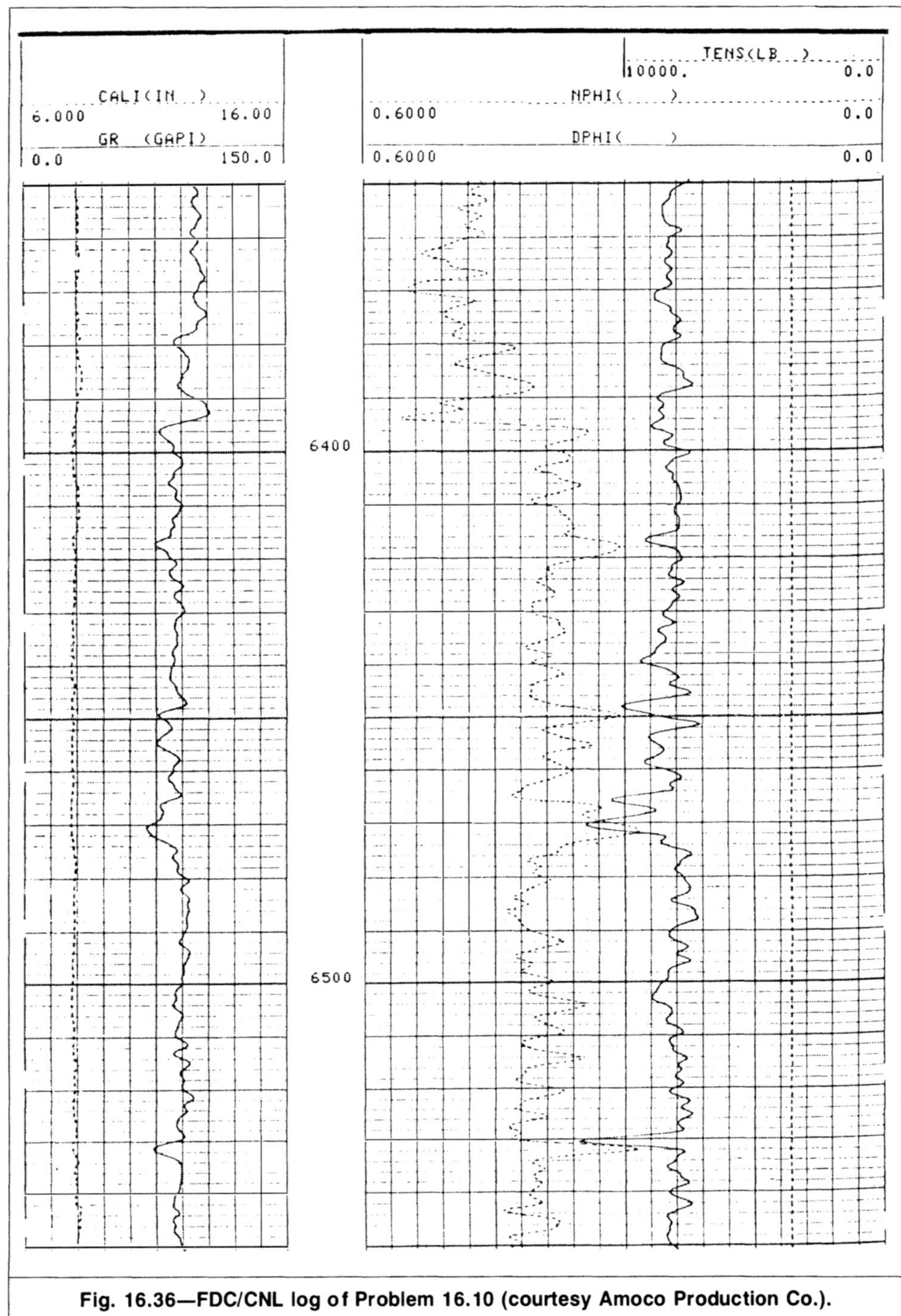

Fig. 16.36—FDC/CNL log of Problem 16.10 (courtesy Amoco Production Co.).

16.11 Interval X in **Figs. 16.37 and 16.38** is clean at the top but becomes shaly as depth increases. The resistivity and R_{wa} curves indicate the presence of hydrocarbons in the interval. The separation at the top of the neutron and density overlay indicates gas. The separation disappears at 9,898 ft. Does a gas/oil contact exist at this depth? Explain.

What is the potential of the shaly formation in Interval Y?

Nomenclature

a = coefficient in Archie's F-ϕ empirical relationship
A = slope
B_{cp} = compaction factor
F = formation factor
g_G = geothermal gradient, °F/100 ft
H = hydrogen
n = fractional salinity
R = resistivity, Ω·m
S = saturation, fraction
S_{wH} = equivalent water saturation based on hydrogen content of pore fluids, fraction
Δt = transit time, μsec/ft
T_f = formation temperature, °F
V_{sh} = shale content, fraction
γ = gamma ray resistivity, API units
ρ = density, g/cm^3
ϕ = porosity, fraction
ϕ_{Dg} = density porosity reading in a clean, gas-bearing formation, fraction
ϕ_{Nc} = neutron log porosity corrected for the excavation effect
ϕ_{Ng} = neutron porosity reading in a clean, gas-bearing formation, fraction
$\Delta\phi_{Nex}$ = excavation effect which should be added to the neutron log porosity reading that does not already contain the correction

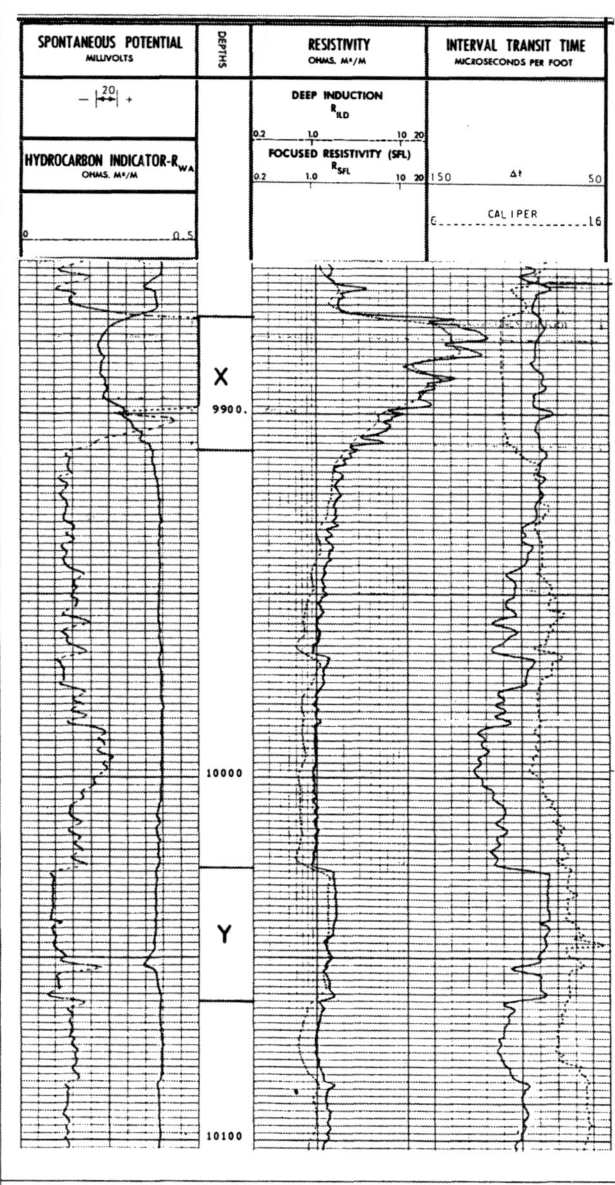

Fig. 16.37—ISF/sonic log of Problem 16.11 (courtesy Schlumberger).

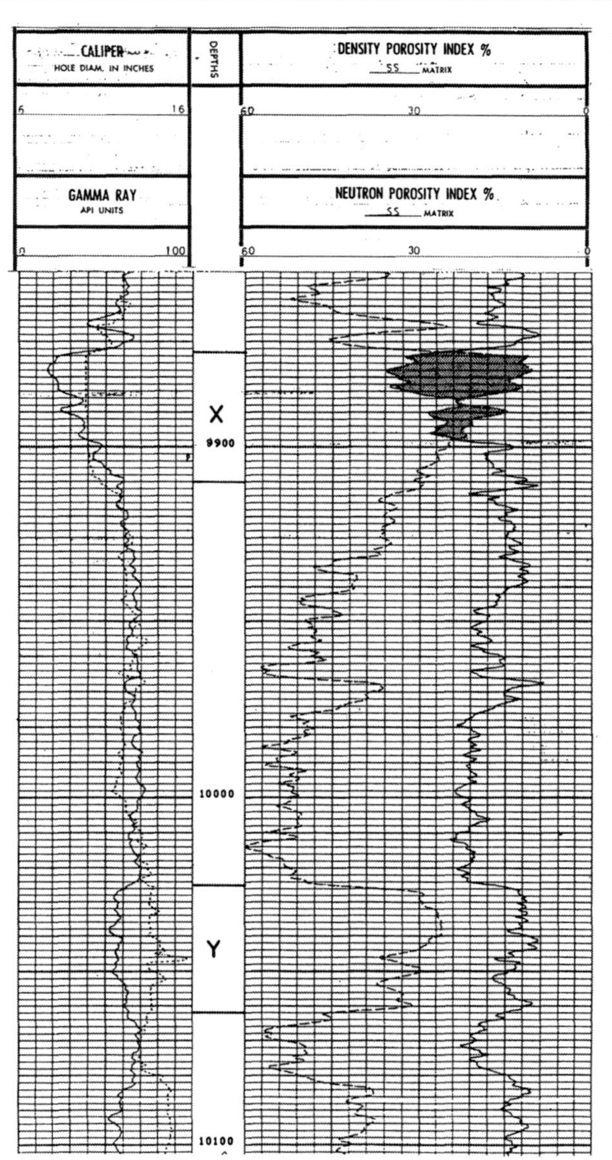

Fig. 16.38—FDC/CNL log of Problem 16.11 (courtesy Schlumberger).

Subscripts

a = apparent
b = bulk
D = density tool
f = fluid
g = gas
m = mud
ma = matrix
mf = mud filtrate
N = neutron tool
o = formation 100% saturated with water (used in R_o only)
o = oil (except when used with resistivity)
S = sonic tool
sh = shale
w = water
xo = flushed zone

References

1. Gardner, G.H.F., Wyllie, M.R.J., and Droschak, D.M.: "Effects of Pressure and Fluid Saturation on the Attenuation of Elastic Waves in Sands," *JPT* (Feb. 1964).
2. Gardner, G.H.F. and Harris, M.H.: "Velocity and Attenuation of Elastic Waves in Sands," paper presented at the Ninth Annual SPWLA Logging Symposium, 1968.
3. Walker, T.: *Acoustic Character of Unconsolidated Sands,* Welex Publication No. L.35.
4. *Log Interpretation—Vol. 1: Principles,* Schlumberger, Houston (1972).
5. Gaymard, R. and Poupon, A.: "Response of Neutron and Formation Density Logs in Hydrocarbon-Bearing Formations," *Log Analyst* (Sept.-Oct. 1968) **IX**, No. 5.
6. Segesman, F.F. and Liu, O.Y.: "The Excavation Effect," paper presented at the 12th Annual SPWLA Logging Symposium, May 1971.
7. Patrickis, A.: "The Determination of Residual Gas Saturation Using Neutron and Density Logs," Petroleum Engineering Dept., Louisiana State U., Baton Rouge (May 1986).
8. Poupon, A., Hoyle, W.R., and Schmidt, A.W.: "Log Analysis in Formations With Complex Lithologies," *JPT* (Aug. 1971) 995–1005.
9. Bettis, F.: "Gas Detection in Sands of High Silt-Clay Content in the Cook Inlet Area," paper presented at the 17th Annual SPWLA Logging Symposium, June 1976.
10. Vohs, J.B.: "Log Analysis in the Shallow Oil Sands of the San Joaquin Valley, California," paper EE presented at the 17th Annual SPWLA Logging Symposium, June 1976.
11. Davis, D.G.: "Histogrammatic Projection of Gas and Heavy Mineral Effects in Shaly Sands," paper Z presented at the 17th Annual SPWLA Logging Symposium, June 1976.
12. Bateman, R.M.: "The Fluid Identification Plot," paper C presented at the 18th Annual SPWLA Logging Symposium, June 1977.

Appendix A
SI Metric Conversion Table

Unit		Factor	Exponent		SI unit
acre	×	4.046 873	E−01	=	ha
acre-ft	×	1.233 489	E+03	=	m^3
aA	×	1.0*	E+01	=	A
aC	×	1.0*	E+01	=	C
aF	×	1.0*	E+09	=	F
aV	×	1.0*	E−08	=	V
a℧	×	1.0*	E+09	=	S
aΩ	×	1.0*	E−09	=	Ω
Å	×	1.0*	E+02	=	m^2
bbl	×	1.589 873	E−01	=	m^3
cp	×	1.0*	E+00	=	mPa·s
curie	×	3.7*	E+10	=	Bq
cycles/sec	×	1.0*	E+00	=	Hz
°C		°C+273.15		=	K
eV	×	1.602 19	E−19	=	J
EMU of capacitance	×	1.0*	E+09	=	F
EMU of current	×	1.0*	E+01	=	A
EMU of electrical potential	×	1.0*	E−08	=	V
EMU of inductance	×	1.0*	E−09	=	H
EMU of resistance	×	1.0*	E−09	=	Ω
ESU of capacitance	×	1.112 650	E−12	=	F
ESU of current	×	3.335 6	E−10	=	A
ESU of electrical potential	×	2.997 9	E+02	=	V
ESU of inductance	×	8.987 554	E+11	=	H
ESU of resistance	×	8.987 554	E+11	=	Ω
ft	×	3.048*	E−01	=	m
ft (U.S. survey)	×	3.048 006	E−01	=	m
ft^2	×	9.290 304*	E−02	=	m^2
ft^3	×	2.831 685	E−02	=	m^3
°F		(°F−32)/1.8		=	°C
°F		(°F+459.67)/1.8		=	K
grain	×	6.479 891*	E−05	=	kg
in.	×	2.54*	E+00	=	cm
$in.^2$	×	6.451 6*	E−04	=	m^2
K		K−273.15		=	°C
lbm/gal (U.K. liquid)	×	9.977 633	E+01	=	kg/m^3
lbm/gal (U.S. liquid)	×	1.198 264	E+02	=	kg/m^3
psi	×	6.894 757	E+00	=	kPa
P	×	1.0*	E−01	=	Pa·s
°R		°R/1.8		=	K
μin.	×	2.54*	E−08	=	m
μsec/ft	×	3.280 840	E+00	=	μs/m
℧	×	1.0*	E+00	=	S

*Conversion factor is exact.

Appendix B
Answers to Selected Problems

Log interpretation requires analysts to determine tool responses, to choose from several possible petrophysical models, and to assume values for missing parameters. Interpretation results vary among analysts but should remain within a reasonable range if each analyst makes judicious choices. The answers to interpretation problems provided in this Appendix are not the only possible answers. However, they are acceptable. Other acceptable answers should be as close as possible.

Chapter 1

Problem 1.1
a. r=600Ω.
b. R=157Ω·cm.
c. R=1.57Ω·m.
d. c=0.00167 ℧.
e. C=0.637 ℧/m and 637 m℧/m.

Problem 1.2
a. $F=\pi/4$.
b. $F=\phi^{-1}$.

Problem 1.3
a. ϕ=47.6%.
b. F=2.64.
c. m=1.3.
d. Unconsolidated sands.

Problem 1.4
a. R=2 Ω·m.
b. F=20.
c. ϕ=19%.

Problem 1.5
a. ϕ=18%.
b. ϕ=20%.
c. r=103.9 Ω.
d. R_o=0.844 Ω·m.
e. S_w=40%.
f. r=520 Ω, 5.28 Ω·m.

Problem 1.6
a. 150 g salt and 950 g water.
b. 50 cm^3.

Problem 1.7
a. 145,000 ppm.
b. 159,400 mg/L.
c. 2.72 g eq/L.
d. 8,456 grains/gal.
e. 0.043 Ω·m and 0.025 Ω·m.

Problem 1.8
a. 170,000 ppm.

Problem 1.10
a. 62,750 ppm.
b. R_w=0.119 Ω·m.
c. 63,100 ppm and 0.119 Ω·m.
d. Yes.

Problem 1.11
a. 3,785 ppm.
b. 1.6 Ω·m.
c. 1,266 ppm.
d. No.

Problem 1.12
m=2.2 and 1.7.

Problem 1.13
From Fig. 1.14, ϕ=18% and the possible range is 15% to 22%.

Problem 1.14
a. ϕ_{15}=14%, F_{15}=32.3.
c. Yes, 0.67.

Problem 1.15
a. ϕ=31%.
b. $n\simeq$ 2.45.

Problem 1.16
$m\simeq1.8$, $n\simeq1.75$.

Problem 1.18
a. R_o=9 Ω·m using a=0.81 and m=2.
b. R_t=0.32 Ω·m.
c. No.

Problem 1.19
$S_o\simeq$80%,

Problem 1.20
$R_o\simeq$2 Ω·m

Problem 1.21
a. $\phi\simeq$6.5%.
b. S_o=50%.

Problem 1.22
a. $\phi\simeq$10%.
b. $S_o\simeq$59%.
c. Tight zone, $\phi\simeq$6%.

Problem 1.23
a. $S_o\simeq$66%.

Problem 1.24
a. S_w=47%.
b. S_w=44% to 55%.

Problem 1.25
a. $(S_o)_{min}$=0%, $(S_o)_{max}$=40%.

Problem 1.26
a. S_w=58%.
b. S_w=46%.
c. S_w=52%.

Problem 1.27
a. S_g=0%.
b. $S_g\cong$ 38%.
c. $S_g\simeq$32%.

Chapter 2

Problem 2.1
938.2 MeV.

Problem 2.3
$f=5\times10^{14}$ Hz, E=2.1 eV.

Problem 2.4
$f=3.75\times10^{20}$ Hz, E=1.55 MeV.

Problem 2.9
109×10^6 years.

Problem 2.15
3×10^{23} electrons/g.

Problem 2.16
5.97×10^{23} electrons/g.

Problem 2.17
a $v=1.96\times10^9$ cm/s.
b. 0, and 1.96×10^9 cm/s.
c. $v=1.66\times10^9$ cm/s, E=1.4 MeV.

Problem 2.18
a. 18 collisions.
b. 115 collisions.

Problem 2.19 $h=0.45$ cm, $\sigma=37$ barns.
Problem 2.20 $\Sigma=22.2$ c.u.
Problem 2.23 b. $\Sigma_w=59.7$ c.u.
Problem 2.25 a. $I=532$ photons/s·cm^2.
b. $I=560$ photons/s·cm^2.
c. $I=926$ photons/s·cm^2.
Problem 2.26 12.5%.
Problem 2.27 a. $C_d=0.035$ day^{-1} and 4×10^{-7}s^{-1}.
b. 25 mCuries, 1.52×10^{-4} mCuries.
c. 4.6×10^{15} atoms, 14×10^{9} atoms.
Problem 2.33 $E=2.6$ MeV.
Problem 2.34 a. $I=228$ photons/s·cm^2.
b. $\alpha=0.227$ cm^{-1}.
c. $I=195$ photons/s·cm^2.
Problem 2.35 a. 6.64×10^{22} atoms/cm^3.
b. 6.81×10^{22} atoms/cm^3.
c. 1.49×10^{21} atoms/cm^3.

Chapter 3

Problem 3.2 a. $\alpha_{pc}=14.9°$.
b. $\alpha_{sc}=26.6°$.
Problem 3.4 $\Delta t_p=384.7$ μs, $\Delta t_s=620.9$ μs.
Problem 3.5 $v_p=2{,}386.3$ m/s, $v_s=1{,}014.5$ m/s.

Chapter 4

Problem 4.2 a. 8.3 in, 12.5 in.
b. ≃0.4 in.
c. different types of clay minerals.
Problem 4.4 c. $T=70e^{8.26\times10^{-5}D}$.
Problem 4.5 a. 3,000 ppm.
b. $R_m=0.08$ Ω·m.
c. $R_{mf}=1.33$ Ω·m at 90°F.
$R_{mf}=0.59$ Ω·m at 180°F.
e. $R_{mc}=0.66$ Ω·m at 180°F.
Problem 4.6 $T_f=180°$F, $R_m=0.18$ Ω·m, $R_{mf}=0.06$ Ω·m, $R_{mc}=0.76$ Ω·m.
Problem 4.7 b. $T_s=90°$F, $m_T=6.8\times10^{-5}$ ft^{-1}.
c. $T_f=460°$F.
Problem 4.8 a. ≃3,250 ppm.
b. 2,681 ft above sea level.
c. $T_f=133°$F at 5,850 ft.
d. $g_G=1.08°$F/100 ft.
e. $T_f=131°$F, $R_m=1.17$ Ω·m, $R_{mf}=1.01$ Ω·m, $R_{mc}=1.05$ Ω·m.
Problem 4.9 $T_f=116°$F.

Chapter 5

Problem 5.1 0.1%.
Problem 5.3 c. $R_{SN}=4.8$ Ω·m, $R_{LN}=3.9$ Ω·m.
d. Zone D is shale.
e. $R_{SN}=3.8$ Ω·m, $R_L=5.1$ Ω·m, $R_L>R_{SN}$ because Zone B is relatively thin.
Problem 5.4 a. 88 ft.
Problem 5.5 50 Ω·m (from the Lateral's plateau).
Problem 5.6 a. 2.937 V.
b. $R_a=30$ Ω·m.
c. $R_a=30$ Ω·m.
d. 0.225 V, 10 Ω·m, 10 Ω·m.
e. Yes.
Problem 5.7 0.3 to 50 Ω·m.
Problem 5.8 2 in.
Problem 5.9 2, 68, 44, and 2 Ω·m.
Problem 5.10 54 and 68 Ω·m.
Problem 5.11 a. 24, 18, and 15 Ω·m.
b. 39, 34, and 24 Ω·m.
Problem 5.12 2.3 and 2 Ω·m.
Problem 5.13 a. $R_{xo}=8$ Ω·m.
b. $\phi=28\%$.
c. $R_{xo}=9.5$ Ω·m.
d. $R_{xo}=9.5$ Ω·m.
e. The Short Normal is affected by R_t.
Problem 5.14 a. $h_{mc}=0.5$ in., $(R_a)_{PL}=65$ Ω·m.
b. $(R_a'')_{PL}=65$ Ω·m.
c. $R_{xo}=55$ Ω·m.
d. $R_{xo}=64$ Ω·m.
Problem 5.15 a. 9.5 and 0.28 Ω·m.
b. Yes.
c. $R_{sh}=0.8$ Ω·m.
d. $h=9$ ft.
e. 9.5 and 0.28 Ω·m.
Problem 5.16 Zone D: $d_i=64$ in., $R_{xo}=29$ Ω·m, $R_t=5$ Ω·m.
Zone E: $d_i=40$ in., $R_{xo}=11.5$ Ω·m, $R_t=2.3$ Ω·m.
Problem 5.17 a. 45 and 45 Ω·m.
b. No invasion; $R_{xo}=R_t=45$ Ω·m.
c. 4.6 and 8 Ω·m.
f. $R_{xo}=10$ Ω·m, $R_t=0.33$ Ω·m.
g. $R_{xo}=14$ Ω·m, $R_t=0.94$ Ω·m.
Problem 5.18 a. $R_a=24$ Ω·m.
b. $R_{\dot{a}}=21$ Ω·m.
c. $R_s=3.7$ Ω·m.
d. $R_{\ddot{a}}=24$ Ω·m.
e. $R_{xo}=2$ Ω·m.
f. $G=0.33$.
g. $R_t=35$ Ω·m.
h. $S_o=75\%$.
i. $S_o=70\%$.

Chapter 6

Problem 6.1 $E_d/E_m=4.8$.
Problem 6.2 a. SSP = −60 mV.
b. SSP = −28 mV.
Problem 6.3 a. SSP = +16 mV.
b. SSP = −32 mV.
Problem 6.4 $R_w=0.03$ Ω·m.
Problem 6.5 90,000 ppm.
Problem 6.6 0.04 Ω·m, 90,000 ppm.

Chapter 7

Problem 7.1 $t=3rc$.
Problem 7.3 c. Zone 1; the relatively high resistivity of the zone suppressed the response of the SP log.
Problem 7.4 80 API units.
Problem 7.6 Yes; the upper sand is responsible for the high produced WOR. Water flow is indicated by the relatively high response of the uranium curve.
Problem 7.7 The high radioactivity of Zone A is caused by a high uranium content. An average shale index of 35% is estimated from the thorium curve.

Chapter 8

Problem 8.2 $(\rho_{log})_{Al}=2.608$ g/cm^3, $(\rho_{log})_{Mg}=1.714$ g/cm^3.
Problem 8.5 $C=1.0078$.
Problem 8.6 Zone X: limestone; $\phi_{avg}=2\%$.
Zone Y: sandstone; $\phi_{avg}=15\%$.
Problem 8.7 $S_g=60\%$ in the bottom part of the formation.
Problem 8.8 a. Not representative because of irregular borehole enlargement.
b. $\phi=33\%$.
c. $\phi=30\%$; shale effect.

Chapter 9

Problem 9.8 a. +3.8%.
Problem 9.9 +3%.
Problem 9.11 a. Gas/oil contact at 9,030 ft.
Oil/water contact at 9,054 ft.
b. The gas and oil saturation is approximately 95%.
Problem 9.12 a. The oil/water contact is at 6,480 ft.
b. $\phi\simeq25\%$, $S_o\simeq50\%$.

Chapter 10

Problem 10.3 ϕ_s=36%.
Problem 10.4 B_{cp}=1.2.
Problem 10.6 a. Cycle skipping.
b. Log porosities of Zones A and B and agrees with core porosities. Log porosities of Zones C and D are affected by the presence of gas in the formation.
Problem 10.7 a. The 3-ft spacing sonic tool.
b. The lower the porosity, the deeper the invasion and the higher the reading of the 16-in. Normal tool.
Problem 10.9 The porosity of Zone 2 is estimated at 13%.
Problem 10.10 The intervals 4,230 to 4,380 ft and 4,460 to 4,480 ft contain zones of altered shales.

Chapter 11

Problem 11.1 a. The FDC log and BHC log are −2 ft and +2 ft offdepth, respectively.
b. Zone 1: R_t=9.8 Ω·m, ρ_b=2.25 g/cm^3, and Δt=95 μsec/ft.
c. No, Δt_{sh}<100 μsec/ft.
d. Zone 1: S_w=18%.
e. Oil/water contact is at 12,378 ft.
Problem 11.3 a. Yes, at 6,968 ft.
b. Approximately 10 ft.
Problem 11.4 a. S_w=100%, 49%, 27%, 19%, and 18%.
Problem 11.6 No, Zones B and C are tight zones—i.e., have extremely low porosity. They probably have a lithology other than sandstone. The relatively high resistivity displayed by the two zones reflects the low porosity rather than the presence of hydrocarbons.
Problem 11.7 ϕ=27%, S_w=60%.
Problem 11.8 Approximately 1.2 million STB.
Problem 11.9 Zone A: ϕ=13%, S_w=43%.
Zone B: ϕ=15%, S_w=57%.
Problem 11.10 Zone A: ϕ=9%, S_w=53%.
Problem 11.11 5.4 ft.
Problem 11.12 a. Change in grain size and sorting.
b. The critical water saturations of Zones A, B, and C are 37%, 55%, and 43%, respectively.
c. S_w<S_{wc}.
Problem 11.13 a. m=1.8, n=1.75.
b. Assuming $\mu_o=\mu_w$,S_{wc}=42%.
c. $(S_w)_{irr}$=22%.
d. 68% of oil in place.
e. ϕ_D=9%, ϕ_{core}=8.5%.
f. S_w=16%.
g. Yes; f_w=0.
i. $(\rho_b)_c$=2.6 g/cm^3.

Chapter 12

Problem 12.1 a. The interval 7,682 to 7,707 ft.
b. 25 ft in gross thickness and 15 ft in net thickness.
c. 7,693 to 7,698 ft, 5 ft.
d. ϕ=26%, S_w=35%.
e. The interpretation is not final; the shaliness of the formation needs to be considered.
Problem 12.2 a. Low-salinity formation water, water chemical analysis indicates that R_w=0.28 Ω·m.
b. Yes; ϕ=13%, R_w=0.32 (from SP log), R_t= 12 Ω·m, and $S_w\simeq$100%.
Problem 12.3 a. Zone 10: R_{wa}=0.50 Ω·m.
b. Zone 10: R_{wa}=0.65 Ω·m.
c. The petrophysical models used to calculate ϕ and F.
d. The zones are very shaly.
e. R_w=0.12 Ω·m.
f. Zones 7, 9, and 10.
Problem 12.4 Zone 10: R_{wa}/R_w=26, R_{mfa}/R_{mf}=1.2.
Problem 12.5 S_w=80% at 5,750 ft.
Problem 12.6 a. Yes. The presence of hydrocarbons is indicated by the separation between the R_t and R_o curves.
b: S_w=30%.
c. ϕ=12%.
d. R_w=0.048 Ω·m.
Problem 12.7 b. The interval of 2,164 to 2,182 ft contains hydrocarbons; net pay = 18 ft, ϕ = 20%, and S_w = 50%.
c. R_w=0.9 Ω·m.
Problem 12.8 a. R_w=0.036 Ω·m.
b. Zones 2 through 5.
c. The F_{xo}/F_s approach confirms the finding in Part b.
Problem 12.9 B=1.48 for 18 in. <d_i<90 in.
Problem 12.10 a. S_w=40%.
b. S_w=40%.
d. S_{xo}=83%.
Problem 12.11 b. Deep invasion leads to optimistic results.
c. Use another method—e.g. R_o log.
Problem 12.12 a. The sand at 4,550 ft.
b. The sand at 4,620 ft.
Problem 12.13 Zone 1: S_{mo}=63%, S_w=27%.
Problem 12.14 N=1.31 million STB.

Chapter 13

Problem 13.3 d. m=2.
Problem 13.4 a. Limestone and sandstone.
b. 0.05 and 0.09 Ω·m.
c. Zone A is more likely to belong to the limestone pattern because of its low porosity. S_o=50%.
Problem 13.5 c. Yes, Zones 5 and 9.
d. R_w=0.06 Ω·m, Δt_{ma}=56 μsec/ft.
e. R_w=0.04 Ω·m, m=2.1.
f. Zones 1, 2, 8, and 11.
Problem 13.6 a. Different lithology.
b. Unconsolidated sands, Δt_{sh}=115 μsec/ft.
c. No, the presence of shale results in water saturation values higher than the true ones.
Problem 13.7 c. Sandstone: m=1.9, R_w=0.16 Ω·m.
Limestone: m=2.23, R_w=0.5 Ω·m.
Problem 13.8 a. m=2.
b. R_w=0.011 Ω·m.
Problem 13.9 a. Yes, tight zones.
b. Yes, the same R_w values are obtained from the two plots.
c. No.

Chapter 14

Problem 14.2.

Zone	Lithology	ϕ (%)
A	Limestone	30
B	Dolomite	5
C	Anhydrite	0
D	Sandstone	30
E	Salt and sulfur	0

Problem 14.3.

Zone	Lithology	ϕ (%)
V	60% sandstone 40% limestone	28
W	80% dolomite 20% limestone	8
X	100% limestone	10
Y	probably anhydrite	0
Z	80% limestone 20% dolomite	20

Problem 14.5 a. $\phi=10\%$, 70% limestone, $\rho_{ma}=2.76$ g/cm^3.
b. $\phi=11\%$, 55% dolomite, $\rho_{ma}=2.77$ g/cm^3.
c. $\phi=11\%$, 60% silica, $\rho_{ma}=2.78$ g/cm^3.
d. $\rho_{ma}=2.77\pm0.1$ g/cm^3, $\phi=10.4\pm0.4\%$.

Problem 14.6 a. $\rho_{ma}=2.71$ g/cm^3.
b. Sandstone.
c. Borehole enlargement.

Problem 14.7 a. 65% limestone, 35% dolomite, 7% porosity.
b. Presence of secondary porosity, 10% total porosity.

Problem 14.8 b. $\Delta t_{ma}=50$ μsec/ft.

Problem 14.9 b. $\rho_b=2.54$ g/cm^3, $\phi_{SNP}=20\%$.
c. $\phi=0\%$.
d. 17% porosity dolomite.

Problem 14.10 42% calcite, 38% anhydrite, 20% dolomite, 5% porosity.

Problem 14.11 b. Sandstone and dolomite.
c. Yes, Zones 1 and 3.

Problem 14.12 b. Limestone, dolomite, and anhydrite.
c. Zones 2, 5, 6, 8, and 12.
d. 5%.

Problem 14.13 Zone 1: 85% dolomite, 15% limestone, 12.5% porosity.

Problem 14.14 Zone A: salt; Zone B: anhydrite.

Problem 14.15 a. Limestone and dolomite.
b. 50% limestone, 50% dolomite, 8% total porosity, 1% secondary porosity index, 90% oil saturation.

Chapter 15

Problem 15.1 a. $V_{sh}=15\%$.
b. The SP response in the clean sand is affected by the relatively high formation resistivity.
c. $\phi=27\%$.
d. $S_w=29\%$.

Chapter 16

Problem 16.1 The gas/oil contact can be placed at about 9,030 ft.

Problem 16.2 A relatively high density porosity of 35% is calculated for the top clean zone of the sand. If the true porosity is that high, then the sand is oil-bearing; otherwise, it is gas-bearing. Additional data—e.g., the core porosity and/or the neutron porosity—are needed to determine the hydrocarbon type.

Problem 16.3 In analyzing this log, one should be aware of the severe borehole enlargement in the interval of 11,860 to 12,020 ft. The borehole size in this interval exceeds the maximum caliper reading of 16 in. Within this interval, the tool is not in contact with the formation. The extremely low readings of the bulk density curve are caused by the borehole size.

Problem 16.5 The zone is a gas-bearing sandstone. The porosity and gas saturation values are 26% and 90%, respectively.

Problem 16.6 The formation is predominantly gas-bearing limestone, displaying a 19% to 20% porosity.

Problem 16.8 No, the separation is caused by the lithology assumed in calculating the neutron and density tool porosities. The formation within this interval is sandstone. Limestone matrix properties were used in tool calibrations.

Problem 16.9 Intervals A through C are gas-bearing; Interval D is oil-bearing; and Interval E is water-bearing.

Problem 16.10 Yes, a positive separation between the neutron and density porosity curves (i.e., $\phi_D > \phi_N$), which is characteristic of a gas-bearing zone, can be seen at 6,470 ft.

Problem 16.11 No, the entire sand Interval X is gas-bearing. The positive separation between the neutron and density curves diminishes and then disappears completely as the shale content increases with depth. Qualitatively, Interval Y could be a hydrocarbon-bearing zone because it displays log responses similar to the bottom few feet of Zone X. Quantitative analysis is required to confirm the qualitative determination and to evaluate the zone's potential.

Appendix C
Abbreviations

API	American Petroleum Inst.
BHC	Borehole Compensated log
BHF	Bradenhead flange
BHT	Bottomhole temperature
CBL	Cement Bond Log
CEC	Cation exchange capacity
CNL	Compensated Neutron Log
DIL	Dual Induction Laterolog
EC	Electron capture
EPT	Electromagnetic Propagation Tool
F-log	Formation Resistivity Factor Log
FDC	Formation Density Compensated
FID	Fluid Identification
GL	Ground Level
GNT	Gamma Neutron Tool
GRS	Gamma Ray Spectrometry
IES	Induction/Electric Survey
ILd	Deep Induction Log
ILm	Medium Induction Log
ISF	Induction Spherically Focused
KB	Rotary kelley bushing
LLd	Deep Laterolog
LLs	Shallow Laterolog
LL3	Laterolog-3
LL7	Laterolog-7
LL8	Laterolog-8
LSS	Long-Spaced Sonic
MID	Matrix identification
ML	Microlog
MLL	Microlaterolog
MOP	Moveable oil plot
MWL	Mean water level
MWD	Measurement while drilling
NLL	Neutron Lifetime Log
OHP	Outer Helmholtz plane
OIP	Oil in place
PL	Proximity log
R_o lo-g	Resistivity of formation 100% saturated with water log
RIL	Resistivity Induction Log
SFL	Spherically Focused Log
SN	Short Normal
SNP	Sidewall Neutron Porosity
SP	Self Potential
SPI	Secondary porosity index
SSP	Static Self-Potential
TD	Total depth
TDS	Total dissolved solids
TDT	Thermal decay time

Author Index

Subject Index

D

E

F

G

H

I

J

K

L

O

P

X

Y

CPSIA information can be obtained at www.ICGtesting.com
Printed in the USA
LVOW091213150413

329015LV00001BA/1/P

9 781555 630560